John George Wood

Animate Creation

OUR LIVING WORLD

John George Wood

Animate Creation
OUR LIVING WORLD

ISBN/EAN: 9783741179419

Manufactured in Europe, USA, Canada, Australia, Japa

Cover: Foto ©Andreas Hilbeck / pixelio.de

Manufactured and distributed by brebook publishing software (www.brebook.com)

John George Wood

Animate Creation

BULL-FROG.

Animate Creation;

POPULAR EDITION OF

"OUR LIVING WORLD,"

A NATURAL HISTORY

BY

THE REV. J. G. WOOD.

REVISED AND ADAPTED TO

AMERICAN ZOOLOGY,

BY

JOSEPH B. HOLDER, M.D.,

Fellow of the New York Academy of Sciences; Member of the Society of Naturalists, E. U. S.; Member of the American Ornithologists' Union; Curator of Vertebrate Zoology, American Museum of Natural History, Central Park, New York.

FULLY ILLUSTRATED WITH SCIENTIFIC ACCURACY.

VOL. III.

NEW YORK:

SELMAR HESS.

COPYRIGHT,
1885,
BY SELMAR HESS.

PREFATORY NOTE.

THE Reptilia and Batrachia form the subjects of this volume. The published Reports of the United States Commission of Fish and Fisheries, and the various Bulletins and Papers of the National Museum and Smithsonian Institution, have been of equal importance in affording the most recent facts and views touching American Zoology.

For the use of the contents of these works we acknowledge the courtesy of the Secretary of the Smithsonian Institution.

The Fishes and Invertebrates are catalogued and described very fully in the above-mentioned publications, all of which are accessible to students or those desiring further technical knowledge, but their contents are too voluminous to be fairly utilized in this volume.

J. B. H.

CONTENTS.

CLASS REPTILES. ... 3
Section SHIELDED REPTILES.—Order CHELARIANS; OR TORTOISES: ... 5
    True Tortoises—*Testudinidae* ... 6
    Terrapins—*Emydidae* ... 9
    Aquatic Tortoises—*Chelididae* ... 17
    Soft Turtles—*Trionycidae* ... 19
    Sea Turtles—*Cheloniadae* ... 21
Order EMYDOSAURI; OR TORTOISE-LIZARDS: ... 28
    Crocodiles—*Crocodilidae* ... 29
    Alligators—*Alligatoridae* ... 35
Order AMPHISBÆNIA: *Amphisbamidae*, Cheirotidae ... 38
Section SQUAMATA; or SCALED REPTILES.—Order SAURA; OR LIZARDS.—Sub-Order SEPTO-GLOSSÆ; or SLENDER-TONGUED LIZARDS,—Tribe CYCLOSAURA ... 40
    Monitors—*Monitoridae* ... 40
    Teguexins—*Teidae* ... 43
    True Lizards—*Lacertinidae* ... 45
    Band-Tailed Lizards—*Zonuridae* ... 50
    Cylindrical-Bodied Lizards—*Chalcidae* ... 54
    Other Families of Lizards—*Anadiadae, Chirocolidae, Cercosauridae, Chamaesauridae* ... 55
Tribe GEISSOSAURA: ... 55
    Gape-Eyed Skinks—*Gymnophthalmidae* ... 57
    Pagopus, Delma—*Pygopidae* ... 57
    Aprasia—*Aprasiadae*.—Lialis—*Lialicidae* ... 58
    Skinks—*Scincoidae* ... 58
    Ophiomore—*Ophiomoridae*,—Seps—*Sepsidae* ... 68
    Javelin Snake—*Acontiadae* ... 69
    Blind Reptiles—*Typhlinidae, Typhlopsidae* ... 70
    Rough-Tailed Lizards—*Uropeltidae* ... 71
Sub-Order PACHYGLOSSAE; or THICK-TONGUED LIZARDS,—Tribe NYCTISAURA: Geckos—*Geckotidae* ... 71
Tribe STROBILOSAURA: True Iguanas—*Iguanidae* ... 75
    Iguanas of the Old World—*Agamidae* ... 84
Tribe DENDROSAURA; or TREE LIZARDS— ... 89
Order OPHIDIA; OR SNAKES: ... 93
    Adders—*Crotalidae* ... 95
    Vipers—*Viperidae* ... 104
    River and Sea Serpents—*Hydridae* ... 115
    Boas—*Boidae* ... 118
    Conocephalus—*Calamaridae*,—Schaap-Sticker with other species—*Coronellidae* ... 125
    Water Snakes—*Natricidae* ... 127
    Racers—*Colubrinae* ... 130
    Tree Serpents—*Dryadidae* ... 134
    Almost-Toothless Serpents—*Dasypeltidae*, or *Rachiodonidae* ... 137
    True Tree Serpents—*Dendrophidae* ... 137
    Wood Snakes—*Dryiophidae* ... 139
    Dipsas—*Dipsadidae*.—Deadly Serpents—*Elapidae* (commencing with Banded Bungarus) ... 140
    Other Families of Deadly Serpents—*Dendraspidae, Atractaspididae* ... 149
Order BATRACHIA; OR FROGS AND TOADS.—Sub-Order BATRACHIA SALIENTIA; or LEAPING BATRACHIANS: ... 149
    A. Aglossa; or Tongueless Batrachians: *Dactylethridae, Pipidae* ... 150
    B. Opisthoglossae; or Tongued Batrachians: True Frogs—*Ranidae* ... 152

    Horned and Land Frog—*Cystignathidae* ... 158
    Painted Frog—*Discoglossidae*.—Nurse Frog—*Alytidae* ... 159
    Toads—*Bufonidae* ... 160
    Tree-Frogs, or Tree-Toads— *Polypedatidae, Hylidae* ... 164
        Pelodryadae (represented by the Blue Frog), *Phyllomedusidae* (represented by the Bicolor Tree-Frog) ... 168
        Hylaplesidae ... 169
    C. Proteroglossa; or Batrachians with Tongue free in front: Rhinophryne—*Rhinophrynidae* ... 169
Sub-Order BATRACHIA GRADIENTIA; or CRAWLING BATRACHIANS: ... 170
    Salamanders— *Salamandridae* ... 170
        Plethodontidae (commencing with Japanese Salamander) ... 175
Order PSEUDOSAURIA; OR FALSE BATRACHIANS:
    Gigantic Salamander, etc.—*Protonopsidae* ... 176
    Congo Snake, etc.—*Amphiumidae* ... 179
Order PSEUDOPHIDIA; OR FALSE SERPENTS (genus Caecilia).—Order PSEUDOICTHYAS; OR FISH-LIKE REPTILES ... 180
Order MEANTIA, comprising old forms of CRAWLING BATRACHIANS: *Proteidae* ... 185
    Mud Eel—*Sirenidae* ... 186

CLASS PISCES; OR FISHES. ... 188
Order CHONDROPTERYGII; OR SOFT-FINNED FISHES.—Sub-Order ELEUTHEROPOMI; or FREE-GILLED FISHES: Sturgeon and Shovel-Fish—*Acipenseridae* ... 190
    Spoon-Bill Sturgeon—*Polyodontidae* ... 192
    Chimaeras—*Chimaeridae* ... 193
Sub-Order TREMATOPNEA; or Fishes with Gills fixed to Bars.—Sub-Section SQUALI, comprising Sharks and Shark-like Fishes: Dog-Fishes—*Scyllidae* ... 196
    Sharks—*Squalidae* ... 196
    Angel-Fish—*Squatinidae* ... 204
Sub-Section RAII; or RAYS: Saw Fish—*Pristinidae*.—True Rays—*Raidae* ... 206
Order ACANTHOPTERYGII; OR SPINE-FINNED FISHES: Sticklebacks—*Gasterosteidae* ... 212
    Fishes with Compressed Body—*Berycidae* ... 216
    Perches—*Percidae* ... 217
    Lip Fishes—*Labridae* ... 221
    Kakaan, Capeuna, Bodian—*Pristipomidae* ... 224
    Mullets—*Mullidae* ... 226
    Braize, Snapper, Gilt-Head, etc.—*Sparidae*. Scaly-Finned Fishes—*Squamipinnes*,—*Chetodonts*—*Chetodontinae* ... 228
    Archer-Fish—*Toxotina*,—Chilodactyle—*Cirrhitidae*.—Gurnards—*Triglidae* ... 231
    Uranoscopus, Sting-Fish, Sillago—*Trachinidae* ... 238
    Tooth-Scaled Fishes—*Sciaenidae* ... 240
    Becuna—*Sphyraenidae* ... 242
    Hair-Tailed Fishes—*Trichiuridae* ... 243
    Mackerels—*Scombridae* ... 244
    Cordonnier, Rudder-Fish, Horse-Mackerel—*Carangidae* ... 249
    Sword Fishes—*Xiphiidae* ... 250
    Gobies—*Gobiidae* ... 251

CONTENTS.

	PAGE
Quoit-Fishes—*Discoboli*	253
Frog-Fishes—*Batrachidae, Pediculati*	255
Blennies—*Blenniidae*	256
Deal-Fish and Gymnetrus—*Trachypteridae*	258
Surgeon and Thorntail—*Acanthuridae*	259
Climbing Perch—*Labyrinthici*	260
Sand Smelt—*Atherinidae*.—Gray Mullet—*Mugilidae*	261
Snake-Head- ⎰ *Ophiocephalidae*	262
ed Fishes. ⎱ *Cepolidae*	263
Spike-Bearing Fishes—*Centriscidae*	263
Snake-Bodied Fishes--*Fistularidae*.—Eel-like Fishes—*Mastacembelidae*	264
Order MALACOPTERYGII; or Fishes having fins with rays closely jointed.—Sub-Order BRACHII: Flat Fishes—*Pleuronectidae*	265
Cod-Fish—*Gadidae*	267
Eels. ⎧ *Ophidini*	270
⎨ *Anguillidae, Congeridae, Muraenidae*.	272
⎨ *Gymnotidae*	273
⎩ *Leptocephalidae*	275
Sub-Order ABDOMINALES: Blind Fish—*Heteropygidae*	275
Herring Tribe—*Clupeidae*	276
Flying Fish, and Pike—*Esocidae*	278
Salmon—*Salmonidae*	279
Stargazer—*Cyprinidontidae*	283
Carp, Barbel, Bream, etc.—*Cyprinidae*	284
Silurus—*Siluridae*	287
Order PLECTOGNATHI; or Fishes with coalescent jaws: Trunk-Fishes—*Sclerodermi*	287
Naked-Toothed Fishes—*Gymnodontes*	289
Order LOPHOBRANCHIATA; OR CREST-GILLED FISHES: Sea-Dragon and Pegasus—*Pegasidae*. Sea-Horse, Great Pipe-Fish, and Phyllopteryx—*Syngnathidae*	290
Order GANOLEPIDOTI: Bony Pike	292
Order CYCLOSTOMI: Lamprey, Lampern, Myxine—*Petromyzonidae*	293
Order LEPTOCARDII: Lancelet	296

DIVISION INVERTEBRATA; OR INVERTEBRATE ANIMALS.	298
Some Early Reminiscences by the American Editor	298

CLASS MOLLUSCA; OR MOLLUSKS. 302

	PAGE
Sub-Class CEPHALOPODA.—Order DIBRANCHIATA.— Section OCTOPODA: Argonaut—*Argonautidae*.	305
Eight-Armed Cuttles—*Octopodidae*	306
Section DECAPODA; or TEN-FOOTED CUTTLES: Calmaries—*Teuthidae*	307
Sepia—*Sepiadae*	310
Spirals—*Spiralidae*	311
Order TETRABRANCHIATA; OR FOUR-GILLED ANIMALS: Chambered Nautilus—*Nautilidae*	311
Order GASTEROPODA.—Sub-Order PROSOBRANCHIATA.— Section SIPHONOSTOMATA: Strombidae	412
Other Families of Shells. ⎰ *Muricidae*	313
⎨ *Buccinidae* (first example: Whelk)	317
⎨ *Conidae, Volutidae*	321
⎩ *Cypraeidae*	322
Section HOLOSTOMATA; or SEA SNAILS:	324
Other Families of Shells. ⎧ *Naticidae, Neritidae*	324
⎨ *Cerithiadae*	325
⎨ *Turritellidae*	326
⎨ *Litorinidae*	327
⎨ *Turbinidae*	328
⎨ *Haliotidae*	329
⎨ *Fissurellidae, Calyptraeidae* (first example: Cup- and Saucer-Limpets)	331
⎩ *Dentalidae, Chitonidae*	332

VOL. III.

	PAGE
Order PULMONIFERA; OR INOPERCULATE AND OPERCULATE GASTEROPODS: Snails—*Helicidae*	333
Slugs—*Limacidae*	335
Apple-Snails—*Paludinidae*	336
Pond-Snails—*Limnaeidae*	337
Section OPERCULATA.—Order OPISTHOBRANCHIATA.—Sub-Order TECTIBRANCHIATA: Bubble-Shells—*Bullidae*	338
Sea-Pigeons—*Aplysiadae*.—Indian Umbrella—*Pleurobranchidae*	339
Sub-Order NUDIBRANCHIATA; or NAKED-GILLED MOLLUSKS: Doris—*Doridae*.—Dendronotus, and Doto—*Tritoniadae*	340
Eolis, and Glencus—*Eolidae*	341
Order NUCLEOBRANCHIATA: Carinaria—*Firotidae*.—Order PTEROPODA; OR WING-FOOTED MOLLUSKS.—Sub-Order THECASOMATA: Hyalea, Cleodora, Spike-Shell, and Cymbulia—*Hyaleidae*	342
Order BRACHIOPODA: Lamp-Shells—*Terebratulidae, Rhynchonellidae, Lingulidae*	344
Order ACEPHALA; OR HEADLESS MOLLUSCS.—Sub-Order ASIPHONIDAE: Oysters—*Astreidae*	345
Wing-Shells—*Aviculidae*	350
Mussels—*Mytillidae*	347
Noah's Ark—*Arcadae*.—Pearl-Bearing Mollusks—*Unionidae*	351
Sub-Order SIPHONIDA: Clams—*Clamidae, Tridacnidae*.—Cockles—*Cardiadae, Cyprinidae*	352
Venus Shells—*Veneridae*.—Through-Shells—*Mactridae*	353
Scrobicularia—*Tellinidae*	354
Razor-Shells—*Solenidae*.— Gaper-Shells—*Myacidae*.—Watering Pot-Shell—*Gastrochaenidae*	355
Piddocks—*Pholodidae*	356
Order TUNICATA: Solitary Tunicates—*Ascidiadae*	358
Social Ascidians—*Clavellinae*.—Compound Ascidians—*Botryllidae*.—Pyrosoma—*Pyrosomidae*	359
Scalpa—*Scalpidae*	360
Sub-Class POLYZOA.—Order INFUNDILULATA.—Sub-Order CHEILOSTOMATA	361
Marine Polyzoa. ⎧ *Catenicellidae, Salicornariadae*	362
⎨ *Cellularidae*	363
⎨ *Scruparidae, Farciminariadae, Gemellaridae, Cabereadae, Bicellariadae*	364
⎨ *Flustradae, Membraniporidae*	365
⎨ *Celleporidae, Escharidae*	366
⎩ *Selenariadae*	367
Sub-Order CYCLOSTOMATA: *Crisiadae, Tubuliporidae*	367
Vesiculariadae (first example: Serialaria lendigeri)	368
Alcyonidiadae, Pedicellinadae	369
Order PHYLACTOLAEMATA.—Sub-Order LOPHOPEA: Fresh-Water ⎰ *Cristatellidae*	369
Polyzoa. ⎱ *Plumatellidae*	371

DIVISION ARTHROPODA.	372

CLASS INSECTA; OR INSECTS. 372

	PAGE
Order COLEOPTERA; OR BEETLES.—Section PENTAMERA.—Sub-Section ADEPHAGA.—Stirps GEODEPHAGA: Tiger Beetles—*Cicindelidae*	373
Ground-Beetles, etc.—*Carabidae*	375
Stirps HYDROPHAGA; or WATER BEETLES: Great Water Beetle—*Dyticidae*.—Whirlwig Beetles—*Gyrinidae*	376
Sub-Section RYPOPHAGA.—Stirps BRACHELYTRA: Rove Beetles—*Staphylinidae*	377

CONTENTS.

Stirps NECROPHAGA; or BURYING BEETLES.—*Section CHILOGNATHOMORPHA.*—*Stirps* LAMELLICORNES:
Cockchaffer—*Melolonthidæ* 377
Stag Beetle—*Lucanidæ.* 378
Watchman Beetle — *Geotrupidæ.* — Scarabaeus—*Scarabeidæ*............................... 379
Chrysophora—*Rutelidæ.*—Hercules Beetle—*Dynastidæ*. 380
Section PRIOCERATA. — *Stirps* MACROSTERNI: Chrysochroa—*Buprestidæ.* — Spring Beetles—*Elateridæ* (first example: the Firefly) 380
Stirps APROSTERNI: Glow-Worm—*Lampyridæ.*—Dermestes—*Dermestidæ.*—Soldiers and Sailors—*Telephoridæ.*—Ptilinus and Death Watch—*Ptinidæ.*......... 381
Section HETEROMERA: Cardinal Beetle—*Pyrochroidæ.*—Ripiphorus—*Mordellidæ.*—Blister Flies—*Cantharidæ.* 382
Section ATRACHELIA: Meal Worm—*Tenebrionidæ* 383
Section PSEUDOTETRAMERA.—*Stirps* RHYNCHOPORA: Weevils—*Bruchidæ, Attelabidæ, Curculionidæ, Scolytidæ* 383
Section LONGICORNES: Xenocerus—*Prionidæ.*— Musk Beetle—*Cerambycidæ.* 385
Section PHYTOPHAGA: Tortoise Beetles—*Cassididæ.*—Chrysomela, and Bloody-Nose Beetle—*Chrysomelidæ.* 386
Section PSEUDOTRIMERA: Ladybirds—*Coccinellidæ.* 386
Order DERMAPTERA; or EARWIGS. 387
Order ORTHOPTERA, comprising Grasshoppers, Locusts, Crickets, etc.—*Section* CURSORIA: Cockroaches—*Blattidæ.* 388
Section SALTATORIA: Crickets—*Achetidæ*, and *Gryttidæ* 389
Locusts—*Locustidæ* 390
Section AMBULATORIA: Walking Sticks and Spectres—*Phasmidæ* 391
Section RAPTORIA: Praying Insects—*Mantidæ* 392
Order THYSANOPTERA; or FRINGE-WINGED INSECTS. — *Order* NEUROPTERA, comprising Termites, Dragon-Flies, etc.—*Section* BIOMORPHOTICA:* Termites—*Termitidæ.* 393
Dragon-Flies—*Libellulidæ* 394
Section SUBNECROMORPHOTICA: Lace-Wing Flies—Hemerobiidæ.—Ant-Lion—*Myrmeleonidæ* ,....... 395
Order TRICHOPTERA; OR CADDIS-FLIES.—*Order* HYMENOPTERA; or FLIES AND BEES.—*Section TEREBRANTIA,* — *Sub-Section* PHYTIPHAGA,—*Tribe* SERRIFERA: Saw-Flies — *Tenthredinidæ, Urocerdæ.*.. 397
Sub-Section EUTOMOPHAGA; or INSECT-EATERS.—*Tribe* SPICULIFERA: Gall Insects — *Cynipidæ.* — Ichneumons—*Ichneumonidæ,* 399
Tribe TUBULIFERA.—*Section* ACULEATA.—*Sub-Section*—INRECTIVORA: Cuckoo Flies—*Crabronidæ.* — Sand and Wood Wasps—*Bembecidæ, Sphegidæ, Scoliidæ.* 400
Large-Headed Mutilla—*Mutillidæ.* 401
Sub-Section DIPLOPTERYGII: Solitary Wasps—*Eumenidæ, Vespidæ.* 401
Sub-Section MELLIFERA: Honey Bees—*Apidæ* 402
Order STREPSIPTERA, comprising Insects parasitic in Bees, Wasps, etc. 404
Order LEPIDOPTERA, comprising Butterflies and Moths 404
Section RHOPALOCERA: 406
Butterflies. { *Papilionidæ.*........................ 406
{ *Heliconiidæ.* 409
{ *Danaidæ, Nymphalidæ.*............... 411
{ *Erycinidæ.* 414
{ *Lycænidæ, Hesperidæ.*............... 415
Section HETEROCERA: 416
Moths. { *Sphingidæ.* 416
{ *Anthroceridæ, Ægeriidæ.* 419
{ *Uraniidæ, Hepialidæ.* 420

Moths. { *Bombycidæ* (Silk-Worm) 421
{ *Arctiidæ.* 422
{ *Lithosiidæ, Noctuidæ.* 424
{ *Geometridæ.* 425
{ *Tortricidæ, Tineidæ,* 426
{ *Alucitidæ* 427
Order HOMOPTERA, comprising the Cicadas, Froghoppers, Plant-Lice, Cochineal Bug, etc.—*Section* TRIMERA: *Cicadæ.*............................... 427
Lantern-Flies—*Fulgoridæ.* 428
Hoppers—*Cercopidæ.* 429
Section MONOMERA: Cochineal Insect—*Coccidæ* 429
Section DIMERA: Grape Phylloxera—*Aphidæ.*.......... 430
Order HETEROPTERA, comprising Water - Beetles, Whirligigs, Skippers, Bed-Bugs. — *Section* HYDROCORISA 430
Back-Swimmers—*Notonectidæ.*—Water Scorpion—*Nepidæ.* 431
Section AUROCORISA: Wheel-Bug—*Reduviidæ.*—Dalader—*Mictidæ* 431
Order APHANIPTERA: Fleas—*Pulicidæ.* 432
Order DIPTERA; OR TWO-WINGED INSECTS.—*Section* CEPHALOTA.—*Stirps* NEMOCERA: Gnat—*Culicidæ.*—Crane-Flies—*Tipulidæ.*................... 433
Army-Worm—*Mycetophilidæ.*........................ 434
Section BRACHOCERA.—*Stirps* NOTACANTHA: Gad-Fly, Isetos, etc.—*Tabanidæ.* 434
Stirps ATHERICERA: Common and Bot-Flies—*Muscidæ, Œstridæ.* 435
Stirps PUPIPARA: Forest, or Horse-Flies—*Hippoboscidæ.*—Lice—*Aptera.* 437

CLASS CRUSTACEA. 438

Section PODOPHTHALMATA; or STALKED-EYED CRUSTACEANS.—*Order* DECAPODA; TEN-LEGGED CRUSTACEANS. — *Sub - Order* BRACHYURA; OR SHORT-TAILED CRUSTACEANS.... 438
Tribe OXYRHYCHITA: 439
Spider { *Macropodiadæ* (commencing with the Stenorhynchus) 439
Crabs. { *Maiidæ.*.............................. 441
{ *Parthenopidæ* 442
Tribe CYCLOMETOPITA: 443
Swimming { *Canceridæ.* 443
Crabs. { *Portunidæ.* 445
Tribe CATOMETOPITA: Land Crabs—*Thelphusidæ, Gecarcinidæ.* 448
Marine Crabs—*Pinnotheridæ* (commencing with Pea Crab), *Myctiridæ.* 449
Swift-Footed Crabs—*Ocypodidæ* 449
Angular Crab—*Gonoplacidæ.*—Painted and Floating Crab—*Grapsidæ.* 451
Tribe OXYSTOMATA: Crested and Armed Crab—*Calappidæ.* 452
Uranis-Crab, Leucosia, and Nut-Crabs—*Leucosidæ* 453
Mask and Polished Crab—*Corystidæ.* 454
Woolly Crab—*Dorippidæ.*—Scallop Crab—*Caphyridæ* 455
Sub-Order ANOMOURA; or ANOMOURAL CRUSTACEANS: 455
Hairy Crab—*Dromiadæ.*—Bearded, Porcupine, and Nodular Crab—*Homolidæ.* 456
Frog Crab—*Raninidæ.*—Extraordinary Forms of Crabs—*Hippidæ.* 457
Hermit Crabs—*Paguridæ.* 458
Porcelain-Crabs—*Porcellanidæ.* 461
Sub-Order MACRURA; or LONG-TAILED CRUSTACEANS: 462
Common Lobster—*Galatheidæ.*—Flat Lobsters—*Scyllaridæ.*—Spiny Lobster—*Palinuridæ.* 463
Burrowing Crabs—*Thalassinidæ.* 464
Fresh-Water Cray-Fish, and Salt-Water Lobster—*Astacidæ* 465
True Shrimps—*Crangonidæ* 466
Varieties of Shrimps—*Alpheidæ.*—Prawns—*Palæmonidæ.* 467
Sword-Shrimp—*Penæidæ.* 469
Order STOMAPODA; OR MOUTH-FOOTED CRUSTACEANS: 469
Opossum-Shrimps—*Mysidæ, Phyllosomidæ, Erichthidæ.* 470
Mantis-Shrimp—*Squillidæ.* 471

CONTENTS.

	PAGE
Sub-Class EDRIOPHTHALMATA; or SESSILE-EYED CRUSTACEANS	472
Order AMPHIPODA; OR AMPHIPODOUS CRUSTACEANS: Jumpers—*Orchestridae*	472
Sand-Screws—*Gammaridae*, *Corophidae*	473
Parasitic Shrimps—*Phronimadae*	474
Caddis- and Fresh-Water Shrimp (These belong to the just mentioned families Corophidae, resp. Gammaridae)	475
Order LAEMODIPODA; OR THROAT-FOOTED CRUSTACEANS: Skeleton-Screw—*Caprellidae*.	476
Whale-Louse—*Cyamidae*	477
Order ISOPODA; OR EQUAL-FOOTED CRUSTACEANS: Arcturus—*Idoteidae*	477
Gribble—*Asotidae*	478
Great Sea-Slater—*Oniscidae*. — Woodlouse—*Porcellionidae*.—Poll-Woodlouse—*Armadillidae*.	479
Order or Sub-Class ENTOMOSTRACA; OR ENTOMOSTRACANS: Gill-Footed Entomostracans—*Branchiopodidae*	480
Order CLADOCERA; BRANCH-HORNS: Water-Flea, and Moina—*Daphniadae*	481
Chydorus—*Lynceidae*, *Polyphemidae*	482
Order OSTRACODA: Cypris—*Cypridae*. — Cythere—*Cytheridae*.—Cypridina—*Cypridinadae*.	483
Order COPEPODA; OR OAR-FOOTED ENTOMOSTRACANS: Cyclops—*Cyclopidae*. — Cetochilus—*Cetochilidae*	484
Legion POECILOPODA; or VARIOUS-FOOTED ENTOMOSTRACANS	485
Order SIPHONOSTOMA; OR TUB-MOUTHED ENTOMOSTRACANS: Argulus—*Argulidae*.—Caligus—*Caligidae*	485
Tribe PACHYCHELA, comprising Entomostracans with broad, shield-shaped heads: Nicothoë, and other ill-Parasites—*Ergasilidae*	485
Order LERNEADA; OR SUCKING FISH-PARASITES.	486
Tribe ANCHORASTOMACHAE; or ANCHOR FISH-PARASITES: Chondracanthus, Lernaeodiscus, Jucculina, etc.—*Chondracanthidae*.	486
Tribe ANCHORACARPACEAE; Anchor-Parasites upon Carps, etc.: Perch-Sucker—*Lernaeopodadae*. — Anchorella—*Anchorelladae*.—Shark-Sucker—*Penelladae*.	486
Carp Sucker—*Lernaeoceradae*	487
Order PYCNOGONIDES, comprising odd forms of Crustaceans. — Order XIPHOSURA; OR SWORD-TAILED CRUSTACEANS.	487
Order CIRRIPEDIA; OR BARNACLES:	488
Goose Mussel—*Lepadidae*	489
Acorn Barnacle—*Balanidae*	491
CLASS ARACHNIDAE, comprising Spiders, Scorpions and Mites	493
Order ARANEIDA; OR TRUE SPIDERS	493
Tribe OCTONOCULINA; or EIGHT-EYED ARACHIDA: Crab-Spider, Mygale, and Trap-Door Spider—*Mygalidae*.	494
Wolf-Spiders—*Lycosidae*	497
Hunting Spiders—*Salticidae*	499
Crab-Spiders—*Thomisidae*	501
Cell-Spiders—*Drassidae*	504
Tube-Weaving Spiders—*Ciniflonidae*	506
Argelena, Tegenaria, Caelotes, Theridion, Linyphia, Tetragnatha—*Agelenidae*, Garden-Spider, and Varieties—*Epeiridae*	509
Tribe BINOCULINA; or TWO-EYED ARACHNIDA (Nops)	511
Tribe SENOCULINA; or SIX-EYED ARACHNIDA	511
Order PSEUDOSCORPIONES; OR FALSE SCORPIONS	512
Order PEDIPALPI; OR TRUE SCORPIONS: Phrynus, *Pedipalpidae*	514
Rock-Scorpion—*Scorpionidae*	515
Order ACARINA; OR MITES: Harvest-Bug—*Gamasidae*—Hippopotamus-Tick—*Ixodidae*	516
Flour-Mite—*Acaridae*	517
Hog- and Dog-Tick—*Pediculidae*. — Deer- and Horse-Tick—*Philopteridae*	518
CLASS MYRIAPODA.	520
Order CHILOPODA; Tribe SCHIZOTARSIA: Cermatia—*Cermatiidae*	520
Tribe HOLOTARSIA: Lithobius—*Lithobiidae*	521
Centipedes—*Scolopendridae*	522
Arthronomalus, Gonibregmatus—*Geophilidae*	523
Order CHILOGNATA; OR DIPLOPODS.—Tribe PENTAZONIA: Zephronia, Glomeris—*Glomeridae*	524
Tribe MONOZONIA: Polydesmidae and Polyzenidae.—Tribe BIZONIA: Julidae	525
CLASS ANNULATA OR ANNELIDA; it includes the Earth-Worms, the Leech, the intestinal Worms and other Worm-like Creatures	527
Order SETIGERA; OR BRISTLE-BEARERS: Serpula, Sabella, Terebella, and Shell-Binder — *Amphitritidae*.	527
Lug-Worms—*Nereidae*	530
Eunice—*Eunicaeae*	531
Cirrhatulus—*Ariciae*.	532
Sea-Mouse—*Aphroditidae*	533
Chaetopterus — *Chaetopteridae*, — Earth-Worm—*Lumbricidae*	534
Order SUCTORIA; OR SUCKERS	535
Sub-Class ENTOZOA; or INTERNAL WORMS. — Order CAELELMINTHA.—Tribe NEMATOIDEA: Ascaris, Guinea Worm, Tricocephalus, Stongylus—*Strongylididae*	538
Order STERELMINTHA: Tape- and Ray-Worm—*Cestoideae*.	539
Trichine—*Trichotrachelidae*	540
CLASS RADIATA; OR ECHINODERMATA.	541
Sub-Class DITREMATA.—Order HOLOTHUROIDEA: Sipunculus—*Sipunculidae*.	541
Priapulus—*Priapulidae*. — Spoon-Worm—*Thalassemadae*.	542
Order HOLOTHUROIDEA: Sea Cucumbers—*Pedidae*.	542
Pentacte—*Pentactidae*	543
Synapta—*Synaptadae*	544
Order ECHINOIDEA: Sea-Urchins—*Cicaridae*	544
Heart-Urchins—*Spatangidae*	546
Cake-Urchin, etc.—*Clypeasteridae*	547
Sub-Class HYPOSTOMATA.—Order ASTEROIDEA; OR STAR-FISHES:	549
Asterias, etc.—*Asteriadae*	551
Brittle Stars—*Ophiuridae*	552
Shetland Argus—*Euryatina*	553
Order CRINOIDEA: Feather Star, and	553
Medusa's Head	554
CLASS ACALEPHA; or Nettles, or Jelly-Fishes, or Hydroids	556
Order SIPHONOPHORA; OR SIPHON-BEARING MOLLUSKS.—Present Classification of Jelly-Fishes:	557
Salloe Man—*Velelladae*	559
Portuguese Man-o'-War—*Physatidae*	560
Diphyes—*Diphyidae*	561
Order CTENOPHORA; OR COMB-BEARERS: Cydippe—*Cattimiridae*	562
Venus' Girdle—*Beroidae*.	563
Order DISCOPHORA; OR DISC-BEARERS.—Tribe GYMNOPHTHALMATA; or NAKED-EYED MEDUSA: Sarsia—*Sarsiadae*.—Eudora, and Aequorea—*Aequoreadae*	564
Tribe STEGANOPHTHALMATA; or COVERED-EYED MEDUSAE:—Chrysaora, Rhizostoma—*Rhizostomadae*.	566
CLASS ZOOPHYTES; OR ANIMAL PLANTS.	568
Order ACTINOIDA; OR RADICATED ZOOPHYTES.—Sub-Order ACTINARIA: Pink Anemones—*Ludernariadae*.	568
Green Anemone—*Antheadae*. — Pearlet Anemone, Pufflets, and Vestlets—*Hyanthidae*. — Plumose Anemone, and Widow—*Sagartiadae*.	569
Warty Anemone—*Bunodidae*.	570
Actinia, and Crambactis—*Actiniadae*.	571
Fungia—*Fungidae*.—Cup-Corals—*Caryophylleadae*.	572
Tree-Corals — *Oculinidae*. — Brain-Coral (Astraea)—*Astraeaceae*	573
True Coral—*Coralladae*.	574
Order ALCYONOIDA, comprising Gorgonias, Sea-Fans, Sea-Whips, etc.:	574
Gorgonia—*Gorgoniadae*	575
Sea Pen—*Pennatuladae*.—Sea-Tiger—*Alcyoniadae*.	576
Order HYDROIDA: *Tubulariadae*.	576
Sertulariadae.—Bell-Zoophytes—*Campanulariadae*,	577
The Coral Reefs of Florida	578
CLASS ROTIFERA; OR WHEEL ANIMALCULES.	581
CLASS RHIZOPODA; OR ROOT-FOOTED PROTOZOANS.	583
Sub-Class FORAMINIFERA	583
Sub-Class POLYCYSTINA	584
CLASS INFUSORIA; OR MICROSCOPIC ANIMALS:	585
Vorticella—*Peritricha*	586
Stentor—*Heterotricha*	587
CLASS PORIFERA; OR SPONGES:	587
Present Classification of Porifera	591

VOL. III.

LIST OF ILLUSTRATIONS.

ILLUSTRATIONS PRINTED IN COLORS.

	PAGE		PAGE		PAGE
Soft Turtles	20	Swarm of Migratory Locusts	390	Hermit-Crabs	458
African Cobra, or Haje, and Gazelles	148	Erycinids	414	Crab-Spider, or Matoudon	492
Bull-Frog	154	Silk-Worm, and Moths	422	Holothurians, and Sea Star	544
Stag-Beetle, and Longicorn Beetle	378	Cicadae, Lantern-Fly, etc.	428	Sea-Anemones	568

FULL-PAGE WOOD ENGRAVINGS.

	PAGE		PAGE		PAGE
Indian Tortoise, or Elephant Tortoise	8	Anakonda	124	Armed Calamary	308
African Crocodiles at Home	28	Cobra Di Capello	142	Hercules Beetle	380
Gavial, or Gangetic Crocodile	30	Archer Fish	230	Dragon-Flies, Laying Eggs	394
Indian Monitor	42	Flying Gurnard	238	May-Fly	396
Elegant Ophiops	50	Sword-Fish	250	Water Boatman, and Water Scorpion	430
Iguana	76	Trout	282	Lobster, and Spiny Lobster	464
Chameleon	90	Burying Beetles, Hornet, Watchman		Red Coral, and Eight-Armed Cuttle	572
Viper, or Adder	110	Beetle, etc.	376		

ILLUSTRATIONS IN THE TEXT.

REPTILES.

	PAGE
Title-Page (Reptiles)	1

TORTOISES.

	PAGE
Gopher Tortoise	6
The Pyxis	9
Lettered Tortoise, and Chicken Tortoise	10
Quaker Tortoise	11
Box Tortoise	12
Mud Tortoise	15
Matamata	17
New Holland Chelodine, or Snake Tortoise	18
Dogania	21
Luth, or Leathery Turtle	22

CROCODILES.

	PAGE
False Gavial	30
Indian Crocodile	32
American Crocodile	34
Margined Crocodile	35

ALLIGATORS.

	PAGE
Alligator	36
Jacaro, or Yacaro	37

AMPHISBÆNIDÆ.

	PAGE
Sooty Amphisbæna	38
Cheirotes, or Hand-Eared Lizard	39

LIZARDS.

	PAGE
White-Throated Regenia	41
Nilotic Monitor	42
Teguexin, or Variegated Lizard	43
Crust Lizard	44
Scaly Lizard	46
Eyed Lizard	47
Green Lizard	48
Rough-Scaled Cordyle	50

PYGOPUS, SKINKS, AND SEPS.

	PAGE
Pygopus	57
Common Skink	59
Hinulia and Mocoa	61
Seps, or Cicigna	68

PACHYGLOSSÆ AND STROBILOSAURA.

	PAGE
Fan-Foot	72
Hatteria	77

	PAGE
Marine Oreocephale	78
Green Caroline Anolis	81
Crowned Tapayaxin	83
Frilled Lizard	85
Spinose Agama	87
Egyptian Mastigure	88
Moloch	89

SNAKES.

	PAGE
Fer-De-Lance	96
Bushmaster	97
Copper-Head Snake	99
Rattlesnake	100
Diamond and Northern Rattlesnake	103

VIPERS.

	PAGE
Tic-Polonga, or Katuka	105
Puff Adder	106
Cerastes, or Horned Viper	108
Horatta Pam	109
Sand-Natter	110

RIVER AND SEA SERPENTS.

	PAGE
Black-Backed Pelamis	116
Acrochodore	117
Carpet Snake	118

ROCK SNAKES AND BOAS.

	PAGE
Rock Snake of India	119
Natal Rock Snake	120
Dog-Headed Boa	122
Boiguagu	123

COLUBRINÆ.

	PAGE
Schaap-Sticker	126
Ringed Snake, or Grass Snake	128
Thunder Snake	131
Chicken Snake	132
Colnber	133
Coach-Whip Snake	135
Boomslange	138
Langaha	139
Banded Bungarus	141
Serpent-Eating Hamadryas	142

FROGS AND TOADS.

	PAGE
Surinam Toad	150

	PAGE
Development of the Egg and of the Tadpole of the Green Frog	156
Savannah Cricket Frog	164
Green Tree-Frog	165
Rhinophryne	169

CRAWLING BATRACHIANS.

	PAGE
Salamander	170
Larva of Axolotl	176
Axolotl	177
Gigantic Salamander	178
Menopome	179

FALSE SERPENTS.

	PAGE
Three-Toed Congo Snake	180
Proteus	186

FISHES.

SOFT-FINNED FISHES.

	PAGE
Sturgeon	191

SHARKS.

	PAGE
Rock Dog-Fish	195
Hammer-Headed Shark	199
Picked Dog-Fish, and Smooth Hound	200
Angel-Fish	205

RAYS.

	PAGE
Common Skate, and Eyed Torpedo	207
Thornback Skate	209

SPINE-FINNED FISHES.

	PAGE
Three-Spined Stickleback, and Fifteen-Spined Stickleback, with Nest	213
Giant Perch, and Common Perch	218
Three-Spotted Wrasse	223
Surmullet	226
Braize, and Young Gilt Head	227

SCALY-FINNED FISHES.

	PAGE
Spotted Scorpion-Fish	232
Bull-Head	234
Sapphirine Gurnard	236
Mediterranean Uranoscopus, and Great Weaver Fish	239
Arapaima	240
Becuna, and Fishing Frog	242
Silvery Hair Tail	244
Mackerel, and Horse Mackerel	245

LIST OF ILLUSTRATIONS.

	PAGE
Tunny	245
Polewig, or Spotted Goby	252
Gemmeous Dragonet	253
Lump-Fish, and Viviparous Blenny	254
Montague's Sucker	255
Sea Wolf	257
Oared Gymnetrus, or Ribbon Fish	259
Climbing Perch	260
Gray Mullet	261
Barca	263

THE COD AND ALLIED SPECIES.

Tobacco-Pipe Fish	264
Haddock, Whiting, and Cod	268

EELS.

Sand Eel, or Hornels	270
Sharp-Nosed Eel	271
Muræna	273
Electric Eel	274

HERRING TRIBE.

Twaite Shad, Sprat, and Herring	277
Flying Fish	279

SALMON, TROUT, GRAYLING, Etc.

Salmon, and Salmon Trout	280
Grayling, and Charr	283
Piraya	284
Star-Gazer	285
Ling, and Sly Silurus	286

PLECTOGNATHI.

Horned Trunk-Fish	288

CREST-GILLED FISHES.

Great Pipe- or Bill-Fish, and Sea-Horse	291
Horse-Like Phyllopteryx	292
Bony Pike	293

LAMPREY, Etc., AND LANCELET.

Lamprey, Lampern, and Sand Pride	294
Lancelet	296

MOLLUSKS.

CEPHALOPODA.

Sepiola	308

CEPHALOPHORA.

Common Woodcock-Shell, and Thorny Woodcock	314
Twisted Triton, Sea Trumpet, and Wrinkled Triton	315
Apple Tun-Shell	319
Helmet-Shell	320
Black Olive	320
Textile Cone	321
Poached Egg, Cowries, Marginella, etc	323
Pelican's Foot	325
Worm Shell	326
Marbled Chiton	333

SLUGS AND SNAILS.

Agate-Shell	335
Pond-Snail	337
Planorbis	337

OPISTHOBRANCHIATA.

Sea Hare	339
Doris	340
Dendronotus	340

NUCLEOBRANCHIATA.

Wing-Footed Mollusk	343

ACEPHALA

Pearl-Oyster	349
Scrobicularia, Razor- and Trough-Shells	354
Gaper Shell	355

POLYZOA.

Net-Pored Animal	366
Group of Polyzoa (Alecto; Tubulipora; Discopora)	368

Vol. III.

	PAGE
Group of Polyzoa (Serialaria; Bowerbankia; Buskia)	368
Group of Polyzoa (Alcyonella; Plumatella; Fredericella; Paludicella)	371

INSECTS.

BEETLES.

Group of Beetles (Tricondyla; Manticora; Harpalus; Lebia; Cicindela; Anthia)	374
Bombardier Beetle	375
Mormolyce	375
Great Water Beetle	376
Whirlwig Beetle	376
Sacred Egyptian Scarabæus	378
Spotted Scarabæus	379
Death Watch	381
Blisters, or Spanish Flies with Larva	382
Nut Weevil	384
Ladybirds	386

CRICKETS.

Field Cricket	389
Mole Cricket	389

ODD FORMS OF ORTHOPTERA.

Walking-Stick Insect	391
Praying Insect	392

FLIES.

Ant-Lion	396
Group of Saw-Flies (Cimbex; Rhyssa; Urocerus; Ichneumon)	398
Turnip-Fly	398

WASPS AND BEES.

Group of Sand Wasps (Crabro; Philanthus)	400
Group of Sand-Burrowing Wasps (Monedula; Pompilus; Scolia)	401
Hive Bee	403

BUTTERFLIES.

Group of Butterflies (Mechanitis; Thecla; Helicopis; Mesosemia; Gynæcia; Papilio; Epicalia; Catagramma; Papilio)	405
Amphrisius	406
Sarpedon and Hector	407
White Butterfly, with Eggs, Caterpillar and Larva	408
Epicharis	409
Phono, Marsæus, Spio, and Erato	409
Midamus	410
Archippus	411
Thyodamas, Thetis, Dido, and Agraulis	412
Peacock Butterfly	412
Neoptolemus	415

MOTHS.

Pine Hawk-Moth	417
Oleander Hawk-Moth	418
Smerinthus, and Humming-Bird Moth	419
Gipsy-Moth	423
Pale Tussock-Moth	424
Clifden Nonpareil	425
Oak-Leaf Roller and Caterpillar	426

SCALE INSECT.

Cochineal Insect	429

HETEROPTERA.

Wheel-Bug with Larva	431

FLEAS AND FLIES.

Flea	432
Gad-Fly	434
Tsetse	435
Horse Bot-Fly	436
Cattle Bot-Fly	437
Sheep Bot-Fly	437
Horse- or Forest-Fly	437

CRUSTACEANS.

TEN-LEGGED CRUSTACEANS.

	PAGE
Sea-Spider	439

CRABS.

Edible Crab	444
Fighting Crab	449
Racing Crab	450
Hairy Crab	455
Porcelain Crab	462
Cray-Fish, or Craw-Fish	466
Edible Prawn	468

MOUTH-FOOTED CRUSTACEANS.

Phyllosome	470
Mantis Shrimb	471

SESSILE-EYED CRUSTACEANS.

Fresh Water Shrimp	475
Mantis Shrimp	476
Whale-Louse	477

BARNACLES.

Goose Mussel	489
Coronet Barnacle	491

ARACHNIDA.

SPIDERS.

Tarantula Spider	497
Wolf Spider	498
Group of Crab-Spiders (Thomisus, and Arkys)	501
Group of various Spiders (Clubiona, Drassus, and Clotho)	504
House Spider	507
Male of the Tetragnathon	509
Female of the Cross-Spider	509
Segestrium	512

PSEUDO-SCORPIONES.

Galeodes	513
Book-Scorpion	514

MYRIAPODA.

Scolopendræ (Centipedes)	521
Polydesmus	525
Millepede	526

ANNULATA.

Serpula	528
Chætopterus	534
Skate-Sucker	537

RADIATA.

Young and Adult Sea-Urchin	545
Heart-Urchin	547
Shield-Urchin	548

NETTLES AND HYDROIDS.

Swimming Sea-Nettles	556
Tubularian Hydroids	558

COMB-BEARERS.

Cydippe	562
Venus' Girdle	563

DISC-BEARERS.

Rhizostoma	566

ANIMAL PLANTS.

ACTINOIDA.

Great Crambactis	571

CORALS.

Cup Coral	572
Madrepora	573
Brain-Coral	574
Organ-Pipe Coral	574
Gorgonia	575
Sea Pen	576

RHIZOPODA.

Amœba	585

INFUSORIA.

Stentor	586

PORIFERA.

Glass Sponge	592
Glass Vase	593

OUR LIVING WORLD.

REPTILES.

THE remarkable beings which are classed together under the general title of Reptiles, or creeping animals, are spread over those portions of the globe where the climate is tolerably warm, and are found in the greatest profusion under the hotter latitudes. Impatient of cold, though capable of sustaining a temperature of such freezing chilliness that any of the higher animals would perish under its severity, and for the most part being lovers of wet and swampy situations, the Reptiles swarm within the regions near the equator, and in the rivers or vast morasses of the tropical countries the very soil appears to teem with their strange and varied forms. Indeed, the number of Reptiles to be found in any country is roughly indicated by the parallels of latitude, the lands near the equator being the most prolific in these creatures, and containing fewer as they recede towards the poles.

Some Reptiles inhabit the dry and burning deserts; but the generality of these creatures are semi-aquatic in their habits, are fitted by their structure for progression on land or in water, and are able to pass a considerable time below the surface without requiring to breathe.

This capacity is mostly the result of the manner in which the circulation and aeration of their blood is effected.

As has been shown in the two volumes on Mammalia and Birds, the heart in these animals is divided into a double set of compartments, technically termed auricles and ventricles, each set having no direct communication with the other. In the Reptiles, however, this structure is considerably modified, the arterial and venous blood finding a communication either within or just outside the two ventricals, so that the blood is never so perfectly aerated as in the higher animals. The blood is consequently much colder than in the creatures where the oxygen obtains a freer access to its particles.

In consequence of this organization the whole character of the Reptiles is widely different from that of the higher animals. Dull sluggishness seems to be the general character of a Reptile, for though there are some species which whisk about with lightning speed, and others, especially the larger lizards, can be lashed into a state of terrific frenzy by love, rage, or hunger, their ordinary movements are inert, their gestures express no feeling, and their eyes, though bright, are stony, cold, and passionless. Their mode of feeding accords with the general habits of their bodies, and the process of digestion is peculiarly slow.

Most of the Reptiles possess four legs, but are not supported wholly upon them, their bellies reaching the ground and being dragged along by the limbs. One or two species can support themselves in the air while passing from one tree to another, much after the fashion of the flying squirrels; and in former days, when Reptiles were apparently the highest race on the surface of the earth, certain species were furnished with wing-like developments of limb and skin, and could apparently flap their way along like the bats of the present time.

Excepting some of the tortoise tribe, the Reptiles are carnivorous beings, and many of them, such as the crocodiles and alligators, are among the most terrible of rapacious creatures. In this class of animals we find the first examples of structures which transmute Nature's harmless gifts into poison, a capacity which is very common in the later orders, such as the spiders and insects, and is developed to a terrible extent in some of the very lowest beings that possess animal life, rendering them most formidable even to man.

The skeleton of a true Reptile, from which class the *Batrachians,* i. e. the frogs, salamanders, and their kin are excluded, for reasons which will presently be given, is composed of well-ossified bones, and is peculiarly valuable to the physiologist. It is well known to all who have studied the rudiments of anatomy, that each bone is formed from several centres, so to speak, consisting of mere cartilaginous substance at its earliest formation, and becoming gradually ossified from several spots.

In the young of the higher animals these centres are only seen during their very earliest stages, and are by degrees so fused together that all trace of them is obliterated. But in the Reptiles it is found that many of the bones either remain in their separate parts, or leave so distinct a mark at the place where they unite, that their shape and dimensions are clearly shown. In the head of the adult crocodile, for example, the frontal bone is composed of five distinct pieces, the temporal of at least five pieces, and each side of the lower jaw-bone is composed of either five or six portions united by sutures.

With the exception of the tortoises, the Reptiles mostly possess a goodly array of teeth, set in the jaw or palate, and as a general fact, being sharp and more or less curved backward. Their bodies are covered with various modifications of the structure termed the dermal, *i. e.* skin skeleton, and are furnished with scales and plates of different forms. In some cases the scales lie overlapping each other like those of the fish, in others they are modified into knobby plates, and in some, of which the tortoises afford well-known examples, they form large flat plates on the back and breast, and scales upon the feet and legs.

The young of Reptiles are produced from eggs, mostly being hatched after they have been laid, but in some cases the young escape from the eggs before they make their appearance in the world. As a general fact, however, the eggs of Reptiles are placed in some convenient spot, where they are hatched by the heat of the sun. Some species are very jealous about their eggs, keeping a strict watch over them, and several of the larger serpents have a curious fashion of laying the eggs in a heap, and then coiling themselves around them in a great hollow cone. The size of the eggs is extremely variable, for, although as a general fact those of the smaller Reptiles are large in proportion to the dimensions of the parent, those of the crocodiles and alligators are wonderfully small, not larger than those of our domestic geese, and in many cases much smaller. They are usually of a dull white color, and in some instances are without a brittle shell, their covering being of a tough leathery consistence.

In form, and often in color, the Reptiles exhibit an inexhaustible variety, and even each order displays a diversity of outward aspect unexampled in the two previous classes of Mammals and Birds. Strange, grotesque, and oftentimes most repulsive in appearance, though sometimes adorned with the brightest tints, the Reptiles excite an instinctive repugnance in the human breast; and whether it be a lizard, a snake, or a tortoise, the sudden and unsuspected contact of one of these beings will cause even the most habituated to recoil from its cold touch. This antipathy may, perhaps, have some connection with the instinctive association of cold with death; but whatever may be the cause, the feeling is deep and universal.

DSI

SHIELDED REPTILES.

TORTOISES.

THE very curious reptiles which are known by the general name of TORTOISES, are remarkable for affording the first example of a skeleton brought to the exterior of the body, a formation which is frequent enough in the lower orders, the crustaceans and insects being familiar examples thereof. In these reptiles the bones of the chest are developed into a curious kind of box, more or less perfect, which contains within itself all the muscles and the viscera, and in most cases can receive into its cavity the head, neck, and limbs; in one genus so effectually, that when the animal has withdrawn its limbs and head, it is contained in a tightly closed case without any apparent opening.

The shell of the Tortoise is divided into two portions, the upper being termed the carapace, and the lower the plastron.

The carapace is formed by a remarkable development of the vertebræ and ribs, which throw out flat processes, and are joined together by sutures like the bones of the skull. The back is therefore incapable of movement, and from the arched shape of the bones is wonderfully strong when resting on the ground. In the Tortoises these bones are united throughout their entire length, but in the Turtles the ends of the ribs retain their original width.

The plastron is similarly formed of the breast-bone, which is thought in these creatures to be developed to the greatest extent of which it is capable. It is composed of nine pieces, each being formed from one of the bony centres already mentioned. These bones are arranged in four pairs, and one in the centre of the front.

As all the limbs have to be worked from the interior of the chest, amid the vital organs and muscles for moving them, they undergo considerable modification. The shoulder-blade, for example, is a curious three-branched bone, quite unique among vertebrate animals, the portion which represents the true shoulder-blade being almost cylindrical, one of the branches flattened, and the other cylindrical, but larger than the real blade-bone. This structure admits of the attachment of powerful muscles, and gives to the fore limbs the great strength which is needed for digging, swimming, climbing, and various modes of exertion. The strong curved bones of the fore limbs bear an evident analogy to the corresponding parts in the mole, with its powerful claws and feet, and its very long blade-bone.

The horny substance commonly termed "tortoise-shell," which is spread in flattened plates on the exterior of the bony case, is thought to be a modification of the scales found on lizards, serpents, etc., and which exist on the legs and other parts of the Tortoises themselves. The row of horny pieces which are found on the edge of the carapace also belong to the "dermal skeleton."

The Tortoises are quite devoid of teeth, the edge of the jaws being sharp and horny, so as to inflict a severe wound; and in many species one or both jaws are sharply hooked at the tip like a falcon's beak. The neck is always rather long, and in many species can be protruded to a considerable extent. Generally, the process of thrusting the neck from the shell is a slow one, but the withdrawal is accomplished with marvellous rapidity, on account of certain long muscles which tie the neck to the back of the carapace. Possibly these muscles, together with their tendons, would, when dried in the baking sunshine, produce musical sounds when touched, and thus give rise to the old poetical legend of the origin of the lyre.

The brain of the Tortoise is very small in proportion to the size of the animal, in the turtle weighing not quite one five-thousandth part of the whole body, and in the land Tortoise about one two-thousandth part. In man the brain is about one-fortieth the weight of the body.

The Tortoises produce their young from eggs, mostly soft and leathery in the texture of their covering, which are laid in some convenient spot, and left to be hatched by heat not derived from the parent. The circulation in the Tortoise is not very complete, but the arterial blood is redder and brighter than the venous.

IN the true TORTOISES the feet are club-shaped and the claws blunt, and the neck can be wholly withdrawn within the shell.

The first example of these creatures is the GOPHER, or MUNGOFA TORTOISE, a native of America. This is a rather pretty, though not brightly colored species, its shell being mostly

GOPHER TORTOISE. — *Testudo gopher*.

brownish-yellow, boldly and variously clouded with rich dark brown. The lower jaw is yellow, and the whole of the plastron is yellow-brown. It is found plentifully in Georgia and Alabama, but, according to Mr. Holbrook, is not seen farther north than South Carolina. When full grown it is a moderately large species, from thirteen to more than fourteen inches in length, and very convex. The following interesting account of its habits is given by Mr. Holbrook in his valuable "North American Herpetology:"—

"They select dry and sandy places, are generally found in troops, and are very abundant in pine-barren countries. They are gentle in their habits, living entirely on vegetable substances. They are fond of the sweet potato (*Convolvulus batatas*), and at times do much injury to gardens by destroying melons, as well as bulbous roots, etc., etc. In the wild state they are represented as nocturnal animals, or as seeking their food by night: when domesticated—and I have kept many of them for years—they may be seen grazing at all hours of the day.

"When first placed in confinement, they chose the lowest part of the garden, where they could most easily burrow. This spot being once overflowed by salt water in a high spring-tide, they migrated to the upper part, nearly eighty yards distant, and prepared anew their habitations. They seldom wandered far from their holes, and generally spent part of the day in their burrows. They delighted in the sun in mild weather, but could not support the intense heat of our summer noons; at those hours they retreated to their holes, or sought shelter from the scorching rays of the sun under the shade of broad-leaved plants. A tanyer (*Arum esculentum*) that grew near their holes was a favorite haunt. They could not endure rain, and retreated hastily to their burrows, or to other shelter, at the coming on of a shower.

"As winter approached, they confined themselves to the immediate neighborhood of their holes, and basked in the sunshine. As the cold increased, they retired to their burrows, where they became torpid; a few warm days, however, even in winter, would again restore them to life and activity.

"The adults are remarkably strong, sustaining and moving with a weight of two hundred pounds or more. The female is generally larger than the male, with the sternum convex; the sternum of the male is concave, especially on its posterior part. The eggs are larger than those of a pigeon, round, with a hard calcareous shell; they are much esteemed as an article of food."

PERHAPS the best known species of these creatures is the COMMON LAND TORTOISE, so frequently exposed for sale in our markets, and so favorite an inhabitant of gardens.

This appears to be the only species that inhabits Europe, and even in that continent it is by no means widely spread, being confined to those countries which border the Mediterranean.

It is one of the vegetable feeders, eating various plants, and being very fond of lettuce leaves, which it crops in a rather curious manner, biting them off sharply when fresh and crisp, but dragging them asunder when stringy, by putting the fore feet upon them, and pulling with the jaws. This Tortoise will drink milk, and does so by opening its mouth, scooping up the milk in its lower jaw, as if with a spoon, and then raising its head to let the liquid run down its throat.

One of these animals, which I kept for some time, displayed a remarkable capacity for climbing, and was very fond of mounting upon various articles of furniture, stools being its favorite resort. It revelled in warmth, and could not be kept away from the hearth-rug, especially delighting to climb upon a footstool that generally lay beside the fender. It used to clamber on the stool in a rather ingenious manner. First it got on its hind legs, rearing itself against the angle formed by the stool and fender. Then it would slowly raise one of its hind legs, hitch the claws into a hole in the fender, and raise itself very gradually, until it could fix the claws of the other hind foot into the thick carpet-work of the stool. A few such steps would bring it to the top of the stool, when it would fall down flat, crawl close to the fender, and there lie motionless. If it were taken off twenty times a day, and carried to the other end of the room, it would always be found in its favorite resort in a few minutes.

This Tortoise had a curious kind of voice, not unlike the mewing of a little kitten. The Common Tortoise is known to live to a great age.

To this genus belongs a very large species, worthy of a passing description. This is the great INDIAN TORTOISE (*Testudo Indica*), a native of the Galapagos. This species is also known scientifically by the name of *Testudo planiceps*. It is seen in the accompanying full-page illustration. Mr. Darwin writes as follows of this animal and its habits: "The Tortoise is very fond of water, drinking large quantities, and wallowing in the mud. The larger islands alone produce springs, and these are always situated toward the central parts, and at a considerable elevation. Hence broad and well-beaten paths radiate in every direction from the wells, even down to the sea-coast; and the Spaniards, by following them up, first discovered the watering-places.

"When landed at Chatham Island, I could not imagine what animal travelled so methodically along the well-beaten tracks. Near the springs it was a curious spectacle to behold many of these great monsters, one set eagerly travelling onwards with outstretched necks, and another set returning, after having drunk their fill. When the Tortoise arrives at the spring, quite regardless of any spectator, it buries its head in the water above its eyes, and greedily swallows great mouthfuls, at the rate of about ten in a minute. The inhabitants say each animal stays three or four days in the neighborhood of the water, and then returns to the lower country.

"For some time after a visit to the springs the bladder is distended with fluid, which is said gradually to decrease in volume, and to become less pure. The inhabitants, when walking in the lower districts, and overcome with thirst, often take advantage of this circumstance by killing a Tortoise, and, if the bladder is full, drinking the contents. In one I saw killed, the fluid was quite limpid, and had only a very slightly bitter taste. The inhabitants, however, always drink first the water in the pericardium, which is described as being best."

The flesh of these Tortoises is very good, and is largely eaten, both fresh and salted. A clear oil is also obtained from the fat. Those who catch these Tortoises do not choose to go through the trouble of cutting up and dressing an animal that is not quite fat, and, as the fitness of its condition cannot be ascertained by the ordinary process, a summary method is employed, viz., cutting a slit through the softer skin near the tail, so as to show the fat under the carapace. Should the Tortoise be in poor condition, it is allowed to go free, and, with the imperturbable temperament of the reptile race, seems to care little for the wound.

Dr. Livingstone mentions a species of Land Tortoise which is remarkable for its love of salt, and the extreme strength of the shell, which, as will be seen, baffles even the teeth of the hyena, which can crush an ox-bone with ease.

"Occasionally we lighted upon Land Tortoises, which, with their unlaid eggs, make a very agreeable dish. We saw many of their trails leading to the salt fountains; they must have come great distances for this health-giving article. In lieu thereof, they often devour woodashes. The young are taken for the sake of their shells, which, when filled with sweet-smelling roots, the women hang around their persons. When taken it is used as food, and the shell converted into a rude basin to hold food or water.

"It owes its continuance neither to speed nor cunning. Its color, yellow and dark-brown, is well adapted, by its similiarity to the surrounding grass and brushwood, to render it undistinguishable; and though it makes an awkward attempt to run on the approach of man, its trust is in its bony covering, from which even the teeth of a hyena glance off foiled.

"When this long-lived creature is about to deposit her eggs, she lets herself into the ground by throwing the earth up around her shell until only the top is visible; then, covering up the eggs, she leaves them until the rains begin to fall, and the fresh herbage appears; the young ones then come out, their shells still quite soft, and unattended by their dam, begin the world for themselves. Their food is tender grass, and a plant named 'thotona,' and they frequently resort to heaps of ashes, and places containing efflorescence of the nitrates for the salts these contain."

THE curious Tortoise which is known only by the comparatively scientific name of PYXIS inhabits several parts of the world, and is not uncommon in some portions of India and Madagascar.

In common with one or two other species, hereafter to be described, the Pyxis has the power of drawing its head, neck, and limbs within the shell and then shutting itself down by means of a lid, formed by the movable front of the sternum. In most of this tribe of reptiles, the sternum is hard and immovable, but in the Pyxis, it moves on a leathery kind of hinge, so as to open when the creature wishes to thrust out its head and limbs, and to close firmly when it withdraws within the shelter of its bony armor.

In order to permit of this total withdrawal into the shell, the carapace is oval and more convex than is usually the case, so as to afford a sufficient space for the reception of the head and limbs. These, too, are rather diminutive in proportion to the size of the animal, and so formed as to be packed into a small compass. The Tortoise employs this curious mode of guarding its vulnerable points whenever it fears danger, and is then so securely locked up in its armor-plates that it is safe from almost every enemy except man. The word Pyxis is Greek, and is very appropriately given to this species, its signification being a box.

The Pyxis is a pretty, but not a large species. The color is extremely variable, scarcely any two individuals being precisely alike, but the general colors are yellow and black. On the

INDIAN TORTOISE, OR ELEPHANT TORTOISE.

carapace the plates are marked with a number of radiating triangular spots, and on the plates which edge the shell there are lines of black. Below, the yellow generally takes a more orange tint, and is diversified with black marks round its edge.

THE PYXIS.—*Pyxis arachnoides.*

WE now come to a group of Tortoises called TERRAPINS.

These creatures are inhabitants of the water, and are mostly found in rivers. They are carnivorous in their diet, and take their food while in the water. They may be known by their flattened heads, covered with skin, sometimes hard, but often of a soft consistency, and their broad feet with the toes webbed as far as the claws.

THE LETTERED TORTOISE is, together with its companion, an American species of the large genus Emys, examples of which are found in various portions of the world, and of which nearly fifty species are known to zoologists. All these creatures have their heads covered with a thin but hard skin.

The Lettered Terrapin is very common in Northern America, and is found in the rivers, ponds, lakes, or even the marshy grounds, where it can obtain an abundant supply of food. It is fond of reptiles, and causes great destruction among the frogs in their earlier stages of existence. It also has a great liking for worms, and, like the green crab of our own coasts, is very apt to take the fisherman's bait, and exasperates him greatly by making him pull up nothing but a little Tortoise when he thought he had caught a fine fish. Regular anglers, therefore, bear an intense hatred to this Tortoise.

It is easily kept in captivity, and will then feed on many substances, preferring those of an animal nature, and being very fond of various reptiles. It will also eat vegetable substances, and one of these Tortoises was fond of purslain (*Portulacea oleracea*).

In color it is very pretty, though rather variable. Generally, it is dark brown above, and the edges are boldly scribbled with broad scarlet marks, something like the letters of some strange language. Below it is yellow, and the head is yellow and black.

The Chicken Tortoise is also found in North America.

It is very common in the ponds, lakes, or marshy grounds, and though very plentiful, and by no means quick in its movements, is not easily caught, owing to its extreme wariness. Hundreds of these Tortoises may be seen reposing on logs, stones, or the branches of fallen trees, where they are apparently an easy prey. But they are very sensitive to the approach of an enemy, and the first that perceives the coming danger tumbles off its perch and falls into the water with a great splash that arouses the fears of all its companions, which go tumbling and splashing into the water in all directions, and in a few seconds not a Tortoise is to be seen where they were so plentiful before they took alarm.

The Chicken Tortoise swims well, but not rapidly, and as it passes along with its head and neck elevated above the surface, it looks so like the dark water-snake of the same country, that at a little distance it might readily be mistaken for that reptile.

LETTERED TORTOISE.—*Emys scripta.*   CHICKEN TORTOISE.—*Emys reticularia.*

It is rather a small species, seldom exceeding ten inches in length. Its flesh is remarkably excellent, very tender and delicately flavored, something like that of a young chicken, so that this Tortoise is in great request as an article of food, and is largely sold in the markets, though not so plentifully as the common salt-water terrapin. Its color is dark brown above, and the plates are scribbled with yellow lines, and wrinkled longitudinally. The neck is long in proportion to the size of the animal, so long, indeed, that the head and neck together are almost as long as the shell. The lower jaw is hooked in front.

An allied species, popularly called the QUAKER TORTOISE, and scientifically *Emys olivacea,* is remarkable for the extreme length of the claws of the fore feet, the three middle claws being elongated in a manner that irresistibly reminds the observer of the nails belonging to a Chinese mandarin of very high rank.

THE SALT-WATER TERRAPIN.

THE SALT-WATER TERRAPIN is a well-known species, living in North and South America, where it is in great request for the table.

The generic name of Malaclemys, or Soft Terrapin, has been given to this species on account of the formation of the head, which is covered with soft, spongy skin. The head is large in proportion to the size of the animal, and flattened above.

This Terrapin lives in the salt-water marshes, where it is very plentiful, and from which it never travels to any great distance. During the warm months of the year it is lively, and constantly searching after prey, but when the cold weather comes on, it burrows a hole in the muddy banks of its native marsh, and there lies buried until the warm sunbeams of spring break its slumbers, and induce it once more to seek the upper earth and resume its former active existence.

QUAKER TORTOISE.—*Emys olivacea.*

It is more active in its movements than is the case with the Tortoises in general, and can not only swim rapidly, but walk with tolerable speed. It is very shy, and discovers approaching peril with a keenness of perception that could scarcely be expected from one of these shielded reptiles, whose dullness and torpidity have long been proverbial.

Mr. Holbrook, in his valuable "North American Herpetology," writes as follows concerning this Terrapin:—

"They are very abundant in the salt marshes around Charleston, and are easily taken when the female is about to deposit her eggs in the spring and early summer months. They are then brought in immense numbers to market; yet, notwithstanding this great destruction, they are so prolific that their number appears undiminished. Their flesh is excellent at all times, but in the northern cities it is most esteemed when the animal has been dug out of the mud in its state of hibernation. The males are smaller than the females, and have the concentric striæ more deeply impressed."

The color of this Salt-water Terrapin is rather variable, but is usually dark greenish-brown on the upper surface, and yellow on the plates which surround the edge of the shell. Below it is yellow, and in many specimens it is marked with variously shaped spots of dark gray. The lower jaw is furnished with a hook, and the sides of the head are dusty white sprinkled with many small black spots.

THE BOX TORTOISE.

VERY many species of Tortoise are extremely variable in their color, but there are few which are so remarkable in this respect as the creature which is appropriately named the Box TORTOISE (*Cistudo carolina*).

This species belongs to America, and is found spread over the whole of the Northern States. It is very plentiful in the localities which it favors, and although so small a creature, is able by means of its wonderful organization to protect itself against almost every foe. Many of the Tortoises can withdraw their limbs and head into their shell, leaving open, however, the apertures through which this movement is achieved, so that the animal might be killed or hooked out by a persevering foe, such as the jaguar, which is known to attack turtles, insinuate its lithe paw within the shell, and scoop out the inhabitant with its sharp curved claws.

But in those instances where the animal has the power of closing the openings through which the legs, tail, and head protrude, there is hardly any mode of getting at the flesh with-

BOX TORTOISE.—*Cistudo carolina.*

out breaking the shell, a feat beyond the power of any animal, except perhaps an elephant, to perform. Certain birds, it is said, are clever enough to soar to a great height with the Tortoise, and break the shell by letting it fall upon a convenient rock, but this story does not seem to be very strongly attested. Several species possess this valuable capability, but none to so perfect a degree as the Box Tortoise, which, according to the Rev. Sydney Smith's felicitous summary, need fear no enemy except man and the boa constrictor, the former taking him home and roasting him, and the latter swallowing him entire and consuming him slowly in its interior, as the Court of Chancery does a large estate.

With regard to this curious propensity, it is evident that there is some analogy between these Tortoises and certain mammalia, which are also able to withdraw themselves within the protection of certain armor with which they are furnished. In the case of the hedgehog, the animal assumes more of an offensive than a defensive character, and relies, not on an impenetrable covering, for the skin is soft, and a pointed weapon can find an easy entrance between the spines, but on the bristly array of bayonet-like spikes that protrude their threatening points in every direction, and bid a tacit defiance to the foe.

The scale-covered manis, again, although guarded with successive layers of broad, horny plates, is, in point of fact, less protected when rolled up than when walking quietly along; for when at rest, the scales overlap each other like the tiles of a house, so that any weapon would glance aside, but when curled up the scales are erected and leave a passage for the arrow or the spear between them.

The real defence of the hedgehog lies in the points of its quills and of the manis in the razor-like edges of its scales, but the defence of the Tortoise is wholly inaggressive, and is more allied to that of the armadillo or perhaps the singular pichiciago (*Chlamydophorus truncatus*), a most remarkable little creature with a curious shelly covering spread over nearly the whole upper surface and down the hind-quarters. A description of this animal may be found in the volume on the Mammalia, page 631. There are again many of the lower animals which have a similar mode of defence, a very familiar example being the well-known pill-woodlouse so common in our gardens, which rolls itself into a round ball when alarmed, and permits itself to be handled and even rolled along the ground without displaying any signs of life.

The Box Tortoise is a terrestrial species, and always keeps to the dry forest-lands, detesting the vicinity of water. It is commonly found in the pine forests, because they are always on thoroughly dry soil, and on account of its fondness for such localities is sometimes known by the popular name of the Pine Terrapin. The negroes call it by the name of Cooter. In the wild state it mostly feeds on insects, and is peculiarly fond of the cricket tribe, but in captivity it will eat almost any food that is offered, taking insects, meat, apples, or even bread.

It is a very little creature, being when adult a very little more than six inches in length. In color it is extremely variable, but is generally yellowish-brown, striped with a brighter hue, and sometimes mottled with black. Of a number of specimens no two were exactly alike, some being yellow, spotted with black, while others exactly reversed these tints, and were black, spotted with yellow. Others again were yellow with black rays, and others olive with yellow rays and streaks. The carapace has a very slight keel along its upper edge.

The upper jaw of this species is furnished with a rather broad hook, and the lower jaw is also hooked, but not so boldly.

This is an interesting species from having its shell so adapted by a hinged cover in front that it shuts itself tightly within. What complete protection is here afforded from any ordinary foe! Tortoises are mostly notable for longevity, and this species seems to be especially favored. We remember to have captured one of them while in a woodland of Worcester County, in Massachusetts, and found the initials of a relative cut on its back. They were recognized as having been cut there thirty or more years previously. This species, from being an inhabitant of dry woods, is more likely than those of ponds and wet places to be found and captured; hence the more frequent selection of this Turtle for such carving purposes.

Other species, found in various parts of the world, seem to have the same curious box-like shell.

The Box Turtle inhabits the United States from Maine and New York to Missouri and southward. A variety called the Three-toed Box Turtle (*C. triunquis*), found in Pennsylvania and southward, is paler in color, and has the hind-feet mostly three-toed. It is called Pine Barren Terrapin, or Cooter, in the South.

THE GOPHER (*Testudo carolina*). This is the common Land Tortoise of the Southern States. It is not known farther north than North Carolina, where among the pine barrens it abounds in great numbers, living entirely on vegetables. The flesh is esteemed a great delicacy. The length of the species is about fourteen inches.

About twenty species of Land Tortoises are known to science, inhabiting both hemispheres. They are all herbivorous, confined to the land, and inhabit the warmer portions of their respective localities. Their special characteristic is the habit of burrowing.

The Pond Turtles, family *Emyidæ*, are represented over the whole world, widely distributed, by about eighty species.

THE WOOD TORTOISE (*Chelopus insculpta*) inhabits the States east of the Ohio, in fields and woods. Its shell is keeled, its plates marked with concentric *striæ*, and radiating lines. A black spot on each scale gives characteristic marking.

MUHLENBERG'S TORTOISE (*C. muhlenbergii*) is the most circumscribed in its *habitat*, being found only in Pennsylvania and New Jersey. It is rare in those places. Its length is about three and a half inches. It inhabits small brooks and streams of running water.

THE SPECKLED TORTOISE (*Chelopus guttatus*) is found in Eastern United States, and as far west as Northern Indiana, where it is abundant. Its main color is black with orange spots. The plastron is yellow, blotched with black. This Tortoise is a favorite pet with the small boy of New England.

BLANDING'S TORTOISE (*Emys meleagris*) inhabits the moist woods and fields in Wisconsin, and eastward to the Alleghanies. Its coloration and markings are somewhat like those of the latter. The shell has no keel.

THE PAINTED TORTOISE (*Chrysemys picta*) is familiarly known in some quarters as the Mud Turtle. It is one of the most common in the Eastern States. It is greenish-black, the plates having a paler margin. The marginal plates are marked with bright red, looking much as if it were freshly painted. The plastron is yellow, blotched with brown. Its length of shell is about six inches. Two varieties are found respectively in Wisconsin and in Western New York.

THE MAP TURTLE (*Malacoclemmys geographicus*) is singular in its markings, suggesting the lines on a map, hence the name. Its locality is the Mississippi River, and northward to New York.

LESUEUR'S MAP TURTLE (*M. lesueri*) is yet another species, found in Wisconsin and Ohio, and from thence southwest. It is much like the preceding, but grayer; the markings are paler and in larger pattern.

THE SALT MARSH TURTLE (*M. palustris*), called also Diamond-back, is of a greenish or dark olive color, with concentric dark stripes along the plates of both shells. It inhabits along the coast from New York to Texas. It is found along the northern shores of Long Island, where it is called Salt Water Terrapin, and is the justly prized and notable luxury of epicures. It frequents low brackish or salt streams near the sea-shore, hibernating in the mud, during the season, from whence it is taken in great numbers for the markets; it is then very fat.

THE SMOOTH TERRAPIN (*Pseudemys terrapin*) is sold in the markets as the same as the preceding. It is said that the two are procured from the same localities. DeKay thinks that the latter inhabits the salt and brackish waters indifferently.

RED-BELLIED TERRAPIN (*Pseudemys rugosa*). This is found in the Middle States, from New Jersey to Virginia. It is a handsome creature—for a reptile—and is easily distinguished by its serrated jaws. As an edible it is somewhat prized. Its length is eleven inches.

HIEROGLYPHIC TURTLE (*P. hyeroglyphica*). Found quite circumscribed in the Eastern States. The shell is smooth, flat, and olive-brown in color, with broad, reticulated, yellowish lines; the lower is pale yellow; the head very small; length of shell, twelve inches.

YELLOW-BELLIED TERRAPIN (*P. troostii*). This species is found in the valley of the Mississippi, and northward to Illinois. Its colors are greenish-black, the side-plates having horn-colored lines and spots; the under shell yellow, with large black blotches; the throat striped green; shell eight inches in length.

THE ELEGANT TERRAPIN (*P. elegans*) is a Western species, being confined to the region east of the Rocky Mountains as far as Illinois. The colors are brown with heavy lines and blotches; a blood-colored band on each side of the neck. The under shell is yellow, with a dusky blotch on each plate.

THE ROUGH TERRAPIN (*P. scabra*) is found along the shore from Virginia to Florida. It is dark brown with yellow stripes; under shell yellow, with small black blotches in front. Length of shell, eight inches.

A number of other species are enumerated as North American, and recorded in the list of North American Reptiles at the close of this volume.

THE SMALL MUD TURTLE (*Cinosternum pennsylvanicum*) is found abundantly in New York, and southward to Florida. The family and generic names indicate the fact that it has a movable sternum. The shell is dusky-brown; the head and neck with light stripes and yellow dots. In some localities it is called Small Box Turtle. It abounds in muddy ponds and pools, living on fish. Length of shell, three and a half inches.

THE MUSK TURTLE (*Aromochelys odorata*). Abundant in eastern United States, and westward to Indiana. Its exceedingly potent musky odor quite distinguishes it. Shell, three and a half inches in length.

THE LITTLE MUSK TURTLE (*A. carinata*) is found in the Mississippi region.

THE common MUD TORTOISE, so called from its mud-haunting propensities, is an example of rather a curious genus of Tortoises, inhabiting America.

It is an odd little creature, being when adult not quite four inches in length, and moving with moderate speed. It is mostly found in ponds and muddy pools, where it feeds upon fish, aquatic insects, and similar diet, catching even the active fish without much difficulty. I lately saw some aquatic Tortoises, which I think belonged to this genus, which had to be ejected from a large basin of a fountain because they killed the newts which inhabited the same locality.

MUD TORTOISE.—*Cinosternum pennsylvanicum.*

Their movements in the water were so deliberate that it was not until they were detected in the very act of biting the newts that their delinquencies were discovered. Their mode of attack was simply to creep under their victim as it balanced itself in the water or swam gently within reach, and then to secure it with a quiet snap of its beak.

Like the lettered Terrapin, already mentioned, it has a vexatious habit of taking the angler's bait, and causes many a fisherman to lose his temper when pulling up a useless little Mud Tortoise instead of the fish on which he had set his heart. It seizes the worm just as it catches the newts, taking it so quietly into its mouth that the float is hardly shaken by the touch. But when the fisherman pulls his line, the Tortoise kicks, pulls and flounces about in so energetic a style that it often deludes the angler into the idea that he has hooked quite a fine fish.

This species has a decided smell of musk, a peculiarity which is found in others of the same genus, one of which (*Cinosternum odoratum*) goes by the appropriate, though not very refined, name of Stink-pot, in consequence of the powerful musky odor which it exudes.

The color of the Mud Tortoise is mostly dusky-brown above, and chestnut below, though this coloring is liable to some variation in different individuals. The tail is thick and pointed, and horny at the tip. The head is large, and there are four large warty appendages on the chin.

THE last example of the Terrapin is that singular animal which is appropriately called the ALLIGATOR TERRAPIN (*Chelydra serpentina*), from the great resemblance which it bears to that reptile.

It is also an American species, and lives mostly in the water. When adult it reaches a large size, often exceeding three feet in length, and as it is very fierce of disposition, lithe of neck, and strong of jaw, it is somewhat dreaded by those who have had a practical acquaintance with its powers. The jaws of this animal are sharp edged, and remarkably strong, cutting like the blades of steel shears. Mr. Bell remarks that he has seen one of these creatures bite asunder a stick of half an inch in diameter. When caught, therefore, the captors always cut off these dangerous heads at once.

Mr. Holbrook gives the following interesting account of the Alligator Terrapin and its habits:—

"It is found in stagnant pools or in streams where the waters are of sluggish motion. Generally they prefer deep water, and live at the bottom of rivers; at times, however, they approach the surface, above which they elevate the tip of their pointed snout, all other parts being concealed, and in this way they float slowly along with the current, but if disturbed, they descend speedily to the bottom.

"They are extremely voracious, feeding on fish, reptiles, or any animal substance that falls in their way. They take the hook readily, whatever may be the bait, though most attracted by pieces of fish; in this way many are caught for market. It is, however, necessary to have strong hooks and tackle, otherwise they would be broken, for the animal puts forth great strength in its struggles to escape, both with its firm jaws and by bringing its anterior extremities across the line. When caught, they always give out an odor of musk, more or less distinct; sometimes in very old animals it is so strong as to be disagreeable.

"Occasionally he leaves the water, and is seen on the banks of rivers or in meadows, even at a distance from his accustomed element. On land, his motions are awkward; he walks slowly, with his head, neck, and long tail extended, elevating himself on his legs like the alligator, which at that time he greatly resembles in his motions. Like the alligator also, after having walked a short distance, he falls on his sternum to rest for a few moments, and then proceeds on his journey.

"In captivity they prefer dark places, and are exceedingly ferocious; they will seize upon and bite severely anything that is offered them, and their grasp upon the object with their strong jaws is so tenacious, that they may even be raised from the ground without loosing their hold.

"In many cities they are brought in numbers to market, and are esteemed excellent food, though I think that they are far inferior to the green turtle, the soft-shelled, or even several of the emydes. They are kept for months in tubs of fresh water, and feed on such offal as may be given them, though they never become fat or increase much in weight."

Though a very valuable and curious reptile, the Alligator Terrapin is far from beautiful, with its little dusky shell, its long, knob-covered tail, its singular legs and feet, and its great, sharply-toothed jaws. On account of its habit of snapping fiercely at its opponents, it is often called by the name of Snapping Turtle, a title, however, which rightly belongs to a species which will shortly be described.

Its head is large, and covered with a hard, wrinkled skin; the neck is long, thick, and furnished with a number of projecting tubercles. Under the chin are two distinct barbels. When adult, the shell is so formed that a depression runs along the centre, leaving a kind of keel at each side of the central line; but when young, the shell forms three distinct keels. It is rather flat, oblong, and at the hinder portion is deeply cleft, so as to form a row of blunt teeth, but while young the teeth are sharp. The tail is stout, long, and is furnished with a series of large, blunt tubercles along its central line.

THE SNAPPING TURTLE is the familiar name of this species in the countryside of New England. It is rather common in all parts of North America, and is found southward as far as Ecuador. Dr. Pickering, of Eastern Massachusetts, records the length of one as "over four feet;" the shell being only about half that in length. This exposure of so large a proportion of its fleshy parts is scarcely paralleled in any other species. Its stout and long tail, and its long neck quite warrant the use of a Southern designation it has—Alligator Tortoise. The savage, snapping habit gives it the common Northern name. The term Alligator Terrapin seems very appropriate, as the Terrapins all have the corrugated backs. In some quarters the flesh is esteemed.

MATAMATA.—*Chelys matamata.*

WE now arrive at another family of Tortoises, termed Chelydes, an example of which is the remarkable MATAMATA, the acknowledge type of its family.

All the Chelydes have broad, flattened heads, long, broad, contractile necks, and when in repose have a curious custom of bending their necks under the side of the carapace. Their feet are webbed, in order to enable them to pass rapidly through the water, and there is always a lobe between the claws. They are aquatic Tortoises, carnivorous, and voracious, and only feed while in the water. When swimming, the whole of the shell is kept below the surface.

The Matamata is certainly the most remarkable of aspect among all the Tortoises, and perhaps may lay claim to be considered one of the oddest-looking animals in the world, far exceeding in its grotesque ungainliness even the wild and weird creations of the middle-age painters.

This Tortoise inhabits Southern America, and is most plentiful in Cayenne. Formerly it was very common, but on account of the excellence of its flesh, it has been subject to such persecution that its numbers have been considerably diminished. It haunts the lakes and rivers, where it swims well and with some speed. As is the case with most aquatic Tortoises, it is carnivorous, and feeds on fish, reptiles, and other creatures, which it captures by a sudden snap of its sharp beak. In general, it appears not to care for chasing the intended prey, but conceals itself among the reeds and herbage of the river-side, and from its hiding-place thrusts

NEW HOLLAND CHELODINE, OR SNAKE TORTOISE.—*Hydromedusa maximiliani.*

out its neck suddenly upon its victims as they pass unsuspectingly within reach of their destroyer. On occasion, however, it will issue from its concealment, dart rapidly through the water and seize a fish, reptile, or even a water-fowl, and then retire with its prey to its former hiding-place.

It is a large and formidable creature, attaining, when adult, to a length of three feet.

The head of the Matamata is most singular in shape, and remarkable for the strange appendages which are placed upon it. The head itself is much flattened, and rather broad, and the snout is prolonged in a most extraordinary manner, so as to form an elongated and flexible double tube.

On the top of the head are two membranous prolongations of the skin, standing boldly from the head, and having much the appearance of ears. From the chin hang two curiously-fringed membranes, and the throat is decorated with four similar membranes, but of larger size and more deeply fringed. The neck is long, and bears upon its upper surface two rows of small, membranous tufts, deeply fringed, and greatly resembling, in every point but that of size, the tufts on the chin and throat. The limbs are powerful, and the tail is short.

The shell of the Matamata is rather convex, broader before than behind, and rather flattened in the middle of the back. The shields are elevated, rather sharp at their tips, and are arranged so as to form three regular keels along the back.

A NEARLY allied species of river Tortoise is figured on page 18. It is the NEW HOLLAND CHELODINE, sometimes called the YELLOW CHELODINE, from the olive-yellow color of the plastron.

This remarkable reptile may almost deserve the name of the Snake Tortoise, its long, flexible neck, and flat, narrow, and pointed head, having a very serpentine aspect. As its name imports, it is an inhabitant of Australia, and is found most commonly in New Holland. It is a water-loving creature, not caring much for rivers and running streams, but haunting the pools, marshes, and stagnant waters, where it lives in the midst of abundance, finding ample food among the fishes and aquatic reptiles which generally swarm in such localities. It is an active animal, traversing the water with considerable speed, and capturing its prey by means of its sharp jaws.

The gape is very large, and the jaws are comparatively slender. The shell is broad, rather flattened, and the shields are thin and smooth, not being elevated as in the preceding species. The general color of the shell is brown above and yellow below, each shield having a black line round its edge.

WE now arrive at another family of the Tortoises, known popularly as Soft Turtles—a rather inaccurate title, inasmuch as they are not turtles, but Tortoises—and scientifically as *Trionycidæ*. The latter title is of Greek origin, signifying three-clawed, in allusion to the fact that, although the species belonging to the family have five toes on each foot, only the three inner toes of each foot are armed with claws.

These Tortoises, represented in the accompanying oleograph, are rather interesting to the careful observer, because the peculiar structure of the external covering permits the formation of the skeleton to be seen without the necessity for separating the shells. In particular, the method in which the breast-bone is developed into the broad, flattened plate which forms the plastron, can clearly be seen through the skin, and even the position of the sutures can be made out without much difficulty.

The head of these creatures is rather oval and flattened, the jaws are horny, but covered with hanging, fleshy lips, and the mouth is lengthened into a cylindrical trunk. The neck is long, and can be contracted, the feet are short, very wide, and the toes are connected together by strong webs. They all live in warm climates, and are found in rivers and lakes.

The typical species is the celebrated FIERCE TRIONYX, or SNAPPING TURTLE, a reptile which derives its former title from the exceeding ferocity of its disposition, and the latter from the method in which it secures its prey or attacks its foes. It is found spread over many parts of North America.

This fierce and determined marauder of the waters is even more formidable than the two previous species, and not only causes terror among the smaller creatures which inhabit the same localities, but is even dreaded by man, whose limbs have often been severely wounded by the bite of these ferocious reptiles. Like the aquatic Tortoises, it is carnivorous in its habits, and is terribly destructive among the fish, smaller quadrupeds, birds, and reptiles. Lurking on the banks, it snatches away many an unfortunate animal as it comes to drink, or seizes the water-fowl that have ventured too close to their terrible neighbor. So fiercely carnivorous is this Tortoise, and so voracious is its appetite, that it will even catch young alligators, and devour them in spite of their teeth and struggles.

The flesh of this species is very delicate, tender, and richly flavored, so that it often meets the doom which it has inflicted on so many other animals. As it is so voracious, it will take almost any kind of bait, provided that it be composed of animal substance, but it prefers fish, and cannot resist a hook so baited.

Its captor's work, however, is not confined to hooking and drawing it ashore, as the Snapping Turtle, when it finds itself with a hook firmly fixed in its jaws, and itself being irresistibly dragged from the water, seems possessed with tenfold ferocity, writhing its long, flexible neck, darting its head furiously at its foes with the rapidity of a serpent's stroke, and snapping sharply with its formidable jaws, one bite of which would shred away the fingers from the hand, or the toes from the feet, as easily as the gardener's scissors sever the twigs and leaves. Such a misfortune has indeed been known to occur. Mr. Bell records an instance where a Snapping Turtle, that was being conveyed to England, contrived to reach the hand of one of the sailors in its fierce struggles, and bit off one of his fingers.

The eggs of the Snapping Turtle are very spherical in form, and brittle of substance. The female lays a large number of these eggs, from fifty to sixty being the usual average, and always deposits them in some dry situation. In order to find a suitable spot for the deposition of her eggs, the female leaves the water, and is often forced to traverse a considerable distance before she can find a spot sufficiently dry for her purpose. Sometimes she will even ascend a very steep acclivity in her anxiety to find a locality that is quite dry, covered with sandy soil, and exposed to the full rays of the sun. She begins her task about May, and the little Tortoises are hatched in July.

The following curious account of the tenacity of life possessed by these creatures has been kindly forwarded to me :—

"As regards the tenacity of life of the Snapping Turtle, and the sympathy (*rapport*) which seems to exist between its severed limbs and main trunk, for some time after the separation has taken place, I witnessed a very curious incident when staying at a farm in Massachusetts.

"When I had brought the animal home, suspended by its tail, I killed it by chopping its head off, yet the head would open and shut the mouth, and roll its eyes. When I held a stick between the open jaws it closed them with violence, and kept hold of it. Meanwhile the headless body was crawling on the ground.

"About a quarter of an hour after having severed the head from the body, my mother had got boiling water, which I threw over the body, placed in a tub, in order to make the horny matter separate from the flesh ; the moment this was done the back heaved and the sides were puffed out as if wind were blown between the skin and flesh, and instantaneously the head, which lay about three or four feet from the tub, on the ground, opened its mouth with a slight hissing sound, let go its hold on the stick, and the part of the neck adhering to the head expanded, as if also wind was blown into it, and both body and head lay motionless and dead. After having taken out thirty-four eggs, I took out the heart, which, strange to say, was still throbbing with life, contracting and expanding. I put it upon a plate, where it kept on beating until about noon the following day."

In this species, the front edge of the carapace is furnished with a great number of tooth-like points, all radiating from the shell. These teeth, or tubercles, distinguish it from two other American species, appropriately termed the Unarmed Trionyx (*Triónyx múticus*), and the Mississippi Snapper (*Macrochelys lacertina*). This species is common in the Gulf States, and as far north as Illinois. It is regarded as one of the strongest and most ferocious of reptiles.

Holbrook records a Turtle under the name of Temminck's Snapper (*Chelonura temmincki*).

BEFORE taking leave of the Soft Turtles, we must cast a casual glance at two rather curious species. The one is the TYRSE (*Tyrse*, or *Triónyx nilóticus*), a native of Africa, as its name imports. This animal is found in the Nile, and other A... rivers, and is a good representation of the American reptile, being very fierce, st... ...racious, and said to devour the young crocodiles, just as the snapping turtle ... ... alligators. The shell of the Tyrse is rather convex, but often is flattened : ... ... line of the vertebræ, and its back is olive-green spotted with yellow or white.

The other species is the DOGANIA (*Dogania subplanus*, or *Triónyx subplanus*). This curious-looking reptile is an Asiatic species, and is found in India. Its neck seems preternaturally long, and supports a very large head, broad behind, and produced into a conical muzzle in front. The shell is rather oval, much flattened, and quite conceals the conical tail. Its color is brown, mottled largely with yellow ; the head is also yellow

SELMAR HESS, PUBLISHER, N.Y.

and brown. The ribs are not fully united together until the animal has attained a rather advanced age.

We now arrive at the TURTLES, a group that can be distinguished by many unmistakable marks. Their feet are very long, those of the fore-limbs being longest, flat, expanded at the end, and often furnished with flattened claws. In fact, the feet are modified into fins or paddles, in order to suit the habits of these reptiles, which only feel themselves at home in the water, and are often met at sea some hundreds of miles from the nearest land. The ribs of the Turtles, instead of being united throughout their length, as in the tortoises, are only wide, flat, and united for part of their length, the remaining portions being free, and radiating like the spokes of a wheel.

DOGANIA.—*Dogania subplanus.*

These reptiles inhabit the seas of the torrid and the temperate zones, and their food is mostly of a vegetable nature, consisting of various seaweeds, but there are a few species which are animal feeders, and eat creatures such as mollusks, star-fish, and other marine inhabitants. Several species are remarkably excellent for food, and caught in great numbers for the table, while others are equally useful in supplying the beautiful translucent substance known by the name of tortoise-shell. Their head is rather globular, and their jaws are naked and horny, and are capable of inflicting a severe wound.

The first example of the true Turtles is the LUTH, or LEATHERY TURTLE (*Dermatochelys coriacea*), so called from the soft leather-like substance with which its shell is covered.

This species is found in the Atlantic, Pacific, and Indian Oceans, where it grows to a very large size, often weighing more than sixteen hundred pounds, and measuring eight feet in length. Being a very good swimmer, owing to the great development of the limbs, especially the fore-legs, it ventures far out to sea, and is occasionally driven to strange countries. Specimens of this reptile have been taken on the coast of France, and on other shores. These individuals were rather large, weighing about seven or eight hundred pounds.

The Leathery Turtle feeds on fish, crustacea, mollusks, radiates, and other animals, and its flesh seems to be hurtful, causing many symptoms of poisoning in those who eat it.

This species is remarkable for having no horny plates, the bones of the carapace and plastron being covered with a strong leathery skin, smooth in the adult animal, but covered with tubercles in the young. Along the back run seven ridges, sharp, and slightly toothed in the full-grown Turtle, but bluntly tubercled in the young. The eye is very curious, as the lids are set vertically instead of horizontally, and when the creature opens and shuts its eyes, have a very singular effect. The jaws are very formidable, being sharply edged, deeply scooped with three rounded notches in the front of the upper jaw, so as to form two curved sharply pointed teeth, and the extremity of the lower jaw is strongly hooked.

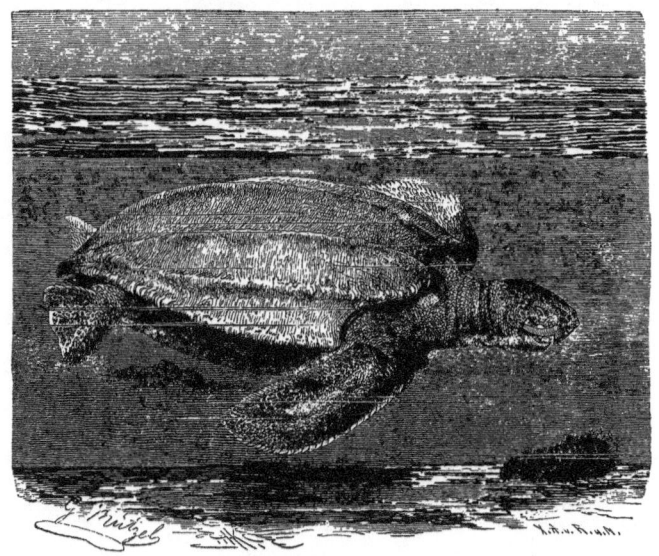

LUTH, OR LEATHERY TURTLE.—*Dermatochelys coriacea.*

The legs of the Leathery Turtle are very long, especially the two fore-limbs, which, in a specimen measuring eight feet in total length, were nearly three feet long, and more than nine inches wide. The feet are not furnished with claws, but the toes have a little horny scale at their tips, which take the place of the claws. The general color of this animal is dark brown, with pale yellow spots, but sometimes the skin is irregularly pied with black and white.

This great creature is essentially a sea-going one, though perhaps not more so than the Hawk's-bill, Green, and Loggerhead species. Its very large flippers rather suggest the above statement.

The editor of this edition has taken the liberty to drop from the original text the statement that this Tortoise resorts to the Tortugas Islands for breeding purposes. This statement has no foundation in fact. The great Loggerheads and the Green Turtles do resort to that group of keys, and breed in considerable numbers, a notice of which will be seen in the text on those species. The breeding-places of the Leathery Tortoise are not known to science.

Our first acquaintance with this creature was during the summer of 1855, when a middle-sized one came ashore on Nahant Beach, near Boston, Mass. A bullet-hole in the neck explained its present condition. Until this specimen came ashore this species was regarded as nearly unknown on the Atlantic coast. The only specimen then known to have been seen and captured near the American Atlantic shores, was an enormous one now preserved in the Boston Museum. It was captured off the mouth of Boston harbor. Its length was eight feet and one inch. In color, jet-black. This was purchased at a large price, and it continues to occupy the same place as one of the most notable of the curiosities of the Museum. For many years this species remained unknown on our shores, excepting as represented by these two specimens. Somewhat later examples came to be more numerous. At the present time it is not an unusual thing to see a Leathery Tortoise when, in summer, cruising some distance off shore.

We saw several in the waters of the Gulf of Mexico, while resident on the Florida Reef. Those we have examined were of a dense black color and rather shiny, like the skin of a porpoise. This is probably the most bulky of living Turtles. The enormous fossil Turtle which was found in one of the Western Territories is allied to the present species.

ANOTHER well-known species of Turtle deserves a passing notice. This is the LOGGERHEAD TURTLE, or CAOUANE (*Thalassochelys caretta*), sometimes called the RHINOCEROS TURTLE.

This fine species has a wide range of locality, being found in the most warm seas. It is extremely powerful, fierce, and voracious, biting with great force, and cutting hard substances without much difficulty. According to Catesby, "the Loggerhead Turtles are the boldest and most voracious of all other Turtles. Their flesh is rank and little sought for, which occasions them to be more numerous than any other kind. They range the ocean over, an instance of which, among many others that I have known, happened in latitude 30° north, when our boat was hoisted out, and a Loggerhead Turtle struck, as it was sleeping on the surface of the water.

"This, by our reckoning, appeared to be midway between the Azores and the Bahama Islands, either of which places being the nearest land it could have come from, or that they are known to frequent, there being none on the north continent of America farther north than Florida. It being amphibious, and yet at so great a distance from land in the breeding-time, makes it the more remarkable. They feed mostly on shell-fish, the great strength of their beaks enabling them to break very large shells." Several other species belong to the same genus.

In general appearance this species is not unlike the common Green Turtle, which will presently be described, but the shell is broader, deeper colored, and has two more plates on the back. The plates along the upper part of the back are six-sided, rather square, and keeled. There are two claws on each foot.

THE LOGGERHEAD TURTLE is so abundant in the waters about the Dry Tortugas, on the Florida Reef, that one of the principal islands or keys is named from it. This creature attains a large size; some measuring quite five feet in length. It is so named from the great comparative size of its head. Considerable difference is seen between this and the Green Turtle in this respect.

The Loggerheads make their appearance in the shoal waters near the keys in early summer. On the first moonlight night they are ready to go on shore to deposit their eggs. On these occasions people living near, mostly wreckers and fishermen, resort to the region and watch for the creatures, to secure their eggs first, and then the carcases. Before leaving the water the reptiles are exceedingly shy and cautious, but once fairly at work digging holes above the high-water mark, they heed nothing until the eggs are all deposited. So intent are they on this business it is a common practice to sit on the creatures' shell and take the eggs as they are deposited. The patient reptiles then carefully draw the sand over the empty hole with as much care as if the complement of eggs was yet there. This accomplished, they hurry off to sea again.

THE well-known CARET, or HAWK'S-BILL TURTLE (*Eretmochelys*), so called from the formation of the mouth, is a native of the warm American and Indian seas, and is common in many of the islands of those oceans.

The Hawk's-bill Turtle is the animal which furnishes the valuable "tortoise-shell" of commerce, and is therefore a creature of great importance. The scales of the back are thirteen in number, and as they overlap each other for about one-third of their length, they are larger than in any other species where the edges only meet. In this species, too, the scales are thicker, stronger, and more beautifully clouded than in any other Turtle. The removal of the plates is a very cruel process, the poor reptiles being exposed to a strong heat which causes the plates to come easily off the back. In many cases the natives are very rough in their mode of conducting this process, and get the plates away by lighting a fire on the back of the animal. This mode of management, however, is injurious to the quality of the tortoise-shell. After the plates have been removed, the Turtle is permitted to go free, as its flesh is not eaten, and after a time it is furnished with a second set of plates. These, however, are of inferior quality, and not so thick as the first set.

When first removed, they are rather crumpled, dirty, opaque, brittle, and quite useless for the purposes of manufacture, and have to undergo certain processes in order that these defects may be corrected. Boiling water and steam are the two principal agents in this part of the manufacture, the plates being boiled and steamed until they are soft and clean, and then pressed between wooden blocks until they are flat. The tortoise-shell possesses the valuable property of uniting together perfectly, if two pieces are thoroughly softened, heated, and then subjected to the action of a powerful press. By this mode of treatment, the tortoise-shell can be formed into pieces of any size or thickness, and can even be forced into moulds, retaining, when cold, a perfect impression of the mould. Even the chippings and scrapings of this valuable substance are collected, and being heated and pressed, are formed into solid cakes fit for the purposes of manufacture.

The uses to which this costly and beautiful substance are put, are innumerable. The most familiar form in which the tortoise-shell is presented to us is the comb, but it is also employed for knife-handles, boxes, and many other articles of ornament or use.

This species is not nearly so large as the green Turtle, and its flesh is not used for food. The eggs, however, are thought to be a great delicacy. It is remarkable that when these eggs are boiled, the albumen, or "white" as it is popularly called, does not become firm. The external membrane is white, flexible, and the eggs are nearly spherical in their form. Their number is very great, and the animal usually lays them in sets at intervals of about three weeks.

The young are generally hatched in about three weeks after the eggs are laid in the sand, the hot rays of the sun being the only means by which they obtain their development. When first excluded from the shell, the young Turtles are very small and soft, not obtaining their hard scaly covering until they have reached a more advanced age. Numberless animals, fish, and birds feed on these little helpless creatures, and multitudes of them are snapped up before they have breathed for more than a few minutes. The rudiments of the scales are perceptible upon the backs of these little creatures, but the only hard portion is the little spot in the centre of each plate, which is technically called the areola, the layers of tortoise-shell being added by degrees from the edges of the plates.

Many birds are always hovering about the islands where Turtles lay their eggs, and as soon as the little things make their appearance from the sand and hurry instinctively towards the sea, they are seized by the many foes that are watching for their prey. Even when they reach the water, their perils are not at an end, for there are marine as well as aërial and terrestrial foes, and as many fall victims on the water as on land. So terrible is the destruction among these reptiles in their early days of life, that were it not for the great number of eggs laid, they would soon be extirpated from the earth.

The shell of the Hawk's-bill Turtle is rather flat, and heart-shaped. When young, the centre of each plate is rather pointed, but in the adult animal the points are worn away and never restored. The plates surrounding the edges of the shell are arranged so as to form strong teeth pointing towards the tail. In the younger specimens, there are two keels running the length of the plastron, but in the older individuals these are worn away like the projections on the back. The jaws are strongly hooked at their tips, and the under jaw shuts within the upper. The tail is very short. The color of this species is yellow richly marbled with deep brown above. The under parts are yellowish-white, splashed with black on the areola in the half-grown and younger individuals, and the head is brown, the plates being often edged with yellow.

The Hawk's-bill Turtle is rather common around the Florida Reef, though large ones are rarely found. The young we have seen among the mangroves in the water-ways of the Everglades. They are highly esteemed as an edible.

THE best known of all the Turtles is the celebrated GREEN TURTLE (*Chelonia mydas*), so called from the green color of its fat.

This useful animal is found in the seas and on the shores of both continents, and is most plentiful about the Island of Ascension and the Antilles, where it is subject to incessant persecution for the sake of its flesh. The shell of this reptile is of very little use, and of small value, but the flesh is remarkably rich and well-flavored, and the green fat has long enjoyed a world-wide and fully deserved reputation.

In Europe the flesh of the Green Turtle is little but an object of luxury, attainable only at great cost, and dressed with sundry accompaniments that increase rather than diminish its natural richness. But in many instances, more especially on board ship, when the sailors have been forced to eat salt provisions until the system becomes deteriorated, and the fearful scourge of scurvy is impending over crew and officers, the Turtle becomes an absolute necessity, and is the means of saving many a noble vessel from destruction, by giving the crew a healthful change of diet, and purifying the blood from the baneful effects of a course of salted provisions.

Landsmen have little notion of the real texture and flavor of "salt junk," their ideas being generally confined to the delicately corned and pinky beef or pork that is served up to table, with the accompaniments of sundry fresh and well-dressed vegetables. Whereas, salt junk is something like rough mahogany in look and hardness, and salted to such a degree as almost to blister the tongue of a landsman. It may easily be imagined how any one who has been condemned to a course of this diet for a lengthened time would welcome fresh meat of any kind whatever, and we need not wonder at the extraordinary relish with which sailors will eat sharks, sea-birds, and various other strangely flavored creatures.

The flesh and fat of the Turtle are valuable in a medicinal point of view, and will supply in a more agreeable, though more costly manner, the various remedies for consumptive tendencies, decline, and similar diseases, of which cod-liver oil is the most familiar and one of the most nauseous examples.

Formerly, before steam power was applied to vessels, the Turtle was extremely scarce and very expensive, but it can now be obtained on much more reasonable terms. Many vessels are now in the habit of bringing over Turtles as part of their cargo, and it is found that these valuable reptiles are easily managed when on board, requiring hardly any attention. The following short account of some captive Turtles has been kindly presented to me by a partaker of their voyage and their flesh :—

"The Island of Ascension is a great resort of Turtle, which are there captured and retained prisoners in some large ponds, from which they are occasionally transferred to ships for 'rations' for the crew. These Turtles may be seen in the ponds, lazily moving along, one above another, sometimes three or four deep. They occasionally come to the surface to take breath, and will splash about at times quite merrily, as though ignorant that their destiny tended towards conversion into soup and cutlets. At the best, however, they are lethargic, awkward creatures.

"About half a dozen fine Turtle were conveyed on board our ship during my stay at the Island of Ascension; they were unwieldly monsters, measuring rather more than four feet six inches in length, and about three feet in breadth. They were allowed to lie either in the boats, or on the after-part of the poop, and seldom disturbed themselves unless the vessel gave an extra roll, or they were stirred up by a pail of water being thrown over them or a wet swab rubbed over their hooked beaks.

"Their tenacity of life was remarkable ; they remained on board ship during upwards of three weeks without any food, and their only refresher was a cold bath, derived from the before-mentioned pail of water, which they usually received with a dreamy lengthy sort of hiss. Even after their three weeks' starvation, they died very hard. One, whose throat was cut in the morning, and from whose body numerous eggs had been extracted, was giving an occasional flap with her fins late in the afternoon ; the fact of her throat having been cut and her body otherwise mutilated appeared merely to produce the effect of ultimately damaging her constitution, and I have grave doubts whether the fact of her ceasing to move was not as much due to the destruction of the various membranes as to the extinction of her reptilian life."

As these animals are large and very powerful, it is not a very easy task to secure and bring them on board. The usual plan is to intercept them as they are traversing the sands, and to turn them over on their backs, where they lie until they can be removed. Many of the tortoise tribe can recover their position when thus overturned, but the Green Turtle is quite unable to restore itself to its proper attitude, and lies helplessly sprawling until it is lifted into the boat and taken on board. In many cases the creature is so enormously heavy that the united strength of the pursuers is inadequate to the task, and they are consequently forced to employ levers and so to tilt it over.

Sometimes the Turtle is fairly chased in the water and struck with a curious kind of harpoon, consisting of an iron head about ten inches in length, and a staff nearly twelve feet long. The head is only loosely slipped into a socket on the staff and the two are connected with a cord. Two men generally unite in this chase, one paddling the canoe and the other wielding the harpoon. They start towards the most likely spots, and look carefully at the bottom of the sea, where it is about six or ten feet in depth, to see whether the expected prey is lying at its ease and does not perceive them.

Sometimes they are forced to give chase to a Turtle on the surface, and sometimes the individual on which they had fixed, takes the alarm, and swims away. In either case they continually pursue the single swimming reptile, until it is fatigued with constant irritation, and sinks to the bottom to rest. No sooner has the Turtle assumed this position than the harpooner lowers his weapon into the water, takes an accurate aim, and then drives the steel spike deep into the shell. Off dashes the Turtle, carrying with it the harpoon. Were it not for the peculiar construction of the harpoon, the weapon would soon be shaken off, and the Turtle escape, but as the shaft slips readily off the head, there is no leverage and the steel head remains fixed, towing after it the long wooden shaft, which soon tires out the poor victim. When thoroughly fatigued, it is drawn to the surface, a rope put around it, and either taken into the boat or hauled ashore.

The food of this Turtle consists of vegetable substances, mostly algæ, which is found in great abundance in those warm climates. This animal grows to a very great size, as may be imagined from the fact that it often requires the united aid of three men to turn it over. A very pure limpid oil is obtained from these species, useful for burning in lamps and other similar purposes. A fat full-grown specimen will sometimes furnish thirty pints of this substance.

The eggs of the Turtle are thought as great delicacies as its flesh, and it is rather a remarkable fact, that although the flesh of the hawk's-bill Turtle is distasteful to all palates and hurtful to many constitutions, the eggs are both agreeable in flavor and perfectly harmless. It is while the female Turtle is visiting shore for the purpose of depositing her eggs that she is usually captured, as these sea-loving reptiles care little for the shore except for this purpose. So admirable an account of the manner in which the Turtle behaves when laying her eggs is written by Audubon, that the description must be given in his own words :—

"On nearing the shore, and mostly on fine, calm moonlight nights, the Turtle raises her head above the water, being still distant thirty or forty yards from the beach, looks around her, and attentively examines the objects on shore. Should she observe nothing likely to disturb her intended operations, she emits a loud, hissing sound, by which such of her enemies as are unaccustomed to it are startled, and apt to remove to another place, although unseen by her.

"Should she hear any more noise, or perceive any indication of danger, she instantly sinks and goes off to a distance; but should everything be quiet, she advances slowly towards the beach, crawls over it, her head raised to the full stretch of her neck, and when she has reached a place fitted for her purpose, she gazes all around in silence. Finding all well, she proceeds to form a hole in the sand, which she effects by removing it from under her body with her hind flappers, scooping it out with so much dexterity, that the sides seldom, if ever, fall in. The sand is raised alternately with each flapper as with a ladle, until it has accumulated behind her, when, supporting herself with her head and fore part on the ground, she, with a spring from each flapper, sends the sand around her, scattering it to the distance of several feet.

"In this manner the hole is dug to the depth of eighteen inches, or sometimes more than two feet. This labor I have seen performed in the short space of nine minutes. The eggs are then dropped one by one, and disposed in regular layers to the number of one hundred and fifty, or sometimes nearly two hundred. The whole time spent in this operation may be about twenty minutes. She now scrapes the loose sand back over the eggs, and so levels and smooths the surface, that few persons, on seeing the spot, would imagine that anything had been done to it. This accomplished to her mind, she retreats to the water with all possible despatch, leaving the hatching of the eggs to the heat of the sand.

"When a Turtle, a loggerhead for example, is in the act of dropping her eggs, she will not move, although one should go up to her, or even seat himself on her back; but the moment it is finished, off she starts, nor would it be possible for one, unless he were as strong as Hercules, to turn her over and secure her."

The Green Turtle is a staple article of commerce in Key West, on the Florida Reef. It is abundant on the waters in the Florida straits, and along the Gulf coasts. The fishermen in the latter regions practise the plugging method of capturing it. In Key West, the Turtles are placed in "crawls," an enclosed space in shallow water, which allows of free circulation of sea water. Steamers plying to New York take on all that are supplied, for the Northern markets. We have seen the young of the Green Turtle in considerable numbers, in the shallow inlets of the mangrove swamps on the southern extremity of Florida. These young are exceedingly good as edibles. The old ones feed in the same localities on the tender algæ, which renders them delicate and fat. The other great sea Turtles are carnivorous, and prove very indifferent as food, though the garrison at Fort Jefferson issued the meat as rations a portion of the year. Its novelty, and change from beef occasionally, made it a welcome article of the commissary.

The aspect of this species is quite in contrast with the loggerhead, the head of the Green Turtle being so much the smaller.

Occasionally, the Green Turtle has been led on, by the influence of the warm waters of the gulf stream, to venture off the entrance of New York harbor—a dangerous locality, one would say, for such a highly prized edible. Fine specimens were kept in the New York Aquarium, where they could be observed with ease. Their peculiar movements in swimming remind one of the flight of a bird.

Dr. Strobel informed Dr. Holbrook, the distinguished author on this subject, that Green Turtles taken at Tortugas Islands were marked by the wreckers, and kept in confinement at Key West, sixty miles distant. Some escaped, and were recaptured while laying eggs, on the same island in the Tortugas group as they were formerly taken from.

CROCODILES AND ALLIGATORS.

The link next to the tortoise tribe is formed of an important group of reptiles, containing the largest of the reptilian order, larger, indeed, than most present inhabitants of the earth, if we except one or two African and Indian animals, and some members of the cetaceous tribe. As is the case with nearly all reptiles, they are carnivorous, and owing to their great size, strength of muscle, voracity of appetite, and the terrible armature of sharp teeth with which their jaws are supplied, they are the dread of the countries which they inhabit, ruling the rivers with a sway as despotic as is exercised by the lion and tiger on land, the eagle in the air, or the shark in the seas.

On account of the peculiar manner in which their bodies are covered with square, keeled, bony plates embedded in the skin, and protecting the body with an armor that effectually guards its upper and more exposed portions from any ordinary weapon, they are separated from the true lizards, and scientifically termed Emydosauri, or Tortoise-lizards, the bony plates being considered to have a certain analogy with those of the shielded reptiles. By some zoological authors these animals are termed Loricata, or Mailed Reptiles, from the Latin word *lorica*, which signifies a coat of mail, or cuirass.

Although these creatures are capable of walking upon land, for which purpose they are furnished with four legs, they are more fitted for the water than its shores, and are swift and graceful in the one, as they are stiff, awkward, and clumsy on the other. Through the water they urge their course with extraordinary speed, their long, flattened, flexible tail answering the double purpose of an oar and a rudder; but on land their bodies are so heavy and their legs are so weak, that they can hardly be said to walk, a term which seems to imply that the body is wholly supported by the legs, but to push or drag themselves along the ground, on which rests a considerable portion of their weight.

The head of these creatures is always rather elongated, and in some species is lengthened into a narrow and prolonged snout. Each jaw is furnished with a row of sharply-pointed and rather conical teeth. These teeth are hollow, mostly grooved on the surface, and are replaced when they fall by new teeth that grow behind them, and in process of time push the old ones out of their sockets.

The nostrils are placed at the very extremity of the skull, and upon a slightly raised prominence, so that the animal is able to breathe by merely exposing an inch or so above the water, and thus can conceal itself from almost any foe, or make an unsuspected approach upon its prey. There is yet another more important use for the position of the nostrils. The Crocodiles feed on fishes and various water-loving creatures, but also are in the habit of lurking by the river-bank, and suddenly seizing upon any unfortunate animal that may come to drink. Suppose, for example, that a calf or a dog is thus dragged into the water, the reptile grasps it across the body, and sinks below the surface, so as to keep the head of the victim below water while itself can breathe by means of the elevated nostrils.

But as during this process the mouth is held widely open, it might be rationally presumed that considerable inconvenience would be caused by the water running down the throat. Such would indeed be the case, were not this difficulty provided for by a simple yet very wonderful contrivance. At the back of the throat, a pair of thin, cartilaginous plates are so arranged, that when the animal opens its mouth the pressure of the water rushing into the mouth immediately closes one upon the other, and effectually prevents the passage of a single drop, the closure being in exact proportion to the volume of water. The structure, indeed, is very like that of the valves of the heart. The channels which lead from the nostrils run very far back through the skull, and open behind the throat-valves, so that respiration is in no way impeded. They cannot, however, swallow their prey while under water, but are obliged to bring it on shore for that purpose. The tongue is small, and fastened down to the lower jaw throughout its length, so that it was formerly thought that the Crocodiles were destitute of that organ.

AFRICAN CROCODILES AT HOME.

There is rather a curious structure in the vertebræ of the neck. These bones are furnished with short, transverse processes like false ribs, which have the effect of preventing the animal from turning its head from side to side. On land, therefore, where its feeble limbs are so inadequate to the support of the long and heavy body, it can easily be avoided by any one of ordinary agility. The eyes are large, and set rather far back upon the head. The ears are carefully guarded from the ingress of water by a pair of tightly-closing valves. Below the throat are a pair of glands which secrete a substance having a strong musky scent which is very disagreeable, and in old individuals taint the whole flesh with its rank odor, and render it uneatable to ordinary palates.

The young of these reptiles are hatched from eggs, which are strangely small in proportion to the large dimensions of the adult animal, the newly-hatched offspring being so small as hardly to be recognized as belonging to the same species as their parents, especially as there are certain differences of shape hereafter to be mentioned.

These great reptiles are divided, or rather fall naturally, into two families, namely, the Crocodiles and the Alligators. All the members of these families can be easily distinguished by the shape of their jaws and teeth, the lower canine teeth of the Crocodiles fitting into a *notch* in the edge of the upper jaw, and those of the Alligators fitting into a *pit* in the upper jaw. This peculiarity causes an obvious difference in the outline of the head, the muzzle of the Crocodiles being narrowed behind the nostrils, while that of the Alligators forms an unbroken line to the extremity. A glance, therefore, at the head will suffice to settle the family to which any species belongs. In the Crocodiles, moreover, the hind legs are fringed behind with a series of compressed scales.

OUR first example of the Crocodiles is the very remarkable GAVIAL, or GANGETIC CROCODILE, sometimes known by the name of NAKOO.

This curious reptile is one of the largest, if not the very largest of its order, sometimes reaching a length of twenty-five feet. As its popular name imports, it is a native of India, and swarms in many of the Indian rivers, the Ganges being greatly infested with its presence. It is a striking animal, the extraordinary length of its muzzle giving it a most singular and rather grotesque aspect.

This prolongation of the head varies considerably according to the age and sex of the individual. In the young Gavial, for example, just hatched from the egg, the head is short and blunt, and only attains its full development when the creature has reached adult age. The males can be distinguished from the other sex by the shape of the muzzle, which is much smaller at the extremity. There are many teeth, the full complement being about one hundred and twenty. They are similar in appearance, and about equal in length.

The color of this species is dark olive-brown, spotted with black. Several species of African Gavials are known to zoologists, besides the Asiatic animal, but on account of the different formation of the head, such as the absence of a swollen muzzle in the male, and some important variations in the plates of the neck and back, they are placed in another genus, and termed False Gavials. Two other Crocodiles are named, BENNETT'S GAVIAL (*Mecistops bennettii*), which is an inhabitant of Western Africa, and the FALSE GAVIAL (*Mecistops cataphractus*). Some naturalists, however, think that these animals are only varieties of the same species. The False Gavial is represented in the engraving on next page.

WE now arrive at the true Crocodiles, in which the jaws are moderately lengthened, wide, flat, tapering, and rather dilated at the extremities. The most peculiar of these reptiles is the long-celebrated CROCODILE of Northern Africa.

This terrible creature is found chiefly in the Nile, where it absolutely swarms, and though a most destructive and greatly dreaded animal, is without doubt as valuable in the water as the hyena and vulture upon the land. Living exclusively on animal food, and rather preferring tainted or even putrefying to fresh meat, it is of great service in devouring the dead animals that would otherwise pollute the waters and surrounding atmosphere.

It also feeds on fish, which it can catch by means of its great swiftness in the water, and is a dangerous foe to cattle or other beasts that come to the river-side for drink. Some persons relate that when its intended victim does not come sufficiently near to be snapped up, the Crocodile crawls to the banks, and with a sweep of its long and powerful tail strikes the poor creature into the water, where it is immediately seized in the Crocodile's ready jaws.

Human beings have a great dread of this terrible reptile. Many instances are known where men have been surprised near the water's edge, or captured when they have fallen into the river. There is, it is said, only one way of escape from the jaws of a Crocodile, and

FALSE GAVIAL.—*Mecistops cataphractus.*

that is to turn boldly upon the scaly foe, and press the thumbs into his eyes, so as to force him to relax his hold, or relinquish the pursuit. Mr. Petherick relates a curious instance, where a man was drawing water, and was chased by a Crocodile into the recess in the earth in which he was standing while working the lever of the "shadoof." The man crouched as far back as he could squeeze himself, and the Crocodile tried to follow him, but got itself so firmly wedged in the narrow channel through which it was endeavoring to force its way, that it could neither reach the man, whose trembling breast was within a span of the reptile's terrible teeth, nor retreat from the strange position into which it had forced itself. After spending some time in terror, the poor man contrived to give the alarm to his comrades, who came running to his assistance, and despatched the Crocodile as it lay helplessly fixed in the crevice.

The plates which cover the skin of the Crocodile are of exceeding hardness, so hard, indeed, that they are employed as armor by some ingenious warriors. A coat of natural scale armor formed from the Crocodile skin may sometimes be seen. Even a rifle ball may be turned by these horny plates, provided that it strikes rather obliquely; and they are impervious to ordinary steel weapons. Modern rifles, however, especially if the ball is hardened with solder or tin, make little account of the plates, but cut their way through them without difficulty.

GAVIAL, OR GANGETIC CROCODILE.

As this reptile is so dangerous and costly a neighbor to the inhabitant of the river banks, many means have been adopted for its destruction. One such method, where a kind of harpoon is employed, is described by Dr. Rüppell: "The most favorable season is either the winter, when the animal usually sleeps on sand-banks, luxuriating in the rays of the sun, or the spring, after the pairing time, when the female regularly watches the sand islands where she has buried her eggs. The native finds out the place, and on the south side of it, that is, to the leeward, he digs a hole in the sand, throwing up the earth to the side which he expects the animal to take. Then he conceals himself, and the Crocodile, should it fail to observe him, comes to the accustomed spot and soon falls asleep.

"The huntsman then darts his harpoon with all his force at the animal, for in order that its stroke may be successful, the iron ought to penetrate to the depth of at least four inches, in order that the barb may be fixed firmly in the flesh. The Crocodile, on being wounded, rushes into the water, and the huntsman retreats into a canoe, with which a companion hastens to his assistance. A piece of wood attached to the harpoon by a long cord, swims on the water, and shows the direction in which the Crocodile is moving. The huntsmen, pulling at this rope, drag the beast to the surface of the water, where it is again pierced by a second harpoon. . . .

"When the animal is struck, it by no means remains inactive; on the contrary, it lashes instantly with its tail, and endeavors to bite the rope asunder. To prevent this, the rope is made of about thirty separate slender lines, not twisted together, but merely placed in juxtaposition, and bound round at intervals of every two feet. The thin lines get between the teeth or become entangled about them."

In spite of the great strength of the reptile, two men can drag a tolerably large one out of the water, tie up his mouth, twist his legs over his back, and kill him by driving a sharp steel spike into the spinal cord just at the back of the skull.

There are many other modes of capturing and killing the Crocodile, such as a hook baited with meat, to which the voracious reptiles are attracted by the cries of a pig, which is pulled by the tail or otherwise maltreated, for the purpose of eliciting those ear-piercing yells which aggrieved swine always produce. The yelping of a dog answers the same purpose, and is used in the same manner. In some cases the negroes are bold enough to engage the Crocodile in its own element, and to attack it with a long knife, which they plunge into the belly.

The eggs of the Crocodile are about as large as those of the goose, and many in number, so that these terrible reptiles would overrun the country, were they not persecuted in the earliest stages by many creatures, who discover and eat the eggs, almost as soon as they are laid. It is curious that the Crocodile is attended by a bird which warns it of danger, just as the rhinoceros has its winged attendant, and the shark its pilot fish. The Crocodile bird is popularly called the ziczac, from its peculiar cry.

SEVERAL other species of Crocodiles are known, among which two species are deserving of a short notice, namely, the INDIAN CROCODILE (*Crocodílus porósus*), and the AMERICAN CROCODILE (*Crocodílus americánus*). As the name of alligator is popularly given to these and other reptiles, there is great confusion respecting the precise animal which is under discussion.

The Indian Crocodile, as its name imports, is an Asiatic species, and is found largely in India. It is sometimes called the DOUBLE-CRESTED CROCODILE, because the head is furnished with two long ridges extending from the front of the eye over the upper jaw. This species is common in Ceylon, and literally swarms in the still waters and tanks, though it is but rarely found in rapid streams, and never except in the low lands, the hill marshes being free from these pests. Respecting this animal, Sir E. Tennent writes as follows:

"The species which inhabit the fresh water is essentially cowardly in its instinct, and hastens to conceal itself on the approach of man. A gentleman who told me the circumstance, when riding in the jungle, overtook a Crocodile evidently roaming in search of water. It fled to a shallow pool almost dried by the sun, and thrusting its head into the mud till it covered up its eyes, it remained unmoved in profound confidence of perfect concealment.

"Some years ago, during the progress of the pearl fishery, Sir Robert Wilmot Horton employed men to drag for Crocodiles in a pond which was infested with them in the immediate vicinity of Aripo. The pool was about fifty yards in length by ten or twelve wide, shallowing gradually to the edge, and not exceeding four or five feet in the deepest part.

"As the party approached the pond, from twenty to thirty reptiles, which had been basking in the sun, rose and fled to the water. A net, specially weighted so as to sink its lower edge to the bottom, was then stretched from bank to bank, and swept to the farther end of the pond, followed by a line of men with poles to drive the Crocodiles forward. So

INDIAN CROCODILE.—*Crocodilus porosus*.

complete was the arrangement, that no individual could avoid the net; yet, to the astonishment of the governor's party, not one was to be found when it was drawn on shore, and no means of escape was apparent or possible, except dashing into the mud at the bottom of the pond."

The extreme tenacity of life possessed by these reptiles is well exemplified, though in a rather painful manner, by an incident which occurred in Ceylon. A fine specimen had been caught by a hook, to all appearance killed, the viscera removed, and the aperture kept open by a stick placed across it. A few hours afterwards the men came to their victim with the intention of cutting off the head, but were much surprised to find the spot vacant. On examination of the locality, it was evident that the creature had recovered itself in some strange manner, crawled away for some distance, and made its escape into the water.

The same author also describes the habits of another species, the MARSH CROCODILE (*Crocodilus palustris*), sometimes known by the names of MUGGER, or GOA; an animal which has a large range of locality, being found in Asia and Australia. Sometimes this species grows to a great length. I have seen a skull twenty-six inches in length, denoting a total length of thirty-three feet.

This animal is in the habit of traversing considerable distances in search of water, but, according to the Singhalese, its feet are sadly cut in passing over the hard, stony ground. If it is baffled in its search, it returns to the exhausted pool, burrows beneath the mud, and there waits until released by the rains. Sir E. Tennent mentions one instance where he saw the recent impress of a Crocodile in the mud from which it had just emerged, and he was told of a curious incident which befell an officer attached to the surveying department. Having pitched his tent, he had retired to rest as usual, but during the night he was disturbed by a movement of the earth below his bed. On the following morning the mystery was solved by the appearance of a Crocodile, which made its way from under the bed.

As is the case with the common Crocodile of Egypt, the young of this reptile are very small when hatched, but so fierce, even in their early days, that they can be caught by pushing a stick towards them, letting them bite it, and pulling them out before they loosen their hold. A gentleman who has resided for eight years in Ceylon told me that one of his friends was so taken with the appearance of these little reptiles, that he captured one, packed it carefully, and took it home. On arriving in his house, he put the Crocodile, then about nine or ten inches long, into a basin of water, and left it. Shortly afterwards a little boy, one of his children, peeped into the basin, and seeing the Crocodile, gave it a push with his finger. The fierce little creature at once snapped at the offending finger, and held it so tightly that the poor child could not shake it off, and ran screaming about the house with the young Crocodile dangling at the end of his finger, until it was removed by an attendant.

ANOTHER well-known species is the AMERICAN CROCODILE, so often and so wrongly termed the alligator. This reptile is found in the tropical and hotter parts of America, and is very common in some localities. When first hatched, the young seem to feed only on living insects, and according to the experiments of M. Bosc, they would not even touch the insects with which they were supplied, until their intended prey began to crawl. During the summer they become lively at night, and make such a hideous bellowing that a person unaccustomed to it has no chance of sleeping. They also have a habit of clattering their jaws together with a loud noise.

This creature is only lately a known resident in North American waters. But few years since, it was supposed that the islands of the West Indies were the most northern range of any species of Crocodile. Dr. Jeffries Wyman, of Boston, discovered a specimen in Bisquine Bay, off the southern extremity of Florida. Three years since, Mr. Ralph Monroe, of Staten Island, N. Y., visited that region, and, while hunting on Virginia Key, some miles from the mainland, discovered several Crocodiles. Two of them he captured, and the preserved skins he presented to the American Museum, in Central Park. Since then, he has killed a specimen of the largest known dimensions, fourteen feet in length, which he has sent to the same institution.

The first comparison with the alligator does not impress one with any considerable sense of difference, but the difference in breadth of the heads, when viewed from above, is very striking. That of the Crocodile is extremely narrow, while that of the alligator is heavy and very wide. The entire "build" of the Crocodile is manifestly favorable to a maritime existence, while that of the alligator is for just such a life as it leads, one of sluggishness and inactivity.

The Crocodile is an active swimmer, and its teeth and jaws are evidently constructed to seize upon fishes while swimming. It is seen mostly in salt-water creeks near the ocean.

Some doubt has been entertained about the identity of this species with that found in the West Indies. It is very natural for this creature, being a sea-going one, to swim across the Florida Straits. It is illustrated on the next page.

ANOTHER species, the MARGINED CROCODILE (*Crocodilus marginátus*), resides in the rivers of Southern Africa. It may be distinguished from the Egyptian species by the great concavity of the forehead, and the strong keels of the dorsal, or back plates. I am indebted to Captain Drayson, author of "Sporting Scenes among the Kaffirs," for the following account of the Margined Crocodile and its habits, from which it appears that the reptile is formidable not only to the creatures on which it usually feeds, but to man himself:—

"About two or three miles from the Bay of Natal there is a river called the Umganie; into this river a lake called the Sea-cow Lake empties itself. The lake was, during my residence at Natal, the retreat of several hippopotami and Crocodiles, both of which were in the habit of *treking* into the Umganie River. Often, when riding round the banks of this lake, I have disturbed two or three Crocodiles, which were stealing amongst the reeds and long grass,

AMERICAN CROCODILE.—*Crocodilus americanus.*

in hope of stalking a fat toad, or a sleepy guana. Sometimes a scaly reptile might be awakened from his doze by the sound of my horse's feet, and would rush through the long reeds towards his retreat. Their movement is much more rapid than would be supposed from their appearance, and they care nothing for a fall head over tail, but almost fling themselves down the steep banks when alarmed.

"On the banks of the Umganie were several Kaffir kraals, in one of which resided a man who had been roughly treated by a Crocodile. This man, seeing me pass his residence, called to me, and asked as a favor that I would watch at a particular part of the river until I shot a rascally Crocodile that had nearly killed him. The Crocodile, he informed me, always made its appearance about sundown, and he hinted that a position might be selected so that the sun would dazzle the Crocodile and prevent him from seeing me. Finding that I was willing to gratify his revenge, he limped out of the inclosure surrounding his huts, and offering me his snuff-gourd, he, at my request, gave me the following account of his escape.

"He had so frequently crossed the stream below his huts at all times of day, and had seen Crocodiles of small dimensions, that he had become, as it were, familiarized to them, and did not imagine that there was any danger to be expected from them. One evening, at about sundown, he was wading across the river, the water of which reached above his waist. Suddenly he felt himself seized by the under part of his thigh, whilst he was at the same instant dragged under water. His wife was following him, and seeing him fall, she scrambled forward to the place where he had disappeared, and thus caused considerable noise and splashing, which (or something else, perhaps the toughness and bad flavor of the Kaffir) had the effect of making the Crocodile quit his hold on the Kaffir, not, however, without tearing off a great portion of the under part of his thigh. The man, with difficulty, escaped to the shore, but he remained a cripple for life, unable to do more than put the toes of his foot on the ground."

MARGINED CROCODILE.—*Crocodilus marginatus.*

We now come to the ALLIGATORS, the second family of those huge reptiles which may be known, as has already been mentioned, by the lower canine teeth fitting into pits in the upper jaw. They are divided into three genera, all of which are inhabitants of the New World. They are indiscriminately called Alligators, Crocodiles, or Caymans, by the natives or the non-zoological traveller, and there is consequently much difficulty in identifying the particular species. The genus Alligator may be known by the partly-webbed toes, the outer toe being free.

The COMMON ALLIGATOR inhabits Northern America, and is plentifully found in the Mississippi, the lakes and rivers of Louisiana and Carolina, and similar localities. It is a fierce and dangerous reptile, in many of its habits bearing a close resemblance to the crocodiles, and the other members of the family.

Unlike the crocodile, however, it avoids the salt water, and is but seldom seen even near the mouths of rivers, where the tide gives a brackish taste to their waters. It is mostly a fish-eater, haunting those portions of the rivers where its prey most abounds, and catching them by diving under a passing shoal, snapping up one or two victims as it passes through them, tossing them in the air for the purpose of ejecting the water which has necessarily filled its mouth, catching them adroitly as they fall, and then swallowing them. Though timid, as are most reptiles as long as their passions are not touched, the Alligator has within it a very mine of furious rage, which, when aroused, knows no fear. Urged by a blind instinct that sees no obstacles, and hardly deserves so intellectual a name as anger, it flings itself upon the assailants, and only ceases its attack as its last breath is drawn.

No easy matter is it to drive the breath out of an Alligator, for its life seems to take a separate hold of every fibre in the creature's body, and though pierced through and through with bullets, crushed by heavy blows, and its body converted into a very pin-cushion, spears taking the place of the pins, it writhes and twists, and struggles with wondrous strength, snapping direfully with its huge jaws, and lashing its muscular tail from side to side with such vigor that it takes a bold man to venture within range of that terrible weapon.

ALLIGATOR.—*Alligator mississipiensis.*

It is fortunate for the assailant that its head is not gifted with mobility equal to that of the tail. The Alligator can only turn its head very slightly indeed, on account of two bony projections, one on each side of the head, which are efficient obstacles to any but the smallest lateral motion. The antagonist may therefore easily escape if on land, by springing aside before the reptile can turn. He must, however, beware of its tail, for the Alligator when angry, sweeps right and left with that powerful member, and deals the most destructive blows with wonderful rapidity. Still, the creature would rather avoid than seek a combat, and does not act in this fashion until driven to despair.

In some parts of America they catch the Alligator in a very ingenious manner. An ordinary hook is said to be of little service against such a quarry, and the natives employ a kind of mixture between a hook and grapnel which very effectually answers their purpose. This so-called hook is made of four sticks of hard tough wood barbed at each end, slightly curving and bound together at one end so as to cause all the upper barbs to radiate from each other.

This apparatus is baited with the flesh of some animal, and suspended just about a foot from the water, the other end of the rope being made fast to a tree or strong stake.

As soon as the Alligator takes this bait and begins to pull at the cord, the barbs begin to make their way into its throat, and it is evident from the construction of the hook that the more the animal pulls, the firmer are the barbs struck into its throat. When thus hooked, its struggles are terrific, and Mr. Waterton, who succeeded in capturing a fine specimen more than ten feet in length, had the greatest difficulty in securing it without damaging its appearance.

The eggs of the Alligator are small and numerous. The parent deposits them in the sand of the river side, scratching a hole with her paws, and placing the eggs in a regular layer therein. She then scrapes some sand, dry leaves, grass and mud over them, smoothes it and deposits a second layer upon them. These eggs are then covered in a similar manner and

JACARE, OR YACARE.—*Jacare sclerops*.

another layer deposited until the mother reptile has laid from fifty to sixty eggs. Although they are hatched by the heat of the sun and the decaying vegetable matter, the mother does not desert her young, but leads them to the water and takes care of them until their limbs are sufficiently strong and their scales sufficiently firm to permit them to roam the waters without assistance.

As is the case with the crocodiles, the young Alligators are terribly persecuted by birds and beasts, and are even in danger of being eaten by the old males of their own species. During the winter months the Alligator buries itself in the mud, but a very little warmth is sufficient to make it quit its retreat and come into the open air again. While lively, especially at night, it is a most noisy animal, bellowing in so loud a tone and in so singular a cadence that even the nightly concert of jaguars and monkeys is hardly heard when the Alligators are roaring.

It sometimes attains to a great size, and is then formidable to man. Mr. Waterton mentions a case when one of these creatures was seen to rush out of the water, seize a man and carry him away in spite of his cries and struggles. The beast plunged into the river with his prey, and neither Alligator nor man were afterwards seen.

The Alligator is a familiar reptile on the Gulf coast, and in the rivers of Florida. Its length is usually about six or seven feet. Specimens are found at times twice this length—fourteen feet being the extreme. A fine example in the Central Park Museum is twelve feet in length. The term Alligator is a corruption of the Spanish *el lagarto*, a lizard. Five species are known in various parts of the world.

THE JACARE, or YACARE (*Jacare sclerops*), also belongs to this family. It inhabits Brazil, and is not uncommon. It may be known by the ridge across the face between the eyes, the scarcely-webbed hind feet and the fleshy eyelids. On account of the aspect of its eyes it is sometimes called the Spectacled Cayman. It is said that, although this reptile attains a very large size, it will not attack a man even in the water, provided that he always keeps in motion. They pass the night in the water and the day on the shore, where they lie sleeping on the sand, dashing into the water if alarmed. It is depicted on the foregoing page.

AMPHISBÆNIDÆ.

WE now leave the crocodiles and alligators, and proceed to another order of reptiles. These creatures are termed Amphisbænidæ, from two Greek words signifying to go both ways, in allusion to the shape of the animal, which looks as if it had a head at each extremity. In former times, indeed, it was thought that not only could these reptiles creep backward and forward with equal ease, but that they absolutely possessed two veritable heads. None of these reptiles are of great size. They are divided into four families, three of which are without external feet, and the members of the other family only possess the front pair of legs very slightly developed. Their eyes are very minute and entirely covered with skin, so that their sight must be of the most limited character. As in the case of the mole, however, this deprivation of sight does not interefere with the welfare of the animal, for it lives mostly beneath the earth, where eyes would be useless.

SOOTY AMPHISBÆNA.—*Amphisbæna americana.*

The SOOTY AMPHISBÆNA is a native of Southern America, being found most plentifully in Brazil and Cayenne. It lives almost wholly underground, boring its way through the soft earth like the common worm, and traversing the soil with considerable address. It feeds upon animal substances, and is very fond of ants, termites, and their young. Indeed, it is no extraordinary occurrence on breaking down a termite's nest, to find an Amphisbæna within, luxuriously curled up in the midst of plenty. Ants' nests below the ground are often penetrated and ransacked by this reptile.

Being too small to injure man by sheer force, and being devoid of poisonous teeth, this creature is quite harmless except to the insects on which it feeds. It is able to crawl in either direction with nearly equal ease and rapidity, and on account of the bluntness of its tail and the almost imperceptible eyes, affords some reason for the popular idea of its possessing two heads.

In speaking of this reptile, Stedman has the following remarks: "This is the snake which, supposed blind, and vulgarly said to be fed by the large ants, is in this country honored with the name of King of the Emmets. The flesh of the Amphisbæna, dried and reduced to a fine powder, is confidently administered as a sovereign and infallible remedy in all cases of dislocation and broken bones, it being very naturally inferred that an animal which has the power of healing an entire amputation in its own case, should at least be able to cure a simple fracture in the case of another."

This process of reasoning alludes to a curious popular error respecting the Amphisbæna. The people of the countries which it inhabits believe that, if one of these reptiles is cut in two, each half, being furnished with a separate head, hastens to its fellow-part, and neatly fitting the severed surfaces, repairs the breach, and is soon restored to its original condition.

It is rather a dull and sluggish animal when exposed to light, crawling slowly upon the ground, twisting itself lazily about, and opening its mouth in a purposeless kind of fashion, without any definite intention of biting or escaping.

The color of the Sooty Amphisbæna is rather variable, but consists of black and white. Its length is about three feet. The White Amphisbæna (*Amphisbæna alba*) belongs also to this genus. It is of a white color, and remarkable for a little pellucid dot in the front edge of each scale.

CLOSELY allied to this creature is another reptile, very appropriately called the CHEIROTES, or HAND-EARED LIZARD (*Cheirôtes lumbricöides*). This is a native of Brazil, and, as far as is known, is of subterranean habits, like the amphisbæna.

The Cheirotes is the only example of all the amphisbænas that possesses external limbs, and even in this instance they are small and but slightly developed. There are no hind legs,

CHEIROTES, OR HAND-EARED LIZARD.—*Cheirotes lumbricoides*

but the two fore legs are set just behind the head; nearly in the place where the ears might be expected to be seen. They are very short, rather flat and strong, and are terminated with five toes, four of which are armed with a tolerably strong claw. The fifth toe is very small and without a claw.

The head of this creature is no larger than the body, the teeth are conical, moderately strong and slightly curved backwards, the muzzle is arched, the tongue horny at the tip, the tail is short, and there is a row of small pores on the under side of the abdomen. In our illustration the animal is shown in its natural size, which varies from eight to ten inches. Its color is yellow, spotted with brown above, and whitish below. This species is the sole representative of its family. The other two families—namely, the Trigonophidæ and the Lepidosternidæ—may easily be distinguished by the fact that in the former the teeth are set in the margin of the jaws, instead of on their inner side as in the other families; and that, in the latter, the scales on the chest are larger and of different shapes, whereas in the other two families they are all squared. Moreover, the pores under the abdomen are absent.

SCALED REPTILES; SQUAMATA.

LIZARDS; OR SAURA.

SLENDER-TONGUED LIZARDS; SEPTOGLOSSÆ.

WE now leave the shielded reptiles and proceed to the Scaled Lizards. These creatures form a very large and important group, and may be distinguished from the previous section by the covering of the body, which is formed of scales either granular or overlapping each other, instead of the straight-edged plates which cover the bodies of the tortoise and crocodiles. The tongue of these animals is rather long, nicked at the tip, and often capable of extension. The young are produced from eggs, sometimes hatched before being deposited, but generally after they have been laid in some suitable spot. The eggs are covered with a rather soft, leathery shell.

The true LIZARDS have four limbs, generally visible, but in a few instances hidden under the skin. Their body is long and rounded, and the tail is tapering and mostly covered with scales set in regular circles or "whorls." The mouth cannot be dilated as in the snakes; because the under jawbones are firmly united in front, instead of being separable as in the serpents. The ear has a very singular appearance, the drum or "tympanum" being mostly distinct and exposed.

There are twenty-four families of true Lizards, and passing by several anatomical and structural distinctions, which will be found at the end of the volume, we will proceed at once to the first family, called the MONITORS. In all these creatures the head is covered with very little, many-sided scales; the tongue is long, slender, and capable of being withdrawn into a sheath at its base; the scales are small, rounded, and arranged in cross rings, those of the side resembling those of the back; the legs are four in number, and each foot has five toes. They are all inhabitants of the Old World, and are seldom, if ever, found far from water.

OUR first example of the true Lizards is the WHITE-THROATED REGENIA, or WHITE-THROATED VARAN, a remarkably fine and powerful species of Lizard, inhabiting Southern Africa. A rather full and accurate description of this Lizard is given by Dr. Smith:—

"It is usually discovered in rocky precipices or on low stony hills, and when surprised seeks concealment in the chinks of the former or the irregular cavities of the latter, and where any irregularities exist on the surface of the stones or rocks, it clasps them so firmly with its toes that it becomes a task of no small difficulty to dislodge it, even though it be easily reached. Under such circumstances the strength of no one man is able to withdraw a full-grown individual, and I have seen two persons required to pull a specimen out of a position it had attained, even with the assistance of a rope tied in front of its hinder legs. The moment it was dislodged it flew with fury at its enemies, who by flight only saved themselves from being bitten. After it was killed, it was discovered that the points of all the nails had been previously broken or at the moment it lost its hold.

WHITE-THROATED REGENIA.—*Regenia albogularis.*

"It feeds upon crabs, frogs, and small quadrupeds, and from its partiality to the two former, it is often found among rocks near running streams, which fact having been observed by the natives, has led them to regard it as sacred, and not to be injured without danger of drought."

This fine Lizard has large, oblique nostrils, a shortish tail with a double keel on its upper surface, and the scales are oblong and have a blunt ridge or keel. The head is short and the scales of the body are large, convex, and surrounded with granulations. The length of the full-grown Regenia is nearly five feet, and its color is dark brown, above variegated with large white spots, and paler beneath, especially under the throat.

THE NILOTIC MONITOR, or VARAN OF THE NILE, as it is sometimes called, is, as its name imports, a native of those parts of Africa through which the Nile, its favorite river, flows.

The natives have a curious idea that this reptile is hatched from crocodile's eggs that have been laid in hot elevated spots, and that in process of time it becomes a crocodile. This odd belief is analogous to the notion so firmly implanted in the minds of our own sea-side population, that the little hermit crab, which is found so plentifully in periwinkle shells, is the young of the lobster before it is big and hard enough to have a shell of its own.

It is almost always found in the water, though it sometimes makes excursions on land in search of prey. To the natives it is a most useful creature, being one of the appointed means for keeping the numbers of the crocodile within due bounds It not only searches on land for the eggs of the crocodile, and thus destroys great numbers before they are hatched, but chases the young in the water, through which it swims with great speed and agility, and devours them unless they can take refuge under the adult of their own species, from whose protection the Monitor will not venture to take them.

NILOTIC MONITOR.—*Monitor niloticus.*

When full grown, the Nilotic Monitor attains a length of five or six feet. The color of this species is olive-gray above, with blackish mottlings. The head is gray, and in the young animal, is marked with concentric rows of white spots. Upon the back of the neck is a series of whitish-yellow bands, of a horse-shoe, or semilunar shape, set crosswise, which, together with the equal-sized scales over the eyes, serve as marks which readily distinguish it from many other species. The under parts are gray, with cross bands of black, and marked with white spots when young.

Specimens belonging to this genus are scattered over the greater part of the world. For example, the INDIAN MONITOR (*Monitor dracæna*) is found in the country from which it takes its name. It is rather a prettily marked animal, being brown with black spots when old, and yellow eye-like marks when young. Another species, GOULD'S MONITOR (*Monitor gouldii*), inhabits Australia, being most commonly found on the western side of the land.

INDIAN MONITOR.

TEGUEXIN, OR VARIEGATED LIZARD.

WE now arrive at another family of Lizards, called from the typical species, the Teguexins. In these reptiles, the head is covered with large, regular, many-sided shields, the sides are flat, and the throat has a double collar.

Our first example is the TEGUEXIN, or VARIEGATED LIZARD, so called on account of the contrasting colors with which it is decorated. It is also known by the name of SAFEGUARD, a title which has been given to it because it is thought to give notice, by hissing, of the approach of the alligator. The monitors derive their name from a similar belief; they being thought to warn human beings of the approach of poisonous serpents.

TEGUEXIN, OR VARIEGATED LIZARD.—*Tejus tejuixin.*

Several species of Teguexin are known, all inhabiting the warmer portions of America, and possessing similar habits. It is said that, although strong and agile, they do not ascend trees, but range at will the hot sandy plains or the dense damp underwood on the margins of lakes and rivers, into which they plunge if alarmed, and remain below the surface until the danger has passed away, their capacious lungs and imperfect circulation permitting them to endure a very long immersion without inconvenience.

The Teguexin is a large and powerful Lizard, exceeding five feet in length when full grown, and extremely active. It feeds mostly, if not entirely, upon animal food, and makes great havoc among snakes, frogs, toads, and other semi-aquatic creatures. It often indulges in diet of a higher nature, and when it can find an opportunity, devours poultry, or breaks and eats their eggs. Sometimes it has been known to eat Lizards of a closely allied species, a fact which has been proved by finding some bones, and other portions of the Ameiva lizard within the stomach of a Teguexin that had been killed. Together with these relics were found the shelly wing-cases of beetles, and the skins of sundry caterpillars.

The teeth of this species are strong, and the reptile can bite with great force. It is a bold and determined combatant when attacked, and if it succeeds in grasping a foe, retains its hold with the pertinacity of the bulldog. The flesh of the Teguexin is eaten, and thought to be excellent. According to Azara, the skin of its tail, when separated into rings, is considered to be a safeguard against paralysis, and worn for that purpose, as well as to remove tumors, another healing power which it is supposed to possess.

The general coloring of the Teguexin is as follows: The upper parts are deep black, with bold mottlings of yellow or green. On the upper part of each side there are two series of white spots, and the under parts are mostly yellow, with black bands. The coloring is, however, extremely variable.

The curious little AMEIVA, which has just been mentioned as falling a victim to the previous species, is closely allied to the Teguexin. It is rather a pretty Lizard, with a very long whip-like tail, and peculiarly elongated toes on the hinder feet. The long tail is covered with a series of scales, arranged in rings, of which about one hundred and twenty have been counted in a perfect specimen. The color of the Ameiva is dark olive, speckled with black on the nape of the neck and front of the back. On the sides are rows or bands of white spots edged with black, from which peculiarity it is sometimes called the Spotted Lizard. There are many species of Ameiva, inhabiting either Central America, or the West Indian Islands.

CRUST LIZARD.—*Heloderma horridum.*

The very odd-looking creature, scientifically termed *Heloderma horridum*, which is seen in the engraving, is an inhabitant of Mexico, where the natives call it Tola-chini. Though looking somewhat like an Ameiva, it forms a separate family, of which it is the only species. It differs from the Ameiva by the formation of its teeth and tail, the latter being thick, and shorter than the body. As the pointed teeth are set as in the deadly snakes, the natives of Mexico believe the reptile's bite to be fatal. This belief, however, is without any foundation, as the reptile really possesses no poisonous fangs. Like some frogs, the *Heloderma* has a penetrating scent, and when disturbed, it ejects an odorous saliva from its mouth. During the day it hides in self-made holes at the foot of trees, and there it lays in a lethargic position until night, when it chases its prey, consisting of beetles, worms and frogs. The *Heloderma* is of an earthy-brown color, the whole body being covered with yellow, white, and brownish-red spots, and the tail with dark scales. It attains a length of nearly three feet three inches.

The Six-lined Taraguira also belongs to the Teguexins. This pretty little Lizard, with its dark green body, and yellow streaks, inhabits North America. Mr. Holbrook makes the following remarks respecting its habits: "This is a very lively, active animal, choosing dry and sandy places for its residence, and is frequently met with in the neighborhood of plantations, or near fences and hedges. Most usually it is seen on the ground in search of insects; its motions are remarkably quick, and it runs with great speed. It is very timid. It feeds on insects, and generally seeks its food towards the close of the day, when they may be seen in corn fields, far from their usual retreat; and not unfrequently I have met male and female in company."

The Six-lined Lizard (*Onemidophorus sexilineatus*), called in the South "Taraguera," inhabits the States from Virginia to Mexico.

A brief notice must also be given of two curious species, also belonging to the same family. The first is the Spurred Centropyx, or Spurred Lizard (*Centropyx calcátus*), so called from two pair of small, sharp, horny spikes, which are set at each side of the base of the tail. The color of this species is olive-green above, with three streaks of a paler hue, and a double series of black spots on the back. Below it is greenish-white.

The other species is the Great Dragon (*Ada guianensis*), a native of tropical America. This fine Lizard is generally from four to nearly six feet in length, and is strong and nimble. It does not appear to be so good a swimmer as some of the preceding species, but runs fast, and can climb trees with great agility. It is generally found among the marshy and low-lying lands, though it spends more time on the land than in the water.

It is a desperate fighter when attacked, and as it has a habit of hiding itself in a deep burrow, and bites fiercely at the hand that is thrust forward to seize it, it is not easily captured. It is, however, much sought after, as its flesh is very good, and the eggs are thought to be great delicacies. There are usually from thirty to forty eggs. The general color of this reptile is olive, yellow beneath, and mottled with brown.

There are twelve genera and about forty species of the family *Teidæ*, or the Teguexins, this name being derived from some local designation. They are all peculiar to the New World.

The true Lizards, or Lacertinidæ, now come before our notice. The tongue of these reptiles is long, flat, can be thrust out to some distance, and very deeply forked. The teeth are hollow at their roots, the scales are keeled, and the sides are flat. They are scattered over the greater part of the globe.

Europe possesses at least two examples of this family, one of which, the Scaly Lizard, is extremely common.

This pretty little reptile is extremely plentiful upon heaths, banks and commons, where it may be seen darting about in its own quick, lively manner, flitting among the grass stalks with a series of sharp, twisting springs, snapping up the unsuspecting flies as they rest on the grass blades, and ever and anon slipping under shelter of a gorse bush, or heather tuft, only to emerge in another moment brisk and lively as ever.

These little creatures are so quick and sharp sighted, that it is not very easy to catch them, especially if they are among gorse bushes, for they twist about so adroitly, that a very smart movement of the hand is required to follow them, and the prickly points of the gorse are always lurking among the grass, to the detriment of a tender skin. They can swim tolerably if thrown into the water, but do not seem to seek that element voluntarily. I have generally found that when flung into water, they lie for a short time quite motionless, with their limbs extended, and tail straight, as if bewildered with the sudden change. They soon, however, get their head towards shore, and then, with a serpentine movement of the tail, scull themselves to land.

This is one of the reptiles that produces living young, the eggs being hatched just before the young Lizards are born. With reptiles, the general plan is to place the eggs in some spot where they are exposed to the heat of the sunbeams; but this Lizard, together with the viper, is in the habit of lying on a sunny bank before her young ones are born, apparently for the purpose of gaining sufficient heat to hatch the eggs. This process is aided by the thinness of the membrane covering the eggs.

The color of this little Lizard is extremely variable, but in general, the upper parts are olive-brown, with a dark brown line along the middle of the back, this line being often broken here and there. Along each side runs a broader band, and between these bands are sundry black spots and splashes. The under parts are orange, spotted with black in the male, and olive-gray in the female. The total length of the Scaly Lizard is about six inches, according to the figure in our illustration.

THE beautiful EYED LIZARD, or GREAT SPOTTED GREEN LIZARD, as it is sometimes called, from the colors with which it is decorated, is a native of Southern Europe, and various other warm portions of the world, being found in Algiers, Senegal, and parts of America.

SCALY LIZARD.—*Zootoca vivipara*.

This creature inhabits dry spots, where the sun has most power, and may be seen among hedges, underwood, or loose stones, running about in search of food, and displaying the gem-like brilliancy of its clothing, as it darts from spot to spot with the agility which characterizes all the species of this genus.

It is of rather a fierce nature, having little fear, and boldly attacking any antagonist that may assail it. If it be irritated with a stick, it will turn sharply upon the offending weapon, and bite it smartly; and if a dog attempts to seize it, the courageous little creature will spring upon its muzzle, and maintain its hold with such pertinacity, that it will suffer itself to be killed rather than relinquish its grasp. In consequence of this combative character, it is greatly respected by the inhabitants of the country where it dwells, and being thought to be poisonous as well as ferocious, is dreaded with a fear quite as keen, though not so reasonable, as would be inspired by a rattlesnake or cobra.

The home of this species is generally made under the roots of trees, if the soil be sufficiently dry and sandy to suit its habits. Otherwise it will excavate a tunnel in the side of a bank or under a hedge, always choosing a southern aspect, so as to ensure the warmth which its nature seems to demand. Sometimes it settles upon a soft sandstone rock for its domicile, and hollows out a deep burrow in the softest part of the rock, mostly choosing the loose, sandy layers that often occur between two tolerably broad strata of rock. Like the rest of the Lizards, it feeds on insects and similar creatures, darting on them with great speed and certainty of aim.

The color of this Lizard is very beautiful, rendering it one of the most lovely of its tribe. The ground color of the body is bright, glittering green, as if covered with an armor of emeralds, upon which are set, along the sides, some rather large, eye-like spots of rich

EYED LIZARD.—*Lacerta ocellata.* (One-half natural size).

azure. A kind of network of black is also spread over the body, sometimes running in well-defined lines, and sometimes composed of rows of black dots. The temples of the Eyed Lizard are covered with unequal, many-sided scales, rather convex in their form. Its length when full grown is about fifteen or sixteen inches, but it is very variable in size as well as in color.

A VERY beautiful species of this genus is common in many parts of Europe, Asia, and Africa. This is the GREEN LIZARD. As its name imports, this reptile is of a green color, and with the exception of the preceding species, is as beautiful a creature as can be seen.

Like the eyed Lizard, it haunts sunny spots, and may be found in orchards, gardens, shrubberies, copses, and similar localities, where it can find plenty of food and obtain concealment when alarmed. Old ruins, too, are greatly haunted by this beautiful Lizard, which flits among the moss-covered stones with singular activity, lying at one moment as if asleep in the sunbeams, or crawling slowly, as if unable to proceed at any smarter pace, and then, when the hand is thrust towards it, disappearing with a rapidity that looks like magic.

Since the great demand for ferneries and vivaria of different descriptions has arisen, this Lizard is used as a beautiful ornament to a glass fern-case, and is sufficiently hardy to be kept alive with a very little care. It seems to revel in the sunshine, and there are few objects more beautiful than the emerald green hues of this Lizard, as the sunbeams flash and glitter on its resplendent surface.

It is susceptible of kindness, and can soon be tamed by those who choose to take the trouble of familiarizing themselves with their bright and lively favorite. Although sufficiently bold and apt to bite if it fancies itself aggrieved, it can be so thoroughly tamed that it will come and take flies out of the hand. In France and other countries this pretty harmless little creature is greatly dreaded, the popular belief attributing to it sundry destructive powers of the same nature as those which our rustic population believe to be exercised by the common newt.

The color of this beautiful creature is rich shining green above, a little blue sometimes appearing upon the head, and the quality of the green being rather variable in different

GREEN LIZARD.—*Lacerta viridis.*

individuals. A multitude of little golden spots are also perceptible on the back, and similar dots of black are not unfrequently sprinkled over the surface. Underneath, the green fades into a yellower hue.

UNTIL comparatively later years, the SAND LIZARD was confounded with the scaly Lizard, which has recently been described.

This reptile is extremely variable in size and coloring, so variable, indeed, that it has often been separated into several species. Two varieties seem to be tolerably permanent, the brown and the green; the former, as it is believed, being found upon sandy heaths where the brown hues of the ground assimilate with those of the reptile, and the green variety on grass and more verdant situations, where the colors of the vegetation agree with those of the body.

Though quick and lively in its movements, it is not so dashingly active as the scaly Lizard, having a touch of deliberation as it runs from one spot to another, while the scaly Lizard seems almost to be acted upon by hidden springs. It does not bear confinement well, and in spite of its diminutive size and feeble powers, will attempt to bite the hand which disturbs it in a place whence it cannot escape. When it finds itself hopelessly imprisoned, it loses all appetite for its food, hides itself in the darkest corner of its strange domicile, and before many days have passed, is generally found lying dead on the ground.

Unlike the scaly Lizard, this species lays its eggs in a convenient spot and then leaves them to be hatched by the warm sunbeams. Sandy banks with a southern aspect are the favored resorts of this reptile, which scoops out certain shallow pits in the sand, deposits her eggs, covers them up, and then leaves them to their fate. Mr. Bell, who has paid great attention to this subject, has remarked that the eggs are probably laid for a considerable period before the young are hatched from them.

As has been already remarked, the coloring of this creature is exceedingly variable in different individuals. Generally it is sandy-brown above, with some faint bands of a darker brown with rows of black spots, which sometimes have a whitish dot in their centre. The sides have a tinge of green more or less distinct, and the under surface is white. In some individuals the green is very distinct. The average length of the Sand Lizard is about seven inches or a little more.

PASSING by a series of genera affording but few interesting points, we come to the curious animal called the CAPE SPINE-FOOT. The generic name Acanthodáctylus, signifies Thorn, or Spine-toed, and is very appropriately given to this animal and the other species of the same genus. All the Spine-foot Lizards are inhabitants of Africa, and most of them are found towards the northern portion of that continent.

According to Dr. Smith, "this Lizard is found on the sandy districts of Great Namaqualand, and where the surface of the country is irregular it is generally met on the highest spots. Where small sand-hills occur, it resorts to them in preference to the other localities, and from the peculiar assistance it derives from the serrated fringes which edge its toes, it runs over the loose sand on the steep surfaces of those slopes with great activity. It feeds on insects."

The color of this Lizard is a very peculiar brown above, changing from yellow-brown to a much warmer hue, partaking of the orange. The top of the head is mottled with dark brown, and the back is freckled with the same hue. From the eyes run two whitish bands on each side, the lower terminating at the hind-leg and the upper reaching some distance along the tail. Between and about these bands are bold brown mottlings in the male, and an orange wash in the female. The upper part of the legs are also mottled with dark brown. The toes are very long, especially those of the hind-foot, and are edged with a fringe composed of sharply pointed scales. The female is larger and more clumsily made than the male.

ANOTHER pretty species of Lizard, termed the NAMAQUA EREMIAS, is found the portion of Africa from which it derives its name. The name Eremias signifies a dweller in a wilderness, and is given to this and several other species because it is always found in hot and arid situations, the sandy flats between Cape Town and Little Namaqua-land being its most favored localities.

It is chiefly remarkable for the great length and slenderness of its tail, which measures five and a half inches in length, although the head and body together are only two inches long. The color of the back and upper parts is delicate brown mottled with a deeper hue, and along the back are drawn four narrow lines of light reddish orange. The sides are cream-yellow, the upper portions of the legs are olive-brown, and the under surface of the animal is yellowish-white. There is a trifling variation in the coloring, according to the age of the individual. Thirteen or fourteen species of this genus are known to zoologists, most of them being natives of Africa.

OUR last example of the true Lizards or Lacertinidæ is the curious little creature termed the ELEGANT OPHIOPS. Two species are known as belonging to this genus, and they can at once be separated from the true Lizards by the character of the eyelids, which are only rudimentary and hardly visible, so as to have gained for their owners the generic title of Ophiops, or Serpent-eyed Lizards.

The Elegant Ophiops inhabits the south-eastern portions of Europe, and the neighboring parts of Asia. The shores of the Mediterranean appear to be favorite localities of the Ophiops, and in those places it is not at all uncommon. It is lively and active in character, and, like the rest of the same family, feeds on insects, which it catches by suddenly springing on them as they repose from their aërial excursions or crawl along the ground. Like most Lizards, it is rather variable in coloring, but the general tints are as follows. The back and upper parts are olive, sometimes deepening into bronze. Along each side run two bands of pale yellow, and between the bands are sundry black spots, also arranged in lines, but varying in form, size, and number, according to the age of the individual. The under parts are white.

QUITTING the true Lizards, we come to another family of reptiles, called the Zonuridæ, or Band-tailed Lizards, because the scales of the tail are arranged in regular series or rings, and by their overlapping cause the edges to stand out boldly in whorls. Along the sides of these reptiles runs a distinct longitudinal fold, covered with little granular scales, and the eyes are furnished with two valvular lids.

THE COMMON ZONURUS, or ROUGH-SCALED CORDYLE, is a native of Southern Africa, and very plentiful at the Cape, where it may be seen among the rocks or in sunny localities flitting

ROUGH-SCALED CORDYLE.—*Zonurus cordylus.* (One-half natural size.)

from spot to spot with some speed, though not exhibiting the singular activity which is possessed by many of the smaller Lizards. It is chiefly remarkable for the curious aspect of the tail, with its whorls of spike-tipped scales, which looks as if a number of thimbles had been deeply notched round their edges and then thrust into one another.

There is a somewhat similar reptile called the COMMON CORDYLE (*Cordylus polygonus*), but it may be distinguished by a peculiarity of structure which has caused it to be placed in a different genus. In the members of the genus Zonúrus, the eyelids are opaque, as is generally the case, but in the genus Cordylus there is a smooth transparent spot in the centre of the lower eyelid.

The form of the Rough-scaled Cordyle is rather stout and flattened, as accords with the comparative slowness of its movements. In color it is variable, but the usual tints are orange-

ELEGANT OPHIOPS.

yellow on the back, sides, and tail, fading into yellow on the head, and white on the under parts. This species may be distinguished from the other Cordyles by the smooth shields of the head and the rhomboidal-shaped scales of the back, which are larger in the centre than on the sides, and decidedly keeled. On the flanks the keels are so long as to become spines, and the sides of the neck are covered with sharp spine-like scales.

THE FALSE CORDYLE is placed in a separate genus, on account of the shape and size of the scales upon the back and sides. Instead of being large and tolerably even in size, as in the preceding genus, they are very small and granular, alternating with bands of larger scales, which are three-sided, convex, and slightly keeled. These scales are largest on the sides of the back. The generic name Microlepidotus signifies small-scaled, and is given to these creatures in allusion to the minute scales of the back and sides.

The habits of this reptile are much like those of the previous species. Dr. A. Smith writes as follows respecting this creature, after describing the singular variations of color to which it is subject:—

"Each of the varieties appeared to be restricted to its own localities, and, so far as my observations extend, no specimens of two varieties are ever found in the same localities. All the varieties inhabit rocky situations; and, when they have a choice, they invariably prefer precipices and the stony walls of difficultly accessible ravines. In this situation they wander carelessly, in search of food or warmth, unless alarmed by what they may regard as enemies. On being closely approached in their retreats, they are with difficulty captured, as, by aid of the prominences on the hinder edge of each temple, they hold on with a tenacity which is quite surprising; and by them they occasionally offer such an effectual resistance to the force applied from behind, that the tail breaks off from the body before the reptile is secured."

As, in Dr. Smith's work, the description of the different varieties occupy nearly five quarto pages of letter-press, it is evidently impossible to give more than a general description in this volume. Suffice it to say, that in one variety, found on the Table Mountain and about Cape Town, the color is ochry-yellow above, banded with dark brown; in another, which inhabits the rocks about Algoa Bay, it is yellow, with bold, black bars along the back; another, which lives on the banks of the Orange River, is brown above, warming into bright chestnut in the male, and olive-green mottled with dusky black in the female; and a fourth variety, which is found in the high, mountainous regions about Natal, is bright green, with an olive-green stripe and short bars of the same tint across the back. The tail is also banded with two shades of green, one a deep olive, and the other having a much yellower hue. The female of this variety is without the bands, and is only mottled with dark olive, and spotted with the same hue along the sides. The length of the False Cordyle is about eighteen inches.

A SMALL group of reptiles is collected under the generic title of Gerrhosauri, or Basket-Lizards, because the arrangement of their scales and coloring has an effect as if the body had been covered with delicate wicker-work, such as is employed to protect glass flasks from injury.

These Lizards are natives of Southern Africa, where they are far from uncommon. They are all rather pretty in form and coloring, but the most pleasing in general appearance is BIBRON'S GERRHOSAURUS (*Gerrhosaurus bibróni*). This animal is found near the Orange River, and may be seen slipping about among the rocky sides of the dark ravines that are so plentiful in that neighborhood. It is a very shy and timid creature, and if it fancies itself watched by an unfriendly eye, or suspects the least shadow of danger, it quietly glides under the heap of dead wood and dried leaves which collect in abundance in such localities, and will not venture out again until it is tolerably sure that the danger has passed away.

As is the case with most of these Lizards, there is considerable variation of coloring, but in general the upper surface is dark brown, and the sides of the head, the throat, and front of the fore limbs are bright scarlet. Along the back run four yellow lines, of which the two central only extend as far as the hind legs, whereas the two outer streaks are continued to the extremity of the tail. It is not a large species, being about ten or eleven inches in length.

THE generic name, SAUROPHIS, which is given to the reptile next in order, is of Greek origin, and signifies Lizard-snake, in allusion to the very serpentine aspect of its body.

This singular creature inhabits Southern Africa, and at first sight might be easily mistaken for a serpent as it crawls about the ground, its four tiny limbs being far too weak to render it any great assistance in progression, which is achieved, as in the serpents, by continual movement of the projecting edges of the scales. Very little is known of its habits.

The head of this reptile is of a somewhat pyramidal shape, and covered with shields, as are both temples. The scales of the back are slightly grooved, and a small keel runs across their length; they are regularly arranged in fourteen series. On the abdomen, the shields are in six rows. There are four very small and feeble limbs, each of which is furnished with four little short and compressed toes, with rather long claws at their extremities. The body is long and cylindrical, and a decided groove runs along each side. Its color is tawny brown, each scale being of a deeper hue at its edge, so as to give a slightly mottled appearance to the creature. The legs and lower edge of the temple are white, spotted with little dots of black.

ON account of the great rapidity of its movements, our next example has received the appropriate title of TACHYDROME, a name derived from the Greek, and signifying a swift runner.

This pretty little Lizard is an Asiatic animal, being mostly found in China, Cochin China, and Java. Although its limbs are much larger and more powerful in proportion to the size of the body than those of the preceding species, its tail is of such great comparative length, and so slender in its proportions, that, quick as is the creature in all its movements, it has much of a serpentine aspect. The tail, indeed, is longer in proportion to the body than is the case with any other of the order, being three times the length of the body and head, and tapers from the body like the thong of a whip from its handle.

The collar of this creature is covered with scales and decidedly toothed. The scales of the back are nearly square in form, slightly overlap each other, and are arranged in four longitudinal series. Each scale has a decided keel along its length. The scales of the sides are small and granular, and those of the abdomen and throat are larger, strongly keeled, and boldly overlap each other, a provision which is evidently intended for the purpose of aiding the creature in progression, and enabling it to hold itself firmly in any cleft into which it may have retreated. The scales of the common snake answer the same purpose, as any one may prove by taking a snake by the tail and drawing it backwards over a carpet, or by allowing itself to insinuate half of its body into a crevice in a rock or old wall, and then endeavoring to draw it out again by pulling at its tail.

The color of this pretty Lizard is dark olive above. On each side a bold, white streak, edged on either side with black, runs from the base of the head to the insertion of the tail. On the sides of the body and neck are a multitude of little black dots, each having a white centre, and between these dots the color is blue, glossed with golden yellow. The abdomen and under parts are pure shining white, and the tail is generally olive, though in some specimens it has something of a metallic or iridescent lustre, and gleams with golden or coppery reflections. Between the nostril and the eye runs a short black line, and on the temples are two similar lines, with a white streak between them. The total length of the Tachydrome is about one foot.

IN the curious snake-like Lizard called the SCHELTOPUSIC, or PSEUDOPUS, the limbs are almost entirely absent, the front pair being altogether wanting, and not even exhibiting a trace of their locality, while the hind pair of legs are only indicated by two slight scale-like appendages at the junction of the tail with the body. It is often the case that with reptiles in which the limbs are externally wanting, their bones, although very small and delicate, are found beneath the skin. But in the Scheltopusic, the only indication of legs is found in a pair of very tiny bones attached to the pelvis, and exhibiting the merest rudiment of the missing limb.

Moreover, the pelvis itself is very small and slight, and is itself scarcely more than rudimentary in its form, though affording one of the needful transition links between the quadrupedal Lizards and the footless snakes, some of which, indeed, possess the rudiments of limbs even in a more doubtful state than is found in the Sheltopusic. In consequence of the absence of limbs, the movements of this reptile are completely those of a serpent, and so snake-like is it in all its gestures, that in the countries where it resides, it is popularly considered as a serpent, as is the case with the blind-worm.

The Scheltopusic is a native of the coast of Northern Africa, and is also found in Dalmatia, the Morea, and parts of Siberia, where it is called by the title under which it is now generally known. It seems to be rather a timid creature, and very mistrustful of strange sights or sounds, always remaining within the vicinity of some familiar spot, whither it seeks an immediate retreat if disturbed.

Thickly wooded valleys, where the underwood is dark and dense, and the vegetation is rank and heavy, are favorite localities of this harmless and weaponless reptile, which has no mode of defence if attacked, and can only retreat from the approach of danger by gliding silently under the brushwood and insinuating itself in some dark crevice, where it lies secure. So watchful is this creature, that although its movements are rather slow, it is not very easily captured, mostly gliding away in so silent a manner that it has reached its haven of safety before its presence is even suspected.

Even if it be seen and followed, it is not readily captured after once it has succeeded in burying itself among the brushwood, for its color is sufficiently sombre so well with the dark soil and dead sticks and leaves among which it resides, that its outline can with difficulty be discerned, even by a practised eye. As is the case with most reptiles, it loves to emerge from its retreat and crawl to some spot where the sunbeams have thoroughly warmed the ground, and there to lie basking in the genial heat. While thus occupied, it is not so wary as at other times, and may be approached and secured before it can make good its retreat.

The whole aspect of this reptile is so serpentine that it has been attacked and killed under the impression that it was a poisonous snake, and great has been the surprise of its slayers to find that they had destroyed, not a venomous serpent, but a harmless Lizard. This creature has been often captured alive and kept in confinement. In its wild state it feeds mostly on insects, the smaller reptiles, and similar creatures, sometimes gliding into a nest of newly hatched birds and swallowing them. This propensity was once exhibited by a captive Scheltopusic; it had fed very contentedly on hard-boiled eggs, until one day it contrived to gain access to a nest full of very young birds, and swallowed the whole brood.

The jaw-teeth of this reptile, although not of a venomous character, are strong, and those of the palate, although small, are probably useful in aiding the creature to secure and swallow its prey. The tongue is thin and covered with a little papillæ of various sizes. Along each side runs a rather deep groove or furrow, which, on a closer inspection, is found to be double. The scales of the back are rather shining and closely set, and there is a slight keel running along the centre of each scale, which is shown more distinctly on the tail than on the body. The keel is shown more distinctly in the young than in the adult.

The color of this reptile is rather variable, but in general the ground color of the body is chestnut, profusely dotted with blackish spots, caused by the dark edges or spots of each scale. These scales are arranged in a regular series of thirteen longitudinal rows. The eye is bright golden-green, and has a very beautiful appearance, as it contrasts well with the chestnut and black of the body and head. The young Scheltopusic is very different from its parent in the coloring, being gray above, with rather obscure bands of grayish-brown, and the under surface is gray, with a whitish lustre. The length of the Scheltopusic is about eighteen inches, the tail occupying about three-fifths of the whole measurement.

In the curious reptile which is appropriately called the GLASS SNAKE, there is not even a vestige of limbs, so that it is even more snake-like than the preceding species. The generic title of Ophisaurus is of Greek origin, signifying Snake-lizard, and is given to the reptile on account of its serpentine aspect. The reader may remember that on page 52 there is an account of the saurophis, a name which is exactly the same as that of the present species, except that the one is called the lizard-snake and the other the snake-lizard, a distinction which, in the present case, is without a difference, so that the two reptiles might exchange titles and yet be appropriately named.

The Glass Snake is indeed so singularly like a serpent that it can only be distinguished from those reptiles by certain anatomical marks, such as the presence of eyelids, which are wanting in the true serpents, the tongue not sheathed at the base, and the solid jaw-bones, which in the serpents are so loosely put together that the parts become widely separated when the mouth of the creature is dilated in the act of swallowing its prey.

The Glass Snake is one of the earliest of the reptile tribe to make its appearance in the spring, shaking off its lethargy and coming out of its home to bask in the sunbeams and look after the early insects, long before the true snakes show themselves. It is generally found in spots where vegetation is abundant, probably because in such localities it finds a plentiful supply of the insects, small reptiles, and other creatures on which it feeds.

It is fond of frequenting the plantations of sweet potato (*Convolvulus batatas*), and during harvest-time is often dug up together with that vegetable. The home of this reptile is made in some very dry locality, and it generally chooses some spot where it can be sheltered by the roots of an old tree, or a crevice in a convenient bank. It moves with tolerable rapidity, and its pursuer must exercise considerable quickness before he can secure it.

To catch a perfect specimen of the Glass Snake is a very difficult business, for when alarmed, it has a remarkable habit of contracting the muscles of its tail with such exceeding force that the member snaps off from the body at a slight touch, and sometimes will break into two or more pieces if struck slightly with a switch, thus earning for itself the appropriate title of Glass Snake. The common blind-worm, which will be described in a future page, possesses a similar capacity, and often uses it in a rather perplexing fashion. Catesby remarks that this separation of the tail into fragments is caused by the construction of the joints, "the muscles being articulated in a singular manner quite through the vertebræ." The tail is more than twice the length of the body, from which it can only be distinguished by a rather close inspection.

The head of the Glass Snake is small in proportion to the body, rather pyramidal in shape. Along each side of the body runs a rather deep double groove. The coloring of this creature is extremely variable, but is generally as follows: The head is mottled above and at the sides with black and green, and the jaws are edged with yellow. The upper part of the body is marked with multitudinous lines of black, green, and yellow, and the abdomen is bright yellow along its length. In the tail there are about one hundred and forty rings of scales. Sometimes the upper surface is black on the sides and neck, and brown on the back, the head being marbled with yellow and black; another variety is chestnut above, with white spots edged with black, and the under parts pale orange; while a third variety is gray mottled with black. The total length of this reptile is from two to three feet.

The Glass Snakes are represented in North America by the *Opheosaurus ventralis*. It is seemingly a serpent, having no external limbs. The tail is very brittle, and the animal has from that fact been regarded as so brittle that a blow will fracture the body. The truth is, there are thin transverse *septi* between the vertebæ, and this is the point where separation takes place so readily. Its range is from Tennessee southward from Kansas.

FOUR small families now follow, containing but very few individuals. The first of these is called the CHALCIDÆ. These reptiles have long cylindrical bodies, with a slight granular groove on the front of each side, and four very short rudimentary limbs. The typical species of this family is the CHALCIS (*Chalcis flavescens*), a native of tropical America, Guiana, and the neighboring parts. The fore-feet have three toes, but the hind-feet are undivided, so as to form a single toe. The scales are squared, and arranged in twenty longitudinal series on the back, and six series on the abdomen.

THE next family, the ANADIADÆ, contains, as far as is known, only one species, the EYED ANADIA (*Anadia ocellata*), thought to inhabit tropical America. In this creature the lower eyelids are pellucid, the scales of the back and sides six-sided and not overlapping each other, while those of the abdomen are squared. The limbs are four in number, and there are five unequal and rather flattened toes on each foot. The color of this species is pale brown, with a bronze gloss, deepening on the sides, and having some white spots edged with black towards the front. Beneath it is shining white.

IN the family of the CHIROCOLIDÆ there is likewise only one species, called the CHIROCOLE (*Heterodactylus imbricatus*), a native of Brazil. This creature has a double collar, and the ears are hidden beneath the skin. The scales of the back, the sides and the tail, are six-sided, rather sharp, arranged in regular rings, and furnished with keels. Those of the abdomen are squared and arranged longitudinally in six rows. There are four short legs, with five toes on each foot, the thumb of the fore-limbs being only rudimentary. The color of the Chirocole is brown, with a pale streak on each side.

THE fourth family is the CERCOSAURIDÆ, containing two genera. These animals have the ears distinct, the throat with a double series of shields, and the collar distinct. On the back and upper part of the tail the scales are large, boldly keeled, and arranged into a regular longitudinal series. The scales of the under portions are squared and flat. There are four limbs, each with five unequal toes. A good type of this family is afforded by the EYED CERCOSAURUS (*Cercosaura ocellata*). The body of this creature is long and rather cylindrical. Its color is black, with four white streaks, the head and the under parts are yellowish, and the sides are sprinkled with green, and variegated with eight or nine white spots edged with black.

OUR last example of the Cyclosaurian reptiles is the ANGUINE LIZARD, or CHAMÆSAURA, the only representative of its family.

The Anguine Lizard is a native of Southern Africa, and is obtained from the Cape of Good Hope. Of its habits there is but little known. It is a curious-looking creature, exceedingly snake-like in general appearance, its four limbs being of the most rudimentary character, small, delicate, feeble, not even separated into toes at the extremity, but ending in a single claw, as if the whole limb were only composed of one small joint. These imperfect limbs are wholly useless for progression, those of the anterior extremity being hardly larger than the long, narrow scales with which the body is covered, and the hinder pair exhibiting but very little more development.

So perfectly serpentine is the form of this creature, that the mark of separation between the tail and body is so slightly defined that the precise line of junction is almost invisible ; whereas, in the common blind-worm, itself a most snake-like reptile, the line of demarcation is plainly shown by a decided diminution in the diameter. The tail is very long and slender, measuring more than twice the length of the body.

The head of the Anguine Lizard is covered with regular, many-sided shields, and the temples, and the whole of the body and tail are clothed with scales, their edges projecting boldly, and arranged in a series of regular rings, or "whorls." Along the back there are six rows of broad scales, and on the sides and abdomen the scales are long, narrow, and with a decided keel running along their central line. There is no groove along the sides, which are rounded. Upon the head the plates are rather long, keeled, and project very slightly over each other. The ears are distinct. The color of the Anguine Lizard is brown, and along each side runs a long yellow streak.

GEISSOSAURI.

A SECOND tribe of Lizards now comes before our notice. These are the GEISSOSAURI, a title derived from two Greek words, the former signifying the eaves of a house, and the latter a Lizard. As in this tribe there are many families, and more than eighty genera, it will be impossible to give more than a very slight account of these reptiles, or even to mention more than a small number selected as types of the large or small groups which they represent.

Indeed, the lower we descend in the scale of creation, the more numerous the species seem to become, and the more perplexing is the task of selecting those species which are worthy of mention on account of their scientific characteristics, and yet possess sufficient individuality to interest the general reader.

To watch the greater number of reptiles in their wild state, is a task simply impossible for any human being to achieve. Many reptiles live in dry and thirsty lands, where no creatures but the white ant and the Lizard seem to acquire moisture, and through which the traveller can only pass with hasty steps, dreading the delay of each minute, lest his precious store of water should fail, and leave him to perish by the most terrible of deaths.

Others reside on the sides of precipitous rocks, over which the enterprising traveller can only pass at hazard of life and limb, and in any case would not be able to watch the proceedings of the shy and timid Lizards that find their home among these craggy recesses, and retreat into them on the slightest alarm. But the chief residence of the reptile race is to be found in hot climates, and in low, swampy ground, where the morasses are ever filled with decaying vegetable matter, and exhale a soft, thick, miasma, as deadly to the white man as the fumes of arsenic, and injurious even to the dark-skinned native, who can breathe unharmed a fetid atmosphere that would smite down his white master as quickly and surely as if he were struck with a bullet, and who only attains his fullest development under these conditions.

In these dread regions, their seething putridity concealed by all the luxuriant vegetation of tropical climes, like a royal mantle flung over a festering corpse, the reptile race abound, the poisoned air being to these creatures the very breath of life, and the surrounding decay the sustaining power of their existence. Indeed, the object of their lives seems to be, by individual transmutation of poisons into living flesh, to destroy by slow but certain degrees the mass of decaying vegetation, and so to prepare an abiding place for beings of a higher order than themselves.

On placing ourselves even in imagination amid such scenes, we seem to be transported back into the former ages of our earth, when man could find no resting-place for his foot, and no atmosphere in which he could breathe and live; when the greater part of the soil was little more than soft mud, the air thick, dank, heavy, and overcharged with decomposition, and the multitude of strange reptiles that bored their slimy way through the deep ooze, crawled lazily upon the slowly hardening banks, or urged their devious course through the turbid waters, were the physically ruling though morally subservient powers of the world.

Little is wanting to complete the illusion, except to give to every object an increase of dimensions; for the vegetation of those days was rank and luxuriant to a degree that is now well indicated, though on a smaller scale, by the foliage of the tropics, and the huge forms of the ancient and now extinct reptile race are closely reproduced by the more familiar inhabitants of the swamp before us.

As the expanse of putrefaction was greater in those epochs, so the miasma destroyers were larger. Frogs and toads as big as calves, reptilian quadrupeds as large as elephants, and reptilian bats expanding leathery wings as wide as those of the pelican, were fit inhabitants of the atmosphere which they breathed, and in which their mission was consummated. Now that the marshy districts are smaller and less poisonous, the reptile race that inhabits them is of smaller dimensions.

The earth has now been so far purified by successive generations and regenerations of life and death, added to human ingenuity and industry, that its harmful districts occupy but a comparatively small portion of its surface, the greater part of the world being suitable for human habitations, the black man settling as a pioneer, a hewer of wood and drawer of water, where the white man cannot yet abide. But in all those localities where the miasmatic exhalations impall the land with their pestilential mantle, and scatter the seeds of death on every breeze, the reptiles may be found luxuriating amid the deadly elements, and thriving in spots where the foot of man dares not tread, and his inquiring eye ventures not to penetrate.

THE PYGOPUS.

THE first family of this tribe is distinguished by the apparent absence of eyelids, those organs being only rudimentary and scarcely visible, so as to give to the eyes a superficial resemblance to those of the serpents. On account of this peculiarity, the reptiles belonging to this family are termed the Gape-eyed Skinks. Their bodies are spindle-shaped, their tongues are scaly, nicked at the tip, their teeth are conical, and their limbs are four in number, and very feeble.

These creatures are found in various parts of the globe, but Australia seems to be their favorite home. The PETE, or AUSTRALIAN TILIQUA (*Crytoblepharus boutonii*), is a good example of the Gape-eyed Skinks, or GYMNOPHTHALMIDÆ, a long name derived from two Greek words signifying naked-eyed. As its name imports, this reptile is a native of Western Australia, but it is also found in other parts of the world, specimens having been taken in Timor and the Mauritius. The color of the Pete is olive, sometimes with a wash of bronze, mottled with brown, and variegated with little black streaks. Sometimes there is a bright yellow streak on each side. Its eyelid is circular and scaly, and the three upper scales are the largest.

THE next family is well represented by the PYGOPUS, or NEW HOLLAND SCHELTOPUSIC, a curious reptile that inhabits Australia.

PYGOPUS.—*Pygopus lepidopus*. (Two-thirds natural size.)

This creature might easily be mistaken for the snake-like Lizard called the Scheltopusic, which has already been described on page 52, as the two fore-legs are entirely absent, and the hinder pair are very small, rudimentary, and set so closely against the body that they would escape a casual glance. They are flattish, covered with scales, and are not even divided into joints or toes, so that they are wholly useless for progression, the Pygopus creeping along after the ordinary fashion of snakes.

If the creature be turned on its back, a curious arrangement of scales is seen. Between the bases of the lower limbs, several large, shield-like scales are seen, and just above them is a row of rather long and arched scales, extending in a semicircular form from one limb to the other, and looking much like the stones that are set upon the summit of an arched doorway. Each of these scales is pierced with a circular pore, so that the general effect is very striking. The whole body of this reptile is very long in proportion to its width, and it has altogether a very serpentine aspect.

The head of the Pygopus is rather short, and is covered above with some rather large shields, that upon the top of the head being equal to any two others in size. The scales of the back are keeled, and its color is coppery gray, with five rows of rather oblong white spots with black centres, and a few black streaks drawn obliquely upon the sides of the neck.

THE DELMA (*Delma fraseri*) is very like the Pygopus, but may be distinguished from it by the scales of the back, which are smooth and without keels, by the shorter hinder limbs, the absence of the pores, and the elliptical shape of the pupil of the eye, that of the Pygopus being circular.

Two more small families of reptiles are worthy of a passing notice. The first is that which is represented by a single species, the APRASIA (*Aprasia pulchella*), and remarkable for being destitute of limbs, and having none of the pores which have just been mentioned. The body is lengthened, and covered with six-sided scales on the upper surface and flanks. The scales are quite smooth, and their color is pale brown, with a dot of dark hue in the centre of each scale, giving a sort of variegated aspect. Along the flanks these dots become longer, so that they almost join each other, and form imperfect streaks on the sides. The lips are yellow. This reptile inhabits Western Australia.

THE next family contains only one genus, which, like the preceding creature, inhabits Australia. In these reptiles the head is long and flattened, the pupil of the eye elliptical and upright, the scales are oval, smooth, and overlap each other, and the curious pores are present, each set in the front edge of a scale. BURTON'S LIALIS (*Lialis burtoni*) may be taken as an example of this family. The color is olive above, with five imperfect brown streaks, and gray below, with large whitish spots.

THE large and important family of the Skinks contains between forty and fifty genera, nearly each of which possesses one or more species, concerning which there is something worthy of notice. In these reptiles the head is rather squared than rounded, and covered regularly with horny shields. The body is mostly spindle-shaped, though sometimes of a cylindrical form, and very much elongated, in which case the legs are generally rudimentary, and sometimes altogether wanting externally. The common blind-worm is a familiar example of this structure. The tail suits the form of the body, being cylindrical in the long-bodied species, and tapering in those of a more spindle-like shape.

The genus in which the COMMON, or OFFICINAL SKINK is placed, is now so restricted, that it only contains a single species; but in the earlier times of zoological science, its rules were so greatly relaxed, that many species were admitted within its limits.

In this genus the muzzle is wedge-shaped, the scales are thin and smooth, and the tail conical and pointed. The toes are rather flattened, and fringed on the side. They eyes are guarded by distinct eyelids, the lower of which is covered with scales. The palate is furnished with teeth, and has a longitudinal groove, and the ears are small, and toothed in front. There are four short and rather stout limbs, tolerably strong, and enabling the creature to make its escape from its enemies by rapidly sinking below the sandy soil on which it is usually found.

The Skink is a native of Northern Africa, and is very common in some localities. Specimens are said to have been found in some portions of Asia, and it seems to be clearly proved to inhabit Syria and several parts of India.

It is a tolerably active little Lizard, not running fast or far, but contenting itself with hanging about the same locality, and feeling itself more secure on the sandy soil of its native districts, than if wandering at large on the plains. Indeed, unless it is alarmed, or except when it is aroused to short exertions by the presence of its prey, the Skink seldom troubles itself to hurry its pace beyond a slow crawl; and not even when most startled, does it attempt to seek safety in flight. No sooner does it perceive the approach of danger, than it slips below the sand with such singular speed and adroitness, that those who have witnessed this performance, say that it seems rather to be gliding into some hole already excavated, than to be engaged in the labor of sinking a tunnel for the purpose of aiding its escape. Several travellers have seen the Skink thus bury itself, and have all carried away the same opinion of its powers.

If quietly approached, it may often be detected sleeping in the hot sunbeams, lying stretched at length upon the stones or rocks, and so far steeped in slumber, that it may be approached quite closely without taking alarm.

The name of Officinal Skink has been given to this reptile on account of the high place which it formerly held among the medical profession, and the extreme value which it was thought to possess when dried, pounded, made up neatly into draughts or boluses, and used as a medicine. There is hardly a disease to which the human race is liable, which was not thought curable by the prepared body of this reptile, certainly not the least repulsive of all the disgusting substances which the early physicians delighted to choose from the animal, vegetable, and mineral kingdom, to fill their multitudinous boxes and bottles, and to inflict upon their patients. Sometimes a physician would even evince his belief in the efficacy of his medicine by taking it himself, and would swallow, with full belief in its

COMMON SKINK.—*Scincus officinalis.*

healing powers, the burnt liver of a hyena, the moss from a dead man's skull, the grated flesh of a mummy, or the remains of a pounded Lizard, together with many other substances too revolting to mention.

Did a warrior receive a wound from a poisoned arrow, or was a woodman bitten by a venomous snake, there was nothing so effectual for the cure as the dried flesh of the Skink, sometimes called El Adda, and sometimes known by the name of Dhab. He who provided himself with this all-powerful medicine was secure against fits of all kinds, which never attacked the system fortified by a dose of powdered Skink, or were speedily driven away if the sufferer had not previously partaken of this panacea. All skin diseases were cured by the Skink, and even the fearful elephantiasis yielded to its potent sway.

Were the system too inexcitable and lethargic, and did the blood course too slowly through the veins, a little Skink powder would restore the natural powers to their full vigor. Or, on the contrary, if the patient happened to be feverish, restless, with a burning forehead, a parched skin, and a hurried pulse, a dose of the same useful medicine would cool the system, cure the headache, and bring the pulse to its normal state. It is an infallible remedy for worms, eradicates cancer, and removes cataract. In fine, a satisfactory estimate of its valuable properties may be gained by perusing, in the daily journals, any advertisement of any patent medicine, together with the list of maladies for which it is a certain remedy.

Even in the present day, this medicine is in great vogue among the sages of the Eastern Hemisphere. Should the reader happen to travel into eastern lands, and fall sick of a fever, be afflicted with a sunstroke, find himself suddenly smarting with a nettle-rash, catch a cold, or suffer from sand-blindness, the remedy which will, in all probability, be offered to him, will consist of this universal panacea. In the time of the ancients, the Skink was in much favor as a medicine, and was imported largely to Rome, ready prepared in white wine. The heads and feet were considered the most efficient portions of the animal, and were relied upon as infallible renovators of a constitution broken by age, or shattered by excess.

Wherever modern civilization has most penetrated, the Skink has, happily for itself, fallen greatly in medical estimation, and in some places is entirely rejected from the pharmacopeia; though there are not wanting some European physicians who assert that the creature really does possess some valuable properties, but that it has fallen into disrepute through the over-estimate which had been formed of its powers, and which naturally created a reaction in the opposite direction.

In Southern Egypt it still commands the firm belief of the people, and is hunted down with the greatest zeal, as it not only can be applied to the personal ailings of the captors, but can be quickly dried in the burning sunbeams, and sent to Cairo and Alexandria, where it commands a ready sale.

In its habits, this Skink much resembles the generality of terrestrial Lizards of its size and locality. As it seeks for safety below the sand, it is generally to be seen upon the hillocks of fine loose sand which are collected by the south wind, at the foot of any tree which may manage to survive in so ungenial a soil, or are blown against the hedges of the more cultivated land. It generally lies quietly upon the sand, but occasionally starts into vigorous action when it perceives an insect passing within easy reach, makes a sudden rush, captures its prey, and subsides again into its former inactive repose. Beetles are its favorite food, and of these insects it will eat a considerable quantity, but can preserve life for a lengthened period without taking any food at all.

Should it be disturbed, it instantly sinks below the sand, with almost magical quickness; and, according to M. Lefebvre, who collected a great number of these Lizards, a few seconds suffice it for constructing and retiring into a burrow several feet in depth. Although armed with tolerably strong teeth and claws, it does not attempt to bite when captured, and any scratch inflicted on the hand of the captor is merely caused by its struggles while endeavoring to effect its escape.

The general color of the Officinal Skink is reddish dun, crossed with bands of a darker hue above. Below and upon both the flanks, it is of a silvery whiteness. It is, however, liable to considerable variations, of which the most important may be briefly denoted as follows: In one variety, the upper parts are yellow, or silver-gray, with seven or eight large brown spots on the sides. In another, the head is yellow, the upper parts are chestnut-brown, profusely sprinkled with little white spots, each scale having two, or even three, white dots upon the surface. The back is marked with a series of broad white bands, generally five or six in number, and having a black patch at either extremity of each band. In another variety, the upper parts are silvery-gray, splashed with pure white, and variegated with irregular brown spots. But however great may be the variations, they are all confined to the upper surface, the abdomen, flanks, and under surface retaining their beautiful silvery whiteness. The banded variety is the most common. The Officinal Skink is by no means a large reptile, seldom exceeding eight inches in length, and being generally about six or seven inches long. The specimen shown in our illustration is drawn of its natural size.

The Skinks form a family of which fifty genera and one hundred and fifty species are enumerated, distributed throughout all parts of the world. Of these the BLUE-TAILED SKINK (*Eumeces fasciatus*) is very abundant in the Southern States, east of the Rocky Mountains. It is black, with fine yellow streaks, the middle one forked on the head. The tail is mostly blue.

THE SCORPION LIZARD.

WESTERN SKINK (*Eumeces septentrionalis*) is found in Nebraska and Minnesota. The COAL SKINK inhabits the Alleghanies, from Pennsylvania southward. The GROUND LIZARD, or SKINK, Mocoa so called, also, is abundant in the Southern States.

THE RED-HEADED SKINK (*Plestiodon erythrocephalus*), according to Dr. Dekay, inhabits Pennsylvania, and extends southward to Florida. Its length is twelve inches. Two other Skinks are recorded by Holbrook, the Five-lined Skink, and the Striped Skink. Both are exceedingly pretty creatures, inhabiting the Gulf States.

PASSING by one or two genera of considerable extent, such as Hinulia and Mocoa, the members of which are mostly found in Australia, though there are species which inhabit

HINULIA AND MOCOA.—*Trachysaurus rugosus.*

China, Java, the Philippines and New Zealand, we come to a reptile very well known by the popular title of the SCORPION LIZARD, and called more scientifically, as well as more correctly, the BROAD-HEADED PLESTIODON.

In spite of the rather alarming name which the terrors of the ignorant have caused them to bestow upon it, the Scorpion Lizard is one of the most harmless, as well as one of the most useful little creatures that inhabit the earth.

It is a native of Northern America, and is spread over a very large tract of country. This curious Lizard is one of the species that delights in trees, and of which we shall see more in a future page. It generally resides in some tree buried in the depths of the forest, and remains at a considerable elevation above the ground, never liking to make its home less than thirty or forty feet above the earth, and often placing itself at a much greater height.

The domicile in which this reptile most delights is the deserted home of a woodpecker, which has brought up her little family, and forsaken the burrow which had taken such time and trouble to hollow from the decaying wood. Here the Scorpion Lizard takes up its residence, and here it remains snugly concealed unless it is alarmed by an enemy at the gate of its wooden fortress, when it runs nimbly to the entrance, and pokes out its red head with so threatening a gesture, that its intending assailant, thinking it must possess a store of poison to assume so resolute an aspect, retreats from the spot and leaves the reptile in quiet possession of its abode.

Happily for the Lizard, the belief in its venomous propensities is widely diffused and deeply engrained in the popular mind, so that without having a single dangerous property except that of undaunted courage when driven to bay, it has established a reputation for ability to avenge itself when injured, which is of no less service to reptiles than men. Not that it is wholly destitute of offensive weapons, for its teeth are strong and sharp, its feet powerful, and its claws are sufficiently pointed to scratch rather deeply.

The Scorpion Lizard is naturally a very timid and retiring creature, and on the approach of danger slips quietly out of the way, wisely preferring flight to combat. But if seized, the captor will have no small struggle before he can fairly secure his small but determined quarry, for the creature bites so fiercely with its sharp teeth, retains its hold with such bull-dog tenacity, and kicks and scratches with such hearty good will, that the non-zoological populace may well be excused for thinking it to be venomous in tooth as well as in temper. The bite, indeed, is so severe, and the creature has such power of jaw, that the wounds inflicted are always exceedingly painful for an hour or two, and might give rise to the idea that the teeth were poisonous like those of the rattlesnake.

The Scorpion Lizard is seldom seen except upon trees, where it can mostly find a sufficiency of food among the insects that always haunt the branches of trees, and of drink in the dewdrops that collect at morning and evening. When, however, it needs a more abundant diet, it descends to the ground for a short visit, but after satisfying its wants, it returns to its tree, runs easily up the trunk, and again establishes itself in its burrow.

The head of the Scorpion Lizard is very broad at the base, and narrows rather suddenly to the snout, which is slightly elongated. The upper part of the head is of a bright red color. The body is olive-brown above, and the throat, abdomen, and whole of the under parts, are yellowish-white. Just in front of the ear is a series of oblong tubercles, and the temples are smooth and covered with rather large shields. The feet are large in proportion to the size of the body, and the toes are rather compressed and exceedingly delicate, in fact almost threadlike in form. The length of the Scorpion Lizard is generally about eleven or twelve inches.

THERE is a closely allied species, also common in North America, popularly called the BLUE-TAIL, and scientifically the FIVE-LINED PLESTIODON (*Plestiodon quinquelineátum*).

Like the preceding species, the Blue-tail inhabits the deepest forests, but is not one of the arboreal reptiles, being always found upon the earth, usually remaining within a short distance of its home, which is made in one of the numerous decaying tree-stumps which are found in these vast forests. Its food consists of insects, which it catches principally upon the ground.

The head of this Lizard is red, like that of the scorpion, but of a much paler quality. The body is olive, with five longitudinal white streaks, the central stripe being forked in front, and with two black bands. The tail is brownish, with a decided wash of blue during the life of the animal, a coloring which has earned for it the popular name of Blue-tail. It is, however, subject to slight variations in the color and shape of the markings. There are several little lobes in front of the ears. The length of the Blue-tail is about eight or ten inches.

A REPTILE which bears some resemblance to the scorpion Lizard is found in Jamaica and the West India Islands, where it seems to take the place of that creature, and to enjoy a reputation almost as bad, with as little cause. The negroes call it by the name of MABOUYA (*Mabouya ágilis*), but as they apply that term to anything which is, or which they consider to be venomous, and as there are very many really poisonous creatures in those countries, and many more which are falsely thought to be so, the word is rather vague in its application.

The Mabouya is a good climber, running up trees with perfect facility, and having a tendency to traverse the huts of the negroes, much to the consternation of the inmates. Its usual habitation, however, is made in the holes of old, decaying trees, and except during the very hot weather, it mostly remains at home. There is another reptile, inhabiting the same country and to which the same title is applied, and which will be mentioned in a future page.

The lower eyelid of the Mabouya is remarkable for a little transparent disc in the centre, the palate is without teeth, and the scales are smooth. Along the back run four black streaks, the two central stripes extending only to the middle of the body, while the two external lines are prolonged nearly to the insertion of the hinder limbs.

THE great family of the Skinks finds a familiar representative in the common BLIND-WORM, or SLOW-WORM, which, from its snake-like form and extreme fragility, might well deserve the title of the glass snake. In this reptile there is no external trace of limbs, the body being uniformly smooth as that of the serpent, and even more so than in some of the snakes, where the presence of the hinder pair of limbs is indicated by a couple of little hook-like appendages. Under the skin, however, the traces of limbs may be discovered, but the bones of the shoulders, the breast, and the pelvis are very small and quite rudimentary.

This elegant little reptile is very common throughout Europe, and is also spread over some portions of Asia, not, however, being found in the north. It is plentiful along hedgerows, heaths, forest lands, and similar situations, where it can find immediate shelter from its few enemies, and be abundantly supplied with food. It may often be seen crawling leisurely over a beaten footpath, and I have once captured it while crossing a wide turnpike road.

Why the name of the Blind-worm should have been given to this creature I cannot even conjecture, for it has a pair of conspicuous though not very large eyes, which shine as brightly as those of any animal, and are capable of good service. Indeed, all animals which prey upon insects, and similar moving things, must of necessity possess well-developed eyes, unless they are gifted with the means of attracting their prey within reach, as is the case with some well-known fishes, or chase it by the senses of hearing and touch, as is done by the mole. Moreover, the chief food of the Blind-worm consists of slugs, which glide so noiselessly that the creature needs the use of its eyes to detect the soft mollusk as it slides over the ground on its slimy course. Speed is not needful for such a chase, and the Blindworm accordingly is slow and deliberate in all its movements except when very young, when it twists and wriggles about in a singular fashion as often as it is touched.

The great fragility of the Blind-worm is well known. By a rather curious structure of the muscles and bones of the spine, the reptile is able to stiffen itself to such a degree, that on a slight pressure, or trifling blow, or even by the voluntary contraction of the body, the tail is snapped away from the body, and on account of its proportionate length, looks just as if the creature had been broken in half. The object of this curious property seems to be to insure the safety of the animal. The severed tail retains, or rather acquires, an extraordinary amount of irritability, and for several minutes after its amputation, leaps and twists about with such violence, that the attention of the foe is drawn to its singular vagaries, and the Blind-worm itself creeps quietly away to some place of shelter.

Even after the movements have ceased, they may be again excited by touching the tail with a stick, or even with the finger, when it will jump about with a vigor apparently undiminished. On frequently repeating the process, however, the movements become perceptibly less active, and after awhile the only sign of movement will be a slight convulsive shiver. Half an hour is, as far as my own experience goes, the limit to which this irritability endures.

I well remember meeting with an incident of this nature. I had come suddenly upon a reptile among the rank grass and underwood, that I at first took for a viper, and at which I aimed a thrust with a little twig of decaying wood, which broke at once. Immediately after the thrust, something began to hop and plunge about most violently just by my feet, and having a very wholesome dread of a viper's fangs, I jumped back a step or two, to the great indignation of a swarm of bees, which had settled themselves in the ruins of an old wooden hut close to the spot. They at once intimated their displeasure in that wing-language so expressive to all bee-owners, so hastily tossing the writhing object to a distance with the shattered remnant of the stick, I got away from the bees, and experimented for some time on the tail of the Blind-worm, as it proved to be. Even the flight through the air, and the heavy fall, seemed to have little or no effect upon the irritability of the severed member, and when I reached it after its fall, I found it hopping about quite merrily.

When the tail of the Blind-worm is thus snapped off, the scales of the body project all round the fractured portion, forming a kind of hollow into which the broken end of the tail can be slipped.

According to popular notions, the Blind-worm is a terribly poisonous creature, and by many persons is thought to be even more venomous than the viper, whereas it is perfectly harmless, having neither the will nor the ability to bite, its temper being as quiet as its movements, and its teeth as innocuous as its jaws are weak. I fancy that the origin of this opinion may be found in the habit of constantly thrusting out its broad, black, flat tongue with its slightly forked tip; for the popular mind considers the tongue to be the sting, imagining it to be both the source of the venom, and the weapon by which it is injected into the body, and so logically classes all creatures with forked tongues under the common denomination of poisonous animals.

It is said that this reptile will bite when handled, but that its minute teeth and feeble jaws can make no impression upon the skin; and also that when it has thus fastened on the hand of its captor, it will not release its hold unless its jaws be forced open. For my own part, and I have handled very many of these reptiles, I never knew them attempt to bite, or even to assume a threatening attitude. They will suddenly curl themselves up tightly, and snap off their tails, but to use their jaws in self-defence is an idea that seldom appears to occur to them.

The pertinacity with which the notion of the Blind-worm's venomous properties is implanted in the rustic mind is really absurd. During the summer of this year, I passed some little time in a forest, and having gone round to the farms in the neighborhood, as distances of several miles are euphuistically called, begged to have all reptiles brought to me that were discovered during the haymaking. In consequence, the supply of vipers and snakes was very large, and on one occasion a laborer came to the house, bare-headed, his red face beaming with delight, and his manner evincing a proud consciousness of deserving valor. Between his hands he held his felt hat tightly crumpled together, and within the hat was discovered, after much careful manœuvring, the head of a Blind-worm emerging from one of its folds.

As I put out my hand to remove the creature, the man fairly screamed with horror, and even when I took it in my hand, and allowed it to play its tongue over the fingers, he could not believe that it was not poisonous. No argument could persuade that worthy man that the reptile was harmless, and nothing could induce him to lay a finger upon it; the prominent idea in his mind being, evidently, not that the Blind-worm had no poison, but that I was poison-proof. To add to his alarm, the creature had snapped off its tail during the rough handling to which it had been subjected—a proceeding which, by his peculiar process of reasoning—only corroborated its venomous properties.

In its wild state the Blind-worm feeds mostly on slugs, but will also eat worms and various insects. Some persons assert that it devours mice and reptiles; but that it should do so is a physical impossibility, owing to the very small dimensions of the mouth and the structure of the jaw, the bones of which are firmly knitted together, and cannot be separated while the prey is being swallowed, as is the case with the snakes.

In captivity it seems to reject almost any food, except slugs; but these molluscs it will eat quite freely. I have kept a specimen in my possession for about four months, which has proved a very interesting creature. After keeping it for a fortnight, I procured six or seven white garden slugs, and placed them in the glass vessel, together with the Blind-worm.

The reptile instantly saw its prey, but did not move from its place, merely following with a slow movement of the head the course of one of the slugs that crawled within an inch or two of its nose. Presently it raised its head very deliberately, and hovered over the slug as it glided along, and, after following it for an inch or two, quickly opened its mouth to the full extent, lowered its head, and grasped the slug just behind the head, squeezing it with some force, and causing a great commotion among the muscles of the foot.

Presently it relaxed its hold a little, again opened its mouth and took a fresh grasp, and after three or four of these movements, it contrived—how, I cannot comprehend, though I have watched the creature over and over again—to get the head of the slug down its throat.

The process of swallowing was then very easy, and, after a few more efforts, the whole of the mollusc had disappeared. After resting for a few minutes, it attacked another slug precisely in the same manner; but I have seldom seen it eat more than two or three at one meal. By degrees it caught and ate all the slugs, and it will finish a dozen in a week or ten days.

After a short time my Blind-worm unexpectedly became the mother of a numerous progeny, nine little Blind-worms having made their appearance in the world during the night. They were remarkably pretty little creatures, and so unlike their parent, that few persons would attribute them to the same species. They are much more serpentine in their general aspect, their heads being considerably wider than their necks, whereas in the adult the head and neck are as nearly as possible of the same width.

Their color is shining creamy-yellow above, and jetty-black below, the line of demarcation running along the flanks, and being very sharply defined. Along the back runs a narrow black line, which upon the head is expanded, and then divides so as to form a letter Y. Just above the nose is another forked, black mark, looking like an inverted V, and both these letters have a notable circular enlargement at the angle. As the creature grows, the V mark becomes gradually uncertain, and finally disappears; but the black line down the back, and its Y-like termination, retain their position through life, though they are not so conspicuous as in the young, owing to the darker coloring of the surface.

How these little things feed I cannot make out. Though the little creatures born in my house had lived for about five weeks, had grown considerably, and had always been very lively, they had taken no food so far as I could discover. For the first three weeks of their life, they lived in a glass jar closed at the top, and with an inch or so of dry earth at the bottom, in which there could be no nourishment. A little milk was poured on the mould now and then; and they perhaps may have licked the moistened earth, and so have obtained some little nourishment, though they were never seen to do so, and indeed appeared perfectly indifferent to the milk.

When I introduced the slugs, the odd little reptiles acted just as their mother was doing, followed the slugs about with their heads, hovered over them, made believe to eat them, and then were quietly walked over by their intended prey, which, being nearly twice as big as themselves, proceeded on its course without paying the least regard to the tiny reptiles, whose bodies were not larger than ordinary knitting-needles, and easily glided over them, or put them to ignominious flight.

After they had been in the jar for some time, I fitted up an old aquarium in a manner intended to imitate as far as possible their natural home, building a bank of earth and stones at either end, laying turf in the middle, and planting ferns upon the banks, with moss round their roots. They enjoyed the change very greatly, immediately proceeded to burrow in all directions through the earth and among the stones, until they established a whole series of tunnels through which they can glide at will, and seem to take great pleasure in permeating their establishment at all hours, especially delighting in pushing their way through the moss and then retreating into their burrows.

On a cold day they bury themselves below the mould; but the first gleam of sunshine that plays among the green fern-leaves brings them from their recesses, and causes them to glide about the moss and turf most merrily. Sometimes, when they are coiled asleep within their home, their bodies are pressed against the glass, and it is curious to see how immovable they will lie, in spite of tapping the glass, but how soon they wake up and brisk they become when the glass is warmed. Even a few warm breaths upon the glass suffice to awake them.

I think that I have discovered another kind of subsistence for the young; but that has only been possible since they have been placed in the aquarium, or rather, the fernery, as it is now. Sundry very minute insects of the dipterous order may be seen flitting about within the glass, probably having been introduced with the turf and ferns; and it is possible that the young Blind-worms may contrive to catch and eat these creatures, and derive some nutriment from them, in spite of their diminutive size.

When wild, the Blind-worm generally retires to its winter-quarters towards the end of August, or even sooner, should the weather be chilly. The localities which it chooses for this purpose are generally dry and warm spots, where the dried leaves and dead twigs of decayed branches have congregated into heaps, so as to afford it a safe refuge. Sometimes it bores its way into masses of rotten wood; and on heathery soils, where the ground slopes considerably, it selects a spot where it will be well sheltered from the winter's rains and snows, and burrows deeply into the dry loose soil.

It is singular to see the creature emerging from the ground when the least touch will soil the fingers, and to see how totally free from earth stains is the bright glittering skin of the reptile, upon which not a particle of mud can cling. I once detected upon the head of my specimen a projection which I thought was a little lump of mud, I having just watered the ferns and turf, greatly to the discomfiture of the Blind-worms, both old and young; but, upon close examination, I found it was only a little pebble which had lodged upon the head, as the reptile came hastily out of its burrow to avoid the water. So quietly did the Blind-worm move, that the stone retained its place upon the head for several minutes, and did not fall off until I startled the creature, and caused it to turn its head rather sharply.

The Blind-worm would be a most useful inhabitant of a garden—not at all repulsive, and, indeed, very seldom seen, its instinct teaching it to remain within some dark recess during the day, and only to come out at night when the slugs leave their earthy hiding-places, and commence feeding. Moreover, it is very prolific, and needs no special appliances, as is the case with the frog and toad, which require the presence of water to produce and hatch their young, and for the little reptiles to come to maturity. Sometimes the number of young is twelve or thirteen, and sometimes there are only seven or eight. The usual average is, however, nine or ten; and they are very hardy little things, requiring no care whatever.

Being one of the earliest to retire into its winter quarters, the Blindworm is one of the first reptiles to leave them, appearing before either the snake or the viper. The reason for this early appearance is simple enough. Neither creature can venture into action when it can find no food, the active powers of the body causing a waste which must be restored with nutriment. The snake feeds upon frogs, and therefore cannot leave its winter's home until it finds the frogs ready for it. The frogs, again, which feed upon insects, must wait until the vegetation has attained sufficient luxuriance to afford food for their insect prey; but the Blind-worm, which finds its nourishment among the mollusks which devour the earliest leaves, is able to leave its winter quarters as soon as the vegetation begins fairly to sprout, and the slugs to devour it.

Even during the winter, a warmer sunbeam than usual will tempt the Blind-worm to come to the mouth of its burrow, poke out its head, and enjoy the temporary, but cheering warmth. My own specimens have not yet made any preparations toward retiring to winter quarters, though the usual time has passed away nearly two months ago, a circumstance which is probably due to the warmth of their home, and the occasional supply of slugs which I now and then put into the case.

Like the snakes, the Blind-worm casts its skin at regular intervals, seeming to effect its object in various modes, sometimes pulling it off in pieces, but usually stripping it away, like the snakes, by turning it inside out, just as an eel is skinned. Some persons, who have witnessed the process, state that this eversion is only extended to the base of the tail, and that the entire tail is drawn out of the skin like a hand out of a glove. Mr. G. Daniel mentions, that a Blind-worm in his possession cast its skin in so many pieces, that the largest portion was only two inches in length. The process began by a split along the abdomen, and the head was the last part extricated from the rejected integument. This mode of shedding the skin was, however, owing, in all probability, to some weakness in the individual, or to the want of the usual aids, such as the stems of grass, heather, and other vegetation, against which the reptile contrives to rub itself, so as to assist its efforts in peeling off the cuticle. The color of the Blind-worm is rather variable. In my own specimen, now crawling over the paper on which I write, and blotting it sadly, the color is dark olive-brown above, with a shining silvery lustre, and diversified with a narrow black line along the back, and a broader black line down each side. The flanks are grayish-white, mottled with black, and the under parts are nearly black, variegated with a little gray. The Y-like mark on the head is still apparent, but there is no trace of the inverted V. On the sides of the head, the mottlings of gray and black are very bold, and round the neck runs a collar of black. This mark, however, may have been caused by the stupidity of the captor, who was so frightened at the contortions of the reptile, that he tied a string round its neck to form a safe handle with which to carry it.

Mr. Bell, in his volume on reptiles, states that the tail is obtuse, but that it rather varies in length, in some cases being not more than half the length of the body, while in others it nearly equals the head and body together. In my own specimen, the tail is by no means obtuse, but very slender and well pointed, and can be so tightly curled at its extremity as not to be removable without damage to the creature. While held in the hand, it generally twists the tip of the tail firmly round one of the fingers, not in a spiral position, but so as to make one complete circle, the extremity of the tail just touching the spot where the circle commences. The total length of this specimen now lying flat against a two-foot rule, towards which I have just succeeded in coaxing it by a judicious arrangement of light and shade, and an occasional touch with the finger, is thirteen inches and a half. The body and head occupy precisely six inches, and the remaining seven inches and a half are given to the tail. The spot where the body ends and the tail begins is very evident, the diameter of the body diminishing slightly but suddenly.

THE family of the Skinks contains so many interesting creatures, that it is difficult to make a satisfactory selection, and impossible to avoid a feeling of regret at the necessity for passing so many species without even a cursory notice. Before, however, proceeding to the next family, we must give a short notice of one or two rather conspicuous species.

The first is the SPINE-BACKED LIZARD of New Guinea (*Tribolonótus novæ guineæ*), a very remarkable creature, notable for the singular formation of the scales which cover the back, and in allusion to which the creature has been placed under the generic name Tribolonotus. This long word is of Greek origin, signifying calthrop-backed—calthrops being certain horrible instruments thrown on the ground to check the advance of cavalry, and consisting of four iron spikes, set round a ball in such a manner, that when flung on the ground, three points rest on the earth, and the other projects perpendicularly into the air.

Though really harmless, the Spine-backed Lizard is a most formidable looking creature, the whole of the back being covered with long and sharply pointed spikes, formed by a modification of the scales, that project boldly in all directions, and fully justify the generic name. Even on the tail the scales, which are arranged in whorls, are long, pointed, and project over each other, so as to give a very formidable aspect to this member. Even the head is armed with these pointed scales, which become larger and larger as they approach the neck. The color of this Lizard is brown above, and grayish-white below.

ANOTHER notable member of this family is the well-known GALLIWASP (*Celestus occiduus*).

This reptile is a native of the West Indian Islands, and is very common in Jamaica, where it is held in great, but groundless dread, by the inhabitants, and especially by the negroes. It generally haunts damp situations, and is mostly found in marshy lands, near water, or hidden under rocks where moisture is retained by the nature of the ground, It is thought that when the Galliwasp is irritated, its bite is as venomous as that of a poisonous snake, and causes immediate death. On account of the dread in which it is held, the negroes call it by the name of Mabouya, in common with the reptile which has already been described on page 62.

The color of the Galliwasp is brown of various tones, diversified with cross bands of blackish brown. It is about one foot in length, There are several species belonging to this genus, all being found in Jamaica.

THE last example of the Skinks which can be mentioned in these pages is SAGRA'S DIPLOGLOSSUS, or DOUBLE-TONGUED LIZARD.

This reptile is a native of Cuba, and is found in localities where the air is cool, and the soil light and moist. It is an active little creature, and moves from place to place with much agility. In this reptile the tongue is rather large, covered with little scale-like papillæ in front, becoming more thread-like behind. The color is gray, with a bronzy lustre, and a black streak runs along each side.

THE next family of Lizards contains only one species, the OPHIOMORE (*Ophiomorus miliáris*), and is separated from the skinks and the sepsidæ on account of a formation of the scales of the head, which seems to place it in an intermediate position between those two families. There are no external limbs, and the whole body and tail are long, cylindrical, tapering, and serpentine in aspect. The color of the Ophiomore is brown above, covered with numerous tiny black dots arranged in regular lines along the body, and being larger upon the sides. The under parts are white, and the sides are gray. It is a native of Northern Africa, and has been brought from Algiers.

IN the SEPSIDÆ, a family which contains seven genera, there are always external limbs, mostly four in number, but in one genus, Scelotes, the front pair of legs are wanting, and the hinder pair are small and divided at the extremity into two toes only.

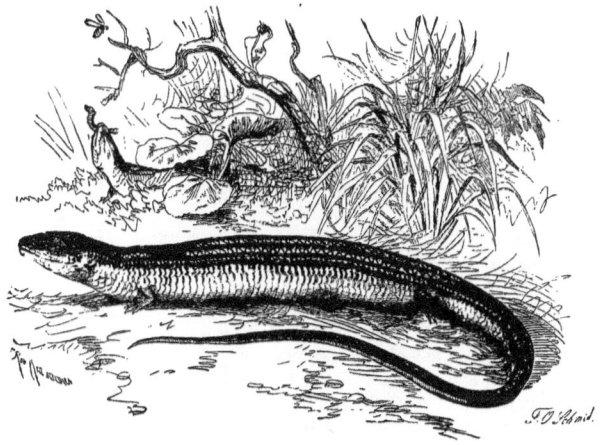

SEPS, OR CICIGNA.—*Seps tridactylus.*

The typical species of this family is the common SEPS, or CICIGNA, a curious snake-like Lizard, found in various parts of the world, and not uncommon in many portions of Europe. Specimens have been taken in the south of France, in Italy, Sardinia, Syria, and the north of Africa. The name of Seps is of Greek origin, and signifies corruption. From ancient times to the present day, this harmless little reptile has been held in great dread by the natives of the country wherein it dwells, being considered as a deadly enemy to cattle, biting them at night during their sleep, and filling their veins with corruption. Horses, and especially mares, were thought to be the most frequent sufferers from the bite of this reptile.

The legs of the Seps are very weak, and are set far apart, so that the creature trusts but little to the limbs for its powers of locomotion, and wriggles itself along after the fashion of the snakes. The food of the Seps consists of worms, small snails, slugs, insects, spiders, and similar creatures, its general habits seeming to resemble those of the blindworm. Like the lizard, when the winter approaches, it burrows deeply in the loose soil, and remains hidden until the succeeding spring.

The teeth of the Seps are small, conical, and simple, and there are no teeth on the palate, which is grooved longitudinally. The eyelids are scaly, and the lower has a transparent disc. The toes are three in number on each foot, and very feeble; the tail is conical and pointed. The color of the common Seps is grey, with four longitudinal brown streaks, which, on a closer inspection, are found to consist of a succession of brown dots.

Two members of the Sepsidæ deserve a passing notice before we pass to the next family. The first is the CAPISTRATED SPHÆNOPS (*Sphænops sepsoides*).

This reptile is a native of Northern Africa, but seems to have a rather restricted range, being seldom, if ever, found out of Egypt. In some parts of that country it is very plentiful, being found in the rice grounds, under hedges, and on the roads where the wheels of passing vehicles have worn deep ruts. Indeed, it appears to have a predilection for ridged ground, over which it passes with considerable speed, and is not to be captured without the exercise of some agility. It is quite harmless, and even when caught, struggles with all its might to escape, but does not attempt to bite the hand that holds it. Like many other reptiles of similar form, it burrows in the ground, but makes its tunnel so near the surface of the ground, and in so horizontal a direction, that the foot of a traveller will often lay open the superficial retreat and render its inmate homeless for a time.

The ancient Egyptians seem to have held this little reptile in religious veneration, as there are several known instances where it has been honored with the ceremony of embalming, and placed in the sacred tombs, together with other creatures formerly reverenced as types of divinity.

The Sphænops has four legs, moderately well shaped, but rather weakly formed, and the feet are divided into four toes, each of which is furnished at the extremity with a claw. The head is wedge-shaped, rounded in front, the palate is without teeth, and the lower eyelid is transparent. The general color of the Sphænops is pale brown, with a longitudinal series of black dots, and a black streak on each side of the muzzle.

OUR last example of this family is the TILIGUGU (*Gongylus ocellátus*), or EYED TILIQUA, another of the numerous reptiles classed under the common title of Mabouya by the ignorant and fearful.

It inhabits the countries bordering the Mediterranean, and is found in Sardinia, Malta, Egypt, and even in Teneriffe. Like the preceding species, it is quick and active in its movements, and when seized does not attempt to bite. It is a lover of dry and elevated spots, where the sand is loose, and there are plenty of stones under which it may hide itself. The food of this reptile consists of insects. Besides the names which have already been mentioned, it is also called LACEPEDE'S GALLIWASP and the OCELLATED SKINK.

In coloring it is one of the most variable of reptiles, but the general tints are gray, with a bronze gloss, diversified by a number of white spots edged with black. It has four legs, the toes are five on each foot, the head is conical, with a rounded muzzle, and the lower eyelid has a transparent disc.

THE ACONTIADÆ form the next family, which contains three genera. The head is small, the upper eyelid is either very small or altogether absent, the body is cylindrical, and the limbs, when present, are very weak and small. In two of the genera, Nessia and Evesia, there are four limbs, in the former with three toes, and in the latter with the feet small, imperfect, and not divided into toes. The upper eyelid is distinct though small.

IN the ACONTIAS, or JAVELIN SNAKE, the limbs are absent externally, and the upper eyelid is rudimentary. The body of this reptile is elongated and cylindrical, not unlike that of the common blindworm. The name Acontias is derived from a Greek word signifying a javelin, and has been given to this creature on account of the shape of the head, which bears some resemblance to the point of a spear. Some writers think that the name is given in allusion to its quick movements when seizing its prey. As in shape, so in habits it resembles the blind-worm, and like that reptile is very common in the fields and under hedges. It is a South African reptile, and is found abundantly at the Cape of Good Hope.

In its coloring this is a very handsome little creature, being sometimes called the Painted Acontias (by the French writers *La Peintade*), in allusion to the variegated tints with which it is bedecked. Like many other reptiles, especially those which are lightly colored, it is susceptible of much variation. Generally, however, it is rich chestnut-brown above, profusely dotted with bright yellow, the spots being arranged in series of varying number, one specimen having eight rows of spots, while another has only six. The scales are smooth, the teeth are conical and rather blunt, the eyes are very small, and the tail is short and rather rounded at the tip.

ANOTHER curious family of reptiles possesses only two limbs at the most, the front pair being always, and the hinder pair sometimes, wanting. There is a curious, cup-like shield on the chin, the body and tail are cylindrical, and both eyes and ears are apparently absent, but may be found hidden under the skin, where the greater part, if not the whole, of their functions must be in abeyance. In consequence of this remarkable privation, they are classed together under the very appropriate name of Typhlinidæ, a term derived from the Greek, and which signifies blindness.

In the typical species, the TYPHLINE, or BLIND ACONTIAS as it is sometimes, but rather erroneously called, the limbs are entirely absent, and the creature looks about as helpless a being as can well be imagined, having no apparent legs, feet, eyes nor ears. The Typhline inhabits Southern Africa, and is found at the Cape of Good Hope. In its coloring it is rather variable, being generally of a brownish hue, with spots of purple upon the hinder part of the scales of the back, and sometimes of a yellowish tint, with violet spots.

THERE are so many reptiles scattered over the world, and they are divided by modern systematic zoologists into so many families, that it is only possible to give a short description of one or two examples of each family, while to supply illustrations would be wholly impracticable without nearly doubling the amount of space that can be allotted to them.

The next family is called by the name of Typhlopsidæ, or Blind Reptiles, a title which has been given to them because their eyes are either very small, or altogether wanting externally. In all these animals the head is broad, rather flattened, and has a large, erect plate near the muzzle. The mouth is small, semilunar in shape, and placed under the muzzle in a manner somewhat resembling that of the sharks. The tail is cylindrical, and has a large shield or plate at the tip, sometimes conical and sometimes spine-shaped.

In the TYPHLOPS, the typical species, the head is nearly covered by a single, very large shield, which is rather bent downwards in front. The tail is very short and tapers suddenly, and the scales of the body are small and uniform. It inhabits India, where it is not uncommon, though, in consequence of its earth-loving habits, it is not very often seen except by those who know its localities, and search purposely for the hidden reptile. It moves over the ground with some rapidity, and burrows easily, penetrating to a depth of three or four feet during the rainy season. At other times it is mostly content with the shelter of large stones and similar places of refuge.

Owing to the small size and the rather remarkable position of the mouth, the Typhlops is unable to act on the offensive, and when captured, although it attempts to glide through the fingers, does not even offer to bite. It is wonderfully tenacious of life, and according to Dr. Russell, will live for some time even when immersed in spirits of wine. The general color of the Typhlops is yellowish-white.

This family contains also the Clawed Snake (*Onychophis*)—so called because the bony shield on the muzzle is erect, keeled, and bent over into a claw-like shape—and the Silver Snakes (*Argyrophis*), a small group of reptiles, deriving their popular name from the silvery lustre of their scales.

The last family of the sub-order Leptoglossæ, or Slender-tongued Lizards, is the group of reptiles termed the Rough-tailed Lizards, or UROPELTIDÆ. In these Lizards the head is rather compressed, flat above, and sharp towards the muzzle. The eyes are of moderate size, and without eyelids, a bony scale answering the purpose. The body is cylindrical, and covered with regular, six-sided scales, sometimes ridged, but mostly smooth. The tail is also cylindrical, and abruptly terminated, as if cut off obliquely. There are no external limbs, and by most systematic naturalists the Rough-tails have been placed among the serpents, which they very closely resemble, except in the arrangement of certain scales, and the short, abruptly truncated tail.

The Rough-tailed Lizards are divided into three genera, separated from each other by the formation of the scales that cover the tail. While moving, the Rough-tails aid themselves by pressing the truncated tail against the ground. As a typical species, we may select the PHILIPPINE SHIELD-TAIL (*Uropeltis philippinus*), a reptile which, as its name imports, inhabits the Philippine Islands. In this creature the tail is rather flattened, and covered above with a curious "flat, roundish, radiating, granular shield." On the lower side of the tail the scales are arranged in six rows. The color of the Philippine Shield-tail is brown above and white beneath, the line of demarcation being very distinct, and regularly waved.

THICK-TONGUED LIZARDS; PACHYGLOSSÆ.

A NEW sub-order now comes before our notice, the members of which are distinguished by the formation of their tongues, which, instead of being flat and comparatively slender, as in the preceding Lizards, are thick, convex, and have a slight nick at the end. On account of this structure, the species of this sub-order are termed PACHYGLOSSÆ, or Thick-tongued Lizards.

These reptiles are divided into sundry groups, the first of which is termed the NYCTISAURA, or Nocturnal Lizards. These creatures have eyes formed for seeing in the dusk, circular eyelids which, however, cannot meet over the eye-ball, and in almost every case the pupil is a long narrow slit like that of the cat. The body is always flattened. The limbs are four in number, tolerably powerful, and are used in progression.

Of these Lizards, the first family is the GECKOTIDÆ, or Geckos, a very curious group of reptiles, common in many hot countries, and looked upon with dread or adoration by the natives, sometimes with both, where the genius of the nation leads them to reverence the object of their fears, and to form no other conception of supreme power than the capability of doing harm.

THE FAN-FOOT, or HOUSE GECKO, is a native of Northern Africa, and is very common in Egypt, and is found, as its name imports, in houses, traversing the floor and walls with astonishing address, in search of its food, which consists of worms, insects, and similar creatures. The natives have a very great dread of this creature, asserting that it is extremely poisonous—the poison not being injected by the teeth, but exuding from the lobules of the toes. The generic title Ptyodactylus, or Toe-spitter, is given to the reptile in allusion to this idea. It is said by Hasselquist, that if a Gecko is taken in the hand, the poisonous matter which is immediately shed over the skin from the feet of the captive, causes an instantaneous eruption, similar to that produced by the sting of a nettle. The same traveller proceeds to relate an incident which is hardly so much in accordance with probability, namely, that two women and a girl were lying at the point of death from having eaten some cheese over which one of these reptiles had walked.

So great is the dread inspired by this creature, that in Cairo it is popularly termed Abouburs, or father of the leprosy. The people fancy that it purposely poisons their provisions, and that it is especially fond of communicating the venom to salted meat of all kinds. In former times the Fan-foot was endowed with even greater powers of offence, its teeth being added to its weapons, and asserted to be capable of leaving their impression even on steel, though in point of fact, the jaws of the Geckos are rather feeble. and their teeth very small, and hardly able to pierce even the human skin.

The Geckos are indebted for their power of traversing perpendicular walls to the formation of their feet, which, although greatly varied in the different genera, have the same essential qualities in all. In this genus the toes are expanded at their extremities, into a round disc, and furnished with claws which are sheathed in a notch cut in the front of the disc. The color of the Fan-foot is reddish brown spotted with white.

FAN-FOOT.—*Ptyodactylus gecko*.

THE COMMON GECKO, or RINGED GECKO, is an Asiatic species, being as common in India as the preceding species in North Africa. It may be easily known from the Fan-foot by the large tubercles upon the back.

This reptile has much the same habits as the Fan-foot, and possesses equally the ability to run over a perpendicular wall. During the day-time it conceals itself in some chink or dark crevice, but in the evening it leaves its retreat, moving rapidly and with such perfectly silent tread that the ignorant natives may well be excused for classing it among supernatural beings.

The Gecko occasionally utters a curious cry, which has been compared to that peculiar clucking sound employed by riders to stimulate their horses, and in some species the cry is very distinct and said to resemble the word Geck-o, the last syllable being given smartly and sharply. On account of this cry, the Geckos are variously called, Spitters, Postilions, and Claqueurs.

During the cold months of the year the Geckos retire to winter quarters, and are thought to retain their condition during this foodless season by means of two fatty masses at the base of the abdomen, which are supposed to nourish them as the camel is nourished by the hump. The male is smaller than the female, and the eggs are very spherical, and covered with a brittle chalky shell. The color of the Gecko is reddish gray with white spots. The scales of the back are flat and smooth, and there is also a series of rather large tubercular projections arranged in twelve rather distinct rows.

CLOSELY allied to these two reptiles is the SPOTTED GECKO, or SPOTTED HEMIDACTYLE, a rather pretty species of Gecko found in various parts of Asia, and tolerably common in India, China and Ceylon. Sir Emerson Tennent, in his valuable work on Ceylon, gives a very interesting account of this little creature, and relates two curious anecdotes, exhibiting the readiness with which even a Gecko can be tamed by kind treatment.

"In a boudoir where the ladies of my family spent their evenings, one of these familiar and amusing little creatures had its hiding place behind a gilt picture-frame, and punctually as the candles were lighted, it made its appearance on the wall to be fed with its accustomed crumb; and if neglected, it reiterated its sharp quick call of *chic-chic-chit*, till attended to. It was of a delicate gray color, tinged with pink, and having by accident fallen on a work-table, it fled, leaving its tail behind it, which, however, it reproduced within less than a month. This faculty of reproduction is doubtless designed to enable the creature to escape from its assailants; the detaching of the limb is evidently its own act.

"In an officer's quarters in the fort of Colombo, a Gecko had been taught to come daily to the dinner-table, and always made its appearance along with the dessert. The family were absent for some months, during which the house underwent extensive repairs, the roof having been raised, the walls stuccoed, and ceilings whitened. It was naturally surmised that so long a suspension of its accustomed habits would have led to the disappearance of the little Lizard, but on the return of its old friends, at their first dinner it made its entrance as usual the instant the cloth had been removed."

ANOTHER rather curious species is the TURNIP-TAILED GECKO (*Thecodáctylus rapicaudus*), so called from the odd shape of its tail, which, when reproduced, is very much swollen at the base, and, with its little conical extremity, has an almost absurd resemblance to a young turnip. It is worthy of mention, that all the Geckos possess the faculty of reproducing their tails when those members have been lost by some accident, and that the second tail is mostly very unlike the original. Before the creature has suffered (if it does suffer) this mutilation, the tail is covered with scales of the same structure and form as those of the back; but when the tail is reproduced, it is generally supplied with little squared scales arranged in cross series. In examining a Gecko therefore, it is necessary to ascertain whether the tail be in its normal condition or only a second and altered edition of that member.

The color of the Turnip-tailed Gecko is brown, mottled boldly with a darker tint, and speckled with tiny dots of dark brown. The scales of the back are six-sided, and on each side of the base of the tail there is a prominent conical tubercle. This species inhabits Tropical America.

A VERY remarkable reptile is the FRINGED TREE GECKO, or SMOOTH-HEADED GECKO. It is a native of Java, and especially worthy of notice on account of the broad membranous expansions which fringe the sides of the head, back, limbs and tail. On the body this membrane is covered with scales, and waved on its edges, but on the tail the waves become suddenly deepened, so as to form bold scollops. The toes are webbed to the tips, and, with the exception of the thumb-joint, are furnished with claws at the swollen extremity. The scales of the back are smooth and flat, and even the membranous fringes are covered with scales.

Formerly this creature was thought to be aquatic in its habits, but it is now known to live on trees, and to employ the membranous expansions in aiding it in its passage from branch to branch, much after the well-known fashion of the flying squirrels. The generic title, Ptychozoön, is composed of two Greek words, the former signifying a fold of a garment, and the latter a living being. The general color of the Fringed Tree Gecko is brown above, with a slight yellowish tinge along the spine, and crossed with small dark brown lines, very narrow and deeply waved. A line of similiar appearance and of a bold zig-zag form encircles the top of the head, looking as if a dark brown string had been tied at the ends, formed into a rude circle, and then pinched at intervals so as to cause deep indentations. Below it is of a whitish gray color.

THE curious and rather interesting little Lizard called the CAPE TARENTOLA, is an inhabitant, as it name signifies, of the Cape of Good Hope, and is found spread over a considerable portion of Southern Africa.

This reptile is of slower habits than the generality of the Geckos, and moves along with deliberate and apparently purposeless steps. It is almost invariably seen upon or near decayed wood, and is frequently found under the bark of dead trees, clinging tightly to the trunk, and shielded by the bark from the unwelcome glare of daylight. In all probability, it finds abundance of food in the same locality, for the space between the bark and wood of a decaying or dead tree, is generally filled with insects of various kinds and in their different states of existence, beside being the chosen home of millipedes, spiders, and similar creatures.

Although a slow mover, the Cape Tarentola can, after the manner of its kin, ascend smooth and perpendicular objects with perfect ease and noiseless motions, and can even traverse and cling to a ceiling or a cross-beam without difficulty, and there remain motionless for hours. Like the generality of the Gecko family, it detests the daylight, and the bright beams of the sun are a torture to this dweller in darkness, which, if overtaken by daylight while out of its refuge, crawls away to the nearest cranny and there buries itself until the evening hours bring with them the desired shades, and restore the animal to its wonted activity. It is extremely shy, and even in the dusk it will avoid the dangerous approach of an intruder by silently slipping under the cover of the loose bark, or hiding itself among the decaying wood.

It is quite a little creature, rarely measuring more than four inches in length, and often not reaching even those moderate dimensions. As is the case with many Lizards, it is liable to certain variations in coloring, but its general tints are as follows: The back and upper portions of the body are yellowish-brown, with a decided yellow wash, and banded with several dark brown bars, rather curved. Scattered over the body are certain protuberant scales of a lighter hue. The tail is a pale brownish-purple with a reddish gloss, and speckled with warm chestnut-brown. The abdomen, and the under portions of the body and limbs are ochry yellow, and the eyes are, although devoid of expression and of a passionless brightness, like polished stone, very shining and of a bright orange-brown. The whole form of this Lizard is rather thick and clumsy.

As this family contains at least forty genera, it is manifestly impossible to mention more than a few species, which can be accepted as types of the family, and serve as links to render the chain of nature complete. Passing, therefore, several series of genera, we will give a short time to one or two species of Gecko before proceeding to the next family.

The WOODSLAVE, as the reptile is popularly termed by the natives of the country where it resides, or the BANDED SPHÆRODACTYLE (*Sphærodáctylus sputátor*), as it is more scientifically called by zoologists, is a small species of Gecko found in most of the American islands, and is spread over many portions of South America; and is held in great dread by the white and dark population. It is generally supposed to possess a store of venomous saliva, causing the part of the body on which it falls to swell grievously, and to eject this poisonous substance from some distance upon those who chance to vex its irascible temper. The specific term *sputátor* signifies a spitter, and has been given to the reptile on account of this supposed propensity. The poisonous saliva is said to be black.

The Woodslave has no claws on its toes, the pupil of the eye is round, and the eyelid circular. The back and tail are covered with small scales. The color is generally black and yellow, arranged in cross bands, and there is a white streak on each side of the head. There are several species belonging to this genus, all inhabiting similar localities.

THE reader will remember that in the turnip-tailed Gecko, mentioned on page 73, the tail is curiously swollen at the base after its reproduction. In the LEAF-TAILED GECKO, otherwise called WHITE'S PHYLLURE (*Phyllúrus platúrus*), the tail is always rather long, flattened considerably, very broad, with a deep notch at its junction with the body, and a shallower double notch in the centre. Along the middle there also runs a shallow groove, and the entire aspect is so quaint, not to say ludicrous, that on seeing a specimen of this odd-looking Lizard, the first impression on the mind is that the tail has been cleverly manufactured and attached to the body by artificial means. This Gecko is a native of New Holland.

Both the scientific names of the Leaf-tailed Gecko refer to the singular formation of its tail, the one signifying Leaf-tail, and the other Broad-tail. The head of this reptile is very broad at the base, very sharp at the snout, and the skin adheres so closely to the bone as to exhibit the form of the skull through its substance. The toes are long, slender, and rather compressed. Along the sides runs a fold of skin, very slight, but sufficiently conspicuous. The tail is very thin and leaf-like; along the edge runs a series of spiny scales, and its surface is covered with rather long conical tubercles arranged in cross rows. The color is brown, and a number of little spiny tubercles are scattered over the back.

In taking leave of the Geckos, we must cast a hasty glance at their feet. In many of their movements the Geckos bear a curious likeness to the common fly, and when one of these reptiles is seen gliding along a perpendicular wall with noiseless step, or clinging with perfect ease to an overhanging beam, quite regardless of the fact that it is hanging with its back downwards, the resemblance is irresistible. And on inspecting the foot and its structure, the resemblance which this member bears in many species to the well-known foot of the fly, is remarkably close and worthy of attention.

STROBILOSAURA.

WE now arrive at an important tribe of Lizards, called by the name of Strobilosaura, a title derived from two Greek words, one signifying a fir-cone and the other a lizard, and given to these creatures because the scales that cover their tails are set in regular whorls, and bear some resemblance to the projecting scales of the fir-cone. In all these reptiles the tongue is thick, short, and very slightly nicked at the tip. The eyes have circular pupils, and are formed for day use.

THE first family of these Lizards consists of those creatures which are grouped together under the general title of IGUANA. This word is employed extremely loosely, as the name of Iguana is applied to many species of Lizards, such as the monitors and the varans, which in reality have little in common with the true Iguanas. These reptiles can mostly be distinguished from the rest of the tribe by the formation of their teeth, which are round at the roots, swollen and rather compressed at the tip, and notched on the edge. There are generally some teeth on the palate. All the true Iguanas inhabit the New World. As the family of Iguanas is extremely large, and contains more than fifty genera, we can only examine a few of the most interesting species, the first of which is the COMMON IGUANA.

This conspicuous, and in spite of its rather repulsive shape, really handsome Lizard, is a native of Brazil, Cayenne, the Bahamas, and neighboring localities, and was at one time very common in Jamaica, from which, however, it seems to be in process of gradual extirpation.

In common with those members of the family which have their body rather compressed, and covered with squared scales, the Iguana is a percher on trees, living almost wholly among the branches, to which it clings with its powerful feet, and on which it finds the greater part of its food. It is almost always to be found on the trees that are in the vicinity of water, and especially favors those that grow upon the banks of a river, where the branches overhang the stream.

Though not one of the aquatic Lizards, the Iguana is quite at home in the water, and if alarmed, will often plunge into the stream, and either dive or swim rapidly away. While swimming, it lays its fore legs against the sides, so as to afford the smallest possible resistance to the water, stretches out its hinder legs, and by a rapid serpentine movement of its long and flexible tail, passes swiftly through the waves. It has considerable power of enduring immersion, as indeed is the case with nearly all reptiles, and has been known to remain under water for an entire hour, and at the end of that time to emerge in perfect vigor.

From the aspect of this long-tailed, dewlapped, scaly, spiny Lizard, most persons would rather recoil than feel attracted, and the idea of eating the flesh of so repulsive a creature would not be likely to occur to them. Yet in truth, the flesh of the Iguana is justly reckoned among one of the delicacies of the country where it resides, being tender, and of a peculiarly delicate flavor, not unlike the breast of a spring chicken. There are various modes of cooking the Iguana, roasting and boiling being the most common. Making it into a fricassee, however, is the mode which has met the largest general approval, and a dish of Iguana cutlets, when properly dressed, takes a very high place among the delicacies of a well-spread table.

The eggs, too, of which the female Iguana lays from four to six dozen, are very well flavored and in high repute. It is rather curious that they contain very little albumen, the yellow filling almost the entire shell. As is the case with the eggs of the turtle, they never harden by boiling, and only assume a little thicker consistence. Some persons of peculiar constitutions cannot eat either the flesh or the eggs of the Iguana, and it is said that this diet is very injurious to some diseases. The eggs are hid by the female Iguana in sandy soil near rivers, lakes, or the sea-coast, and after covering them with sand, she leaves them to be hatched by the heat of the sun.

In consequence of the excellence of the flesh and eggs, the Iguana is greatly persecuted by mankind, and its numbers considerably thinned. Those who hunt the animal for sport, or merely to supply their own homes, generally employ a noose for the purpose, which they cast dexterously round the neck of the reptile as it sits on a branch, and then by a sudden and sharp jerk loosen its hold, and secure it. The creature is very bold, having but little idea of running away, and in general is so confident of its capability of frightening away its antagonist by puffing up its long dewlap, and looking ferocious, that it is captured before it discovers its mistake. Even when caught, it has no notion of yielding without a struggle, but bites so fiercely with its sharp, leaf-like teeth, and lashes so vigorously with its long whip-like tail, that it is not secured without some trouble and risk. It is also very tenacious of life, and does not readily die even from repeated blows with heavy sticks, so that the spear or the pistol are often employed to kill it.

Those, however, who hunt the Iguana for sale, are obliged to have recourse to other expedients, such as nets, and dogs, the latter being trained to secure the Iguana without killing it. Many persons set out on regular expeditions of this sort, embarking in a little vessel and visiting numbers of different islands and inlets in chase of the Iguana. Those which they can succeed in taking alive, have their mouths carefully secured to prevent them from biting, and are then stowed away in the hold, where they will live for a considerable time without requiring any nourishment. Those which are killed, they either eat on the spot, or salt them down in barrels for winter consumption. Were the Iguanas quick of foot, they would seldom be captured, but, fortunately for the hunters, they cannot run fast, and according to the quaint language of Catesby, who visited the Bahamas about 1740, "their holes are a greater security to them than their heels."

The food of the Iguana seems to consist almost entirely of fruits, fungi, and other vegetable substances, and it is known that in captivity it feeds upon various leaves and flowers. Yet it has been said by some persons, who have seen the Iguana in its native state, that it eats eggs, insects, and various animal substances. Perhaps these creatures were not the true Iguanas, but belonged to the monitors, varans, or similar carnivorous Lizards.

The Iguana is capable of domestication, and can be tamed without much difficulty by those who are kind to it and accustom it to their presence. It will even permit itself to be carried about in its owner's arms, though it will not permit a stranger to approach.

The general aspect of the Iguana is most remarkable, and can perhaps be better understood by reference to the illustration than by any lengthened description. Suffice it to say that the head is rather large, and covered above with large scales. The mouth is enormously wide, and studded around the edge with those singularly shaped teeth which have already been described. About the angles of the jaw there are generally some large, solitary, rounded scales. The chin is furnished with a kind of dewlap, large, baggy, and capable of being inflated at the will of the animal, scaly, and edged in front with a row of bold, tooth-like projections. The sides of the neck are covered with tubercles. The tail is extremely long, and very thin and tapering. The usual color of the Iguana is dark olive-green, but is

IGUANA.

rather variable even in the same individual, being affected by change of weather, or locality, or temper. On the sides a few brown bands are generally seen, and the tail is marked with brown and green of various tones, the two colors being arranged in alternate rings. The average length of the Iguana is about four feet, but it often attains a much greater size, reaching a length of six feet or a little more.

The NAKED-NECKED IGUANA was long confounded with the preceding species, bearing a great resemblance to that reptile in color, form, and habit, and being found in the same localities. It can, however, be readily distinguished from the common Iguana by the absence of tubercles upon the sides of the neck. Along each side of the lower jaw runs a series of

HATTERIA.—*Hatteria punctata.* (⅙ Natural size. See next page.)

large strong scales. The general color of this species is bluish-green, darker on the back than on the abdomen. Its flesh is esteeemed equally with that of the preceding species.

BESIDES these Iguanas, there are one or two which deserve a short notice. One of these animals is the MARBLED IGUANA or CAMALEAO (*Polychrus marmorátus*), also a native of Brazil and Central America. This species has the throat compressed into a small dewlap, and the scales of the back and sides equal. There is no crest upon the back and tail. Its color is brown, mottled with bold marblings and diverging lines of a darker hue, and sometimes having a slight purple gloss.

The APLONOTE (*Aloponótus ricardi*) is another species of Iguana, having its head covered with small equal many-sided plates, and its throat dilated into a small pouch without the toothed projections in front. A shallow crest runs along the back and tail, and the back is without scales, but covered with multitudinous granular tubercles of a very small size. The tail is compressed. The color of this species is blackish-brown, variegated with many spots of tawny brown.

Another curious species is appropriately called the HORNED IGUANA (*Metopóceros cornútus*), deriving its name from the horn-like projections upon its head. Upon the forehead there is a large horn-like tubercle, and two pairs of large horny plates between the nostrils. There is a crest upon the back, but it is very low between the shoulders, and upon the loins it is not continuous. It inhabits St. Domingo.

THE next family, termed Rhynchocephalia, which is represented in the illustration on page 77, contains only one species, the *Hatteria punctata*. This reptile inhabits New Zealand where the natives regard it with fear, though without any reason, as the animal is quite inoffensive. They nevertheless like the flesh of the "Guana," "Tuatera," or "Narara," as they call this great Lizard. A specimen caught in Wellington, New Zealand, was brought to Europe and has lived there in captivity for many years. It has fed on meal-worms and other scaled insects.

The general color of the HATTERIA is a dark olive-green, the sides and limbs are variegated with many yellow sprinkles. There is a conspicuous crest of sharp scales which runs along the head and the back, while the tail shows rather flattened projections. The scales of the head and back are of a yellow color, those of the tail being brown.

IT has already been mentioned that the Iguana possesses the power of swimming to a large extent, and that it is capable of sustaining a long submersion without suffering any injury.

MARINE OREOCEPHALE. - *Oreocephalus cristatus*.

There is a curious species of Iguana, the MARINE OREOCEPHALE, which exists upon the seashore, and passes a considerable portion of its time in the water. This creature was first made known to science by Mr. Darwin, who found it on the coasts of the Galapagos islands, and describes its habits in the following words :—

"It is a hideous-looking creature, of a dirty-black color, stupid and sluggish in its movements. The usual length of a full-grown one is about a yard, but there are some even four feet long. I have seen a large one which weighed twenty pounds. These lizards are occasionally seen some hundred yards from the shore swimming about, and Captain Collnett in his voyage says that they go out to sea in shoals to catch fish. With respect to the object I believe he is mistaken, but the facts stated on such good authority cannot be doubted.

When in the water, the animal swims with perfect ease and quickness by a serpentine movement of its body and flattened tail, the legs during this time being perfectly motionless and closely collapsed on its sides. A seaman on board sunk one with a heavy weight attached to it, thinking thus to kill it directly, but when, an hour afterwards, he drew up the line, the Lizard was quite active. Their limbs and strong claws are admirably adapted for crawling over the rugged and fissured masses of lava which everywhere form the coast. In such situations, a group of six or seven of these hideous reptiles may oftentimes be seen on the black rocks, a few feet above the surf, basking in the sun with outstretched legs."

In this reptile the throat is not formed into a pendent pouch, but the skin is much crumpled, so that the animal can dilate it at will. The whole body is covered with sharp, rough, tubercular scales, and a crest of longer scales runs along the back. The teeth are sharp and three-lobed, and although, when the wide mouth is opened, they present a very formidable array of weapons, the creature is quite harmless, and feeds on vegetable diet, seaweeds forming the chief part of its subsistence. The middle toes are united by a strong web, and the claws are large. There is some difference in the aspect of the young and adult, this distinction being most obvious in the head, where the scales are rather convex in the young, but in the adult are enlarged into unequal and rather high tubercular shields.

Of the family *Iguanidæ* there are about sixty genera, and one hundred and fifty species, all of North and South America and the Antilles. According to Holbrook, four genera of this family are known in the United States.

IN the earlier ages of science, when a few facts were struggling their way through the superincumbent mass of fiction that had so long caused Natural History to be little more than a collection of moral fables, the BASILISC was a creature upon whose wondrous properties the inventive pens of successive narrators were never tired of dilating. Crowned with a royal diadem, emblematical of its sovereign rule, the Basilisc held supreme sway over the reptile race, and derives its name of Basilisc, or kinglike, "because he seemeth to be the King of Serpents, not for his magnitude or greatness. For there are many serpents bigger than he, as there be many four-footed beasts bigger than the lyon, but because of his stately face and magnanimous minde."

The Basilisc was thought to be an occasional *lusus naturæ*, having during his life no companion of his own kind, and to derive his existence from an egg laid by a cock when he was very old, and sat upon by a snake. Some scientific writers, however, better informed than the more popular zoologists, said that the egg was not incubated by a snake, but by a toad.

Before the Basilisc all living creatures but one were forced to fly, and even man would fall dead from the glance of the kingly reptile's eye. "This poyson," says Topsel, "infecteth the air, and the air so infected killeth all living things, and likewise all green things, fruits and plants of the earth: it burneth up the grasse whereupon it goeth or creepeth, and the fowls of the air fall down dead when they come near his den or lodging. Sometimes he biteth a man or beast, and by that wound the blood turneth into choler, and so the whole body becometh yellow or gold, presently killing all that touch it or come near it." Even a horseman who had taken into his hand a spear which had been thrust through a Basilisc, "did not only draw the poyson of it into his own body and so dyed, but also killed his horse thereby."

The only creature that could stand before the Basilisc and live, was said to be the cock, whose shrill clarion the bird-reptile held in such terror, that on hearing the sound it fled into the depths of the desert and there concealed itself. Travelers, therefore, who were forced to pass through the sandy deserts of Libya, were advised always to carry with them a supply of strong, lively, loud-voiced cocks, by whose vigorous crowings they would be protected from the Basiliscs haunting those parts.

There is an old proverb, "No smoke without fire," and this saying is verified in the present case. In some parts of Tropical America there is a perfectly harmless Lizard of no great dimensions, belonging to the family of the Iguanas, and having a bold crest on the back of its head. It is probable that one of these reptiles was imported into the Old World at some time now forgotten, and that its rather odd shapé and the crest on its head were seized upon by the first describers, and reported with continually increasing exaggerations by succeeding writers.

Like the rest of the Iguanas, this animal is a good climber of trees, it can swim well, and its food consists apparently of insects and the various little creatures which frequent the water and the foliage of its banks.

Although quite innocuous, it certainly is rather forbidding, and when it obtains its greatest length of three feet, presents a sufficiently formidable appearance to warrant in some degree the wild and fabulous tales which were deduced from its strange shape. Along the back, instead of the row of pointed spines which generally cross the back of the Iguanas, runs a broad crest-like membrane, another broad membrane occupying the upper surface of the tail. These curious appendages are supported by a series of slender bones, formed by elongations of the vertebræ of the back and tail, so that the animal looks exactly as if the fins of a fish had been grafted on the body of a reptile. There is a slight pouch on the throat, and the palate is toothed.

MANY species of the Lizard tribe are called by the name of Anolis, but are divided by systematic zoologists of the present day into several distinct genera. The CRESTED ANOLIS inhabits some of the hotter portions of America and the neighboring islands.

The chief point of interest in this Lizard is the curiously expansile throat, which, in common with others of the same genus, it is able to expand at will. When terrified, it tries to escape, but if it finds itself deprived of all means of eluding its antagonist, it turns to bay, and by puffing out the throat until it assumes a very great size, endeavors thereby to intimidate the foe. While thus engaged, the creature has the faculty of continually altering its color; the hues of the body to a certain degree, but more especially those of the throat, changing with a rapidity that is said even to surpass the famed powers of the chameleon.

It is an active little creature, traversing perpendicular objects with nearly as much ease as the Gecko, and to aid it in these movements the last joint but one of the toes is swollen, so as to form a pad, and is covered below with cross ridges, so as to enable the creature to take a firm hold of the object to which it is clinging. The food of the Anolis consists chiefly of insects, which are captured by means of singular address on the part of the Lizard. The Anolis can run up and down trees, walls, or rocks, with such rapidity, and leap so boldly from one spot to another, that at a little distance its movements might easily be mistaken for those of a bird.

Though not aquatic in its habits, and apparently not taking willingly to the water, the Anolis is mostly to be found in the woods and thickets that are in the close neighborhood of a stream or lake. It is a timid, yet a restlessly inquisitive animal; for although it hides itself with instinctive caution on hearing the approach of a footstep, it is of so curious a nature that it must needs poke its head out of its hiding-place, and so betray itself in spite of its timidity. So absorbed, indeed, is the Anolis in gratifying its curiosity, that it will allow itself to be captured in a noose, and often falls a victim to the rude and inartificial snares made by children. Its voice is a little sharp chirruping sound; and by imitating these notes, the children decoy it within reach of the fatal noose.

The usual resting-place of the Crested Anolis is within the hollow of some decaying tree, where also the female deposits her eggs.

The color of the Crested Anolis is dark, ashen blue, a blackish spot being apparent on each side. Along the nape of the neck and the back runs a series of long compressed scales, forming a toothed crest, and on the basal half of the tail is a fin-like crest, strengthened by bony rays. The throat-pouch is extremely large, and when inflated gives to the reptile quite an ungainly appearance. The greatest known length of the Crested Anolis is about eighteen inches, but the other species are generally of much smaller dimensions. The name Xiphosurus is of Greek origin, and signifies Sword-tail.

Of the restricted genus Anolis, we take two examples. In this genus the back and nape of the neck are either smooth, or have a low crest formed by two series of short scales. The scaly plate at end of the muzzle is erect. All these Lizards are very active, inhabiting trees, and jumping about from branch to branch with wonderful skill, and clinging even to the pendent leaves by means of their curiously formed feet.

GREEN CAROLINA ANOLIS.—*Anolis carolina.*

This GREEN CAROLINA ANOLIS is, as its name imports, a native of North America, where it is tolerably common. It is a pretty lively little creature, specially brisk and active in its movements.

This Lizard is, according to Holbrook, "a bold and daring animal, haunting outhouses and garden fences, and in new settlements it even enters the houses, walking over the tables and other articles of furniture in search of flies. It is very active, climbing trees with great rapidity, and leaping with ease from branch to branch and from tree to tree, securing itself even on the leaves by means of the oval disks of the fingers and toes, which enable it also to walk easily on glass, and on the sides and ceilings of rooms. It feeds on insects, and destroys great numbers, seizing them suddenly and devouring them, unrestrained even by the presence of man."

Towards the spring, the Green Anolis becomes quarrelsome, and is so exceedingly pugnacious, that the adult males hardly ever meet without a fight, the vanquished usually coming off with the loss of his tail—a misfortune, however, that sometimes occurs to both the combatants. This Lizard is seldom seen in all its beauty except when engaging in battle, for at the sight of its antagonist it remains stationary for a moment, nods its head up and down two or three times, as if to work itself into a proper state of fury, puffs out its dewlap, which then becomes of a light scarlet, and having gone through all these preliminaries, it leaps on its foe, and the struggle begins. As the summer draws on the irascibility of its temper diminishes, and during the whole summer and early autumn these pretty Lizards may be seen amicably associating together. They are fond of basking in the sun, and will then dilate their dewlaps, at the same time assuming the most brilliant emerald hues.

The color of this reptile is extremely variable, altering even in the same individual according to the season of the year, the temperature, the health, or even the present state of the creature's temper. Generally the whole upper surface is beautiful golden green, and the abdomen white, with a tinge of green. The dewlap, or throat-pouch, is white, with a few little spots and five bars of red, which color, when the pouch is inflated, spreads over its whole surface. The total length of this reptile is, according to the figure in our illustration, nearly seven inches.

THE GREEN LIZARD (*Anolis principalis*), also called CHAMELEON, is an attractive creature, quite in contrast to the latter-named. It is of very graceful shape and movements, and is a beautiful green in color. It inhabits along the Gulf and Atlantic shores southward; length, six to eight inches.

THE second species, the RED-THROATED ANOLIS, is a native of America and the neighboring isles.

It is a brisk and lively little creature, darting about the ground, over rocks, among the branches, or upon the leaves, with equal address. It is, perhaps, a little too fond of fighting, and terribly apt to quarrel with others of its own kind. Those who have witnessed a combat between two of these Lizards say that it is remarkable for ferocity, courage, and endurance. They face each other with expanded throats and glaring eyes, their skin changing its lustrous coloring, and their whole being instinct with fury.

As during each combat one or two females are generally spectators of the fight, it is probable they may be the cause of war, and that the victor may receive his reward from one of the female witnesses of his prowess. So furious do they become, that the conqueror is said to devour the vanquished, who, however, sometimes runs away as fast as he can, and escapes with the loss of his tail, which is left writhing in the victor's mouth and soon swallowed. Those who have thus lost their tails seem to be greatly affected by the mutilation, and are timid and languishing afterwards.

The inflated throat part of the angry animal has a very curious effect, as it becomes of a bright cherry-red, due probably to the excited state of the creature.

Mr. Bell, in his work on reptiles, mentions a curious anecdote of one of these Lizards which was worsted in combat with a common garden-spider. "The activity of the smaller insectivorous Lizards, when in pursuit of their food, is exceedingly curious and interesting. They watch with all the caution of a cat, and dart upon their prey with the quickness of lightning.

"In the act of seizing their food, however, they must necessarily be exposed to some danger from the noxious qualities of the insects which they indiscriminately attack. The following fact would seem to indicate that, even in our own temperate climate, an insect not generally recognized as poisonous may inflict a fatal injury on its saurian enemy.

"Some years since, I had in my possession two living specimens of the beautiful little green Anolis of the West Indies, a Lizard about the size of our smallest species. I was in the habit of feeding them with flies and other insects; and, having one day placed in the cage with them a very large garden-spider (*Epeira diadema*), one of the Lizards darted at it, but seized it only by the leg. The spider instantly ran round and round the creature's mouth, weaving a very thick web round both jaws, and then gave it a severe bite on the lip, just as this species of spider usually does with any large insect which it has taken. The Lizard was greatly distressed; and I removed the spider and rubbed off the web, the confinement of which appeared to give it great annoyance, but in a few days it died, though previously in as perfect health as its companion, which lived for a long time afterwards."

With regard to the injury produced by the bite of the spider, I can say from personal experience that even to human beings, especially those who are tender-skinned, the bite of the common garden-spider is extremely painful. I have suffered for some hours from the bite of one of these creatures, and I have seen the arm of a young lady flushed and swollen, because a garden-spider had bitten the back of her hand. The pain is something like that produced by the sting of a wasp, but more dull, and seeming to throb with the pulse.

The color of the Red-throated Anolis is greenish blue, excepting on the throat when the creature is excited. There is no crest on the nape and back, but the tail is slightly toothed above. When full-grown, it is about the size of our sand Lizard.

Our last example of this large and interesting family is the CROWNED TAPAYAXIN, one of the singular North American reptiles which are popularly known by the name of Horned Toads, their general form and mode of sitting being extremely toad-like.

This animal is not at all uncommon in California, and is said when at liberty in its wild state to move with much rapidity over the ground, in search of its insect prey. Its habits in

CROWNED TAPAYAXIN.—*Phrynosoma orbiculare.*

confinement, however, do not carry out this statement, as it is then sluggish to a degree, remaining for many consecutive hours in precisely the same attitude, heedless of the falling rain or the burning rays of the sun, and scarcely changing its position even when pushed with the finger. It is quite harmless, in spite of its very formidable looks, and does not attempt to avenge itself upon its captor, however roughly it may be handled. After a while it can be made to know its owner, and will even take flies and other insects out of his hand. Little red ants seem to be its favorite food; but it lives on beetles and insects of various kinds.

The head of this curious reptile is armed with long, pointed, conical spines, set around its edge and directed backward. Shorter and stouter spines, but of a triangular shape, are scattered over the back, and extend even over the odd, short, and pointed tail. Each edge of the tail is armed with a strong row of spines, giving it a regularly toothed appearance. The general color of the Crowned Tapayaxin is gray, variegated with several irregular bands of rich chestnut-brown. The head is light brown, blotched with a darker hue, and the under parts are ochry-yellow, marked with sundry blotches of dark gray.

THE HORNED LIZARD (*Phrynosoma douglassi*). This strangely armed creature is found in Central America, and in western portions of the United States. Holbrook records three other species, which inhabit the region about the Columbia River.

THE family which comes next in order is that in which are included the AGAMAS, a group of Lizards which have been appropriately termed the Iguanas of the Old World. In the members of this family the teeth are set upon the edge of the jaws, and not upon their inner side, as in the true Iguanas of the New World. Between thirty and forty genera are contained in this family, and some of the species are interesting as well as peculiar beings.

PERHAPS the most curious of all this family, if not, indeed, the most curious of all the reptiles, is the little Lizard which is well known under the title of the FLYING DRAGON.

This singular reptile is a native of Java, Borneo, the Philippines, and neighboring islands, and is tolerably common. Some writers believe that this creature was the original source from which the many fables respecting the formidable dragon of ancient and modern mythology were derived. Perhaps, however, the real clue to the various fables that were once so common respecting the formidable dragon may be found in one of the huge saurians of the ancient days, which had survived its comrades, and preserved its existence upon the earth after man had been placed upon this planet.

The most conspicuous characteristic of this reptile is the singularly developed membranous lobes on either side, which are strengthened by certain slender processes from the first six false ribs, and serve to support the animal during its bold leaps from branch to branch. Many of the previously mentioned Lizards are admirable leapers, but they are all outdone by the Dragon, which is able, by means of the membranous parachute with which it is furnished, to sweep through distances of thirty paces, the so-called flight being almost identical with that of the flying squirrels and flying fish.

When the Dragon is at rest, or even when traversing the branches of trees, the parachute lies in folds along the sides; but when it prepares to leap from one bough to another, it spreads its winged sides, launches boldly into the air, and sails easily, with a slight fluttering of the wings, towards the point on which it had fixed, looking almost like a stray leaf blown by the breeze. As if in order to make itself still more buoyant, it inflates the three membranous sacs that depend from its throat, suffering them to collapse again when it has settled upon the branch. It is a perfectly harmless creature, and can be handled with impunity. The food of the Flying Dragon consists of insects.

The color of this reptile is variable, but is usually as follows: The upper surface is gray, with a tinge of olive, and daubed or mottled with brown. Several stripes of grayish-white are sometimes seen upon the wings, which are also ornamented with an angular network of dark, blackish-brown. Sometimes the black is rather plentiful upon the wings, forming four or five oblique bands near the edge. It is a small creature, measuring only a few inches in length.

THE FRINGED DRAGON is mostly found in Sumatra, where it seems to be tolerably common. In habits, and in general appearance, this reptile bears a great resemblance to the preceding species, from which, however, it may be known by the conspicuous black spots on its wings, each spot being surrounded with a ring of white. The head is grayish-white, covered with an irregular network of dark brown, and on the throat are a number of circular specks covered with granular scales. Upon the under parts of the male, the scales are rather large and keeled, and upon the wing are a number of rather short, white dashes of a partly triangular shape. Along the sides runs a series of small, triangular, keeled scales.

Besides these species there are several other flying Dragons, all inhabiting similar localities. They are divided into genera on account of the different structure of the ear, and the position of the nostrils. The tail of all the Dragon Lizards is extremely long, and very slenderly formed.

A VERY curious reptile of this family deserves a passing notice. This is the TIGER LIZARD, or GONYOCEPHALE (*Gonyocéphalus chameleontina*), a native of Java. This creature is remarkable for the high and deeply-toothed crest which runs along the nape of the neck, like the crest of an ancient helmet, and far overtops the head, although the upper part of the skull is much raised by an enlargement of the orbits. A large but compressed pouch hangs from the lower jaw and throat, and is prolonged so as to form an angular fold just before the shoulder. A toothed crest runs along the back, but is barely one quarter the height of that which passes over the nape, and the tail is long and compressed. The color of this Lizard is green, with variable streaks and scribblings of black, and the legs are deeply banded. The Tiger Lizard sometimes attains a length of three feet.

THE Lizards of this family are remarkable for the extraordinary modifications of form which they exhibit. In one species, such as the tiger Lizard, a row of long, spike-like scales is raised upon the neck, in the dragons the skin of the sides is dilated to an enormous extent, and even the ribs are drawn out like wire and turned out of their usual course to support the membranous expansion, and in the FRILLED LIZARD the neck is furnished with a large, plaited

FRILLED LIZARD.—*Chlamydosaurus kingii.*

membrane on each side, forming a most remarkable appendage to the animal without any apparent object.

The Frilled Lizard is a native of Australia, and, like most of the family, is generally found on trees, which it can traverse with great address. It seems to be a bold and courageous animal, trusting to its formidable teeth and generally ferocious aspect as a means of defence. "As we were pursuing our walk in the afternoon," writes Captain Gray, "we fell in with a specimen of the remarkable Frilled Lizard. It lives principally in trees, though it can run very swiftly along the ground. When not provoked or disturbed, it moves quietly about, with its frill lying back in plaits upon the body; but it is very irascible, and directly it is frightened, it elevates the frill or ruff, and makes for a tree, where, if overtaken, it throws itself upon its stern, raising its head and chest as high as it can upon the fore-legs; then, doubling its tail underneath the body, and displaying a very formidable set of teeth from the concavity of its large frill, it boldly faces an opponent, biting furiously whatever is presented to it, and even venturing so far in its rage as to fairly make a charge at its enemy.

"We repeatedly tried the courage of this Lizard, and it certainly fought bravely whenever attacked. From the animal making so much use of its frills as a covering and means of defence for its body, this is probably one of the uses to which nature intended the appendage should be applied."

This remarkable Lizard was discovered by Mr. Allan Cunningham, who caught the first specimen as it was perching on the stem of a small decayed tree.

The general color of the Frilled Lizard is yellow-brown mottled with black, and it is remarkable that the tongue and the inside of the mouth are also yellow. The frill, which forms so conspicuous an ornament to this creature, is covered with scales, and toothed on the edge. It does not come to its full size until the animal has attained maturity, and increases in regular proportion to the age of its owner. In the young the frill does not even reach the base of the fore limbs, while in the adult it extends well beyond them. The head is somewhat pyramidal in shape, and four-sided. There is no pouch on the throat. A small crest runs along the nape of the neck, but does not extend to the back. The tail is long and tapering, and like the back, is devoid of a crest. The eyes are rather prominent during the life of the reptile, and the tongue is thick, short and nicked at the end. It is rather a large species, measuring when full grown nearly a yard in total length.

IN the genus Grammatophora, the head is three-sided, and rather flattened, with a sharpish muzzle. There is no throat-pouch, but the skin of the chest is folded into a kind of cross plait. The tail is long, conical, rather flattened at the base, and covered with overlapping keeled scales. All the members of this genus inhabit Australia.

THE MURICATED LIZARD, or GRAMMATOPHORE, is a native of New Holland. It is almost arboreal in its habits, being seldom if ever seen except on trees, which it traverses with remarkable agility, being quick, sharp, and dashing in its movements. It feeds on insects, and is enabled to catch them as they settle on the leaves or branches. It also eats caterpillars, grubs, and other larvæ, which it can find in profusion among the boughs.

The coloring of this Lizard is rather variable. Generally the back is brownish-gray, traversed by sundry brownish bars, running longitudinally on the body and transversely upon the legs and tail. Upon the nape of the neck and the back run a crest composed of triangular compressed scales, having two or three similar rows of pointed scales at each side. Upon the sides of the nape are rows of triangular keeled scales, and the sides are covered with little compressed scales intermixed with large keeled shields. The toes are long, and the two central ones are much longer than the others. This is a small Lizard, only measuring when full grown about fourteen inches.

THE STELLIO, sometimes called the HARDIM by the Arabs, is a well-known Lizard inhabiting Northern Africa, Syria, and Greece.

It is a very active little creature, haunting the ruins of ancient dwellings, heaps of stones, rocks, and similar localities, among which it flits from spot to spot with ceaseless activity. It has a curious habit of bending or nodding its head downwards, a movement which is greatly resented by the stricter Mahometans, who are pleased to consider the Lizard as offering an insult to their religion by imitating them in their peculiar actions of prayer. The more religious among them, therefore, take every opportunity of killing the Stellio, blending amusement, piety, and destructiveness with a happy appreciation of their several merits, earning a good position in Paradise on easy terms, and consoling themselves for the present dearth of infidel heads by slicing off those of the unbelieving Lizards.

The Stellio lives almost entirely on the various insects that flit about the sand, and its quick, rapid movements are needed to secure its prey. A kind of cosmetic was anciently made from this reptile, and even at the present day the Turks employ it in the offices of the toilet.

The color of the Stellio is olive-green above, clouded with black, and the under parts are yellow, sometimes tinged with green. There is no crest upon the nape of the neck, and the scales of the tail are rather large, and arranged in distinct whorls. There is no decided throat-pouch, but the skin of the throat is loose and plaited into a single cross fold towards its base. The body is rather flattened, and there is a longitudinal plait on each side. The tail is round and conical.

IN the restricted genus AGAMA—a word, by the way, which is not derived from any classical source, but is simply the popular name among the natives of Jamaica—the scales of the back are flat and keeled, and the third and fourth toes are nearly equal in length. The throat is marked with one longitudinal fold, and one, or sometimes two transverse folds towards its base. Upon the sides of the neck and near the ears are curious groups of spiny scales. There is a slight crest along the back, the body is rather flattened, and the tail is long, tapering, and is covered with whorls of boldly projecting scales.

In a very old work on natural history, it is stated that the Lizards which have their tails thus armed with sharp, spiny scales, make use of them in a rather singular fashion. They feed, according to these old writers, on cattle and other animals, and judging that from their small size they cannot bring an ox or a cow home after they have killed it, they jump on its back, cling tightly there with their feet, and by judicious lashing of the sharp tail, guide the animal to their home, where they give the fatal bite.

SPINOSE AGAMA.—*Agama colonorum.*

THE SPINOSE AGAMA (*Agama colonorum*) is a well-known example of this genus, residing in Northern Africa, and plentiful in Egypt. The color of this reptile is brown; the scales on the sides of the neck are very long and sharp, and those of the back are broad, boldly keeled, and sharply pointed, so that the creature presents rather a formidable appearance. The tail is long and powerful.

THERE is a very remarkable Lizard belonging to this family, called the EARED MEGALO-CHILE, or sometimes, though wrongly, the EARED AGAMA.

This curious creature is found in Russia. In this genus, containing, as far as is at present known, only one species, the head is flat and round, the eyes large, and the ears sunken and concealed under the skin. On the angle of the mouth at each side is placed a large membranous fold of skin, curved so as to bear a close resemblance to a large external ear, and boldly toothed on its edge. The neck is rather contracted, as if pinched, and has a cross fold below. The back has no crest, the tail is much flattened throughout its length, and the toes are long and very strongly toothed on the edge. The color of this reptile is gray and brown, with a slight green wash upon the top of the head.

The EGYPTIAN MASTIGURE, or SPINE-FOOTED STELLIO, is a native of Northern Africa, and was said, though wrongly, to be the reptile spoken of by the ancients as the land-crocodile. Our figure of this creature is of one-third natural size.

This species attains a rather large size, a full-grown specimen sometimes measuring a yard in length. It is an inhabitant of desert spots, preferring old ruins, rocky ground, and similar

EGYPTIAN MASTIGURE.—*Uromastix spinipes.*

localities, where it can obtain instant refuge in case of alarm. The color of this reptile is bright grass-green during life, but, as is generally the case with all these animals, the brilliant colors fade soon after death, and change to dingy blackish-brown if the skin be stuffed, or to mottled grays, browns, and blacks, if preserved in spirits. The head of this creature is rounded, the back without a crest, the skin of the throat so folded as partly to cover the ears, and the ears themselves are oblong, and toothed in front. The tail is rather flattened, and furnished with transverse rows of large scales, boldly keeled, and sharply pointed. A few conical spines are scattered upon the upper part of the thigh, the sides, and loins.

THE last example of the Agamidæ which can be figured in these pages, is the most ferocious-looking of the whole family, and were its dimensions much enlarged, would be universally allowed to be the most terrible-looking creature on the face of the earth. Many reptiles are spiny in different parts of their bodies, but this creature, appropriately termed the MOLOCH, bristles like a hedgehog with sharp spikes, which project both above and below in such profusion, that this Lizard almost seems to have been formed for the purpose of testing the number of effective spikes that can be planted on a given space. The creature is all spikes, and thorns, and projections. Upon the top of the head two very large spikes are seen, projecting from each eyebrow, and on the back of the neck is a large rounded protuberance, covered with little spiny scales, and having one long projecting spine on each side. On the back, the arrangement is very curious. A number of long spines are scattered at intervals

MOLOCH.—*Moloch horridus.*

over the surface, each of which is surrounded by a circle of lesser spines. It is worthy of notice that these large spines are hollow, and fit upon protuberances of the skin much in the same way that a cow's horn is sheathed on its core. The whole head and limbs are covered with spines similar in formation, but smaller in size. The tail is covered with long, sharp, spiny scales, arranged in whorls, and boldly radiating from their centre; and even the toes are covered as far as the long, sharp claws, with boldly keeled scales. The general color of this reptile is palish yellow, spotted regularly with brown above, and below with dark red blotches edged with black. The Moloch is a native of Australia. The natural size of this creature is given in our engraving.

TREE LIZARDS; DENDROSAURA.

THE last tribe of the Lizards contains but one genus and very few species. From their habit of constantly living on trees, these creatures are called DENDROSAURA, or TREE LIZARDS. In these, the scales of the whole body are small and granular, and arranged in circular bands. The tongue is very curious, being cylindrical and greatly extensile, reminding the observer of a common earth-worm, and swollen at the tip. The eyes are as peculiar as the tongue, being very large, globular, and projecting, and the ball is closely covered with a circular lid, through which a little round hole is pierced, much like the wooden snow-spectacles of the Esquimaux. The body is rather compressed, the ears are concealed under the skin, and the toes are separated into two opposable groups, so that the creature can hold very firmly upon the boughs. All the Dendrosaura are inhabitants of the Old World. The tail is very long and prehensile, and is almost invariably seen coiled round the bough on which the reptile is standing.

The most familiar example of the Dendrosaura is the common CHAMELEON, a reptile which is found both in Africa and Asia.

This singular reptile has long been famous for its power of changing color, a property, however, which has been greatly exaggerated, as will be presently seen. Nearly all the Lizards are constitutionally torpid, though some of them are gifted with great rapidity of movement during certain seasons of the year. The Chameleon, however, carries this sluggishness to an extreme, its only change being from total immobility to the slightest imaginable degree of activity. No one ever saw a Chameleon even walk, as we understand that word, while running is a feat that no Chameleon ever dreamed of.

When it moves along the branch upon which it is clinging, the reptile first raises one foot very slowly indeed, and will sometimes remain foot in air for a considerable time, as if it had gone to sleep in the interim. It then puts the foot as slowly forward, and takes a good grasp of the branch. Having satisfied itself that it is firmly secured, it leisurely unwinds its tail, which has been tightly twisted round the branch, shifts it a little forward, coils it round again, and then rests for a while. With the same elaborate precaution, each foot is successively lifted and advanced, so that the forward movements seem but little faster than the hour-hand of a watch.

The extreme slowness and general habits of this animal are well depicted in an account of a tame Chameleon, kindly presented to me by Captain Drayson :—

"I once owned a Chameleon, which was a very quaint creature. He had been captured by some Kaffir boys, whom I found laughing immoderately at the animal, a practice which I found very common amongst these people whenever they saw one of these reptiles. For a trifle the creature became my property, and I carried him to a little wattle and daub house in which I then resided. Being anxious to watch the private habits of my visitor, I drove a stick into the wall, and placed him upon it. The stick was about four feet in length, and half an inch in diameter, so that the locomotion of the Chameleon was rather limited.

"The first peculiarity I remarked about him was the very slow, methodical way in which he moved. To turn to the right about would occupy him several minutes, whilst to move from one end of the stick to the other was a recreation of which he was sparing, a whole day being devoted to this performance. There was something rather antique in his general appearance, both as regards his form and movements; the long, independent-moving, swivel eyes, giving him the characteristics of an Egyptian production, whilst the habit of puffing himself out occasionally, and of hissing, made him seem old-fashioned in the extreme.

"I was disappointed when I found how slight was the variation in his color. I had been led to believe that if placed on a scarlet, blue, or black ground-work, he would soon assume the same hue; this I found was a delusion. His usual color was a light yellowish-green, and this he could alter to a dark blue, or brown-green, and he could make several dark brown spots become very prominent on his skin.

"The method I used to adopt to make him show off, was to rub his side with my finger. He objected to this treatment, and used to puff away pompously, and vary his tints, as it appeared to me, by means of contracting or expanding his muscles under the skin. He looked very lantern-like, as though he were merely skin and ribs, and he was never found guilty of eating anything. Sometimes I saw flies settle upon him, a liberty which he did not resent. He merely turned one of his swivel eyes towards the delinquent and squinted calmly at it. Occasionally I put a fly in his mouth, and forced him to keep it there; he took the affront very coolly, and the fly was seen no more. So hollow did he appear, that I frequently listened to hear if the flies were buzzing about inside him, but all was

CHAMELEON.

quiet. He stayed on the stick during two months. I then gave him a run out of doors, but having left him a few minutes, he took advantage of my absence and levanted, after which I saw him no more."

The food of the Chameleon consists of insects, mostly flies, but, like many other reptiles, the Chameleon is able to live for some months without taking food at all. This capacity for fasting, together with the singular manner in which the reptile takes its prey, gave rise to the absurd fable that the Chameleon lived only upon air. To judge by external appearance, there never was an animal less fitted than the Chameleon for capturing the winged and active flies. But when we come to examine its structure, we find that it is even better fitted for this purpose than many of the more active insect-eating Lizards.

The tongue is the instrument by which the fly is captured, being darted out with such singular velocity that it is hardly perceptible, and a fly seems to leap into the mouth of the reptile as if attracted by magnetism. This member is very muscular, and is furnished at the tip with a kind of viscid secretion which causes the fly to adhere to it. A lady who kept a Chameleon for some time, told me that her pet died, and when they came to examine it, they found that its tongue had in some strange way got down its throat, an accident which they took to be the cause of its death. Its mouth is well furnished with teeth, which are set firmly into its jaw, and enable it to bruise the insects after getting them into its mouth by means of the tongue.

The eyes have a most singular appearance, and are worked quite independently of each other, one rolling backwards while the other is directed forwards or upwards. In connection with this subject some very curious and valuable remarks will be found on the next page. There is not the least spark of expression in the eye of the Chameleon, which looks about as intellectual as a green pea with a dot of ink upon it.

Owing to the exceeding slowness of its movements, it has no way of escaping when once discovered, and as a French writer well says, "un Caméléon aperçu est un Caméléon perdu." Great numbers of these creatures fall victims to enemies of every kind, and were it not that their color assimilates so well with the foliage on which they dwell, and their movements are so slow as to give no aid to the searching eye of their foes, the race would soon be extinct. The Chameleon has an odd habit of puffing out its body for some unexplained reason, and inflating itself until it swells to nearly twice its usual size. In this curious state it will remain for several hours, sometimes allowing itself to collapse a little, and then reinflating its skin until it becomes as tense as a drum and looks as hollow as a balloon.

The Chameleon is readily tamed, if such a word can be applied to the imperturbable nonchalance with which it behaves under every change of circumstance. It can be handled without danger, and although its teeth are strong, will not attempt to bite the hand that holds it. It is, however, rather quarrelsome with its own kind, and the only excitement under which it has been seen to labor is when it takes to fighting with a neighbor. Not that even then it hurries itself particularly, or does much harm to its opponent, the combatants contenting themselves with knocking their tails together in a grave and systematic manner.

A few words on the change of color will not be out of place. The usual color of the Chameleon when in its wild state is green, from which it passes through the shades of violet, blue and yellow, of which the green consists. In moderate climates, however, it rarely retains the bright green hue, the color fading into yellowish-gray, or the kind of tint which is known as *feuille-morte*. One of the best and most philosophical disquisitions on this phenomenon is that of Dr. Weissenbaum, published in the "Magazine of Natural History." The writer had a living Chameleon for some time, and gives the result of his observations in the following words:—

"The remote cause of the difference of color in the two lateral folds of the body, may be distinctly referred to the manner in which the light acts upon the animal. The statement of Murray that the side turned towards the light is always of a darker color, is perfectly true; this rule holds good with reference to the direct and diffused light of the sun and moon as to artificial light. Even when the animal was moving in the walks of my garden, and happened to come near enough to the border to be shaded by the box edging, that side so shaded would instantly become less darkly colored than the other.

"Now the light in this way seldom illumines *exactly* one half of the animal in a more powerful manner than the other, and as the middle line is constantly the line of demarcation between the two different shades of color, we must evidently refer the different effects to two different centres, from which the nervous currents can only radiate, under such circumstances, towards the organs respectively situated on each side of the mesial line. Over these centres, without doubt, the organ of vision immediately presides; and, indeed, we ought not to wonder that the action of light has such powerful effects on the highly irritable organization of the Chameleon, considering that the eye is most highly developed. The lungs are but secondarily affected, but they are likewise more strongly excited on the *darker side, which is constantly more convex than the other.*

"Many other circumstances may be brought forward in favor of the opinion that the nervous currents in one half of the Chameleon are going on independently of those in the other; and that the animal has two lateral centres of perception, sensation and motion, besides the common one in which must reside the faculty of concentration.

"Notwithstanding the strictly symmetrical construction of the Chameleon as to its two halves, the eyes move independently of each other, and convey different impressions to their different centres of perception; the consequence is, that when the animal is agitated, its movements appear like those of two animals glued together. Each half wishes to move its own way, and there is no concordance of action. The Chameleon, therefore, is not able to swim like other animals; it is so frightened if put into water, that the faculty of concentration is lost, and it tumbles about as if in a state of intoxication.

"On the other hand, when the creature is undisturbed, the eye which receives the strongest impression propagates it to the common centre, and prevails on the other eye to follow that impression, and direct itself to the same object. The Chameleon, moreover, may be asleep on one side and awake on the other. When cautiously approaching my specimen at night with a candle, so as not to awake the whole animal by the shaking of the room, the eye turned toward the flame would open and begin to move, and the corresponding side to change color, whereas the other side would remain for several seconds longer in its torpid and changeable state, with its eye shut."

It seems probable that the change of color may be directly owing to the greater or less rapidity of the circulation, which may turn the Chameleon from green to yellow, just as in ourselves an emotion of the mind can tinge the cheek with scarlet, or leave it pallid and death-like. Mr. Milne Edwards thinks that it is due to two layers of pigment cells in the skin, arranged so as to be movable upon each other, and so produce the different effects.

The young of the Chameleon are produced from eggs, which are very spherical, white in color, and covered with a chalky and very porous shell. They are placed on the ground under leaves, and there left to hatch by the heat of the sun, and the warmth produced by the decomposition of the leaves. The two sexes can be distinguished from each other by the shape of the tail, which in the male is thick and swollen at the base.

THERE are nearly twenty species of Chameleons known to zoologists at the present day, all presenting some peculiarity of form or structure. One of the most remarkable species is the LARGE-NAPED CHAMELEON, or Fork-nosed Chameleon, as it is sometimes called.

This creature inhabits Madagascar, that land which nourishes so many strange forms of animal life. It is also found in India, the Moluccas, and Australia. When full grown, the muzzle of the male is very deeply cleft, or forked, the two branches diverging from each other. The female has no horns, and in the male they are short and blunt while the creature is young, not obtaining their full length and sharpness until it has attained full age. These curious forked projections belong to the skull, and are not merely a pair of prolonged scales or tubercles.

SNAKES; OPHIDIA.

The large and important order at which we now arrive, consists of reptiles which are popularly known as SNAKES, or more scientifically as OPHIDIA, and to which all the true serpents are to be referred.

Almost every order is bordered, so to speak, with creatures so equally balanced between the characteristics of the orders that precede and follow it, that they can be with difficulty referred to their right position. Such, indeed, is the case with the Ophidia, from which are excluded, by the most recent systematic zoologists, the amphisbænians and many other footless reptiles, now classed among the lizards. The greater number of the Snakes are without any vestige of limbs, but in one or two species, such as the pythons, the hinder pair of limbs are represented by a pair of little horny spurs placed just at the base of the tail, and are supported by tiny bones that are the undeveloped commencements of hinder limbs. Indeed, several of the true lizards, the common blind-worm, for example, are not so well supplied with limbs as these true Snakes.

The movements of the serpent tribe are, in consequence, performed without the aid of limbs, and are, as a general rule, achieved by means of the ribs and the large cross scales that cover the lower surface. Each of these scales overlaps its successor, leaving a bold horny ridge whenever it is partially erected by the action of the muscles. The reader will easily see that a reptile so constructed can move with some rapidity by successively thrusting each scale a little forward, hitching the projecting edge on any rough substance, and drawing itself forward until it can repeat the process with the next scale. These movements are consequently very quiet and gliding, and the creature is able to pursue its way under circumstances of considerable difficulty.

Oftentimes the Snake uses these scales in self-defence, offering a passive resistance to its foe when it is incapable of acting on the offensive. Any one may easily try this experiment by taking a common field Snake, letting it glide among the stubble or into the interstices of rocky ground, and then trying to pull it out by the tail. He will find that even if the reptile be only half concealed, it cannot be dragged backward without doing it considerable damage, for on feeling the grasp, it erects all the scales and opposes their edges so effectually to the pull that it mostly succeeds in gliding through from the hand that holds it. I have often lost Snakes by allowing them to insinuate themselves into crevices, and have been fain to let them escape rather than subject them to the pain, if not absolute damage, which they must have suffered in being dragged back by main force.

The tongue of the Snake is long, black, and deeply forked at its extremity, and when at rest is drawn into a sheath in the lower jaw. In these days it is perhaps hardly necessary to state that the tongue is perfectly harmless, even in a poisonous serpent, and that the popular idea of the "sting" is entirely erroneous. The Snakes all seem to employ the tongue largely as a feeler, and may be seen to touch gently with the forked extremities the objects over which they are about to crawl or which they desire to examine. The external organs of hearing are absent.

The vertebral column is most wonderfully formed, and is constructed with a special view to the peculiar movements of the serpent tribe. Each vertebra is rather elongated, and is furnished at one end with a ball and at the other with a corresponding socket, into which the ball of the succeeding vertebra exactly fits, thus enabling the creature to writhe and twine in all directions without danger of dislocating its spine. This ball-and-socket principle extends even to the ribs, which are jointed to certain rounded projections of the vertebræ in a manner almost identical with the articulation of the vertebræ upon each other, and as they are moved by very powerful muscles, perform most important functions in the economy of the creature to which they belong.

Sometimes the Snakes advance by a series of undulations, either vertical or horizontal, according to the species, and when they proceed through water, where the scales of the abdomen would have no hold of the yielding element, their movements are always of this undulatory description. The number of vertebræ, and consequently of ribs, varies much in different species, in some Snakes being about three hundred.

The jaws of the serpents are very wonderful examples of animal mechanics, and may be cited among the innumerable instances where the existing construction of living beings has long preceded the inventions of man. We have already seen the invaluable mechanic invention of the ball-and-socket joint exhibited in the vertebræ of the Snakes, and it may be mentioned that in the spot where the limbs of almost all animals, man included, are joined to the trunk, the ball-and-socket principle is employed, though in a less perfect manner than in the Snakes. It is by means of this beautiful form of joint that posture-masters and mountebanks are able to contort their bodies and limbs into so many wonderful shapes, the muscles and tendons yielding by constant use and enabling the bones to work in their sockets without hindrance. Indeed, a master of the art of posturing is really an useful member of society, at all events to the eye of the physiologist, as showing the perfection of the human form, and the wonderful capabilities of man, even when considered from the mere animal point of view.

In the jaw of the serpents, we shall find more than one curious example of the manner in which human inventions have succeeded, if, indeed, they have not been borrowed from some animal structure.

All the Snakes are well supplied with teeth; but their number, form, and structure differ considerably in the various species. Those Snakes that are not possessed of venomous fangs have the bones of the palate as well as the jaws furnished with teeth, which are of moderate size, simple in form, and all point backward, so as to prevent any animal from escaping which has ever been grasped, and acting as valves which permit of motion in one direction only.

The bones of the jaw are, as has already been mentioned, very loosely constructed, their different portions being separable, and giving way when the creature exerts its wonderful powers of swallowing. The great python Snakes are well known to swallow animals of great proportionate size, and any one may witness the singular process by taking a common field Snake, keeping it without food for a month or so, and then giving it a large frog. As it seizes its prey, the idea of getting so stout an animal down that slender neck and through those little jaws appears too absurd to be entertained for a moment, and even the leg which it has grasped appears to be several times too large to be passed through the throat. But by slow degrees the frog disappears, the mouth of the Snake gradually widening, until the bones separate from each other to some distance, and are only held by the ligaments, and the whole jaw becoming dislocated, until the head and neck of the Snake look as if the skin had been stripped from the reptile, spread thin and flat, and drawn like a glove over the frog.

No sooner, however, has the frog fairly descended into the stomach, than the head begins to assume its former appearance; the elastic ligaments contract and draw the bones into their places, the scales, which had been far separated from each other, resume their ordinary position, and no one would imagine, from looking at the reptile, to what extent the jaws and neck have recently been distended. As many of the Snakes swallow their prey alive—the frog, for example, having been heard to squeak while in the stomach of its destroyer—the struggles of the internal victim would often cause its escape, were it not for the array of recurved teeth, which act so effectually, that even if the Snake wished to disgorge its prey it could not do so. Mr. Bell had in his collection a small Snake which had tried to swallow a mouse too large even for the expansile powers of a Snake's throat, and which had literally burst through the skin and muscles of the neck.

The lower jaw, moreover, is not jointed directly to the skull, but to a most singular development of the temporal bone, which throws out two elongated processes at right angles with each other, like the letter L laid horizontally ⌐, so that a curious double lever is obtained, precisely after the fashion of the well-known "throwing-stick" of the aboriginal Australians, which enables those savages to fling their spears with deadly effect to a distance of a hundred yards.

The teeth of the venomous Serpents will be described in connection with one of the species.

The Serpents, in common with other reptiles, have their bodies covered by a delicate epidermis, popularly called the skin, which lies over the scales, and is renewed at tolerably regular intervals. Towards the time of changing its skin, the Snake becomes dull and sluggish, the eyes look white and blind, owing to the thickening of the epidermis that covers them, and the bright colors become dim and ill-defined. Presently, however, the skin splits upon the back, mostly near the head, and the Snake contrives to wriggle itself out of the old integument, usually turning it inside out in the process. This shed skin is transparent, having the shape of each scale impressed upon it, being fine and delicate as goldbeater's-skin, and being applicable to many of the same uses, such as shielding a small wound from the external air. In two very fine specimens of cast skins, formerly belonging to a viper and boa-constrictor, now lying before me, the structure of each scale is so well shown, that the characteristics of the two reptiles can be distinguished as readily as if the creatures were present from whose bodies they were shed. Even the transparent scale that covers the eyes is drawn off entire, and the large elongated hexagonal scales that are arranged along the abdomen, and aid the animal in its progress, are exhibited so boldly that they will resist the movement of a finger drawn over them from tail to head.

THE first sub-order of Snakes consists of those Serpents which are classed under the name of VIPERINA. All these reptiles are devoid of teeth in the upper jaw except two long, poison-bearing fangs, set one at each side, and near the muzzle. The lower jaw is well furnished with teeth, and both jaws are feeble. The scales of the abdomen are bold, broad, and arranged like overlapping bands. The head is large in proportion to the neck, and very wide behind, so that the head of these Snakes has been well compared to an ace of spades. The hinder limbs are not seen.

In the first family of the Viperine Snakes, called the CROTALIDÆ, the face is marked with a large pit or depression on each side, between the eye and the nostril. The celebrated and dreaded FER-DE-LANCE belongs to this family.

This terrible reptile is a native of Brazil, and in some parts is very common, owing to its exceeding fecundity and its habit of constant concealment. It has an especial liking for the sugar plantations, and a field of canes is seldom cut without the discovery of seventy or eighty of these venomous creatures. Martinique and St. Lucia are terribly haunted by this Snake, which is held in great dread by the natives and settlers. In general, the Serpents, even those of a poisonous character, avoid the presence of man, but the Fer-de-Lance frequently takes the initiative, and leaping from its concealment, fastens upon the passenger whose presence has disturbed its irritable temper, and inflicts a wound that is almost invariably fatal within a few hours.

Even in those cases where the sufferer recovers for the time, the system is terribly injured, and the latent virulence of the poison can hardly be eliminated from the frame, even at the cost of painful boils and ulcerations which last for many years. The nervous system is also much affected, as giddiness and paralysis are among the usual consequences of the strong venom which this reptile extracts, by some inexplicable chemistry, from perfectly harmless food. Convulsions, severe pain at the heart, together with distressing nausea, are among the many symptoms produced by this poison.

To escape this creature in its chosen haunts is a matter of very great difficulty, as it is either concealed under dead leaves, among the heavy foliage of parasitic plants, or coiled up in the nest of some poor bird whose eggs or young it has devoured, and from this spot of vantage makes its stroke, swift and straight as a fencer's thrust, and without the least warning by hiss or rattle to indicate its purpose.

All animals dread the Fer-de-Lance; the horse prances and snorts in terror on approaching its hiding-place, his whole frame trembles with fear, and he cannot be induced by spur or whip to pass within striking distance of this formidable reptile. Birds of all kinds have a horror of its presence, and will pursue it from place to place, or hover near the spot on which it is resting, fluttering their wings, stretching their necks, and uttering hoarse cries of mingled rage and terror. The honey guide is especially fearful of this Serpent, and has often guided a man, not as he supposed, to the vicinity of a hive of wild bees, but to the resting-place of this venomous Snake. The pig, when in good condition, is said to be the only animal that can resist the poison, the thick coating of fat which covers the body preventing the venom from mingling with the blood. It is said, indeed, that a fat hog cares nothing for Fer-de-Lance or rattlesnake, but receives their stroke with contemptuous indifference, charges at them fearlessly, tramples upon them until they are disabled, and then quietly eats them.

Against the effects of this poison there seems to be no certain remedy; but the copious use of spirits has lately appeared to neutralize in some measure the full virulence of a Snake

FER-DE-LANCE.—*Craspedocephalus lanceolatus.* (One-sixth natural size).

bite. The amount of strong spirits which can be drunk under such circumstances is almost incredible, its whole force seeming to be employed in arming the nerves against the enfeebling power of the poison. Some recent and valuable experiments have shown, that if a man, bitten by a venomous Serpent, can be kept in a state of semi-intoxication through the use of spirituous liquors, this rather strange process will give him almost his only hope of escape.

Yet nothing is made in vain, and terrible as is this creature to man, it is of no small use to him even in the localities where it is most dreaded. But for the presence of the Fer-de-Lance and one or two other Serpents closely allied to it, the sugar plantations would be devastated by the rats which crowd to such fertile spots, and on which this Snake chiefly feeds.

As is the case with many Serpents, the color of the Fer-de-Lance is rather variable. Its usual tints are olive above with dark cross bands, and whitish gray below, covered with very minute dark dots. The head is brown. This reptile attains a considerable size, being generally five or six feet long, and occasionally reaching a length of seven or eight feet. The tail ends in a horny spine which scrapes harshly against rough objects, but does not rattle.

CLOSELY allied to the Fer-de-Lance is another poisonous Serpent of Southern America, remarkable for the very large size to which it attains, and the glowing radiance of its fearful beauty. This is the CURUCUCU, more familiarly known by the popular title of BUSHMASTER, (*Láchesis mutus*.)

Mr. Waterton, who has incidentally mentioned this Snake in his "Wanderings," has kindly sent me the following information about this terrible creature: "The Bushmaster will sometimes reach fourteen feet in length. The Dutch gave it the name of Bushmaster on account of its powers of destruction, and being the largest poisonous Snake discovered. It still continues to have the same name among the colonists of British Guiana. Its Indian name is COUANACOUCHI. It is a beautiful Serpent, displaying all the prismatic colors when alive, but they disappear after death. All these three species (the Bushmaster, Labarri, and

BUSHMASTER.—*Lachesis muta*. (One-sixth natural size.)

Coulacanara) inhabit the trees as well as the ground, but as far as I could perceive, they never mount the trees with a full stomach."

THE WATER MOCCASIN (*Ancistrodon piscivorus*). This reptile is restricted to the region between the Carolinas and the Gulf, and the valleys of the Mississippi River. This is emphatically a Water Snake. This reptile is, perhaps, the most dreaded of any in this country. It has the reputation of attacking unprovoked any one that may be in reach—a circumstance that is true of very few animals throughout the world. The Southern negroes are much exposed to its venom in the wet rice lands, where it abounds. It is very stout, and in color and markings very forbidding; the length being about nineteen inches.

Another species is recorded as a native in Indianola, Texas, called *A. pugnax*. The Black Moccasin (*A. atrofuscus*) is found in the mountains of North Carolina.

The name of WATER VIPER (*Ancistrodon piscivorum*) is appropriately given to the creature now before us, in consequence of its water-loving habits.

It is a native of many parts of America, and is never seen at any great distance from water, being found plentifully in the neighborhood of rivers, marshes, and in swampy lands. It is a good climber of trees, and may be seen entwined in great numbers on the branches that overhang the water. On the least alarm, the reptile glides from the branch, drops into the water, and wriggles its way into a place of safety. The object of climbing the trees seems to be that the creature delights to bask in the sun, and takes that method of gratifying its inclination where the whole of the soil is wet and marshy. But in those localities where it can find dry banks and rising grounds, the Water Viper contents itself with ascending them and lying upon the dry surface enjoying the genial warmth.

It is a most poisonous reptile, and is even more dreaded by the negroes than the rattlesnake, as, like the fer-de-lance, it will make the first attack, erecting itself boldly, opening its mouth for a second or two, and then darting forward with a rapid spring. At all times it seems to be of an aggressive character, and has been known to chase and bite other Snakes put into the same cage, the poor creatures fleeing before it and endeavoring to escape by clinging to the sides of the cage. But when several other individuals of the same species were admitted, the very Snake that had before been so ferocious, became quite calm, and a box containing four or five specimens has been sent on a journey of many miles without any quarrels ensuing among the inmates.

The food of the Water Viper consists of fishes, which it can procure by its great rapidity of movement and excellent swimming powers, of reptiles and even of birds. Mr. T. W. Wood has favored me with an account of the manner in which a Water Viper devoured the prey that was put before it:—

"A short time ago I had the good fortune to be present when some captured reptiles of this species were fed. Some sparrows and titlarks were put into the apartment containing several specimens of the Water Viper. The sparrows seemed very much terrified, and soon huddled together in a corner, afraid, as I suppose, of the spectators.

"One of the titlarks, however, bolder than the rest, ran about as if at home. One of the Water Vipers perceiving it quiet for a moment, seemed to fix its eye upon the poor little creature. The reptile commenced moving towards the bird slowly but surely, their eyes being intently fixed upon each other. When the Serpent had approached within about half an inch, it opened its mouth and seized the bird by the side, its left wing being grasped in the Snake's mouth. The ill-fated bird instantly gave two or three convulsive struggles, the head then dropped, the eyes closed, and all was over; a drop of blood oozed slowly out of the bird's bill. The reptile did not release the bird after it was bitten, but began to swallow it almost immediately.

"Another titlark was then introduced by the keeper. This bird was, when I approached, lying on its side as if dead. Another Water Viper seized its head and commenced swallowing it, the bird struggling violently; at each effort of deglutition the venomous fangs were seen to move forward. In this case the poison did not take such rapid effect, as the bird was evidently alive when it disappeared down the reptile's throat."

The color of the Water Viper is greenish brown, taking a yellowish tone along the sides, and banded with blackish brown. It seldom exceeds two feet in length. This serpent is also known by the popular names of COTTON-MOUTH and WATER MOCCASIN SNAKE.

THE COPPER-HEAD SNAKE of the same country is closely allied to it. An illustration of it is to be found on next page. This is the dreaded Cotton-mouth of the Southern negroes. It inhabits rather low ground, and extends along the Catskill range as far as the Gulf States. Its color is a hazel-brown, with a light coppery hue upon its head. Its length is about two feet. It is justly dreaded as a most vicious and venomous reptile. Though differing from the preceding in some respects, particularly in having no rattles, it has poison fangs that are quite deadly in application to man or beast. The names Dumb Rattle, Red Adder, Red Viper, Deaf Adder, and Chunk-head, are applied to it in various sections of country.

THE RATTLESNAKE.

THE well-known and terrible RATTLESNAKE now comes before us.

This dreaded reptile is a native of North America, and is remarkable for the singular termination to the tail, from which it derives its popular name. It has already been mentioned that the fer-de-lance has a long, horny scale at the tip of its tail, and in the Rattlesnake this appendage is developed into a rather complicated apparatus of sound.

At the extremity of the tail are a number of curious loose horny structures, formed of the same substance as the scales, and varying greatly in number according to the size of the individual. It is now generally considered that the number of joints on the "rattle" is an indication of the reptile's age, a fresh joint being gained each year immediately after it changes its skin and before it goes into winter quarters. There is, however, another opinion prevalent among the less educated, which gives to the Rattlesnake the vindictive spirit of the North American Indian, and asserts that it adds a new joint to its rattle whenever it has slain a human being, thus bearing on its tail the fearful trophies of its prowess, just as the Indians wear the scalps of their slain foes.

COPPER-HEAD SNAKE.—*Ancistrodon contortrix.*

The joints of this remarkable apparatus are arranged in a very curious manner, each being of a somewhat pyramidal shape, but rounded at the edges, and being slipped within its predecessor as far as a protuberant ring which runs round the edge. In fact, a very good idea of the structure of the rattle may be formed by slipping a number of thimbles loosely into each other. The last joint is smaller than the rest, and rounded. As was lately mentioned, the number of these joints is variable, but the average number is from five or six to fourteen or fifteen. There are occasional specimens found that possess more than twenty joints in the rattle, but such examples are very rare.

When in repose the Rattlesnake usually lies coiled in some suitable spot, with its head lying flat, and the tip of its tail elevated in the middle of the coil. Should it be irritated by a passenger, or feel annoyed or alarmed, it instantly communicates a quivering movement to the tail, which causes the joints of the rattle to shake against each other, with a peculiar skirring ruffle, not easily described, but never to be forgotten when once heard. All animals, even those which have never seen a Rattlesnake, tremble at this sound, and try to get out of the way. Even a horse newly brought from Europe is just as frightened as the animal that has been bred in the same country with this dread Serpent, and at the sound of the rattle will prance, plunge, and snort in deadly fear, and cannot be induced to pass within striking distance of the angry Snake.

It has already been mentioned that swine are comparatively indifferent to the Rattlesnake, and will trample it to death and eat it afterwards. It is certain that they will eat a dead Rattlesnake, though almost any other animal will flee from the lifeless carcase nearly as swiftly as from the living reptile. Perhaps the thick coating of fat that clothes the body of the well-fed swine may neutralize the poison of the venomed teeth, and so enable the hog to receive the stroke with comparative impunity. The peccary is also said to kill and devour the Rattlesnake without injury, and deer are reported to jump upon it and kick its life out with their sharp hoofs.

Fortunately for the human inhabitants of the same land, the Rattlesnake is slow and torpid in its movements, and seldom attempts to bite unless it is provoked, even suffering itself to be handled without avenging itself. Mr. Waterton tells me in connection with these

RATTLESNAKE.—*Orotalus durissus.*

reptiles: "I never feared the bite of a Snake, relying entirely on my own movements. Thus, in presence of several professional gentlemen, I once transferred twenty seven Rattlesnakes from one apartment to another, with my hand alone. They hissed and rattled when I meddled with them, but they did not offer to bite me." The fer-de-lance Snake is, as has already been mentioned, most fierce and irritable in character, taking the initiative, and attacking without reason. But the Rattlesnake always gives notice of its deadly intentions, and never strikes without going through the usual preliminaries. When about to inflict the fatal blow, the reptile seems to swell with anger, its throat dilating, and its whole body rising and sinking as if inflated by bellows. The tail is agitated with increasing vehemence, the rattle sounds its threatening war-note with sharper ruffle, the head becomes flattened as it is drawn back ready for the stroke, and the whole creature seems a very incarnation of deadly rage. Yet, even in such moments, if the intruder withdraw, the reptile will gradually lay aside its angry aspect, the coils settle down in their place, the flashing eyes lose their lustre, the rattle becomes stationary, and the Serpent sinks back into its previous state of lethargy.

It is rather curious that the Rattlesnake varies much in its powers of venom and its irritability of temper, according to the season of the year. During the months of spring it will seldom attempt to bite, and if it does strike a foe, the poison is comparatively mild in its effects. But after August, and before it seeks its winter quarters, the Rattlesnake is not only more fierce than at any other time of the year, but the venom seems to be of more fearful intensity, inflicting wounds from which nothing escapes with life.

The rapidity of the effects depends necessarily on the part which is bitten. Should the points of the teeth wound a moderately large vein or an artery, the venom courses swiftly through the blood, and the victim dies in a few minutes. But if, perchance, the tooth should pierce some fleshy and muscular part of the body, the poison does not have such rapid effect, and the injured person may be saved by the timely administration of powerful remedies. There seems, indeed, to be no one specific for the bite of this reptile, as the effects vary according to the individual who happens to be bitten, and the state of health in which the sufferer may be at the time. Immediate suction, however, and the unsparing use of the knife appear to be the most efficacious means of neutralizing the poison, and strong ammonia and oil have been employed with good results. Catesby, in writing about this reptile, remarks that he has known instances where death has occurred within two minutes after the infliction of the bite.

The food of the Rattlesnake consists of rats, mice, reptiles, and small birds, the latter of which creatures it is said to obtain by the exercise of a mysterious power termed fascination, the victim being held, as it were, by the gaze of its destroyer, and compelled to remain in the same spot until the Serpent can approach sufficiently near to seize it. It is even said that the Rattlesnake can coil itself at the foot of a tree, and by the mere power of its gaze, force a squirrel or bird to descend and fling itself into the open mouth waiting to receive it.

These phenomena have been strongly asserted by persons who say that they have seen them, and are violently denied by other persons who have never witnessed the process, and therefore believe that the circumstances could not have happened. For my own part I certainly incline to the theory of fascination, thinking that the power exists, and is occasionally employed, but under peculiar conditions. That any creature may be suddenly paralyzed by fear at the sight of a deadly foe is too well known to require argument, and it is therefore highly probable that a bird or squirrel, which could easily escape from the Serpent's jaws by its superior agility, might be so struck with sudden dread on seeing its worst enemy, that it would be unable to move until the reptile had seized it.

Birds, especially, are most sensitive in their nature, and can be fascinated in a manner by any one who chooses to try the experiment. Let any bird be taken, laid on its back, and the finger pointed at its eyes. The whole frame of the creature will begin to stiffen, the legs will be drawn up, and if the hand be gently removed, the bird will lie motionless on its back for any length of time. I always employ this method of managing my canaries when I give them their periodical dressing of insect-destroying powder. I shake the powder well into their feathers, pour a small heap of it on a sheet of paper, lay the bird in the powder, hold my finger over its eyes for a moment, and leave it lying there while I catch and prepare another bird for the same process. There is another way of fascinating the bird, equally simple. Put it on a slate or dark board, draw a white chalk line on the board, set the bird longitudinally upon the line, put its beak on the white mark, and you may go away for hours, and when you return the bird will be found fixed in the same position, there held by some subtle and mysterious influence which is as yet unexplained.

Thus far there is no difficulty in accepting the theory of fascination, but the idea of a moral compulsion on the part of the Snake, and a perforced obedience on the part of its victim, is so strange that it has met with very great incredulity. Still, although strange, it is not quite incredible. We all know how the immediate presence of danger causes a reckless desire to see and do the worst, regardless of the consequences, and heeding only the overpowering impulse that seems to move the body without the volition of the mind. There are many persons who cannot stand on any elevated spot without feeling so irresistible a desire of flinging themselves into the depths below, that they dare not even stand near an open window or walk near the edge of a cliff. It may be that the squirrel or bird, seeing its deadly enemy below, is so mentally overbalanced that it is forced to approach the foe against its own will, and is drawn nearer to those deadly fangs by the very same impulse that would urge a human being to jump over the edge of a precipice or from the top of a lofty building.

Every squirrel or every bird may not succumb to the same influence, just as every human being does not yield to the insane desire of jumping from heights, and it is probable that a Rattlesnake may coil itself under a tree and look all day at the squirrels sporting upon the branches, or the birds flitting among the boughs, without inducing one of them to become an involuntary victim. Yet it is possible that out of the many hundreds that could see the Serpent, one would be weak-minded enough to yield to the subtle influence, and, instead of running away, find itself forced to approach nearer and nearer the fearful reptile.

Some persons acknowledge the fact that the bird approaches the Snake, and is then snapped up, but explain it in a different manner. They say that the bird is engaged in mobbing or threatening the Snake, just as it might follow and buffet a hawk, an owl, or a raven, and in its eagerness approaches so closely that the Snake is able to secure it by a sudden dart. Such is very likely to be the case in many instances, as the little birds will often hover about a poisonous Snake, and, by their fluttering wings and shrieking cries, call attention to the venomous reptile. But the many descriptions of the fascinating process are too precise to allow of such a supposition in the particular instances which are mentioned.

Even common Snakes can exercise a similar power. I have seen one of these Snakes in chase of a frog, and the intended victim, although a large and powerful specimen of its race, fully able to escape by a succession of leaps such as it would employ if chased by a human being, was only crawling slowly and painfully like a toad, its actions reminding one of those horrid visions of the night, when the dreamer finds himself running or fighting for his life, and cannot move faster than a walk, or strike a blow that would break a cobweb. In such cases, the victim may be taken from the pursuer, but unless it is carried to a considerable distance, it will soon be in the jaws of the Serpent a second time.

It is worthy of notice that in all such instances, a sudden sound will seem to break the spell and snap the invisible chain that binds the victim to its destroyer. If birds are spellbound by finger or chalk line, as has already been described, a quick movement or a heavy footstep will release them from their bonds, and a sudden shout will in a similar manner enable a bird to break away from the Serpent into whose jaws it was on the point of falling. One of my friends when in Canada saw a little bird lying on the ground, fluttering about as if dusting itself, but in a rather strange manner, and on his nearer approach, a Snake glided from the spot, and the bird gathered its wings together and flew away. The Snake was one of the harmless kind, and being taken to the house of the person who had interrupted it in its meal, served to keep the premises clear of rats and mice. The Serpent is not the only creature to which this singular power is attributed, for the natives of Northern Africa assert that the lion is also gifted with this influence, and can induce certain hapless men and women to leave their homes and follow him into the woods. This, however, is only a popular tradition among the natives, and has met with no corroboration.

The Rattlesnake retires to its winter quarters as soon as the increasing coldness of the weather gives it warning to seek a home where it can find protection against the frosts. Sometimes the Snake chooses a convenient hole or crevice for this purpose, but in general it prefers the neighborhood of marshy ground, and harbors under the heavy masses of a certain long-stemmed moss (*sphagnum palustre*) which grows plentifully in such situations. In such localities the Rattlesnake may be found during the winter, either coiled up in masses containing six or seven individuals, or creeping slowly about beneath the protecting moss. Many of these fearful Snakes are killed during the cold months by persons who are acquainted with their habits, and surprise them in their winter quarters.

The general color of the Rattlesnake is pale brown. A dark streak runs along the temples from the back of the eye, and expands at the corner of the mouth into a large spot. A series of irregular dark brown bands are drawn across the back, a number of round spots of the same hue are scattered along the sides, upon the nape of the neck and back of the head.

The Rattlesnakes are peculiar to America, embraced in the family *Crotalidæ*, the latter term meaning, in the Greek, rattlers, referring to the characteristic habit of some of the species. They have two fangs on the upper jaw, which are grooved, and suited to deliver the liquid poison which lies in a sac at the roots. Eighteen species of Rattlesnakes are now known in North America.

The Northern Rattlesnake (*Crotalus horridus*), called also the Banded Rattlesnake, is the more common of the few species of this dreaded family of reptiles. It is illustrated together with the *Crotalus adamanteus*, another American Rattlesnake. The Banded Rattlesnake is found in rocky places on dry soil, reaching in its range as far north as the middle of New England and New York State, west as far as the Rocky Mountains, and south to the Gulf States. Along the shores of Lake Champlain it is particularly abundant. Dr. DeKay, the eminent zoologist of the State of New York, gives the following from a local newspaper of the day:—

"Two men in three days killed eleven hundred and four Rattlesnakes on Tongue Mountain, in the town of Bolton, New York."

THE DIAMOND AND THE NORTHERN RATTLESNAKE.—*Crotalus adamanteus* and *Crotalus horridus*. (One-tenth natural size.)

The popular belief that a rattle is added yearly is not correct. Dr. Holbrook, the author on American Reptiles, says he has known one to add two rattles in a year, and Dr. Bachman observed four added in the same period. Mr. Peale, of the Museum in Philadelphia, kept a Rattlesnake fourteen years. It had, when first confined, eleven rattles. Several were lost annually, and new ones took their place. At its death there were but eleven rattles, though it had increased in length four inches. Holbrook saw one having twenty-one rattles. Accounts are occasionally given of a more numerous series. We have an example of one bearing twenty-four rattles. This is probably about the limit. The pretended powers of "charming" are not credited by naturalists.

The Diamond Rattlesnake is strictly a Southern species, being confined to the seaboard below the Carolinas. Its habits differ, in so far that this one inhabits damp, shady places; hence the local name, Water Rattle. In size it exceeds the Banded species, some specimens attaining the length of eight feet. The common name is suggested by the elegant diamond or lattice-work markings of its body. Several smaller species are enumerated as North American: The *C. atrox*, of Texas; *C. lucifer*, Oregon; *C. confluentus*, Texas; and *C. molossus*, New Mexico.

THE SOUTHERN GROUND RATTLESNAKE (*Caudisona miliaria*), called also the SMALL RATTLESNAKE, is about thirteen inches in length, with a small button, or what appears to be an aborted rattle, on the tail. It ranges from the Carolinas to the Gulf States, and is particularly abundant on the prairies of the Western Territories and States. It is venomous, but its small size is thought to render its poison less potent. This serpent is thought to be even more dangerous than either of the preceding reptiles, because its dimensions are so small that a passenger is liable to disturb it before he sees the deadly creature in his path, and the sound of the rattle is so feeble that it is inaudible at the distance of two or three paces, and can only be heard when special attention is paid to it. It is a prolific species, and still maintains its numbers, in spite of the constant persecution to which it is subjected.

The food of the Miliary Rattlesnake consists of mice, frogs, insects, and similar creatures, which it mostly obtains by darting suddenly upon them as they pass near the spot where the reptile is lying. This serpent is fond of coiling itself on the fallen trunks of trees, decaying stumps, or similar situations. Fortunately, it is very easily killed, a smart blow dealing instant death even from a very small stick. The color of this reptile is brownish olive, darker upon the cheeks, which are diversified by a narrow white streak from the back of the eye. A series of brown spots runs along the centre of the back, and the sides are ornamented with two rows of brown spots, each spot corresponding with a space in the other row. The abdomen is sooty black, marbled with a darker and rather more polished hue. An irregular, dark brown band runs along each side of the nape and the crown of the head.

THE VIPERS.

WE now come to the second great family of poisonous Serpents, namely the VIPERS, or VIPERIDÆ. All the members of this family may be distinguished by the absence of the pit between the eyes and the nostrils. There are no teeth in the upper jaw except the two poison-fangs.

A rather celebrated species of these Snakes is the TIC-POLONGA, or KATUKA (*Daboia elegans*), a native of Asia, and perhaps of Brazil. This Serpent is much dreaded, its poison being of a very deadly character. A chicken that was bitten by a Tic-polonga died in thirty-six seconds, and a dog bitten by the same creature was dead in twenty-six minutes after receiving the injury. It is tolerably common in India and Ceylon, but is not so familiarly known as the cobra and other species, because it is not employed for public exhibition as is the case with those Serpents.

Sir Emerson Tennent, in his well-known "Natural History of Ceylon," writes thus of the Tic-polonga: "These formidable Serpents so infested the official residence of the District Judge of Trincomalie, as to compel his family to abandon it. In another instance, a friend of mine, going hastily to take a supply of wafers from an open tin case which stood in his office, drew back his hand on finding the box occupied by a Tic-polonga coiled within it."

The word Tic-polonga signifies Spotted-polonga, the latter word being a kind of generic title given by the natives to many Serpents, no less than eight species being classed under this common title. It is said that the Tic-polonga and the cobra bear a mortal hatred towards each other, and to say that two people hate each other like the Tic-polonga and cobra is equivalent to our proverb respecting the cat and dog. The Tic-polonga is said always to be the aggressor, to find the cobra in its hiding-place, and to provoke it to fight. There are many native legends in Ceylon respecting the ferocity of this Snake.

Its general color is brown; there are two dark brown spots on each side of the back of the head, and a yellow streak runs between them. Upon the body are three rows of oblong brown spots, edged with white.

TIC-POLONGA, OR KATUKA.—*Daboia elegans.*

The terrible PUFF ADDER is closely allied to the preceding species.

This reptile is a native of Southern Africa, and is one of the commonest, as well as one of the most deadly, of poisonous Snakes. It is slow and apparently torpid in all its movements, except when it is going to strike, and the colonists say that it is able to leap backwards so as to bite a person who is standing by its tail. Captain Drayson, who has seen much of this reptile and its habits, has kindly forwarded to me the following short account of this creature:—

"This formidable looking reptile is more dreaded than any other of the numerous poisonous Snakes in Africa, a fact which mainly results from its indolent nature. Whilst other and more active Snakes will move rapidly away upon the approach of man, the Puff Adder will frequently lie still, either too lazy to move, or dozing beneath the warm sun of the south. This reptile attains a length of four feet, or four feet six inches, and, some specimens may be found even longer; its circumference is as much as that of a man's arm. Its whole appearance is decidedly indicative of venom. Its broad ace-of-clubs-shaped head, its thick body, and suddenly tapered tail, and its chequered back, are all evidences of its poisonous nature. It derives its popular name from its practice of puffing out or swelling the body when irritated.

Vol. III.—14.

"In a country so infested with poisonous Snakes as are some portions of South Africa, it is surprising that there are not more instances of lives having been lost by this means. It is, however, as rare to hear of a person having been bitten and dying from the bite of a poisonous Snake in South Africa as it is to hear of a death in civilized countries from the bite of a mad dog. The fact, however, is that all Snakes will, if possible, make their escape when man approaches them, and it is merely when they are trodden upon, or are oppressed by their own superabundant poison, that they are disposed to bite an animal unsuited for their food.

"An infuriated Puff Adder presents a very unprepossessing appearance. I once saw a female of this species in a most excited state. She had been disturbed in her retreat under an old stump by some Kaffirs, who were widening the highroad through the Berea bush at Natal. She had several young ones with her, and showed fight immediately she was discovered. The Kaffirs were determined to kill the whole family, but were fearful of approaching

PUFF ADDER.—*Vipera arietans.*

her. Happening to pass at the time of the discovery, I organized a ring, and, procuring some large stones, directed the Kaffirs to open fire. After a few minutes the excited lady was killed, and she and her young were carefully buried in a retired locality, lest some bare-footed Kaffir might tread upon her head, and thus meet his death."

There is certainly in nature no more fearful an object than a full-grown Puff Adder. The creature grovels on the sand, winding its body so as to bury itself almost wholly in the tawny soil, just leaving its flat, cruel-looking head lying on the ground and free from sand. The steady, malignant, stony glare of those eyes is absolutely freezing as the creature lies motionless, confident in its deadly powers, and when roused by the approach of a passenger, merely exhibiting its annoyance by raising its head an inch or two, and uttering a sharp angry hiss. Even horses have been bitten by this reptile, and died within a few hours after the injury was inflicted. The peculiar attitude which is exhibited in the illustration is taken from life, one of the Puff Adders in a collection having been purposely irritated.

It is rather curious that the juice of tobacco is an instant poison to these creatures, even more suddenly deadly to them than their poison to the human beings who can absorb the tobacco juice with impunity. The Hottentots will often kill the Puff Adder by spitting in its face the juice of chewed tobacco, or making it bite the end of a stick which has been rubbed in the tobacco oil found in all pipes that have been long used without being cleaned.

The Bushmen are in the habit of procuring from the teeth of this serpent the poison with which they arm their tiny but most fearful arrows. In the capture of the Puff Adder they display very great courage and address. Taking advantage of the reptile's sluggish habits, they plant their bare feet upon its neck before it has quite made up its reptilian mind to action, and, holding it firmly down, cut off its head and extract the poison at their leisure. In order to make it adhesive to the arrow point, it is mixed with the glutinous juice of the amaryllis.

There seems to be no certain remedies for the bite of the Puff Adder. Ammonia appears to be the least inefficacious substance for that purpose, and the natives occasionally attempt to heal the injury by splitting a living fowl across the breast, and applying the still palpitating halves to the wound. There is a kind of seed called the "gentleman bean," which is said to have a beneficial effect. If one of these beans be placed on the recently inflicted wound, it adheres with great firmness, and is said to absorb the poison from the system, and to fall off as soon as this object is achieved. The Bushmen are in the habit of swallowing the poison whenever they kill a Puff Adder and do not need its venomous store for their arrows, hoping thereby to render themselves proof against its effects. When examined under the microscope, the poison resolves itself into minute crystalline spiculæ, not unlike those of Epsom salts, which must be kept perfectly dry or they will soon vanish from the glass on which they are placed.

The color of the Puff Adder is brown, chequered with dark brown and white, and with a reddish band between the eyes. The under parts are paler than the upper.

SEVERAL other deadly serpents of the same country are closely allied to the puff adder. The first is the DAS ADDER, or RIVER JACK (*Clotho nasicornis*) of the colonists, remarkable for the long curved horn or spine upon the nose, formed by the peculiar development of the scales over the nostril. This curious structure is only found in the male. In color it is much darker than the puff adder, being black, marbled with a paler hue, and decorated with sundry lozenge-shaped spots along the back.

THE BERG ADDER (*Clotho átropos*) is another of these fearful reptiles. As its name denotes, it is found more among the hills and stony ranges than on the plains, but is not unfrequently found upon the flats, and will sometimes intrude into very awkward positions, such as the floor of a hut, or even the bed upon which some wearied man is about to cast himself. It is not quite so poisonous as the puff adder, though its looks are quite as unprepossessing, and it never bites unless purposely irritated or trodden upon.

It is an ugly, thick-bodied, slow-crawling creature, with a suddenly tapering tail and a most evil looking head. It is not a large reptile, its average length being about eighteen inches. Its color is olive-gray, marbled on the sides, and decorated along the back with four rows of dark squared spots.

YET one more species of this genus deserves a passing notice. This is the HORNED ADDER (*Clotho cornúta*), sometimes, but erroneously, called the Cerastes, a term that is rightly applied to another Serpent shortly to be described. It sometimes goes by the popular name of HORNSMAN. It derives its name of Horned Adder from the groups of little thread-like horns that are seen on the head, one group appearing above each eye. In some works of Natural History, it is called the PLUMED VIPER, in allusion to these curious groups. It is not very graceful in form, being decidedly short, squat, and puffy in shape, but is very prettily marked, its body being richly marbled with chestnut, covered with a multitude of minute dots, and variegated with four rows of dark spots along the back, two rows running on each side of the vertebral line.

THE CERASTES, OR HORNED VIPER.

THE true CERASTES, or HORNED VIPER, is a native of Northern Africa, and divides with the cobra of the same country the questionable honor of being the "worm of Nile," to whose venomous tooth Cleopatra's death was due.

The bite of this most ungainly looking Serpent is extremely dangerous, though, perhaps, not quite so deadly as that of the cobra, and the creature is therefore not quite so much dreaded as might be imagined. The Cerastes has a most curious appearance, owing to a rather large horn-like scale which projects over each eye, and which, according to the natives, is possessed of wonderful virtues. They fancy that one of the so-called horns contains the supply of poison for the teeth, and that the other, if pounded and the powder rubbed over the eyelids, will enable the fortunate experimenter to see all the wealth of the earth—a privilege which, according to the peculiar cast of the Oriental mind, is of nearly as much value as the actual possession. The reader may remember a tale in the "Arabian Nights," in which a similar story is narrated.

The Cerastes has, according to Bruce, an awkward habit of crawling until it is alongside of the creature whom it is about to attack, and then making a sidelong leap at its victim. He

CERASTES, OR HORNED VIPER.—*Vipera cerastes.*

relates an instance where he saw a Cerastes perform a certainly curious feat: "I saw one of them at Cairo crawl up the side of a box in which there were many, and there lie still as if hiding himself, till one of the people who brought them to us came near him, and though in a very disadvantageous position, sticking, as it were, perpendicularly to the side of the box, he leaped near the distance of three feet, and fastened between the man's forefinger and thumb, so as to bring the blood."

The man who was thus bitten happened to be one of the men who profess Serpent charming, and avow themselves to be proof against the bite of any poisonous Snake. In this instance no ill effects followed the hurt, although Bruce proved that the poison-fangs had not been extracted, by making the reptile bite a pelican, which died in about thirteen minutes. Some persons have suggested that in this, as well as in other similar instance, the man was a clever juggler, who substituted a really venomous specimen for a Snake whose poison-fangs had been extracted. But in any case it would be necessary to handle the really poisonous reptile for the purpose of effecting the exchange, and, in my opinion, the necessary rough handling of the creature would be a matter of no small danger. Bruce enters into this subject at some length, and records the result of a long series of experiments in a form which, though very interesting, is now so familiar as to need no quotation.

That in many instances the poison-teeth of venomous Serpents have been extracted, in order to allow the performer to play his tricks with them without harm, is very well known, but the fact of acknowledged and detected imposture does not invalidate the reality which is clumsily imitated by pretenders, any more than a forgery disproves the existence of a genuine document. More will be said on this subject when we come to the different species of cobra.

The Cerastes usually lives in the driest and hottest parts of Northern Africa, and lies half-buried in the sand until its prey shall come within reach. Like many Serpents, it can

HORATTA PAM.—*Echis carinata.*

endure a very prolonged frost without appearing to suffer any inconvenience; those kept by Bruce lived for two years in a glass jar without partaking of food, and seemed perfectly brisk and lively, casting their skins as usual, and not even becoming torpid during the winter.

The color of the Cerastes is pale brownish white, covered irregularly with brown spots. Its length is about two feet.

PASSING to another genus of venomous Snakes, we come to a rather pretty little Serpent, an inhabitant of India, and called by the natives HORATTA PAM (*Echis carináta*). It is said to be very dangerous in spite of its small dimensions, and to require a double dose of Serpent medicine in order to counteract the effects of its poison. Its color is grayish brown, darkening into rather deep brown on the head, and variegated with angular white streaks on the body, and large oblong spots on the head, edged with a deeper hue. Its length is about fifteen or sixteen inches.

THE common ASP, or CHERSÆA (*Vipera aspis*) is nearly allied to the preceding species.

This Snake is common in many parts of Europe, and is plentiful in Sweden and the neighboring countries, besides being distributed over nearly the whole continent. It is much dreaded, and with reason, for its bite is very severe, and in some cases will cause death. As is the case with other venomous reptiles, the Asp is most dangerous during the hottest months of the year, and it has well been remarked that there is probably some connection between the electrical state of the atmosphere and the venom of Serpents, as the poison is always most deadly and the creatures most fierce when the electrical conditions of the atmosphere are disturbed, and the thunder-clouds are flying quickly through the air. When a person is bitten in one of his limbs, he quickly digs a hole and buries the injured part below the surface of the earth, as the fresh mould is thought to be very efficacious in alleviating the ill effects of the poison. Should the injury be in a toe or a finger, the rougher but more effectual remedy of instant amputation is generally employed.

The color of this reptile is olive above, with four rows of black spots. The two middle rows are often placed so closely together, that they coalesce and form a continous chain of black spots along the spine, very like the well-known markings of the common viper.

SAND-NATTER.— *Vipera ammodytes.*

ANOTHER venomous Snake, the AMMODYTE, or SAND-NATTER (*Vipera ammodytes*), belongs to the same genus as the asp.

This reptile inhabits southern Europe, and is generally found in rocky localities. The bite of this creature is very dangerous, and the remedies employed are generally of little efficacy. Enlarging the wound with a thorn, and squeezing a garlic upon the part bitten, is the general mode of alleviating the pain, but is of little use to the injured person. Its color is olive above, with a broad oblique dark streak on each temple, two similar streaks on each side of the head, and a wavy dark line along the crown of the spine.

THE common VIPER, or ADDER, is very well known in many parts of Europe, but in some localities is very plentiful, while in others it is never seen from one year's end to another.

Many persons mistake the common grass Snake for the Viper, and dread it accordingly. They may, however, always distinguish the poisonous reptile from the innocuous, by the

VIPER, OR ADDER.

chain of dark spots that runs along the spine, and forms an unfailing guide to its identification. It is the only poisonous reptile inhabiting some European countries, the variously-colored specimens being nothing more than varieties of the same species.

Like most reptiles, whether poisonous or not, the Viper is a very timid creature, always preferring to glide away from a foe rather than to attack, and only biting when driven to do so under great provocation.

The following interesting account of a Viper's bite and its consequences, has been kindly forwarded to me by Mr. W. C. Coleman:—

"Several years ago, in my school-boy days, I had an experience with a Viper, which may possibly interest such of your readers as have not enjoyed a similar intimacy with the creature, especially as it places the Viper character in a somewhat more amiable light than it is usually represented.

"One cold, damp day in the beginning of May, I was out in the country on a foraging expedition; birds' nests and objects of natural history in general being the objects of search. Entering, in the course of exploration, a likely coppice, I descried a blackbird's nest perched among some tangled stems of underwood three or four feet from the ground. A glance at the interior, however, soon showed that some other marauder had forestalled me, as the sole occupants of the nest were some crushed and empty egg-shells, and scanty remains of the fluid contents spilt about. 'A weasel,' thought I, but wrongfully, as it happened, for on turning away in dudgeon, a rustling movement among the herbage on the ground a couple of yards off, attracted my eyes and ears; and there I saw the undoubted spoiler of the nest, a large Viper, moving away briskly with his tail in the direction of the nest.

"A little knowledge is a dangerous thing, and my slight natural history reading, assisted by bad engravings, had helped me to fancy that I knew the Viper from the common Snake well enough; and so, deciding that this was only a common harmless Snake, I made a plunge at the creature and apprehended him with my unprotected hand. Receiving no bite, I was now confirmed in my idea of the beast's perfect innocence (except in the bird's-nest matter), and decided on adopting him as a pet. So presently set off home, a distance of more than two miles, taking my serpentine friend in my hand. Not always in my hand, however, for to beguile the homeward journey I proceeded to try sundry experiments on the supple backbone and easy temper of the animal, occasionally tying him round my neck, and so wearing him for a considerable distance; then twining him round my wrist into a fancy bracelet, and weaving him into various knots and devices according to taste, all this with perfect impunity on my part, and the utmost apparent good humor on his.

"On the road, a kind farmer of my acquaintance, whose natural history lore was more practical than my own, endeavored to convince me that I was 'harboring a Viper in my bosom,' but I was not going to hear my good-tempered playmate called bad names; put my finger into the Adder's very mouth to prove he had no idea of biting, and so passed on, in much conceit with myself as an accomplished herpetologist.

"We thus reached home in perfect safety and amity. My brothers and sisters greeted the stranger with some little instinctive horror at first, but got over that feeling when they heard of his innocent nature and amusing capabilities, in proof of which I repeated the necktie experiment, etc. About this stage, however, I must mention that he exhibited a somewhat unpleasant phenomenon common to the Snake tribe in general, who can relieve themselves of the torpor consequent on a heavy meal, by disgorging the same when irritated and requiring restoration of their usual activity. The rejectamenta in this case consisted of portions of unhatched young birds, thus confirming the nest robbery.

"Being thus lightened, and perhaps stimulated by the warmth of a fire in the room, he was now lively enough, unhappily for me, for on essaying to continue my experiments, by tying him into a double knot, his endurance was at an end; one dart at my finger and a sharp puncture told me that the thing was done. Then, too late, I recollected that the 'Adder is distinguished by a zigzag chain of dark markings down the back,' and sure enough the vile creature before me had those very marks. In a rage, I battered his life out with a stick, lest he should do more damage, and then settled down to watch the progress of the poison within my system

"It was not slow to take effect; first the wound looked and felt like a nettle sting, then like a wasp sting, and in the course of a few minutes the whole joint was swollen, with much pain. At this juncture my father, a medical man, arrived from a country journey, and set the approved antidotes to work, ammonia, oil and lunar caustic, to the wound, having previously made incisions about the punctured spot, and with paternal affection attempted to draw out the poison by suction; but nothing availed, and all sorts of horrid symptoms set in, fainting, sickness, delirium, and fever; the hand and whole arm to the shoulder greatly swollen and discolored, with most intense pain. This state of things lasted for several days. I forget the exact time, but I was not fully restored for more than a fortnight after the bite.

"Since that day I have taken care to put my acquaintance with Serpents on such a footing as to be able at a glance to tell the species of any of the common Snakes; a piece of useful knowledge most easily gained, and well worth the acquirement."

It was a most providential circumstance that the reptile did not bite him immediately after its capture, and that the wound was inflicted on the finger and not on the neck, as in the one case he could hardly have reached his home, and in the other, the great swelling might have caused suffocation, as is known to be the case with persons bitten in the neck by other poisonous Serpents.

A FEW words will not be out of place respecting the alleged capability of the Viper of receiving its progeny into its mouth when in danger.

A long-standing controversy on this subject has elicited a vast amount of correspondence, the whole of which seems to resolve itself into two divisions, namely, communications from a great number of persons who assert that they have seen the young Vipers crawl into their parent's open mouth, and letters from two or three persons who say that they did not do so, because such a proceeding is impossible, and contrary to the laws of nature.

One of the most learned of the objectors remarks, that no amount of testimony can prevail against reason, and that the persons who assert that they have seen the young Vipers crawl into their mother's mouth, have fallen into the dangerous fallacy of believing what they saw. Now this argument, novel though it may be to the scientific world in general, is perfectly familiar to theologians as being the sheet-anchor of a certain school of controversialists, who deny the credibility of the miraculous events narrated in the Scriptures. It has been repeatedly exploded in polemical controversy, and long abandoned by impartial thinkers, inasmuch as it assumes a knowledge of all the laws of nature, and contracts the power of the Divine Creator of the Universe within the narrow limits of the individual idiosyncracy and mental capacities of the disputant.

It has ever been conceded that, in all ages, the testimony of credible witnesses has been the surest mode of confuting false reasoning and thereby eliciting truth; so that when any unprejudiced reasoner finds that a favorite theory is contradicted by the testimony of even one trustworthy observer, much more when the united accounts of many competent judges all tend to the same point, he feels that it is time for him to reflect whether, however perfect may be the form of his syllogism, there may not be something wrong with his premises. Reasoning is more liable to falsity than the senses to deception. It is easy enough to talk of a flagrant violation of the laws of nature, but before we venture to do so it is as well to be quite certain that we are sure of the full extent of those laws. Who is there, even among the most learned, that can define the full working of even a single known law and its ever-varying action under different circumstances? And who can venture to say that some hitherto unrecognized law may not be in existence, which, if known and acknowledged, would account for the circumstances which at present seem so unaccountable?

In the second place, if we are not to depend upon the testimony of our acknowledged senses, on what are we to depend for the whole of natural philosophy, astronomy, or, indeed, any other established science? It is simply on the testimony of our senses that all existing sciences are founded, and even analogous reasoning is not admitted as valid proof of an asserted fact. There is hardly any new discovery which does not destroy some old and respectable theory, and give entirely a new idea of the law of nature on which it depends.

The operation of the senses is in itself one of the known laws of nature, by which we discover facts and through which we are enabled to exercise our reasoning faculties. A human being without the senses of sight, hearing, and touch, would be the dullest animal on the face of the earth, and as long as the privation lasted, would hold a lower place than a sponge or a medusa. If we once acknowledge that the evidence of the senses is not to be believed, we must reject the whole of the physical sciences. Astronomical observations, chemical experiments, geological surveys, anatomical researches, and the whole of natural history, must be at once thrown aside if such a theory is to be consistently carried out; and for the same reason, the courts of law must be abolished, depending as they do on the personal observations of human beings, mostly illiterate, and often ignorant to a degree. Repeated observations are the only method of ascertaining the laws of nature, and if they show that certain events, however strange they may appear, have really occurred, they surely prove, not that the senses of the witnesses were deceived, but that another law of nature has been discovered.

Were the Viper the only creature of whom such an act is related, the phenomenon would be less worthy of belief; but there is hardly a poisonous Snake of any country by whom the same act is not said to be performed, the narrators not being professed naturalists with a theory, but travellers, hunters, and settlers, casually noting the result of their personal experience. I cannot but think that the accumulated testimony of many trustworthy persons, acting independently of each other, accustomed to observation, and mostly unaware of the importance that would be afterwards attached to their words, is entitled to some respect, and affords legitimate grounds to the truth-seeker, not for contemptuous denial, but for further investigation.

Several observant inhabitants assert that both sexes assume this protective habit, the male as well as the female receiving the young into the mouth in cases of sudden danger. In those localities, the head of the Viper is always chopped off as soon as the reptile is killed, and the Viper-catchers say that in such cases the young Vipers frequently are seen crawling out of the severed neck.

I certainly never saw the Viper act in this manner, but I have had very few opportunities of watching this reptile in a wild state and noting its habits; whereas those who spend their lives in the forests, and especially those men who add to their income by catching or killing these reptiles, speak of the reception of the young into the mouth of the parent, as a fact too well known to be disputed.

It has been objected that the young would be consumed by the gastric juice of the parent —one of the most sensible objections that has been made. But this assertion has been invalidated by the researches of able anatomists and experimentalists, such as Mr. F. T. Buckland, etc., who have discovered by careful dissection, two facts; the one, that the young may be concealed within the expansile body of the parent without entering the true stomach at all, the œsophagus or gullet forming a highly expansile antechamber between the throat and the actual stomach ; and the other, that if they should happen to do so, the gastric juice would not hurt them. Incredible, therefore, as the possibility of such an act may seem, it can but be acknowledged that the weight of practical testimony is wholly in its favor. Moreover, the various suggestions offered to account for the deception practised by the Viper upon the eyes of observers, just as if it had been a professed conjurer performing before an audience, are really puerile in the extreme, and if they happen to affect the written testimony of one person, they are contradicted by the written testimony of another. It is to be hoped that if the Viper really does act in the manner stated, a specimen may be obtained with the young still within her body, and attested in such a manner that no objector may invalidate the proof by saying that the old one had been captured and the young pushed down her throat by force.

The head of the Viper affords a very good example of the venomous apparatus of the poisonous Serpents, and is well worthy of dissection, which is better accomplished under water than in air. The poison-fangs lie on the sides of the upper jaw, folded back and almost undistinguishable until lifted with a needle. They are singularly fine and delicate, hardly larger than a lady's needle, and are covered almost to their tips with a muscular envelope through which the points just peer. The poison-secreting glands and the reservoir in which the venom is stored are found at the back and sides of the head, and give to the venomous Serpents that peculiar width of head which is so unfailing a characteristic. The color of the poison is a very pale yellow, and its consistence is very like that of salad oil, which, indeed, it much resembles both in look and taste. There is but little in each individual; and it is possible that the superior power of the larger venomous Snakes of other lands, especially those under the tropics, may be due as much to its quantity as its absolute intensity. In a full-grown rattlesnake, for example, there are six or eight drops of this poison, whereas the Viper has hardly a twentieth part of that amount.

On examining carefully the poison-fangs of a Viper, the structure by which the venom is injected into the wound will be easily understood. On removing the lower jaw, the two fangs are seen in the upper jaw, folded down in a kind of groove between the teeth of the palate and the skin of the head, so as to allow any food to slide over them without being pierced by their points. The ends of the teeth reach about half-way from the nose to the angle of the jaw, just behind the corner of the eye.

Only the tips of the fangs are seen, and they glisten bright, smooth and translucent, as if they were curved needles made from isinglass, and almost as fine as a bee's sting. On raising them with a needle or the point of the forceps, a large mass of muscular tissue comes into view, enveloping the tooth for the greater part of its length, and being, in fact, the means by which the fang is elevated or depressed. When the creature draws back its head and opens its mouth to strike, the depressing muscles are relaxed, the opposite series are contracted, and the two deadly fangs spring up with their points ready for action. It is needful, while dissecting the head, to be exceedingly careful, as the fangs are so sharp that they penetrate the skin with a very slight touch, and their poisonous distilment does not lose its potency, even after the lapse of time.

The next process is to remove one of the teeth, place it under a tolerably good magnifier and examine its structure, when it will be seen to be hollow, and, as it were, perforated by a channel. This channel is, however, seen, on closer examination, to be formed by a groove along the tooth, which is closed, except at the one end whence the poison exudes and the other at which it enters the tooth. If the tooth be carefully removed, and the fleshy substance pushed away from its root, the entrance can be seen quite plainly by the aid of a pocket lens. The external aperture is in the form of a very narrow slit upon the concave side of the fang, so very narrow, indeed, that it seems too small for the passage of any liquid.

There are generally several of the fangs in each jaw, lying one below the other in regular succession. From the specimen which has just been described I removed four teeth on each side, varying in length from half to one-eighth the dimensions of the poison-fangs.

The Viper seems to be well aware of the power of its fangs, and to discriminate between animate and inanimate antagonists. I have tried in vain to make a Viper bite a stick with which I was irritating it; but no sooner did a kitten approach, than the reptile drew back its head and made its lightning-like dart at the little creature with such rapidity, that it would have gained its point, had not its back been so much injured as to deprive it of its natural powers.

The ordinary food of the Viper is much the same as that of the common Snake, and consists of mice, birds, frogs, and similar creatures. It is, however, less partial to frogs than the common Snake, and seems to prefer the smaller mammalia to any other prey. The young of the Viper enter the world in a living state, having been hatched just before they are born. The fat of the Viper was once in high estimation as a drug, and the older apothecaries were accustomed to purchase these reptiles in considerable numbers. Even now this substance is in some repute in many agricultural districts, being employed as a remedy for cuts, sprains, or bruises, and especially as a means of alleviating the painful symptoms of a Viper's bite.

The color of the Viper is rather variable; but the series of very dark marks down the back is an unfailing sign of the species, and is permanent in all the varieties. Generally, the ground color is grayish-olive, brown, or brownish-yellow; along the back runs a chain of zigzag blackish markings, and a series of little triangular spots is found upon each side. The largest specimen I have yet seen in a wild state was one of the yellow varieties. Sometimes the ground is brick-red, and now and then a nearly black specimen is found. Mr. Bell mentions an example where the ground color was grayish-white, and the markings jetty-black.

The reptile that is called by the significant title of DEATH ADDER, or DEATH VIPER, is a native of Australia, where its poisonous fangs render it an object of much fear. A very excellent, though short description of this Snake is given by Mr. Bennett in his "Wanderings in New South Wales."

"The most deadly Snake in appearance, and I believe also in effect, is one of hideous aspect, called by the colonists the Death Adder, and by the Yas natives 'Tammin,' from having a small, curved process at the extremity of the tail; or, more correctly, the tail terminating suddenly in a small, curved extremity, bearing some resemblance to a sting. It is considered, by popular rumor, to inflict a deadly sting with it.

"This hideous reptile is thick in proportion to its length; the eye is vivid yellow, with a black longitudinal pupil. The color of the body is difficult to be described, being a complication of dull colors, with narrow, blackish bands shaded off into the colors which compose the back; abdomen slightly tinged with red; head broad, thick and flattened. The specimen I examined measured two feet two inches in length, and five inches in circumference. A dog that was bitten by one died in less than an hour. The specimen I examined was found coiled up near the banks of the Murrumbidgee river; and being of a torpid disposition, did not move when approached, but quietly reposed in the pathway, with its head turned beneath its belly."

The generic title of Acanthophis, or Thorny-Snake, is given to this species on account of the structure of the tail, which is furnished at its extremity with a recurved horny spine.

RIVER OR SEA SERPENTS.

WE now arrive at a very remarkable family of Snakes, which pass their lives in water, either fresh or salt, and are river or sea Serpents as the case may be. In order to enable them to pass through the waters without injury to the organs of respiration, the nostrils are furnished with a valve so as to prevent the ingress of water while the creature is below the surface.

A good example of these marine Serpents is the BLACK-BACKED PELAMIS (*Pélamis bícolor*) the Nalla Whallagee Pam of the Indian fishermen. This Snake is found only at sea, and is said seldom if ever to approach the shore, except for the purpose of depositing its eggs, which are laid on the beach sufficiently near high-water mark for the young Snakes to seek their congenial element as soon as they are hatched. The Black-backed Pelamis is frequently found sleeping on the surface of the sea, and is then caught without much difficulty, as it is forced to throw itself on its back before it can dive. It has been suggested that this movement is intended to expel the air in the ample lungs. Sometimes it is unwillingly captured by the fishermen in their nets, and is an object of considerable dread to them on account of the formidable character of its teeth. In these Serpents the fangs are but little larger than the other teeth of the jaw, but can be distinguished by their slightly superior size and the groove that runs along their front edge. The average length is about one yard.

THE SHOOTER SUN (*Hydrophis obscúra*) is another of the sea Serpents. This reptile is also one of the Indian species, and inhabits the sea or the saline waters of the river-mouths, not being able to exist in fresh water. It is an admirable swimmer, but is very awkward on dry land, and cannot survive for any length of time unless it has access to salt water. The outline of this Serpent is most remarkable. The head and neck are almost absurdly minute in proportion to the wide thick body, bearing about the same proportion as the tip of the little finger does to the wrist. The tail is also very wide, extremely blunt, and compressed.

The markings of this reptile are rather curious. The ground color is black. There is a large yellow spot on each side of the head, a series of pale, gray-brown spots runs on each side of the neck, and a row of large rounded white marks is arranged along the back so as to form a richly variegated pattern of boldly contrasted colors.

THE CHITTUL (*Hydrophis sublævis*) is another of these marine Snakes, and is found in India and Ceylon. It is rather a large species, sometimes exceeding five feet in length, and is handsomely colored. It is extremely venomous, a fowl that had been bitten by a Chittul dying within five minutes after receiving the injury. The ground color of this Snake is yellow, and the body is covered with an irregular row of black rings. Some black bands also cross the neck.

In the ACROCHORDE, sometimes called the Oular Carron, the tail, instead of being flattened, is rounded, conical, and very short, diminishing in diameter in a very sudden manner. It is a native of Java, and is said to be wholly vegetarian in its diet, the stomach

BLACK-BACKED PELAMIS.—*Pelamis bicolor.*

having been found to contain nothing but half-digested fruit. The flesh of the Acrochorde is said to be excellent.

Upon the head are a number of little scales, each of which is divided into three ridges. The creature is in the habit of distending its body with air to a very great extent, and when it so acts the scales separate from each other and make the head and body look as if they were covered with tubercles. The general color is brown in the adult, and brown banded and streaked with a darker hue in the young.

THE CHERSYDRUS (*Chersydrus granulátus*) is a rather curious aquatic Serpent, found in Asia and most common in Java. It is sometimes called the Banded Acrochorde, but wrongly so, as its tail, instead of being round and conical, is flat, compressed, and sword-like in shape.

It inhabits the bottoms of marine creeks and the mouths of rivers. The Javanese call it Oular Limpe. The body of this reptile is covered with small scales, each boldly keeled in the centre, and its color is black and white arranged in alternate rings.

THE ERPETON, or HERPETON, as the name is sometimes written, is a truly curious reptile, of no great size, but bearing a pair of appendages on the head that seem to serve no recognized purpose save to bewilder zoologists. The muzzle of this creature is covered with scales, and on each side of it rises a curious appendage. This remarkable organ is soft, but completely covered with scales and defended by them. Of the habits of the Erpeton nothing appears to be known, and even its country is dubious. Its color is pale brown streaked with white.

ACROCHORDE.—*Achrochordus javanicus.*

THE sombre and rather unsightly CERBERUS, better known by its native name of KAROO BOKADAM, is an Asiatic reptile, being found in India, the Philippines, Ceylon, Borneo, and similar countries. It is an ugly looking Serpent, but is not much dreaded, and is thought to be practically non-venomous. It is a stout, thick-bodied Snake, with a very large head in proportion to the size of its neck, though small in comparison with the body. The mouth is not large, and the teeth are small, regular, and set rather closely together. The nostrils of this Serpent are very small, and placed close to each other almost on the very tip of the muzzle. The eyes are small, round, and projecting as if squeezed out of the head, and are surrounded by a curious circle of nearly triangular scales, much as a circular window in a brick wall is edged with wedge-shaped bricks.

The general color of this Serpent is grayish-brown above, covered with narrow bands of black set rather closely together. The abdomen is black mottled with yellow, the sides are white with spots of pale brown, and the lips and throat are of the same tint, but spotted with black. The tail is nearly black. The usual length of this Serpent is about three feet six inches.

WE now arrive at a very important family of serpents, including the largest species found in the order. These Snakes are known by the popular title of Boas, and scientifically as Boidæ, and are all remarkable, not only for their great size and curious mode of taking their prey, but for the partial development of the hinder limbs, which are externally visible as a pair of horny spurs, set one on each side at the base of the tail, and moderately well developed under the skin, consisting of several bones jointed together. In most of the species the tail is

CARPET SNAKE.—*Morelia variegata.*

rather short and strongly prehensile. The peculiar habits of these enormous Snakes will be mentioned in connection with the various species. The first of these creatures is the DIAMOND SNAKE of Australia (*Morelia spilótes*), a very handsome species and tolerably common. It is called the Diamond Snake on account of the pattern of the colors, which are generally blue, black, and yellow, arranged so as to produce a series of diamonds along the back. The CARPET SNAKE (*Morelia variegata*), of the same country, is closely allied to it. Both these reptiles are variable in their coloring.

THE members of the restricted genus PYTHON are remarkable for their habit of depositing the eggs together and coiling their bodies round them, so as to form a large conical heap. The common grass Snake is said to perform the same feat. The true Pythons are inhabitants of

Asia, and are generally found in India. The common ROCK SNAKE of India (*Python molúrus*) is a good example of this genus. The natives believe that the little spurs are useful in fighting, and therefore cut them off whenever they capture the reptile. It is the Pedda-Poda of the Hindoos. It is not one of the largest of its kind, usually attaining a length of ten or eleven feet, and not being held in much dread. A fowl that was inclosed in a cage with one of these Serpents, soon obtained the mastery over her terrible companion, and was seen quietly pecking at its head.

One of these reptiles that was kept at the gardens of the Zoological Society, once made a curious mistake while being fed, and had well-nigh sacrificed the life of its keeper. The man had approached the reptile with a fowl in his hand and presented it as usual to the Snake. The

ROCK SNAKE OF INDIA.—*Python molurus.*

Serpent darted at the bird, but as it was just then shedding its skin and nearly blind, it missed its aim, and instead of seizing the bird, grasped the keeper's left thumb, and instinctively flung its coils around his arms and neck, as is customary when the animal seized is of considerable size.

The keeper tried to force the Snake's head from its hold, but could not reach it, as he was bound in the folds of the Snake. He then cast himself on the ground in order to battle to the greatest advantage, but would probably have succumbed to the fearful pressure, had not two keepers providentially entered the room, and by breaking away the Serpent's teeth released the man from his terrible assailant. Except the fright and a few wounds from the Serpent's teeth, no evil results ensued. The representation in our picture is one-tenth of the actual size of the specimen from which it was drawn.

THE NATAL ROCK SNAKE, OR PORT NATAL PYTHON.

ANOTHER species of Indian Rock Snake, called by the natives ULAR SAWA (*Hypsirhina aër*), is tolerably common, and in its habits resembles the preceding species. It often attains to a very considerable size, and is said when full-grown to be about thirty feet in length. This terrible Snake has been known to kill mankind, crushing the body in its numerous folds until nearly every bone was broken. In one such instance, the man had been caught by the right wrist, as was seen by the marks of the Serpent's teeth.

NATAL ROCK SNAKE.—*Hortatla natalensis*. (One-eighth natural size.)

THE handsome NATAL ROCK SNAKE, or PORT NATAL PYTHON, as it is sometimes called, now comes under our notice. It is a fine, handsome species, sometimes attaining a great length, and being most beautifully colored. During life and when in full health and in the enjoyment of liberty, this, in common with many other Snakes, has a beautiful rich bloom upon its scales, not unlike the purple bloom of a plum or grape. Should, however, the Snake be in ill health, this bloom fades away, and in consequence, we seldom if ever see it on the scales of the Serpents which are kept in glass cases.

The dimensions of this reptile are often very great. Dr. A. Smith has seen a specimen measuring twenty-five feet in length, exclusive of a portion of the tail which was missing. Flat skins of this creature are, however, very deceptive, and cannot be relied upon, as they stretch almost as readily as India rubber, and during the process of drying are often extended several feet beyond the length which they occupied while surrounding the body of their quondam owner.

The teeth of this Serpent are tolerably large, but not venomous, and although of no insignificant size, are really of small dimensions when compared with the size and weight of their owner. Few persons have any idea of the exceeding heaviness of a large Snake, and unless the reptile has been fairly lifted and carried about, its easy gliding movements have the effect of making it appear as if it were as light as it is graceful.

Both jaws are thickly studded with these teeth, and their use is to seize the prey and hold it while the huge folds of the body are flung round the victim, and its life crushed out of its frame by the contracting coils. In order to secure its prey, the Rock Snake acts after the manner of all this family. It waits in some spot where it knows that its victim will pass, coils its tail round some object, such as a tree or a stone, so as to give it a firm hold, and then, rapidly darting at the prey, it draws back its head, carrying the poor victim into the fatal grasp of its folds. It usually seizes by the throat, and retains its hold until the crushed animal is quite dead.

The following interesting account of the Rock Snake of Natal has been kindly forwarded to me by Captain Drayson:—

"The Rock Snake is somewhat rare, even in the least populous districts, and, in consequence of its retired habits and silent method of moving, it is not frequently seen. Although on an average I traversed the forests and plains near my various stations at least five times a week, I saw but seven Rock Snakes during a period of nearly three years. This Snake retreats into rocky crevices, or amongst the most tangled brushwood, after it has devoured its prey, which consists of toads, frogs, lizards, such as guanas, etc., birds of any size, and even small bucks. Its bite is quite harmless compared to that of the poisonous Snakes, and it destroys its victims by pressure.

"So cautious is this Snake to remain quite quiet if it thinks itself unseen, that on one occasion I nearly rode over a rather large Boa, which lay on a small path along which I was riding. On each side of this path there was a dense jungle, and there was merely room for one animal to travel along it. I happened to 'pull up' my pony to examine the surrounding bush, when I noticed that his erected ears indicated that he had seen game, he being a most accomplished shooting pony. Upon looking on the path before me I observed a very large Snake, lying perfectly still, and looking at me in a very suspicious manner. The reptile being partly concealed by the long grass, I could not see whether or not it was a poisonous Snake, so I quietly 'reined back' about a yard, and shot the creature through the body. The coils and contortions were something terrific to see, as the monster fought hard for his life; but even the bone and muscle of a Boa has but a poor chance against gunpowder and lead. A charge of buck shot in the head settled the business, and cleared the path of a very disagreeable *vis-à-vis*. This Snake measured about sixteen feet in length, and was in very fair condition, having a fine bloom on his skin. He had resided about a hundred yards from a long *vlei* (lagoon), in which frogs and lizards abounded.

"A much larger Rock Snake was shot by me some time after this, and measured upwards of seven yards. I once had an opportunity which rarely occurs to many men, viz., that of trying my speed with a young Boa-constrictor. Upon returning from shooting one afternoon I crossed the Umbilo River near Natal, and shortly after observed a *coran* flying up and down in a very singular manner. This bird being very good eating, I dismounted, and commenced stalking him, and approached within a few yards of him without being discovered. I then noticed a Snake creeping towards the coran, which merely flew on a few feet and then settled again. The Snake again approached the bird, which, however, seeing me, became disenchanted, and was making its escape when I shot it, and then turned my attention to the Snake, which remained quite still. I soon saw that the animal was a young Rock Snake about twelve feet long, and, being desirous to obtain a live specimen of this reptile, I ran to my pony, where on the saddle I had a long leather strap, with which I hoped to noose the young Boa.

"Upon returning to the scene of the coran's death, I found the Snake making off as fast as he could towards a clump of thick bush. Immediately starting after him, I headed him after a race of about sixty yards, when he turned and tried another direction. I failed in noosing him, and, finding that he would probably escape into the bush, I was compelled to knock him on the head with a dead branch which happened to be near me. I believed him to have been killed outright; but on conducting a naturalist to the scene on the following morning the Snake had vanished, a fact which, combined with subsequent experience of the Snake nature, induces me to believe that he was merely stunned by the blow, and became refreshed during the cool of the evening, after which he retreated to his stronghold."

The color of the Natal Rock Snake is olive, variegated with yellow cross-bands and spots, edged with deep black. The head is marked with an arrow-headed spot, and a dark streak runs from the back of the eye. The under parts and the sides of the face are yellow.

There are several other species inhabiting Africa, resembling the preceding creature in general habits and appearance.

THE splendid RINGED BOA of America, sometimes called the ABOMA, has been celebrated for its destructive powers, and in ancient times was worshipped by the Mexicans and propitiated with human sacrifices. Naturally, the people of the country would feel disposed to awe in the presence of the mighty Snake whose prowess was so well known by many fatal experiences; and this disposition was fostered by the priests of the Serpent deity, who had

DOG-HEADED BOA.—*Xiphosoma caninum.*

succeeded in taming several of these giant Snakes, and teaching them to glide over and around them, as if extending their protection to men endowed with such supernatural powers.

This Serpent destroys its prey, after the fashion of its family, merely by squeezing it to death between its folds. While thus engaged, the reptile does not coil itself spirally round the victim, but wraps fold over fold, to increase its power, just as we aid the grasping strength of one hand by placing the other over it. It is said that the Snake can be removed from its prey by seizing it by the tail, and thus unwinding it. Moreover, a heavy blow on the tail, or cutting off a few feet of the extremity, is the best way of disabling the monster for the time.

This creature is rather variable in its coloring, the locality having probably some influence in this respect. Generally, it is rich chocolate-brown, with five dark streaks on the top and sides of the head, a series of large and rather narrow dark rings along the back, and two rows of dark spots on the sides. Sometimes a number of large spots are seen on the back, and white streaks on the sides. In all the members of this genus, the hinder limbs or "spurs" of the male are larger and stronger than in the female.

Another American species, the Dog-headed Boa, or Bojobi (*Xiphosóma canínum*), is notable for the formidable armament of teeth which line the mouth, and the beautiful green color of its skin. As is the case with all the Boidæ, this species is only found in the hottest parts of the country, and is most plentiful in Brazil. It may be known from the other species, partly by its green color, partly by the deep pits on the plates that edge the lips, and partly by the regular ring of scales that surrounds the eye. This Snake is sometimes called the Araramboya.

BOIGUACU.—*Boa constrictor.*

We now come to the Boiguacu, or true Boa Constrictor, a title which is indifferently applied to all the family, and with some degree of appropriateness, inasmuch as they all kill their prey by pressure or constriction.

This magnificent reptile is a native of Southern and Tropical America, and is one of those Serpents that were formerly held sacred and worshipped with divine honors. It attains a very large size, often exceeding twenty feet in length, and being said to reach thirty feet in some cases. It is worthy of mention, that, before swallowing their prey, the Boas do not cover it with saliva, as has been asserted. Indeed, the very narrow and slender-forked tongue of the Serpent is about the worst possible implement for such a purpose. A very large amount of this substance is certainly secreted by the reptile while in the act of swallowing, and is of great use in lubricating the prey, so as to aid it in its passage down the throat and into the body; but it is only poured upon the victim during the act of swallowing, and is not prepared and applied beforehand.

The dilating powers of the Boa are wonderful. The skin stretches to a degree which seems absolutely impossible; and the comparison between the diameter of the prey and that of the mouth through which it has to pass, and the throat down which it has to glide, is almost ludicrous in its apparent impracticability, and, unless proved by frequent experience, would seem more like the prelude to a juggler's trick than an event of every-day occurrence. To such an extent is the body dilatable, that the shape of the animal swallowed can often be traced through the skin, and the very fur is visible through the translucent eyes, as the dead victim passes through the jaws and down the throat.

There is a popular idea among the inhabitants of the country in which the Boa lives, that, if it attacks a man in a forest, he may possibly escape by slipping round a tree in such a manner that the Serpent may squeeze the trunk of the tree, mistaking it for the body of the man, and so burst itself asunder by the violence of its efforts. Whether any one has escaped by this rather transparent device is not mentioned.

The color of the Boa Constrictor is rich brown, and along its back runs a broad chain of large blackish spots of a somewhat hexagonal shape, and of pale white spots scooped at each end. These dark and pale spots are arranged alternately, and form a really pretty pattern; and, should the colors be faded, as is always the case when the skin has been renewed, the species may be recognized by the arrangement of the scales round the eyes, which are set in a circle, are thirty in number, and are separated from the scales of the lips by two rows of smaller scales.

An equally celebrated Snake, the ANACONDA, is figured in the accompanying full-page illustration.

This gigantic serpent is a native of tropical America, where it is known under several names, La Culebra de Agua, or Water Serpent, and El Traga Venado, or Deer-Swallower, being the most familiar. The flesh of this Serpent, although firm and white, is seldom if ever eaten by the natives, although the flesh of Serpents is considered a delicacy by many nations. Within the body is a large amount of fat from which can be obtained a very considerable quantity of oil. This oil is thought to be a specific for many complaints, especially for rheumatism, strains, and bruises. Seven or eight gallons of fine oil can be extracted from one of these reptiles; but the process of draining off the oil is generally performed in so careless a manner, that half of the amount is usually wasted.

Sir R. Ker Porter has some curious remarks on the Anaconda: "This Serpent is not venomous nor known to injure men (at least not in this part of the New World); however, the natives stand in great fear of it, never bathing in waters where it is known to exist. Its common haunt, or rather domicile, is invariably near lakes, swamps and rivers; likewise close to wet ravines produced by inundations of the periodical rains; hence, from its aquatic habits, its first appellation (i. e. Water Serpent). Fish, and those animals which repair there to drink, are the objects of its prey. The creature lurks watchfully under cover of the water, and while the unsuspecting animal is drinking, suddenly makes a dart at the nose, and with a grip of its back-reclining double range of teeth, never fails to secure the terrified beast beyond the power of escape."

Compression is the only method employed by the Anaconda for killing its prey, and the pestilent breath which has been attributed to this reptile is wholly fabulous. Indeed, it is doubtful whether any Snake whatever possesses a fetid breath, and Mr. Waterton, who has handled Snakes, both poisonous and inoffensive, as much as most living persons, utterly

ANACONDA.

denies the existence of any perceptible odor in the Snake's breath. It is very possible that the pestilent and most horrible odor which can be emitted by many Snakes when they are irritated, may have been mistaken for the scent of the breath. This evil odor, however, is produced from a substance secreted in certain glands near the tail, and has no connection with the breath.

The color of the Anaconda is rich brown; two rows of large round black spots run along the back, and each side is decorated with a series of light golden yellow rings edged with deep black.

ONE or two members of this family are worthy of a passing notice. The well-known YELLOW SNAKE of Jamaica (*Chilabothrus inornátus*) is allied rather closely to the boa and the anaconda. It is a rather handsome reptile, being of an olive-green upon the head and front part of the body, covered with a multitude of little black lines, drawn obliquely across the body. The hinder part of the body is black, spotted with yellowish olive.

ANOTHER member of this family, the CORAL SNAKE (*Tortrix scytale*) is a well-known inhabitant of Tropical America, and is feared or petted by the natives, according to the locality in which it happens to reside. In some parts of the country, the native women, knowing it to be perfectly harmless, and being pleased with the bold contrast of black and pale gold which decorate its surface, are in the habit of taming it and of placing it round their necks in lieu of a necklace. In other parts of the country, however, the natives believe it to be terribly poisonous, and flee from its presence with terror.

It lives chiefly on insects, worms, and caterpillars, and is very timid. This creature does not taper so gradually from the middle of the body to the tail as is usual in most Serpents, but is nearly of the same cylindrical form throughout its length. The ground color of this Serpent is pale yellow, decorated with jetty-black rings, about sixty in number, that are drawn irregularly over its surface. The Coral Snake never grows to any great size, and seldom reaches two feet and a half in length.

COLUBRINÆ.

WE now come to another section of the Serpents, termed COLUBRINÆ, the members of which are known by the broad, band-like plates of the abdomen, the shielded head, the conical tail, and the teeth of both jaws. Some of them are harmless and unfurnished with fangs, whereas some are extremely venomous and are furnished with poison-fangs in the upper jaw. These, however, do not fold down like those of the viper and rattlesnake, but remain perfectly erect. The formation of the fangs again differs in the various species. In some the fang is grooved for the introduction of poison into the wound, whereas in others it is perforated nearly throughout its length.

As an example of the first family of these Serpents, we may take the common BROWN SNAKE of America (*Conocéphalus striátus*).

This reptile is quite harmless, and is plentiful in many portions of America, having rather a wide range of locality. Although common, it is not conspicuous, for its small dimensions, its sombre hue, and its retiring habits serve to conceal it from the general gaze. It is usually found hiding under the bark of trees, in stone heaps, or among the crevices of rocky ground, choosing those localities because it feeds principally on insect prey, and can find abundance of food in such places. Its color is grayish brown above and white below. It is a small species, rarely reaching eleven inches in length.

THE large family of the Coronellidæ contains many curious Serpents, among which may be mentioned the well-known SCHAAP-STICKER of Southern Africa.

This Snake has a rather wide range of country, being spread over nearly the whole of Southern Africa, and very common at the Cape of Good Hope. It is a handsome little reptile, prettily marked, and brisk and lively in its movements, as is required for the purpose of catching the agile prey on which it feeds. The Schaap-sticker lives mostly on insects and small lizards, and darts upon them with great swiftness of movement. It is generally found crawling among heaps of dead leaves, or trailing its variegated form over grassy banks, where it finds the prey on which it subsists.

The color of this Serpent is extremely variable, and decidedly different in the old and young. In the young specimen, the spots that ornament the back are darker than in the adult, and there is generally a little wash of green over the surface. The general color of this Snake is brown, with a grayish or golden tint according to the individual. Along the back run several rows, usually three or four in number, of dusky spots, generally of a somewhat oval or rhombic form, and edged with deep black. In one specimen the spots have coalesced so as to form three continuous bands running along the body. The length of the Schaap-sticker is about two feet.

ANOTHER species belonging to this family is the *Coronella Austriaca*. It is rather remarkable, that where the Snake is tolerably common, the sand lizard (*Lacerta stirpium*) is

SCHAAP-STICKER.—*Psammophylax rhombeatus.* (One-half natural size.)

also generally found. In general appearance, this Snake is not unlike the viper, and is about the same size, attaining a length of two feet when adult. It may, however, easily be known from the viper, by the absence of the chain of dark lozenge-shaped marks upon the back, for which is substituted a double series of small dark spots, one row at each side of the spine. There is a dark patch upon the shoulder and head, and under the eyes runs a blackish streak. The body is generally brown, but the depth and tone of the ground color and the markings are extremely variable, but are almost always darker towards the head. Below, the color is light brown, often marbled with black. The neck is large, being scarcely smaller than the body.

THE BLACK SNAKE, or ZWARTE SLANG (*Coronella cana*), of Southern Africa, belongs to the same genus.

This reptile is common throughout Southern Africa, but is not very often seen, on account of its timid habit of hiding itself in some crevice, except when in search of food, or when coiled up in repose enjoying the hot beams of the sun. When young, it frequents little hillocks covered with stones, but when it reaches adult age, it takes to the plains, preferring those that are of a sandy nature, interspersed with little shrubs. It is a shy reptile, and mostly runs away when alarmed. Sometimes, however, it will turn upon the pursuer, and if grasped, will coil itself round the arm and squeeze so tightly, that the hand becomes numbed and unable to retain its hold.

Many Snakes are variable in their coloring, but the Black Snake is, perhaps, the most remarkable among them for this peculiarity. Usually, as its name imports, it is black, but sometimes it is bright chestnut. Many specimens are gray, mottled with black, while others are chestnut, marbled with deep rich brown. When full grown, it attains a length of seven feet.

THE common GRASS SNAKE, or RINGED SNAKE, is a good example of the Natricidæ.

It is extremely plentiful throughout Europe, being found in almost every wood, copse, or hedgerow, where it may be seen during the warm months of the year, sunning itself on the banks, or gently gliding along in search of prey, always, however, betraying itself to the initiated ear by a peculiar rustling among the herbage. Sometimes it may be witnessed while in the act of creeping up a perpendicular trunk or stem, a feat which it accomplishes, not by a spiral movement, as is generally represented by artists, but by pressing itself firmly against the object, so as to render its body flatter and wider, and crawling up by the movement of the large banded scales of the belly, the body being straight and rigid as a stick, and ascending in a manner that seems almost inexplicable.

The Ringed Snake is perfectly harmless, having no venomous fangs, and all its teeth being of so small a size that even if the creature were to snap at the hand, the skin would not be injured. Harmless though the Serpent be, it will occasionally assume so defiant an air, and put on so threatening an aspect, that it would terrify those who were not well acquainted with its habits. I have kept numbers of these Snakes, and have often known them, when irritated, draw back their heads and strike at the hand in true viperine fashion. Indeed, the venomous look of the attitude is so strong, that I never could resist the instinctive movement of withdrawing the hand when the Snake made its stroke, although I knew full well that no injury could ensue.

The food of the Ringed Snake consists mostly of insects and reptiles, frogs being the favorite prey. I have known Snakes to eat the common newt, and in such cases the victim was invariably swallowed head first, whereas the frog is eaten in just the opposite direction. Usually, the frog, when pursued by the Serpent, seems to lose all its energy, and instead of jumping away, as it would do if chased by a human being, crawls slowly like a toad, dragging itself painfully along as if paralyzed. The Snake, on coming up with its prey, stretches out its neck and quietly grasps one hind foot of the frog, which thenceforward delivers itself up to its destroyer an unresisting victim.

The whole process of swallowing a frog is very curious, as the creature is greatly wider than the mouth of the Snake, and in many cases, when the frog is very large and the Snake rather small, the neck of the Serpent is hardly as wide as a single hind leg of the frog, while the body is so utterly disproportioned, that its reception seems wholly impossible. Moreover, the Snake generally swallows one leg first, the other leg kicking freely in the air. However, the Serpent contrives to catch either the knee or the foot in its mouth during these convulsive struggles, and by slow degrees swallows both legs. The limbs seem to act as a kind of a wedge, making the body follow easily, and in half an hour or so the frog has disappeared from sight, but its exact position in the body of the Snake is accurately defined by the swollen abdomen. Should the frog be small, it is snapped up by the side and swallowed without more ado.

In captivity, this Snake will eat bread and milk, and insects of various kinds, such as the cockroach, meal-worm, or any beetle that may be found running about under stones and leaves. It always, however, prefers frogs to any other food, and seems to thrive best on such a diet.

The skin or slough of the Ringed Snake is often found in the hedgerows or on waste grounds, entangled among the grass stems and furze through which the creature had crawled with the intention of rubbing off the slough against such objects. In some countries the rejected slough is thought to be a specific against the headache, and is tied tightly round the forehead when employed for alleviating pain.

The Ringed Snake is fond of water, and is a good swimmer, sometimes diving with great ease and remaining below the surface for a considerable length of time, and sometimes swimming boldly for a distance that seems very great for a terrestrial creature to undertake. This reptile will even take to the sea.

I have often seen tame Snakes taken to an old deserted stone-quarry for a bath in the clear water which had collected there. Generally the Snake would swim quietly from one side to another, and might then be recaptured, but on sundry occasions it preferred diving to the very bottom, and there lay among the stones, heedless of all the pelting to which it was subjected, and impassive as if perfectly acquainted with the harmless nature of stones projected into water. Nothing would induce the Snake to move but a push with a stick, and as the water was rather deep and the quarry wide, a stick of sufficient length was not readily found. The motions of the Snake while in the water are peculiarly graceful, and the rapid progress is achieved by a beautifully serpentine movement of the body and tail.

This Snake is susceptible of kindness, and if properly treated, soon learns to know its owner, and to suffer him to handle it without displaying any mark of irritation. Though harmless and incapable of doing any hurt by its bite, the Snake is not without other

RINGED SNAKE, OR GRASS SNAKE.—*Tropidonotus natrix.*

means of defence, its surest weapon being a most abominable and penetrating odor, which it is capable of discharging when irritated, and which, like that of the skunk, adheres so closely to the skin or the clothes, that it can hardly be removed even by repeated washings. Moreover, it is of so penetrating a nature that it cannot be hidden under artificial essences, being obtrusively perceptible through the most powerful perfumes, and rather increasing than diminishing in offensiveness by the mixture. The reptile will, however, soon learn to distinguish those who behave kindly to it, and will suffer itself to be handled without ejecting this horrible odor.

The young of the Ringed Snake are hatched from eggs, which are laid in strings in some warm spot and left to be hatched by the heat of the weather or other natural means. Dunghills are favorite localities for these eggs, as the heat evolved from the decaying vegetable matter is most useful in aiding their development, and it often happens that a female Snake obtains access into a hothouse and there deposits her eggs. Some persons say that the mother is sometimes known to remain near the eggs, and to coil herself round them as has already been related of the boa. The eggs are soft, as if made of parchment, and whitish.

They are found in chains containing fifteen or twenty, and are cemented together by a kind of glutinous substance.

During the winter the Snake retires to some sheltered spot, where it remains until the warm days of spring call it again to action. The localities which it chooses for its winter quarters are always in some well sheltered spot, generally under the gnarled roots of ancient trees, under heaps of dry brushwood, or deep crevices. In these places the Snakes will congregate in great numbers, more than a hundred having been taken from one hollow. A few years ago I saw a hole from which a great number of Ringed Snakes had been taken; it was situated in a bank, at some depth. The color of the Ringed Snake is grayish-green above and blue-black below, often mottled with deep black. Behind the head is a collar of golden yellow, often broken in the middle so as to look like two patches of yellow. Behind the yellow collar is another of black, sometimes broken in the middle also. Along the back run two rows of small dark spots, and a row of large, oblong spots is arranged down each side. Both the color and the shape of the spots are very variable.

The length of this reptile is generally about a yard, but it sometimes attains a length of four feet. The female is always larger than the male. The generic title *Tropidonotus* is formed from two Greek words signifying keel-backed, and is given to these Serpents because the scales of the back are keeled.

THE HOG-NOSE SNAKE is so called from the odd formation of the muzzle, which is rather blunt, and slightly turned up at the tip, something like the snout of a hog. It generally frequents moist and marshy localities, as the edges of rivers and ponds, where it finds a plentiful subsistence among the toads, frogs, lizards, and insects which swarm in such spots. It is an inhabitant of Northern America.

Although as harmless as our ringed Snake, and of similar dimensions, so that it need not be feared on account of its bodily strength, the Hog-nose Snake is rather feared by those who are not acquainted with its structure and habits. If it be irritated in any way, it assumes a most threatening attitude, coils itself like a rattlesnake, flattening its head after the fashion of venomous Serpents, utters a furious hiss, and strikes at the foe with the rapidity of lightning. Yet all this flourish of defiance is without the least foundation, and although it might serve to intimidate the ignorant, only raises the mirth of the better instructed. For the Serpent does not even open its mouth when it strikes, but darts its closed jaws at the foe, without even inflicting the trifling wounds which might be caused by its small but needle-like teeth. Even if pushed about with a stick, and handled in the roughest manner, it never bites, but contents itself with its impotent personation of the venomous Snakes.

Sometimes it tries other arts, and instead of stimulating envenomed rage, pretends to be dead and lies motionless, hoping to escape as soon as the enemy has gone away. So perfectly does it assume the semblance of death, suffering itself to be tossed about without displaying the least sign of life, the muscles relaxed and the body hanging loosely and heavily in the hand, that experienced naturalists have been repeatedly deceived, and only discovered the deception by seeing the reptile make its escape after they had left it lying apparently dead upon the ground.

The color of the Hog-nose Snake is rather variable, but is generally of a darker or lighter brown above, with a row of large blotches of a different shade of brown running along the sides. Sometimes these blotches are so large, that they unite across the back and form broad bands. There is a dark band between the eyes. The average length of this reptile is about three feet.

THE species called BLOWING VIPER (*Heterodon platyrhynchus*), and Buckwheat-nose, is a most vicious appearing reptile, yet wholly harmless. Its habit of inflating its head and throat renders it unusually forbidding. It is common in the Middle and New England States west of the Connecticut River. *H. simus* is common in the Western States. Five other species are known in North America.

The sombre BLACK VIPER belongs to the same genus as the preceding species, and is very similar to that reptile in many of its habits. It is also an inhabitant of Northern America. Like the hog-nose Snake, it is much dreaded from its fierce aspect, but without the least reason. It is a very ugly and ungraceful-looking Snake, with a neck of great width, and a head very narrow in front and very wide behind, and is by no means a pleasing object to the eye. It does not frequent the marshy localities so constantly as the hog-nose, but prefers the more elevated and drier situations, having a great fondness for the pine-barren districts where the soil is dry and the fallen leaves afford it a shelter and a hunting-ground. It feeds mostly on little mammalia, certain reptiles and insects. Like the hog-nose Snake, it hisses and strikes with fangless jaws when irritated, and on account of its thick body, flat, wide head, and little glittering eyes, has so venomous an aspect, that it terrifies almost any antagonist for the moment, and then glides away before he has recovered from the instinctive shock to the nerves.

The color of the Black Viper is wholly black above, without any spots, though on the living Snake there are indications here and there of a deeper tint. The under parts are blackish-slate, and the throat takes a whiter hue. It is but a little Serpent, in spite of all its airs, being seldom more than twenty inches in length.

THE PINE SNAKE (*Pityophis melanoleucus*), called also Bull Snake, is found in the pine-barrens of New Jersey, and southwards in such localities to Georgia. Six other species of this genus are recorded as North American.

ALLEGHANY BLACK SNAKE (*Coluber obsoletus*). This was first discovered on a summit of the Blue Ridge Mountains, in Virginia. Specimens have since been found in the Highlands of the Hudson River. It resembles the common Black Snake, but has carinated scales, which readily distinguish it. It is credited with an exceedingly mild disposition, quite in contrast with the latter reptile.

Holbrook's specimen measured five feet three inches.

DeKay calls it the Racer, and Pilot Black Snake. Yet it surely cannot be the Racer that is so often referred to by observers, who report a long and large Black Snake, which runs along the tops of bushes, and well justifies the popular designation.

THE FOX SNAKE (*C. vulpinus*) inhabits from Massachusetts westward to Kansas and northward.

The family of the Colubrinæ is represented in most parts of the world, North America possessing a large number of examples.

The CORN-SNAKE of America may be reckoned among the most handsome of its tribe. This pretty reptile is extremely common in many parts of America, although it is not very frequently seen, owing to its dislike of daylight. As long as the sun is above the horizon, the Corn-Snake conceals itself in some hiding-place, and issues from its home as soon as the shades of evening begin to approach. It is fearless after its fashion, and has an instinctive liking for the habitations of mankind, haunting farms and houses, where it does considerable service by devouring rats and mice. Occasionally it takes toll in the form of a chicken, but its services most certainly outbalance its little perquisites. It will even enter houses, and can be tamed and made quite familiar. Sometimes it takes a fancy to frequent the roadside, and may be seen quietly coiled and at rest, or trailing its beautiful scales out of the reach of wheels or hoofs.

The colors of this Serpent are brilliant, and arranged in a bold and striking manner. The general color is rich chestnut-red, and along each side runs a series of large patches of a brighter, but deeper red, each patch being edged with jetty-black. There is also a row on each side of much smaller spots of an oval shape, just outside the larger row, and arranged alternately with them. These spots are golden-yellow, and are also edged with black. There are some similar spots on the head, and a streak is generally found over each temple. The under parts are silvery white, boldly checkered with black. The length of the Corn-Snake varies from five to six feet.

The SPOTTED RACER is another name for the Corn-Snake of the South, its northern limit being the Carolinas.

ANOTHER example of this genus is the THUNDER SNAKE, so called from the threatening black and white of its body, which seems to have a lowering aspect, and to menace poison as the thunder-cloud augurs lightning. Sometimes it is known by the name of KING SNAKE, or CHAIN SNAKE, the latter title being given because the black and white markings of the body are arranged alternately in a chain-like fashion.

The Thunder Snake is mostly found in moist and shady places, where it feeds upon small quadrupeds, reptiles, and birds if it can catch them. The portentous aspect of this Snake is fully carried out by its character, which is fierce, quarrelsome, and aggressive to a degree

THUNDER SNAKE.—*Ophibolus getulus.*

seldom found even in poisonous Serpents, and in a fangless Snake not at all to be expected. If put in a box with other Serpents, it always quarrels and fights with them; and in one instance, when a Thunder Snake had been introduced into a cage where a miliary rattlesnake was residing, it attacked the venomous reptile in spite of its poisonous weapons, overpowered, killed, and ate it. Some persons think that a deadly feud always rages between the Thunder Snake and rattlesnake, but the truth of this supposition is somewhat dubious. In the instance just mentioned, the creature would probably have treated a Serpent of any species in precisely the same manner.

The Thunder Snake is colored after a very peculiar fashion. All along the body run alternate bands of jetty-black and pure white, the black being very broad and the white very narrow, and not reaching completely across the body. The head is also mottled and scribbled with black upon white after a curious and most complicated fashion. The full length of this Serpent is about four feet.

THE SCARLET KING SNAKE (*Ophibolus doliatus*) inhabits Florida, and extends northward only as far as North Carolina. Its length is three feet six inches. Five other varieties of this genus are recorded as North American, found in the Southwestern States.

The CHICKEN SNAKE (*Coluber quadrivittatus*) derives its name from its habit of entering farms and houses and stealing chickens from the roost. As, however, it feeds largely on rats and mice, its services in this respect may in all probability counterbalance the loss caused by its thefts. Like the corn-Snake, it is soon tamed, and will become very familiar. In color it is a very delicate looking reptile, being of a soft bright golden-brown, and having four narrow stripes upon a rich dark brown running the whole length of the body. In length it is usually about four feet six inches, though a few specimens attain the length of six and even seven feet. This is also a Northern American reptile.

CHICKEN SNAKE.—*Coluber quadrivittatus.*

SAY'S SNAKE is a most attractive creature, having a bluish-black body, with round milk-white spots, thickly bespattered over the entire upper surface. It measures from three to four feet in length.

Its *habitat* is throughout the Gulf States.

THE MILK SNAKE, or HOUSE SNAKE (*Ophibolus triangulus*), is common in many parts of North America, and has derived its popular names from its habit of entering houses and its fondness for milk, which some persons fancy it obtains from the cows. Its general food consists of mice and insects, and, like the preceding species, it is probably of some use to the farm where it takes up its residence, and worthy of the encouragement which it sometimes receives.

In the general arrangement of the markings, it is not unlike the corn-Snake, with which it has often been confounded, especially after the fresh beauty of its colors has been dimmed by death, or extracted and changed by spirits. There are similar rows of patches along the sides, but in this species the spots are much broader, often coalescing over the back and forming bands, and the general hue of the body is a beautiful blue tinge. The under parts are silver-white, boldly tesselated with oblong and sharply defined marks of black.

The length of the Milk Snake is generally about four feet. It inhabits as far north as Maine. In Massachusetts it is called Checkered Adder; in New York it is Sachem Snake and Sand King, and Spotted Adder. In Arkansas and Georgia two species are found, respectively.

Kennicott's Chain Snake (*Ophibolus calligaster*) is a species found from Illinois to Kansas.

The Indigo Snake (*Spilotes couperi*), called also Gopher Snake, is a dark indigo-blue in color, much resembling the Black Snake in the bluish-black color. It is stouter in body, and from that fact and a fancied courageousness, the negroes regard it as an enemy and victor of the rattlesnake. Its habit of frequenting the holes of the Gopher suggests the local name.

Species belonging to the genus Coluber are found in Australia, India, Japan, China, and Europe, the latter (*Coluber æsculapii*) being the Serpent which is represented by the ancients as twined round the staff of Æsculapius and the caduceus of Mercury.

The Black Snake of America (*Bascanium constrictor*) is perhaps the best known of the numerous Serpents, which, happening to be black or dark brown, have been called by the same title.

This Snake is common in Northern America, where it is sometimes known under the name of Racer, on account of its great speed. It is a perfectly harmless, but highly irascible reptile, especially during the breeding-season, when it seems to become endowed with an unreasoning ferocity, which, happily for the world, is seldom found in reptiles better provided with offensive weapons.

COLUBER.—*Coluber æsculapii.*

It has a curious habit of rustling its tail among the herbage in such a manner as to resemble the whirr of the dreaded rattlesnake, and then darts at the object of its rage and inflicts a tolerably severe bite, thereby inducing great terror on the part of the sufferer, who, in the hurry of the moment, naturally believes that he has been bitten by the rattlesnake itself.

It is fond of climbing trees in search of young birds, eggs, and similar dainties, and even in that position, is of so tetchy a disposition, that when irritated, it will descend in order to attack its foe. Even if confined with other Snakes, it becomes quarrelsome, fights with them, and if possible will kill them.

The haunts of the Black Snake are usually to be found along the edges of streams and ponds or lakes, and the reptile is mostly to be seen in shady spots, well sheltered by brushwood. Sometimes, however, it goes farther a-field, and wanders over the free country, traverses rocky soil, or glides along the roadside.

It is a most useful reptile, being very fond of rats, and able from its great agility to climb over walls or buildings in search of its prey, and to insinuate its black length into their holes.

It also feeds much on birds, especially when they are young, and is consequently an object of detestation to the feathered tribes. It often happens that the locality of the Black Snake is indicated by the proceedings of the little birds, which collect above their hated enemy, scold with harsh cries, flutter their wings noisily, and by dint of continual annoyance will often drive the reptile away from the locality. It has been thought that this Serpent was in the habit of killing its prey by pressure, after the fashion of the boas, but this statement has not been satisfactorily confirmed.

The color of this Snake is blue-black above, and ashen slate below, becoming rather whiter upon the throat. In some specimens a number of spots are observed upon the back of a deeper and duller hue than the general tint. In length the Black Snake generally reaches from five to six feet.

This familiar Serpent of our country-side is the *beau ideal* of its race; expressing the most slender and graceful form, with an extreme length of body. The tail is prolonged gradually to a mere point, and becomes highly prehensile. The head is graceful, and the steel-bluish, uniform color, with the beautifully tessellated arrangement of scales, all tend to render the creature attractive in spite of its being a Snake, usually the embodiment of the unsightly. Add to this, the Black Snake is the most active of its order; and even the most powerful.

This is widely distributed over the United States. A species found in California is dedicated to General Fremont. It has a stouter body and a larger head than the preceding. Nine other species will be found enumerated in the catalogue at the end of this volume.

BEAD SNAKE (*Elaps fulvius*). Inhabits Virginia and southward to Alabama. It is also called Harlequin, from its curiously marked body. Though possessing poison-fangs, it is very gentle and mild in disposition. It is jet-black, with seventeen broad crimson rings, each bordered with yellow. Two other species are recorded, *E. tenere*, and *E. tristis*, of Texas.

RIBBON SNAKE (*Eutænia saurita*). Called also Swift Garter Snake. The markings are slender and ribbon-like. Inhabits east of the Alleghanies and southward to Georgia.

FAIRIES GARTER SNAKE (*E. faireyi*). Inhabits the Mississippi valley, and northward to Michigan. SAY'S GARTER SNAKE has the same *habitat*. HAY'S GARTER SNAKE inhabits from Lake Michigan, westward to Oregon.

COMMON GARTER SNAKE (*E. sirtalis*), is the familiar striped Snake of eastern New England, and is our most common species. Several varieties are known. Ten other species are also enumerated as inhabiting the United States.

WATER ADDER (*Tropidonotus sipedon*). This is an exceedingly common aquatic Snake, indigenous to the Eastern United States.

A variety, called the RED-BELLIED WATER SNAKE, is common in Michigan. Twelve distinct species are known as North American.

THE small, but interesting family of the Dryadidæ contains a number of Serpents remarkable for the slender elegance of their form, the delicate beauty of their coloring, and the singular swiftness of their movements.

The well-known COACH-WHIP SNAKE, of North America, is a useful example of this family.

This remarkable reptile has not earned its popular name without good reason, for the resemblance between one of these Serpents and a leather whip-thong is almost incredibly close.

The creature is very long in proportion to its width, the neck and head are very small, the body gradually swells towards the middle and then as gradually diminishes to the tail, which ends in a small point. The large smooth scales are arranged in such a manner that they just resemble the plaited leather of the whip, and the polished brown-black of the surface is exactly like that of a well-worn thong.

The movements of this Snake are wonderfully quick, and when chasing its prey, it seems to fly over the ground. The mode of attack is very remarkable. Seizing the doomed creature in its mouth, it leaps forward, flings itself over the victim, envelops it with coil upon coil of its lithe body, so as to entangle the limbs and bind them to the body, and, in fact, makes itself into a living lasso. One of these Snakes was seen engaged in battle with a hawk, and would apparently have conquered in the seemingly unequal combat had not the foes been separated. It had grasped the hawk by one wing, had dragged it to the ground, and had succeeded in disabling the terrible claws from striking, when the sudden approach of the narrator alarmed the Snake, which released its hold, darted into the bushes, and permitted the rescued hawk to fly away in peace.

COACH-WHIP SNAKE.—*Bascanium flagelliforme.*

The color of this Serpent is rather variable. Generally it is shining black above and lighter beneath, with splashes of purple-brown. Sometimes, however, it is cream or clay-colored, and occasionally has been seen almost white. But, whatever color may be the body, the portion near the head is always raven-black. The length of this Snake is about five or six feet.

The Coach-Whip Snake (*Bascanium flagelliforme*) is a rare species, inhabiting the Gulf States. As its name suggests, the body is long, slender, and graceful; and it is a rapid runner.

Other species are from Texas, from the great Salt Lake, and from California. A genus, *Salvadora*, has a species found in Mexico.

ANOTHER very slender Snake, also a native of America, is closely allied to the preceding species. This is the GREEN SNAKE, well known for its grass-green color and its singular activity.

The Green Snake is fond of climbing trees, traversing the boughs in search of food with marvellous celerity, and darting at its insect prey through considerable distances. So slender is this Serpent, that a specimen which measures three feet in length, will barely reach one-third of an inch in thickness at its widest part. Partly owing to this extreme delicacy of form, and partly on account of the leaf-green color of its body, the Green Snake is not easily seen among the foliage, and in many cases would be undiscovered but for its rapid and energetic movements. The food of this Snake consists mostly of insects. It is very readily tamed, and many persons are fond of carrying the beautiful creature about them, tying it round their throats as a necklace, or as a bracelet on the wrist. The eye corresponds in beauty to the rest of the person, being very large and of a beautiful topaz-yellow.

The color of the Green Snake is delicate grass-green above, and silvery-white below. Its average length is about three feet. Its shape is much like the Black Snake, but it is smaller.

Its *habitat* is in the Southern States.

Another species is found in Texas and Arkansas.

THE common GREEN SNAKE, called also Grass Snake, in the Northern States, inhabits from Massachusetts to Pennsylvania.

Contia, a genus of Baird and Giraud, has three species, *C. mitis*, of California.

THE RING-NECKED SNAKE (*Diadophis punctatus*), called also the Little Black and Red Snake, inhabits from Maine to Florida. Another species, *D. amabilis*, inhabits California; *D. docilis*, Texas; *D. pulchellus*, California; *D. regalis*, Mexico.

Lodia is a genus of Baird and Giraud. *L. tenuis* is the species found in Puget Sound region.

Sonora is a genus of same authorities; species *semi-annalata*, found in Mexico.

THE SCARLET SNAKE (*Cemophora coccinea*) inhabits the Gulf States. It is an exceedingly handsome reptile; richly colored.

Rhinochilus lecontei is a form discovered by Lecont in San Diego, California.

THE BROWN SNAKE (*Haldea striatula*) inhabits from Virginia to the Gulf States.

THE HORN SNAKE (*Farancia abacura*), called also Red-bellied Snake, inhabits the Gulf States. Two other allied genera, *Abaster* and *Virginia*, with one species each, are known in the Southern States.

THE WORM SNAKE, called also GROUND SNAKE (*Carphophiops amœnus*), inhabits from Pennsylvania to Gulf States.

The genus *Tantilla* embraces two species, each of the Southern States.

THE small BROWN SNAKE (*Storeria dekayi*) found rather common from New York State to the Gulf States. It is a small gray form, with minute spots of black along its upper parts.

Another species is STORER'S SNAKE (*S. occipito maculata*).

A FAMILY of North American Reptiles named *Boidæ*, is characterized by the individuals having rudimentary hinder limbs, or spur-like appendages, situated near the anas. The LEAD-COLORED WENONA (*Charina plumbea*) inhabits Puget Sound. *C. bottæ* is another species, found in the same region.

BRAZIL possesses a most lovely example of these Serpents, the EMERALD WHIP SNAKE (*Philodryas viridissimus*).

Dr. Wucherer, of Bahia, writes as follows concerning this pretty species in a letter quoted by Sir J. E. Tennent, in his "Natural History of Ceylon": "I am always delighted when I find that another tree-Snake has settled in my garden. You look for a bird's nest: the young ones have gone, but you find their bed occupied by one of these beautiful creatures, which will coil up its body of two foot in length within a space not larger than the hollow of your hand.

"They appear to be always watchful, for at the instant you discover one, the quick playing of the long, black, forked tongue, will show you that you, too, are observed. On perceiving the slightest sign of your intention to disturb it, the Snake will dart upwards through the branches and over the leaves, which scarcely seem to bend beneath the weight. A moment more, and you have lost sight of it. Whenever I return to Europe, you may be sure that in my hothouse these harmless lovely creatures shall not be missing."

The green color of this species is paler than above.

THE GRAY SNAKE of Jamaica (*Dromicus ater*) is another instance of this family. It is often called the BLACK SNAKE, but as that title has already been employed, it is better to use the popular name which is first mentioned.

This reptile is extremely plentiful in Jamaica, where it is mostly found haunting heaps of dead leaves, rocks and buildings. It is especially fond of the crevices found in old walls, and will lie for hours with its head and neck hanging out of some cranny partially awaiting the approach of any miserable lizard which may come within reach while searching after flies. It is rather a savage ophidian, darting fiercely at its adversary if irritated, and inflicting a wound which, though not dangerous, is very unpleasant, and causes the limb to swell and ache for some time. It is said, that if it is attacked by a dog, it strikes at the eyes, and can blind the poor creature. While preparing to strike, it dilates its neck, and flattens its head, so as to look as like a venomous Serpent as its limited means will permit.

The color of the Gray Snake is exceedingly variable. Mostly, it is uniformly black, with a tinge of brown; but it often happens, that the former tint is subservient to the latter, and in many cases the color is gray, sometimes of a uniform tint, and sometimes variegated with large dark spots. The length of this Snake is rather more than three feet.

THE little family of the Dasypeltidæ possesses but one genus, but is remarkable for the formation of the teeth and their use. The teeth of the jaws are very minute and scanty, being at the most only six or seven in number; but some sharp and strong processes issue from the hinder vertebræ of the neck, through holes in the membranes, and form a series of tooth-like projections in the gullet.

The most familiar example of this family is the ROUGH ANODON, of Southern Africa. The name Anodon is of Greek origin, and signifies toothless. This reptile lives almost wholly upon eggs, which it eats after a curious fashion. When it finds a nest, it takes the eggs into its mouth, where they lie unharmed, on account of the absence of teeth, so that the shell is not broken, and the liquid contents are preserved. When, however, the reptile swallows the egg, it passes into the throat, and meets the saw-like row of vertebral teeth which have just been mentioned. In its passage, the shell is cut open by these teeth, and the muscular contraction of the gullet then crushes the eggs, and enables the contents to flow down the Snake's throat. These bony processes are tipped with enamel like real teeth.

The color of this remarkable Serpent is brown, with a row of black marks along the back, sometimes coalescing into a continuous chain, a series of smaller spots upon each side, and some arrow-head marks upon the head of a jetty-black.

THE next family is composed of the Tree-Serpents, or Dendrophidæ, so called from the habit of residing among the branches of trees.

Our first example of this family is the well-known BOOMSLANGE, of Southern Africa. In pronouncing this word, which is of Dutch or German origin, and signifies Tree-Snake, the reader must remember that it is a word of three syllables. The Boomslange is a native of Southern Africa, and is among the most variable of Serpents in coloring, being green, olive, or brown; of such different colors, that it has often been separated into several distinct species.

Dr. A. Smith has given the following valuable description of the Boomslange and its habits:—

"The natives of South Africa regard the Boomslange as poisonous; but in their opinion we cannot concur, as we have not been able to discover the existence of any gland manifestly organized for the secretion of poison. The fangs are inclosed in a soft, pulpy sheath, the inner surface of which is commonly coated with a thin, glairy secretion. This secretion possibly may have something acrid and irritating in its quality, which may, when it enters a wound, occasion pain and swelling, but nothing of greater importance.

"The Boomslange is generally found on trees, to which it resorts for the purpose of catching birds, upon which it delights to feed. The presence of a specimen in a tree is generally soon discovered by the birds of the neighborhood, who collect around it, and fly to and fro, uttering the most piercing cries, until some one, more terror-struck than the rest, actually scans its lips, and, almost without resistance, becomes a meal for its enemy. During such a

BOOMSLANGE.—*Bucephalus capensis.*

proceeding, the Snake is generally observed with its head raised about ten or twelve inches above the branch, round which its body and tail are entwined, with its mouth open and its neck inflated as if anxiously endeavoring to increase the terror which, it would almost appear it was aware, would sooner or later bring within its grasp some one of the feathered group.

"Whatever may be said in ridicule of fascination, it is nevertheless true, that birds, and even quadrupeds also, are, under certain circumstances, unable to retire from the presence of certain of their enemies; and what is even more extraordinary, unable to resist the propensity to advance from a situation of actual safety into one of the most imminent danger. This I have often seen exemplified in the case of birds and Snakes; and I have heard of instances equally curious, in which antelopes and other quadrupeds have been so bewildered by the sudden appearance of crocodiles, and by the grimaces and contortions they practised, as to be unable to fly or even move from the spot towards which they were approaching to seize them."

The beautiful BOIGA, sometimes called the AHÆTULLA, also belongs to the family of Tree-Serpents. This pretty and graceful creature inhabits Borneo, and, on account of the extreme gentleness of its disposition and the ease with which it is tamed, the children are in the habit of considering it as a kind of living toy, and allow it to twine around their bodies, or carry it about in their little hands, without the least alarm. It is a most active Serpent, living in trees, and darting its lithe form from branch to branch with arrow-like celerity, leaping, as it were, from the coiled folds in which it prepares itself for the spring, and passing through the boughs as if shot from a bow, its glittering scales flashing an emerald or sapphirine radiance, as it glances through the sunbeams.

The head of the Boiga is long and slender, as beseems the delicate body; the eye is very full and round, and the gape very wide. The upper part of its body is rich, shining blue, shot with sparkling green; and three bright, golden stripes run along the body, one traversing the spinal line, and another passing along each side. Behind each eye is a bold jetty-black streak, and immediately below the black line runs a stripe of pure white.

The specific name ought properly to be spelled leiocercus. It is of Greek origin, and signifies smooth-tail, in allusion to the smooth-surfaced scales of the back and tail.

THE family of the Wood-Snakes, or Dryiophidæ, as they are learnedly called, contains some interesting and rather curious reptiles.

The GOLDEN TREE-SNAKE, which is a native of Mexico, is a most lovely species, and of a most singular length, looking more like the thong of a "gig whip" than a living reptile. It lives in trees, and in many respects resembles the preceding species. It is not so gorgeously decorated as the boiga, but its colors are beautifully soft and delicate. The general tint of this Serpent is gray, tinged with yellow, and having a golden reflection in certain lights, and being decidedly iridescent in others. The body is profusely covered with minute dottings of black.

LANGAHA.—*Langaha nasuta.* (Two-thirds natural size.)

THE accompanying illustration represents the LANGAHA, one of the Serpents of Madagascar, remarkable for the singular appendage to the head. The muzzle is extremely elongated, and is furnished with a fleshy projection, about one-third as long as the head, and covered with small scales. There is another species, the COCK'S-COMB LANGAHA (*Langaha crista-galli*), also a native of Madagascar, which is known from the ordinary species by the form of the appendage, which is toothed something like the comb upon a cock's head. The color of the Langaha is reddish-brown.

A VERY beautiful example of the Wood-Snakes is found in Ceylon. This is the BROWN WOOD-SNAKE (*Passerita mycterizans*). Like the langaha, the snout of this Serpent is furnished with an appendage, which is pointed, and covered with scales, and is about one-fourth as long as the head. This appendage is conspicuous, but its use is not very plain. It lives almost wholly in trees, and is nocturnal in its habits, traversing the boughs at night for the purpose of catching the small birds as they sleep, taking their young out of the nest, and preying upon the lizards and geckos which also prowl about the trees by night in search of their insect food. There are two varieties of this beautiful Serpent, one being bright green above, with a yellow stripe down each side, and paler below; while the other is brown, glossed with purple, and without the yellow stripe. This variety is rare. The length of these Snakes rarely exceeds three feet.

THE DIPSAS and its congeners may be known from the preceding Snakes, which they much resemble in general form, by the large size of the head compared with the extremely delicate and slender neck. The body, too, is much wider in the centre, causing the neck and tail to appear disproportionately small. This Snake is a native of many parts of Asia, and is found in the Philippines. The name Dipsas is derived from a Greek word, signifying thirst, and is given to this Snake because the ancients believed that it was eternally drinking water and eternally thirsty; and that, to allay in some degree the raging drought, it lay coiled in the scanty springs that rendered the deserts passable. As they considered almost all Serpents to be venomous, and, according to the custom of human nature, feared most the creatures of which they knew least, they fancied that the waters were poisoned by the presence of this dreaded Snake. Lucan, in the Pharsalia, alludes to this idea:—

> "And now with fiercer heat the desert glows,
> And mid-day gleamings aggravate their woes;
> When lo! a spring amid the sandy plain
> Shows its clear mouth to cheer the fainting train.
> But round the guarded brink, in thick array
> Dire aspics rolled their congregated way,
> And thirsting in the midst the horrid Dipsas lay.
> Blank horror seized their veins, and at the view,
> Back from the fount the troops recoiling flew."

The ancient writers also averred that the bite of the Dipsas inoculated the sufferer with its own insatiate thirst, so that the victim either died miserably from drought, or killed himself by continually drinking water.

The colors of the Dipsas are not brilliant, but are soft and pleasing. The general tint is gray, banded with brown of different shades, sometimes deepening into black. The top of the head is variegated with brown, and a dark streak runs from the eye to the corner of the mouth.

THE BANDED BUNGARUS is a native of India, where, from its habits, it is sometimes called the Rock Serpent. The name Bungarus is a most barbarous Latinization of the native word Bungarum-Pamma, which, though not euphonious, has at all events the advantage of being indigenous, and might have been spared the further distortion of being wrested into a sham classical form. In this reptile the head is rather flat and short, and the muzzle is rounded. The upper jaws are furnished with grooved fangs.

The color of the Banded Bungarus is very variable, but always consists of some light hue, relieved by bands or rings of jetty-black along its length.

AN allied species, the SERPENT-EATING HAMADRYAS (*Hamadryas elaps*), is noted for the peculiarity from which it derives its name. It feeds almost wholly on reptiles, devouring the lizards that inhabit the same country, and also living largely on Snakes. Dr. Cantor says of this Serpent that it cannot bear starvation nearly so well as most reptiles, requiring to be fed at least once a month. "Two specimens in my possession were regularly fed by giving them a Serpent, no matter whether venomous or not, every fortnight. As soon as this food is brought near, the Serpent begins to hiss loudly, and expanding its hood, rises two or three feet, and retaining this attitude, as if to take a sure aim, watching the movements of the prey, darts upon it in the same manner as the *naja tripudians* (*i. e.* the cobra) does. When the victim is killed by poison, and by degrees swallowed, the act is followed by a lethargic state, lasting for about twelve hours.

The Hamadryas is fond of water, will drink, and likes to pass the tongue rapidly through water as if to moisten that member. It is a fierce and dangerous reptile, not only resisting when attacked, but even pursuing the foe should he retreat, a proceeding contrary to the general rule among Serpents. The poison of this creature is virulent and active, a fowl dying in fourteen minutes, and a dog in less than three hours, after receiving the fatal bite, although the experiments were made in the cold season, when the poison of venomous Snakes is always rather inactive. The poisonous secretion reddens litmus paper very slightly, and, as is the case with most Serpent poisons, loses its efficacy by being exposed to the air. The native name of the Hamadryas is Sunkr Choar.

BANDED BUNGARUS.—*Bungarus fasciatus.*

The color of this Snake is generally of an olive hue, auburn and pale below, but there is a variety marked with cross-bands of white. It is a large species, varying from four to six feet in length, while some specimens are said to reach ten feet.

We now come to some of the most deadly of the Serpent tribe, the first of which is the well-known Cobra di Capello, or Hooded Cobra of India.

This celebrated Serpent has long been famous, not only for the deadly power of its venom, but for the singular performances in which it takes part. The Cobra inhabits many parts of Asia, and in almost every place where it is found, certain daring men take upon themselves the profession of Serpent-charmers, and handle these fearful reptiles with impunity, cause them to move in time to certain musical sounds, and assert that they bear a life charmed against the bite of these reptilian playmates. One of these men will take a Cobra in his bare hands, toss it about with perfect nonchalance, allow it to twine about his naked breast, tie it round his neck, and treat it with as little ceremony as if it were an earth-worm. He will then take the same Serpent—or apparently the same—make it bite a fowl, which soon dies from the poison, and will then renew his performances.

Some persons say that the whole affair is but an exhibition of that jugglery in which the Indians are such wondrous adepts; that the Serpents with which the man plays are harmless, having been deprived of their fangs, and that a really venomous specimen is adroitly substituted for the purpose of killing the fowl. It is moreover said, and truly, that a Snake, thought to have been rendered innocuous by the deprivation of its fangs, has bitten one of its masters and killed him, thus proving the imposture.

Still, neither of these explanations will entirely disprove the mastery of man over a venomous Serpent. In the first instance, it is surely as perilous an action to substitute a venomous Serpent as to play with it. Where was it hidden, why did it not bite the man

SERPENT-EATING HAMADRYAS.—*Hamadryas elaps.* (See page 140.)

instead of the fowl, and how did the juggler prevent it from using its teeth, while he was conveying it away? And, in the second instance, the detection of an impostor is by no means a proof that all who pretend to the same powers are likewise impostors.

The following narrative of Mr. H. E. Reyne, quoted by Sir J. E. Tennent in his "Natural History of Ceylon," seems to be a sufficient proof that the man did possess sufficient power to induce a truly poisonous Serpent to leave its hole and to perform certain antics at his command: "A Snake-charmer came to my bungalow, requesting me to allow him to show me his Snake dancing. As I had frequently seen them, I told him I would give him a rupee if he would accompany me to the jungle and catch a Cobra that I knew frequented the place.

"He was willing, and as I was anxious to test the truth of the charm, I counted his tame Snakes, and put a watch over them until I returned with him. Before going I examined the

COBRA DI CAPELLO.

man, and satisfied myself he had no Snake about his person. When we arrived at the spot, he played upon a small pipe, and after persevering for some time, out came a large Cobra from an ant-hill which I knew it occupied. On seeing the man, it tried to escape, but he caught it by the tail and kept swinging it round until we reached the bungalow. He then made it dance, but before long it bit him above the knee. He immediately bandaged the leg above the bite, and applied a Snake-stone to the wound to extract the poison. He was in great pain for a few minutes, but after that it gradually went away, the stone falling off just before he was relieved.

"When he recovered, he held up a cloth, at which the Snake flew, and caught its fangs in it. While in that position, the man passed his hand up its back, and, having seized it by the throat, he extracted the fangs in my presence and gave them to me. He then squeezed out the poison on to a leaf. It was a clear oily substance, and when rubbed on the hand, produced a fine lather. I carefully watched the whole operation, which was also witnessed by my clerk and two or three other persons."

With regard to the so-called charming of Serpents, there is no need of imagining these men to be possessed of any superhuman powers; for these, and most of the venomous Serpents, are peculiarly indolent, and averse to using the terrible weapons which they wield; in proof of which assertion, the reader may recollect that Mr. Waterton, though not pretending to be a Snake-charmer, carried a number of rattlesnakes in his bare hand without being bitten for his meddling. Not that I would positively assert that the Snake-charmers do not possess some means of rendering themselves comparatively proof against the Serpent's bite; for it is reasonable to conclude that, just as a secretion of a cow will, when it has been suffered to pervade the system, render it proof against the poison of the small-pox, there may be some substance which, by a kind of inoculation, can guard the recipient against the poison of the Cobra. In the last century, the one was quite as irremediable as the other.

Another fact is yet to be mentioned. In almost every instance where a poison, vegetable or animal, is likely to gain access to human beings, Nature supplies a remedy at no great distance, just as, to take a familiar instance, the dock is always to be found near the nettle. There certainly are many poisons for which no sure remedy has been discovered, and, until lately, the venom of the Cobra ranked among that number. Recently, however, some important discoveries have been made, which seem to prove that the bite of the Cobra may be cured in two methods, viz., the external application of certain substances to the wound, and the internal administration of others. As the general character of the Cobra is almost precisely the same as that of many other venomous Serpents, and has long been familiar to the public, I shall devote the greater portion of the space, not to the creature itself, but to the remedies for its bite.

The first of these remedies is a plant belonging to the group of birth-worts, and known to botanists by the name of *Aristolochia indica*.

This plant has long been considered as a valuable remedy for the bite of the Cobra, but the accounts of its use and mode of operation have mostly been vague and scarcely trustworthy. I have, however, been fortunate enough to obtain much valuable information on this subject from R. Lowther, formerly Commissioner in India, who was accustomed to employ this plant very largely in cases of Cobra-bites, and has kindly forwarded the following communication on the subject:—

"According to your request I have the pleasure of inclosing a statement of one out of at least twenty cases of Snake-bites, in which the exhibition of the *Aristolochia indica* was attended with complete success, on patients who were brought to my house on a litter, in a perfect state of coma from the bites of venomous Snakes.

"The *Aristolochia indica* is noticed by medical writers as a powerful stimulant, much extolled as a remedy for Snake-bites, in support of which I need only refer you to my detailed statement, as also to the circumstances under which the plant was transferred to my garden at Allahabad. The gentleman from whom I received it (Mr. Breton, Deputy Collector of Customs) gave me the following account of it;

"A Cobra, to the great alarm of his servants, had taken up its abode in a mound of earth, formed by white ants, in the vicinity of his house. A party of Snake-catchers having one day made their appearance in the village, Mr. Breton was afforded the opportunity of getting rid of the reptile by having it dug out of its lodgment. After having reached a considerable depth, the man at work used his finger for the purpose of ascertaining the direction of the hole. This seemed to have been its termination, or nearly so, as the Snake caught hold of his finger. His companion immediately ran of to the bank of a stream near at hand, and brought back some leaves, which, having bruised with a stone, he administered to his friend's relief. Mr. Breton requested the man to take him to the plant, which he forthwith removed to his own garden. The Snake-catcher informed him the plant was a specific, and that they usually carried the dried root about with them in case of need.

"Mr. Breton, having been subsequently appointed to Allahabad, brought the plant away, and was successful in the treatment of numerous cases. On being removed to a distant station, he transferred the plant to me. The plant is a creeper, and sheds its leaves at that season when Snakes, for the most part, are lying inert in their holes. I should have mentioned, that the Cobra above referred to was killed in the hole.

"There are several species of *Aristolochia*, all of them I believe stimulant; but the *Indica* is that which I refer to—it is intensely bitter and strongly aromatic.

"In one bad case which came under my treatment, in which large doses had been exhibited, I gave an additional leaf to the patient to take home, but to be used only in case of relapse. Her husband informed me that, although quite recovered, she took the extra dose at one o'clock in the morning, and became so giddy that in attempting to move she reeled about like a drunken creature.

"A young Hindoo woman was brought to my door in a 'charpoy,' or litter, in a state so apparently lifeless from a Snake-bite, that I had no hesitation in refusing to prescribe. An officer, who was on a visit at my house at the time, considered the woman beyond the power of human relief, and advised me to send her away, as my failure would bring discredit on a remedy which was attracting public notice. In this instance the patient was as cold as marble; there was no pulsation; countenance death-like.

"The woman's husband manifested great distress at my refusal, at the same time urging that as the remedy had been prepared, I might, at any rate, give his wife the chance of recovery. I explained to him my motives, and my firm belief that his wife was dead long before he had reached my door. However, rather than add to his distress by persisting in my refusal, I forced her jaws open, and poured down her throat three medium-sized leaves of the *Aristolochia Indica*, reduced to a pulp, with ten black peppercorns, diluted with a graduated ounce of water. The remedy having flowed into her stomach, I directed her body to be raised and supported in a sitting posture, and with some anxiety, though without the slightest prospect of success.

"I attentively watched her features, and in the course of eight or ten minutes I observed a slight pulsation on her under lip. I instantly directed her husband, with the aid of my own servants, to drag her about for the purpose, if possible, of increasing the circulation. Supported by two men, holding her up by the waist and arms, she was moved about, her feet helplessly dragging after her. After the lapse of a few minutes, I perceived an attempt on the part of the patient to use her feet. I accordingly directed them to raise her body sufficiently high to admit of the soles of her feet being placed on a level with the ground. In a few minutes she gave a deep inspiration, accompanied with a kind of shriek, manifesting the return of consciousness. This was followed by an exclamation, 'A fire is consuming my vitals!' At this time her chest and arms were *deadly cold*. I immediately gave her the pulp of one leaf in an ounce of water, which greatly alleviated the burning sensation in the stomach.

"She was then enabled to explain the position of the wound in her instep, which had the appearance of a small speck of ink, surrounded by a light-colored circle. I had the part well rubbed with the *Aristolochia*, after which she was able to walk without assistance. I kept her walking up and down for at least a couple of hours. Having expressed herself entirely recovered, I allowed her to depart. She called on the following morning to show herself.

"The Snake unfortunately escaped, but the woman described it as a 'Kala Samp,' which is the term ordinarily used for the Kobra Kapelle.

"I have written the above entirely from memory, the case having occurred eight or nine years ago.

"A middle-aged woman was brought to my door in the early part of the rainy season, who had been bitten by a Snake at daybreak, while stooping down for the purpose of sweeping the floor. She called out to the people of the house that a rat had bitten her, and nothing more was thought of it, as her attention was directed to her infant, who became fractious for the breast. She accordingly went to bed to give the child sustenance, and not long afterwards complained of giddiness. It was suggested to her that a Snake might have bitten her, but she referred to a hole in the mud-wall from which the rat must have darted out.

"Nothing further transpired until the household were alarmed on finding her in a state of insensibility, foaming at the mouth, and the infant at her breast. They were then convinced that a Snake must have done the mischief, and immediately carried her off to the charmer! After detaining the woman for a full hour, the fellow coolly told her friends to take her off to the Commissioner, who would prescribe for her. The poor woman had been dead for some time before the incantations were finished. On arriving at my house, I found the deceased in a state of incipient decomposition, and, having heard the statement of her friends, directed them to take the body away for the performance of funeral rites, and to lose no time in bringing her infant, who was said to be suffering from the effects of the poison.

"The poor thing reached my house in a state of insensibility, though not dead. Its head was hanging on its shoulder, and when raised beyond the perpendicular would fall on the opposite shoulder. The body was not cold, and that was the only indication that death had not supervened. I selected one of the smallest of the leaves of the *Aristolochia*, and pounded one-third of it, and, with a small table-spoonful of water, poured the solution into the stomach. After the lapse of four or five minutes the child heaved a deep sigh, opened its eyes wildly, gave a loud scream, and afterwards became quite composed. The child was brought to me on the following morning quite well."

As this plant is so valuable, and seems likely to become an acknowledged remedy, a few lines may be spared for a short description of the species, and the mode of its action.

The *Aristolochia Indica* is one species of a rather large genus, inhabiting many parts of the world, but being most plentiful in the hotter regions. It is a creeping plant, and the specimens grown by Mr. Lowther were trained upon a trellis-work, which they clothed with their narrow, abruptly pointed leaves. Another species of this group of plants, the *Aristolochia serpentina*, is not uncommmon in parts of North America, where it is known under the title of the Virginian Snake-root. An infusion of this plant is used as a specific against ague and liver affections.

The fresh leaf of the *Aristolochia Indica* is, when tasted, very bitter and aromatic, bearing some resemblance to quinine in the clear searching quality of the bitter. It is remarkable that when persons are suffering from the poison of the Cobra, they describe it as being sweet. There is certainly a kind of sweetness in the leaf, for on chewing a dried leaf of this plant, kindly sent me by Sir W. Hooker, from the collection in the botanical gardens at Kew, I find it to be rather, but not very bitter, with a pungent aroma, something like that of the common ivy, and a faint, though decided sweetness as an after-flavor.

It is not a universal specific, for when experiments were tried by getting some dogs bitten by the Cobra, and treating them with this leaf, they died to all appearance sooner than if they had been entirely neglected. Mr. Lowther has made rather a curious series of experiments on the Cobra's poison and the mode of its action, and has found that while human beings become cold as marble under the influence of the venom, dogs are affected in precisely an opposite manner, being thrown into a high fever, from which they die. The body of a dog killed by a Cobra's bite, will remain quite hot for some ten hours. The *Aristolochia*, therefore, which is a powerful stimulant, rather aids than counteracts the operation of the poison.

In the case of a human being, however, the effect of this remedy seems to be infallible, and Mr. Lowther informs me that he always kept a mortar and pestle by the plant, so that no time should be lost in bruising the leaf, and mixing it thoroughly with water, before pouring it down the throat of the sufferer. The admixture of water was necessary, because, in most instances, the patient was insensible, and the jaws stiffened, so that the mouth needed to be opened forcibly, and the preparation poured down the throat.

THE second mode of cure employed by the natives of India, Ceylon, and even of some parts of Africa, is the now celebrated Snake-stone, so carefully described by Sir J. E. Tennent in his "Natural History of Ceylon." On being bitten by a Cobra, the sufferer applies one of these remarkable objects to each puncture, where they adhere strongly for a variable space of time, five or six minutes appearing to be the usual average. They seem to absorb the blood as it flows from the wound, and at last fall off without being touched, when the danger is considered to be over. This mode of application is general throughout all parts of the world where the Snake-stone is known.

Through the kindness of Sir J. E. Tennent, I have been enabled to make a careful inspection of these objects, and to peruse the original letters relating to their use. They are flattish, shaped something like the half of an almond with squared ends, rather light, bearing a very high polish, and of an intense black—in fact, looking much as if they were rudely cut from common jet. The value of these singular objects is placed beyond doubt by the carefully accredited narratives lately published.

In one case, a native was seen to dart into the wood, and return, bearing a Cobra, about six feet in length, grasping it by the neck with the right hand and by the tail with the left. The Serpent was powerful, and struggled so hard, that its captor was forced to call for assistance. As, however, he held the reptile awkwardly, it contrived to get its head round, and to the horror of the spectators, fastened on his hand, retaining its hold for several seconds. The white bystanders at once gave up the man for lost, but his companion speedily produced from his waistband two Snake-stones, one of which he applied to each puncture. They clung firmly, seemed to absorb the flowing blood, and in a minute or two relieved the extreme pain which the man was already suffering. Presently both Snake-stones dropped simultaneously, and the man declared that the danger had then passed away.

Another native then took from his stores a little piece of white wood, passed it over the head of the Cobra, grasped it by the neck and put it into his basket, averring that when armed with this weapon, a man could handle any kind of Snake without being bitten.

A similar instance is related by Mr. Lavalliere, formerly District Judge of Kandy, and forwarded to Sir J. E. Tennent by the writer, together with the materials employed. The woody substances will presently be described; at present our business is with the Snake-stone, or Painboo-Kaloo as the natives call it.

The formation of these objects has long been a mystery, and they have been made into a very profitable article of commerce by those who possess the secret. The monks of Manilla are said to be the chief makers of Snake-stones, and to supply the merchants, by whom they are distributed throughout so many countries.

One of these stones was sent for analysis to Mr. Faraday, who pronounced it to be made of charred bone, and in all probability to have been filled with blood, and again charred. "Evidence of this is afforded, as well by the apertures of cells or tubes on its surface, as by the fact that it yields and breaks under pressure, and exhibits an organic structure within. When heated slightly, water rises from it and also a little ammonia, and if heated still more highly in the air, carbon burns away, and a bulky white ash is left, retaining the shape and size of the stone." This ash is composed of phosphate of lime, and Sir J. E. Tennent remarks, with much judgment, that the blood discovered by Mr. Faraday was probably that of the native to whom the Snake-stone was applied.

Another light has been thrown on the subject by Mr. R. W. H. Hardy, who states that the Snake-stone is in use in Mexico, and that it is formed by cutting a piece of stag's-horn into the proper shape, wrapping it tightly in grass or hay, folding it in sheet copper so as to exclude the air, and calcining it in a charcoal fire.

Being desirous of testing the truth of this recipe, I procured a piece of stag's-horn, cut it into proper shape, and exposed it to the heat of a fierce charcoal fire for an hour and a half. On removing it from the copper, the hay had been fused into a black mass, easily broken, and forming a complete cast of the inclosed horn, which fell out like an almond from its shell.

On comparing the charred horn with the veritable Snake-stones, I find them to be identical except in the polish. The fracture of both is the same, and when exposed to a white heat in the air, my own specimen burned away, leaving a white ash precisely as related of the real specimen, and the ashes of both are exactly alike, saving that my own is of a purer white than that specimen calcined by Mr. Faraday, which has a slight tinge of pink, possibly from the absorbed blood. On throwing it into water it gave out a vast amount of air from its pores, making the water look for a few seconds as if it were newly opened champagne, a peculiarity which agrees with Thunberg's description of the Snake-stone used at the Cape, and imported at a high price from Malabar. The rather high polish of the Cingalese Snake-stone I could not rightly impart to my own specimen, probably for want of patience. I found, however, that by rendering the surface very smooth with a file, and afterwards with emery paper, *before* exposing it to the fire, it could be burnished afterwards by rubbing it with polished steel. Even in the original objects, the polish is not universal, the plane side being much rougher than the convex.

We will now pass to the little pieces of woody substance, by which the natives assert that they hold dominion over the Serpent tribe. It has already been mentioned that the native who produced the Snake-stones, employed a small piece of wood as a charm to render the Snake harmless while he handled it. Mr. Lavalliere, in the course of his narration, remarks that the man who was bitten proceeded to bandage his leg above the wound, and to stroke it downwards with a piece of some root. I have also inspected the identical substances used in the two cases just narrated, and have come to the conclusion that no virtue resides in the particular plant from which the charm is taken, but the whole of its value lies in the confidence with which the possessor is inspired.

There are three specimens of charmed woods, all belonging to different plants. One is apparently a part of an aristolochia, another is so small and shrivelled that it cannot be identified, while the third, on being cut and tasted, proves to be nothing more or less than a piece of common ginger. This fact serves to establish the theory of Mr. Waterton, that there is no particular secret in Snake-charming, except the possession of confidence and unhesitating resolution.

ONE notable peculiarity in the Cobra is the expansion of the neck, popularly called the hood. This phenomenon is attributable, not only to the skin and muscles, but to the skeleton. About twenty pairs of the ribs of the neck and fore part of the back are flat instead of curved, and increase gradually from the head to the eleventh or twelfth pair, from which they decrease until they are merged into the ordinary curved ribs of the body. When the Snake is excited, it brings these ribs forward so as to spread the skin, and then displays the oval hood to best advantage. In this species, the back of the hood is ornamented with two large eye-like spots, united by a curved black stripe, so formed that the whole mark bears a singular resemblance to a pair of spectacles.

The native Indians have a curious legend respecting the origin of this mark, and their reverence for the reptile. One day when Buddha was lying asleep in the sun, a Cobra came and raised its body between him and the burning beams, spreading its hood so as to shade his face. The grateful deity promised to repay the favor, but forgot to do so. In those days the Brahminny kite used to prey largely on the Cobras, and worked such devastation among them, that the individual who had done Buddha the forgotten service ventured to remind him of his promise, and to beg relief from the attacks of the kite. Buddha immediately granted the request by placing the spectacles on the Snake's hood, thereby frightening the kite so much that it has never since ventured to attack a Cobra.

It is rather curious that many persons fancy that the Cobra loses a joint of its tail every time that it sheds its poison, this belief being exactly opposite to the popular notion that the rattlesnake gains a new joint to its rattle for every being which it has killed.

The color of this Serpent is singularly uncertain, and in the museums may be seen several specimens of each variety. In some cases the body is brownish-olive, and the spectacles are white, edged with black. Another variety is also brownish-olive, but covered with irregular cross-bands of black. The spectacles are remarkably bold, white, edged with black. Other specimens are olive, marbled richly with brown below. The spectacles are like those of the last variety. Sometimes a few specimens are found of a uniform brownish-olive without any spectacles; others are black with white spectacles, and others, again, black without spectacles. Even the number of rows in which the scales are disposed is as variable as the color. The specimens without spectacles seem to come from Borneo, Java, the Philippines, and other islands. The length of the Cobra di Capello is usually between three and four feet.

THE AFRICAN COBRA, or HAJE, is equally poisonous with its Asiatic relative. It is sometimes called SPUUGH-SLANGE, or Spitting-Snake, on account of its power of projecting the poisonous secretion to a distance. It effects this object by a sudden and violent expiration of the breath, and, if aided by the wind, will strike an object at the distance of several feet. Gordon Cumming mentions an instance of his suffering from the poison of this Serpent: "A horrid Snake, which Kleinberg had tried to kill with his loading-rod, flew up at my eye and spat poison in it. I endured great pain all night; the next day the eye came all right again." This short narrative was much ridiculed when the work first appeared, familiar as the existence of the Spitting-Snake has been to naturalists for many years.

The Haje is one of the fiercest among poison-bearing Snakes, seldom running from an adversary, but generally turning to fight, and not unfrequently beginning the attack. Generally, it moves slowly, but when angry, it darts at its foe, and strikes and spits with such rapid energy, that the antagonist stands in need of a quick hand and eye to conquer the furious reptile. It is a good climber, and is in the habit of ascending trees in search of prey. It is fond of water, and will enter that element voluntarily. While immersed, it swims well, but slowly, scarcely elevating its head above the surface.

In coloring it is one of the most variable of Snakes. Sometimes it is light yellow-brown, either of a uniform tint, or covered with irregular blotches. This is the variety shown in the colored illustration. Other specimens are black when adult, having, when young, a series of broad yellow bands on the fore part of the body. Another variety is black, with a grayish-white spectacle-like mark on the neck, and the fore part of the abdomen yellow, with some broad cross-bands. It is rather curious that the hood of the black specimens is not so wide as in the yellow and brown varieties. The length of the Haje is about five or six feet.

ONE of the brightest and loveliest of Serpents is the BEAD SNAKE of North America.

This beautiful little reptile inhabits the cultivated grounds, especially frequenting the sweet-potato plantations, and burrowing in the earth, close to the roots of the plants, so that it is often dug up by the negroes while getting in the harvest. It possesses poison-fangs, but is apparently never known to use them, permitting itself to be handled in the roughest manner, without attempting to bite the hand that holds it.

The colors of this Snake are bright, pure, and arranged in a manner so as to contrast boldly with each other. The muzzle and part of the head are black, the remainder of the head is golden-yellow, and the front of the neck jetty-black. A narrow band of golden-yellow with undulating edges comes next the black, and is followed by a broad band of the lightest carmine. From this point the whole of the body and tail are covered with narrow rings of golden-yellow, alternating with broad bands of carmine and jetty-black. Towards the tail the

AFRICAN COBRA, OR HAJE, AND GAZELLES.

carmine bands become paler and more of a vermilion hue, and for the last four inches there are no red bands, the black and yellow alternating equally. The extreme tip of the tail is yellow. The Bead Snake never attains any great size, seldom exceeding two feet in length.

It is very remarkable that the terrible LABARRI Snake of South America (*Elaps lemniscátus*) should be closely allied to and belong to the same genus as the bead Snake of the Northern States. Mr. Waterton states that this Serpent is fond of lying coiled on a stump of a tree or some bare spot of ground, where it can hardly be distinguished from the object on which it is reposing. The same writer remarks in a letter to me, that "the Labarri Snake has fangs, and is mortally poisonous when adult. It exhibits the colors of the rainbow when alive, but these colors fade in death. I have killed Labarri Snakes eight feet long."

We now arrive at a most curious family, known by the possession of very long poison-fangs, perforated, and permanently erect. They only include one genus, of which the best known species is the NARROW-HEADED DENDRASPIS (*Dendraspis angústiceps*).

This Serpent is very long, slender, and unusually active and a good climber, exceeding the haje in this accomplishment. It is found in Southern Africa, and is tolerably common at Natal. Its color is olive-brown washed with green above, and a paler green below. It is rather a large though very slender Snake, sometimes reaching the length of six feet.

The last example of the Serpent tribe is the ATRACTASPIS of Southern Africa (*Atractaspis irreguláris*). The fangs of this Snake are longer in proportion than those of any other known Serpent, reaching nearly to the angle of the mouth. They are so long, indeed, that Dr. Smith is of opinion that the creature cannot open its mouth sufficiently wide to erect the fangs fully, so that the poison-teeth are always directed backwards. They still, however, serve an important purpose; for when the Atractaspis seizes its prey, the poison-fangs necessarily pierce the skin, so as to inject the venom into the body of the victim, and from their shape act as grapnels, by which all attempts at escape are foiled. Very little is known of the habits of this Snake, but it is thought to burrow in loose ground.

The color of the Atractaspis is blackish-green above, shaded with orange-brown, and orange-buff below. It is a small Serpent, rarely measuring more than two feet in length.

THE BATRACHIANS.

FROGS AND TOADS.

The BATRACHIANS are separated from the true reptiles on account of their peculiar development, which gives them a strong likeness to the fishes, and affords a good ground for considering these animals to form a distinct order. On their extrusion from the egg, they bear no resemblance to their parents, but are in a kind of intermediate existence, closely analogous to the caterpillar or larval state of insects, and called by the same name. Like the fish, they exist wholly in the water, and breathe through gills instead of lungs, obtaining the needful oxygen from the water which washes the delicate gill-membranes. At this early period they have no external limbs, moving by the rapid vibration of the flat and fan-like tail with which they are supplied. While in this state, they are popularly called tadpoles, those of the frog sometimes bearing the provincial name of pollywogs. The skin of the Batrachians is not scaly, and in most instances is smooth and soft. Further peculiarities will be mentioned in connection with the different species.

These creatures fall naturally into two sub-orders—the leaping or tail-less Batrachians, and the crawling Batrachians. The leaping Batrachians, comprising the frogs and tóads, are familiar in almost all lands.

The tongue plays an important part in separating the frogs and toads into groups; and in the first group the tongue is altogether absent, these creatures being, in consequence, called Aglossa, or tongueless Batrachians.

The first of these creatures, the XÉNOPUS of Western and Southern Africa (*Dactylethra lævis*), is remarkable for possessing nails on its feet, the first three toes being tipped with a sharply-pointed claw or nail. The family is very small, comprising only one genus, and, as far as is known, two species. The color of the Xenopus is ashy-brown, veined with blackish-brown. It is rather a large species.

The celebrated SURINAM TOAD has long attracted attention, not for its beauty, as it is one of the most unprepossessing of beings, but for the extraordinary way in which the development of the young is conducted.

When the eggs are laid, the male takes them in his broad paws, and contrives to place them on the back of his mate, where they adhere by means of a certain glutinous secretion,

SURINAM TOAD.—*Pipa americana.*

and by degrees become embedded in a series of curious cells formed for them in the skin. When the process is completed, the cells are closed by a kind of membrane, and the back of the female Toad bears a strong resemblance to a piece of dark honey-comb, when the cells are filled and closed. Here the eggs are hatched; and in these strange receptacles the young pass through their first stages of life, not emerging until they have attained their limbs, and can move about on the ground.

The skin of this, as well as of other Batrachians, is separated from the muscles of the back, and allows room for the formation of the cells, being nearly half an inch thick. The full-sized cells are much deeper than long, and each would about hold a common horse-bean, thrust into it endways. The mouths of the cells assume an irregularly hexagonal form, probably because their original shape would be cylindrical, were they not squeezed against each other.

When the young have attained their perfect state, they break their way through the cover of the cells, and present a most singular aspect as they struggle from the skin, their heads and paws projecting in all directions. After the whole brood have left their mother's back, the cells begin to fill up again, closing from below as well as from above, and becoming irregularly puckered on the floors. The cells in the middle of the back are the first developed; the whole process occupies rather more than eighty days.

As its name implies, this singular creature inhabits Surinam, but is also found in various parts of Central America. In spite of its repulsive aspect, the negroes are said to eat its flesh.

The color of the Surinam Toad is brownish-olive above, and whitish below. The skin is covered with a large number of tiny and very hard granules, among which are interspersed some horny tubercular projections. The snout is of a very curious shape, the nostrils being lengthened into a kind of leathery tube. The throat of the male is furnished with a very large bony apparatus, of a triangular, box-like shape; and within are two movable pieces by which the voice is modulated.

In the illustration the animal appears one-half of its natural size.

WE now come to the Batrachians with tongues. In the greater number of these creatures, the tongue is fastened to the front of the mouth, and free behind, the tip pointing down the throat. The prey is taken by the rapid throwing forward of this tongue, and its equally rapid withdrawal into the mouth, carrying the doomed creature on its tip, with such celerity, that the eye can hardly follow the movement.

The skeleton of the adult Frog is worthy of a short notice before we proceed to the further investigation of these remarkable creatures. The first remarkable point is the shape of the head, and the enormous size of the orbits of the eyes, which are so large, that, when the skull is placed flat upon an open book, several words can be read through the orifices. Very little room is left for the brain, and, in consequence, the intellectual powers of the Frog are but slender.

The vertebræ are furnished with projections at each side, but the ribs are totally wanting. On account of this deficiency, the process of respiration cannot be maintained as is usual among the better developed beings, but is similar to that which is employed by the tortoises. The needful movements are made not by the sides but by the throat, so that if a quiescent Frog be watched, it appears to be continually gulping something down its throat, as is indeed the case, the material being air, which is thus forced into the beautifully formed lungs.

The hind-legs are extremely long, and the toes so much lengthened, that in the common Frog the middle toe occupies about three-fifths of the length of the entire body, and in some species is even more produced. Owing to the peculiar shape of the limbs, the Frog when reposing sits almost upright, and is at once ready for the extraordinarily long leaps which it can take when alarmed. The usual mode of progression is by a series of jumps, though of short range, but the creature will often crawl after the fashion of the toad—the presence of a snake seeming almost always to have the effect of causing the change of action.

The skin of the Frog is very porous, and is capable of absorbing and exuding water with wonderful rapidity. If a Frog, for example, be kept for some time in a perfectly dry spot, it loses its fine, sleek condition, becomes thin and apparently emaciated, and assumes a very pitable appearance. But if it be then placed merely on wet blotting-paper, its thirsty skin drinks the needful moisture, and it soon becomes quite plump and fresh. A familiar proof of the extreme porosity of the skin is afforded by the dead Frogs which are often found on the highroad or dry paths in the middle of summer, and which are dried into a shrivelled, horny mass, which would be shapeless but for the bones of the skeleton around which the skin and muscles contract.

The whole of these creatures are most tenacious of life, suffering the severest wounds without appearing to be much injured at the time, and bearing the extremes of cold and hunger with singular endurance. Heat, however, is always distasteful to the Frog, and when carried to any extreme becomes fatal. In the hot countries, where Frogs of various species exist, they all unite in the one habit of avoiding the hot beams of the sun by hiding in burrows or crevices during the day, and only emerging from their refuge in the night-time, or during rainy weather. Many species even dive below the muddy soil of pools as soon as the water has nearly disappeared, and there remain moist, torpid, and content until the next rains refill their home with the needful waters.

Most of the Frogs have a power of changing the color of the skin, which is often found to lose its brightest tints and become dark brown or nearly black in a very short space of time. Any sudden alarm will often produce this change, the presence of a snake being an almost unfailing means of effecting this object; and it is known that the color of the Frogs is greatly affected by the locality in which they are at the time placed. The Tree-Frogs are more subject to this change of color than the ordinary species; but even the common Frog is well known to alter from yellow to brownish-black in a very short space of time. This change is produced by some mental emotion acting upon certain masses of pigment or coloring matter in the skin; and for a further elucidation of the subject, I must refer the reader to my "Common Objects for the Microscope," where the pigment masses are drawn as seen through the microscope, and their peculiar action explained.

ONE of the most singular members of this group of animals is the PARADOXICAL FROG (*Pseudis paradoxa*).

This curious creature is a native of Surinam and South America, and is remarkable for the enormous size of the larva, or tadpole. As a general rule, and indeed, as might be expected, the generality of the batrachians are smaller in their larval than in their adult state; the tadpole of the common Frog being a good example. But the Paradoxical Frog exhibits a phenomenon which is perhaps found in none of the higher animals, though common enough among the non-vertebrated beings, and is less in its adult state than in its preliminary form of tadpole.

The tail of this tadpole is exceedingly voluminous, and the body has other envelopes or appendages, which, when thrown off as it proceeds to its perfect state, reduce the bulk so greatly that the earlier observers thought that the creature reversed the usual order of nature, and from a Frog became a tadpole. Some persons went even further, and said that it was changed from a Frog into a fish. The appropriate title of Paradoxical was given to it in allusion to this opinion.

Strange, however, as this phenomenon may appear, and remarkable as it undoubtedly is, it finds abundant parallels in the insects, where the larva is often of greater bulk than the perfect insect, or imago, as it is technically called. We may take for example the common silk-worm, where the caterpillar is extremely large when compared with the moth into which it afterwards changes; or that great, fat, bulky, subterranean grub which eats continually for three years, becomes so obese that it is forced to lie on its side, and afterwards turns into the neat, compact, and active little cockchaffer.

The color of the Paradoxical Frog is greenish, spotted with brown, and streaked irregularly with brown along its legs and thighs. The snout is tapering and rather pointed in front.

OUR next example of the Ranidæ is the AFRICAN BULL-FROG.

This fine species is spread over the whole of Southern Africa, but is found most plentifully towards the eastern coast, where it always frequents springs, pools, or the vicinity of fresh water. It is most impatient of drought, and when a more than usually dry season has parched the ground and rendered the hot soil uncomfortable for the delicate skin of the creature's feet and abdomen, these Frogs are said to congregate in the pools in great numbers, and just before the water has quite dried up, to burrow deeply into the soft mud and there lie until the next rains bring the welcome moisture.

Fifty of these large Frogs have been seen gathered together in one little pool, far from any other water. It is, moreover, evident that they must have some place of concealment, for they are sure to appear in great numbers after a few heavy rains, and it is quite consistent with probability that they should possess a simple and obvious method of preserving their lives during the frequent droughts of the climate in which they reside.

THE SHAD-FROG.

Dr. Livingstone mentions this fine species in his well-known work on Southern Africa, as follows:—

"Another article of which our children partook with eagerness was a very large Frog, called 'Matlamétlo.'

"These enormous Frogs, which, when cooked, look like chickens, are supposed by the natives to fall down from the thunder-clouds, because after a heavy shower the pools which are filled, and retain water a few days, become instantly alive with this loud croaking pugnacious game. This phenomenon takes place in the driest parts of the desert, and in places where to an ordinary observer there is not a sign of life.

"Having been once benighted in a district of the Kalahari, where there was no prospect of getting water for our cattle for a day or two, I was surprised to hear in the fine, still evening the croaking of Frogs. Walking out until I was certain that the musicians were between me and our fire, I found that they could be merry on nothing else but a prospect of rain.

"From the bushmen I afterwards learned that the Matlamétlo makes a hole at the root of certain bushes, and there ensconces himself during the months of drought. As he seldom emerges, a large variety of spider takes advantage of the hole, and makes its web across the orifice. He is thus furnished with a window and screen gratis, and no one but a bushman would think of searching beneath a spider's web for a Frog. They completely eluded any search on the occasion referred to; and as they rush forth into the hollows filled by the thunder-showers when the rain is actually falling, and the Bechuanas are cowering under their skin garments, the sudden chorus struck up simultaneously from all sides seems to indicate a descent from the clouds.

"The presence of these Matlamétlo in the desert in a time of drought was rather a disappointment, for I had been accustomed to suppose that the note was always emitted by them when they were chin-deep in water. Their music was always regarded in other spots as the most pleasant sound that met the ear after crossing portions of the sandy desert; and I could fully appreciate the sympathy for these animals shown by Æsop, himself an African."

It is a large and handsome species, but becomes duller in color as it increases in age. The young, however, are very lightly tinted. The general color is greenish-brown above, with a decided rusty wash, variegated with mottlings of reddish-brown, and streaked and spotted with yellow. The green takes a brighter and purer hue along the sides of the head and legs. The abdomen is yellow, mottled with orange, and the chin is striped and splashed with brown. The eyes are very curious and beautiful, being of a rich chestnut hue, covered with a profusion of little golden-white dots, which shine with a metallic lustre.

When young, the yellow lines on the body are edged with jetty-black, and the legs are covered with bold black bars. The head is stout and rather flat, and the skin of the body is puckered into longitudinal folds. The lower jaw is remarkable for two large, bony, tooth-like projections in front. The ordinary length of a full-grown specimen is about six inches.

WE now come to the very large genus of which the common Frog is so familiar an example, and which finds representatives in all except cold latitudes. The very handsome SHAD-FROG derives its popular name from its habit of making its appearance on land at the same time that the shads visit the shore. The specific title *halecína* also alludes to this circumstance, as the Indian word for a shad is halec.

This Frog requires much moisture, and is seldom seen at any distance from the banks of rivers or pools of fresh water. Sometimes, however, when the dew lies very heavily on the grass, the Shad-Frog makes its way over the fields to spots far from the water-side, but takes care to return before the hot sunbeams have dried up the grateful moisture of the herbage. The food of this reptile consists chiefly of insects. It is a very active creature, and ever lively, making leaps of eight or ten feet in length,

It is thought by many persons to rank among the handsomest of the froggish tribe. The general color is light golden-green, variegated with four rows of olive spots, edged with rich gold. One regular row of these spots runs along each side of the spine, and the others are scattered rather vaguely along the sides. The throat is white with a silvery lustre, and the abdomen whitish-yellow. The aural vesicles are brown, with a circular centre of azure-blue, and look like two little targets on the side of the head. The eyes are very large, of a beautiful golden lustre, and with a bold black streak drawn horizontally through their centre. The legs are exceedingly long in proportion to the size of the body, being five inches in length, whereas the body measures scarcely three inches. This length of limb and lightness of body adds greatly to the leaping powers, for which this creature is so celebrated.

The Shad-Frog is called also the Leopard Frog, as well as Shad-Frog in Pennsylvania. It inhabits North America generally.

THE GREEN FROG (*Rana clamitans*), called also Spring Frog, inhabits the United States east of the Rocky Mountains. It is a handsome species—bright green, spotted with black.

THE WOOD FROG (*Rana temporaria*) is a European species, common in all eastern portions of the United States. Two varieties are recognized.

THE SOLITARY SPADE-FOOT (*Scaphiopus*) is a curious form, inhabiting sparsely the eastern United States.

THE common TREE TOAD (*Hyla versicolor*) is very abundant and familiar in the eastern Middle States. Its length is about two inches. *H. pickeringii*, Pickering's Frog, is less common. *H. andersonii*, Anderson's Frog, is quite rare. Pickering's is about one inch in length.

THE TREE FROG (*Chorophilus triseriatus*) inhabits the eastern portion of the United States.

THE CRICKET FROG (*Acris gryllus*) has the same *habitat*. A variety, *crepetans*, is enumerated as a North American form—called the Western Cricket.

ANOTHER very common and very pretty Frog is abundant in the eastern United States. This is the PICKEREL FROG, so called because it enjoys a sad pre-eminence among anglers as a bait for pike, too fortunate if it can be snapped up at once by the voracious fish, instead of dangling for a season in mid-water, with a hook delicately inserted under its skin so as to keep it lively as long as possible, and prevent it from losing by death its attractive appearance.

It is mostly found in or near the salt marshes, and is remarkable for possessing a powerful and extremely disagreeable odor. In spite, however, of this seeming drawback, its flesh is said to be very delicate, and to be quite as good as that of the edible Frog of Europe.

The coloring of this species is very striking, on account of its irregularly squared aspect. The ground tint is pale brown above, covered with moderately large square spots of dark brown arranged like the stones of a tesselated pavement, and producing a somewhat regular pattern. A bright yellow line, not raised above the general surface, runs from behind each eye, and the under parts are yellowish-white. It is quite a little Frog, being under three inches in length.

UPON the accompanying oleograph is presented the figure of the celebrated BULL-FROG of America (*Rana catesbiana*), one of the largest and most conspicuous of its kind.

This enormous batrachian is perhaps the best swimmer among the Frog race, having been known to live for several years in water without any support for its feet. It leads a solitary life for the greater part of its existence, living in a hole near the water, and seldom leaving its domicile by day unless when suddenly alarmed. If frightened by an unknown sound or sight, the Bull-Frog leaps at once into the water, and instead of diving to the bottom immediately, skims along the surface for a few yards before it disappears.

During the breeding season, these huge Frogs assemble together in great multitudes, congregating to the amount of four or five hundred in some pool or marsh, sitting with their bodies half submerged, and making night hideous with their horrid bellowing cries. Few persons, except those who have had personal experience, and who have lost night after night of needful sleep by the ceaseless noise, can imagine the loudness of voice and variety of tone possessed by the different species of Frogs. And travellers who lie awake at night, unwilling hearers of the nocturnal concerts, are disposed to envy the happy ignorance of those whose calmer lot is cast in countries where the drummings, bellowings, chatterings, and pipings of the Frog race are practically unknown. Among these nightly musicians the Bull-Frog is the loudest and most pertinacious; mostly remaining quiet by day, but sometimes exulting in a black cloud or a heavy shower, and raising its horrid din even in the hours of daylight.

It is a most voracious creature, feeding mostly on snails and similar prey, which it catches on its nocturnal excursions from its domicile, but often devouring animals of a larger size, such as crayfish, two of which crustaceans have been found in the stomach of a single Bull-Frog, and even gobbling down an occasional chicken or duckling. Taking advantage of its voracity, the inhabitants of the country are in the habit of catching it by means of a rod and line. The hook is generally baited with an insect, and gently drawn along the ground near the Frog, which leaps upon it, seizes it, and is hooked without difficulty. It is rather curious that the Frog will not touch the insect as long as it is allowed to rest quietly on the ground, but as soon as the line is pulled, so as to make the insect move, it is at once pounced upon. The common Frogs and toads have the same custom.

The flesh of the Bull-Frog is very delicately flavored, and in some places the creature is kept in captivity and fed for table.

This species is exceedingly active, making leaps of eight or ten feet in length and five feet in height. There is a well-known story of a race between a Bull-Frog and an Indian, the former to have three jumps in advance, and the distance about forty yards, to a pond from which the Frog had been taken. When the parties were ready to start, the glowing tip of a burning stick was applied to the Bull-Frog, which set off at such a rate, and made such astonishing leaps to get into the welcome water, that its human opponent was vanquished in the race.

In some places this creature is never disturbed, as it is supposed, perhaps with some justice, to aid in keeping the water pure. The popular name of Bull-Frog is derived from its cry, which is said to resemble the bellowing of the animal whose name it bears. Several species of Frog have been classed under the same popular name.

The color of the Bull-Frog is brown, mottled with black above, and taking a greener hue upon the head. The abdomen is grayish-white, and the throat is white dotted with green. The length of the head and body of the large species is rather more than six inches, and a fine specimen will sometimes measure nineteen or twenty inches from the nose to the extremity of its feet. The skin of the back is smooth, and without any longitudinal fold.

THERE is another tolerably common species inhabiting the same country, which is also popularly called the Bull-Frog. It may be readily distinguished from the bull-Frog, which it otherwise greatly resembles, by the presence of a glandular fold on each side of the back. It is a very noisy creature, with a sharper and more yelping cry than the preceding species. When disturbed, it shoots at once into the water, and there sets up its peculiar cry. It is more active than the common bull-Frog, and if once released, is almost certain to escape, from the great length and rapidity of its leaps, the creature never seeming to pause between two jumps, but springing off the earth with an instantaneous rebound not unlike the flying leaps of the jerboa or kangaroo. It is a moisture-loving species, and is never found far from water.

WE now come to the best known of all the batrachians, the COMMON FROG.

The general form and appearance of this creature are too well known to need much description. It is found plentifully in all parts of Europe and America, wandering to considerable distances from water, and sometimes getting into pits, cellars, and similar localities, where it lives for years without ever seeing water. The food of the adult Frog is wholly of an animal character, and consists of slugs, possibly worms, and insects of nearly every kind, the wire-worm being a favorite article of diet. A little colony of Frogs is most useful in a garden, as they will do more to keep down the various insect vermin that injure the garden, than can be achieved by the constant labor of a human being.

The chief interest of the Frog lies in the curious changes which it undergoes before it attains its perfect condition. Every one is familiar with the huge masses of transparent jelly-like substance, profusely and regularly dotted with black spots, which lie in the shallows of a river or the ordinary ditches that intersect the fields. Each of these little black spots is the egg of a Frog, and is surrounded with a globular gelatinous envelope about a quarter of an inch in diameter. According to gipsy lore, rheumatism may be cured by plunging into a bath filled with Frog spawn.

On comparing these huge masses with the dimensions of the parent Frog, the observer is disposed to think that so bulky a substance must be the aggregated work of a host of Frogs. Such, however, is not the case, although the mass of spawn is forty or fifty times

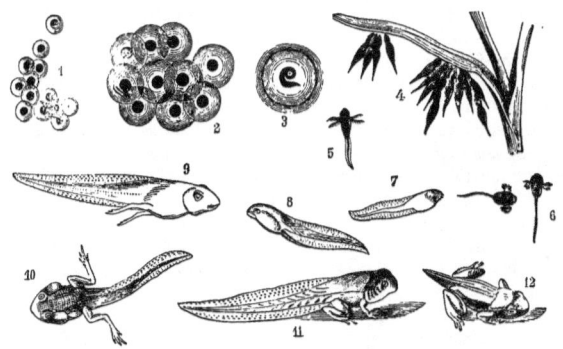

DEVELOPMENT OF THE EGG AND OF THE TADPOLE OF THE GREEN FROG.—1. Eggs just laid. 2. The same, a short while later. 3. Tadpole in the Egg. 4 and 5. Tadpoles just out of Egg. 6 to 12. Further development of the Tadpole, up to its last transformation.

larger than the creature which laid it. The process is as follows: The eggs are always laid under water, and when first deposited, are covered with a very slight but firm membranous envelope, so as to take up very little space. No sooner, however, are they left to develop, than the envelope begins to absorb water with astonishing rapidity, and in a short time the eggs are inclosed in the centre of their jelly-like globes, and thus kept well apart from each other.

In process of time, certain various changes take place in the egg, and at the proper period the form of the young Frog begins to become apparent. In this state it is a black grub-like creature, with a large head and a flattened tail. By degrees it gains strength, and at last fairly breaks its way through the egg and is launched upon a world of dangers, under the various names of tadpole, pollywog, toe-biter, or horsenail.

As it is intended for the present to lead an aquatic life, its breathing apparatus is formed on the same principle as the gills of a fish, but is visible externally, and when fully developed consists of a double tuft of finger-like appendages on each side of the head. The tadpole, with the fully developed branchiæ, is shown at Fig. 6, on the accompanying illustration. No sooner, however, have these organs attained their size than they begin again to diminish, the shape of the body and head being at the same time much altered. In a short time they entirely disappear, being drawn into the cavity of the chest, and guarded externally by a kind of gill cover. This is is shown in Figs. 7 and 8.

Other changes are taking place meanwhile. Just behind the head two little projections appear through the skin, which soon develop into legs, which, however, are not at all employed for progression, as the tadpole wriggles its way through the water with that quick undulation of the flat tail which is so familiar to us all. The creature then bears the appearance represented in Fig. 10.

Presently another pair of legs make their appearance in front, the tail is gradually absorbed into the body—not falling off, according to the popular belief—the branchiæ vanish, and the lungs are developed. Figs. 11 and 12 represent a young Frog in a state absorbing the tail.

The internal changes are as marvellous as the external. When first hatched, the young tadpole is to all intents and purposes a fish, has fish-like bones, fish-like gills, and a heart composed of only two chambers, one auricle and one ventricle. But in proportion to its age, these organs receive corresponding modifications, a third chamber for the heart being formed by the expansion of one of the large arteries, the vessels of the branchiæ becoming gradually suppressed, and their place supplied by beautifully cellular lungs, formed by a development of certain membranous sacs that appear to be analogous to the air-bladders of the fishes.

The Frog, contracted as are its intellectual powers, is yet susceptible to human influence, and can be tamed by kind treatment. Mr. Bell mentions a curious instance where one of these creatures became so completely domesticated, that it used to come nightly from a hole in the skirting-boards where it had established itself, partake of food offered to it by the members of the family, and even jump upon the hearth-rug in winter in order to enjoy the warmth of the fire. A favorite cat, which inhabited the same house, took a strange fancy to the Frog, and these seemingly incongruous companions were to be constantly seen sitting together on the hearth-rug, the Frog nestling under the soft warm fur of the cat. The Frog was, however, more than a year an inmate of the house before it became domesticated, and for many months would retreat to its stronghold when approached.

Stories of so-called "showers of Frogs" are often seen in the papers, and as a general rule are little to be credited, the solution of the supposed phenomenon being merely that a shower of rain has induced the creatures to come simultaneously from their retreats. There are, however, instances where credible spectators have seen them fall, and in such cases the little creatures were probably sucked up by a waterspout, or even by a brisk whirlwind, together with the water in which they were disporting, carried away for some distance, and at last dropped on the ground, as is sometimes the case with sticks, stones, and leaves, picked up by a passing whirlwind.

The general color of the common Frog is greenish-yellow, or brown, the same individual often passing through all these colors in a few days. A long patch of blackish-brown or warm brown is placed behind each of the eyes, and it is yellowish-white below. There are no teeth in the lower jaw, and only a single row of very tiny teeth in the upper jaw and on the palate. The ordinary length of the Frog is rather less than three inches, and the total length of the hinder leg is about four inches.

A VERY pretty species of this genus is found in Southern Africa. This is the BANDED FROG, remarkable for the beautiful stripes which adorn its body, and the inordinate length of the second toe of the hind-foot.

This pretty creature is not very plentiful in any one locality, but is spread widely throughout the Cape district and the whole of Southern Africa. It is very active, being a good leaper, and brisk in all its movements. The second toe of the hind-foot is truly remarkable. The whole of the toes are but slightly webbed, and project boldly beyond the connecting membrane; but the second toe is nearly as long as the whole body, which is longer than in the generality of Frogs. The object of this exceeding development is not very clear.

The general color of this species is wood-brown, upon which are drawn six dark streaks, the two centre stripes running nearly the entire length of the body. The hinder part of the thigh is orange-brown, and the under parts are yellowish-white. The length of the head and body is nearly two inches.

THE celebrated EDIBLE FROG, or GREEN FROG of Europe (*Rana esculenta*), also belongs to this large genus. This handsome species is common in all the warmer parts of that Continent, but in the vicinity of large cities is seldom seen, except in the ponds where it is preserved, and whence issues a horrid nocturnal concert in the breeding-time. The proprietors of these froggeries supply the market regularly, and draw out the Frogs with large wooden rakes as they are wanted. In Paris these creatures are sold at a rather high price for the table, and as only the hind-legs are eaten, a dish of Frogs is rather an expensive article of diet.

It is needful to make a very early visit to the market, four or five A. M. being about the best time, to see the manner in which the Frogs are brought to market. They are generally sold by women, each of whom has by her side two tubs or barrels, one containing living Frogs, and the other having a leather band nailed to the side, in which is stuck a sharp, broad-bladed knife. When the purchaser has bargained for a certain number, the seller plunges her left hand into the one barrel, brings out a Frog by its legs, lays it across the edge of the second barrel, and with a single cut of the knife, severs the hind-legs just above the pelvis, leaving the whole of the body and fore-quarters to fall into the tub. The hind-legs are then carefully skinned, and dressed in various ways, that with white sauce seeming to be the best, at all events according to my own taste. They require considerable cooking, but when properly dressed have a most delicate and peculiar flavor, which has been compared, but not very happily, to the wing of a chicken. I would suggest that a mixture of the smelt and the breast of the spring chicken would convey a good idea of the Edible Frog when cooked.

Poachers are very apt to invade the froggeries, and without entering the boundaries often contrive to kidnap a goodly number of the inmates by a very curious mode of angling, something like "bobbing" for eels. They get a very long fishing-rod, tie a line of sufficient length to the tip, and at the end of the line they fasten, in place of a hook and bait, a simple piece of scarlet cloth. Thus prepared, they push the rod over the fence, let the scarlet rag just touch the surface of the water, and shake the rod so as to make the rag quiver and jump about. The Frog, thinking that it has found a very savory morsel, leaps at the rag, closes its mouth firmly upon it, and is neatly tossed out of the water and over the hedge before it can make up its mind to loosen its hold.

The color of this species is bright green spotted with black, and having three bold yellow stripes along the back. The under parts are yellowish. In size it is rather larger than the common species.

THE remarkable HORNED FROG is one of the quaintest species among the Frog tribe.

There are several species belonging to this genus, all inhabiting Southern America, and all notable for the singular development of the upper eyelids, which are prolonged into hard, horn-like points. In the present species the back is furnished with a bony shield, and the prominences over the eyes are bold and well defined. The body is short, stout, and squat, the skin covered with tubercles and folds, and the opening of the mouth enormous. It is a large and voracious species, one specimen when opened being found to have swallowed a full-grown land-Frog (*Cystignathus fuscus*), belonging to the same genus as our next example. The toes are long, powerful, and with hardly a vestige of web except just at the base.

THE little ORNATE LAND-FROG affords a remarkable contrast to the last-mentioned species on account of its small dimensions, the activity of its movements, and the beauty of its coloring.

It is found in Georgia and South Carolina, and is always seen on land and dry spots, its thirsty frame being amply supplied by the dews and casual rains without needing immersion in water. Indeed, this Frog is so little conversant with the element usually so familiar to all its tribe, that if thrown into water, it makes no attempt to swim, but lies helplessly sprawling on the surface. On land, however, it displays wonderful activity, being of an extremely lively nature, and making long and bold leaps in rapid succession, so that it is not to be captured without considerable difficulty.

The color of this species is rather variable, but is generally of a soft dove tint, on which are placed several oblong marks of deep rich brown edged with golden-yellow. Below it is silvery-white granulated with gray. It is a very little species, measuring only one inch and a quarter when full-grown.

ANOTHER species of this genus, the SENEGAL LAND-FROG (*Cystignathus senegalensis*), inhabits Southern Africa.

It resides in burrows in the ground, and is tolerably quiet, except before rain or on a dull day, when it begins to pipe, and continues its curious cry for several hours together. The voice of this Frog is a sharp piping whistle several times repeated. Dr. A. Smith relates that he was greatly puzzled on hearing this strange whistling sound, and made many a fruitless search after the utterer. At last one of the Hottentots showed him the animal in its burrow, and after that time he was able to procure as many as were desired.

The head and body of this species are short, puffy, and smooth, and the color is yellowish-gray, with three longitudinal bands. Below, it is yellowish-white without any mottlings. Its length is about two inches.

THE pretty PAINTED FROG is a European species, being found in Greece, Sicily and Sardinia. It has a rather wide range of locality, as it is not uncommon in Northern Africa, along the banks of the Nile, and is tolerably plentiful along the shores of the Mediterranean.

It is fond of water, but seems careless whether it be salt or fresh, and is found indifferently in rivers, streams, lakes, and the saline morasses. The common esculent Frog possesses similar habits, and the two species are often seen in company. The food of the Painted Frog consists of insects, spiders, slugs, and snails, both terrestrial and aquatic. There is a difference in the web of the toes in the sexes, those of the female being scarcely webbed at all, while in the male the membrane extends to half their length. The thumb is quite rudimentary, and its place is indicated by a small tubercular projection.

The color and general aspect of the skin are extremely variable, the difference seeming to be quite capricious, and not depending on sex or locality. The ground color is usually yellowish-green or olive, decorated with spots and having several white longitudinal streaks. In some specimens the skin is smooth, while in others it is covered with tubercles, and the spots are seldom alike in two individuals, sometimes running together so as to form continuous bands. The white lines too are often partially, and sometimes wholly absent. In this species the male does not possess any vocal sacs.

THE reader will remember that in the description of the Surinam Toad, on page 150, mention was made of the curious manner in which the female carries her eggs upon her back until they have passed through their preliminary stages of existence. A noteworthy analogy, close in some respects, but failing singularly in others, is to be found in the NURSE FROG of Europe (*Alytes obstétricans*).

In this species it is the male that undergoes the anxieties of watching over the young offspring, his mate being comparatively free from that duty.

When the eggs, about sixty in number, are laid, he takes possession of them, and fastens them to his legs by means of a glutinous substance, and carries them about with him wherever he goes. In process of time, the eggs swell, and become so transparent that the black eyes of the future young are seen through their envelopes. Their careful parent then proceeds to some spot where he can find still water, deposits them, and departs, rejoicing in his freedom. The young soon burst their way through the envelopes in which they had been surrounded, and swim off merrily.

Except at such times, the Nurse Frog is seldom seen in the vicinity of water, and even at that season, the creature does not care to swim about, or even to enter the water. The color of this species is olive-brown with small dark spots.

THE very odd-looking species which is popularly and appropriately termed the SOLITARY FROG is a native of North America, and is remarkable for several peculiarities of form, the eye and the foot being chiefly notable.

It is a land-loving species, never seen in or near water except during the breeding-season. During the greater part of the year it resides in holes which it scoops in the sandy soil, and at the bottom of which it sits watching for prey, much like a gigantic ant-lion. In order to assist it in digging, the animal is furnished with a flat, sharp-edged spur, with which it scoops out the loose soil. Sometimes, however, it wedges itself into the sand, tail foremost, and shovels its way downwards much after the fashion of the crab. The hole is about six inches in depth.

Quick though it is in this labor, it is but a sluggish and inactive creature when compared with most of its kin, being a very poor leaper, and slow in most of its movements. It is generally to be seen in the month of March, just after the spring rains, and is a very hardy species, caring little for cold, and traversing the snow without apparent inconvenience.

The eye of the Solitary Frog is very beautiful, and at the same time most remarkable. It is large, full, and of a rich topaz hue, and across its centre run two bold black lines at right angles to each other, so as to form a cross very like that which is seen upon starch grains when viewed by polarized light.

Altogether, the aspect of this species is very unique. It looks much more like a toad than a frog, and has a remarkably blunt snout. Its general color is olive, mottled with brown above, and covered with tubercles. Along each side of the spine runs a line of "king's yellow," and the under parts are yellowish-white. The average length of the Solitary Frog rather exceeds two inches.

THE last of the true Frogs which can be mentioned in this work is the BOMBARDIER (*Bombinátor igneus*), a native of many parts of Europe, and common in France.

It is fond of water, and seldom found in very dry localities. When disturbed, it has the power of emitting a strong and very unpleasant odor of garlic, which serves it as a means of defence, like the penetrating scent of the common ringed snake. It is active, and can both swim and leap well. The eggs are laid in long strings, and the tadpole is of a very large size when compared with the earliest state of its perfect existence, and, like the paradoxical Frog already described, is larger in the tadpole state than after it has assumed its perfect form.

The color of the Bombardier is grayish-brown above, and orange below, marbled or spotted with blue-black.

WE now arrive at another section of Batrachians, including those creatures which are known under the title of TOADS, and of which the COMMON TOAD of Europe is so familiar an example. The members of this section may be known by the absence of teeth in the jaws and the well-developed ears.

The general aspect and habits of this creature are too well known to require more than a cursory notice. Few creatures, perhaps, have been more reviled and maligned than the Toad, and none with less reason. In the olden days, the Toad was held to be the very compendium of poison, and to have so deadly an effect upon human beings, that two persons were related to have died from eating the leaf of a sage bush under which a Toad had burrowed. Still, even in those times, it was held to possess two virtues, the one being the celebrated jewel supposed to be found in its head, and the other the power of curing bleeding at the nose.

This jewel could not be procured by dissection, but must be obtained by causing the owner to eject it. "But the art," says one of the quaint old writers, "is in taking of it out, for they say it must be taken out of the head alive before the Toad be dead, with a piece of cloth of the color of red Scarlet, wherewithal they are much delighted, so that while they stretch out themselves as it were in sport upon that cloth, they cast out the stone of their head, but instantly they sup it up again, unless it be taken from them through some secret hole in the said cloth, whereby it falleth into a cistern or vessel of water into which the Toad dareth not enter, by reason of the coldnesse of the water. . . . . The probation of this Stone is by laying of it to a live Toad, and if she lift up her head against it, it is good, but if she run away from it, it is a counterfeit."

The same writer gives, in his own racy language, an account of the use to which even so venomous an animal as a Toad may be put by those who know how to employ the worst things for the best purposes. "Frederic, the Duke of Saxony, was wont to practis in this manner. He had ever a Toad pierced through with a piece of wood, which Toad was dryed in the smoak or shadow, this he rowled in a linnen cloth; and when he came to a man bleeding at the nose, he caused him to hold it fast in his hand until it waxed hot, and then would the bloud be stayed. Whereof the Physitians could never give any reason, except horrour and fear constrained the bloud to run into his proper place, through fear of a Beast so contrary to humane nature. The powder also of a Toad is said to have the same vertue."

For these and other similar opinions too numerous for mention, there is some little foundation. The skin of the Toad's back is covered thickly with little glands, and some larger glands are gathered into two sets, one at each side of the back of the head, and secrete a liquid substance, with sufficient acridity to make the eyes smart should they be touched with this fluid, and to force a dog to loose his hold, if he should pick up a Toad in his mouth, and run away with open jaws and foaming mouth. The glands at the back of the head secrete a large quantity of liquid, and if pressed, will eject it in little streams to the distance of a few inches.

In France, this poor creature is shamefully persecuted, the idea of its venomous and spiteful nature being widely disseminated and deeply rooted. The popular notion is that the Toad is poisonous throughout its life, but that after the age of fifty years it acquires venomous fangs like those of the serpents. I once succeeded, but with great difficulty, in saving the life of a fine fat Toad that was leisurely strolling in the Forest of Meudon and had got into a rut too deep for escape. I had stooped down to remove the poor creature from danger, but was dragged away by the by-standers, who quite expected to see me mortally bitten, and who proceeded to slaughter the Toad on the spot. "Every one kills Toads in France," said they.

Hearing from them, however, that tobacco was instantaneously fatal to Toads, I made a compromise that they might kill it by putting tobacco on it, but in no other way. The experiment was accordingly tried, and I had the pleasure of seeing the creature walk away with the tobacco on its back, quite unconscious that it ought to have been dead. One of the spectators not only insisted upon the quinquegenarian fangs, but averred that he had a pair at home in a box. However, I never could induce him to show them to me.

In point of fact, the Toad is a most useful animal, devouring all kinds of insect vermin, and making its rounds by night when the slugs, caterpillars, earwigs, and other creatures are abroad on their destructive mission. Many of the market-gardeners are so well aware of the extreme value of the Toad's services, that they purchase Toads at a certain sum per dozen, and turn them out in their grounds.

Dull and apathetic as the Toad may seem, it has in it an affectionate and observant nature, being tamed with wonderful ease, and soon learning to know its benefactors and to come at their call. Mr. Bell had one of these creatures, which was accustomed to sit on one hand and take its food out of the other. Many persons have possessed tame Toads, which would leave their hiding-place at the sound of a whistle or a call, and come hastily up to receive a fly, spider, or beetle. Toads can be rendered useful even in a house, for they will wage unceasing war against cockroaches, crickets, moths, flies, and other insect pests.

It is worthy of notice, that the Toad will never catch an insect or any other prey as long as it is stationary, but on the slightest movement, the wonderful tongue is flung forward, picks up the fly on the tip, and returns to the throat, placing the morsel just in the spot where it can be seized by the muscles of the neck, and passed into the stomach. So rapidly is the act performed, that Mr. Bell has seen the sides of a Toad twitching convulsively from the struggles of a beetle just swallowed, and kicking vigorously in the stomach.

Entomologists sometimes make a curious use of the Toad. Going into the fields soon after daybreak, they catch all the Toads they can find, kill them, and turn the contents of their stomachs into water. On examining the mass of insects that are found in the stomach, and which are floated apart in the water, there are almost always some specimens of valuable insects, generally beetles, which from their nocturnal habits, small dimensions, and sober coloring, cannot readily be detected by human eyes.

The Toad will also eat worms, and in swallowing them it finds its fore-feet of great use. The worm is seized by the middle, and writhes itself frantically into such contortions that the Toad would not be able to swallow it but by the aid of the fore-feet, which it uses as if they were hands. Sitting quietly down with the worm in its mouth, the Toad pushes it further between the jaws, first with one paw and then with another, until it succeeds by alternate gulps and pushes to force the worm fairly down its throat.

These paws are also useful in aiding it to rid itself of its cuticle, which is shed at intervals, as is the case with many reptiles and Batrachians. The process is so singular, and so admirably described by Mr. Bell, that it must be given in his own words:—

"I one day observed a large Toad, the skin of which was particularly dry and dull in its color, with a light streak down the mesial line of its back; and on examining further, I discovered a corresponding line along its belly. This proved to arise from an entire slit in the old cuticle, which exposed to view the new and brighter skin underneath. Finding, therefore, what was going to happen, I watched the whole detail of this curious process.

"I soon observed that the two halves of the skin, thus completely divided, continued to recede farther and farther from the centre, and became folded and rugose; and after a short space, by means of the continued twitching of the animal's body, it was brought down in folds on the sides. The hinder leg, first on one side and then on the other, was brought forward under the arm, which was pressed down upon it, and on the hinder limb being withdrawn, its cuticle was left inserted under the arm, and that of the anterior extremity was now loosened, and at length drawn off by the assistance of the mouth. The whole cuticle was thus detached, and was now pushed by the two hands into the mouth in a little ball, and swallowed at a single gulp. I afterwards had repeated opportunities of watching this curious process, which did not materially vary in any instance."

Though apparently unfit for food, the Toad is eaten by some nations, and certainly is not more unprepossessing than the iguana. The Chinese, however, are in the habit of eating a species of Toad for the purpose of increasing their bodily powers, thinking that the flesh of this creature has the property of strengthening bone and sinew.

This animal is extremely tenacious of life, and is said to possess the power of retaining life for an unlimited period if shut up in a completely air-tight cell. Many accounts are in existence of Toads which have been discovered in blocks of stone when split open, and the inference has been drawn that they were inclosed in the stone while it was still in the liquid state, some hundreds of thousands of years ago, according to the particular geological period, and had remained without food or air until the stroke of the pick brought them once more to the light of day.

Such an account appears at once to be so opposed to all probability as to challenge a doubt; but if there had been sufficient testimony, even to one such fact, an unprejudiced thinker would be justified in placing it among the wonderful but veritable occurrences that occasionally startle mankind. But there really seems to be no account which is sufficiently accurate to permit of such a conclusion. In more than one case, the whole story has proved to be nothing more than an imposition; and in others, there is hardly sufficient evidence to show that some crevice did not exist, which would supply the inclosed animal with sufficient air for its narrow wants, and permit many minute insects to crawl into the cavity which held the imprisoned Toad.

There is no doubt that in many cases a little Toad has crept into a rocky crevice after prey or in search of a hiding-place, and by reason of its rapid increase in size been unable to make its exit. As, moreover, the creature is very long lived, it would, by frequent movements, give a polish to the walls of its cell in a few years; a circumstance that has been employed as a proof of the antiquity of the Toad and its residence. Similar instances are known where the animal has been found inclosed in timber. Here, however, is less difficulty in accounting for the fact, because the growth of wood over a wounded part is often extremely rapid, and has been known to cause the inclosure of nails, tools, and even birds' nests with their eggs. Even in such a case, there is not sufficient evidence to prove that the closure was absolutely perfect, and that the Toad was hermetically sealed in the wooden walls of its cell.

Dr. Buckland made some experiments on this supposed property of the Toad, and inclosed a number of these creatures in artificial chambers, made to represent as nearly as possible the rock and wood in which the imprisoned Toads have been found. None of these experiments met with success; and in those cases where the Toads lived longest, the plaster was found imperfect. Some of the Toads whose cells were really air-tight died in a month or two.

It may, however, be reasonably urged that such experiments do not fairly represent the original conditions under which an animal could survive for so long a period, and that in order to carry out the experiment in a consistent manner, the Toads ought to have been procured when very young, inclosed in a chamber with a moderate aperture, and that aperture lessened gradually, so as to prepare the creature by degrees for its long fast and deprivation of air. For a good summary of this subject and a collection of almost every narrative, I may refer the reader to Mr. Gosse's "Romance of Natural History," second series.

The development of the Toad is much like that of the Frog, except that the eggs are not laid in masses, but in long strings, containing a double series of eggs placed alternately. These chains are about three or four feet in length, and one-eighth of an inch in diameter. They are deposited rather later than those of the Frog, and the reptiles, which are smaller and blacker than the Frog larvæ, do not assume their perfect form until August or September. The general color of the Toad is blackish-gray with an olive tinge, and the tubercles which stud the surface are brown. Beneath, it is yellowish-white, tinged with gray, and in some specimens spotted with black. The full size of the Toad is not well ascertained, as it seems to have almost unlimited capacities for increasing in size together with years. The length of a very large specimen is about three inches and a half.

The American Toad (*Bufo lentiginosus*) is exceedingly common in most parts of the United States. This species is peculiar in that it varies in several respects. An average specimen is about three inches in length, and two in breadth. Its general appearance is sufficiently familiar.

This humble appearing, and to some vicious, but perfectly harmless reptile, or batrachian, is regarded by American agriculturists as a valuable agent in suppressing certain damaging insects. Five varieties are known. Sixty-seven species and varieties of Frogs and Toads are known in North America.

ANOTHER species of Toad, the NATTERJACK, is found in many parts of Europe. It may be known from the common species by the short hind-legs, the more prominent eyes, the less webbed feet, the yellow line along the middle of the back, and the black bands on the legs. It is not so aquatic as the common Toad, haunting dry places, and seldom approaching water except during the breeding season. Its ordinary length is about three inches.

THE GREEN or VARIABLE TOAD (*Bufo viridis*, or *variábilis*) is rather a handsome species, and is found plentifully in the South of France. It derives its popular names from the large spots of deep green with which its upper surface is adorned. Many of the Batrachians possess the capability of changing their hues according to locality or through mental emotion, and the Green Toad is extremely conspicuous in this respect, wearing different colors in light and shade, sleep and wakefulness.

THE WARTY TOAD of Fernando Po (*Bufo tuberósus*) is a singular looking species, remarkable for the extreme development of the hard tubercles on the back, and being among Batrachians analogous to the moloch among lizards, or the porcupine among mammalia. The whole upper surface of the body is thickly covered with large tubercles, each having a horny spine in the centre. The glands on the back of the head are large and very conspicuous. Even the under parts are covered with tubercles, but without the spine in the centre. Above each eyelid is a group of horny tubercles, so that the creature presents a most remarkable appearance. Its length is about three inches.

Our last example of these creatures is the large AGUA TOAD of America (*Bufo agua*).
This large species digs holes in the ground, and resides therein. It is one of the noisiest of its tribe, uttering a loud snoring kind of bellow by night and sometimes by day, and being so fond of its own voice that even if taken captive it begins its croak as soon as it is placed on the ground. It is very voracious, and as it is thought to devour rats, has been imported in large numbers from Barbadoes into Jamaica, in order to keep down the swarm of rats that devastate the plantations. When these creatures were first set loose in their new home, they began to croak with such unanimous good-will that they frightened the inhabitants sadly, and caused many anxious householders to sit up all night.

This Toad grows to a great size, often obtaining a length of seven inches, and nearly the same measurement in breadth. It may be recognized by the great enlargement of the bone over the eyes, and the enormous dimensions of the glands behind the head. Its color is extremely variable.

We now come to the Tree-Frogs, or Tree-Toads, so called from their habits of climbing trees, and attaching themselves to the branches or leaves by means of certain discs on the toes, like those of the geckos. In the first family the toes are webbed, and the processes of the vertebræ are cylindrical. A good example will be found in the SAVANNAH CRICKET FROG of America.

SAVANNAH CRICKET FROG.—*Acris gryllus.*

This species is very common in its own country, and is found throughout a very large range of territories, specimens having been taken from several Northern and Southern States of America. It is a light, merry little animal, uttering its cricket-like chirp with continual reiteration, even in captivity. Should it be silent, an event sometimes greatly to be wished, it can at any time be roused to utterance by sprinkling it with water. It is easily tamed, learns to know its owner, and will take flies from his hand.

This species frequents the borders of stagnant pools, and is frequently found on the leaves of aquatic plants and of shrubs that overhang the water. It is not, however, possessed of such strongly adhesive powers as the true Tree-Frogs, and is unable to sustain itself on the under side of a leaf. It is very active, as may be surmised from the slender body and very long hind-legs, and, when frightened, can take considerable leaps for the purpose of avoiding the object of its terror.

The color of this species is greenish-brown above, diversified by several large oblong spots edged with white, and a streak of green, or sometimes chestnut, which runs along the spine and divides at the back of the head, sending off a branch to each eye. The legs are banded with dark-brown, and the under surface is yellowish-gray with a slight tinge of pink. It is but a little creature, measuring only an inch and a half in length.

Another species (*Hyla carolinensis*) is sometimes called by the same popular title, because its voice, like that of the preceding species, bears some resemblance to that of a cricket. Being one of the true Tree-Frogs, it is not a frequenter of the water, but proceeds to the topmost branches of trees, and there chirps during the night.

ANOTHER family, containing the well-known Tree-Frog of Europe, has the toes webbed, and the processes of the vertebræ flattened. The best-known species is the common GREEN

Tree-Frog of Europe, now so familiar from its frequent introduction into fern-cases and terrestrial vivaria.

This pretty creature is mostly found upon trees, clinging either to their branches or leaves, and being generally in the habit of attaching itself to the under side of the leaves, which it resembles so strongly in color, that it is almost invisible even when its situation is pointed out. When kept in a fern-case, it is fond of ascending the perpendicular glass sides, and there sticking firmly and motionless, its legs drawn closely to the body, and its abdomen flattened against the glass.

The food of the Tree-Frog consists almost entirely of insects, worms, and similar creatures, which are captured as they pass near the leaf whereto their green foe is adhering. It is seldom seen on the ground except during the breeding season, when it seeks the water, and there deposits its eggs much in the same manner as the common Frog. The tadpole is hatched rather late in the season, and does not attain its perfect form until two full months have elapsed. Like the Toad, the Tree-Frog swallows its skin after the change. The common Tree-Frog is wonderfully tenacious of life, suffering the severest wounds without seeming to be much distressed, and having even been frozen quite stiff in a mass of ice without perishing.

GREEN TREE-FROG.—*Hyla arborea*.

The following interesting account of a young Tree-Frog is by Mr. G. S. Ullathorne :—

"My acquaintance with this interesting reptile (which had already passed through all the stages of the tadpole state) began in the following manner :—

"I was at school in Hanover at the time, and used frequently to take walks in the neighboring woods, with a companion. During one of these walks we came across three Green Frogs (or rather they came across our path). Guessing at once they were Tree-Frogs, and thinking that they were just the things to keep, we were 'down upon them,' and tied them up in our handkerchiefs. I contented myself with one, and let my companion have the others. When I arrived safely at my journey's end with my Frog, I procured for him a good-sized glass jar, put a little water in the bottom, a branched stick for him to climb up (though he generally preferred the sides of the glass), covered the top of the jar with a piece of muslin, and installed him on a shelf with a salamander (*Salamandra maculosa*), a ring snake (*Natrix torquata*), and various other 'pets.'

"My great amusement was to watch the little creature eat. When I put a fly into his jar, as long as the fly remained quiet, the Frog took no notice of it, but directly the fly began buzzing about, the Frog would wake up from his lethargic state, and on a suitable opportunity would make a leap at the poor fly, adroitly catch it in his mouth (though he sometimes missed his mark), and, I need hardly add, swallow it. On one occasion, I gave my little favorite a very large 'blue-bottle,' almost as large as himself, but nothing daunted, he caught it in his mouth and endeavored to swallow it, though in vain, for had I not been there I verily believe he would have been choked.

"Before he changed his skin, which he did now and then, his color became much darker and looked more dirty, and he went into quite a torpid state, but when the event was over, he appeared greener and livelier than ever. One day, after I had had him some time, I was playing upon the pianoforte, when I was astonished by an extraordinary sound, but on looking round I discovered the cause of the great noise, for there was my Frog swollen to an immense extent under the chin, and croaking in a very excited manner, making quite a loud noise. I mention this circumstance because it has been imagined that a *solitary* Tree-Frog will not croak, but mine certainly proved to the contrary, for though the first croaking was evidently the effect of the piano, yet he would frequently croak after that time without being excited by any apparent noise whatever. I may here mention that the noise of a quantity of Frogs croaking and nightingales singing, has frequently kept me awake for a considerable time during a spring night.

"And now comes the most melancholy part of my story. Leaving my Frog carelessly on the window-sill, I went to school; when I came back there was the glass certainly, and the Frog also, but oh! distressingly melancholy to relate, the water was quite hot from the intense heat of the sun, and the poor Frog was scorched, or rather boiled to death—he was quite discolored, being instead of green, a sort of yellow. And thus ends my tale."

The color of this species is green above, sometimes spotted with olive, and a grayish-yellow streak runs through each eye towards the sides, where it becomes gradually fainter, and is at last lost in the green color of the skin. In some specimens there is a grayish spot on the loins. Below, it is of a paler hue, and a black streak runs along the side, dividing the vivid green of the back from the white hue of the abdomen.

THE CHANGEABLE TREE-TOAD is a native of many parts of America, being found as far north as Canada, and as far south as Mexico. It is a common species, but owing to its faculty of assimilating its color to the tints of the object on which it happens to be sitting, it escapes observation, and is often passed unnoticed in spots where it exists in great numbers.

This is a curious and noteworthy species, as it possesses the capability of changing its tints to so great an extent that its true colors cannot be described. It is usually found on the trunks of trees and old moss-grown stones, which it so nearly resembles in color, that it can hardly be detected, even when specially sought. The skin of this creature will, in a short time, pass from white through every intermediate shade to dark-brown, and it is not an uncommon event to find a cross-shaped mark of dark-brown between the shoulders. Old and decaying plum-trees seem to be its favorite resting-places, probably because the insects congregate on such trees,

It is a noisy creature, especially before rain, and has a curious liquid note, like the letter *l* frequently repeated, and then ending with a sharp, short monosyllable. During the breeding season, this Frog leaves the trees and retires to the pools, where it may be heard late in the evening. In the winter it burrows beneath the damp soil, and there remains until the spring. The contour of this species is very toad-like in shape and general appearance, and this resemblance is increased by the skin glands, which secrete a peculiarly acrid fluid.

The upper surface of this creature is, as has already been remarked, too variable for description. There is always, however, a little bright yellow on the flanks, and the under surface is yellowish-white, covered with large granulations. The length of this species is about two inches.

IN the POUCHED FROG we find a most singular example of structure, the female being furnished with a pouch on her back, in which the eggs are placed when hatched, and carried about for a considerable period.

This pouch is clearly analogous to the living cradle of the marsupial animals. It is not merely developed when wanted, as is the case with the cells on the back of the Surinam Toad, but is permanent, and lined with skin like that of the back. The pouch does not attain its full development until the creature is of mature age, and the male does not possess it at all. When filled with eggs the pouch is much dilated, and extends over the whole back nearly as far as the back of the head. The opening is not easily seen without careful examination, being very narrow, and hidden in folds of the skin.

Its color is very variable, but green has the predominance. It is found in Mexico, but many specimens have been brought from the Andes of Ecuador.

A VERY curious species, called the LICHENED TREE-TOAD (*Trachycéphalus lichenátus*), inhabits Jamaica, and is described by Mr. Gosse in his "Naturalist's Sojourn" in that island.

It derives its name from the aspect of the head, which looks as if it was overgrown with lichens. It is generally found among the wild pine trees, and is very active, being able to take considerable leaps. Sometimes it puffs out its body, and causes a kind of frothy moisture to exhale from the skin. This moisture adheres to the fingers like gum, and causes the Frog to leave a trail behind it like that of a snail or slug.

The color of the Lichened Tree-Toad is pale red mottled with brown, and having a large patch of the same color between the shoulders. The muzzle and sides are pale green, spotted with dark reddish-brown, and below it is whitish-gray, the chin being speckled with reddish-brown. The head is flattened, sharply pointed at the muzzle, and studded with sharp bony ridges. Its ordinary length is about four inches.

ANOTHER species of the same genus, the MARBLED TREE-TOAD (*Trachycéphalus marmorátus*), is described by the same writer:—

"One of them was taken in a bedroom at Savannah-le-Mar, one night in October, having probably hopped in at the open window from the branches of a mango tree only a few feet distant. I was surprised at its change of color, in this respect resembling the chameleon and anoles, or still nearer, the geckos.

"When I obtained it, the whole upper parts were of a rich deep amber-brown, with indistinct black bands. On looking at it at night, to my surprise I saw a great alteration of hue. It was paler on the head and back, though least altered there; on the rump and on the fore and hind legs it was become a sort of semi-pellucid drab, marked with minute close-set dark specks. When disturbed, it presently became slightly paler still, but in a few minutes it had recovered its original depth of tint. In the course of half an hour it displayed again the speckled dark hue, and now uniformly so, save a black irregular patch or two on the head, and a dark patch between the mouth and each eye. The belly, which was very regularly shagreened, was of a dull buff, not susceptible of change. Its eyes retained their proverbial beauty, for the irides were of a golden-brown tint, like sun-rays shining through tortoise-shell.

"This specimen was about as large as a middling English Frog, being two inches and a quarter in length.

"While in captivity, if unmolested, it spent a good deal of time motionless, squatting flat and close, with shut eyes, as if sleeping, but sometimes it was active. I kept it in a basin covered with a pane of glass, for facility of observation. It would keep its face opposite the window, altering its position pertinaciously if the basin were turned, though ever so gently. It took no notice of cockroaches, nor of a large flesh-fly which buzzed about it, and even crawled over its nose. If taken in the hand, it struggled vigorously, so as to be with difficulty held; once or twice, while thus struggling, it uttered a feeble squeak; but if still retained, it would at length inflate the abdomen with air, apparently a sign of anger. It leaped, but not far."

A VERY odd-looking species is the BLUE FROG. It is the sole representative of a family, remarkable for having webbed toes, flattened processes of the vertebræ, and glands at the back of the head.

The Blue Frog, as it is called from its hue, inhabits Australia, and is not uncommon at Port Essington, whence several specimens have been brought to Europe. The head of this species is broader than long, the muzzle short and rounded, and the gape very large. The secreting glands at the back of the head are large, and extend in a curve over the ear as far as the shoulder. They are pierced with a large number of pores, and by their shape and dimensions give to the creature a very singular aspect. The discs of the fore-feet are extremely large, and the toes of the hind-feet are about three-quarters webbed. The color of the Blue Frog is light, uniform blue above, and below silvery-white. Its length is about three inches and a half.

THE large and handsome BICOLORED TREE-FROG is the only species at present known as belonging to the family.

In this creature the toes are not webbed, but in other respects the form resembles that of the preceding family, except, perhaps, that the processes of the vertebræ are wider in proportion to their volume. The Bicolored Tree-Frog inhabits South America, Brazil, and Guiana, and seems to be tolerably common. Possibly its bright and boldly contrasting colors render it more conspicuous than its green and olive relatives. The popular name of this creature is very appropriate, as the whole of the upper parts are intense azure, and the under parts pure white, or white tinged with rose. The thighs and sides are spotted with the same hue as the abdomen.

PASSING over the small section of Frogs (*Micrhylina*) distinguishable by their toothed jaws and imperfect ears, and represented by a single species, we come to the third section of these animals (*Hylaplesúra*), known by their toothless jaws and perfectly developed ears. Of this section, the TWO-STRIPED FROG affords a good example.

This species is a native of Southern Africa, and is chiefly found in the eastern and northeastern parts of the colony of Cape Town. It lives almost entirely upon or in trees, and may be seen either in the cavities of a decaying trunk, or clinging to the bark in close proximity to one of these holes.

In Dr. A. Smith's "Illustrations of the Zoology of Southern Africa," there is so curious and important an account of the imprisonment of this species in the bole of a tree, that it must be given in his own words :—

."On the banks of the Limpopo River, close to the tropic of Capricorn, a massive tree was cut down to obtain wood to repair a wagon. The workman, while sawing the trunk longitudinally, nearly along its centre, remarked on reaching a certain point—'It is hollow, and will not answer the purpose for which it is wanted.'

"He persevered, however, and when a division into equal halves was effected, it was discovered that the saw in its course had crossed a large hole, in which were five specimens of the species just described, each about an inch in length. Every exertion was made to discover a means of communication between the external air and the cavity, but without success. Every point of the latter was probed with the utmost care, and water was left in each half for a considerable time, without any passing into the wood. The inner surface of the cavity was black, as if charred, and so was likewise the adjoining wood for half an inch from the cavity.

"The tree, at the part where the latter existed, was nineteen inches in diameter, the length of the trunk was eighteen feet; the age, which was observed at the time, I regret to say, does not appear to have been noted. When the Batrachia above mentioned were discovered, they appeared inanimate, but the influence of a warm sun, to which they were subjected, soon imparted to them a moderate degree of vigor. In a few hours from the time they were liberated, they were tolerably active, and able to move from place to place, apparently with great ease."

The color of this species is deep liver-brown above, with two longitudinal yellow stripes, beginning at the eyes and extending as far as the base of the hind-legs. A forked yellow mark appears between these stripes just where they end, and the limbs are liver-brown, spotted with yellow. The under parts are very pale brownish-red profusely variegated with pale yellow spots In length it is nearly two inches. The generic name Brachymerus is derived from two Greek words, signifying short-thighed.

THE TINGEING FROG of Southern America (*Hylaplesia tinctória*) is worthy of a casual notice.

This creature is so called because the Indians are said to employ it for imparting a different tinge to the plumage of the green parrot. They pluck out the feathers on the spots where they desire to give the bird a different colored robe, and then rub the wounded skin with the blood of this Frog. The new feathers that supply the places of those that have been removed, are said to be of a fine red or yellow hue.

It is found in various parts of Southern America, and is common in Surinam, where it mostly inhabits the woods, traversing the branches and leaves by day, and at night concealing itself under the loose bark. Like the common Tree-Frog of Europe, it seldom visits the water except during the breeding-season, for the purpose of depositing its eggs.

In color it is extremely variable. Some specimens are black, with a white spot on the top of the head, and two stripes of the same color running from the head along each side. In certain individuals there are cross bands of white between the stripes. Other examples are gray above and black below; some are wholly black, spotted with large round white marks; others are black; others are gray, spotted with black; while a few specimens are brown, with a large white spot on each side, and two white bands on the fore limbs.

THE RHINOPHRYNE is remarkable as being the only known example among the Frogs where the tongue has its free end pointing forward, instead of being directed towards the throat.

This curious species inhabits Mexico, and can easily be recognized by the peculiar form of its head, which is rounded, merged into the body, and has the muzzle abruptly truncated, so as to form a small circular disc in front. The gape is extremely small, and the head would, if separated, be hardly recognizable as having belonged to a Frog. There are two glands by the ears, but although they are of considerable dimensions, they are scarcely apparent externally, being concealed under the skin. The legs are very short and thick, and the feet are half-webbed. Each hind-foot is furnished with a flat, oval, horny spur formed by the development of one of the bones. There are no teeth in the jaws, and the

RHINOPHRYNE. *Rhinophryne dorsalis.*

ear is imperfect. The color of the Rhinophryne is slate-gray, with yellow spots on the sides and a row of similar spots along the back. Sometimes these latter spots unite so as to form a jagged line down the back.

THE CRAWLING BATRACHIANS.

WE now arrive at the Crawling Batrachians, technically called Amphibia Gradientia. All these creatures have a much elongated body, a tail which is never thrown off as in the frogs and toads, and limbs nearly equal in development, but never very powerful. Like the preceding sub-order, the young are hatched from eggs, pass through the preliminary or tadpole state, and, except in a very few instances, the gills are lost when the animal attains its perfect form. Both jaws are furnished with teeth, and the palate is toothed in some species. The skin is without scales, and either smooth or covered with wart-like excrescences. There is no true breast-bone, but some species have ribs.

The development of the young from the egg is not quite the same as that of the tailless Batrachians. Instead of being deposited in masses or long strings, the eggs are laid singly, and are hatched in succession. When the young are first hatched they bear some resemblance to the tadpole of the frog, the gills being very conspicuous. In these creatures, however, the fore-legs make their appearance first, and are soon followed by the hinder pair, whereas in the frogs the hind-legs are seen for some time before the fore-limbs are visible externally. Further remarks will be made on this subject when we come to the well-known representative of this sub-order, the common newt or eft.

SALAMANDER.—*Salamandra maculosa.*

THE celebrated SALAMANDER, the subject of so many strange fables, is a species found in many parts of the continent of Europe.

This creature was formerly thought to be able to withstand the action of fire, and to quench even the most glowing furnace with its icy body. It is singular how such ideas should have been so long promulgated, for although Aristotle repeated the tale on hearsay, Pliny tried the experiment, by putting a Salamander into the fire, and remarks, with evident surprise, that it was burned to a powder. A piece of cloth dipped in the blood of a Salamander was said to be unhurt by fire, and certain persons had in their possession a fire-proof fabric made, as they stated, of Salamander's wool, but which proved to be asbestos.

Another fable related of this creature still holds its ground, though perhaps with little reason. I have already mentioned one or two instances of the prejudices which are so deeply ingrained in the rustic mind, and given a short account of the superstitions prevalent in France regarding toads. The Salamander there suffers an equally evil reputation with the toad, as may be seen by the following graphic and spirited letter:—

"Returning homeward a few evenings ago from a country walk in the environs of D——, I discovered in my path a strange-looking reptile, which, after regarding me steadfastly for a few moments, walked slowly to the side of the road, and commenced very deliberately clambering up the wall. Never having seen a similar animal, I was rather doubtful as to its properties; but, reassured by its tranquil demeanor, I put my pocket-handkerchief over it, and it suffered itself to be taken up without resistance, and was thus carried to my domicile. On arriving *chez moi*, I opened the basket to show my captive to the servants (French), when, to my surprise and consternation, they set up such a screaming and hullabaloo, that I thought they would have gone into fits.

"'Oh! la, la, la, la, la!—Oh! la, la, la, la, la!' and then a succession of screams, in altissimo, which woke up the children, and brought out the neighbors to see what could be the matter.

"'Oh, monsieur a rapporté un sourd!'

"'Un sourd!' cried one.

"'Un sourd!' echoed another.

"'UN S-O-U-R-D!!!' cried they all in chorus; and then followed a succession of shrieks.

"When they calmed down into a mild sample of hysterics, they began to explain that I had brought home the most venomous animal in creation.

"'Oh! le vilain bête!' cried Phyllis.

"'Oh! le méchant!' chimed in Abigail; 'he kills everybody that comes near him; I have known fifty people die of his bite, and no remedy in the world can save them. As soon as they are bitten they *gonflent, gonflent*, and keep on swelling till they burst, and are dead in a quarter of an hour.'

"Here I transferred my curiosity from the basket to a glass jar, and put a saucer on the top to keep it safe.

"'Oh! Monsieur, don't leave him so; if he put himself in a rage, nothing can hold him. He has got such force, that he can jump up to the ceiling; and wherever he fastens himself he sticks like death.'

"'Ah! it's all true,' cried my landlady, joining the circle of gapers. 'Oh! la, la! Ça me fait peur; ça me fait tr-r-r-r-embler!'

"'Once I saw a man in a hay-cart try to kill one, and the *bête* jumped right off the ground at a bound and fasten itself on the man's face, when he stood on the hay-cart, and nothing could detach it till the man fell dead.'

"'Ah! c'est bien vrai,' cried Abigail; 'they ought to have fetched a mirror and held it up to the *bête*, and then it would have left the man and jumped at its *image*.'

"The end of all this commotion was that, while I went to inquire of a scientific friend whether there was any truth in these tissue of *bêtises*, the whole household was in an uproar, *tout en émoi*, and they sent for a *commissionnaire* and an ostler with a spade and mattock, and threw out my poor *bête* into the road, and foully murdered it, chopping it into a dozen pieces by the light of a stable lantern; and then they declared that they could sleep in peace!—*les misérables!*

"But there were sundry misgivings as to my fate, and as with the Apostle, 'they looked when I should have swollen or fallen down dead suddenly; and next morning the maids came stealthily and peeped into my room to see whether I was alive or dead, and were not a little surprised that I was not even *gonflé*, or any the worse for my *rencontre* with a *sourd*.

"And so it turned out that my poor little *bête* that had caused such a disturbance was nothing more nor less than a Salamander—a poor, inoffensive, harmless reptile, declared on competent authority to be no ways venomous; but whose unfortunate appearance and somewhat Santanic livery have exposed it to obloquy and persecution."

This notion of the poisonous character of the Salamader is of very old date, as the reader may see by referring to any ancient work on Natural History. One of the old writers advises any one who is bitten by a Salamander to betake himself to the coffin and winding-sheet, and remarks that a sufferer from the bite of this animal needs as many physicians as the Salamander has spots. If the Salamander crawled upon the stem of an apple-tree, all the crop of fruit was supposed to be withered by its deadly presence, and if the heel of a man should come in contact with the liquid that exudes from the skin, all the hair of his head and face would fall off.

There is certainly an infinitesimally minute atom of truth in all this mass of absurdities, for the Salamander does secrete a liquid from certain pores in its surface, which, for the moment, would enable it to pass through a moderate fire, and this secretion is sufficiently acrid to affect the eyes painfully, and to injure small animals if taken into the mouth.

The Salamander is a terrestrial species, only frequenting the water for the purpose of depositing its young, which leave the egg before they enter into independent existence. It is a slow and timid animal, generally hiding itself in some convenient crevice during the day, and seldom venturing out except at night or in rainy weather. It feeds on slugs, insects, and similar creatures. During the cold months it retires into winter-quarters, generally the hollow of some decaying tree, or beneath mossy stones, and does not reappear until the spring.

The ground color of this species is black, and the spots are light yellow. Along the sides are scattered numerous small tubercles.

THE YELLOW SALAMANDER (*Amblystoma xiphias*), called also Desmognath, inhabits the Alleghanies. The Dusky Salamander (*A. obscurum*) inhabits from Ohio to Massachusetts, and southward, and is one of the commonest species in our springs and brooks. The Black Salamander is the largest of the Eastern species, inhabiting from Pennsylvania southward.

RED-BACKED SALAMANDER (*Plethodon cinereus*) is common in the Eastern States. A variety is noticed with no red dorsal band. The Viscid Salamander (*P. glutinosus*) is chiefly terrestrial; like the preceding, inhabits the same localities.

TWO-STRIPED SALAMANDER (*Spelerpes bilineatus*), called Cave Salamander and Green's, inhabits from Maine to Wisconsin, and southward. *S. longicaudus* abounds in the caves from Maine to Kentucky. The Red Triton (*S. ruber*), inhabits from Maine to Nebraska, and southward.

THE PURPLE SALAMANDER (*Gyrinophilus porphyriticus*), a large aquatic species, inhabits the Alleghany Mountain region. It is said to be the only Salamander that exhibits any attempt at self-defense, the others being too sluggish.

THE common NEWT, ASKER, EFFET, EFT, or EVAT, as it is indifferently termed, is well known throughout Europe. At least two species of Newt inhabit the northern parts of Europe, and some authors consider that the number of species is still greater. According to the system employed in this work, we accept only two species, the others being merely noted as varieties.

THE CRESTED NEWT derives its popular name from the membranous crest which appears on the back and upper edge of the tail during the breeding-season, and which adds so much to the beauty of the adult male.

This creature is found plentifully in ponds and ditches, during the warm months of the year, and may be captured without difficulty. It is tolerably hardy in confinement, being easily reared even from a very tender age, so that its habits can be carefully noted.

I had some of these animals in a large slate tank through which water was constantly running, and which was paved with pebbles, and furnished with vallisneria and other aquatic plants, for the purpose of imitating as nearly as possible the natural condition of the water from which the creatures had been taken. Here they lived for some time, and here the eggs were hatched and the young developed.

It was a very curious sight to watch the clever manner in which the female Newts secured their eggs; for which purpose they used chiefly to employ the vallisneria, its long slender blades being exactly the leaves best suited for that purpose. They deposited an egg on one of the leaves, and then, by dexterous management of the feet, twisted the leaf round the egg, so as to conceal it, and contrived to fasten it so firmly that the twist always retained its form. The apparent shape of the egg is oval, and semi-transparent; but on looking more closely, it is seen to be nearly spherical, of a very pale yellow-brown, and inclosed within an oval envelope of gelatinous substance.

When the young Newt is hatched, it much resembles the common tadpole, but is of a lighter color, and its gills are more developed. It rapidly increases in size, until it has attained a length of nearly two inches, the fore-legs being then tolerably strong, and the hinder pair very small and weak. The gills are at this time most beautiful objects; and if the young creature be properly arranged under the microscope, the circulation of the blood, as seen through their transparent walls, is one of the most exquisite sights that the microscope can afford.

The legs now attain greater strength, the gills become gradually more opaque and slowly lessen in size, being at last entirely absorbed into the body. In exact proportion to the diminution of the gills, the lungs increase in size; and the animal undergoes exactly the same metamorphosis as has already been related of the frog, being changed, in point of fact, from a fish into a batrachian. The tail, however remains, and is made the principal, if, indeed, not the only means by which the Newt propels itself through the water.

When it has passed through its changes, the Newt is no longer able to lead a sub-aquatic life, but is forced to breathe atmospheric air. For this purpose it rises to the surface at tolerably regular intervals, puts its snout just out of the water, and, with a peculiar little popping sound, ejects the used air from its lungs and takes in a fresh supply.

Towards the breeding-season, the male changes sensibly in appearance; his colors are brighter, and his movements more brisk. The beautiful waving crest now begins to show itself, and grows with great rapidity, until it assumes an appearance not unlike that of a very thin cock's comb, extending from the head to the insertion of the hinder limbs, and being deeply toothed at the edge. The tail is also furnished with a crest, but with smooth edges. When the animal leaves the water, this crest is hardly visible, because it is so delicate that it folds upon the body and is confounded with the skin; but when supported by the water, it waves with every movement of its owner, and has a most graceful aspect.

After the breeding-season, the crest diminishes as rapidly as it arose, and in a short time is almost wholly absorbed. Some remnants of it, however, always remain, so that the male may be known, even in the winter, by the line of irregular excrescences along the back. The use of this crest is not known, but it evidently bears a close analogy to the gorgeous nuptial plumage of many birds, which at other times are dressed in quite sober garments.

The Newt feeds upon small worms, insects, and similar creatures, and may be captured by the simple process of tying a worm on a thread by the middle, so as to allow both ends to hang down, and then angling as if for fish. The Newt is a ravenous creature, and when it catches a worm, closes its mouth so firmly that it may be neatly landed before it looses its hold. Some writers recommend a hook; but I can assert, from much practical experience, that the hook is quite needless, and that the Newt may be captured by the simple worm and thread, not even a rod being required.

It is curious to see the Newt eat a worm. It seizes it by the middle with a sudden snap, as if the jaws were moved by springs, and remains quiet for a few seconds, when it makes another snap, which causes the worm to pass farther into its mouth. Six or seven such bites are usually required before the worm finally disappears.

The skin or epidermis of the Newt is very delicate, and is frequently changed, coming off in the water in flakes. I found that my own specimens always changed their skin as often as

I changed the water; and it was very curious to see them swimming about with the flakes of transparent membrane clinging to their sides. The skin of the paws is drawn off just like a glove, every finger being perfect, and even the little wrinkles in the palms being marked. These gloves look very pretty as they float in the water, but if removed they collapse into a shapeless lump.

The food of the Newt consists of worms, insects, and even the young of aquatic reptiles. I have seen a large male Crested Newt make a savage dart at a younger individual of the same species, but it did not succeed in eating the intended victim.

This creature is very tenacious of life, and the muscular irritability of the body seems to endure for a long time after the creature is dead. One of these animals, that had been dead for some time, whose heart and lungs had been removed, and whose limbs had been pinned out ready for dissection, was so retentive of this singular irritability, that when the tail was touched with the point of a scalpel, the body and limbs writhed so actively as to free the limbs from their attachments. On repeating the experiment, it was found that this susceptibility gradually departed, lingering longest towards the body. The eel possesses an even greater degree of this muscular irritability, as is well known by all who have made an eel-pie or seen it prepared. The tail of the blind-worm, too, which has already been described, is equally irritable when separated from the body.

The color of the Crested Newt is blackish or olive-brown, with darker circular spots, and the under parts are rich orange-red, sprinkled with black spots. Along the sides are a number of white dots, and the sides of the tail are pearly-white, becoming brighter in the spring. The length of a large specimen is nearly six inches, of which the tail occupies rather more than two inches and a half.

The STRAIGHT-LIPPED NEWT of Mr. Bell (*Triton bibronii*) is only ranked as a variety of this species. In this variety the upper lip does not overhang the lower, and the skin is more tubercular than in the ordinary examples.

The MARBLED NEWT (*Triton marmorátus*) is a continental species, and is found plentifully in the southern parts of France.

It is a much larger species than the preceding, often attaining the length of eight or nine inches. It mostly lives in the water, but will leave that element voluntarily when the weather is stormy, or even if the hot sunbeams are too powerful to please its constitution. A rather powerful and not very pleasant odor is exhaled from this creature. During the winter it leaves the water, seeks for some hole in a decaying tree, and there remains until the following spring. The color of the Marbled Newt is olive-brown above, marbled with gray and dotted with white on the back. The head is gray, with black dots and spots. Along the centre of the back runs a streak of white and orange, and the under parts are dotted with white.

The SMOOTH NEWT is more terrestrial in its habits than the crested species, and is often seen at considerable distances from water.

By the rustics this most harmless creature is dreaded as much as the salamander in France, and the tales related of its venom and spite are almost equal to those already mentioned. During a residence of some years in a small village, I was told some very odd stories about this Newt, and my own powers of handling these terrible creatures without injury was evidently thought rather supernatural. Poison was the least of its crimes, for it was a general opinion among the rustics in charge of the farm-yard that my poor Newts killed a calf at one end of a farm-yard, through the mediumship of its mother, who saw them in a water-trough at the other end; and that one of these creatures bit a man on his thumb as he was cutting grass in the church-yard, and inflicted great damage on that member.

The worst charge, however, was one which I heard from the same person. A woman, he told me, had gone to the brook to draw water, when an Effert, as he called it, jumped out of the water, fastened on her arm, bit out a piece of flesh, and spat fire into the wound, so that she afterwards lost her arm.

All the Newts possess singular powers of reproducing lost or injured members, this faculty proving them to hold a rather low place in the scale of creation. The Smooth Newt has been known to reproduce the tail, and even the limbs ; and in one case an eye was removed entirely, and reproduced in a perfect state by the end of the year.

This species may be known by its smooth and non-tubercular skin, and its small size. During the breeding-season the male wears a crest, which runs continuously from the head to the end of the tail, and is not so deeply cleft as that of the crested species.

This ornament is very delicate and beautiful, and at the height of the season is often edged with beautiful carmine or violet. The color is brownish-gray above and bright orange below, covered with round spots of black. In the autumn and during the winter, the abdomen becomes much paler. The length of this species is about three inches and a half.

THE PALMATED WATER NEWT of Mr. Bell (*Lissotriton pálmipes*) is held to be merely a variety of this species.

WE now arrive at another family, known by the curious manner in which the teeth of the palate form a broken cross-series.

The first example is the JAPANESE SALAMANDER (*Onychodáctylus japónicus*), remarkable for having, during the larval state and in the breeding-season, claws upon the toes. Its color is purplish-black, variegated irregularly with white, and the claws are black. It is thought by the natives to possess medical properties, and they employ its flesh in sundry ailments, killing, and drying it in the sun for better preservation.

ANOTHER example of this family is the AMBLYSTOME, or SPOTTED EFT, of North America.

This species is not uncommon in the countries which it inhabits, and is found in some numbers in Pennsylvania. The eggs of this creature are not deposited singly and in the water, as is the case with the newts, but are laid in small packets, and placed beneath damp stones. The head of the Amblystome is thick, convex, and with the muzzle rounded. Its color is deep violet-black above, and purple-black below, with a row of circular or oval yellow spots along the sides. These spots are large in proportion to the dimensions of the individual, and have a very bold effect. The genus is rather large, containing about eleven acknowledged species. One of them, *Amblystoma talpoideum*, or Mole-like Amblystome, derives its name from its habit of burrowing in the ground after the fashion of the mole. It lives in South Carolina, and is found on the sea-islands. The fore-limbs are peculiarly short and stout, and the body is rather thick and clumsily made.

It is found northward as far as Illinois. There are eight other species enumerated, found, respectively, in Ohio, New Jersey, and the Southern States. Specimens kept in the New York Aquarium during the year 1878, passed through the usual and various stages of transformation. The tail was first noticed to be growing gradually smaller, and on absorption of the branchiæ, the transformation was seen to be complete. The more delicate and comely Salamander, with proper lungs, and its body prettily decorated with round spots, was the perfect and permanent form. This process of change was clearly visible in the well-arranged tanks of the Aquarium. The larva state of one species is the celebrated Axolotl of Mexico, and Lake Como one of the western territories.

One genus and nineteen species are recorded as embraced under the family *Amblystomidæ*, all found in North America. They are particularly abundant in the South and West.

WE now come to a very remarkable creature, the AXOLOTL, which is presumed to be but the larva or tadpole state of some very large batrachian. Like many other enigmatical animals, it has been bandied about considerably in the course of investigation, and, according to the latest observations, the original opinion seems to be correct, namely, that it is not an adult crawling batrachian with perpetual gills, but that it is in its preliminary or tadpole stage of existence. Mr. Baird makes the following sensible remarks on this subject:—

"It so much resembles the larva of *Amblystoma punctatum*, in both external form and internal structure, that I cannot but believe it to be the larva of some gigantic species of this genus. It differs from all other perennibranchiates in possessing the larval character of the gular or opercular flap, this being unattached to the adjacent integuments, and free to the extremity of the chin. The non-discovery of the adult is no argument against its existence. I had caught hundreds of the very remarkable larva of *Pseudotriton salmoneus* before I found an adult. Until then I knew nowhere to refer the animal, supposing this species to exist no nearer than the mountains of New York and Vermont."

LARVA OF AXOLOTL.

As may be seen from the illustrations, the gills or branchiæ are quite as large in proportion as those of the newt in its larval state. They are furnished with fringes.

The Axolotl inhabits Mexico, where it is tolerably plentiful, and in some places is found in such numbers that it is sold in the markets for the table. It frequents the lake surrounding the city of Mexico, and, according to Humboldt, is also found in the cold waters of certain mountain lakes at a considerable elevation above the sea.

The color of this remarkable creature is rather dark grayish-brown, covered thickly with black spots. The length varies from eight to ten inches.

ANOTHER small order now comes before us, containing a few species, and only two very small families. In all these creatures the body is long and lizard-like, the legs four and feeble, and the gills internal, but permanent throughout life.

OUR first example of this family is the now celebrated GIGANTIC SALAMANDER.

This is undoubtedly one of the least attractive of the vertebrate animals, being dull in habits, sombre in color, with a sort of half-finished look about it, and not possessing even that savage ugliness which makes many a hideous creature attractive in spite of its uncomeliness. It is a native of Japan, and even in that country seems to be rare, a large sum being asked for it by the seller. It lives in the lakes and pools that exist in the basaltic mountain ranges of Japan.

Dr. Von Siebold brought the first living specimen to Europe, and placed it in a tank at Leyden, where it was living when the last accounts were heard, having thus passed a period of many years in captivity. Its length is about a yard. Two specimens were brought over at the same time, being of different sexes, but on the passage, the male unfortunately killed and ate his intended bride, leaving himself to pass the remainder of his life in celibacy. It fed chiefly on fish, but would eat other animal substances.

Another fine specimen attracted much notice in spite of its ugliness and almost total want of observable habits. It is very sluggish and retiring, hating the light, and always squeezing itself into the darkest corner of its tank, where it so closely resembles in color the rock-work near which it shelters itself, that many persons look at the tank without even discovering its presence. The length of this specimen is about thirty-three inches, and if it survives, it may possibly attain even a larger size. The specimen shown in the engraving on next page is reduced to one-fifth of its natural size.

AXOLOTL.—*Axolotles guttatus.*

The head of this creature is large, flattened, and very toad-like in general aspect, except that it is not furnished with the beautiful eyes which redeem the otherwise repulsive expression of the toad. The head is about four inches wide at the broadest part, and is covered with innumerable warty excrescences. The eyes are extremely small, placed on the fore part of the head, and without the least approach to expression, looking more like small glass beads than eyes.

The whole upper part of the body is covered thickly with excrescences, and even the under part of the rounded toes are studded with little tubercles, which can be plainly seen with a magnifying lens as the creature presses its feet against the glass wall of its tank. Despite of its sluggish nature, it is quite able to obtain its own subsistence by catching the fish on which it feeds, and the keeper told me that even in captivity it easily catches the fish that are put into its tank. On the journey, it was mostly fed upon eels, and at the present time it eats eels as well as other fish, provided they are rather small.

It is well to mention casually in this place that the human-looking skeleton, discovered at Œningen in 1726, and long supposed to be the fossil skeleton of a man who had perished in the deluge, is nothing more than the bones of a huge Salamander, closely allied to the present species. The color of the Gigantic Salamander is a very dark brown, with a tinge of chocolate, and taking a lighter and more yellowish hue upon the under surface of the feet.

THE MENOPOME.

THE great MENOPOME of America (*Menopoma alleghaniense*) has been honored with a large array of names, among which are TWEEG, HELLBENDER, MUD DEVIL, and GROUND PUPPY, the first being an Indian name, and the others given to the creature in allusion to its mud-loving habits or the ferocity of its disposition.

The Menopome inhabits the Ohio and Alleghany rivers, and it is a fierce and voracious animal, so dangerous a foe to fish and other living beings that it is in some places known by

GIGANTIC SALAMANDER.—*Cryptobranchus maximus.*

the name of Young Alligator. It is very ugly, and rather revolting in appearance, so that the fishermen stand in great awe of the fierce, active beast, and think it to be venomous as well as voracious. The teeth, however, are very small in proportion to the size of the creature. Its color is slaty-gray, with dark spots, and a dark streak runs through the eye. Its length is about two feet.

THE SIREN.   179

It is also called Big Water Lizard by the inhabitants along the Ohio and other interior portions. This as well as the other members of the group is harmless, though seemingly ferocious and venomous. Specimens were kept in the New York Aquarium, and much additional knowledge was thereby gained of its habits.

THE second family of this order is represented by its typical species, the CONGO SNAKE.

This curious creature is a native of America, and is found rather plentifully near New Orleans, in Florida, Georgia, and South Carolina. It is fond of burrowing in mud, and will often descend to a depth of three feet below the surface of the soil, acting indeed more like an earth-worm than a vertebrate animal. Many of these creatures have been accidentally dug out while deepening or clearing ditches. The negroes are much afraid of the Congo Snake, and think it to be poisonous, a belief which has its only foundation in fear, generated by ignorance.

The legs are extremely small and feeble, and there are only two toes on each foot. Its color is dark blackish-gray above, and lighter beneath. Another species, the THREE-TOED

MENOPOME.—*Protonopsis horrida.*

CONGO SNAKE (*Murænopsis tridáctylus*), is much like the common Congo Snake, from which it may be distinguished by possessing three toes on each foot instead of two. The length of both these creatures is from two to three feet. These two species constitute the whole of the family to which they belong.

THE TAILED BATRACHIANS (*Proteidæ*) are now regarded as differing sufficiently from near forms to belong to a distinct order. The family *Proteidæ*—MUD PUPPIES—embraces one genus and one species.

THE *Necturus* is called in the Middle States MUD PUPPY, WATER DOG, MENOBRANCHUS, and DOG-FISH. It is common north and west of the Alleghanies, and is abundant in the Great Lake region.

THE great SIREN (*Siren lacertina*) is a species consisting of the entire family *Sirenidæ*. This creature has a most remarkably long, eel-like form.

AMONG these remarkable animals, the orders multiply themselves rapidly. The Pseudophidia, or False Serpents, include some very curious species, whose position remained long unsettled. There is but one family, and all its members have very long and cylindrical bodies, no limbs, a very short tail, and a smooth wrinkled skin, in which are embedded a multitude of minute scales. The two worm-like creatures, the White-bellied Cæcilia and the Slender Cæcilia, are good examples of this very remarkable family.

The name Cæcilia is derived from a Latin word signifying blindness, and is given to the creature because the eyes are always minute, and in some species are hidden under the skin. The WHITE-BELLIED CÆCILIA inhabits Southern America, and, like the rest of its kin, burrows under the ground after the fashion of the earth-worm, to which it bears so strong an external resemblance, preferring wet and marshy ground to dry soil. Its body is rather thick and cylindrical, and is surrounded by about one hundred and fifty incomplete rings. The muzzle is rounded and so is the tail. There are teeth in the jaws and on the palate, all of which are short, strong, and conical; the tongue has a curiously velvety feel to the touch. Below each nostril there is a small pit, sometimes taken for a second nostril.

The color of the White-bellied Cæcilia is blackish, marbled with white along the under surface.

THREE-TOED CONGO SNAKE.—*Murænopsis tridactyla.*

THE SLENDER CÆCILIA derives its name from its slight form. In this species the body is smooth throughout the greater part of its length, but towards the tail the skin is gathered into fifteen circular folds pressed closely together. The muzzle is rather broad and rounded. The body of the Slender Cæcilia is extremely elongated, being about two feet in length, and not thicker than an ordinary goose-quill. Its color is almost wholly black.

THE small but very remarkable order of animals which stands next in our list, has proved an insoluble enigma to the systematic zoologists, who not only are unable to decide upon any order to which it may belong, or in what precise relation it stands to other reptiles, but are not even able to announce positively its class, or to say whether it is a reptile or a fish. The three species which comprise this order—if indeed they do not form a separate class—are so fish-like in most parts of their anatomy and their general habits, that they might be regarded as belonging to the fishes, were not they allied to the reptiles by one or two peculiarities of their structure. Some accurate and experienced anatomists accordingly place these creatures among the fishes, while others, equally experienced, consider them as belonging to the reptiles.

In fact, the position in which these creatures are placed depends wholly on the amount of importance given to the reptilian or piscine characters.

The species known by the name of LEPIDOSIREN, or MUD-FISH, is found in Africa, inhabiting the beds of muddy rivers.

The habits of this creature are very remarkable. Living in localities where the sun attains a heat so terrific during a long period of the year that the waters are dried and even their muddy beds baked into a hard and stony flooring, these animals would be soon extirpated unless they had some means of securing themselves against this periodical infliction, and obtaining throughout the year some proportion of that moisture for lack of which they would soon die. The mode of self-preservation during the hot season is very like that which has already been mentioned in the case of certain frogs and other similar creatures, but is marked by several curious modifications.

When the hot season has fairly commenced, and the waters have begun to lessen in volume, the Lepidosiren wriggles its way deeply into the mud, its eyes being so constructed that the wet soil cannot injure them, and the external nostrils being merely two shallow blind sacs. After it has arrived at a suitable depth, it curls itself round, with its tail wrapped partly over the head, not unlike the peculiar attitude assumed by fried whitings, except that its flexible spine enables it to squeeze the two sides closer together than can be accomplished in that fish, and in that position awaits the coming rains. It will lie in a torpid condition for a very considerable space of time, depending entirely on the advent of rain for the re-assumption of vitality.

After it has curled itself up and resigned itself to the exigencies of its condition, a large amount of a slimy substance is secreted from the body, which has the effect of making the walls of its cell very smooth, and probably aids in binding the muddy particles together. When the rains fall, the moisture penetrates rapidly through the fissures of the earth, cracked in all directions by the constant heat, reaches the cell of the Lepidosiren, dissolves its walls, and restores the inhabitant to life and energy.

Several specimens have been brought to Europe, most of which I have had opportunities of seeing while alive, as well as of examining parts of their structure after death.

While retained in an ordinary aquarium, it passes much of its time in an apparently semi-torpid condition at the bottom of the tank, generally seeking the darkest corner and squeezing itself along one of the perpendicular angles of the case. It was found, however, that whenever the surface of the water was disturbed, the creature woke up, as it were, and rose to see what was the matter. In this way it could be induced to come at a signal to take the food on which it lived.

Further investigations and experiments on a larger scale, afforded a considerable insight into the habits of this singular creature.

Several batches of these animals have been kept alive, all of which have died, some after a life of only a few weeks, and others after surviving for three years. It will, however, be useless to follow the fortunes of each separate individual, and we will therefore only examine the general habits which seem to be common to all.

The Lepidosirens, or Mud-fish as they are popularly called, were sent while still in their muddy nests, or "cocoons," according to the technical term, and, in one instance, three specimens were inclosed in a single lump of hard mud, weighing when dry about twenty pounds.

One of the cocoons is now lying before me, together with the dried and shrivelled body of its former inhabitant, still curled up in the singular fashion already mentioned. The walls of the cocoon are composed of a thick, grayish clay, quite hard and dry, and intermixed here and there with remnants of vegetable matter. The hollow in which the Lepidosiren resided is quite smooth in the interior, but gives no idea of the real shape of the inhabitant, the cell seeming to be somewhat large, most probably on account of the coat of mucous substance with which it was lined, and part of which is to be seen still adhering, like flakes of dry membrane, to the sides of the cell.

By rapidly tearing this membranous substance with an oblique bearing, it can be in some places split like a scrap of paper under similar circumstances; but when placed under the microscope, it shows no signs of organization, being of a light brown color, irregularly mottled with black. When burned, it rapidly takes fire and bursts into flame, giving out a very nauseous odor, like that which is perceived on burning the wing-case of a beetle, and leaves a firm black ash, of nearly the same shape and form as before the light was applied to it.

The remainder of this substance is found loosely adhering to the body of the former inhabitant, and can be easily stripped off.

On being immersed in water, the earthy cocoons fell to pieces as if they had been made of sugar, and the imprisoned creatures were thus released. At first they were exceedingly sluggish, and hardly stirred, but after the lapse of an hour or two they became tolerably alert.

One of these specimens died after it had been kept about six weeks, and a good plaster-cast of it is now before me. Its length is ten inches, and the circumference of the head, just in front of the fore pair of limbs, is exactly three inches. The scales are tolerably well marked, and are shown even in the plaster-cast, though in the living animal there is hardly a trace of them. They are also very evident after the creature has been immersed in spirits for some time. In taking a cast of the Lepidosiren, the mucous secretion with which the body is covered affords a serious obstacle to the correctness of the image, as it is apt to adhere to the plaster, and pull away with it some portions of the skin.

A fellow-specimen, that floated dead from its cocoon, is also before me, bent on itself in the manner usual among these creatures, and with its mouth widely open, showing the peculiar teeth.

Finding, as has already been mentioned, that the Lepidosiren would rise to the surface of the water when a splashing was made, the attendants used to feed it by paddling about with the finger, and then holding a piece of raw beef in the spot where the disturbance had been made. The creature used to rise deliberately, snatch the meat away, and, with a peculiarly graceful turn of the body, descend to its former resting-place for the purpose of eating its food.

The mode of eating was very remarkable. Taking the extreme tip of the meat between its sharp and strongly formed teeth, it would bite very severely, the whole of the head seeming to participate in the movement, just as the temporal muscles of the human face move when we bite anything hard or tough. It then seemed to suck the meat a very little farther into its mouth and gave another bite, proceeding in this fashion until it had subjected the entire morsel to the same treatment. It then suddenly shot out the meat, caught it as before by the tip, and repeated the same process. After a third such manœuvre, it swallowed the morsel with a quick jerk. The animal always went through this curious series of operations, never swallowing the meat until after the third time of masticating.

After a while, it was thought that the water in which it lived was not sufficiently warm to represent the tepid streams of its native land, and its tank was consequently sunk in a basin, where the water is kept at a tepid heat for the purpose of nourishing the tropical plants which grow in it. Here the creature remained for some time, but at last contrived to wriggle itself over the side of its tank, and roam about in the large basin quite at liberty.

It remained here for some time, and being deprived of its ordinary supply of raw beef, took to foraging for itself. The gold-fish with which the basin is stocked became its victims, and it was quite as destructive as an otter would have been. It had quite a fancy for attacking the largest fish; and though apparently slow in its movements, could catch any fish on which it had set its wishes. As the fish was quietly swimming about, suspecting no evil, the Lepidosiren would rise very quietly beneath it until quite close to its victim, just as the terrible ground-shark rises to take its prey. It then made a quick dart with open mouth, seized the luckless fish just by the pectoral fins, and with a single effort bit entirely through skin, scales, flesh, and bone, taking out a piece exactly the shape of its mouth, and then sinking to the bed of the basin with its plunder. The poor fish was never chased, but was suffered to float about in a half-dead state, and numbers of mutilated gold-fish were taken out of the basin.

I have several times seen the creature while swimming about in search of a dinner, and have been much struck with the exceeding grace of its movements, which, indeed, very strongly resemble those of the otter.

At last its depredations were checked, for when the basin was cleansed, according to custom, a portion was fenced off, so that the Lepidosiren could not get out, and the gold-fish could not get in.

Not choosing to supply a succession of gold-fish, out of each of which the fastidious creature would only take one bite, the superintendent bethought himself of frogs, and fed the animal regularly with these batrachians. But having been warned, by the effects on the gold-fish, not to trust his fingers within reach of the teeth that could inflict such very effective bites, he got a long stick, cleft one end of it, put one hind-foot of the frog into the cleft, and held it on the surface of the water, so that the struggles of the intended victim should agitate the surface, and warn the Lepidosiren that its dinner was ready. No sooner did the frog begin to splash, than the Lepidosiren rose rapidly beneath it, seized it in its mouth, dragged it off the stick like a pike striking at a roach, and sunk to the bottom with its prey. Not a vestige of the frog was ever seen afterwards; and Mr. Wilson naturally conjectures that the poor victim was gradually chewed up, like the beef with which the creature was formerly fed.

Under this regimen the Lepidosiren grew apace, and in three years had increased from ten inches in length and a few ounces in weight, to thirty inches long, and weighing six pounds and a quarter. The rapidity of its growth may be accounted for by the fact, that it had fed throughout the entire year, instead of lying dormant for want of water during half its existence, and its size was apparently larger than it would be likely to attain in its native state.

Thinking that perhaps the creature might need its accustomed season of repose—happily called æstivation, in opposition to the term hibernation—it was well supplied with clay similar to that from which its cocoon had been formed, but without any result, the animal evincing no disposition to avail itself of the stores so thoughtfully collected in its behalf. This is, I think, a very interesting example of the manner in which nature accommodates herself to circumstances, and is paralleled by many other instances in the several departments of Natural History. Bees, for example, on finding themselves within easy distance of a sugar plantation, have been known to decline honey making; and the same result has occurred when they were transported to fertile localities where the honey-bearing flowers are in blossom throughout the year.

As an example of a similar phenomenon occurring in the vegetable kingdom, I may instance some Australian flowers brought over by Mr. Howitt, and planted in his garden. These plants were at first sadly puzzled by the seasons, wanting to blossom just as our winter had set in, but in the course of a few years they grew gradually later in blossoming, until they had found the proper season, and then were content to put forth their leaves and flowers at the same time as the indigenous plants.

The cause of this specimen's regretted death was rather curious. In the winter time, when the basins were cleaned, the animal was removed from one basin to another, while the former was being emptied. Unfortunately, the fires which warmed the water were suffered to expire during the night, and in the morning the poor Lepidosiren was found chilled to death.

The history of this creature is not only interesting, but is valuable as it shows the comparative advantages of watching the habits of animals in large and small habitations. Had, for example, the creature lived from the first in the large basin, its remarkable mode of eating its food could not have been observed, as it always seeks the bottom of its prison for that purpose; while, had it been always kept in the glass tank, its graceful movements and fish-eating propensities would never have been discovered.

The bones of the Lepidosiren are, when first taken from the body, of a bright green color, and so gelatinous in structure, that if left in the water they would probably dissolve. After a time, however, the green color fades, though traces of it can still be discerned. The bones of the head are, however, of a firmer character, as is needful for the management of the sharp and powerful teeth; and in the skull of the above-mentioned specimen, the green tint still lingers on several of the bones.

The teeth are most remarkable, looking as if they were made from a ribbon of enamel-covered bone, plaited in a series of very deep undulations in front, and sweeping off at each side with a bold curve. Those of the palate and lower jaw are so made that they lock into each other, the folds exactly corresponding, and fitting into each other with such exactness, that no creature when seized could hope to escape without much detriment. The edges of this continuous tooth-ribbon, if I may so call it, are very sharp, and armed with small saw-like teeth, rather worn away in front, but very perceptible on the sides. In the very front of the upper jaw are two little pointed teeth, set apparently loosely in the soft parts of the nose, and quite useless for biting. When, however, the skull is removed from the body, and cleared of muscle and other soft parts, these teeth retain their place, and by the hardening of their attachments become tightly fixed in the skull.

During life the points of these teeth project very slightly through those two little holes just inside the upper lip, which are considered as the internal nostrils. While the creature is alive, the teeth cannot be seen even when the mouth is open, being covered by a very soft and yielding substance, through which they seem to cut when in use.

The external aspect of this creature is very singular, the chief characteristics being its eel-like form, and the four long slender projections which stand in the place of limbs, and are analogous to similar structures in certain reptiles already described and figured. These are not true limbs, and the cartilaginous ray by which they are supported has no joint. They are quite soft and flexible, as if they were made of leather, and are of very trifling use in locomotion. The two fore-limbs are set at the shoulders, just behind the head, and widely separated from each other, while the hinder pair are quite close together at their bases. In the species just described, two short tubercular appendages, about an inch in length, accompany the larger limb-like projections, and, except in dimensions, bear a close resemblance to those organs. I may take this opportunity of remarking that the creature is not known to leave the water and to crawl on land.

Another specimen has not attained to any great size, being scarcely half as large as the individual just described, though it has lived in captivity for three years. The tank in which it resided was small, and may have probably accounted for the slight increase in dimensions. It was interesting to watch this creature move about its prison, as the peculiar screw-like or spiral movement of the limbs was well exhibited. The whole body was covered with rather large scales, embedded deeply in the skin, and not easily to be seen in living specimens.

The name of Lepidosiren, or Scaly Siren, is given to this creature on account of its scaly covering. At about one-third of the distance from the head to the tip of the tail a rather narrow and fin-like membrane arises, which runs completely round the tail until it is terminated close to the bases of the hind pair of limbs. It is strengthened throughout by a series of soft jointed rays.

The flesh of the Lepidosiren is very soft and white, and is thought to be excellent for the table, so that in its native country it is dug up from its muddy bed and used for food. It usually burrows to a depth of eighteen inches. This creature possesses both lungs and gills, the latter organs being twofold, the external gills being tufted on the under side, and the internal gills being placed on the edge of the divisions between the gill openings on the side of the neck. The heart is more reptilian than piscine, having three compartments, two auricles and one ventricle, and affords one of the strongest reasons for ranking the creature among the former class.

There are several species of Lepidosiren, divided into two genera, distinguished from each other by the number of ribs. The species which is found in Southern America, and is there known under the popular name of CARAMURU (*Lepidosiren paradoxa*), has fifty-five pairs of ribs, whereas the African species has only thirty-six pairs. The color of the Lepidosiren is darkish brown with a wash of gray.

The next order of Crawling Batrachians is called by the name of Meantia, and contains a very few but very remarkable species. In all these creatures the body is long and smooth, without scales, and the gills are very conspicuous, retaining their position throughout the life of the animal. There are always two or four limbs, furnished with toes, but these members are very weak, and indeed rudimentary, and both the palate and the lower jaw are toothed.

The first example of this order is the celebrated PROTEUS, discovered by the Baron de Zois, in the extraordinary locality in which it dwells.

At Adelsberg, in the duchy of Carniola, is a most wonderful cavern, called the Grotto of the Maddalena, extending many hundred feet below the surface of the earth, and consequently buried in the profoundest darkness. In this cavern exists a little lake, roofed with stalactites, surrounded with masses of rock, and floored with a bed of soft mud, upon which the Proteus may be seen crawling uneasily, as if endeavoring to avoid the unwelcome light by which its presence is known. These creatures are not always to be found in the lake, though after heavy rains they are tolerably abundant, and the road by which they gain admission is at present a mystery.

The theory of Sir H. Davy is, "that their natural residence is a deep subterraneous lake, from which in great floods they are sometimes forced through the crevices of the rocks into the places where they are found; and it does not appear to me impossible, when the peculiar nature of the country is considered, that the same great cavity may furnish the individuals which have been found at Adelsberg and at Sittich."

Whatever may be the solution of the problem, the discovery of this animal is extremely valuable, not only as an aid to the science of comparative anatomy, but as affording another instance of the strange and wondrous forms of animal life which still survive in hidden and unsuspected nooks of the earth.

Many of these animals have been brought in a living state to England, and have survived for a considerable time when their owners have taken pains to accommodate their condition as nearly as possible to that of their native waters. I have had many opportunities of seeing some fine specimens, brought by Dr. Lionel Beale from the cave at Adelsberg. They could hardly be said to have any habits, and their only custom seemed to be the systematic avoidance of light. Dr. Beale has kindly forwarded to me the following account of these curious creatures:—

"One of the Proteuses I brought over from Adelsberg lived for five years, and, what is very interesting, passed four years of his life in the same water, a little fresh being added from time to time to make up for the loss by evaporation. He lived in about a quart of water, which was placed in a large globe, this being kept dark by an outer covering of green baize. Perhaps half a pint of water may have been added during two years.

"He was not once fed while he was in confinement, and one of his companions died soon after taking a worm before he had been two years in this country.

"The one I kept was very active, and his movements were as rapid as those of an eel. He was thinner just before death than when he was brought from the cave, but the loss of substance was so very slow as not to be perceptible from year to year, and to the last he retained the power of performing very active muscular movements.

"His external gills always contracted when a strong light was thrown upon them. The circulation of the blood in the vessels of these organs was very often exhibited; the animal being placed in a long tube with a flat extremity, provided with an arrangement for the constant supply of water, and on several occasions some of the large blood corpuscles were removed for the purpose of microscopical examination, so that the animal was not placed under the most favorable circumstances for living without food.

"There are probably very few more striking examples of very slow death from starvation than this, and it is probable that the ultimately fatal results were as much caused by confinement, change of air and temperature, and occasional exposure to light for some hours, as from mere starvation. It is well known, for example, that, as a general rule, the Batrachia endure starvation most remarkably."

The gills of the Proteus are very apparent, and of a reddish color, on account of the blood that circulates through them. I have often witnessed this phenomena by means of the ingenious arrangement invented by Dr. Beale, by which the creature was held firmly in its place while a stream of water was kept constantly flowing through the tube in which it was confined. The blood discs of this animal are of extraordinary size; so large, indeed, that they can be distinguished with a common pocket magnifier, even while passing through the vessels. Some of the blood corpuscles of the specimen described above, are now in my possession, and, together with those of the lepidosiren, form a singular contrast to the blood corpuscles of man, the former exceeding the latter in dimensions as an ostrich egg exceeds that of a pigeon.

The color of the Proteus is pale faded flesh tint, with a wash of gray. The eyes are quite useless, and are hidden beneath the skin, those organs being needless in the dark recesses where the Proteus lives. Its length is about a foot. What are the natural habits of this strange animal, what is its food, of what nature is its development, and what is its use, are a series of problems at present unanswered. By some writers it has been thought to be merely

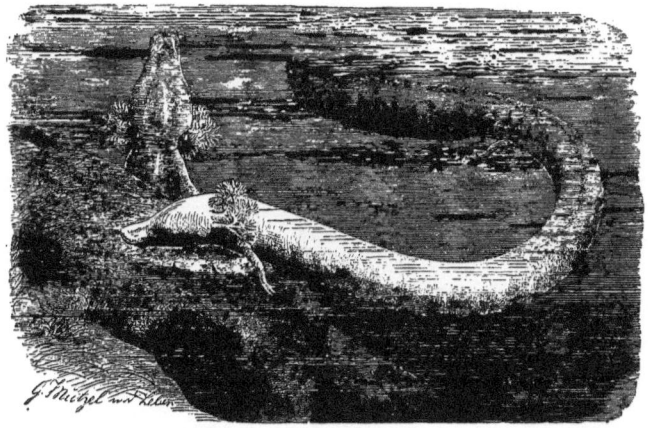

PROTEUS. *Proteus anguineus.*

the larval state of some large Batrachian at present unknown; but the anatomical investigations that have been made into its structure seem to confirm the idea that it is a perfect being, and one of those species which carry the gills throughout their whole existence.

IN the NECTURUS, the head is much broader and flatter and the tail shorter than in the preceding species. This animal belongs to the same family as the proteus, but is a native of America, being found in the Mississippi and several of the lakes. It is rather a large animal, attaining, when adult, a length of two or three feet, and being of a thick and sturdy make. The gills of this creature are large and well tufted, and the limbs are furnished with four toes on each foot, but without claws.

The general color of this creature is olive-brown above, dotted with black, and with a black streak from the nostril through the eye, and along each side to the tail. Below it is blackish-brown with olive spots.

OUR last example of the Batrachians is the curious SIREN, or MUD-EEL, as it is sometimes called, on account of its elongated eel-like form and its mud-loving habits.

It is a native of several parts of America, and is found most plentifully in Carolina, where it haunts the low-lying and marshy situations. The rice-grounds seem to be its most favored localities, the muddy soil being the substance best adapted for its means of progression. Its food seems to consist almost entirely of worms and various insects, of which it will consume a considerable quantity every day. A fine specimen used to feed upon earth-worms, of which it would devour about eighteen or twenty every two days. This individual passed the greater part of its time beneath the thick stratum of soft mud with which the bed of the basin was profusely covered. This was a very long specimen, and by an uninitiated observer would probably have been taken for an eel.

The head of the Siren is small in proportion to the size of the animal, the eye is very small, and the gill tufts are three in number on each side, and beautifully plumed. It has only one pair of legs, the hinder set being wanting, and the front pair are extremely small, and of no practical use in progression. It has only three toes on each foot. The color is dark blackish-brown, and the length of a fine specimen is about three feet.

FISHES; PISCES.

IN the FISHES, the last class of vertebrated animals, the chief and most obvious distinction lies in their adaptation to a sub-aqueous existence, and their unfitness for life upon dry land.

There are many vertebrate animals which pass the whole of their lives in the water, and would die if transferred to the land, such as the whales and the whole of the cetacean tribe, an account of which may be found in Vol. I., page 418. But these creatures are generally incapable of passing their life beneath the waters, as their lungs are formed like those of the mammalia, and they are forced to breathe atmospheric air at the surface of the waves. And though they would die if left upon land, their death would occur from hunger and inability to move about in search of food, and in almost every case a submersion of two continuous hours would drown the longest breathed whale that swims the seas.

The Fishes, on the contrary, are expressly formed for aquatic existence; and the beautiful respiratory organs, which we know by the popular term of "gills," are so constructed that they can supply sufficient oxygen for the aeration of the blood. They have not the power, as is sometimes imagined, of separating the oxygen, which, in its combination with certain proportions of hydrogen, compose the element in which they live, but are able to take advantage of the atmospheric air which is contained in the water.

Any reader who happens to possess a globe with gold-Fish can prove, and doubtlessly has proved, the truth of this assertion. It often happens that when the supply of water is insufficient, or the mouth of the vessel too small to permit the air to be absorbed by the water in sufficient volume, the Fish come gasping to the surface, and there swim with gaping mouths, sucking in the air with audible gulps. But if a little water be taken up in a cup or spoon, and dashed back from a little height, so as to cause a sharp splash, or, better still, if a syringe be employed for the same purpose, so as to drive a quantity of atmospheric air into the water, the Fish soon become contented, their anxious restlessness abates, and they quietly swim backward and forward, without displaying any more signs of uneasiness.

The reason that Fishes die when removed from the water, is not because the air is poisonous to them, as some seem to fancy, but because the delicate gill membranes become dry and collapse against each other, so that the circulation of the blood is stopped, and the oxygen of the atmosphere can no longer act upon it. It necessarily follows, that those Fish whose gills can longest retain moisture will live longest on dry land, and that those whose gills dry most rapidly will die the soonest. The herring, for example, where the delicate membranes are not sufficiently guarded from the effects of heat and evaporation, dies almost immediately it is taken out of the water; whereas the carp, a fish whose gill-covers can retain much moisture, will survive for an astonishingly long time upon dry land, and the anabas, or climbing perch, is actually able to travel from one pool to another, ascending the banks, and even traversing hot and dusty roads.

The entire shape of these creatures, subjected though it be to manifold variations, is always subservient to the great object of passing rapidly through the ponderous liquid in which they swim, so as to enable them to secure their prey or avoid their enemies. Even in creatures of such different shapes as the sharks, the eels, the salmon tribe, and the flat fish, the capacity for speed is really wonderful, and is in all effected by simple and beautiful modifications of one mechanical principle, that of the inclined plane or screw.

In all Fishes, the power of progession lies in the wonderfully muscular tail with its appended fin, and the creature drives itself forward by repeated strokes of this organ in exactly the same manner that a sailor urges a boat through the water by the backward and forward movements of a single oar in the stern.

To show the power of this principle, I will mention that, being on one occasion left with a party of friends on board a fishing-barge in a small lake, and deserted by an ill-conditioned boatman, who refused either to put us ashore or take us to a better fishing-ground, and so went misanthropically home to his dinner, I called to mind the progression of the Fishes, and straightway became independent of the boatman. After hauling up the anchor, I inserted the butt end of the largest fishing-rod into the head of the rudder so as to form an extempore tiller, and by moving the rudder gently to and fro I was able to propel the barge in any direction and to any distance. We thus traversed the lake at our pleasure, drove the barge ashore at its further extremity, and left the boatman to find it and take it back as he could.

Even the eels and the flat Fishes, with their gracefully serpentine movements, adopt this mode of progression, though it is not so apparent as in the Fish whose bodies are less flexible, and accordingly employ more force in the tail itself.

The fins are scarcely employed at all in progression, but are usually used as balancers, and occasionally to check an onward movement. Before proceeding further, I may mention that all the fins of a Fish are distinguished by appropriate names. As they are extremely important in determining the species and even the genus of the individual, and as these members will be repeatedly mentioned in the following pages, I will briefly describe them.

Beginning at the head and following the line of the back, we come upon a fin, called from its position the "dorsal" fin. In very many species there are two such fins, called, from their relative positions, the first and the second dorsal fins. The extremity of the body is furnished with another fin, popularly called the tail, but more correctly the caudal fin. The fins which are set on that part of the body which corresponds to the shoulders are termed the "pectoral" fins; that which is found on the under surface and in front of the vent is called the abdominal fin, and that which is also on the lower surface, and between the vent and the tail, is known by the name of the "anal" fin. All these fins vary extremely in shape, size, and position.

The gill-cover, or operculum as it is technically called, is separated into four portions, and is so extensively used in determining the genus and species that a brief description must be given. The front portion, which starts immediately below the eye, is called the "præ-operculum," and immediately behind it comes the "operculum." Below the latter is another piece, termed, from its position, the "sub-operculum," and the lowest piece, which touches all the three above it, is called the "inter-operculum." Below the chin and reaching to the sub-operculum, are the slender bones, termed the "branchiostegous rays," which differ in shape and number according to the kind of Fish.

The scales with which most of the Fish are covered are very beautiful in structure, and are formed by successive laminæ, increasing therefore in size according to the age of the Fish. They are attached to the skin by one edge, and they overlap each other in such a manner as to allow the creature to pass through the water with the least possible resistance. The precise mode of overlapping varies materially in different genera. Along each side of the Fish runs a series of pores, through which passes a mucous secretion formed in some glands beneath. In order to permit this secretion to reach the outer surface of the body, each scale upon the row which comes upon the pores is pierced with a little tubular aperture, which is very perceptible on the exterior, and constitutes the "lateral line." The shape and position of this line are also used in determining the precise position held by any species. In comparing the scales taken from different Fishes, it is always better to take those from the lateral line.

The heart of the Fish is very simple, consisting of two chambers only, one auricle and one ventricle. The blood is in consequence cold.

The hearing of Fishes appears in most cases to be dull, and some persons have asserted that they are totally destitute of this faculty. It is now, however, known that many species have been proved capable of hearing sounds, and that carp and other fish can be taught to come for their food at the sound of a bell or whistle. The internal structure of the ear is moderately developed, and there are some curious little bones found within the cavity, technically called otoliths.

The sense of touch seems to have its chief residence in the mouth and surrounding parts, the scaly covering rendering the surface of the body necessarily obtuse to sensation. The smell seems to be strongly developed, if it be possible to pronounce an opinion from the size and distribution of the nasal nerves. The brain is very small in these creatures, and from its shape, as well as its dimensions, denotes a low degree of intelligence.

In the anatomy of the Fishes there are many other interesting structures, which will be described when treating of the particular species in which they are best developed.

SOFT-FINNED FISHES; CHONDROPTERYGII.

The fishes comprised in the first order are called by the rather harshly-sounding title of Chondropterygii, a term derived from two Greek words, the former signifying cartilage and the latter a fin, and given to these creatures because their bones contain a very large amount of cartilaginous substance, and are consequently soft and flexible. The bones of the head are rather harder than those of the body and fins.

It is necessary, before entering into any description of the different species, to premise that the arrangement of the fishes is a most difficult and complicated subject, in which no two systematic naturalists seem to agree entirely. I have, therefore, followed the course which has been adopted throughout the whole of this work.

The cartilaginous fishes are again subdivided into groups, in the first of which the gills are quite free, and the members of this group are accordingly called by the name of Eleutheropómi, or free-gilled fishes. What quality in the fishes should give birth to such polysyllabic and harsh-sounding names, is not easy to say; but the fact is patent that not even in botany is the scientific terminology so repulsive as in the fishes. I shall endeavor, as far as possible, to avoid this technical language, and to throw the scientific descriptions to the end of the work, as in the two former volumes; and the reader may feel sure when his attention is struck by a long and difficult name, that it is only used in consequence of the exigencies of the occasion.

The first family, of which the common STURGEON is a good and familiar example, are at once known by the cartilaginous or bony shields with which the head and body are at intervals covered.

In this remarkable fish the mouth is placed well under the head, and in fact seems to be set almost in the throat, the long snout appearing to be entirely a superfluous ornament. The mouth projects downwards like a short and wide tube, much wider than long, and on looking into this tube no teeth are to be seen. Between the mouth and the extremity of the snout is a row of fleshy finger-like appendages, four in number, and apparently organs of touch.

One or two species of Sturgeon are important in commerce, as two valuable articles, namely, isinglass and caviare, are made from them. The former substance is too well known to need a description, and the mode of preparing it for use is briefly as follows: The air-bladder is removed from the fish, washed carefully in fresh water, and then hung up in the air for a day or two so as to stiffen. The outer coat or membrane is then peeled off, and the remainder is cut up into strips of greater or lesser length, technically called staples, the long staples being the most valuable. This substance affords so large a quantity of gelatinous matter, that one part of isinglass dissolved in a hundred parts of boiling water will form a stiff jelly when cold.

Caviare is made from the roe of this fish, and as nearly three millions of eggs have been taken from a single fish, the amount of caviare that one Sturgeon can afford is rather large. It is made by removing all the membranes, and then washing the roe carefully with vinegar or white wine. It is next dried thoroughly in the air, well salted, subjected to strong pressure in order to force out all moisture caused by the wet-absorbing properties of the salt, and is lastly packed in little barrels for sale. The caviare made on the Caspian is considered the best. In Russia it forms a large item in the national consumption, probably on account of the great number of fasts observed by the Greek Church. The roes of several other fish are employed in the same manner, and in Italy, a substance called "botargo" is prepared from the roe of a species of mullet.

The common Sturgeon has sometimes, but not very often, been found in English rivers, and whenever it is captured in the Thames within the jurisdiction of the Lord Mayor, it is

STURGEON.—*Acipenser attilus.*

termed a royal fish, and becomes the property of the Crown. It is not unfrequently taken near the English shores, more especially on the eastern coast, and most persons are familiar with the occasional appearance of one of these fine fish on a fishmonger's stall. The flesh of the Sturgeon is held in some estimation; and in the olden English days, it was always reserved for the table of the king. Some very fine specimens have sometimes been caught in English rivers, the largest on record having weighed four hundred and sixty pounds. The size of this specimen may be imagined from its weight, as another individual which weighed only one hundred and ninety pounds measured eight feet in length.

The body of the Sturgeon is elongated, and slightly five-sided from the head to the tail. Along the body run five rows of flattened bony plates, each plate being marked with slight grooves in a radiating fashion, and having a pointed and partly conical spine on each plate, the points being directed towards the tail. The plates along the summit of the back are the largest.

THERE are many species of Sturgeons, and among the most remarkable are the Shovel-fish and the Spoon-bill Sturgeon, both being natives of the rivers of North America.

The SHOVEL-FISH derives its name from the curious form of its head, which is flattened, rounded, and really not unlike the implement after which it receives its popular title. The adult and young differ somewhat in their aspect. Both are of large size and show a conspicuous arrangement of the bony scales along the body.

THE SHOVEL-NOSE STURGEON (*Scaphiorhynchops platyrrhynchus*). About four species inhabit the waters of Central and Eastern Asia, and the United States. One only is found in the latter region.

The term White Sturgeon is also used to designate this species. It inhabits the Mississippi Valley and the streams of the Western and Southern States. Both of the long technical terms literally mean spade-snout, and flat-snout.

THE family of Sturgeons is regarded as one very strongly marked; there being little danger of confusion as to the rights of membership. Though the Sturgeon is, in most portions of the United States, rather an unfamiliar fish, yet the characters are such, that once the species are seen they are quickly and correctly placed in a natural classification. It has no near allies, excepting those lying entombed in rocks of past geological ages. The skeleton is cartilaginous. Species are found in all north temperate portions of the globe. They all breed in fresh water; but some reside in the sea during a part of the season. Others are permanent residents in the fresh waters of the Great Lakes and rivers.

THE SHARP-NOSED STURGEON (*Acipenser sturio*). From Cape Cod to Florida this fish finds its *habitat*. It is also found in European waters. The Short-nosed Sturgeon has the same American range.

According to Jordan and Gilbert, there are two genera and twenty species of Sturgeons. There are seven species inhabiting the waters of North America. Most of the species are migratory, like the Salmon which are found in the same waters.

A variety of this species is very common in waters between Massachusetts and Florida.

THE WHITE STURGEON is native to the Columbia River. It is also called SACRAMENTO STURGEON, as it inhabits along the Pacific Coast to Monterey. It reaches a weight of from three hundred to six hundred pounds, and is used extensively for food.

THE GREEN STURGEON is reputed as unfit for food, and, indeed, it has the merited reputation of being poisonous. It is smaller in size than the preceding.

THE LAKE STURGEON, called also Ohio, Black, Stone, and Rock Sturgeon, inhabits the Mississippi River and northward to the Great Lakes. Its weight is from fifty to one hundred pounds. This is the common Fresh-water Sturgeon, which usually does not descend to the sea.

THE SHORT-NOSE STURGEON is found from Cape Cod to Florida.

THE SPOON-BILL STURGEON is, in allusion to the singular shape of the head, sometimes called the PADDLE-FISH. This creature is remarkable for several reasons. In the first place, the uncommonly elongated and flattened snout is sufficiently conspicuous to arrest the attention of even the most casual observer, and in the second place, the body is quite smooth, and wants those bony plates which generally form so characteristic an adornment of the Sturgeon. This remarkable fish is frequently found in the Ohio and Mississippi.

THE DUCK-BILLED CAT. This curiously endowed fish is represented in this country by two species, in two genera—being equally distributed in the fresh waters here and in China. They are embraced in the sub-class, Sturgeons, and in a separate order and a single family.

THE very singular family of the Chimæridæ contains a few but remarkable species.

Both these creatures are sufficiently quaint and ungainly in aspect. The NORTHERN CHIMÆRA is also known by the title of RABBIT-FISH, probably on account of its general aspect, and KING OF THE HERRINGS, because it follows the shoals of those fishes during their wonderful migrations, and makes great havoc among their numbers. The appendage to the top of the head is also looked upon by the Norwegians in the light of a kingly crown, and has contributed towards its royal title. It is known in some localities under the name of SEA CAT.

This species is mostly found in the Northern seas, and is, when living, a most beautiful creature, its body glowing with golden-brown variegations upon a white ground. The title of Gold and Silver Fish is sometimes given to the Northern Chimæra in consequence of this gorgeous coloring. The pupil of the eye is green, and the iris is white. It feeds mostly upon the smaller fish, but finds much of its subsistence among the various mollusks, crustaceans, and other inhabitants of the ocean. The flesh is not considered good, being hard and coarse.

The form of this fish is very peculiar, the body being tolerably large and rounded towards the point, and the tail tapering rapidly until it ends in an elongated thong, almost like the lash of a whip. The second dorsal fin commences immediately behind the first, and extends along the tail nearly to the extremity of its lengthened filamentary termination. The sexes may readily be distinguished from each other, both by the shape of the head and first dorsal fin, and by a pair of bony appendages close to the ventral fins. It is not a large species, seldom exceeding a yard in length.

IN the seas of the southern hemisphere, there is another species of Chimæra, called from its locality, the SOUTHERN CHIMÆRA (*Callorhynchus antárctica*) or ELEPHANT-FISH, the latter title being given to it on account of the extraordinary prolongation of the snout. The Araucanian name for this species is CHALGUA ACHAGUAL. The snout of this fish is developed into a strange cartilaginous prolongation, which is bent backwards in a hook-like form, and is thought by some persons to bear a resemblance to a common hoe.

The tail of this species does not correspond in oddity with its head, being without the long filament that gives so strange an aspect to its Northern relative. The color is satiny-white mottled with brown, and the size is about the same as that of the Northern Chimæra.

The Chimæras are so manifestly different from fishes more or less allied on each side, they are regarded as forming naturally a sub-class. They are all embraced under one order, and one family. The extraordinary appearance of these creatures quite justifies the titles given them. Two species are enumerated, one called Rat-fish, and the other Elephant-fish. The former is found in the Atlantic Ocean, from Cape Cod northward, in deep water; the other inhabits the Pacific, from Monterey northward, and is very abundant.

THE SHARKS.

THE fishes belonging to the next sub-order have their gills fixed by their outer edge to the divisions in the gill-openings at the side of the neck. This sub-order includes the Sharks and the Rays.

The first family of this large and important group is known by the name of Scyllidæ, and its members can be recognized by several distinguishing characteristics. They have spout-holes on the head, and the gill-openings are five in number on each side. Sometimes there only seems to be four openings, but on closer examination the fourth and fifth are found set closely together, the opening of the fifth appearing within that of the fourth. The teeth are sharp and pointed, and the tail is long, notched on the outer side, and is not furnished with a fin.

One of the commonest species is the LITTLE DOG-FISH, called by several other names, as is usual with a familiar species that is found in many localities. Among such names are SMALL SPOTTED DOG-FISH, LESSER SPOTTED SHARK, MORGAY, and ROBIN HUSS.

This fish is plentiful on the northern coasts of Europe, and is often thought a great nuisance by fishermen, whose bait it takes instead of the more valuable fish for which the hook was set. It generally remains near the bottom of the water, and is a voracious creature, feeding upon crustaceans and small fish. It often follows the shoals of migrating fish, and on account of that custom is called the Dog-fish.

Generally its flesh is neglected, but when properly dressed it is by no means unpalatable, and is said to be sometimes trimmed and dressed in fraudulent imitation of more valuable fish.

The skin of this and other similar species is rough and file-like, and is employed for many purposes. The handles of swords, where a firm hold is required, are sometimes bound with this substance; and joiners use it in polishing the surface of fine woods so as to bring out the grain. It is also employed instead of sand-paper upon match boxes.

The egg of this species is very curious in form and structure, and is often found on the sea-shore, flung up by the waves, especially after a storm. These objects are familiar to all observant wanderers by the sea-shore, under the name of mermaid's purses, sailor's purses, or sea purses. Their form is oblong with curved sides, and at each angle there is a long tendril-like appendage, having a strong curl, and in form not unlike the tendrils of the vine. The use of these appendages is to enable the egg to cling to the growing sea-weed at the bottom of the ocean, and is to prevent it from being washed away by the tide. After a storm, however, when the agitated waves have torn up the beds of marine wrack and other sea-weeds that usually lie in still calmness beneath their sheltering waters, and especially during the time of low tide, these objects may be found lying upon the uncovered and dripping shore, their strong but delicate tendrils entwined in almost inextricable complexity among the salt-loving vegetation of the ocean, and their tiny inmates as yet imperfect and unborn.

Water, which to these creatures contains the breath of life, gains access to the imprisoned sharkling through two slight, longitudinal apertures, one towards each end of the egg; and it is a very remarkable fact that in these waters the undeveloped young are furnished with small external gills, which are afterwards absorbed into the system—a phenomenon curiously analogous to the structure of the tadpole.

The substance of the egg-shell, if such a term can be applied to the envelope which contains the young, is of a moderately stiff, horny character, becoming harder when dry, and of a semi-transparent, yellowish hue, not very unlike, though not so clear as the yellow portions of tortoise-shell.

For the escape of the young Shark, when strong enough to make its own way in the wider world of waters, an outlet is provided in the opened end of the envelope, which opens when pushed from within, and permits the little creature to make its way out, though it effectually bars the entrance against any external foe. When it first leaves its horny home, the neophyte Shark bears with it a capsule, containing a portion of the nutrimental principle of the egg, as is seen in the chicken of the common fowl, and is enabled to exist upon this substance until it has attained the power of foraging for itself, when the small remainder of the capsule is absorbed into the abdomen.

The head of the Little Dog-fish is rather flat upon the top, there is a little spiracle or blow-hole behind each eye, and the shape of the mouth is somewhat like a horse-shoe.

The general color of the body is pale reddish on the upper parts, covered with many little spots of dark reddish-brown; below it is yellowish-white.

The length of this species is about eighteen inches.

THE ROCK DOG-FISH derives its name from the fact that it is often found on rocky coasts. From its superior size, it is also known by the name of LARGE SPOTTED DOG-FISH, and on several coasts it goes by the curt and not euphonious name of BOUNCE.

The habits of this fish are so like those of the preceding species, that they need no description.

It may readily be distinguished from the little dog-fish by the large size and fewer number of the spots, as well as by the shape of the ventral fins, which in this species are nearly squared at the end, whereas in the former they are of a diamond-like form. The color of the Rock Dog-fish is brownish-gray above, without the red tinge of the little dog-fish, and covered rather sparingly with large patches of blackish-brown. Below it is whitish. The length of a fine specimen will sometimes be nearly a yard.

ANOTHER species of Dog-fish, namely, the BLACK-MOUTHED DOG-FISH, or the EYED DOG-FISH (*Pristidúrus melanóstomus*), is mentioned by Mr. Yarrell among the list of European fishes. It may be at once distinguished from either of the preceding species by its large snout, and a row of small, flat, and sharp-edged prickles, arranged in saw-like fashion on the upper rim of the tail fin. The generic title Pristidurus, or Saw-tail, is given to the fish in allusion to this peculiarity.

Its color is light brown on the upper surface, sprinkled with spots, the smaller of which are scattered irregularly, and the larger arranged in four rows, two on each side.

Its length is between two and three feet.

ROCK DOG-FISH.—*Scyllium catulus*.

The Dog-fish family includes six or more genera. The species number fifteen,—rather small sharks, chiefly of the Atlantic. The Black Dog-fishes are represented by one species, found lately off Gloucester, Massachusetts, by the naturalists of the United States Fishery Commission. It is a native in the Greenland seas.

THE COMMON DOG-FISH, or PICKED or PIKED DOG-FISH; also called BONE-DOG, from its potent bony weapons; also SKITTLE-DOG, and HOE. Its range in the Atlantic is very wide, being very abundant on the shores of the Northern and Middle States. Its oil, from the liver, is prized, and it forms an important item of commerce among the fishermen.

A SPECIMEN of the other genus, *Centrocymnus*, was taken near Gloucester, Massachusetts. Its range extends to Portugal, on the opposite shores of the Atlantic.

THE Nurse Shark family includes two genera and four species. These Sharks are not at all the same as the so-called Nurse of the fishermen of the Northern States. It is a large, small-mouthed, harmless Shark, seen in shoals in the warm waters of the sub-tropical and tropical regions. We have seen shoals of this Shark, numbering scores, feeding in the shallow lagoons of the Tortugas reef. Their mouths are situated beneath the snout, as is the case with most Sharks; but in this species they are somewhat like those of the sturgeons, and are not armed with teeth of any considerable size; consequently, their prey consists of small stuff, as the mollusca and crustacea of the shoals. The wide-spread lagoons on the reef at Tortugas are rich feeding-grounds for this Shark. Its rather clumsy form and sluggish, harmless habits render it a tempting source of sport for the youngsters resident there. To sail with a fair wind into a drove of a hundred, more or less, and harpoon a sizeable one that would tow the boat over the reef to their hearts' content, was a privilege our boys highly appreciated, though, perhaps, savoring of the ruder class of sport.

The length of this species is from six to ten feet. Those we were accustomed to see were about seven feet on an average.

Several times we found the young of this species ensconced in some crevice of the broken coral rocks.

Under the family designated as "True Sharks" are embraced twenty or more genera, and about sixty species, found in all seas.

THE SMOOTH HOUND, or DOG SHARK (*Mustelus hinnulus*). This is the smallest of the Sharks of the American waters, and is identical with that of Venice. Another of this genus frequents the coast of California. In the same waters a kind nearly allied is

THE COMMON DOG SHARK (*Triacis semifasciatus*). It is distinguished by a row of rounded black spots along the sides of the body, alternating with the interdorsal cross bars. Another, called *T. henlei*, inhabits the same waters.

THE TIGER SHARK (*Galeocerdo tigrinus*) is the sole representative of this genus in America, being found in waters near Cape Cod. It also inhabits the Indian Ocean. It is rather large, and peculiar for the variegated appearance of coloration.

Another genus, embracing species called Smooth-toothed Sharks, is represented in American waters by THE SMOOTH-TOOTHED SHARK (*Aprionodon punctatus*).

OBLIQUE-TOOTHED SHARK (*Scoliodon terræ-novæ*), called also SHARP-NOSE, is found from Newfoundland to South America. Rather small in size.

The family of Hammer-heads embraces three genera and five species, which inhabit most seas. They are large Sharks, known at once by their most singularly-shaped heads.

SHOVEL-HEAD, or BONNET-HEAD (*Reniceps tibur*), inhabits the Atlantic southward, and extends to China.

Under the term Sharp-nosed Sharks are enumerated several large species, that live in tropical seas, of which there is one species in our waters.

THE SPOTTED-FIN SHARK (*Isogomphodon limbatus*). A stray specimen was found at Wood's Hill, Massachusetts.

THE BLUE SHARK, so called from the fine slaty-blue color of its skin, is a not unfrequent visitor of the shores of Northern Europe, and is the object of the deadliest hatred to the fishermen, who are sometimes doomed to see their fish stolen, their nets cut to pieces, and their lines hopelessly ruined by this fish, without the least power of checking its depredations.

About the month of June, according to Mr. Couch's observations, this Shark makes its appearance on the coasts, and has sometimes been so plentiful that nine or ten have been taken by the fishing boats in a single day. As the fishermen are hauling up their lines with the fish upon the hooks, the Blue Shark will follow the fish as it is drawn upwards, seize upon it, and hook itself for its trouble. Exasperated by the unsuspected check upon its maraudings, it tries to bite the line asunder, a feat easily performed by its lancet-like teeth with their notched edges.

Sometimes, however, it takes to another stratagem, and as soon as it feels the hook, rolls itself round so rapidly on its axis, that it winds the line round its body into a mass of inextricable entanglement. So effectually is this feat achieved, that, in spite of the value of the line, the fishermen have been known to give up any attempt to unravel its knotty convolutions. This fish has another fashion of biting the line asunder without any apparent reason.

Perhaps, however, it never is so thoroughly destructive as in the pilchard season, when it follows the vast shoals of these fish to the continental shores, and devours them wholesale. Even when they are inclosed in the net, the Blue Shark is not to be baffled or deprived of its expected banquet; for, swimming along the whole length of the net, it bites at the inclosed fish, caring nothing for the meshes, and taking out large mouthfuls of mingled net and pilchards, swallows them together.

The sailors have an idea that this voracious fish is able to succor her young, when in danger, by opening her mouth and letting them swim down her throat. It is undoubtedly true, that living young have been found in the stomach of large sharks; but whether they had been swallowed as a means of protection, is by no means proved. The reader will doubtlessly remember the similar stories that have been told of the viper and other poisonous snakes.

The skull of a Shark shows the terrible teeth with which it is armed. They lie in several rows, ready to take the place of those which are broken or cast off when their work is done. From these teeth, which cut like broken glass, the natives of many savage lands make tools and weapons of war, by ingeniously fixing them into wooden handles.

The voracity and dullness of nerve belonging to the Shark is really wonderful. One of my friends was fishing after a large Shark that was following the vessel, and, after a little time, succeeded in inducing the fish to take the great hook that had been nicely baited with pork to suit his palate. Too sudden a jerk, however, having been given to the line, the hook tore its way through the side of the cheek, setting the Shark free. The wound was a terrible one, and bled profusely; but the Shark seemed to care little or nothing about it, still hovered about the bait, as if unable to resist its attractions, and after a little while was hooked a second time and hauled safely on board.

The capture of a Shark is always an event on board ship, especially if she be a sailing vessel and the wind has fallen. A hook made for the purpose is secured to a fathom or so of iron chain, the Shark being capable of biting through a rope in an instant, and in no way so particular in its diet as to need fine tackle. Indeed, as in the last-mentioned instance, the creature seems to be perfectly aware of the danger, but to be incapable of resisting the tempting morsel. The other end of the chain is firmly lashed to a stout rope, and the latter secured to the vessel, as one rush of a powerful Shark would pull half a dozen men overboard.

All things being ready, a good large piece of pork is fixed tightly on the hook, and allowed to tow overboard. The Shark, being to the full as inquisitive as the cat, comes up with true feline curiosity, and sniffs at the bait with an air of deliberate scrutiny. Sometimes, having perhaps lately partaken of a good meal, it is very coy about taking the bait, and keeps the anxious anglers above in a state of tantalized impatience for an hour or more. Generally, however, it dashes at the bait at once, and has even been known to leap from the water and hook itself before the bait had even reached the surface.

Now begins a mighty struggle, and all is eager excitement. The Shark knows no wiles, but uses all its great strength to tear away from the hook by sheer force, having apparently but slight sense of pain, and in many cases would do so were not a check put upon its efforts by a rope knotted into a bowline and dexterously slipped over its tail. Being now held by both extremities, it is shorn of its strength, like Samson without his locks, and lifted on deck by both lines. Sometimes a trident-like harpoon, technically called a "grains," the handle of which is heavily loaded with lead to make it fall with greater force, is dropped upon the struggling fish.

Being brought on deck, however, the struggles of the creature recommence with tenfold violence. Twisting with marvellous agility, snapping right and left with its murderous teeth, and dealing heavy blows with its terrible tail, it makes the deck tremble under its strokes, until some experienced sailor runs in with an axe, and, with a blow across the tail, reduces the creature to malignant impotence. The muscles of the Shark are endowed with astonishing irritability, and long after the body has been cut to pieces and parts of it cooked and eaten, the flesh will quiver if pricked with a knife-point; the separated heart will beat steadily while lying on the bare boards, and the jaws of the severed head will snap with frightful vehemence if any object be put between the teeth.

Sailors generally make high festival at the dismemberment of a Shark, and have great delight in opening the creature, for the purpose of finding out the articles which it had swallowed. For a Shark, when following a vessel, will eat anything that falls overboard. The contents of a lady's workbox, a cow's hide entire, knives, hats, boots, and all kind of miscellanea have been found in the interior of a Shark; while, on one occasion, were discovered the papers of a slaver, which had been flung overboard when the vessel was overhauled, and, by means of which papers so strangely recovered, the vessel was identified and condemned.

The color of this species is beautiful slate-blue above, and white below.

The Blue Sharks are represented in our waters by four species, being very numerous in species in other and tropical seas.

THE GREAT BLUE SHARK (*Charcharinus glaucus*) is a large form of the tropics, occasionally found in our American seas. Mr. Couch, the British naturalist, says: "This Shark is so plentiful about the month of June, that nine or ten have been taken in one day." It is a constant and serious trouble to the fishermen. This Shark is one of the kinds that are so frequently taken on ocean-going vessels.

THE DUSKY SHARK (*C. obscurus*) is a large one, reaching the length of nine or ten feet. It is very common off the American Atlantic shores.

THE SMALLER BLUE SHARK (*C. milberti*) ranges from Cape Cod to Florida, and is also found in the Mediterranean Sea. De Kay says: "This is taken frequently along our shores, and as far north as New Hampshire."

A SPECIES, *C. lamia*, was identified by Prof. Putnam from a tooth which is large enough to represent a Shark thirteen feet in length.

THE remarkable fish depicted in the accompanying illustration affords a striking instance of the wild and wondrous modifications of form assumed by certain creatures, without any ascertained purpose being gained thereby. We know by analogous reasoning that some wise and beautiful purpose is served by this astonishing variation in form; but as far as is yet known, there is nothing in the habits of this species that accounts for the necessity of this strange shape.

The shape of the body is not unlike that of the generality of Sharks, but it is upon the head that the attention is at once rivetted. As may be seen from the figure, the head is expanded laterally in a most singular manner, bearing, indeed, no small resemblance to the head of a hammer. The eyes are placed at either end of the projecting extremities, and the mouth is set quite below, its corners just coinciding with a line drawn through the two projecting lobes of the head. It is worthy of notice, that several of the commonest insects— those beautiful dragon-flies belonging to the genus Agrion—have heads modelled on a very similar principle, and there are some exotic insects where this singular shape is even more exaggerated, the eyes being set quite at the end of long lateral footstalks.

This species attains to a considerable size, seven or eight feet being a common measurement, and specimens of eleven or twelve feet having been known. Its flesh is said to be almost uneatable, being hard, coarse, and ill-flavored.

The HAMMER-HEADED SHARK produces living young, and from the interior of a very fine specimen captured near Tenby, and measuring more than ten feet in length, were taken no less than thirty-nine young, all perfectly formed, and averaging nineteen inches in length.

Several species of Hammer-headed Sharks are known, among which the Heart-headed Shark (*Sphyrnias tiburo*), has the best developed head, and the Broad-headed Shark (*Sphyrnias láticeps*), the most so. Another species, the Tudes (*Sphyrnias tudes*), thought to inhabit the Mediterranean, and the shores of Southern America, is intermediate between the two extremes.

The general color of this species is grayish-brown above, and grayish-white below.

HAMMER-HEAD (*Sphyrna zygæna*). This is a large Shark, found in most seas. It is common on the American coast from Cape Cod southward. The width of head is about twice its length.

In Cuba this is called Cornuda. Dr. Mitchell says "the voracity of this Shark may be judged from the following occurrence at Sag Harbor, in September, 1805. Three of this species were taken in a net by Mr. Joshua Terry; the largest was eleven feet in length. On

HAMMER-HEADED SHARK.—*Spyrnius zygæna*. (One-fifteenth natural size.)

opening him many detached parts of a man were disclosed, with portions of clothing." DeKay says it is much dreaded by the Long Island fishermen, for its boldness. Some have been seen in "Hell Gate" four feet in length. The Hammer-head is equally well known on both sides of the Atlantic. Its range is from the coast of Brazil northward, but is not known to pass Cape Cod.

The extraordinary shape of this creature's head is, seemingly, a deformity; yet we are not justified in so believing—for Nature doeth all things well, and for a purpose. This Shark brings forth living young, from thirty to forty in number, all perfectly formed, and averaging nineteen inches in length.

THE destructive and voracious fish, which is indiscriminately known by the names of TOPE, PENNY DOG, or MILLER'S DOG, according to the particular coast near which it is found, is another familiar representative of the great Shark family.

The Tope is commoner towards the southern than the northern coast, but wherever it is found, it is an intolerable nuisance, behaving itself much after the example set by the blue Shark, and being, in proportion to its dimensions, quite as injurious to the fishing interest. Like the last-mentioned species, it produces living young, the number of a single family being about thirty. They are born in May and June, and mostly remain on the coasts through their first winter, not retiring into deep water till they have entered their second year.

Like the blue Shark, the Tope is fond of robbing the fishermen's hooks, and will in like manner endeavor to free itself when hooked, biting through the line, or rolling round with

PICKED DOG-FISH AND SMOOTH HOUND.—*Acanthias vulgaris et Mustelus vulgaris.*

such rapidity that it winds the long cord about its body into tangled knots. The upper surface of the Tope is slaty-gray, becoming lighter towards the abdomen, which is nearly white.

TOPE (*Galeorhinus galeus*). The common name of this Shark is local in the tropical countries. PENNY DOG and MILLER'S DOG are names applied to it in Europe. It is one of the species that brings its young forth alive. They are born in May, and the brood is said to be thirty in number. San Francisco is recorded as one locality it inhabits. It seems to be the only species yet known in America.

THE prettily marked and curiously toothed SMOOTH HOUND is also known under the titles of SKATE-TOOTHED SHARK and RAY-TOOTHED DOG, the two latter titles being appropriately given it on account of its curious and beautifully formed teeth, which resemble in form the cylinders of a crushing mill, and are used for a similar purpose.

The jaws, instead of being studded with rows of sharp and knife-like teeth, are supplied with two rounded projections on which the flat-topped teeth are set closely together like the stones of a mosaic, and which are so formed that they roll over each other as the jaws are closed, producing a crushing effect of enormous power. These curious teeth are rendered needful by the food on which the Smooth Hound lives, namely, the hard-shelled crustaceans, whose armor of proof is nevertheless soon comminuted under the bony rollers.

As may be inferred from the character of its food, the Smooth Hound is not destructive to the fisheries, and may be allowed to live in harmless security. Its flesh is said to be tolerably well-flavored, and even moderately tender. It produces its young in a living state, but is not very prolific, the number at a birth rarely exceeding ten or twelve. Almost as soon as born they retire into deep water, so that, though a tolerably plentiful species, it is not seen so often as those which live in shallow waters.

The color of the Smooth Hound is pearly-gray, and above the lateral line, which in this species is very strongly marked, the body is decorated with small round white spots, very conspicuous while the creature is young, but becoming fainter when it attains maturity. The under parts are whitish-yellow.

BEFORE noticing some of the larger and more terrible species, we must not omit the PORBEAGLE, sometimes called the BEAUMARIS SHARK (*Isúrus cornúbicus*), a fish of a wonderfully mild aspect for a Shark, and notable for a very porpoise-like aspect. The name of Porbeagle is in fact owing to this resemblance. This species feeds on fish of various kinds, three full-grown hakes having been found in the stomach of one individual, and derives some of its subsistence from the larger mollusks. It attains a rather large size, five or six feet being a common length. Its color is uniform grayish-black above, and white below.

THE dreadful WHITE SHARK, the finny pirate of the ocean, is one of the large species that range the ocean, and in some seas are so numerous that they are the terror of sailors and natives. One individual, whose jaws are still preserved, was said to have measured thirty-seven feet in length; and when we take into consideration the many instances where the leg of a man has been bitten off through flesh and bone as easily as if it had been a carrot, and even the body of a boy or woman severed at a single bite, this great length will not seem to be exaggerated.

Many portions of this fish are used in commerce. The sailors are fond of cleaning and preparing the skull, which, when brought ashore, is sure of a ready sale, either for a public museum, or to private individuals who are struck with its strange form and terrible armature. The spine, too, is frequently taken from this fish, and when dried, it passes into the hands of walking-stick makers, who polish it neatly, fit it with a gold handle, and sell it at a very high price. One of these sticks will sometimes fetch thirty or forty dollars. There is also a large amount of oil in the Shark, which is thought rather valuable, so that in Ceylon and other places a regular trade in this commodity is carried on.

The fins are very rich in gelatine, and in China are, as is said, employed largely in the manufacture of that gelatinous soup in which the soul of a Chinese epicure delights, and of which the turtle soup is thought by Chinese judges to be a faint penumbra or distant imitation. The flesh is eaten by the natives of many Pacific islands; and in some places the liver is looked upon as a royal luxury, being hung on boards in the sun until all the contained oil has drained away, and then carefully wrapped up in leaves and reserved as a delicacy.

These islanders have a very quaint method of catching the Shark—absurdly impotent in theory, but strangely efficacious in practice. They cut a large log of wood into the rude resemblance of a canoe, tie a rope round the middle, form the end of the rope into a noose, and then set it afloat, leaving the noose to dangle in the water. Whether induced by curiosity, or by what strange impulse urged, is not very clear, but the fact is patent that before the noose has been floating very long, a Shark is sure to push its head through it, and on backing as soon as it feels the obstruction, is caught by the tightening of the noose. The natives then go off in their canoes, chasing the bewildered Shark, who is unable to dive on account of the floating log, and who is so lustily battered about the head with the heavy clubs so admirably made by those ingenious natives, that it is soon killed and hauled ashore in triumph.

The color of the White Shark is ashen-brown above, and white below.

The Great White Shark is named in America Man-eater, and Atwood's Shark, though the latter term, applied in honor of the sea captain of Provincetown, Mass., who assisted greatly in contributing to Dr. Storer's report on the Fishes of Massachusetts, is now dropped, as the species is regarded as identical with the European. It is found in all temperate and tropical seas, and is one of the largest of Sharks, reaching a length of fifteen feet, and the weight of nearly a ton.

This Shark is especially abundant on the Florida Reef. While resident there, at Fort Jefferson, our lads of the garrison had much sport in capturing them. On one occasion, one measuring ten feet in length, and very bulky, as this species is, was hooked off shore. The boys had one of the flat pontoon boats of the engineers, otherwise they would have been taken off seaward. As it was, they had enough to do. Several soldiers went to their assistance, and rowed the party in, while the prize, hauled "short up" to the gunwale, made savage resistance. On gaining the moat of the fort planks were laid, and the monster hauled over into the confined water of the ditch. Here he remained, restlessly swimming its length of waters, steadily refusing any food. In about two months he died, seemingly exhausted.

THE Basking Sharks, or Whale-nosed, are represented in American waters by the GREAT BASKING SHARK (*Cetorhinus maximus*.) It is thought to be the largest existing Shark, being nearly forty feet in length. It is quite notable, among other things in being a resident in Arctic waters, from whence it strays as far as the latitude of Portugal, and of Virginia. It is also strange in being so enormous in bulk, and at the same time quite harmless, as it has but a small mouth and quite insignificant armament of teeth. It is called in Europe Sun-fish, Sail-fish, and Hoe-mother, from the habit of basking in the sun on the surface of the sea; from the sail-like appearance of its high dorsal fin when near the surface, and from the fact that the fishermen affect to believe it to be the Mother of the Hoe, or Piked Dog-fish, respectively.

It seems to be of a rather dull and listless character, allowing itself to be approached quite closely by a boat, without giving any signs of alarm until the bow of the boat actually touches its person.

The gill apertures of the Basking Shark are extremely long, reaching almost across the neck. The head is conical, the muzzle short, and the eyes near the snout. The skin is very rough to the touch, whether the hand be passed from head to tail or *vice versâ*, and the color is blackish-brown, glossed with a bluish tint.

Among our fishermen it is known as Bone Shark and Elephant Shark. A specimen drifted ashore at Provincetown, Mass., which afforded six barrels of oil, taken from the liver alone, which sold for $103. The food of this Shark is probably small mollusks and crustaceans, and, according to some authors, marine algæ.

A WELL-KNOWN species, familiar under the names of THRESHER (*Alopias vulpes*), FOX SHARK, SEA FOX, SEA APE, SWINGLE-TAIL, LONG-TAIL, etc., is at once to be recognized by the peculiar form of the head and the wonderfully long upper lobe of the tail, which equals in length the body from the tip of the snout to the base of the tail. The lower lobe is quite short, and in no way conspicuous.

This fish is appropriately called the Thresher on account of its habit of using its long and flexible tail after the fashion of a quarter-staff, and dealing the most tremendous blows on or near any object that may excite its ire. Sometimes it seems to employ its tail in playing off a practical joke or frightening away dolphins or other creatures that are disporting themselves in apparent security. The following short account by Captain Crow will give a good idea of the powers of this tremendous weapon when wielded by the iron muscles of the Thresher:—

"One morning during a calm, when near the Hebrides, all hands were called up at three A. M. to witness a battle between several of the fish called Threshers or Fox Sharks and some swordfish on the one side and an enormous whale on the other. It was in the middle of summer, and the weather being clear and the fish close to the vessel, we had a fine opportunity of witnessing the contest. As soon as the whale's back appeared above the water, the Threshers, springing several yards into the air, descended with great violence upon the object of their rancor, and inflicted upon him the most severe slaps with their long tails, the sounds of which resembled the reports of muskets fired at a distance.

"The swordfish in their turn attacked the distressed whale, striking from below, and thus beset on all sides, and wounded, where the poor creature appeared, the water around him was dyed with blood. In this manner they continued tormenting and wounding him for many hours, until we lost sight of him, and I have no doubt that they in the end completed his destruction." This strange alliance of two different fish against a marine mammal is a truly curious circumstance, and may have a deeper meaning than appears on the surface.

The food of the Thresher consists mostly of fish, and in the stomach of one of these creatures taken off the coast of Cornwall were found a quantity of young herrings. The color of the Thresher is dark slaty-blue above, and the same color, but mottled with white, below.

It abounds in all warm seas, and in summer is one of the most abundant kinds on our New England coast. It is also occasionally taken on the Pacific coast. The Thresher is the only representative of its family.

THE family of Sand Sharks includes one genus and three species, of which *Carcharias americanus* is the more familiar form. It is a small voracious Shark, rather common on our Atlantic shores.

The family of Porbeagles is well known through its very familiar representative, the Mackerel Shark (*Isurus glaucus*).

A species, allied of the genus *Lamna*, was lately discovered at Wood's Holl, Mass. It is a large and fierce creature.

A large spotted species of the Whale-shark family inhabits the California waters.

The family which embraces the Port Jackson Sharks is represented in California seas by *Cestracion francesci*.

Another family in this connection is known as the Cow-shark family, having one species which ranges from Cape of Good Hope to California. It is named "Perlous," and is of the genus Heptangus.

AMONG some other Sharks, the PICKED DOG-FISH deserves notice, on account of the curious weapons from which it derives its name.

In front of each dorsal fin is placed a strong and sharply pointed spine, or pike, which has caused the fish to receive its popular name in most parts of the coast. The word is a dissyllable, and pronounced Pick-ed. On some of the shores it is called the BONE DOG, and on others it is known by the name of the HOE.

These spines form aggressive weapons of a rather formidable character, the fish having the capability of directing a blow with wonderful accuracy. Mr. Couch says, that he has known the Picked Dog-fish able to pierce a finger if laid on its head, and never to miss its aim. When about to strike, it bends its body like a bow, and suddenly lashes out in the intended direction. It is a very common species, especially during the herring season, as it follows the shoals of those fish for the purpose of feeding on them. Even the tiny, quarter-grown young, not half the size of their intended prey, instinctively follow the herrings, though it is manifestly impossible that they should be able to eat them.

The Picked Dog-fish is destructive to the fishing trade, not only on account of its large appetite and the number of fish it consumes, but because it cuts the hooks away from the lines with its sharp teeth. As, moreover, it is extremely plentiful, some twenty thousand having been captured at one haul of a seine net, the destruction which it causes can be readily imagined. Sometimes this fish assembles in large shoals, and then the fishermen avenge themselves of their injuries, by shooting their nets around them, and capturing them by boats' loads at a time. Their flesh is tolerably good, a useful oil is obtained plentifully from the liver, while the refuse portions are most valuable as manure, and are strewed in unfragrant richness over the fields, warning the nostrils at a considerable distance that the next year's crop is likely to be successful, and that a nearer approach is undesirable except to the farmer and the entomologist.

The color of the Picked Dog-fish is slaty-gray above, diversified, when young, with a few white spots, and the under parts are yellowish-white. The skin is rough if stroked from the tail to the head, and smooth when rubbed in the reverse direction. The average length of this species is about eighteen inches. It is illustrated, together with the Smooth Hound, on page 200.

THE GREENLAND, or NORTHERN SHARK (*Dalátias boreális*), must receive a brief notice, as it is frequently mentioned in accounts of whaling voyages.

This species is remarkable for the very small proportionate size of the fins, and for the manner in which the points of the teeth diverge from the centre of the jaw. It is a great foe to the whale and whalers, and is so heedless of danger when intent on satisfying its hunger, that it will follow a dead whale to the ship, mix boldly with the men who are engaged in cutting the blubber, thrust its head boldly among them, and at every bite scoop out lumps as large as a man's head.

So deeply engaged is the creature in this interesting occupation, that even if a man should slip into the water from the smooth oily skin of the whale, the Greenland Sharks take no notice of him, but continue their depredations on the whale. Even after the long whaling knife has been thrust through its body, it will dart off for the moment on feeling the wound, but will soon return to the same spot and continue its banquet. It also feeds on crustaceans and small fishes. Many specimens are nearly if not wholly blinded by a parasitic animal technically called *Lernæa elongata*, some three inches in length, which fastens upon the corner of the eye, and lives upon its fluids.

The color of this species is brown with a shade of deep blue. Its length, when full-grown, is about fourteen feet.

ANOTHER curious species of Shark, called appropriately the SPINOUS SHARK (*Echinorhínus spinósus*), is notable for the spine-topped bony tubercles which are scattered over the surface of the body. The greater number of these spinous projections are boldly hooked, in a manner not unlike the thorns of the common bramble, and the points are directed backwards; others, however, are quite straight and stand upright. The object of these curious spines is not clearly known. They are very small in proportion to the size of the fish, and it is said that the males are more thickly studded with them than the females.

The color is dark leaden gray on the head and back as far as the first dorsal fin, the remainder being reddish-yellow with mottlings and cloudings of purple and brown. On the abdomen are irregular spots of vermilion. The chin and sides of the mouth are white. The average length of a full-grown specimen seems to be about seven or eight feet. In most, if not in all, of these creatures, the female is larger than the male, as is the case with the birds of prey.

THE dark-skinned, wide-mouthed, leather-finned, and thorn-backed fish which is shown in the illustration, is popularly known throughout many parts of Europe by the name of the ANGEL-FISH, a term singularly inappropriate except on the well-known principle "lucus a non lucendo," or perchance as leaving the spectator the option of choosing the kind of angel which the creature is thought to resemble.

Sooth to say, it is as hideous a fish as is to be found in the waters, and from all accounts is as unprepossessing to the inhabitants of the sea as to those of the land, being voracious to a degree, and attaining a size that causes it to be a most formidable foe to the many fishes on which it feeds. It is also known by the name of MONK-FISH, in allusion to the rounded head, which was thought to bear some resemblance to the shaven crown of a monk; and in some places is called the SHARK RAY because it seems to be one of the connecting links between the sharks and the rays, and has many of the characteristics of both. On some parts of the English coasts it is known as the KINGSTON.

It has many of the habits of the flat-fishes, keeping near the bottom, and even wriggling its way into the muddy sand of the sea-bed so as to conceal its entire body. As in the course of these movements it disturbs many soles, plaice, flounders, and other flat-fishes that inhabit the same localities, it snaps them up as they endeavor to escape, and devours great quantities of them, so that it is really a destructive fish upon a coast.

It is most common upon the southern shores, and has there been taken of considerable size, attaining a weight of a hundred pounds. Unfortunately, the flesh is now thought to be

ANGEL-FISH.—*Squatina vulgaris.*

too coarse for the table, though it was formerly in some estimation, so that the creature is useless to the fisherman, who can only avenge himself for his losses by killing the destructive creature, but cannot repay himself by eating or selling it. The skin, however, being rough, is of some small use in the arts, being dried and employed, like that of the dog-fish, for polishing joiner's work, and it is in some places manufactured into a sort of shagreen.

The eyes are set rather far back on the upper part of the head, and a little behind each eye is the temporal orifice, very large, in proportion to the dimensions of the fish, very long, and set transversely on the head. The wide mouth, which opens in front of the head and not below as in the sharks, is furnished with rather long and sharply-pointed teeth. The color of the upper parts is dark chocolate-brown mottled with a darker hue, and very rough. Along the back runs a row of short, sharp spines, their points directed backwards, and the under parts are smooth and of a dull brownish-white. The length of an adult specimen is seven or eight feet.

THE RAYS.

This group of fishes forms a separate order, in which are seven families.

The Saw-fish (*Pristis antiquorum*) is a familiar form, sometimes reaching the length of fifteen feet; having a saw-like snout four or five feet in length. It is found in nearly all the warmer seas, and even in the colder regions. It is reported that this Ray swings itself sidewise with rapidity and thereby cuts down fishes by the double-edged saw, which proves a most effective weapon.

The color of the Saw-fish is dark gray above, nearly black in some individuals, the sides are ashen, and the abdomen white. It often attains a great size, measuring fifteen or eighteen feet in length, including the saw.

The Tentaculated Saw-fish (*Pristióphorus cirrátus*) is worthy of notice as forming a transition link between the sharks and the true Saw-fish. In this creature, the snout is lengthened and armed with spines; but these structures are of different lengths, hooked, and only attached to the skin, and not implanted in the bone, as is the case with the true Saw-fish.

In the true Rays, or Raidæ, the fore part of the body is flattened and formed into a disc-like shape, by the conjunction of the breast fins with the snout.

Our first example of the Rays is the Torpedo, a fish long celebrated for its power of emitting at will electrical shocks of considerable intensity. In consequence of this property, it is sometimes called the Cramp-fish, Cramp Ray, Electric Ray, or Numb-fish.

The object of this strange power seems to be twofold, namely, to defend itself from the attacks of foes, and to benumb the swift and active fish on which it feeds, and which its slow movements would not permit it to catch in fair chase. It does not always deliver the electric shock when touched, though it is generally rather prodigal of exercising its potent though invisible arms, but will allow itself to be touched, and even handled, without inflicting a shock. But if the creature be continually annoyed, the shock is sure to come at last, and in such cases with double violence. It has been observed, moreover, that the fish depresses its eyes just before giving its shock.

The power of the shock varies greatly in different individuals, with some being so strong as to cause the recipient to fall to the ground as if shot, and, with others, so feeble that it is hardly perceived. According to M. de Quatrefages, the fishermen are sometimes unpleasantly made aware that they have captured a Torpedo in their meshes, by the sudden shock through their arms and breast as they are hauling in their net. Anglers, too, are sometimes struck by means of the line which they are holding; and I presume that in either case the line must be wet, or it would not act as a conductor of the electrical fluid.

One of these fishes was placed in a vessel of water, and a duck was forced to swim about in the same vessel. The Torpedo soon became excited, and in a few hours the duck was dead. Fish, also, of different kinds are killed by this remarkable influence; and it is plausibly suggested by one writer, that this mode of destruction would render them liable to rapid decomposition, and would aid the organs of digestion in a creature like the Torpedo, where they are but imperfectly developed.

The shocks of this fish were once used as remedies for gout and fevers. In the first case, the patient had to lay his foot on the Torpedo, and bravely hold it in its place, despite of all the shocks sent by the angry fish through the sensitive limb of the aggressor; and in the latter case the Torpedo was used, as it were, to frighten the fever out of the system,

The patient was stripped, and the Torpedo placed successively to the joints, trunk, and extremities, so that the whole of the body and limbs were permeated, in their turn, by the electric shock.

That the stroke of the Torpedo is veritable electricity is a fact which was once much disputed, but is now conclusively proved by a host of experiments. Needles have been magnetized by it just as if the shock had been that of a galvanic battery, the electrometer showed decided proofs of the nature of the fluid that had been sent through it, and even the electric spark has been obtained from the Torpedo—very small, it is true, but still recognizably apparent. It is rather curious, that in the course of the experiments it was discovered that the upper surface of the Torpedo corresponded with the copper plate of a battery, and the lower surface with the zinc plate.

COMMON SKATE AND EYED TORPEDO.—*Raja batis et Torpedo oculatus.*

The structure of the electrical organ is far too complex to be fully described in this work, as it would require at least forty or fifty pages, and a large number of illustrations. I will, however, give a brief summary of the strange organ by which such wonderful results are obtained, and any of my readers who would like to examine it more in detail, will find ample information in an article on the subject by Dr. Coldstream, in the "Cyclopædia of Anatomy and Physiology."

Briefly, then, this organ is duplex, and consists of a great number of columns, placed closely against each other, each inclosed in a very thin membrane. These columns are again built up, as it were, of flat discs, separated by a delicate membrane, which seems to contain fluid. This structure may be roughly imitated by piling a number of coins upon each other, with a bladder between each coin and its successor—in fact, a kind of voltaic pile. The length of the columns, and consequently the number of discs, varies according to their position in the body. The columns extend quite through the creature, from the skin of the back to that of the abdomen, and are clearly visible on both sides, so that those of the middle are necessarily the longest, and those at either end become gradually shorter. In many large specimens, more than eleven hundred columns were counted, and the number of discs is on an average a hundred to the inch. It seems, from the best researches, that the growth of this organ is produced, not by the increase of each column, but by a continual addition to their number. A vast amount of blood-vessels pass through the electric organ, and it is permeated with nerves in every direction.

How the electrical effect is produced is a very deep mystery. In fact, we know scarcely aught of this marvellous power, save the knowledge that it pervades all nature, and even in its external manifestations is one of the most ethereal and most potent of the second means through which the will of the Creator guides His universe. That the same electrical principle exists in all animals is familiarly known, and also that it is far more intense in some individuals than in others of the same species. It is known that the contact of two different kinds of flesh, such as the muscle of a fish and an ox, both newly killed, will produce similar effects; and that it exists so largely in human beings, that no two individuals can place themselves on isolated stools, and join their hands, without emitting so much electricity by that slight contact, that the instrument will record its presence. But the origin of this wonderful power eludes our mental grasp like the receding waters of the mirage, and the increase of our knowledge serves but to betray the extent of our ignorance.

I cannot but think that this subtle and potent emanation, which is able to strike the victim through an intervening space of the fluid common to both aggressor and sufferer, has some affinity with the still more subtle and equally mysterious influence by which certain of the serpent race are enabled to paralyze or attract the creatures which they could not secure by actual contact. It may possibly be that the electric powers of the Torpedo, which need water or some other conducting substance for their exercise, are, after all, but a more concentrated and palpable manifestation of that force, which enables the rattlesnake to arrest an animal not in physical contact with itself, the pointed finger to lay a bird motionless on its back until released by a sudden sound or touch, and one human being to influence his fellow without the use of words, and to attract or repel him by an irresistible though invisible agency.

It is rather remarkable that even the Torpedo, gifted with such puissant arms, dealing pain and death around at will, should find at all events one foe insensible to the electric stroke, and perhaps even needing its exciting influence to preserve it in health. This is a parasitic creature, termed scientifically the Branchellion, which clings to the Torpedo and feeds upon its juices, quite indifferent to all the shocks which its victim dispenses. It generally measures from an inch to an inch and a half in length.

This fish is found in the Mediterranean, and the Indian and Pacific Oceans, and occasionally off the Cape. Happily, the Torpedo does not attain a very great size, one of the largest specimens being about four feet long, and weighing sixty or seventy pounds.

Of the Torpedo family there are three species known to American waters. The Numbfish, or Cramp-fish, Torpedo, was formerly common off Cape Cod, Mass., ranging southward to Florida.

A specimen weighing sixty pounds was sent to Boston, where Dr. Storer examined it. It was powerful enough to give an all sufficient shock to an average man.

A species is known in California waters. The Ray family proper, embracing the forms called Skates, has four genera and about forty species. Those in American waters are: the Common Skate, Ocellated Ray, Starry Ray, Brier Ray, Barn-door Skate, Granulated Skate. On the Pacific coast are four kinds. The Skates are well known to all who go down to the sea "to fish."

The Butterfly Ray inhabits the same waters. The young of the Skates are produced from eggs, called barrows, from their resemblance to hand barrows. Many have seen those black or brownish oblong objects on the beaches, having four arms, which cling to seaweeds sometimes. These are the eggs. Fine examples will be seen on page 195, and in the plate near the close of this volume, entitled *Gorgonia verrucosa;* the "barrows" are seen coiling on the branches.

The Eagle Rays form another family, having three genera and twenty species.

The Cow-nose Rays are members of this family. Some of them are very handsome— really looking like large butterflies while swimming, being about a foot wide, with pure white bellies and delicate gray backs.

THE Rays are well represented by several large and curious species. One of the commonest examples is the THORNBACK SKATE or RAY, so called from the large number

THORNBACK SKATE.—*Raja clavata.*

of thorny projections which are scattered over its back, and especially along the spine. This species is represented by the illustration.

The Thornback is one of the common Rays, and is taken plentifully on the shores of northern Europe. As is the case with many of the same genus, the flesh is considered rather good, and is eaten both when fresh and when salted for consumption during stormy weather. Autumn and winter are the best seasons for procuring this fish, as the flesh is then firm and white, while during the rest of the year it is rather liable to become flabby. Thornbacks taken in November are thought to be the best.

This species, like the rest of the Rays, feeds on crustacea, flat-fish, and mollusks, and as many of these creatures possess very hard shells, the Rays are furnished with a crushing mill of teeth, which roll on each other in such a way that even the stony shell of a crab is broken up under the pressure. It is notable that the teeth differ in the two sexes when adult. Those of the female are flat on the top, but those of the male throw out a strong angular projection, which is so arranged that the projections of one jaw exactly fit into the interstices of the other, and the roller-like arrays of teeth bear a wonderful resemblance to the well-known clod-crushing machine.

The young of this and other Skates are produced from eggs, whose form is familiar to every visitor to the sea-shore, where they go by the popular name of Skate-barrows. Their color is black, their texture leathery, thin, and tough, and their form wonderfully like a common hand-barrow, the body of the barrow being represented by the middle of the egg, and the handles by the four projections at the angles. The empty cases are continually thrown on the beach, but it is seldom that the young are found inclosed, except after a violent storm, or when obtained by means of the dredge.

This species is notable for certain thorny appendages to the skin, which are profusely sown over the back and whole upper surface, and among which stand out conspicuously a few very large tubercular spines, with broad, oval, bony bases, and curved, sharp-pointed projections. Fifteen or sixteen of these bony thorns are found on the back. Along the spine runs a single row of similiar spines, and at the commencement of the tail it is accompanied by another row on either side, making that member a very formidable instrument of offence. In point of fact, the tail is as formidable a weapon as can be met with, and the manner in which this living quarter-staff is wielded adds in no slight degree to its power. When angered, the Skate bends its body into a bow-like form, so that the tail nearly touches the snout, and then, with a sudden fling, lashes out with the tail in the direction of the offender, never failing to inflict a most painful stroke if the blow should happen to take effect.

The color of the Thorn-back Skate is brown, diversified with many spots of brownish-gray, and the under parts are pure white.

THE family of "Sea Devils" embraces the most remarkable forms of any.

The SEA DEVIL, DEVIL-FISH, MANTA (*Manta birostris*), inhabits from the tropical waters northward to the Carolinas, reaching the length, or width, properly, of twenty feet. Dr. Mitchell records one that required the strength of three yoke of oxen to drag it. It was estimated to weigh over four tons. Singular instances have occurred of this creature becoming entangled in the anchor gear of small vessels, and actually towing them some distance before the cause of the unusual movement was discerned.

We have seen several of these monsters, feeding apparently on shoals of small fishes, in the Gulf of Mexico. As they turned in the course of their feeding, the great pectoral flaps were thrown upward out of the water, exposing a white under surface, and creating a great commotion in the sea.

THE COMMON SKATE, sometimes called the TINKER, is so well known that only a very short description is needed.

This fish is found in great plenty, and sometimes attains to a really large size, a fine specimen having been known to weigh two hundred pounds. The fishermen have a custom of calling the female Skate a Maid, and the male, in consequence of the two elongated appendages at the base of the tail, is called the Three-Tailed Skate. It is a very voracious creature, eating various kinds of fish, crustaceans, and other inhabitants of the deep.

The color of this species is grayish-brown on the upper surface, and a little reddish-brown and black-brown are found on the edges of the broad fins. Below, it is grayish-white, over which divers darker lines are drawn, and upon which are scattered a great number of bluish spots with small sharp points among them. It is illustrated on page 207, together with the Eyed Torpedo.

A FAMILY called STING RAY has seven species. Some of them are of great size. Terrible as is the armed tail of the thorn-back skate, and severe as are the wounds that can be inflicted by it, the Sting Ray is furnished with a weapon even more to be dreaded, and capable of causing a still more serious injury.

The tail itself of this species is long, flexible, whip-like, and smooth, so that were it unaided by any additional armature, it could only inflict a sharp and stinging blow, which, however painful, would do no more damage than the cut of a horsewhip. The tail is further armed with a projecting bony spine, very sharp at the point, and furnished along both edges with sharp cutting teeth. . When attacked or irritated, the Sting Ray suddenly strikes its whip-like tail around the offender in lasso fashion, and holding him tightly against the barbed spine, wields the latter with such strength and rapidity that it lacerates the flesh to a frightful and dangerous extent, in some cases even causing the death of the victim.

Along the coast, where the offensive powers of this fish are familiarly and practically known, an opinion prevails that the bony spine is supplied with poison. This notion, however, is one of the many popular errors on similar subjects, having been founded on the aggravated inflammation that sometimes follows the wounds caused by the Sting Ray. There is no poison whatever in this bone, and any such symptoms are due, not to the inherent venom of the weapon, but to the unsound constitution of the sufferer.

The reader will at once perceive the exact resemblance between the spine of the Sting Ray and the many-barbed spears used by the savage inhabitants of the Pacific islands. In fact, this spine not only furnished them with the original idea of those cruel weapons, but is constantly taken from the fish and affixed to the shaft of a lance. In their eyes, its great merit—and one which they imitate in their manufactured weapons—is that when the spear is struck into the body of a foe, the jagged blade is sure to snap asunder at the point where it enters the body, leaving several barbs fixed in the wound without any handle by which they may be withdrawn.

It is found that in the Sting Ray, a second spine exists below the first, which is provided in order to supply the place of the first in case it should be broken off or dragged out.

The Sting Ray is in some places called the FIRE FLAIRE, probably on account of the very red color of the flesh when cut open. This fish is not approved for the table, being rank and disagreeable in flavor.

The color of the Sting Ray is grayish-yellow above, taking a slaty-blue tint towards the middle of the body, and spotted with brown when the creature is young. Below, it is white. The eyes are golden color, the temporal orifice behind each eye is extremely large, and the tail is very thick and muscular at the base. The spine is set about one-third of its length from the base. The mouth and teeth are small.

The Sting Ray of our waters is the same as that of Europe. It ranges from Cape Cod to Florida. In various places it is called Whip Ray, Clam-cracker, and Sting-a-ree, or Whip-sting Ray. A large form is common off Long Island, with tail of five feet in length.

IN some respects, such as the long tail and double-barbed spine with which it is armed, the EAGLE RAY (*Myliobatis aquila*) bears some resemblance to the preceding species, but must be distinguished from that fish by the projecting head, the bluntness of the snout, the very great length and comparative tenuity of the tail, the shortness of the spine, and the diminutive size of the temporal apertures. In some places this fish is called the Whip Ray, in allusion to the extreme length of the slender tail.

The flesh of the Eagle Ray is not eaten, being hard, rank, and disagreeable, but the liver is thought to be eatable, and a large quantity of good oil is obtained from it. It sometimes attains to a very large size, weighing as much as eight hundred pounds. Its color is dark brown above, deepening towards the edges, and grayish-white below.

The Eagle Ray is a Mediterranean species, found occasionally in English waters. Its great wing-like sides give it a resemblance to a bird. Its long tail and double barbed spine at its base give it a most singular and vicious aspect.

BEFORE quitting these fish entirely, a short notice must be given of several interesting species, of which figures cannot be inserted for want of space.

The first is the HORNED RAY (*Cephaloptera johnii*), sometimes called, from its huge dimensions, horned head, dark body, and lowering aspect, the SEA DEVIL. There are, however, several species which are popularly called by the latter title.

This enormous creature is found in the Mediterranean and the warmer seas in general, and has been taken in the nets together with the tunny. The flesh is not eaten except by the very poor, but the supply of oil from the liver is abundant and valuable. There seem to be hardly any bounds to the size which this creature will attain. M. Le Vaillant saw three of these huge fish sporting round the ship in lat. 10° 15′ N. long. 35° W. and, after some persuasion, induced the crew to attempt their capture. They secured the smallest of the three, and when it was brought on board, it was found to measure twenty-eight feet in width, twenty feet in length, to weigh a full ton, and to have a mouth large enough to swallow a man.

This gigantic Ray feeds almost wholly on fishes and mollusks. On account of their horned heads, the Italian fishermen call the old ones cows and the young calves. A strong attachment seems to exist between the male and female, for it has more than once happened that when one fish has been harpooned or otherwise captured, its mate has hung about the boat until it shared the same fate with its deceased partner; and in one instance, where the female had been caught in a tunny net, the male was seen wandering about the net for several days, and at last was found dead in the same partition where his mate had been captured. So, in common justice, the name of Sea Devil ought not to be applied to so loving and faithful a creature.

The color of the Horned Ray is very dark black-blue above, and gray-white beneath. The jaws and mouth are proportionately greater than is generally the case with these fishes. The tail is long, thin, and smooth for the first quarter of its length, after which it is furnished with tubercles. At its base there is a sharp, flattened spine, armed, like that of the preceding species, with a double row of barbs.

THERE are several other Rays, among which may be briefly mentioned the LONG-NOSED SKATE (*Raia salviani*), remarkable for the great length of the snout; the FLIPPER SKATE (*Raia intermedia*), notable for the olive-green color of the upper surface, and the numerous white spots with which it is covered; the BORDERED RAY (*Raia marginata*), which may be known by the dark edge to the side fins, or wings as they are generally called, and the three rows of sharp spines on the tail; and lastly, the HOMELYN RAY (*Raia miraletus*), which may be distinguished by the large size of the eyes and temporal orifices, and the bold dark spots on the sides.

SPINE-FINNED FISHES; ACANTHOPTERYGII.

WE now arrive at the vast order of the SPINE-FINNED FISHES, known scientifically as the ACANTHOPTERYGII. In all these fishes, the skeleton is entirely bony, and part of the rays of the dorsal, anal, and ventral fins are formed into spines, in some species very short, and in others of extraordinary length.

Without devoting more time or space to the purely scientific and anatomical characteristics, which will be separately described at the end of the volume, we will proceed at once to the various species of this vast and important order. I may here mention, that, whenever possible, I have selected examples of the various common genera, employing only those foreign species that are needful to fill up the links of the chain, or that are worthy of notice from some remarkable points in their form or their habits.

THE family of Sticklebacks comprises eight genera, and about twenty species of small, active, and exceedingly pugnacious fishes. They are very destructive to spawn and fry of other fishes. It is scarcely to be conceived how damaging these little creatures are, and how greatly detrimental they are to the increase of all the fishes among which they live.

Most of them build nests quite elaborately—which the male defends with great spirit. They inhabit the fresh waters and arms of the sea in Northern Europe and America.

THE NINE-SPINED STICKLEBACK (*Gasterosteus pungitius*) is a form equally known in the northern parts of Europe and America, found in both fresh and salt water. Eight other species are known in American waters. The following account of European species discloses the habits of nest-building, and other habits that apply equally to American forms.

THE THREE-SPINED STICKLEBACK, a very common fish, is also known under the names of TITTLEBAT, PRICKLEFISH, and SHARPLIN.

THREE-SPINED STICKLEBACK AND FIFTEEN-SPINED STICKLEBACK, WITH NEST.—*Gasterosteus aculeatus* and *Gasterosteus spinachia*. (Natural Size.)

It is a most bold and lively little fish, hardly knowing fear, pugnacious to an absurd degree, and remarkably interesting in its habits. Even more voracious than the perch, it renders great service to mankind in keeping within due bounds the many aquatic and terrestrial insects, which, although performing their indispensable duties in the world, are so extremely prolific, that they would render the country uninhabitable were they allowed to increase without some check.

So voracious and fearless indeed is this little creature that it always forms the earliest game of the juvenile angler, who need not trouble himself in the least about the temper of his hooks, the fineness of his tackle, or the delicate balance of his float. Any one can catch a Stickleback without rod, float, or even hook. All that is needful is to repair to the nearest streamlet, armed with a yard or two of thread and a walking-stick. Thin twine will answer very well instead of the thread, and even the stick is not absolutely needed. Having proceeded thus equipped to the bank of the stream, a worm may be picked out of the ground, tied by the middle to the thread, and thrown quite at random into the water.

The Sticklebacks will not be in the least frightened by the splash, but rather rejoice in it as calling their attention to food. In a moment the worm will be the centre of a contending mass of little fishes, rolling over and over, struggling to the utmost of their power, and entirely hiding the worm from sight. Now let the angler quickly lift the bait out of the water, swing it on shore, and he will almost certainly find that he has captured two Sticklebacks, one hanging to each end of the worm, and retaining its hold so perseveringly that it can hardly be induced to relinquish its gripe. This process may be repeated at pleasure, and as the Sticklebacks never seem to learn wisdom, a large store may soon be accumulated. This is a good way of stocking an aquarium, as the strongest and liveliest fish are sure to be caught first.

I have caught them by hundreds in a common butterfly-net, by the simple stratagem of lowering the net into the water, dangling the worm over the ring, and by degrees lowering the worm and raising the net until I had the whole flock within the meshes.

Should the reader be disposed to place his newly-captured specimens in an aquarium, he must make up his mind that they will fight desperately at first, and until they have satisfactorily settled the championship of the tank, their intercourse will be of the most aggressive character. Never were such creatures to fight as the Sticklebacks, for they will even go out of their way to attack anything which they think may possibly offend them, and they have no more hesitation in charging at a human being than at one of their own species. I have known one of these belligerent fish make repeated dashes at my walking-stick, knocking his nose so hard against his inanimate antagonist, that he inflicted a perceptible jar upon it, and in spite of the blows which his nose must have suffered, returning to the combat time after time with undiminished spirit.

These combats are, however, most common about the breeding season, when every adult Stickleback challenges every other of his own sex, and they do little but fight from morning to evening. They are as jealous as they are courageous, and will not allow another fish to pass within a certain distance of their home without darting out and offering battle.

Any one may see these spirited little combats by quietly watching the inhabitants of a clear streamlet on a summer day. The two antagonists dart at each other with spears in rest, snap at each other's gills or head, and retain their grasp with the tenacity of a bull-dog. They whirl round and round in the water, they drop, feint, attack, and retreat, with astonishing quickness, until one confesses itself beaten, and makes off for shelter, the conqueror snapping at its tail, and inflicting a parting bite.

Then is the time to see the triumphant little creature in all the glory of his radiant apparel; for with his conquest he assumes the victor's crown; his back glows with shining green, his sides and head are glorious with gold and scarlet, and his belly is silvery-white. It is a little creature certainly, but even among the brilliant inhabitants of the southern seas, a more gorgeously colored fish can hardly be found. If the conqueror Stickleback could only be enlarged to the size of a full-grown perch or roach, it would excite the greatest admiration. It is curious, that the vanquished antagonist loses in brilliance as much as the conqueror has gained; he sneaks off ignominiously after his defeat, and hides himself, dull and sombre, until the time comes when he, too, may conquer in fight, and proudly wear the gold and scarlet insignia of victory.

These struggles are not only for mastery, but are in so far praiseworthy, that they are waged in defence of home and family.

The Stickleback is one of the very few fish who build houses for their young, as a defence against the many foes which are ever lying in wait for the destruction of the eggs or the newly-hatched young. These nests are built of various vegetable substances, and their structure is admirably described in the following passage extracted from an educational magazine:—

"In a large dock for shipping, thousands of Prickle-fish were bred some years ago, and I have often amused myself for hours by observing them. While multitudes have been enjoying themselves near the shore in the warm sunshine, others have been busily engaged making their nests, if a nest it can be called. It consisted of the very minutest pieces of straw or sticks, the exact color of the ground at the bottom of the water on which it was laid, so that it was next to an impossibility for any one to discover the nest, unless they saw the fish at work, or observed the eggs.

"The nest is somewhat larger than a twenty-five-cent piece, and has a top or cover, with a hole in the centre, about the size of a very small nut, in which are deposited the eggs or spawn. This opening is frequently concealed by drawing small fragments over it, but this is not always the case. Many times have I taken up the nest, and thrown the eggs to the multitude around, which they instantly devoured with the greatest voracity. These eggs are about the size of poppy seeds, and of a bright yellow color; but I have at times seen them almost black, which, I suppose, is an indication that they are approaching to life.

"In making the nest, I observed that the fish used an unusual degree of force when conveying the material to its destination. When the fish was about an inch from the nest, it suddenly darted at the spot, and left the tiny fragment in its place, after which it would be engaged for half a minute in adjusting it. The nest, when taken up, did not separate, but hung together like a piece of wool."

This interesting little account is doubly valuable, as not being the work of a professed naturalist, but of an observant lover of nature, who saw some curious phenomena, and recorded them in simple and unpretending language. The fifteen-spined Stickleback, a marine species, also makes a nest, though hardly of so careful a construction.

The Three-spined Stickleback is very fond of inhabiting the mouths of rivers where they empty themselves into the sea, the brackish water appearing to suit its constitution. It can therefore be easily acclimatized to new conditions, and a specimen that has been taken from an inland stream can soon be brought to inhabiting the water of a marine aquarium, though such water is usually, in consequence of evaporation, more salt than that of the sea.

As a general fact, the flesh of the Stickleback is despised as an article of food, and in my opinion wrongly so. I have often partaken of these little fish fried, or even baked, and think them decidedly palatable—delicate, crisp, and well-flavored, with the slightest possible dash of bitter that gives a unique piquancy to the dish. At all events, the young of the Stickleback and the minnow frequently do duty as whitebait, and the guests never discover the deception. Yet there is hardly any place in Europe where people, except the starving poor, will condescend to eat this delicate and nutritious little fish, which can be scooped by thousands out of any streamlet, and does not require more trouble in cooking than the red herring. The only use that at present seems to be made of this fish is to spread it over the ground as manure, an office which it certainly fulfils admirably, but might, in all probability, be better employed in feeding man than manuring his fields. An oil is sometimes expressed from them, and the refuse carted off to the fields, but the value of the oil seems hardly to repay the trouble of procuring it.

Mr. Yarrell mentions a considerable number of Sticklebacks; but Dr. Günther, in his elaborate catalogue of Acanthopterygian fishes, comprises several species together, as only varieties and not different species. For example, the QUARTER-ARMED STICKLEBACK (*Gasterosteus gymnúrus*), or SMOOTH-TAILED STICKLEBACK, known by its four or five scaly plates above the pectoral fin; the HALF-ARMED STICKLEBACK (*Gasterosteus semiarmatus*), where the plates extend throughout half the length of the body; the HALF-MAILED STICKLEBACK (*Gasterosteus semiloricatus*), where they extend still farther; and the NEW YORK STICKLEBACK (*Gasterosteus noveboracensis*), are all considered as being only varieties of the species which has just been denoted.

THE TEN-SPINED STICKLEBACK is nearly, if not quite, as plentiful as the three-spined species, and is perhaps the smallest of our river-fish.

It may be readily distinguished by the nine or ten spines upon the back, all in front of the dorsal fin, and by the absence of plates upon the sides. All the Sticklebacks are voracious little creatures, and I am told by an angler friend that they destroy quantities of the spawn of other fish, and seize upon the young as soon as they are hatched. He also informs me that they are extremely capricious in their choice of locality. For example, at the head of a millstream they may be found by thousands, while at the tail of the same stream not a single Stickleback can be found. There are parts of rivers where they are so plentiful that the fisher is entirely baffled in his sport by these little creatures eating his bait before it sinks to its full depth, while the middle of the stream might be quite free from them.

The Ten-spined Stickleback does not like salt water, and cannot be acclimatized to the marine aquarium like its three-spined relative. All the Sticklebacks are remarkable for the comparative nakedness of the skin, which for the most part bears no scales, as in the generality of fish, and in the Ten-spined species is wholly naked. The place of the scales is supplied by certain bony or scaly plates upon the side, and it is the nakedness of the skin which permits the colors of these little fish to glow with such bright and changeful hues.

The color of this species is green upon the back, and on the abdomen and sides silvery-white spotted minutely with black. The fins are very slightly tinged with yellow. The length of the Ten-spined Stickleback is variable, but rarely exceeds two inches.

The FIFTEEN-SPINED STICKLEBACK, SEA-ADDER, or BISMORE, is wholly a marine species, and is nearly as common as its companion on the picture on page 213.

It is remarkably elongated in proportion to its width, and this formation, together with its armature of sharp, tooth-like spines, has gained it the name of Sea-Adder. It is a voracious creature, feeding on all sorts of marine animals, mollusks, worms, eggs, and fry, and minute crustaceans. Mr. Yarrell advises the collector of marine crustaceans to examine carefully the stomachs of the shore-frequenting fishes, and especially of this species, as he will be likely to discover some curious species of those animals, too active or too small to lodge in his net, but unable to avoid the quick eye and ready jaws of the Stickleback. The same writer mentions that on one occasion, when a Fifteen-spined Stickleback had been caught with a net and placed in water together with a small eel, three inches in length, the voracious creature seized on the eel in a very short time, and contrived to swallow it. The eel, however, was too long to be wholly accommodated in the stomach of the Stickleback, and after a while was disgorged, only partly digested.

As in the case of the frog, the color of the Stickleback varies with singular rapidity, being dull or bright according to the mental emotions of the individual.

WE now arrive at another family of fishes, in which the body is rather compressed—*i. e.*, flattened sideways—the eyes are large, and the mouth oblique. It is scientifically known by the name of Berycidæ, and all its members are inhabitants of the tropical and temperate seas.

Our first example of this family is the JAPANESE SINGLE-THORN.

In all the fishes of this genus, the scales are rather large, very strong, and so closely compacted together, that they form a strong, mailed covering to the body. The name of Monocentris, or Single-thorn, is derived from the curious modification of the ventral fins, which are devoid of membrane, and are reduced to a single, very strong, and rather lengthened spine, and a few very short rays. In the place of the dorsal fin are four or five thick spines, and the shield-like scales of the body are rough, projecting, and keeled.

The Japanese Single-thorn is an inhabitant, as its name imports, of the seas of Japan, and is almost, if not quite, the only species of its genus. It is chiefly remarkable for the size of its head, the strong, thorn-like spines, and the mailed suit of hard and projecting scales. It is of a tolerably uniform color, its whole body being silvery-white, and its length is about six or seven inches.

THE large-eyed and deep-bodied fish, HOPLOSTETHUS, or Armed-breast, derives its name from the strong and sharp spines which are placed on the scapular bone and the angle of the præoperculum. Like the last-mentioned species, it seems to be the only member of its genus.

This fish is found in the Mediterranean, and is not uncommon on the coast of Madeira.

It is remarkable not only for the offensive weapons with which it is armed, but for the large, full eye, the saw-like series of notches on the abdomen, and the beautiful rosy hue of its scales. The dorsal fin of this fish is single, but is composed of two distinct portions, the one being supported by strong, spinous rays, and the other by soft and flexible rays. The muzzle is very short, rounded, and does not protrude; the tail is deeply forked, and the serrated portion of the abdomen consists of eleven, twelve, or thirteen scales. The body is very deep, in proportion to its length.

BEFORE proceeding to the next family, we must casually notice two large genera belonging to this family. Of the first genus the MURDJAN PERCH (*Myripristis murdjan*) is a good example.

This handsome fish is found off the coasts of India and in the Red Sea, and can be easily recognized by its beautiful coloring, its large scales, short muzzle, and prominent chin. The general color of this splendid fish is bright rose-pink, beautifully mottled by a rich violet edge to each scale. The soft portions of the dorsal, ventral, and anal fins are boldly margined with white, and the front rays have a cross band of violet-brown. The tail fin is edged with white, and a longitudinal stripe of violet-brown traverses each lobe. About fourteen or fifteen species of this genus are known.

OF the next genus, the SCARLET PERCH (*Holocentrum rubrum*) is rather a striking example.

This fine fish inhabits the Asiatic seas, and there are specimens in the museums from the Red Sea, Amboyna, Louisiade Archipelago, the Philippines, Japan, and China. On the operculum are two strong spines, the upper being the larger. The color of this fish is shining red, diversified with eight bands of grayish-white. The outer edges of the tail fin are black, and there is a patch of the same color on the ventral fins. This genus contains many very handsome species, and in almost every case the prevailing colors are red and violet.

PERCH, MULLET, BRAIZE.

WE now come to the large and important family of the Perches, which comprises many of the handsomest and most valuable fishes. The members of this family are found in all parts of the globe.

THE COMMON PERCH is well known as one of the handsomest river-fish, and, on account of its boldness and the voracious manner in which it takes the bait, and the active strength with which it struggles against its captor, is a great favorite with many anglers.

Moreover, when captured, and placed in an aquarium, it very soon learns to distinguish the hand that feeds it, and will come to the surface and take food from the fingers. It has a fashion of seizing its food with a rather sharp jerk, and then snatches it away with such violence, that, when it takes the hook, it will drag a stout cork float several inches below the surface, and, by the force of its own stroke, will mostly hook itself, without any exertion on the part of the angler. Bold-biting, however, as is its reputation, there are some seasons of the year when it is almost impossible to catch a Perch, and even the shy and gently-nibbling roach is an easier prey.

The flesh of the Perch is white, firm, well-flavored, and is thought to be both delicate and nutritious.

The Perch is not a large fish; from two to three pounds being considered rather a heavy weight. Individuals, however, of much greater dimensions have been, though rarely, captured.

The color of the Perch is rich greenish-brown above, passing gradually into golden-white below. Upon the sides is a row of dark transverse bands, generally from five to seven in number. The first dorsal fin is brown, with a little black between two or three of the first and last rays; the second dorsal and the pectoral are pale brown, and the tail and other fins are bright red.

THE fine fish so well known under the name of BASS, or SEA-DACE, or SEA-PERCH, is common on many coasts, and is considered by anglers as affording good sport.

It seems, from the accounts of practical sportsmen, to bite with readiness at a bait, but to be a difficult fish to secure, on account of its tender mouth, its ingenious stratagems, and its great strength. When hooked, it leaps, plunges, and swims with such force and swiftness, that the captor is forced to exercise the greatest skill in preventing it from breaking away. One of its favorite ruses is to double back under the boat, in hopes of cutting the

GIANT PERCH AND COMMON PERCH.—*Lucioperca sandra et Perca fluviatilis.*

line against the keel, or gaining a fixed point by which it may be able to drag the hook from its mouth.

Even when fairly tired out, and drawn to the edge of the boat, it is by no means secured, for its scales are so hard that a very sharp blow of the gaff is needed to fix the hook in its side, and its gills and fins are so formidably armed, that it cannot be grasped with impunity. The spines of the dorsal fin, in particular, are strong and sharp as packing-needles, and the various portions of the operculum are edged with projecting teeth that cut like lancets. Many are the wounds that have been inflicted by the sudden twist and wriggle of the Bass, when grasped in a careless manner. When lifted into the boat, the hook is not to be taken from the mouth without some risk.

It is a voracious fish, and derives its name of "lupus," or wolf, in consequence of its insatiate appetite. It feeds upon other fish and various inhabitants of the sea. Mr. Couch states that it is very fond of wood-lice, and is bold enough to venture among rocks in a tempest for the sake of snapping up these creatures, as they are washed by the waves and beaten by the winds from their places of concealment among the stones.

The flesh of the Basse is very excellent, and is thought to be in best condition when the fish is small, measuring about eighteen inches in length. The color of this fish is dark dusky blue on the back, and silvery white on the abdomen; the fins are brown.

THE pretty little RUFFE (*Acerina cérnua*) is common in many rivers, where it is sometimes known under the name of POPE, the reason for the latter title not being very clear.

In general appearance the Ruffe bears some resemblance to the perch, the shape of its body and the thorny fins being not unlike those of that handsome fish. It may, however, be immediately distinguished from the perch by its spotted fins, and the absence of the dark band over the sides. Moreover, the dorsal fin is single. It is a tolerably bold biter, and takes a hook readily when baited with a little bright-red worm.

The color of the Ruffe is light olive-brown above, and silver-white on the abdomen; the flanks are yellowish-brown. The back, dorsal fin, and tail, are covered with little brown spots, set so closely in the tail as to resemble bars, and upon the gill-covers there is a little pearly-green. The length of this fish seldom exceeds six or seven inches.

One of the two creatures figured in the engraving on page 218 is the GIANT PERCH. It is a remarkably fine fish, which is found in many of the rivers and lakes of Germany and Eastern Europe.

This handsome species derives its name of Lucioperca, or Pike-Perch, from the resemblance which it bears to both these fishes, having the lengthened body of the one and the spine-armed fins of the other. It has, however, nothing to do with the pike, and is closely allied to its companion on the engraving, the perch, belonging, indeed, to the same family. The teeth are rather large, and are thought to resemble those of the pike in length and sharpness.

The color of the Giant Perch is greenish-olive above, banded with brown. Below, it is white. It is a very fine fish, attaining, when full-grown, to a length of three or four feet. There are several species belonging to the same genus.

A very handsome fish, that is popularly but erroneously called the AMERICAN PIKE, has derived its name from the elongated and somewhat pike-like form of its body. The teeth, however, are even, and bear no resemblance to those of the real pike.

The flesh of this fish is thought to be good for the table, and as the dimensions to which the creature attains are often considerable, it is really one of the valuable inhabitants of the American waters. It is one of the sea-loving species, and is mostly found on the Atlantic shores of tropical America. Many specimens now in the museums were taken in the West Indies, others off the coast of Guiana, some from Bahia, and others from Surinam. The general color of the American Pike is silvery-white, tinged on the back with green, and becoming a pure, shining white on the abdomen. The dorsal fins are two in number, the first being shortish, and having eight very strong and sharp spines. The second spine of the anal fin is very long and sharp, and the præoperculum is armed with two sharply-toothed edges.

The PIKE, or PIKE-PERCHES, so called in America, are equally a European form. They are large carnivorous fishes, living in fresh waters. Two strongly marked species are on each hemisphere.

The WALL-EYED PIKE (*Stidostedium vitreum*), DORY, GLASS-EYE, YELLOW PIKE, BLUE PIKE, and JACK SALMON, are names common to this species. It is found in the Great Lakes, and the Upper Mississippi, and some of the Atlantic streams far north; is an abundant and valuable food fish, reaching nearly three feet in length, and a weight of twenty pounds. Another species, called SANGER, SAND-PIKE, GRAY PIKE, and HORN-FISH, is smaller, and is found in similar regions.

The family *Percidæ*, the Perches, in which this fish belongs, embraces twenty-two genera, and from ninety to one hundred species. They are inhabitants of fresh waters of cool regions, most of them being American, and nearly all belonging to the fauna of the United States. A great majority belong to the Darters, all of which are American. They are among the most singular and interesting of our fishes. Their colors are generally brilliant; in some of the genera are some of the most brilliantly colored of any known fishes. The sexual differences are very great in some species, the females being more sombre in color and more speckled or barred than the males. Most prefer clear running water, where they lie concealed under stones, darting from under with great velocity when hungry. They all can turn the head from side to side, and lie with the head turned on one side at times, the body supported on the expanded ventrals. The Ammocrypts are fond of lying concealed under the sand, the eyes only exposed. They are carnivorous, feeding greatly on larvæ of *Diptera*. All are small in size, being at most eight inches in length. The group *Microperca* are the smallest spiny-rayed fishes known, being only about one and a half inch. The general aspects of this numerous family are here indicated. The species would require volumes to describe. They are interesting and beautiful, but of no account commercially, and scarcely large enough for food.

The well-known BLACK BASS of America (*Centropristis atrarius*) inhabits the rivers and lakes of North America.

This fine fish is a really valuable species, on account of its large dimensions and the excellence of its flesh, and the attention of scientific men has lately been turned towards its preservation. In the Patent Reports upon some Black Bass that were transferred to Waramang Lake, Connecticut, some years previous, it is said that they multiplied very rapidly, grew at the average rate of one pound per annum, and ordinarily attained a weight of five pounds or a little more. They are very hardy, and can be taken from one locality to another if placed in a tub of water covered with a wet canvas. So rapid, indeed, is its increase, that although less than a hundred were originally placed in the lake, they have probably increased to several millions in a space of seven years.

It is a marvellously bold-biting fish, and affords good sport to all anglers, whether they only fish for the sake of the amusement, using a fly or other delicate bait, or whether they merely seek to take their prey as a matter of business, and employ small fish as a bait, or the obstruction "spoon," whose treacherous glitter the Black Bass is seldom able to withstand. It is an active and powerful fish, and when hooked struggles so long and so fiercely, that it tests all the angler's skill before it can be safely landed.

The color of the Black Bass is brown, washed with golden-green, and mottled with dark spots on the centre of each scale, darker on the back, and becoming nearly white on the abdomen. When newly caught, the body is traversed with several dark bands. It is a very fine fish, specimens having been known to weigh nearly twenty pounds.

BLACK SEA BASS (*Serranses atranius*). This fish ranges from Cape Cod to Florida. It is called Black Perch in Massachusetts; Black Bass and Black Fish in New Jersey.

The STRIPED BASS (*Roccus lineatus*) is one of the most notable of our sea fishes—one of the largest and finest of our game fishes, reaching four feet in length. Dr. Storer records one that weighed eighty-four pounds. We have taken small ones off Nahant rocks by angling, while fishing for Tautog. It is found in our markets nearly all the year.

CLOSELY allied to these fish is an enormous genus, containing about one hundred and forty known species, from which the OUATALIBI, or RUDDY SERRANUS (*Serránus ouatalibi*), is selected as an example.

This beautiful fish inhabits the warm Caribbean Sea, and is plentiful upon the West Indian coasts. Its color is bright red, and the head, body, and sometimes the dorsal fin, are profusely powdered with small blue spots, edged with black. Just by the joint of the lower jaw there is a pair of largish black spots, and on the back of the tail, immediately behind the dorsal fin, is another black spot. Of its habits nothing interesting is told.

THE STONE BASS is an inhabitant of the European seas. It is otherwise known as COUCH'S POLYPRION, in honor of the eminent naturalist who first made it known as an own species, and as JEW-FISH and WRECK-FISH—the last title being given to it on account of its habit of frequenting drifting timbers, apparently for the purpose of feeding upon the various marine creatures that swarm about such localities. In Madeira it is called CHERNE, when full grown, and CHERNOTTE when young.

Barnacle-laden timber seems to have great attractions for the Stone Bass, and it is mentioned by Mr. Yarrell that a becalmed vessel was surrounded for a fortnight with these fish, probably on account of the trailing barnacles with which her planking was covered. Their presence was most valuable, as they were caught in great numbers, and the men fed almost wholly upon them for twelve or fourteen days.

From examination of the stomach, the Stone Bass seems to feed mostly on small fish of various kinds, sardines having been found in its interior in large quantities. Mollusks also form part of its food. It lives mostly in the deeper waters, preferring a rocky bottom, and generally remaining deeply immersed, unless attracted to the surface by the presence of its food.

When following floating timbers, it is a remarkably bold fish. Mr. Couch remarks thus upon its habits : "When a piece of timber, covered with barnacles, is brought by the currents from the more southern regions which these fishes inhabit, considerable numbers of them sometimes accompany it. In the alacrity of their exertions, they pass over the wreck in pursuit of each other, and sometimes for a short space are left dry on the top, until a succeeding wave bears them off again. From the circumstance of their being usually found near floating wood covered with barnacles, it might be supposed that this shell-fish forms their food ; but this does not appear to be the case, since, in many that were opened, nothing was found but small fishes. Perhaps the young fishes follow the floating wood for the sake of the insects that accompany it, and thus draw the Stone Bass after them."

The color of the Stone Bass is dark purple-brown above, and silvery-white below. The fin-membranes are brown, and the tail is tipped with white. When young, it is mottled with darker and lighter brown. The lower jaw is larger than the upper, and over the operculum runs horizontally a bold, bony ridge, ending in a sharp point directed backwards. There is also a row of short, sharp spines over the eye, and the first ray of the ventral fins and the first three rays of the anal fin are furnished with strong, thorny spines, so that the fish is armed at all points, and when struggling violently is likely to inflict rather severe wounds on the hand that grasps it incautiously.

THE great and important family of the Labridæ, or Lip-fishes, so called in allusion to the thick and fleshy lip with which their mouths are furnished, are spread over the greater portion of the globe, few coasts being without several representatives of the group.

These fishes are not only remarkable for the full fleshiness of their evidently sensitive lips, but for the endless variety of rich and vivid tints with which their bodies are decorated ; hues pure as the bright patterns of cathedral windows, and often arranged with a symmetrical regularity of outline and a daring harmony of contrasting colors that, when seen on the healthy and living fish, appear as if flung on its scales direct from the kaleidoscope. Of the three primary colors, red seems to retain its purity of tone more perfectly than either the blue or the yellow, the former usually being mingled with yellow, and forming greens of varying qualities, while the latter frequently takes a slight tinge of red, and becomes warm orange. These tints are extremely variable, ranging through every tone of the secondary colors, and even in different individuals of the same species the color is so uncertain that no dependence can be placed upon it as a means of determining the particular species.

The Labridæ are most lovely creatures, but it is in the tropical and warmer seas of the world that they are to be seen in their full brilliancy. No artist can transfer to paper the radiant hues that glow on these favored members of the finny race, and no pen can do justice to their wondrous splendor, as they dart through the crystalline water like living meteors, or leisurely traverse the forests of moving algæ, balancing themselves among the submarine foliage like humming-birds of the sea.

THE CHECKERED SWINE-FISH is one of the singular species which are so frequent in the hotter seas, and which exhibit a surface at once colored with the brightest hues, and decorated with a pattern as geometrically regular as if drawn by the aid of rule and compass. This fish is found on the coasts of Ceylon. Its color is bright green on the back, gradually changing to golden-yellow towards the abdomen. The checkered marks on the body are purple and gray, and the stripes and spots on the head are rich, glowing orange. In allusion to the wicker-like markings and the brilliant colors of the fish, the natives call it by the name of Hembili Girawah ; the former word signifying the little basket in which betel is carried, and the latter a parrot.

Though edible, this fish is not regularly captured for the markets. It generally frequents the rocky portions of the coast. It never attains any great dimensions, none seeming to measure more than fifteen inches in length.

THE curious fish, SLY EPIBULUS, has derived its popular and scientific titles from the crafty manner in which it obtains its food.

In this fish, the mouth and jaws exhibit a very remarkable modification of structure, which enables the creature to protrude its mouth with great rapidity, and to such a degree, that when pushed forward to its greatest extent it is even longer than the head. Aided by this apparatus, the Sly Epibulus captures its prey as follows : It feeds mostly upon small fishes, and instead of alarming them by charging among their ranks, and so giving itself a tedious and possibly an unsuccessful chase, it quietly withdraws itself to some sheltered spot, and waits, still and motionless as the watchful kingfisher, and no less dangerous to the smaller members of the finny tribe. No sooner does an unwary fish pass near the cunning enemy, deceived by its apparent harmlessness, than the movable mouth is suddenly projected with the rapidity of a serpent's stroke, and the victim is seized and swallowed in a moment.

THE TILE-FISH (*Lopholatilus chamæleonticeps, Goode and Bean*) is a newly-discovered food-fish, the value of which is thought to be very considerable as an addition to our list of food-fishes. It is abundant in deep water off Massachusetts Bay ; it is a fine large fish and highly colored, and belongs to family *Latilidæ*.

FROM the many species of the genus Labrus, the BALLAN WRASSE has been selected. This fish, otherwise known by the popular names of ANCIENT WRASSE, or OLD WIFE, is one of those species which is mostly found haunting the rocky portions of the shore.

It is not in any great estimation as an article of food, the flesh being too soft, and not possessing any particularly good flavor. It generally frequents the deep rocky gullies, where the water is tolerably tranquil, except when the waves are beaten into foam by a storm. Here it may be seen swimming about in the clear element, concealing itself among masses of seaweed, and ever and anon darting forth to secure some tempting morsel, such as a passing crab or prawn.

Mr. Couch remarks concerning this species, that "it takes a bait freely, and fishermen remark that when they first fish in a place, they take but few, and those of a large size ; but on trying the same spot a few days after, they catch a greater number, and those smaller ; from whence they conclude that the large fish assume the dominion of a district, and keep the younger at a distance."

The Wrasse deposits its spawn in spring or towards the beginning of summer, and, as is usual with most fishes, assumes its brightest apparel previous to performing that duty. There is much uncertainty about the coloring of the Ballan Wrasse, but in general the color is red above and on the sides, variegated with oval spots of rich bluish-green ; the fins are green, sparely spotted with red, and the abdomen is pale orange. It does not attain to any great size, seldom having been known to exceed sixteen or eighteen inches in length, and two or three pounds in weight.

OF the family *Labridæ*, the Cunners are important as food-fishes. The CHOGSETT or CUNNER (*Ctenolabrus adspersa*) is, perhaps, the most familiar sea-fish on our New England coast. During the summer months it is the principal source of angling sport. It is a singular fact that in Boston this fish is called Cunner, and by no other name, while in Lynn, Salem, and Nahant, adjoining towns, it is universally known as Nipper and Chogsett, the latter the aboriginal term. It is sometimes called Perch or Sea Perch.

From the first settlement of the country this fish has been esteemed, and parties have been accustomed to visit the shore yearly to enjoy a day's nippering, and feast on fried Nippers and haddock chowder. The Nipper is taken by hook from the rocks, and affords much sport, as it is very "gamey."

Like many other fishes, this is better and larger as the water is colder; therefore, in Portland harbor the Nipper is much larger and proves an excellent food-fish.

THE TAUTOG (*Tautoga onitis*) is of this family. It much resembles the Cunner, but is quite black. In New England it is called Black-fish by some. Its range is from Maine to South Carolina. No ocean fish is more "gamey" than this. With small live-crab bait it affords sharp sport with the rod. This fish abounds in the vicinity of Long Island. But a few years since it was not known north of Cape Cod, but is now quite abundant.

Dr. Smith, in his "Fishes of Massachusetts," an early work, says: "Within the recollection of a gentleman now living (1833), the Tautog was unknown in Boston harbor." Its weight reaches twelve pounds, though it averages but about half that. The blossoming of the dogwood (*Cornus florida*) early in April, or the chestnut-trees, is understood to denote the time of baiting Black-fish:

> "When chestnut leaves are big as thumb-nail,
> Then bite Black-fish without fail;
> But when chestnut leaves are long as a span,
> Then, catch Black-fish if you can!"

This is an old verse recorded in Mitchell's interesting book.

This fish is related to the Wrasse of Europe.

THREE-SPOTTED WRASSE.—*Labrus trimaculatus.*

SEVERAL other species of the same genus are known, such as the GREEN STREAKED-WRASSE, or GREEN-FISH (*Labrus donovani*), a rather rare but very beautifully colored fish, almost wholly green and slightly streaked. Some naturalists think that this is only the young of the preceding species. The most curiously decorated species is, however, the THREE-SPOTTED WRASSE (*Labrus trimaculatus*). This fish is decorated with a rich ruddy orange over the greater part of its body, becoming slightly paler on the sides, and changing to golden yellow on the abdomen, with the slightest possible dash of red. On the upper part of the back, and occupying portions of the dorsal fish, are three large spots of deep rich purple, between which are placed four similarly shaped spots of pale rose. These spots, however, are rather variable in number.

THE RED GROUPER (*Epinephilus morio*) is an important food-fish, forming much of the fish cargoes gathered by the fishermen on the Florida coast for the Havana market.

IN the course of the preceding pages our notice has been drawn to many remarkable forms of fishes, some terrible in their fearful armatures of spines and teeth, some repulsive from their slimy exterior and coldly malignant aspect, and others almost bordering on the grotesque from the odd and eccentric manner in which various parts of their structure are modified. The TESSELATED PARROT-FISH of the Ceylonese seas, though not strikingly unique in its external appearance, as many of the species already described, is, when closely examined, one of the most wonderfully colored fish in the world. The whole body is covered with a beautifully drawn pattern of elongated hexagons, as perfect and regular as those of a honeycomb.

The colors of this remarkable fish are as follow: The general hue of the Tesselated Parrot-fish is azure-blue, covered with a hexagonal network of golden-yellow. The oddly shaped head is bright yellow, streaked and spotted with blue. The dorsal and anal fins are brown edged with green, and the pectorals and ventrals are brown with the front rays green. The tail fin is wholly green. The natives call this fish by the name of Laboo Girawah, the former term being the name of a certain gourd or pumpkin which is marked in a somewhat similar fashion.

The species belonging to this genus are very numerous, and have received their rather appropriate title of Parrot-fishes from the rich beauty of their colors and the peculiar form of their jaws, which are very strong, covered with great numbers of mosaic-like teeth, and curved in a manner that greatly resembles the beak of a parrot. As the fish wears out the teeth rapidly while crushing the corallines and other hard substances on which it feeds, a provision is made for insuring a continual supply of new teeth to replace those which are worn away and rendered useless. The young teeth are perpetually being developed towards the base of the jaws, and by a beautiful yet simple adaptation of existing parts, which cannot be made intelligible without the use of diagrams, advance in orderly succesion towards the front, and take their places with unfailing certainty in the densely compacted mosaic-work which arms the jaws.

BEFORE leaving this family we must briefly examine another very large genus, here represented by the BANDED MULLET (*Apógon fasciátus*).

This fish is found off the Fejee Islands, upon the coast of Mozambique, and in the Australian and Moluccan seas. The genus to which it belongs comprises about sixty species, all inhabiting the warmer waters, and some entering the mouths of rivers. They are most plentiful in the Indian and Australian seas, but are never seen in the colder waters of the northern and southern regions. The scales of these fish are large, and fall off almost at a touch. The gill-cover is rather formidably armed, the operculum bearing spines, and the præoperculum having a double-notched ridge.

The coloring of the Banded Mullet is bold and striking. The general tint of the body is a glowing rose, and a series of broad dark bands are drawn along the body, four or five on each side, and one on the back. At the base of the tail fin is a large round black spot, and a black band runs across the root of the second dorsal and anal fins.

THE next family, termed the Pristipomidæ, after the typical genus, forms a large and somewhat important group of fishes. They are all carnivorous, *i.e.*, they feed upon fish in preference to other diet; they have no molar or cutting teeth, and all inhabit the waters of the warm and temperate regions of the globe. The greater number of the species are marine, but a few are found in the rivers.

As an example of the typical genus, we will take the KAKAAN (*Pristipóma hasta*), a species found in the "Red Sea, along the east coast of Africa, through all the Indian seas to the northern shores of Australia."

In this prettily marked species, the dorsal fins are separated by a notch, rather variable in depth, and the fourth dorsal spine is much elongated, being indeed equal to half the length of the head. The second spine of the anal fin is also long and sharp. The coloring of the Kakaan is seldom precisely the same in any two individuals, but the body is always covered with a great number of brown spots, arranged with some degree of regularity. Sometimes these spots fall into horizontal lines, so as to look at a little distance like a series of brown bars drawn along the body, while in other specimens the spots are gathered into vertical bands. There are also several series of circular brown spots on both the dorsal fins.

THE CAPEUNA, or FOUR-STREAKED RED-THROAT, is a remarkably pretty fish, and a good example of the genus to which it belongs.

The generic title of Hæmulon is given to these fishes on account of the bright ruddy color of that part of the lower jaw which is concealed when the jaws are shut. The French call this genus Rougegueule. The profile of their rather elongated head is thought to bear some resemblance to that of a pig. The Capeuna is most beautifully colored, as will be seen when the description is compared with the figure. The spines of the dorsal fins are tolerably firm, but cannot be termed strong or formidable, and the same may be said of the lengthened second spine of the anal fin. The eye is large and full, and the tail is deeply forked. A rich brown band runs along the whole of the body just above the dorsal line, and a corresponding band is drawn immediately below it. Between the upper band and the spinous portion of the dorsal fin, a short brown streak is drawn, looking as if dashed hastily with one sweep of a brush, and a still shorter stripe of the same color runs along each side of the head just above the eye. From the eyes are drawn two wider stripes of rich golden-yellow, which pass beneath the lateral line, and run to a considerable distance, the lower streak being continued as far as the tail fin, and the upper reaching to the middle of the soft portion of the dorsal fin, where it turns slightly upwards.

ONE of those remarkably colored species for which the warmer seas are so famous, and whose vivid coloring and striking forms put to shame the comparatively sober inhabitants of the northern waters, is the BODIAN, or CUVIER'S BODIAN, as it is generally called.

What connection there may be between colors and caloric is one of the unsolved enigmas of creation, and though it is most evident that such a connection exists, its principles and even its results are at present shrouded in mystery.

The tints which decorate the finny inhabitants of these tepid waters are brilliant beyond all power of description, and the most glowing colors of the artist, though painted on a ground of burnished gold, fail to convey more than a dim idea of the wondrous chromatic effects produced by the living creatures. Even the patterns in which these colors are arranged are as unexpected as they are effective, and the art student would gain no slight knowledge of that most difficult science of color, were he to visit the tropical seas, and study the fishes as they swim calmly in the crystalline water, amid the forests of waving seaweeds or branching corals.

The harmony of the tints is not less remarkable than their brilliancy, for the brightest and most glowing colors are flung boldly together in kaleidoscopic profusion, and, in defiance of all the conventional rules by which artists like to govern themselves and others, are so exquisitely harmonious that not a tint could be altered or removed without destroying the entire chromatic effect. Examples of some of these fish will be given in the course of the succeeding pages, and the reader will see that, even when laboring in this instance under the disadvantage of substituting plain black and white for their natural colors, they must be truly the humming-birds of the ocean.

The Cuvier's Bodian is a species spread over the greater part of the Indian seas, and caught, though it appears but rarely, on the coasts of Ceylon, being most frequently captured on the southern shores and upon rocky ground. The Cingalese name is Deweeboraloowah. In color it is a remarkably handsome fish, though not of such pure primary tints as others which will presently be mentioned. The color of this fish is yellowish-brown on the back, changing gradually to reddish-gray on the sides, and fading to simple gray on the abdomen. The head, tail, and fins are bright golden-yellow, and the bars and patches of darker color are deep chocolate-brown. Its average length is from eighteen to twenty inches.

THE next family, the Mullidæ, finds a well-known representative in the common SUR-MULLET, sometimes called the STRIPED RED MULLET, on account of the yellow longitudinal stripes that are drawn along the body.

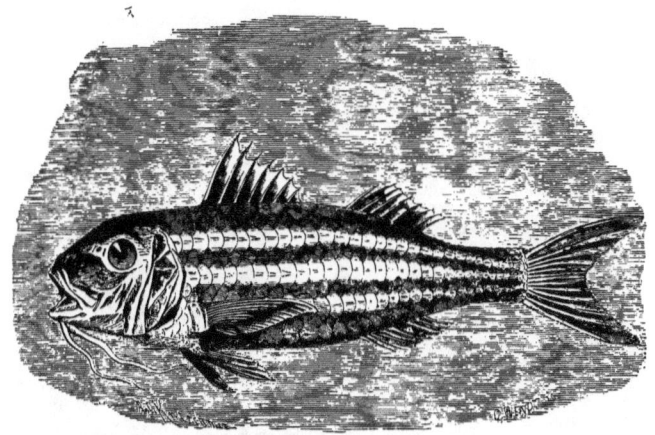

SURMULLET.—*Mullus surmuletus.*

This fish is celebrated for the excellence of its flesh, and in the time of the ancients was one of the most costly luxuries that the wealthy epicure could place upon his table, from two hundred to three hundred dollars being paid for a fish weighing six or seven pounds. These dimensions are but rarely reached, and never, as it is believed, on cold shores. The liver is held to be the best part of this fish, but the whole of its flesh is firm, white, and delicately flavored. Its value in the market is extremely variable, owing to its migratory habits, being at one time caught by hundreds in the trawl or mackerel nets, while at other times there is not a single individual to be found. There seems, however, to be one definite rule in its migrations, namely, that it approaches the shore in the summer time, and in the winter retires into deep water, whence it can only be taken in the trawl net.

ANOTHER species of this genus is the PLAIN RED MULLET (*Mullus barbátus*).

In general habits it closely resembles the preceding species, but may be distinguished from that fish by the almost vertical line of the head, which rises abruptly from the muzzle to the eyes, and by the different coloring. In the Plain Red Mullet the back is light pink, the sides and part of the abdomen dark red, and there is a single yellow streak below the lateral line.

A RATHER extensive genus belonging to the present family cannot be passed over without some notice, as it contains many fish which are remarkable for their form and coloring, if not for their habits or utilities.

The THREE-BANDED MULLET is a native of the Indian and Polynesian Seas, and has been taken off the coasts of China, Amboyna, Celebes, Ceylon, and India.

THE family *Mullidæ* is represented in American waters by five genera, containing thirty-five species. They are called collectively Surmullets, and inhabit all tropical seas; some species straying northward. The Goat-fishes belong to this family.

THE family of the *Sparidæ* is represented by the BRAIZE, otherwise known as the BECKER, PANDORA, and KING OF THE SEA-BREAM.
This is a common fish in the Mediterranean.

BRAIZE AND YOUNG GILT HEAD.—*Pagrus vulgaris et Chrysophrys aurata.*

The family *Sparidæ* is a very large one, embracing four hundred and fifty species, in fifty-five genera; abounding in temperate and tropical seas. The Snappers (*Lutjanus*) include several very notable table fishes. The Gray Snapper and the Red Snapper are important. The latter, *L. blackfordii*, is a late introduction in our markets. It is named in honor of the notable dealer in fish at Fulton Market, who adds to an exceptional reputation as a business man a scientific spirit which is highly commendable. The Hog-fish, or Sailors' Choice (*Pomadasys*), is a food-fish of some note, found along the Atlantic from New York southward. The various species of Grunts are classified here. The Scup, Scuppung, or Porgee (*Diplodus*) is an abundant and valuable food-fish. The Sheepshead (*Diplodus probatocephalus*) is regarded as equal if not the superior of any of our fishes as a table luxury. It is abundant from Cape Cod to Texas; though it is not so common north of Virginia, where it is prized very highly. Its flesh is compared to the English turbot. Its name is derived from the appearance of the mouth, which resembles that of a sheep. Its weight is occasionally sixteen pounds. It is wary and timid, and is very difficult to take with a hook; though they are captured in numbers by the seine.

The well-known Common Sea-Bream (*Pagellus centrodontus*) is a handsome fish, notable for its large round eyes, and the reddish-gray hue of its body. It is sometimes called the GILT-HEAD, because part of the head looks as if it were silvered, and when young, it goes by the name of CHAD. The general color is reddish with a tinge of gray, becoming light on the sides, and fading into white below. A few very faint bands are drawn along the sides.

As allusion has been made to the term GILT-HEAD as one of the popular names of the sea-bream, it is as well to mention that the title rightly belongs to a closely allied species, *Chrysophrys aurata*, a fish that properly inhabits the Mediterranean.

This fish derives its name from a semilunar golden spot over the eye. At the upper part of the edge of the operculum there is a violet patch. The back is blue, fading delicately into silver-gray, and the sides are longitudinally banded with golden streaks. The fins are grayish-blue, and at the bases of the dorsal and anal fins the scales are so raised at each side, that the fin looks as if it were set in a groove. This arrangement is seen in many of the fish belonging to this family. It is represented through the lower figure in the illustration on page 227.

SCALY-FINNED FISHES; SQUAMIPINNES.

We now arrive at a large family, containing a series of fishes remarkable for their extraordinary shape, their bold and eccentric coloring, and their curious habits. In Dr. Günther's elaborate arrangement of the Acanthopterygian fishes, this family is called by the name of Squamipinnes, or scaly-finned fishes, because "the vertical fins are more or less densely covered with small scales;" the spinous portions sometimes not scaly. They are nearly all carnivorous fishes, and for the most part are exclusively inhabitants of the tropical seas or rivers. Their bodies are very much compressed and extremely deep in proportion to their length, and the mouth is usually small and placed in front of the snout.

CHÆTODONTINA.

The large family of *Chætodontidæ*—the Chetodonts—so called from the Greek, meaning bristle-tooth, embraces one hundred and seventy species, in about five genera. They are carnivorous fishes; most of them belonging to the genera *Pomacanthus* and *Chætodon*. They are remarkable for their extraordinary shape, bold and eccentric coloring, and curious habits. One special characteristic is that the body is deep, often extremely so, and very thin or compressed, comparatively. The mouth is usually very small and placed in front of the snout.

The ANGEL-FISH (*Pomacanthus ciliaris*), called also ISABELITA in the West Indies, is found on our southern coast, and is quite abundant in the waters of the Florida Reef. It is one of the most beautiful of fishes, and has been eagerly sought for aquaria. Mr. Barnum, who first put in operation a sea-water aquarium, sent some assistants to the Florida Reef in the winter of 1859-60. The editor of this edition was then resident of Fort Jefferson, Tortugas, where the party ultimately arrived. In order to fit up Mr. Barnum's aquarium in New York City with the beautiful fishes and marine objects that are so abundant in the waters of the Florida Reef, we, who felt much interest in desiring our northern people to enjoy some of the beauties of the coral reefs, gave them all assistance. We encircled a lot of old roots that were lying in shallow water, the most favorable places for finding the Angel-fishes and many rare forms. When the seines were ready to haul, the roots were turned over, and the fishes and other forms being temporarily disturbed were quickly captured. The party had secured the services of a fishing-crew and their smack. The latter had the usual "well," which is used to keep the "fare" of fish alive within while waiting a market. Into this well the fishes, etc., were placed. The sea-water playing in and out through the bottom was comparatively pure. After securing many specimens of great value, as many as could safely be bestowed, the smack set sail for the colder waters of the north. Here in the Florida Straits the sea is, even in winter, warmer than that around the northern coast in summer. Some of the choicest specimens and duplicates were judiciously placed in glass globes, and kept in the cabin. The almost inevitable thing happened—the cold water killed everything before the party had reached Hatteras. The specimens in the globes were saved, and the "only Angel-fish ever exhibited in northern waters" was continued a long time a pleasing object to the many visitors of the aquarium, and a nearly sole consolation for the loss of the entire cargo of living tropical fishes. One of the most beautiful of the objects saved was a peculiar form of Sea-Anemone, which we had captive in our own aquarium, and which we added to the New York collection. This lived a long time, and was a constant source of admiration. It will be described in the proper place at the end of this volume.

The Wandering Chætodon is an example of a very large genus, comprising about seventy species, all of which are striking from their shape and color. Some of them are almost circular or disc-like in the general contour of their figure, and the arrangement of the markings is very conspicuous. The muzzle is moderate in length, and the scales are rather large in proportion to the dimensions of the body.

The Wandering Chætodon is a native of the waters extending from the Red Sea to Polynesia, and is one of the common fishes of the Ceylonese coasts.

The colors of this fish are very beautiful, and are arranged after a very curious fashion. The ground color of the body is golden-yellow, on which a number of purplish-brown lines are drawn. Some which start from the upper edge of the gill-cover are drawn obliquely towards the centre of the dorsal fin, and from the last of these lines a number of streaks issue nearly at right angles, take a slight sweep downwards, and then converge towards the tail. From the upper part of the head a broad black band descends to the angle of the interoperculum, and envelops the eye in its progress. The dorsal fin has a narrow black edge, and a black band extends along the soft portion of the same fin, crosses the tail, and is continued on the anal fin, which has a black and white edge. Two bold black bands are drawn across the tail. It is not a large species, rarely exceeding one foot in length.

A most remarkable species is called, from the form of its mouth, the Beaked Chætodon.

The curiously elongated muzzle is employed by this fish in a rather unexpected manner, being used as a gun or bow, a drop of water taking the place of the arrow or bullet. Perhaps the closest analogy is with the celebrated "sumpitan," or blow-gun, of the Macoushi Indians, a tube through which an arrow is driven by the force of the breath. The Beaked Chætodon feeds largely on flies and other insects, but is not forced to depend, as is the case with nearly every other fish, on the accidental fall of its prey into the water. If it sees a fly or other insect resting on a twig or grass-blade that overhangs the water, the Chætodon approaches very quietly, the greater part of its body submerged, and its nose just showing itself above the surface, the point directed towards the victim. Suddenly, it shoots a drop of water at the fly with such accuracy of aim, that the unsuspecting insect is knocked off its perch, and is snapped up by the fish as soon as it touches the surface of the water.

This habit it continues even in captivity, and is in consequence in great estimation as a houshold pet by the Japanese. They keep the fish in a large bowl of water, and amuse themselves by holding towards it a fly upon the end of a slender rod, and seeing the finny archer strike its prey into the water. Another fish, which will be described in the following pages, possesses the same faculty, but is not so remarkable for its eccentric form and the bold beauty of its tints.

The Beaked Chætodon inhabits the Indian and Polynesian seas, and has been taken off the west coast of Australia, where it is usually found in or near the mouths of rivers. Over the head and body of this species are drawn five brownish cross-bands edged with darker brown and white, and in the middle of the soft dorsal fin there is a rather large circular black spot edged with white.

Several other species of this genus are recognized, one of which, the LONG-BEAKED CHÆTODON (*Chelmo longirostris*), is truly remarkable for the exceeding development of the snout, which considerably exceeds half the length of the head. This species is also notable for a large triangular patch of jetty-black, which covers the upper surface of the head, the neck, and the side of the head as far as the lower edge of the eye. There is also a circular spot of the same hue on the anal fin. This species is a native of Amboyna.

A VERY remarkable fish adds to the singular shape of all the group the peculiarly elongated dorsal spine from which it has received its name of LONG-SPINED CHÆTODON, or CHARIOTEER. It also well exhibits the scale-covered fins, a structure which is indicative of the large family to which it belongs. Both scientific names are of Greek origin, the former signifying a charioteer, the long slender spine representing the whip; and the latter signifies "single-horned," in allusion to the same peculiarity.

The fourth dorsal spine of this species is enormously elongated and whip-like; its use not being as yet ascertained or even conjectured with any show of reason. Over each eye is a conical projection, not easily distinguished, on account of the deep black hue with which it is colored, and a similar protuberance arises on that part of the fish which is by courtesy termed the nape of the neck. Three very broad black bands are drawn across the body; their edges are sharply defined, as if a painter had drawn them with black varnish. The foremost band commences at the first dorsal spine, and sweeps over the neck, upper part of the head, snout, and chin, the eye being imbedded, as it were, in the black ground, and shining with great vividness on account of the contrast. The second band passes from the fifth to the seventh dorsal spines to the abdomen, being rather narrow at the top and widening as it passes downwards below, but not comprising the pectoral fin. The third band starts a little below the central streak, and is drawn rather obliquely over the body, through the hinder portion of the anal fin.

THE members of the curious genus to which the SEMILUNAR HOLOCANTHUS belongs are remarkable for a very strong, sharp-pointed, thorny spine with which the præoperculum is armed. These curious fish are found in almost all tropical seas.

Nearly forty species of this genus are now known, all of which possess some remarkable peculiarity in coloring. There is, for example, the RINGED HOLOCANTHUS (*Holocanthus annularis*), where the shoulder is decorated with a blue ring, and the body is marked with six or seven arched blue stripes, all radiating from the base of the pectoral fin. The SPOTTED HOLOCANTHUS (*Holocanthus maculosus*) has a number of black, semilunar spots on the fore part of the body; the CILIATED HOLOCANTHUS (*Holocanthus ciliaris*) is marked with an azure ring on the nape of the neck, and a number of blue spots and streaks about the head; the EMPEROR HOLOCANTHUS (*Holocanthus imperator*) has a number of blue lines upon the head, chest, and anal fin, a large black spot on the shoulder, and the body decorated with many waved, orange-colored streaks; and lastly, the ARCHED HOLOCANTHUS (*Holocanthus arcuatus*), though not so brightly clad, is quite as striking a species as any that has been mentioned, simply on account of the single arched stripe that is drawn along the body, from the eye to the end of the dorsal fin, taking a slight upward curve like a bent bow.

It is said of one of the species, LAMARCK'S HOLOCANTHUS (*Holocanthus lamarckii*), that the attachment between the sexes is very strongly developed, and that, if one individual be captured, its mate will haunt the fatal spot, and even fling itself ashore or into the net, in the eagerness of its search.

ARCHER FISH.

OTHER SCALE-FINNED FISHES.

OF another group or sub-family of the scale-finned fishes the ARCHER FISH is a good example. It is depicted on the accompanying full-page illustration.

This curious species is a native of the East Indian and Polynesian seas, and possesses the power of shooting water at its prey with even more force than the beaked chætodon. So powerful, indeed, is the projectile force, and so marvellously accurate is the aim, that it will strike a fly with certainty at a distance of three or even four feet. In general appearance, there is little to attract attention about this fish, the only remarkable point in its form being the greatly elongated lower jaw, which may possibly aid it in directing the liquid missile, on which it partially depends for its subsistence as does a hunter on the accuracy of his rifle. The general color of the Archer Fish is greenish, and the short, wide bands across the back are dark brown with a shade of green. Two species of this genus are known.

As an example of the next family, the Cirrhitidæ, we take the BANDED CHILODACTYLE. The family to which this fish belongs is a very small one, containing only eight genera, none of which comprise many species. Altogether, this family is not larger than many single genera. The members of which it is composed are all exotic species, inhabiting the "seas of the tropical regions and the southern temperate parts of the Pacific."

Perplexing as is the task of ascertaining the habitation of migrating birds, the difficulty of fixing the range of fishes is far less easy to overcome, as the transition from the tropical to the temperate, and from them to the colder seas, is so extremely gentle, that a fish of errant disposition, or one that has been caught in a long-lasting storm might be, and has been often, driven into strange waters which it does not know, and from which it can find no retreat.

THE large and important family of the Triglidæ, or Gurnards, is represented by several European fishes. This family contains a great number of species, many of which are most remarkable, not only for their beautiful colors, which alone are sufficient to attract attention, but also for the strange and wild shape, and large development of the fins. They are carnivorous fish, mostly inhabiting the seas, a very few species being able to exist in fresh water. They are not swift or strong swimmers, and therefore remain, for the most part, in deep water. Some, however, are able, by means of their largely developed pectoral fins, to raise themselves into the air, and for a brief space to sustain themselves in the thinner element. The mouth is mostly large; and in some cases the gape is so wide, and the head and jaws so strangely shaped, that the general aspect is most repulsive.

On account of its fiery color and ungainly aspect, the RED SCORPION-FISH has long been supposed to possess qualities as dangerous as its appearance is repulsive, and has been termed the SEA SCORPION and SEA DEVIL, from the supposed venom of its spines and frowardness of its temper. It is, however, a harmless fish enough, not capable of inflicting such severe injuries as several species that have already been described. When captured, it certainly plunges and struggles violently, in its endeavors to escape, and if handled incautiously it will probably inflict some painful injuries with its bony spears. This result, however, is attributable to the carelessness of the captor and to the natural desire for liberty, and not to any malevolent propensities innate in its being.

Another species, the SPOTTED SCORPION-FISH (*Scorpæna porcus*), represented in the fine engraving on next page, has similar habits. Both these fish are extremely voracious, as may be inferred from their wide mouth and general aspect, feeding on the smaller fish and similar creatures. They have a habit of lying in ambush, under overhanging tufts of sea-weed, and thence issuing in chase of any unfortunate little fish that may happen to pass near the fatal spot. All the fish of this genus are remarkable for their large head, with its armature of spines and odd skinny flaps, and the curious naked groove that runs along its summit. The pectoral fins are always large and rounded, and the body is mostly decorated with sundry skinny appendages. Examples of this genus are found in all the tropical seas, extending as far north as the Mediterranean, and to the Atlantic shores of Northern America.

The general color is brownish-red, marbled with dark brown and dotted with black. In some individuals the dots are arranged in lines around the dark marblings.

The extraordinary creature which is known by the name of RED FIRE-FISH, and to the natives of Ceylon by the title of GINI-MAHA, inhabits the greater part of the tropical seas, from Eastern Africa, through the Indian seas, to Australia.

This fish is remarkable for the singular development of the dorsal and pectoral fins, the latter being of such vast proportionate size, that they were formerly supposed to act like the

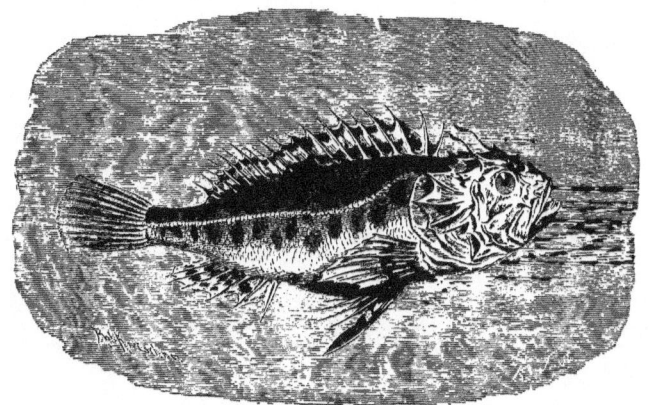

SPOTTED SCORPION-FISH.—*Scorpæna porcus.*

corresponding organs in the flying fish, and to raise the creature out of the water into the air. Such, however, is not the case; for the rays which carry the connecting membrane are not supported by a corresponding strength of bone as in the true flying fishes, and are far too weak to serve that purpose. Indeed, the object of this remarkable development is one of the many mysteries with which the inquiring zoologist is surrounded, and which make his task so exhaustlessly fascinating.

The structure of the entire skeleton is very interesting to comparative anatomists, but is too complicated, and requires too many technical terms to be described in these pages.

The Red Fire-fish is common off the Ceylonese coast, and is said to be rather valuable as an article of food, its flesh being very white, firm, and nutritious. The native fishermen hold this species in some dread, thinking that it can inflict an incurable wound with the sharp spines which arm its person and stand out so boldly in every direction. This idea, however, is without any foundation; for, although the thorny spines may prick the hand deeply and painfully, they carry no poison, and inflict no venomed hurt.

ONE or two notable fishes require a cursory notice.

The SEA LOCUST (*Apistos israelitórum*) is a native of the Red Sea, and is remarkable as being the only flying-fish of those strange waters. It is particularly plentiful on that part of the coast near which the Israelites were forced to wander for a space of forty years, and on that account has received its specific title. Ehrenberg has noticed that it is very abundant near Tor, and that several specimens fell into his boat almost every time that the sea was agitated. He further throws out a suggestion, that the quails to which allusion is made in the sacred volume are really the Sea Locusts, but this conjecture seems to be entirely gratuitous, and is unsupported by facts.

Another curious fish is the SEEPAARD of the Dutch (*Agriopus torvus*), a native of the seas around the Cape of Good Hope.

It is a rather powerfully armed species, on account of the strong, sharp, and recurved spines of the dorsal fin, but its head is not supplied with the thorny projections that render the preceding fish so perilous to handle. The dorsal fin of the Seepaard is single, and the spinous portion is greatly developed, rising in a bold curve over the shoulders and back like the crest of an ancient helmet, and being continued almost as far as the tail. Very little is known of this fish, though it is far from uncommon, and is eaten by the Dutch colonists of the Cape.

Its color is brown, mostly marbled with black, and the skin is smooth.

The strange and quaintly decorated fish, called YELLOW SCORPÆNA, is an inhabitant of the American coast, being found on the Atlantic shores of Northern America.

This odd-looking species frequents the same localities as the cod, and is often taken at the same time as that fish. The skin of the Yellow Scorpæna is devoid of scales, and the ventral and pectoral fins are enveloped in thick skin. The head is depressed, naked, and is covered with a series of loose, skinny appendages, that flap and wave about in the water without any apparent purpose. It is also armed with a number of rather sharp spines. There are two dorsal fins, the first being so deeply scooped that at one time the fish was described as possessing three dorsals. The first four spines of the dorsal fin are very long, and the membrane is deeply scooped between the fourth and fifth spines. The general color of this fish is yellow, tinged more or less with red, and in some specimens marbled with brown. The length of a very fine specimen is about two feet, but the ordinary average is from fourteen to eighteen inches.

The FILAMENTOUS GURNARD affords another example of this apparent capriciousness of grotesque formation. It is found on the coasts of the Isle of France. It appears to feed mostly upon crustaceans and mollusks, and the bony remnants of certain cuttle-fish have been found in its stomach. Its color is grayish-brown, marbled with a deeper hue of the same tint, and covered with minute spots of white.

There is another species of this genus which is colored in a rather bold and pleasing manner. This is the SPOTTED PELOR (*Pelor maculatum*), which derives it name from the manner in which the black hue of the skin is variegated with white. In this species there are three large white patches on the back, and three more on the dorsal fin. Some circular white spots are scattered on the head, and a white ring encircles the eyes. The pectoral fins are decorated with a bold white band, and the tail fin is marked with two white bands alternating with the same number of black stripes.

The odd-looking fish which is known by the name of the THREE-LOBED BLEPSIAS, is one of those species to which the ancient naturalists had affixed certain names without any apparent motive for so doing. There is no particular meaning in the word, and the sum of information obtainable from lexicons is, that it signifies a certain fish.

The members of this genus are found on the coasts of Kamschatka, and some fine specimens were obtained from the New Orcas Islands, in the Gulf of Georgia. This species is not very common, but may easily be known from its congener, the TWO-LOBED BLEPSIAS (*Blepsias bilobus*), by the peculiar manner in which the spiny portion of the dorsal fin is notched so as to form the whole fin into three distinct lobes. In the second species this structure is not seen. In both, the soft portion of the dorsal fin is greatly developed, and the body and fins are boldly marked with dark streaks upon a lighter surface. The body is entirely covered with prickles.

FAMILY *Triglidae*—the Gurnards. This group has about forty species, included in five genera. They are singular-looking fishes, resembling the *Cottidæ*, their allies. The Sea Robins are among them, and the curious Flying Gurnards.

THE SCULPINS—family *Cottidæ*. The genera are very numerous, being forty in number, and the species about one hundred and fifty. They are mostly confined to the rock pools and shores of northern regions. Many are found in fresh waters. Most are of small size and singular aspect, and none are valued as food. The Sea Ravens are included in this family. Several species called Sculpin are familiar to the fishermen and anglers of our northern coast; more particularly the tyro, who allows his line to lie loosely, and the bait to drag over the sea-weeds near bottom, where lurk these exceedingly odd-looking creatures.

BULL-HEAD.—*Cottus gobio*. (Natural size.)

WE now come to a very familiar and not very prepossessing fish—the well-known BULL-HEAD, or MILLER'S THUMB, sometimes called by the name of TOMMY LOGGE.

This large-headed and odd-looking fish is very common in European brooks and streams, where it is generally found under loose stones, and affords great sport to the juvenile fisherman.

The name of Miller's Thumb is derived from the peculiarly wide and flattened head, which is thought to bear some resemblance to the object whence its name is taken. A miller judges of the quality of the meal by rubbing it with his thumb over his fingers as it is shot from the spout, and by the continual use of this custom, the thumb becomes gradually widened and flattened at its extremity. The name of Bull-head also alludes to the same width and flatness of the skull. It is but a small fish, averaging four, and seldom exceeding five inches in length.

SEVERAL other species of this genus inhabit Europe. There is the SHORT-SPINED COTTUS, or SEA SCORPION (*Cottus scorpius*), which, as its name denotes, is one of the marine species. It is a very common fish, being found plentifully under heavy sea-weeds and stones, in the pools that are left above low-water mark by the retreating tide. The name of Scorpion is given to it on account of the sharp spines with which its head is armed, no less than eight sharp and four rather blunt prickles being found on the head. The rays of the dorsal and pectoral fins are also sharply pointed, so that it must be cautiously handled by those who wish to escape wounded fingers. This is a much prettier species than the preceding fish, its body being rich purple-brown, mottled with a warm red hue, and in the adult male there are some stripes of red on the pectoral fins, and the abdomen is brightly decorated with some snowy-white circular spots on glowing scarlet. Its extreme length seldom exceeds eight inches.

ANOTHER and much more formidable species is the well-known FATHER LASHER, LONG-SPINED COTTUS, or LUCKY PROACH (*Cottus bubalis*). In color this species is very like the sea scorpion, but it may be readily distinguished from that fish by the array of long and sharply-pointed spines with which its head is armed. It is a rather large species, measuring ten inches in length. It is common on European coasts, and like the preceding species, may be taken in the rock pools at low water.

THE FOUR-HORNED COTTUS (*Cottus quadricornis*) may be easily known by the four bony protuberances on the crown of the head. There are four spines on the præoperculum. Its general color is brown above, and grayish-white below, the sides being yellow. The lateral line is marked with rough points.

THE generic name of Platycephalus, which is appropriately given to this and the other fish placed in the same group, is of Greek origin, and signifies Broad-head.

The head is of great width, but also of very considerable flatness. It is even wider in proportion than that of the bull-head, but is narrower towards the snout and not so rounded. The body is also flattened in front, but assumes a more cylindrical form towards the tail.

The ARMED PLATYCEPHALUS is remarkable for the great length of the lower spine which proceeds from the præoperculum, and which reaches almost to the edge of the elongated operculum. It is also very wide and strong, being, indeed, about four times as large as the spine immediately above it. There are three little spines in front of the eye. Its color is brown, mottled and spotted on the fins with deeper and lighter shades of the same color.

WE now come to the typical genus of this family, which is represented by several well-known species.

The SAPPHIRINE GURNARD, so called from the fine, deep blue which tints the inner surface of the pectoral fins, is of tolerably common occurrence.

This seems to be the most valuable of nine species, being, like all the others, excellent for the table, but exceeding them considerably in size. The name of Hirundo, or swallow, has been given to this fish on account of the great size of the pectoral fins, which are almost as proportionately large to the dimensions of the fish as the wings of the swallow to the bird. Putting aside the great development of these members and their rich blue color, the Sapphirine Gurnard may be distinguished from the other species by the extreme smoothness of the lateral line, which may be rubbed with the finger in either direction without exhibiting the spiny roughness which is found in other Gurnards. In consequence of this structure, the fishermen sometimes call the fish the Smoothside Gurnard.

All the scales of this species are very small. The large head is armed with spines, some springing from just before the eye, and others from the operculum and the shoulder.

Of several other species of Gurnard, may be mentioned the CUCKOO GURNARD (*Trigla pini*), sometimes called the RED GURNARD from the color of its body. This is a very common species, and when young may be found in the rock pools at low water, measuring only a few inches in length, but perfectly exhibiting the characters of its genus. The specific title of "pini," or belonging to the pine-tree, is given to the Cuckoo Gurnard on account of the peculiar aspect of the lateral line, which is crossed with numerous short, straight, narrow, and elevated lines, which have been compared by some writers to the needle-shaped leaves of the pine. The name of Cuckoo Gurnard is given to it, because when it is first taken out of the water it emits a sound which bears a distant resemblance to the cuckoo's cry. The curious soft rays which project from the base of the pectoral fin in this and other Gurnards are evidently organs of touch, being plentifully supplied with nerves and movable at the will of the owner.

The color of this fish is bright rosy-red above, and silvery-white on the sides and abdomen. These colors soon fade after the fish has been removed from the water.

THE GRAY GURNARD (*Trigla gurnardus*) is also tolerably common, and is readily to be known by its short pectoral fins and the greenish-brown body, spotted with white above the lateral line. On account of the peculiar sound which it utters, it is popularly known in Scotland by the name of CROONER, and in Ireland it is called the NOWD.

SAPPHIRINE GURNARD.—*Trigla hirundo*. (One-fifth natural size.)

ANOTHER curious and remarkable species, the SHINING or LONG-FINNED GURNARD (*Trigla obscúra*), is at once known by the great length of the second spine of the dorsal fin, which is nearly double the length of the other spines, and projects boldly with a slight curve towards the tail. It is a handsomely colored fish, the head and upper part of the body being vermilion-red, and the abdomen white, tinged with red. The flanks are shining silvery-white, and have given cause for the name of Shining Gurnard. The fins are all bright red, with the exception of the pectorals, which are deep blue.

THE LYRIE, or ARMED BULL-HEAD, is known by a great variety of names, such as POGGE, SEA POACHER, and NOBLE.

It is a curious-looking fish, with its bony armor-plates and shielded head. It is most commonly taken near the mouths of rivers, though it is sometimes captured far out at sea. Its flesh is firm and good, but its small size and bony shields render it scarcely serviceable for the table. It feeds mostly on aquatic animals.

The body of the Lyrie is covered by eight rows of bony plates, strongly reminding the observer of the sturgeon, and the head, gill-cover, and shoulders are strongly armed with spines.

The general color of the Lyrie is brown above, crossed with several broad bands of dark brown, and the abdomen is white, with a trifling tinge of brown.

IN the remarkable genus which now comes before our notice, the body is covered with bony plates, like ancient armor, and the front part of the head is formed into a deeply cleft fork on account of the development of certain bones of the skull.

The ORIENTAL GURNARD is found in the Japanese seas, and is a good example of the genus to which it belongs, the bony plates being very large, and the forked processes of the head well developed. Between the ventral fins, each bony plate is just three times as long as it is broad. The præoperculum is furnished with a strong spine, crossed by a projecting ridge from its angle.

A very curious species belonging to this genus is known by the name of MAILED GURNARD (*Peristethus cataphractum*).

In this fish, the bony plates between the ventral fins are twice as long as they are broad. It mostly prefers rather deep water over rocky ground, but approaches the shallows for the purpose of spawning. Its food consists of the softer crustaceans, medusæ, and similar creatures. It is a swift swimmer, but seems to be rather reckless, as it not unfrequently strikes its forked snout against the stones, and breaks off one or both points. The flesh of the Mailed Gurnard is tolerably good, but requires some care in cooking, besides costing some little trouble in freeing it from the hard, bony plates in which the body is so securely enveloped. In order to clear away these defences, the fish must be soaked in warm water, and the scales stripped off from the tail upward. In some places, such as the coasts of Spain, it is held in considerable estimation, and is especially sought by fishermen. Its color is like that of the Red Gurnard. Nearly all the rays of the first dorsal fin are extremely elongated, and, together with the mailed body, the armed head, and the double snout, give to the fish a most singular aspect. The total length of the Mailed Gurnard is about two feet.

THE Flying Gurnards are extraordinary and beautiful fishes, remarkable not only for the very great development of the pectoral fins, their muscles and attachments, but for the unexpected use to which those members are occasionally subservient.

These fishes, together with one or two other species, hereafter to be described, possess the power of darting from the water into the air, and by the mingled force of the impetus with which they spring from the surface, and the widely spread wing-like fins, to sustain themselves for a short space in the thinner element, and usurp for a time the privileges of the winged beings whose trackless path is through the air.

The object of exercising these strange powers seems to be, not the pleasure of the fish, but the hope of escaping from the jaws of some voracious monster of the deep, whose sub-aquatic speed is greater than that of the intended victim, but whose limited powers are incapable of raising it into the air. Foremost among these persecutors is the coryphene, often called the dolphin by sailors, and which is the so-called "dolphin" whose colors glow with such changeful beauty during its death-pangs.

Little, however, do the powers of flight avail the unfortunate fish, for winged foes, known by the name of albatross, frigate-bird, and similar titles, are hovering above in waiting for their prey, and no sooner does the Gurnard launch itself fairly into the air, and so escape the open jaws of the pursuer coryphene, than the albatross swoops down with extended wings, snatches up the fish in its beak, and without altering the bold and graceful curve in which it has made the swoop, sweeps up again into its airy height, where it wheels on steady wing awaiting another victim.

Between the hungry coryphene below and the voracious albatross above, the poor Flying Gurnard leads no very happy life, and its intermediate existence, persecuted on either side, has been often employed as a type of those unfortunate persons who are ashamed of the more lowly society in which they were born, and aspire to ascend to an elevated condition for which they are not fitted by nature.

While passing through the air, the Flying Gurnard is able slightly to change its direction, but cannot prolong its flight, by flapping its finny wings. In fact, its elevation into the air may be readily imitated by throwing an oyster-shell in a horizontal direction, taking care to throw it in such a manner that the concavity is downwards and the convexity upwards. The flight is closely analogous to that of the flying squirrels, rats, and mice among mammalia, and of the flying dragon among reptiles.

The COMMON FLYING GURNARD, represented in the accompanying full-page illustration, is brown above, passing into a beautiful rose-color below. The fins are black, variegated with blue spots, and on the tail fin the spots run together so as to resemble continuous bands. Its length varies from ten to fifteen inches. It is a native of the Mediterranean and warmer parts of the Atlantic, and in many parts of those seas is very common.

The second species, the INDIAN FLYING GURNARD, is found throughout the Indian Ocean and Archipelago, and on account of its habits, its singular and striking form, and its lovely coloring, has always attracted the attention of voyagers, even though they have possessed no skill in natural history.

This beautiful fish is notable for the two long detached filaments that are planted between the head and the dorsal fin, the first being extremely elongated and the second much shorter. The first spine of the dorsal fin is solitary, and at first sight looks like another isolated filament. In all the members of this genus, the præoperculum is armed with long, sharp, and powerful spines, the scales of the body are strongly keeled, and there is no appearance of a lateral line. Four species of Flying Gurnards are known, the two which have been selected affording excellent types of their general form. In the Indian Flying Gurnard, the pectorals are covered with brown spots, and dotted rather profusely with bluish white.

WE now arrive at a moderately large family of fishes, called, from the typical genus, Trachinidæ. In these creatures the body is long and rather flattened, the gill-covers are wide, and the teeth are arranged in bands.

OUR first example of these fishes is the very remarkable MEDITERRANEAN URANOSCOPUS, a word which requires some little explanation before examining the form and habits of the species. The generic title is derived from two Greek words, literally signifying sky-gazer, and is given to the fish on account of the peculiar position of the eyes, which are set so singularly on the upper part of the head, that they look upwards instead of sideways, as is the usual custom among the finny inhabitants of the waters. It is illustrated on next page.

This species lives mostly at the bottom of deep seas, and is said to angle for the smaller fish, on which it feeds, by agitating a slender filamentary appendage of its mouth in such a manner as to resemble a worm, and to pounce on the deluded victims when they hurry to the spot in hopes of a meal. Though a fish of rather repulsive aspect, its flesh is tolerably good, and is eaten in many parts of Europe and along the shores of the Mediterranean.

Its head is very large and broad, and is partially covered with bony plates, and the opening of the mouth is nearly vertical. The slender filament which has already been mentioned is set before and below the tongue, and the shoulders and gill-covers are armed with an array of strong sharp spines.

THE STAR-GAZERS, or family *Uranoscopidæ*, are divided among seven genera, and twenty species are known. They are carnivorous fishes, of singular appearance, living on the sea bottoms in most warm regions. The great protruding eyes are placed upon the surface of the face, and near each other, that they may be observant when buried, as they are much of the

FLYING GURNARD.

time, in the sand. It is, therefore, a fancy about star-gazing. Their goggle eyes, directed upward, suggested the name. Two species only are known to North American waters.

The typical genus of this family is represented by several species, of which the GREAT WEAVER is one of the most familiar.

This species is the dread of fishermen, the wounds occasioned by the sharp spine of the gill-cover, and those of the first dorsal fin, being extremely painful, and said to resemble the sting of a hornet, the evil effects extending from the hand up the arm, and even reaching the shoulder. On the first infliction of the injury, it gives little more pain than the prick of a pin or needle, but in a short time, a dull hot pain creeps up the arm, and increases in intensity for several hours. Fishermen, taught by experience, are very cautious in handling this dangerous fish, and before they place it in their basket they cut off the whole of the first dorsal fin and the hinder part of the gill-cover. In France, this precaution is rendered compulsory by law.

MEDITERRANEAN URANOSCOPUS—*Uranoscopus scaber.*   GREAT WEAVER-FISH.—*Trachinus draco.*

THE curious fish called the INDIAN SILLAGO is a good example of a moderately large genus which is spread over many seas, being found on various shores from the Red Sea to the coast of Australia.

The Indian Sillago is easily recognized by the extraordinary length of the second dorsal spine, which, in a good specimen, is developed to such an extent that it equals the length of the body. The use of this structure is very obscure. As its name imports, this fish inhabits the Indian seas, and is found in the Bay of Bengal and near the mouth of the Ganges. It is held in some estimation for the table, as its flesh is light, digestible, and well flavored. The color of the Indian Sillago is brown.

THE BRAZILIAN PERCOPHIS is found upon the coasts of Brazil, and is apparently the sole representative of the genus in which it has been placed. The first dorsal fin is very small in proportion to the second, and the space between them is about equal to the length of the first dorsal. The ventral fins are set very far forward, being placed under the throat. The lower jaw projects considerably beyond the upper, and the cleft of the mouth is horizontal. The canine teeth are very large in proportion to the dimensions of the fish.

THE ARAPAIMA (*Sadis gigas*), also called PIRARUCU, is one of the most remarkable fishes known, as to size especially. Specimens have been caught measuring fifteen feet in length, and of 410 pounds weight. In our illustration the gigantic fish is shown one-twentieth of its

ARAPAIMA.—*Sadis gigas*.

natural size. The body is entirely covered by large scales. The remarkable colors add to its singularity, as large fishes are usually plainly decorated. This fish has the tail so small, it appears to have been shorn of a large part. The color of the body and base of fins is a mixture of bluish and crimson lake, with a terminal bar of blue along the fins and tail. It is abundant in the Amazon, where it is prized as an edible.

ANOTHER family, the *Sciænidæ*, now come before us. The members of this family are clothed with ctenoid, or toothed scales; the mouth is set in front of the snout, the teeth are arranged in bands, and the gill-covers are either unarmed or furnished with feeble spines.

The first example of this family is the BELTED HORSEMAN, a striking and boldly marked species.

This fish is found upon the Atlantic coasts of tropical America, and is, perhaps, the most striking of the limited genus to which it belongs. The body is oblong, and the nape of the neck is very high, its elevated line being continued by the first dorsal fin, which is short, high, and pointed, its height being just equal to the depth of the body. The second dorsal fin is long, rather low, and is covered with very thin scales. The tail fin is covered in like manner. The scales of the body are of moderate size.

Its general color is grayish yellow, diversified with three broad brown belts, edged with whitish gray.

Another species of the same genus, the SPOTTED HORSEMAN (*Eques punctatus*), is nearly, though not quite as remarkable a fish, and is notable for the bluish-white spots which decorate the dorsal, ventral, and anal fins. The general color of this fish is brown, with two vertical bands running over the side of the head, and some curved bands passing along the body from the back to the tail. This fish is found in the Caribbean seas.

CLOSELY allied to these creatures is a rather remarkable fish, called scientifically *Pogonias chromis*, and more popularly known by the name of BEARDED DRUM-FISH.
This title is given to the fish on account of the peculiar sounds produced by the fish, which are thought to bear some resemblance to the beating of a drum. The sound is apparently produced both while the fish is immersed and after its removal from the water, and probably on account of the sound-conducting powers of the water, the hearer finds great difficulty in referring the strange noises to any particular spot. These fish do not seem to thrive well in fresh water, as the drumming was invariably found to cease as soon as the boat in which the observers were sitting had left the sea-coast and entered a river. It is a native of the North American coasts, and is known to extend as far south as Florida.

ANOTHER noisy fish is well known under the title of MAIGRE, the strange sounds produced by this species having been heard from a depth of one hundred and twenty feet.
In one instance, perhaps in many others, the novel accomplishment has led to the destruction of its possessors, the fishermen having been directed by the sounds to the whereabouts of the utterers, and inclosed them in their nets. The flesh of the Maigre is thought to be peculiarly excellent, the head and shoulders being held in the greatest estimation.
It is a rather large fish, seldom measuring less than a yard in length, and often attaining nearly double those dimensions, and is in consequence extremely valuable to the fisherman. Although at one time it might be captured with tolerable frequency on the coast of France, and now and then on more northern European shores, it is now very scarce, having shifted its localities, and being found most plentifully on the southern shores of the Mediterranean. There it seems to be hatched and to remain until it attains nearly adult age, when it crosses to the northern side of that sea, and is there found to be of considerable dimensions.

THE BLACK CORVINA of the Mediterranean (*Corvina nigra*) is allied to the maigre, and is scarcely less celebrated than that fish for the excellence of its flesh.
This fish is not exclusively a marine species, but frequents salt lakes and ponds, and, though it hovers about the mouths of large rivers, probably for the purpose of feeding on the many animal and vegetable substances which are borne by their currents into the sea, does not appear to ascend their streams. In general appearance it is not unlike the maigre, and is often sold in the markets under that name.

THE SQUETEAGUE, or WEAK-FISH of North America (*Cynoscion regale*) is another of the noisy fishes, producing dull sounds like those of a drum. It is plentiful about New York, and is captured in large quantities for the table. The name of Weak-fish is attributable to two causes, the one that when hooked it makes but a feeble resistance, and the other that its flesh is popularly supposed to be weakening to those who habitually live upon it. It is a useful species, for it not only affords delicate food, but its swimming-bladder can be made into isinglass which is said to be in no way inferior to that of the sturgeon. On account of its spotted skin, the French of New Orleans call it by the name of Trout.
It is a member of the family *Sciænidæ*, the Croakers, which number one hundred and twenty-five species, included in twenty-five genera. The species are all carnivorous; and most reach a large size.
A genus embraces what are called Fresh-water Drums.
The BIG DRUM (*Pogonias chromis*) inhabits from Cape Cod to Florida. A peculiar drumming sound is heard from it. Some fine ones, three feet in length, were kept in Mr. Coup's Aquarium, and proved of great interest to visitors.
LAFAYETTE FISH (*Stromateus triacanthus*). This is a species which made its appearance, as was supposed, for the first time, in great numbers in the Long Island Sound and adjacent waters, at the time Lafayette made his last visit to the country, in 1824. Its habit is to reap-

pear in large shoals after long intervals, and though then known to science, it was new to the public. It is called Goody at Cape May, and Chub and Roach in Virginia.

THE KING-FISH (*Menticirrus nebulosus*) is a large silvery fish, much esteemed in Key West and other Southern cities as a table fish. It resembles the mackerel or blue-fish in its flesh.

This fish affords much occupation to the fishermen of the Florida Keys; the Havana market being supplied therefrom. It is found as far north as Cape Cod. It is called in some quarters Bermuda Whiting. Its excellence as a food-fish induced the early settlers to name it King-fish in token of its superiority. Barb is another name for it.

BECUNA.—*Sphyræna vulgaris*.   FISHING FROG.—*Lophius piscatorius*.

ONE example of the Sphyrænidæ, the family next in order, is the BECUNA, a rather large and tolerably ferocious fish, inhabiting the Mediterranean and many parts of the Atlantic Ocean.

This long-bodied, deep-mouthed, and sharp-toothed fish bears some resemblance to the pike both in general appearance and in habits, and is hardly less voracious than the veritable pike of our own country. It is said that from the scales of the Becuna are washed those minute crystalline spiculæ, which are so useful in the preparation of artificial pearls, and which, when mixed and prepared for commerce, are termed *essence d'orient*. Some parts of the air-bladder are also used in the manufacture of this substance. The flesh of the Becuna is well flavored and is often brought to table; being capable of being dressed in a fresh state and after salting.

On the back, the color of this fish is leaden-blue with a wash of green, and on the abdomen it is white. The sides are in many specimens marked with dark cross-bars of the same green color as the back. When young it is spotted with brown.

THE BARRACOUDAS, family *Sphyrænidæ*, consist of one genus and about fifteen species. They are voracious, pike-like fishes, inhabiting nearly all temperate and tropical seas. A species is found on the Pacific coast, which proves a valuable food-fish, measuring about three feet.

These fishes resemble strongly the pickerels or pikes, and their habits are certainly much the same. We have often seen the fishermen on the Florida Reef strike some great specimen of the Barracouda—the *S. picuda*, we think, which abounds there. The fisherman stands in his "dingy" and sculls with his back to the sun, just outside the shallow waters of the reef. The Barracouda is curious about the oar-blade, and follows it. The sun is in *his* eye. A well-directed aim transfixes the victim.

WE now arrive at a small family of fishes, termed Trichiuridæ, or Hair-tailed fishes, in consequence of the delicate filamentary finlets which decorate the tail in some species. In all these fishes, the body is long and compressed, almost like a riband, and indeed is not at all unlike those flat "snakes" that are sold in the toy shops, and which dart in all directions when held by the tail.

The first example of these curious creatures is the SCABBARD-FISH, so called because in shape it bears some resemblance to the sheath of a sword.

On account of its shape and bright silvery whiteness, it is a most striking inhabitant of the ocean, and when writhing its way through the translucent water, in elegant undulations, it looks like a broad riband of burnished silver winding through the waves. This shining brilliancy is caused by a thin epidermis, which covers the body in place of scales, and which can be easily rubbed off by the fingers, to which it adheres, transmitting to them a portion of the metallic whiteness which it imparts to its proper owner.

In spite of the exquisite beauty of this fish, it is captured for the sake of its flesh, which is highly esteemed, and is generally sought in the months of April and May when it approaches the coasts. The drag-net is the usual instrument of capture. It seems to be a solitary fish, and lives at a considerable depth. The rapid undulations of the body are capable of propelling the creature through the water with great velocity, but, from all appearances, it is not able to make much way against a rapid tide, or to overcome the dashing waves raised by a tempestuous wind. Along its back runs a single dorsal fin, and the ventral fins are only represented by a pair of scales, a structure which has gained for the fish the generic name of Lepidopus, or Scale-fin.

ANOTHER fish that much resembles the preceding species is sometimes, but very rarely seen. This is the SILVERY HAIR-TAIL (*Trichiúrus leptúrus*), a species that may easily be distinguished from the scabbard-fish by the shape of the tail, which has no fin at its extremity, but tapers into a long and gradually diminishing point. This species is common in many parts of the Atlantic Ocean, and by the Spanish inhabitants of Cuba it is termed the Sabre-fish.

Another species, the SAVALA (*Trichiúrus savala*), is found in the East Indian seas, and is sufficiently plentiful to form a recognized article of diet, and to be sold in the markets. The Savala bears salting well, and is much used for consumption when the inclemency of the weather will not permit fishing-boats to put to sea. When fresh, however, it does not suit the taste of Europeans, though in Malabar the salted fish is esteemed both by the native inhabitants and the European colonists.

BEFORE leaving this small but curious family, the ATUN (*Thyrsites atun*) deserves a passing notice.

This elegant and useful fish is found on the coasts of Southern Africa and part of Australia, and is much valued for the flakey whiteness and pleasant flavor of its flesh, which bears some resemblance to that of the cod, but is even superior in delicacy. It feeds mostly upon the cuttle-fish, the calamary being its favorite prey. So voracious is this creature that it is readily caught by making a sham calamary out of lead and leather, dressing it with projecting hooks, and flinging it into the sea. The fishermen throw this bait to some distance, and then draw it rapidly through the water, when the Atun takes it for the real calamary darting along after its usual fashion, dashes at it and is immediately hooked. In default of this bait, a strip of red cloth stuck on a hook is often a sufficient lure for this voracious fish.

THE MACKERELS, family *Scombridæ*, include seventeen genera and about seventy species of highly brilliant and metallic-tinted fishes, found in the high seas. Many of them are cosmopolitan, and all have a wide range.

The COMMON MACKEREL (*Scomber scombrus*), the well-known food-fish, is abundant along the whole coast of North America, occasionally straying to the Pacific ocean.

The notable SPANISH MACKEREL is a common article in our New York market. It is not frequently seen above that.

The Bonito is another ally, of considerable repute as a food-fish,—occasional on our coast.

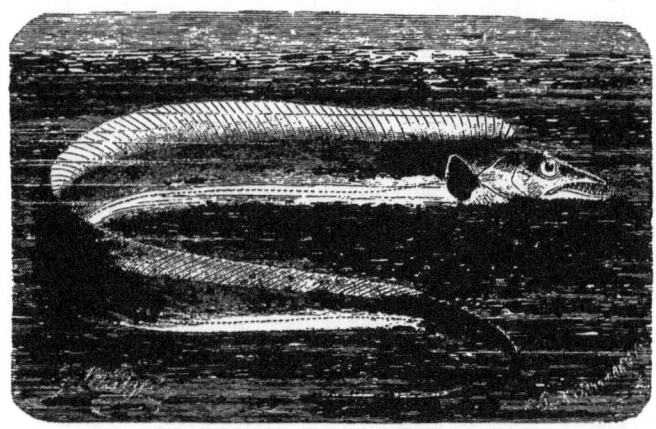

SILVERY HAIR-TAIL.—*Trichiurus lepturus.*

The Tunnies of this family are wonderful for their size. The COMMON TUNNY, or HORSE MACKEREL, is a notable creature, reaching the length of ten feet, and weighing a half ton. It makes its appearance in the summer months, sometimes being taken in the seines. Though large in the anterior half, its terminal portion has all the beauty of the shape of the Mackerel. The small of the body and the sharply-defined crescent tail render it a graceful fish. It is one of the well-known ancient fishes, being abundant in the Mediterranean Sea from the earliest time. A single specimen has yielded twenty gallons of oil. So much like the Mackerel is its flesh, it is captured for the market, and its flesh sold as third-rate mackerel.

The LITTLE TUNNY, or ALBICORE, is an active, graceful fish, running in schools of a hundred or more. We have seen them leaping out of water, and gambolling around Egg Rock, at Nahant, Massachusetts.

The Mackerel is well known for the exceeding beauty of its colors and the peculiar flavor of its flesh. This is one of the species that are forced by the irresistible impulse of instinct to migrate in vast shoals at certain times of the year, directing their course towards the shores, and as a general rule frequenting the same or neighboring localities from year to year. The time of their advent is rather variable, and in consequence the price of this fish varies with the scarcity or abundance.

The flesh of the Mackerel is very excellent, and it possesses a rather powerful and unique flavor that has caused fennel to be looked upon as a necessary corrective in the sauce with which the fish is served. Unfortunately, it must be eaten while quite fresh, as it becomes

MACKEREL.—*Scomber scombrus.*   HORSE MACKEREL.—*Trachurus saurus.*

unfit for consumption in a very short time after being taken out of the water; and in consequence of this property, the London costermongers are permitted to hawk it about the streets on Sundays, much to the discomfort of peaceable householders who long for repose and do not want Mackerel.

TUNNY.—*Oroynus thynnus.*

THE TUNNY does not visit the European coasts in sufficient numbers to be of any commercial importance; but on the shores of the Mediterranean, where it is found in very great abundance, it forms one of the chief sources of wealth of the sea-side population.

In May and June, the Tunnies move in vast shoals along the shores, seeking for suitable spots wherein to deposit their spawn. As soon as they are seen on the move, notice is given by a sentinel, who is constantly watching from some lofty eminence, and the whole population is at once astir, preparing nets for the capture, and salt and tubs for the curing of the expected fish. There are two modes of catching the Tunny, one by the seine-net and the other by the "madrague." The mode of using the seine is identical with that which has already been described when treating of the mackerel, but the madrague is much more complicated in its structure and management.

The general shape of the Tunny is like that of the Mackerel, but in size it is vastly superior, generally averaging four feet in length and sometimes attaining the dimensions of six or seven feet.

Of an allied species, the PACIFIC ALBACORE (*Orcynus alalonga*), Mr. F. D. Bennett writes as follows, in his well-known "Whaling Voyage." "Ships, when cruising slowly in the Pacific Ocean, are usually attended by myriads of this fish, for many successive months. A few days' rapid sailing is nevertheless sufficient to get rid of them, however numerous they may be; for they seldom pay more than very transient visits to vessels making a quick passage. When a ship is sailing with a fresh breeze, they swim pertinaciously by her side and take the hook greedily; but should she be lying motionless or becalmed, they go off to some distance in search of prey, and cannot be prevailed upon to take the most tempting bait that the sailor can devise."

The BONITO (*Sarda mediterranea*) is a very pretty and common species that is found in the Mediterranean and many parts of the Atlantic.

This is a smaller species than the albacore, not exceeding two feet and a half in length. The flesh of this fish is eaten both fresh and when pickled, but in a fresh state is not held in very high estimation. At some seasons, it appears to contract an unwholesome quality, which is injurious to certain constitutions, causing rather a painful rash to break out on the face and body, though others can eat it with impunity. The flesh is very red in color, and looks very like butcher's meat.

Like the albacore, the Bonito is a determined foe of that much persecuted creature, the flying-fish, and is often taken by means of a hook dressed with feathers so as to resemble its natural prey. It is a truly beautiful species, deserving fully the popular name of Bonito, which may be freely translated as Little Beauty. The back is deep indigo-blue, mottled with a lighter shade of the same hue, and when young a number of dark streaks are drawn across the back. The abdomen is silvery-white, and the cheeks and gill-covers are of the same brilliant hue.

BONITO (*Sarda chiliensis*), SKIP-JACK, and TUNA, so called, and the *S. mediterranea* are food-fishes of something the same quality as the mackerel. They are especially "sea-going" species.

Another species, the STRIPED BONITO (*Auxis rochei*), inhabits the same localities, and is nearly as plentiful as the preceding fish. It may readily be known from the plain Bonito by the four dark lines which extend along each side of the abdomen and end at the tail.

THE prettily-marked PILOT-FISH is frequently seen off the American coasts, but seems to be rather shy, and is not very often captured.

This little fish has long been supposed to act the part of the shark's provider, and to perform in the ocean the same actions that were once attributed to the jackal on land. Many modern writers, however, deny the truth of the statement, by saying that the Pilot-fish only follows the shark for the sake of the scraps that the larger fish is likely to leave, and that it would probably be snapped up by the shark but for its watchfulness and agility.

As is usual in such a disputation, the evidence is very conflicting, and many accounts have been published tending to throw discredit on the one side or the other, according to the particular circumstances under which the observations were made. One well-known naturalist, for example, mentions an instance where a shark was directed towards a baited hook by two Pilot-fish that accompanied him; but, on the other hand, another accomplished observer narrates an interesting anecdote of a shark being continually warned of a baited hook by his little friends, who struck their noses against his snout whenever he turned towards the bait.

At last, however, he dashed at the tempting morsel and was captured, to the sorrow of the Pilot-fishes, who swam about for some time in search of their friend, and then darted down into the depths of the sea.

BLUE-FISH (*Pomatomus saltator*), called in some quarters GREEN-FISH and SKIP-JACK. This is valued generally as one of the choicest of ocean fishes, being much the same as the mackerel as a food-fish, but rather preferred. In the first quarter of this century, Blue-fishes were unknown in New England above Cape Cod. About 1850, single individuals were taken at Nahant, and for several years a few were taken, and valued very highly. Soon we heard of their abundance to such an extent that they were hauled on to the land as manure, on the coast above Cape Cod. With a good breeze and trolling lines, this fish affords much sport. Its range is remarkable; it is found in the Mediterranean and Indian Ocean, and near New Holland. Blue-fish are very destructive to the mackerel fishery. They are voracious and make havoc in the shoals of those fishes.

THE POMPANO (*Trachynotus carolinus*) is a much prized fish of the Southern waters, ranking ahead of all others. It is put down in salt for the market, and is always in great request. In South Carolina it is called CREVALLI. Its range is from Cape Cod to Florida.

SPANISH MACKEREL (*Scomberomorus maculatus*). A "sea-going" fish, but frequently exposed for sale in New York markets.

EVERY one has heard of the SUCKING-FISH, and there are few who are not acquainted with the wild and fabulous tales narrated of its powers.

This little fish was reported to adhere to the bottom of ships, and to arrest their progress as suddenly and firmly as if they had struck upon a rock. The winds might blow, the sails might fill, and the masts creak, but the unseen fish below could hold the vessel by its single force, and confine her to the same spot as if at anchor. It is wonderful how fully this fable was received, and how many years were needed to root the belief out of prejudiced minds. Both scientific names refer to this so-called property, *echeneis* signifying "shipholder," and *rémora* meaning delay.

That the Sucking-fish is able to adhere strongly to smooth surfaces is a well-known fact, the process being accomplished by means of the curious shield or disc upon the upper surface of the head and shoulders. This disc is composed of a number of flat, bony laminæ, arranged parallel to each other in a manner resembling the common wooden window-blind, and capable of being raised or depressed at will. It is found by anatomical investigation, that these laminæ are formed by modifications of the spinous dorsal fin, the number of laminæ corresponding to that of the spines. They are moved by a series of muscles set obliquely; and when the fish presses the soft edge of the disc against any smooth object and then depresses the laminæ, a vacuum is formed, causing the fish to adhere tightly to the spot upon which the disc is placed.

When the creature has once fixed itself, it cannot be detached without much difficulty; and the only method of removing it, without tearing the body or disc, is to slide it forwards in a direction corresponding with the set of the laminæ. In the opposite direction it cannot be moved; and the fish, therefore, when adhering to a moving body, takes care to fix itself in such a manner that it cannot be washed off by the water through which it is drawn. Even after death, or when the disc is separated from the body, this curious organ can be applied to any smooth object, and will hold with tolerable firmness. In order to accommodate the disc, the upper part of the skull is flattened and rather widened.

The Sucking-fish will attach itself to many moving objects, and has been found adhering to the plankings of ships and boats, to turtles, to whales, and to fishes of various kinds. Even the albacore, which eats the Sucking-fish whenever it can catch it, is occasionally honored by its adhesion, and in the British seas a specimen has been captured while sticking to a cod-fish. The shark, however, is its favorite companion; and it often happens that one of these voracious creatures is attended by quite a little train of Sucking-fishes. What object is fulfilled by this capability of adhesion, is a problem as yet unsolved. The Remora is perfectly organized and capable of procuring food for itself, and, though not a swift swimmer, is able to proceed through the water with tolerable rapidity. Its mouth is moderately large; and that the creature has no difficulty in seeking a subsistence is proved by the fact that its stomach usually contains remnants of small crustacea and mollusks.

The color of this species is dusky brown, darker on the back than on the abdomen. The fins are darker than the body, and are of a dense leathery consistence. The length of this fish seldom exceeds eight inches.

There are about ten species of Sucking-fishes known, of which the SHIELDED SUCKING-FISH (*Echeneis scutáta*) is perhaps the most remarkable. This species may be at once recognized by the very great size of the disc, and its length being nearly one-half that of the body. At the hinder portion of the disc the laminæ are wanting, and its surface is smooth. This species attains to considerable dimensions, sometimes to nearly two feet in length.

The family of Remoras (*Echeneididæ*) is made up of the species of two genera, found in all seas, all having a long range. The COMMON REMORA is found attached to large sharks in the warmer waters, from the Atlantic to the Pacific. Another species is found north as far as Massachusetts. We have seen several Remoras drop from the Blue Shark, of the Gulf of Mexico, when taken from the water.

THE well-known JOHN DORY, so dear to epicures, is frequently seen in the fishmongers' shops, where its peculiar shape seldom fails of attracting attention even from those who are not likely to purchase it, or even to have seen it on the table.

The name of John Dory is thought to be a corruption of the French name *jaune dorée*, a title given to the fish on account of the gilded yellow which decorates its body. It was called Zeus by the ancients, because they considered it to be the king of eatable fish; and the name of Faber, or blacksmith, has probably been earned by the smoky tints which cloud its back. The dark and conspicuous spots on the side are thought in many places to be imprinted on the fish as a memorial of the honor conferred upon its ancestor in times past, when St. Peter took the tribute-money from the mouth of the Dory, and left the print of his finger and thumb as a perpetual remembrance of the event. Some persons, however, contend that the marks are due, not to St. Peter, but to St. Christopher; and the Greeks, who hold to the latter tradition, call the fish Christophoron.

The Dories (family *Zenidæ*) inhabit the warmer seas. Five genera and about ten species are known. One species only is recorded as familiar to our coasts, the *Z. ocellatus*, though it is oceanic, approaching our shores near Cape Cod.

WE now come to a most beautiful and interesting fish, the CORYPHENE, so often erroneously spoken of as the dolphin.

This splendid fish is found in many of the warmer seas, inhabiting the Mediterranean Sea, and the Indian, Pacific, and Atlantic Oceans. The reader has, in all probability, heard the old story respecting the lovely and changeful colors of the dying dolphin, and is quite aware that in the shining black and gray skin of the true dolphin no such changes take place. There is, however, more truth than usual in this tale; for the dolphin in question is really the Coryphene, whose colors are always most brilliant, and glow with changeful beauty during the death struggle. A similar phenomenon occurs in several other fishes, of which the common red mullet is a familiar example.

The Dolphins (family *Coryphænidæ*) are embraced in one genus and six or eight species. They are very large fishes, inhabiting the high seas in warm regions, well known through their representative which is so often seen by the ship's sides in the warm waters of the tropics. The term *Coryphænas* would be more appropriate for these fishes instead of the Dolphin. The latter was applied by the ancients to the small whale-like creature which resembles our porpoise. The beautiful form so frequently used in sculpture was borrowed from the cetacean, although it is true that the Coryphene also has the graceful protuberance on the head which characterizes the Dolphin of the ancients. Ancient authors do not give much attention to the Coryphene; hence, to the cetacean rightfully belongs the place in art. Poets have celebrated the beauties of Coryphene. The colors are beyond description, and the changeable hues are surprisingly beautiful. As seen from the side of a vessel, as the Coryphene playfully accompanies it, the gleam of golden and silvery lights, changing now and again to rich metallic tints—emerald, sapphire, and many gorgeous colors—it is a never-failing source of wonder and enjoyment. Two species are seen near our shores in the warmer seasons, though they are essentially pelagic.

A VERY remarkable fish is allied rather closely to the preceding species, in spite of the great difference in form, and by some writers was placed in the same genus as that fish.

The EYED PTERACLIS is a good example of the curious genus to which it belongs, and which can always be recognized by the extreme depth of the dorsal and anal fins, and their delicate tenuity of structure. The dorsal fin is, moreover, remarkable for the bold sweep of its extent, passing in an unbroken curve from the forehead to the tail. Owing to the development of the anal fin, the two ventrals are placed very far forward, and are seen under the throat. The members of this genus are spread over the Indian Ocean, the Sea of Marmora, and some of the American coasts.

The Eyed Pteraclis is found on the Mozambique coast. It is a very beautiful fish, the general color being shining white, as if made of polished silver, with a wash of gold upon the pectoral and tail fins, and a deepish tint of blue-gray upon the others. On the dorsal fin there is a round spot of dark blue. It seems to be a small species. About four members of this genus are known to naturalists.

These fishes form a small family called *Bramidæ*, the Bramoids, included in four genera and about ten species. *Pteraclis carolina*, a small fish, inhabits the waters of the Carolinas.

BEFORE quitting this family, we must briefly notice the handsome OPAH, or KING-FISH (*Lampris luna*).

This beautiful species seems to be the sole representative of its genus, it having been separated from the genus Zeus, in which it had formerly been placed, in consequence of its single dorsal fin. It sometimes attains to a considerable size, a specimen having been taken which measured five feet in total length, and weighed about one hundred and fifty pounds. The flesh of this fish is red, very good, and is said to resemble that of the salmon.

The color of the Opah is bright green on the upper part of the back and sides, with reflections of purple and gold in certain lights. The fins and eyes are scarlet, and a number of round spots of pale gold are scattered upon the sides.

The Opahs (family *Lamprididæ*) are fishes of large size and gorgeous coloration, inhabiting the open Atlantic. A single genus is known. *Lampris guttata* is sometimes seen off Newfoundland. It is estimated as one of the most gorgeously colored fishes known. Sun-fish is a local name.

WE now arrive at a rather large family of fishes, which has been separated from the mackerels on account of certain anatomical variations, which will be mentioned at the end of the volume.

The CORDONNIER, or COBBLER-FISH, has derived its popular name from the long sharp spines of the dorsal and anal fins, which are thought to resemble the awl and bristles employed by cobblers in their trade. This fish is a good example of the large genus to which it belongs, and in which no less than seventy species have been classed. It is found in various localities, from the Red Sea throughout all the Indian seas, and is tolerably common. The form of this fish is sufficiently curious to render it a conspicuous species, and it may be easily distinguished from its many congeners by the oblong spot on the operculum, and the six black bands that are drawn across the body and reach nearly to the abdomen.

ANOTHER species of this genius is the RUDDER-FISH, (*Caranx carangus*), so called because it is fond of hovering about the rudders of vessels, apparently for the sake of picking up the refuse food that is thrown overboard. It is rather a pretty fish, the general color being silvery white and blue. The lateral line is covered, near the tail, with a row of spinous plates. It is somewhat remarkable that this fish, when hooked, emits a rather loud chattering kind of noise, thought to proceed from the passage of air through the gills. The flesh of the Rudder-fish is rather coarse, but is digestible and nourishing. Another fish (*Pammelas perciformis*), found in the seas of Northern America, is sometimes called by the name of Rudder-fish.

The family *Carangidæ*—from *Caranx*, the generic name of several of the groups—embraces twenty-five genera, and one hundred and eighty species, called, collectively, Pilot Fishes. Most are widely distributed, and are excellent food. They abound in warm waters, and move northward in summer. The familiar Horse Crevallé is one of the group; and the Mackerel Scads. The Cobbler-fishes, Moon-fishes, and the notable Pompano, also are included. The Common Pompano (*Trachynotus carolinus*) is one of the choicest food-fishes in the south. It ranges northward to Cape Cod, though is not taken so far north in quantities.

CLOSELY allied to these fishes is the well-known HORSE MACKEREL (*Trachurus saurus*), sometimes known by the popular name of SCAD. Its picture is to be seen on page 245.

This species is common in the European seas, and occasionally appears in enormous shoals, almost rivalling in numbers those of the common mackerel, and crowding so closely against each other that they cannot escape if threatened by danger, and may be taken out of the sea by hand or dipped out in buckets. The flesh of the Horse Mackerel is rather coarse, and when fresh is held in very slight esteem. However, it readily takes salt, and is then much eaten, especially during the winter months.

The color of the Horse Mackerel is dusky olive on the upper part of the back, changing in certain lights to resplendent green, which descends down the sides, and is variegated by wavy bands of blue. The sides of the head and the abdomen are silvery-white. The lateral line is furnished with a row of strong and deeply keeled bony plates, which give to the hinder part of the body a somewhat squared outline.

THE well-known SWORD-FISH, represented in the accompanying full-page illustration, derives its popular name from the curious development of the snout, which projects forward, and is greatly prolonged, into a shape somewhat resembling a sword-blade. The "sword" is formed by the extension of certain bones belonging to the upper part of the head.

This fine fish is found in the Mediterranean Sea, and also in the Atlantic Ocean, and in the former locality is often very plentiful. The Sicilian fishermen are accustomed to pursue the Sword-fish in boats, and mostly employ the harpoon in its capture. The weapon is not very heavy, and by a strong and practised hand can be hurled to some distance.

The fishermen are accustomed to chant a kind of song, set to words which no one can understand, but which are supposed to be the more efficacious for their incomprehensibility. This song is thought by some writers to be a corruption of some old Greek verses, and the fishermen believe that the Sword-fish is so fond of this song that it follows the boat in which it is sung. They will not venture to speak one word of Italian, thinking that the Sword-fish would understand what they were saying, learn that they contemplated its death, and then dive and make its escape. No bait of any kind is employed, the unintelligible chant being thought to be far more efficacious than any material aid.

The flesh of the Sword-fish is always eatable and nourishing, and in small specimens is white and well-flavored.

The use of the "sword" is not clearly ascertained. In all probability the fish employs this curious weapon in gaining its subsistence, but the precise mode of so doing is not known. It is an ascertained fact that the Sword-fish will sometimes attack whales, and stab them deeply with its sharp beak; and it is also known that this fish has several times driven its beak so deeply into a ship that the weapon has been broken off by the shock. In such cases, the blow is so severe, that the sailors have fancied that their vessel has struck upon a rock. Several museums possess examples of pierced planks and beams, but it is possible that the

SWORD-FISH.

fish may have struck them by accident, and not in a deliberate charge. The Sword-fish generally go in pairs.

The food of this creature is rather varied, consisting of cuttle-fish, especially the squid, and of small fishes, neither of which animals would in any way fall victims to the sword. It certainly has been said that the weapon is used for transfixing the flat fish as they lie on the bed of the sea, but this assertion does not appear to be worthy of credit.

The young and adult specimens are very different from each other. In the young, the body is covered with projecting tubercles, which gradually disappear as it increases in size, and when it has attained the length of three feet, they are seldom to be seen. Those on the abdomen remain longer than the others. The dorsal fin extends in the young specimens from the back of the head to the root of the tail, but the membranes and spines of its centre are so extremely delicate, that they are soon rubbed away, and the adult specimen then appears to have two dorsal fins.

The color of the Sword-fish is bluish-black above, and silvery-white below. The whole body is rough, and the lateral line is almost invisible. The usual length of the Sword-fish is from ten to twelve feet, but specimens have been seen which much exceed those dimensions. A few examples of the Sword-fish have been captured that measured seven feet in length.

The Sword-fishes, family *Xiphiidæ*, have three genera and about five species. They are large, strong fishes, and all good for food. Off Portland, Me., they frequent in considerable numbers. The fishermen here find it profitable to fit out for their capture. The vessels are provided with resting bars on the bow-sprit, and a lance is always at hand on the bar. When the Sword-fish is seen the fisherman hastens to his bar, and, leaning over it, to make all firm, he hurls the spear, and usually secures his prize. The handle slips out of the iron spear, and the line which is fast to the spear-head, serves to haul the fish on board. The flesh of the Sword-fish is very excellent; rather dry, but the union of the flavor of mackerel and halibut renders it quite a good food-fish.

THE SAILOR SWORD-FISH is still of much more curious aspect. It is a representative of a genus of Sword-fishes that have been separated from the previous genus on account of the very great height of the dorsal fin.

The Sailor-Sword fish is sometimes called the FAN-FISH or SAIL-FISH, and is said to possess the power of raising or lowering the enormous dorsal fin just as a lady opens or closes her fan. Sir J. Emerson Tennent mentions this fish in the following terms: "In the seas around Ceylon, Sword-fishes sometimes attain to the length of twenty feet, and are distinguished by the unusual height of the dorsal fin. Those both of the Atlantic and Mediterranean possess this fin in its full proportions only during the earlier stages of their growth. Its dimensions even then are much smaller than in the Indian species; and it is a curious fact, that it gradually decreases as the fish approaches to maturity; whereas in the seas around Ceylon, it retains its full size throughout the entire period of life. They raise it above the water while dashing along the surface in their rapid course, and there is no reason to doubt that it occasionally acts as a sail."

In this genus the ventral fins are reduced to one, two, or three spines, which in the present species are two in number. The tail is very deeply forked, and the enormous dorsal fin is a uniform deep blue.

WE now arrive at the large family of the Gobies, which include many curious fish.

The BLACK GOBY, sometimes known as the ROCK-FISH, is a moderately common example of the enormous genus to which it belongs, and which contains more than a hundred and fifty authenticated species. The members of this genus may easily be recognized by the peculiar form of the ventral fins, which are united together so as to form a hollow disc, by which they can attach themselves to rocks or stones at pleasure. In fact, this disc, although differing in shape, acts on exactly the same principle as that of the sucking-fish.

The Black Goby prefers the rocky to the sandy coasts, and may be found in the pools left by the retreating tide. Some naturalists deny that the disc is used for adhesion, but I have caught and kept many Gobies, and have frequently seen them sticking to the sides of the vessel in which they were confined. The adhesion was achieved with astonishing rapidity, and the little fish contrived to hold itself with wonderful tenacity. The surface of the Black Goby is very slippery, owing to the abundant mucous secretion which is poured from the appropriate glands, but after it has been in spirits for some time, the edges of the scales begin to project through the mucous, and are exceedingly rough to the touch.

Several species of Goby inhabit the American shores, such as the POLEWIG, or SPOTTED GOBY (*Gobius minutus*), a rather pretty little fish, transparent golden-gray, with a multitude of tiny black dots upon the back, and generally marked with some darkish blotches upon the sides, and a black spot on the dorsal fin. The TWO-SPOT GOBY (*Gobius Ruthen sparii*) is another species, and may be distinguished by the two deep brown spots on either side, one just above the root of the pectoral fin, and another on the side of the tail.

In some places along the sea-coast, the Gobies are known by the popular appellation of Bull-routs, and are rather feared on account of the sharp bite which their strong jaws and pointed teeth can inflict upon the bare hand.

POLEWIG, OR SPOTTED GOBY.—*Gobius minutus*.

The general color of this fish is blackish-brown above, changing to white along the abdomen and under the chin. The length of this species seldom exceeds five or six inches.

The Gobies, family *Gobidæ*, are carnivorous fishes, mostly of small size, living on the bottoms near the shores in warm regions. Some inhabit fresh waters, and others live indiscriminately in either fresh or salt water. There are sixty to seventy genera, and nearly four hundred species.

THE pretty GEMMEOUS DRAGONET, FOX-FISH, SOULPIN, or GOWDIE, can easily be distinguished from any other species, on account of its very remarkable shape

It is not a very uncommon fish, and is captured either with the hook or in a net, the latter being the ordinary method of securing it. It is rather a voracious fish, and feeds chiefly on mollusks and marine worms. The flesh of this species is firm, white, and well-flavored, and in spite of its small size the Dragonet repays the trouble taken in its capture. It generally remains near the bottom of the sea, and does not often enter shallow water except when young, when it approaches the shore, and sometimes is taken in the net of the shrimper.

It is a lovely fish, well deserving its name of Gemmeous Dragonet, as its scales glitter as if set with gems, and of Gowdie, or golden, on account of the gilded lustre of its exterior. The name of Dragonet is given to it on account of the dragon-like aspect of the body and fins.

The color of this beautiful fish is golden-yellow of different shades, variegated with spots and streaks of sapphire upon the head and sides. The under surface is white. The first dorsal fin consists of four rays, the first being enormously lengthened, and reaching, if depressed, to the base of the tail. The succeeding rays rapidly diminish in length, the fourth being extremely short, barely an inch in length. The pectorals are rounded and triangular, the central ray being the largest. The length of the Gemmeous Dragonet is about ten or eleven inches.

GEMMEOUS DRAGONET.—*Callionymus lyra.*

More than twenty species of Dragonets are known, spread over a very large portion of the globe, and inhabiting the temperate seas of the Old World, and the Indian Ocean from Mozambique to the Western Pacific Islands. They are marine fishes, and inhabit the bottom of the sea at no great distance from the shore.

WE now come to a very small, but curious family, termed Discoboli, or Quoit-fishes, because the spines of the ventral fins are modified into a flattened disc, something like the quoits of the ancients. This disc has a soft, leathery margin, and enables them to attach themselves to rocks or stones, after the manner of the gobies.

A very good example of these curious fishes may be found in the LUMP-SUCKER, otherwise called the LUMP-FISH, SEA-OWL, and COCK-PAIDLE, the latter name being given to it on account of the elevated ridge along the back, which is covered with a notched and tuberculated skin not unlike the comb of the cock.

The sucker or disc of this fish is capable of very powerful adhesion, retaining its hold with such tenacity, that on one occasion, when a Lump-fish was placed in a pail containing several gallons of water, it immediately affixed itself to the bottom, and held so firmly, that when grasped by the tail and lifted, it raised the vessel in which it was placed, notwithstanding the combined weight of the water and pail.

The Lump-fish is said to make a kind of home, and to hover about the spot where the eggs are placed, for the purpose of guarding them from foes. When thus engaged, it is a brave and combative fish, permitting no other finny inhabitant of the water to pass within a certain distance of its charge, and, in cases of necessity, biting fiercely with its short but sharp teeth. It is said that after the young have attained some little size, they attach themselves to their careful parent, who conveys the young family into deep water.

The dimensions of this fish are variable, but the average length is about sixteen inches.

The Lump-suckers (*Cyclopteridæ*) are included in two genera, four species being known. *Cyclopterus lumpus* is rather common off the coasts of both Europe and America, though never abundant. A species is found in the North Pacific.

LUMP-FISH.—*Cyclopterus lumpus.* VIVIPAROUS BLENNY.- *Zoarces viviparus.*

There are only two genera in this small family, and both find examples in the seas.

Of the second genus, the Unctuous Sucker, or Sea-Snail (*Liparis vulgaris*), is a good illustration.

This species appears to be less common in the south than in the north. It derives its names of Unctuous Sucker and Sea-Snail from the soft and slime-covered surface of its body. It seems to prefer the rocky coasts, and may be found in the water-pools at low tide. The color of this fish is pale brown streaked irregularly with a darker tint. Both the dorsal and anal fins are low, long, and reach to the commencement of the tail fin. It is a little fish, seldom exceeding four or five inches in length.

Montague's Sucker (*Liparis montagui*) is remarkable for its habit of adhering to a stone or rock by the disc, and then curving its body to such an extent that the tail and the head almost meet. Even when merely lying at rest, and not employing the sucker, it assumes this remarkable attitude. It is smaller than the last species, rarely exceeding three inches in length. Its color is rather dull orange above, with bluish reflections, and white below. The fins are of a rather deep orange hue.

ANOTHER small family now comes before us, called the Frog-fishes, from the froggish aspect of the body, and especially of the head.

The TOAD-FISH is a very curious-looking creature, with its flattened and wide head, gaping mouth, and spacious gill-cover. All the members of this genus are carnivorous fishes, and are spread through the coasts of the tropical regions, where they are mostly found on the bottom and partially buried in the sand or mud, in hope of surprising the active prey on which they feed. Some species, however, are found even in the temperate seas.

The Toad-fish inhabits the East Indian seas, and has been taken at the mouth of the Ganges. Its color is brown, marked with a much darker tint, and the fins are streaked and blotched with similar colors. The body is without scales.

MONTAGUE'S SUCKER. *Liparis montagui.*

The Toad-fishes, family *Batrachidæ*, are included, twelve species in five genera. They are carnivorous coast fishes, mostly of the warmer seas. The young of some fasten themselves upon rocks by means of ventral discs, which, however, disappear. Common names of them are Oyster-fish and Sarpo.

THE FISHING-FROG, ANGLER-FISH, or WIDE-GAB, which is shown in the lower figure of the illustration on page 242, has long been famous for the habit from which it has derived its popular name.

The first dorsal fin is almost wholly wanting, its place being occupied merely by three spines, movable by means of certain muscles. The manner in which these spines are connected with the body is truly marvellous. The first, which is furnished at its tip with a loose shining slip of membrane, is developed at its base into a ring, through which passes a staple of bone that proceeds from the head. The reader may obtain a very perfect idea of this beautiful piece of mechanism by taking a common iron skewer, slipping a staple through its ring and driving the staple into a board. It will then be seen that the skewer is capable of free motion in every direction.

The second spine is arranged after a somewhat similar fashion, but is only capable of being moved backwards and forwards. Fishing-Frogs are somtimes found in the shops, and the inquiring reader will find himself amply repaid if he purchases one of these fishes and dissects its head, merely for the purpose of seeing the beautiful structure which has been briefly described.

The use of these spines is no less remarkable than their form.

The Fishing-Frog is not a rapid swimmer, and would have but little success if it were to chase the swift and active fishes on which it feeds. It, therefore, buries itself in the muddy sand, and continually waves the long filaments with their glittering tips. The neighboring fish, following the instincts of their inquisitive nature, come to examine the curious object, and are suddenly snapped up in the wide jaws of their hidden foe. Many fishes can be attracted by any glittering object moved gently in the water, and it is well known by anglers how deadly a bait is formed of a spoon-shaped piece of polished metal, furnished with hooks, and drawn quickly through the water.

The arrangement of the spines in this fish—which is equally well known in our American waters—as our author says, will well repay the examination. We have frequently seen these fishes in the market, brought there as curiosities, but have seen very large specimens on Nahant beaches. One example was about five feet in length, the head being about one-third as much in width. The gape of such a head is enormous, and the creature had partially swallowed a cod-fish of the largest kind, which, with its head protruding, was heavy enough to weigh the Angler to the bottom, when the heavy seas threw it with its prey to the shore. The first free spine on the top of the head is about nine inches in length, and with its bit of membranes as bait, is a veritable fishing-rod. The creature is sluggish, and, lying on the soft bottom, partly covered, it moves this rod gently, and thereby attracts the luckless fishes that form its food.

THE very odd-looking creature called the WALKING-FISH, is one of the strange and wild forms that sometimes occur in nature, and which are so entirely opposed to all preconceived ideas, that they appear rather to be the composition of human ingenuity than beings actually existing. The traveller who first discovered this remarkable fish would certainly have been disbelieved if he had contented himself with making a drawing of it, and had not satisfied the rigid scrutiny of scientific men by bringing home a preserved specimen.

THE TOAD-FISH (*Antennarius histrio*) is the curious little creature that is seen at times floating on the surface, evidently distressed on account of its body being unduly inflated. Its habit is to inflate itself, but often it seems to be helpless in this state. Its curious nest, made among floating algæ on the ocean, is familiar to readers of popular books on Natural History. This fish is so decorated by algæ-like excrescences it becomes a complete piece of deception. It is difficult to distinguish the difference between the fish and the surrounding sea-weed.

THE BAT-FISH (*Malthea vespertilio*) is a sluggish fish, found in the warmer waters. Its whole appearance is that of a creature adapted to live on the bottom. The pectoral fins are developed into feet-like organs, and it actually crawls like a reptile. Its more interesting feature lies in the development over its mouth of an erectile club shaped fleshy process, which protrudes from a concave locality just over its mouth. Lying in the mud secure in its protective resemblance to the surrounding bottom, it causes this erectile organ to turn slowly in imitation of a worm, which it resembles. Any inquisitive body that comes within reach is taken in below, the great mouth being quite ready, and capable of swallowing anything near its own size.

THE important family of the Blennies comes next in order. They are all carnivorous fishes, many being extremely voracious, and are spread over the shores of every sea on the globe. They mostly reside on or near the bottom.

The SEA WOLF, SEA CAT, or SWINE-FISH, is one of the fiercest and most formidable of the finny tribes that are found on our coast, and has well earned the popular names by which it is known.

The general color of the Sea Wolf is brownish-gray, with a series of brown vertical stripes and spots over the upper parts; the under parts are white. On European shores it attains a length of six or seven feet, but in the northern seas, where it thrives best, it greatly exceeds those dimensions. There is an American variety where the vertical streaks are modified into round spots of blackish-brown.

The Sea Wolf is taken by the fishermen of Swampscott, and along the shores of New England, in winter frequently, and once was considered a nuisance, as it interfered with cod-fishing. It has been found to be a valuable food-fish, since the considerable advance in the price of cod-fish and haddock. Blennies are numerous among the rocks on the eastern coast—the Butter-fish among them.

The typical genus of this family is represented by several specimens, of which the EYED BLENNY is one of the most conspicuous.

This pretty fish is not very common. From the elevated dorsal fin, and the bold dark brown spot that decorates it, this Blenny has sometimes been called the Butterfly-fish. In the Mediterranean it is tolerably common, and lives mostly among the seaweed, where it finds abundance of the smaller crustacea and mollusks.

The dorsal fin of this fish is very large, being greatly elevated and extending from the back of the head almost to the tail. The dark spot is placed between the sixth and eighth rays.

SEA WOLF.—*Anarrhichas lupus.*

The color of the Eyed Blenny is pale brown, patched here and there with a darker tint. The dark spot on the fin is mostly edged with white or very pale yellow. The length of this fish is seldom more than three inches.

Among other species of Blenny the SHANNY, or SHAN (*Blénnius pholis*) is tolerably common in European seas.

PASSING by the remaining Blennies, all of which are very similar in habits and general appearance, we must pause for a short space to examine a very curious species belonging to the same family, called the JUMPER-FISH (*Salárias tridáctylus*).

This odd little fish offers no remarkable beauties of color or form, being of a simple dark brown, and without any salient points of external structure; but it is possessed of a wonderful power of suddenly leaping out of the water, darting over the wet stones and rocks and snapping up flies and other insects with the nimble agility of the lizard. It can scramble up a nearly perpendicular face of rock, and is so wary and agile, that on the least attempt to seize it, the little creature darts towards the sea and is nearly certain to make its escape. While engaged in this pursuit, the Jumper-fish adheres so tightly to the rock, that 't is not detached even by the shock of repeated waves. It is quite a little fish, not more than four inches in length. Its residence is in the seas of the East Indian Archipelago. At least fifty species of the Salarias are known to zoologists.

THE BUTTER-FISH, SWORDICK, or SPOTTED GUNNEL (*Centronôtus gunellus*), belongs to this family, and is evidently one of the transitional species between the true blennies and those which are placed at the end of the family.

This fish is frequently captured, especially on the rocky shores, and is mostly found hidden under stones and sea-weeds in the rock-pools left by the receding tide. The name of Butter-fish is very appropriate, and is given to it on account of the plentiful mucous secretion which is poured over its body, and which renders it so slippery that it can with difficulty be retained in the hand. It is quick and agile in its movements, and even if confined within the limits of the rocky pool is not easily captured.

The body of this fish is much elongated and somewhat eel-shaped, the head is small, the muzzle blunt, and the dorsal fin is low and long, extending the whole length of the back. The ventral fins are very small. The color of the Swordick is brown, in some specimens with a purple and in others with a golden wash. Along the base of the dorsal fin, and in some individuals upon the fin itself, are a number of bold, black spots, each with a white streak on either side. A dark brown stripe is also drawn from the eye to the lower jaw. The length of the Butter-fish is about six inches.

OUR last example of this family is the well-known VIVIPAROUS BLENNY, called also by the popular names of EEL-POUT, LUMPER, GUFFER, and GREENBONE, the last-mentioned title being given to it because, when boiled, the bones have a green hue. It is illustrated, with the Lump-fish, on page 254.

As its name imports, the Viviparous Blenny lays no spawn, but produces its young alive, and able to shift for themselves. In one case, where a female fish of about fifteen inches in length was taken, the young were about four inches long. It is a very curious fact, that the size of the new-born young seems to depend upon that of their parent, the offspring of a Blenny of seven inches in length measuring only one inch and a half.

The flesh of this fish is tolerably good, but is not in very great repute, so that it is but seldom to be seen in the markets. It generally hides itself under stones or sea-weed, preferring the large, heavy algæ, called tang.

The body of this fish tapers gradually from the shoulders to the tail, in thickness as well as in depth, and when examined with a pocket magnifier, the surface appears to be studded with circular depressions. Its general color is pale brown, and its length varies between six and sixteen inches.

PASSING by several small families, we come to a very curious fish, denominated the RIBAND-SHAPED VAAGMAR, sometimes called the DEAL-FISH (*Trachypterus árcticus*).

This singular fish is remarkable for the extreme compression of the body, a specimen three feet in length not being thicker than an ivory paper-knife. The dorsal fin of this fish extends completely along the back; there is no anal fin, and the tail fin stands boldly erect, like the closed tail-feathers of a fan-tail pigeon. The general color of the Vaagmär is silvery-white, and the body is covered with very small scales. The dorsal fin is bright orange, sometimes being of a blood-red, and the tail fin is of the same hue. On each side are two oval spots of blackish-gray, set obliquely on the body. The length of this fish often reaches six feet.

It is one of the northern fishes.

A SPECIES even still more remarkable is, on very rare occasions, obtained on the North Sea; but, owing to the extreme fragility of its structure, it is mostly deficient in some of its parts. Our picture of this creature is remarkably true to nature.

The OARED GYMNETRUS, or RIBBON-FISH (*Regalecus banksii*), as it is called, is also greatly compressed throughout its length, and is equally delicate with the last-mentioned species. It is chiefly notable for the very odd structure of the ventral fins, which are reduced to long, slender filaments, much resembling in shape the long tail-feathers of the racket-tail humming-bird. This fish sometimes attains very great dimensions; specimens have been taken measuring twelve feet in length. Its color is silvery-gray, mottled with dusky spots of varying depth, which are most conspicuous towards the head. The whole surface of the skin

OARED GYMNETRUS, OR RIBBON-FISH.—*Regalecus banksii*. (One-twelfth natural size.)

is plentifully studded with bony tubercles, and on the line of the abdomen each tubercle is furnished with a hooked point directed backwards. Along the lateral line runs a row of elongated flat scales.

In the next family, the tail is mostly armed with one or more bony spines or plates, small in the young, but increasing in size with the dimensions of the fish.

The SEA SURGEON (*Acanthurus chirurgus*) is a good type of these fishes, and derives its popular name from the sharply-pointed and keen-edged spine on the side of the tail, which cuts and wounds like a surgeon's lancet. The generic name, signifying Thorn-tail, is given to it in consequence of this structure. This species is found on the Atlantic coasts of tropical America and Africa, and is tolerably plentiful in the Caribbean seas. The scales of this fish are very small, and the single spine on each side of the tail is movable and set in a longitudinal groove. Its food is of a vegetable nature.

In color it is rather variable, but the ground tint is usually of a brownish hue, and the operculum has a black edge. In some specimens the end of the tail is marked with a white band, which encroaches on part of the tail fin, and there is also a narrow white edge to that fin. There are in certain individuals a few darkish streaks drawn across the body, some black longitudinal stripes on the dorsal and anal fins, and in the young the sides are marked with darkish waving lines. This fish sometimes attains a rather large size, a specimen being nineteen inches in length. The genus is rather comprehensive, containing between forty and fifty known species.

This remarkable fish is common on the Florida Reef. We have kept it in our aquarium, and found it wonderfully lively and interesting. It is one of the few fishes that seem to have strong combative impulses. It will bear no trifling, but strikes powerfully with its tail, the sharp lance-like weapon proving dangerous to its foes. Pass a rod lightly enough towards it, and the Surgeon instantly turns, and, facing the object, makes a thrust with savage celerity.

NEARLY allied to the surgeon-fish is a very curious species, called the UNICORN THORN-TAIL (*Naseus unicornis*), on account of the singular structure of the forehead, which is modified in front into a long and horn-like protuberance, rather conical in shape, and projecting forwards in a line with the body. This horn is not to be seen in the young fish, and only attains its full dimensions when its owner has reached adult age. Sometimes the horn is longer than the snout, but in most specimens it is slightly shorter. Each side of the tail is furnished with two lancet-bearing plates, which are not movable.

CLIMBING PERCH.—*Anabas scandens*. (One-half natural size.)

This species is found from the Red Sea to Japan and Polynesia. Its color is brownish-gray, and the dorsal and anal fins are marked with longitudinal blue stripes. The largest specimen I have known of measures twenty-two inches in length, and its horn is three inches long.

THE extraordinary fish called, from its habits, the CLIMBING PERCH, is a native of Asia, and is remarkable for its apparent disregard of certain natural laws.

This singular creature has long been celebrated for its powers of voluntarily leaving the failing streams, ascending the banks, and proceeding over dry land towards some spot where its unerring instinct warns it that water is yet to be found. There are several fish which are known to have this power; the common eel, for example, which has frequently been observed crossing the fields in its passage from one stream to another. I have even seen the eels creeping over rocks, and contriving, in some mysterious manner, to crawl along the flat horizontal surface of an overhanging rock as easily as a fly walks on the ceiling. But I believe that the eel only passes over moist ground, whereas the Anabas seems quite indifferent to such considerations, and takes its journey over hard, dry, and dusty roads, heated with the burning beams of the noonday sun, without appearing to feel much inconvenience from the strange nature of the transit.

Several species, of which the Anabas Scandens has been chosen as the best example, possess this singular property of walking over dry ground, so that the old proverb of a fish out of water is, in these cases, quite inapplicable.

It is known of the Climbing Perch that the fishermen of the Ganges, who subsist largely on these fishes, are accustomed to put them into an earthen pan or chatty as soon as caught; and although no water is supplied to them, they exist very well without it, and live this strange life for five or six days.

On opening the head of this fish, the curious structure which enables it to perform such marvellous feats is clearly seen. Just within the sides of the head, the "pharyngeal" bones, *i. e.*, the bones that support the orifice between the mouth and gullet, are much enlarged, and modified into a series of labyrinthine cells and duplications, so that they retain a large amount of water in the interstices, and prevent the gill-membranes from becoming dry. Some writers say that this fish is capable of climbing up the rough stems of palm-trees, in search of the water that lodges between the bases of the dead leaves and the stem, but this account is now held unworthy of belief. In the Tamoule language it called Paneiri, or Tree-climber.

THE small genus Atherinidæ has a representative in the SAND SMELT (*Atherina presbyter*), a pretty little fish, and one that is of great use to fishermen, both for sale and for bait.

It is extremely plentiful here in America, as well as in Europe, and in many places is sold as the true smelt, which it somewhat resembles in flavor and the peculiar odor as of cucumber.

GRAY MULLET. *Mugil capito.*

Owing to the small size of this fish the net is the usual mode of capture, the fashion of which varies according to the locality. On some coasts the net is about ninety feet in length and eighteen in depth, and is drawn along the sands by the united aid of one party in a boat and the other on the shore. In other places, however, it is circular and supported on an iron hoop. It is then baited with broken crustacea and lowered into the water. At intervals it is raised smartly to the surface, and the entrapped Sand Smelts removed.

The color of the Sand Smelt is the palest pink, diversified with a broad belt of shining silvery-white, which is drawn along the side. The cheeks, gill-covers, and the base of each pectoral fin are of the same white hue. Upon the upper part of the back and head are a great number of little black spots. The length of the fish is from six to seven inches.

WE now come to the important family of the Mugilidæ, of which the common GRAY MULLET is a good example. In all these fish there are two dorsal fins, the first having four stiff spines. They are spread over all sea-coasts and fresh waters of the temperate and tropical regions. The mode of feeding is rather curious. These fish live chiefly on the soft organic substances that are found mixed with weed and sand, and in swallowing the food a considerable amount of sand is taken into the mouth. The fish, however, is furnished with a kind of self-filtering apparatus, by means of which the heterogeneous mass is raked and sifted, as it were, and the indigestible portions rejected.

The Gray Mullet deserves notice as being one of the most daring and ingenious of the finny race, and is, in fact, a very fox for artfulness. The idea of constraint is most obnoxious to it, and its instincts of freedom are so strongly developed that it endeavors to recover its liberty in the most extraordinary ways.

If, for example, it has been inclosed in a net, it will at once dart to the side and try to leap over the head-rope into the open sea. Moreover, if one fish succeeds in the attempt, the remainder immediately follow their leader, like a flock of sheep jumping over a hurdle. If the net is raised so high that the leap is impracticable, the fish tries to creep under it; and if that mode of escape be cut off, it examines every mesh, in hopes of finding some defective spot through which it may insinuate itself. Mr. Couch mentions that he has seen a Gray Mullet, after trying all other modes of escape, deliberately retire to the greatest possible distance from the wall of net, and then dash furiously at the meshes, as if to break through them.

The genus Mugil is very large, containing between sixty and seventy species.

Mullets of the Florida waters are numerous. A novel method of taking them we witnessed at Punta Rassa, on the Gulf coast. They are about eighteen inches in length, and have very wide backs. The shoals are few in number. Negro boys took them in this manner: common "grains," or spear, secured to a long handle by a line, the latter is held upright in the palm, the line retained; the pole is tossed upwards to return spear first directly over the broad backs of the fishes, and, as a rule, it strikes home.

THE fishes belonging to the family of the Ophiocephalidæ, or snake-headed fishes, are able to leave the water for a time and to crawl upon land, deriving their power from a curious structure of the breathing organs. It has already been stated that a fish can breathe as long as the delicate membranes of the gills are wet; and that in those fishes which are able to live out of water for any length of time, a peculiar modification of the breathing organs is requisite in order to supply the needful moisture. In the family to which the climbing perch belongs, a series of thin laminated plates are arranged in a cavity above the gills, thus retaining a sufficient supply of water between the laminæ. In the present genus, however, there are none of these laminæ, but the water is retained in a simple cavity which communicates with the gills.

Of this family the CORA-MOTA, or GACHUA (*Ophiocephalus gachua*), is a good example.

This fish is a native of the fresh waters of Eastern India and its archipelago, and in its general shape and movements is so very snake-like that Europeans will seldom eat it. The Cora-mota is common in the ponds and dykes of Bengal; and is one of the fish popularly supposed to be rained from the clouds, as it is generally to be found on the grass after a heavy shower. However this may be in other instances, it is tolerably clear that the Cora-mota has been in concealment during the drought, and ventures into the fresh wet grass as a welcome change from the muddy ditches in which it has been forced to reside. It can also find a plentiful supply of food on the moist herbage; and as on account of its peculiar formation it is able to move on land with considerable ease, its migrations will often extend to considerable distances.

The Cora-mota is remarkably tenacious of life, and can survive the severest wounds for a wonderfully long period. The natives of India take advantage of this peculiarity, and with the disregard of inflicting torture that seems to be inherent in the Oriental mind, are in the habit of selling the fish piecemeal, and cutting it up for sale while still living. Indeed, the *habitués* of the market will not pay the best price if the fish does not flinch from the knife.

The color of this species is brown crossed with several dark bars. Its length seldom exceeds a foot.

Another species of this genus, the BARCA (*Ophiocephalus barca*), is a much handsomer fish, attains a considerable size, and is considered to be useful for the table. This fish is one of the mud-lovers, living for the most part in holes excavated in the banks of Indian rivers, and only putting out its head in search of prey.

The color of this species is violet spotted profusely with black, and the fins are marked with sundry bold bars and dots. In length it often attains three feet.

THE remarkable BAND-FISH, or SNAKE-FISH (*Cépola rubescens*), is an example of a curious family, consisting of one genus only, and about seven species.

The Band-fish is not uncommon in the Mediterranean, though it is seldom taken off the English coasts. Its body is long and much compressed, like that of the vaagmär, already described; and when winding its way through the translucent water, its carmine body with the glittering scaly mail have earned for it the popular names of FIRE-FLAME and RED RIBAND.

Little is known of its habits, except that it is a shore-loving fish, delighting to bask under the heavy masses of sheltering sea-weed, and that it feed smostly on mollusks and crustacea. Several specimens of this fish have been found on the beach after a storm; and Mr. Yarrell

BARCA.—*Ophiocephalus barca.*

remarks, with some acumen, that all the fish formed after this pattern, with their compressed bodies affording little resistance to the water, and their length preventing the concentration of muscular force upon a single centre of motion, are ill fitted for combating tempestuous waters, and are flung about at the mercy of the waves.

The head of the Band-fish is small, and the eye is full and very large, its diameter being nearly half the depth of the head. The body is greatly compressed, slender, and very smooth; the scales being minute and glittering in the sunbeams. The dorsal fin extends from the top of the head to the end of the tail, and the anal fin is nearly as long. Its color is rather variable, shades of purple and orange exhibiting themselves in certain specimens. In all examples, however, red is the predominant hue. The length of the adult Band-fish is usually about fifteen or twenty inches.

IN the curious species which belong to the genus Centriscidæ, or spike-bearing fishes, the body is much compressed, and one of the spines of the first dorsal fin is long, sharp, and powerful. The bones which form the front of the head are greatly prolonged, and are modified into a kind of long tube, at the end of which is placed the narrow mouth. It is thought that the fish obtains its food by sucking it along the tube, the needful vacuum being formed by the dilatation of the throat.

The BELLOWS-FISH, sometimes called the TRUMPET-FISH and the SEA SNIPE, is most common in the Mediterranean. It prefers to reside in moderately deep water, and is mostly found where the bottom of the sea is muddy. Its food is not precisely known, but is thought to consist of minute marine animals. The first spine of the dorsal fin is enormously large, strong, sharply pointed, and armed on its under surface with a row of saw-like teeth, that must render it a very efficient weapon of offence. The spine is also movable. The flesh of this fish is eatable; but as the head occupies so large a portion, the amount of flesh is rather small when compared with the size of the fish.

The family which now comes before our notice is in many ways remarkable, and deserves some little attention before proceeding to the remaining fishes.

In the Fistularidæ the snout is greatly prolonged, as in the preceding family, and bears the mouth at the end of a bony tube. The body, however, is extremely long and snake-like, and there is no long spine to the dorsal fin. There are only two genera in this family, the one being covered with scales, and the other destitute of these appendages.

The TOBACCO-PIPE FISH is found in several parts of the tropical Atlantic, and is notable for its very peculiar form. The body is without scales, and the tail-fin is deeply forked, the

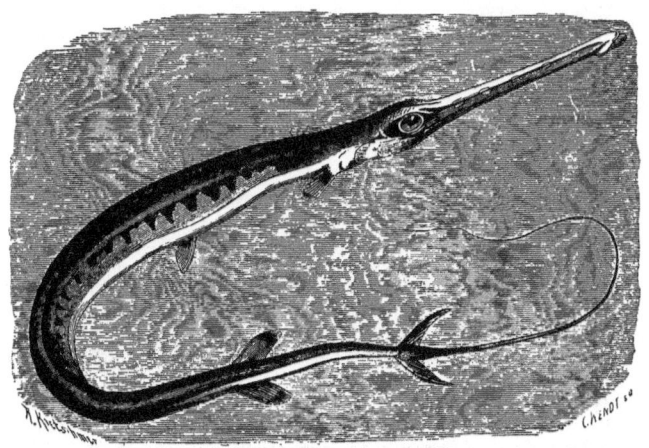

TOBACCO-PIPE FISH.—*Fistularia tabaccaria.*

two central rays being sometimes united and prolonged into a lengthened filament, and at others separate, but still elongated. The outer edge of the tube is either smooth or very slightly notched. The color of this fish is greenish olive, and the upper parts of the body are marked with blue streaks and spots. In some specimens, the back takes a reddish brown hue.

THERE is a curious family of fishes, termed the Mastacembelidæ, in which the body is long and eel-like, covered with little scales, and remarkable for the odd-looking snout and its appendage.

In these strange-looking fishes, of which the SPOTTED MASTACEMBELUS (*Mastacémbelus maculátus*) is a good example, the dorsal fin is very long, its front portion consisting of a number of short free spines. The anal fin is also furnished with similar spines, and the ventral fins are altogether wanting. The gill-openings are reduced to a narrow slit, and the movable appendage of the upper jaw is smooth on its under side. The jaws are furnished with minute teeth, and the lower jaw is but slightly movable. In all the species of this genus, with the exception of the Spotted Mastacembelus, the præoperculum is armed at its angle with small teeth.

This species is found in the fresh waters of Java and Sumatra. The dorsal fin joins that of the tail, which is again joined by the long anal fin. The color of the fish is brown, diversified with darker blotches, and the fins are edged with yellow.

FLAT FISHES; PLEURONECTIDÆ.

The Flat Fishes, as they are popularly called, or the Pleuronectidæ, as they are named scientifically, are among the most remarkable of the finny tribe. The latter name is of Greek origin, and signifies side-swimmer, in allusion to the mode of progression usually adopted by these fishes.

The Common Sole is one of the most familiar of the flat fishes.

The Sole can be taken by the line, but the fishermen always use the trawl-net, a kind of huge dredge, with a mouth that often exceeds thirty feet in width. As these nets are drawn along the bed of the sea, the great beam which edges the mouth scrapes the mud and sand, and alarms the fish to such an extent that they dash wildly about, and mostly dart into the net, whence they never escape. Vast numbers of Soles are taken by this method of fishing, and as the trawls bring to the surface enormous quantities of crustaceans, mollusks, zoophytes, and other marine inhabitants, the energetic naturalist cannot employ his time better than in taking a sail in one of these boats, and enduring a few hours' inconvenience for the sake of the rich harvest which he is sure to reap.

It is a hardy fish, and can soon be acclimatized to live in fresh water; and it is said that under such circumstances the fish can be readily fattened, and becomes nearly twice as thick as when bred in the sea. Sometimes the Soles venture into the mouths of rivers, passing about four or five miles into the fresh water, and depositing their multitudinous eggs in such localities.

The Zebra Sole is a native of Japanese waters, and is remarkable for the waving dark streaks with which its body is covered, and which bear a great resemblance to the stripes upon the zebra's hide. In habits it appears to resemble the common species.

The Lemon Sole, or French Sole (*Solea pegusa*), derives the former of these titles from the lemon-yellow color of its upper surface, and the latter from the localities in which it is most commonly found. It is found generally about sixteen miles off the English coasts. The color of this fish is orange, mixed with light brown, and mottled with little round spots of wood-brown. It is wider in proportion to its length than the common Sole. Another species, the Variegated Sole (*Solea variegata*), may be known by the reddish-brown color, clouded with dark brown. The body is rather thick in proportion to its length.

The Solenette, or Little Sole (*Monochirus linguátulus*), is seldom more than five inches long, and of a reddish-brown color, without cloudings.

Perhaps the most remarkable of these fishes is the Transparent Sole (*Achirus pellúcidus*).

This rare and interesting fish is a native of the Pacific Ocean, and is notable for the extreme pellucidity of its body, which is so marvellously transparent, that when swimming in a vase of water, or lying on the bottom, the algæ or stones can be distinctly seen through its structures. It is quite colorless, except a very slender and very delicate pink streak on the edge of the back, and several similar lines upon the sides; the perfect but glass-like skeleton is hardly to be detected, and even the viscera are almost invisible. It is a very little fish, appearing not to exceed two inches in length; but its width is proportionately great, so that the fish assumes a nearly circular form. The eyes are silvery-white, and the pectoral fins are wholly absent.

The American Sole (*Achirus lineatus*). This fish is found from Cape Cod southward. At one time a notion prevailed that the flesh was not eatable, but the truth is, it is not only wholesome, but very delicate. It is called Calico and Hoe-choke Cover Clip in New Jersey, and Spotted Sole in Massachusetts.

The well-known Turbot, so widely and so worthily celebrated for the firm delicacy of its flesh, inhabits many of the European coasts, and is generally found in tolerable abundance. Like all flat fishes, it mostly haunts the sandy bed of the sea, but will sometimes swim boldly to the surface of the water. It is a restless and wandering fish, traversing considerable distances as it feeds, and generally moving in small companies.

The SPOTTED TURBOT (*Bothus maculatus*) is a small species, not very familiar; called in New Jersey Window Pane, and in New York Sand Flounder. Its range is from Cape Cod to Hatteras. Mitchell described it as the New York Plaice (*Pleuronectus maculatus*), and it is also called Watery Flounder. It has been sold in New England as English Turbot, and is nearly, if not quite equally good a food-fish, as the latter. Its common name, Sand Flounder, associated with that of the miserable Flounders of our harbors, does not help its reputation as an edible. A species called Smooth Plaice is common along the coast from Maine southward. Several other species of Flounders are known.

ANOTHER flat fish, the BRILL (*Pleuronectes rhombus*), called in Scotland the BONNET FLEUK, and in other places known by the names of KITE and BRETT, is held in much estimation for the sake of its flesh, which is but little inferior to that of the turbot, and is, indeed, sometimes fraudulently substituted for that fish. The Brill resembles the turbot in food and habits as well as in appearance, but does not attain the same dimensions, seldom exceeding seven or eight pounds in weight. The skin of the dark side is devoid of the bony tubercles which are found in the turbot. Its color is reddish-brown, mottled with a darker tint of the same color, and variegated with numerous round white spots of a pearly lustre. On account of these spots the Brill is sometimes called the PEARL. When young, the pale reddish-brown is covered with spots of black or very dark brown.

PASSING by the two species of Topknots, we come to the PLAICE, so well known by the bright red spots which are scattered over its dark side.

This is one of the commonest of the flat fishes, and, happily for the poor, is taken in such quantities that it supplies nutritious aliment at a very low rate of purchase. It is taken chiefly with the trawl-net, but can be captured with the line, as it bites freely at a bait, generally the common lugworm, and is one of the fish that is most usually caught by amateur sea-fishers. Even the shrimpers take large quantities of small Plaice in their nets; and along the coast this fish is so numerous, that at low water it may be seen in great numbers darting over the sandy flats, the white surface glittering in the light as the little creatures dash wildly along in their terror of the approaching enemy.

THE FLOUNDER, MAYOCK FLEUK, or BUTT, is quite as common as the plaice, and is found in salt, brackish, or fresh water, sometimes living in the sea, sometimes inhabiting the mouths of rivers, and sometimes passing up the stream for many miles. As this fish is capable of living in fresh water, it has often been transferred to ponds, and will there fatten rapidly.

The color of the Flounder is usually brown, taking a darker or lighter shade, according to the nature of the ground on which the fish rests, those that inhabit the muddy shores being nearly black, and those which prefer the sand taking a yellower hue. Generally, the eyes and the color are on the right side, but reversed specimens are very common, and in some instances the fish has been entirely white or wholly brown. The average weight of the Flounder is three or four pounds.

ONE or two other examples of the flat fishes deserve a passing notice.

The COMMON DAB (*Platessa limanda*) is plentiful upon sandy coasts, and may at once be recognized by the roughness of its surface, or structure, which has gained for it the specific title of Limanda, or file-back—the Latin word *lima* signifying a file. Its flesh is very good, and is thought to be in best condition from the end of January to April. Its color is pale brown, and its length seldom exceeds eight inches.

A VERY large species of flat fish is called the HALIBUT (*Hippoglossus vulgáris*). The flesh is tolerably good, but is rather dry and without much flavor. It is rather longer in proportion to its width than is generally the case among flat fishes. Its color is brown of different shades, and the surface smooth, the small, oval-shaped scales which cover it being soft and without projections. This fish attains a large size, specimens of five feet in length not being uncommon.

The largest example on record measured above seven feet in length, and weighed more than three hundred pounds.

The Halibut is found in all northern seas, south, to France and San Francisco. It reaches a weight of 400 pounds. Dr. Storer records an example, on the authority of Mr. Newcomb, a noted fishmonger of Boston Market, as weighing 420 pounds, after the head and bowels were removed. The Halibut fishing of Grand and George's Banks is an important industry.

As an edible it ranks high. Great numbers are taken in Massachusetts Bay. Dr. Storer adds, in relation to weight: "The largest specimen of which I have any certain knowledge, was taken at New Ledge, near Portland, Maine, in 1807, and weighed upwards of 600 pounds."

At Nantucket there were once employed eighty vessels, of from 60 to 80 tons burthen each, in this fishery. Ancient names of this fish are *Fler*, and *Helbut*.

A species called Greenland Halibut is found in the northern seas.

THE COD.

THE well-known COD-FISH is a native of many seas, and in some localities is found in countless legions.

This most useful fish is captured in vast numbers at certain seasons of the year, and is always taken with the hook and line. The lines are of two descriptions, namely, the long lines to which a great number of short lines are attached, and the simple hand-lines which are held by the fishermen. The long lines sometimes run to an extraordinary length, and shorter lines, technically called snoods, are affixed to the long line at definite distances. Whatever may be the length of the snoods, they are fastened at intervals of double their length, so as to guard against the entanglement of the hooks. For example, if the snoods are six feet long, they are placed twelve feet apart on the line; if four feet long, eight feet apart, and so on.

To the end of each snood is attached a baited hook, and as the sharp teeth of the fish might sever a single line, the portion of the snood which is near the hook is composed of a number of separate threads fastened loosely together, so as to permit the teeth to pass between the strands. At each end of the long line is fastened a float or buoy, and when the hooks have been baited with sand launce, limpets, whelks, and similar substances, the line is ready for action.

The boat, in which the line is ready coiled, makes for the fishing-place, lowers a grapnel or small anchor, to which is attached the buoy at one end of the line, and the vessel then sails off, paying out the line as it proceeds, and always "shooting" the line across the tide, so as to prevent the hooks from being washed against each other, or twisted round the line, which is usually shot in the interval between the ebb and flow of the tide, and hauled in at the end of about six hours.

As soon as the long line has been fairly shot, and both ends firmly affixed to the grapnels, the fishermen improve the next six hours by angling with short lines, one of which is held in each hand. They thus capture not only Cod-fish, but haddock, whiting, hake, pollack, and various kinds of flat fishes. On favorable occasions, the quantity of fish captured by a single boat is very great, one man having taken more than four hundred Cod alone in ten hours.

The Cod is a most uncertain fish in its habits, sometimes haunting the same locality for a number of successive years, and then suddenly leaving it and repairing to some spot where not a fish might be found on the preceding year. New fishing-grounds are frequently discovered, and it sometimes happens that the fishermen are fortunate enough to alight on a spot hitherto untouched, where, to use the graphic description of a sailor, the Cod are "as big as donkeys, and as common as blackberries."

Rockall, for instance, is one of the discoveries of this nature. It is a sandbank in the North Atlantic, about 136 miles from St. Kilda, and only distinguishable by a small rock like a rude haystack. The Cod are there so plentiful and so large that each fishing-boat sold her five days' catch for $700; and after due preparation, the fish were disposed of at nearly double that price.

A great part of the estimation in which this fish is held depends upon the perfect manner in which it takes salt, and the length of time during which it can be preserved in an eatable state. Salted Cod is to many persons a great dainty, but to others, among whom I must be reckoned, it is insufferably offensive, and even with all the additions of sauce and condiment is barely eatable.

The Cod is sometimes sent away in a fresh state, but is often split and salted on the spot, packed in flats on board, and afterwards washed and dried on the rocks. In this state it is called Klip-fish or Rock-fish. The liver produces a most valuable oil, which is now in great favor for the purpose of affording strength to persons afflicted with delicate lungs or who show symptoms of decline. The best oil is that which drains naturally from the livers as they are thrown into a vessel which is placed in a pan filled with boiling water. The oil is then carefully strained through flannel, and is ready for sale.

HADDOCK.—*Morrhua æglefinus.*   WHITING.—*Merlangus vulgaris.*   COD.—*Gadus æglefinus.*

The roe of the Cod is useful for bait, the sardine in particular being very partial to that substance. Much of the roe is stupidly wasted by the fisherman, who carelessly flings into the sea a commodity of which he can sell any amount, and for which he can obtain two dollars and a half per hundredweight. In Norway, the dried heads of the Cod are used as fodder for cows, and, strange to say, the graminivorous quadrupeds are very fond of this aliment.

Like several other marine fish, the Cod can be kept in a pond, provided the water be salt; and if the pond should communicate with the sea, these fishes can be readily fattened for the table. Several such ponds are in existence, and it is the custom to transfer to them the liveliest specimens that have been caught during the day's fishery, the dead or dying being either sold or cut up as food for their imprisoned relatives. These fishes are extremely voracious, and will eat not only the flesh of their kinsmen, but that of whelks and other mollusks, which are abundantly thrown to them. It is found that under this treatment the Cod is firmer, thicker, and heavier in proportion to its length, than if it had been suffered to roam at large in the sea.

The color of the Cod is ashen-green, rather mottled with deeper tints, and the abdomen is white. The head is very large, there is a long, fleshy barbule on the chin, and the pupil of the eye is blue. Varieties in color and even in form are not uncommon, and in some cases are thought to be produced by difference of diet and locality. The average length of an adult Cod-fish is about three feet, and its weight twelve pounds.

The Common Cod (*Gadus callarias*, L.), having a range extending from the northern seas to Virginia, and from Oregon to Japan, is the most important of all food-fishes. It is taken along the coast of Massachusetts during the whole year, leaving the vicinity of land in February and going into deeper water. During the preparation of the State Reports of Massachusetts on Natural History, Dr. Storer, of Boston, had occasion, in his task of writing up the histories of the fishes, to consult several well-known and reliable authorities. Among them, Jonathan Johnson, of Nahant, is prominent. We had the pleasure of his acquaintance, and have personal knowledge of the great services he, as well as some others in the trade, extended to scientific observers. He states that the largest Cod he has seen taken weighed eighty-eight pounds. Mr. Holbrook, a fishmonger in Boston Market, reported to Dr. Storer that he "saw taken, in the spring of 1807, at New Ledge, near Portland, Me., a Cod that weighed one hundred and seven pounds, which had barnacles on its head as large as one's thumb."

The AMERICAN COD is very voracious, attacking and feeding on smaller fishes, crustaceans, and marine shell-fishes. During the winter months the Cod-fishes have their stomachs full of small mollusks, crustaceans, worms, etc., which are obtained on the rocks in deep water. Boston is supplied largely with fresh Cod and haddock by small vessels sent out from Nahant and Swampscott. These vessels average about fifty tons, and are built much after the models of the pilot-boats, being staunch and fleet. The writer once accepted an invitation to take a place on board and accompany the crew on one of their winter day trips. "Green's Harbor ground" is a favorite place for Cod and haddock fishing at this season. There we came to anchor after three hours' sail; having started from home at the early hour of 2 A. M. This early start brings them on the grounds at daylight, when the fishes bite more briskly. Of the six men, each has his boat on deck. They are put out at distances from each other, and after four or five hours' fishing are ready to be picked up; the vessel lying to during the day until then. We were clothed, like the others, in heavy woolens, and an oil-cloth suit over all, with "Sou-Wester" hat. The boots are immensely heavy, and being soaked with tar, become impervious to wet. Heavy woolen mittens for the hands, most singularly afford complete comfort during the intense cold by frequent immersion in a bucket of sea-water, the mittens being soaked in it. So long as this was done the hands remained warm. Green's Harbor is directly opposite Daniel Webster's estate, at Marshfield, Mass., and is a favorite resort for fishermen for Boston market. Cod-fish that live around the shore, among the algæ, become delicate, and are often quite brilliant in color. Their flesh is tender and sweeter than those of the deep sea.

The TOM COD is a miniature of the Cod-fish, reaching in the colder waters of Maine about twelve inches. But farther south it is usually about eight inches in length. In the fall, when the first cold weather comes, this is called Frost-fish, and is taken by the hook from our wharves. It is a savory fish. Dr. Mitchell says it has been taken in great numbers in the creeks by a common hoe, the fishes being so abundant. Its range is from Newfoundland to Hatteras.

The Codlings, of genus *Phycis*, are represented by several species in our American waters north of Hatteras. One is called Squirrel Hake, and Chuss in New York; American Hake in New England. It is also called Ling, a picture of which will be seen on same cut with Sly Silurus, on a following page. These fishes are caught at night. During the bright summer nights, off Nahant, the light dancing dories of the Swampscott fishermen may be seen manned by busily engaged fishers for the local markets. This fish varies from ten to thirty pounds weight. Cusk is another variety allied to the preceding.

Of several other species of this genus we may mention the DORSE (*Morrhua callarias*), the HADDOCK (*Morrhua æglefinus*), a well-known and very valuable fish, which is represented in the previous illustration, and the WHITING, POUT, SMELTIE, or KLEG (*Morrhua lusca*), so often manufactured into whitings by the simple process of slicing off certain parts of the fish, skinning it, and pushing its tail through the head. In this state it is sold and consumed as whiting; and as one fish is just as good as the other, the consumer suffers no injury, and the enterprising vendor is recompensed for his trouble. The Pout is graphically termed by the fishermen the Stinkalive, because it becomes putrid so soon after death. While living, various iridescent colors play over the surface of the fish, but as soon as it is dead the colors and the dark bands disappear, and the whole upper surface becomes of a dull yellow-brown, the abdomen being whitish with a tinge of blue-gray.

THE common WHITING (*Merlangus vulgaris*), which is also figured in the previous illustration, is closely allied to the fishes of the preceding genus, and is too well known to need description. The COAL-FISH (*Merlangus carbonarius*), and the POLLACK (*Merlangus pollachius*), belong to the same genus as the whiting; and the HAKE (*Merlucius vulgaris*) is closely allied to them.

THE EELS.

IN the large and important group of fishes to which our attention is now drawn, the ventral fins are wholly wanting, the body is long, snake-like, smooth, and slimy

SAND EEL, OR HORNELS.—*Ammodytes tobianus.*

on the exterior, and in many cases covered with very little scales hidden in the thick, soft skin.

OUR first example is the SAND LAUNCE, a very common fish on many coasts, and usually found wherever the shore is of a sandy character. The generic name Ammodytes signifies sand-diver, and is given to this fish in consequence of its habit of burying itself in the wet sand, where it remains hidden and secure from marine foes.

THE FIERASFERS are small fishes of tropical waters, parasitic or commensal in echinoderms and mollusks; allied near the eel-pouts and sand eels.

While resident on the Florida Reef, at Fort Jefferson, we discovered a Fierasfer living in a large Holothuria. At this time, 1859, this was a novelty to naturalists. Since then species have been found in other parts of the world, and in various objects.

The large Holothuria, or Sea Cucumber of the Reef, is often eighteen inches in length; and it is abundantly spread over the reef in shoal water. The visitor, in sailing leisurely in these waters, may reach one easily from his seat in the boat. Lift the creature into a bucket of water: he soon exhausts the oxygen of so small amount of water, and there peeps out from his mouth another creature that requires more oxygen than is left, and thus asserts his rights by leaving. It is a fish, a Fierasfer, that so appears, and a delicate transparent one it is, of eight inches in length. Its name Fierasfer is derived from the Greek, meaning sleek and

SHARP-NOSED EEL.—*Anguilla acutirostris.*

shining, in allusion to its great delicacy. Its habit of living protected within the halls of the great echinoderm renders its exterior tissues delicate, if indeed it be not originally made so. Who can tell? These cases of parasitic or, more properly, commensal life are, indeed, puzzling. This species is *Fierasfer dubius* (Putnam), and its locality is recorded as Florida Keys to Cuba.

ANOTHER species of this genus, the SAND EEL, or HORNELS (*Ammodytes tobianus*), is sometimes mistaken for the preceding species, from which, however, it may be distinguished by its greater size, its larger head, the farther setting back of the dorsal fin, the browner color, and more opaque body. When full-grown, the Sand Eel will reach the length of a foot or thirteen inches.

THE MURÆNA.

THE SHARP-NOSED EEL, represented in the engraving on page 271, derives its name from the shape of its head, and by that structure may be distinguished from the second species. In their habits the Eels are so similar, that the present species will be taken as an example of the whole genus.

Eels are found in almost all warm and temperate countries, and grow to a very large size in tropical regions. They are, however, impatient of cold, and in the extreme northern or southern parts of the world are not to be found. In many of the Pacific islands these fish are held in great estimation, being preserved in ponds and fed by hand, and in New Zealand they afford one of the staple articles of consumption. In some parts of the world, however, a strong prejudice exists against Eels, probably on account of their resemblance to snakes, and even a hungry man will not eat one of these wholesome and nutritious fish.

The Eel is one of the most mysterious river fishes, and although much is now known that formerly was involved in obscurity, there is still much to learn respecting its habits, and, more especially, its mode of reproduction. It is probable, that difference of locality may influence the Eel and cause difference of habit; but it is certain, that, if a number of practical observers set themselves to watch the Eel and its customs, their accounts would vary in the most perplexing manner, and to build a theory upon so unsafe a basis is quite impossible.

The Common Eel (*Anguilla rostrata*) is abundant along the Atlantic coast, from Maine to Mexico, ascending all the streams and resident in the Mississippi valley.

The BROAD-NOSED EEL is at once to be distinguished by the greater breadth of its head, bluntness of its nose, and soft unctuousness of its body. It does not seem to attain so great a size as its sharp-nosed relative. Besides these species, I mention a third Eel, the SNIG, which is known by its olive-green back and the golden-yellow of the under parts. The Grig is a term applied by fishermen to any Eel of a small size, and even the name of Snig is employed in a very vague fashion.

THE well-known CONGER EEL is a marine species, very common in our seas, and being most usually found on the rocky portion of the coast.

This useful fish has, of late years, come into more general use than formerly, and its good qualities are more appreciated. The flesh, though not very palatable if dressed unskilfully, is now held in some estimation, and for the manufacture of soup is thought to be almost unrivalled. The fishermen can now always obtain a ready sale for the Congers; and those which are not purchased for the table are mostly bought up and made into isinglass. It often attains to a very great size, measuring ten feet in length, and weighing more than a hundred pounds.

The Conger Eel (*Conger niger*) is the same in species, found in Europe and East Indies, and on our Atlantic coast. In Europe it grows to a large size, and appears to be more plentiful, weighing one hundred pounds and measuring ten feet in length. A strong diversity of opinion exists concerning its value as food. In the fourteenth century it was prized, and was reserved by the nobility. A special preparation of this fish formed an established trade in the west of England in the time of King John. This preparation was dried Conger, called *Conger doust*, or Sweet Conger, which was exported to Spain.

THE beautifully mottled MURÆNA is tolerably common in the Mediterranean, but is scarce towards the northern coasts.

In former days the Muræna was held in great distinction by epicures; and the wealthy were accustomed to preserve them in ponds built for that special purpose. In these ponds the Muræna were fattened, and several of the aristocrats labored under the imputation of feeding them with an occasional slave, whenever an ill-fated domestic had the misfortune to offend them. The flesh is very white in color, and of a peculiar and very delicate flavor. This fish can live either in salt or fresh water, but appears to prefer the sea.

The color is golden-yellow in front and purple towards the tail; and the whole body is covered with bands, irregular rings, and spots of deep and pale gold, purple, and brown.

the dorsal fin begins a little behind the head and runs to the tail, where it is united with the anal fin. Both these fins are, however, low and fleshy, and not at all conspicuous. The length of this fish is extremely variable; one captured specimen measured four feet four inches in length.

The Muræna, or Moray (*Muræna melanotis*), is found from Charleston, S. C., to Florida. Its usual length is about twenty inches. Its shape is somewhat like that of the Blennies. Its propensity to bite, and general appearance, are suggestive of snake.

THE ELECTRIC EEL is even more remarkable for its capability of delivering powerful electric shocks than the torpedo.

MURÆNA.—*Muræna helena.*

The Electric Eel is a native of Southern America, and inhabits the rivers of that warm and verdant country. The organs which enable it to produce such wonderful effects are double, and lie along the body, the one upon the other.

The reader will remember that in the torpedo the electric effect was produced by a number of little columns; in the Electric Eel, the corresponding organ consists of a great number of divisions, technically called "septa," which are again subdivided by lesser transverse membranes. One organ is always larger than the other; and it was found that in a fish measuring about two feet four inches in length, there were thirty-four septa in the larger organ and fourteen in the smaller. On an average two hundred and forty transverse membranes are packed in each inch, thereby giving a vast extent of electricity-producing surface. It was calculated by Lacepéde, that the expanse of this organ in an Electric Eel of four feet in length is equivalent to one hundred and twenty-three square feet, while that of a large torpedo only equals fifty-eight feet.

In the native country of these fishes they are captured by an ingenious but somewhat cruel process. A herd of wild horses are driven to the spot and urged into the water. The alarmed Gymnoti, finding their domains thus invaded, call forth all the terrors of their invisible artillery to repel the intruders, and discharge their pent-up lightnings with fearful rapidity and force. Gliding under the bellies of the frightened horses, they press themselves against their bodies, as if to economize all the electrical fluid, and by shock after shock generally succeed in drowning several of the poor quadrupeds.

Horses, however, are of but slight value in that country, hardly, indeed, so much valued as pigeons in North America, and as fast as they emerge from the water in frantic terror, are driven back among their dread enemies. Presently the shocks become less powerful, for the Gymnotus soon exhausts its store of electricity, and when the fishes are thoroughly fatigued they are captured with impunity by the native hunters. A most interesting account of this process is given by Humboldt, but is too long to be inserted in these pages.

ELECTRIC EEL.—*Gymnotus electricus.*

Several of these wonderful fish have been brought to foreign countries in a living state. I well remember a fine Gymnotus that lived in captivity. Numbers of experimenters were accustomed daily to test its powers; and the fatal, or at all events the numbing, power of the stroke was evident when the creature was supplied with the fish on which it fed. Though blind, it was accustomed to turn its head towards the spot designated by the splashing of the attendant's finger, and as soon as a fish was allowed to fall into the water the Gymnotus would curve itself slightly, seemed to stiffen its muscles, and the victim turned over on its back, struck as if dead by the violence of the shock.

When full-grown, the Electric Eel will attain a length of five or six feet, and is then a truly formidable creature. The body is rounded, and the scales small and barely visible. According to Marcgrave, the native name for this fish is Carapo.

WE have already seen some examples of fishes where the body is extremely transparent, and now come to an entire family where this peculiarity is the chief and most obvious characteristic.

The skeleton of the Leptocephalidæ, or Glass Eels as they are termed, from their Eel-like shape and singular translucency, is very imperfect, merely consisting of cartilage, and so slight that even in the head, where the greatest strength is required, the brain can be seen through the translucent skull in which it lies. Their bodies are always extremely compressed and mostly leaf-like, so transparent that when lying in a vessel containing water they would hardly be noticed, and the lateral line is formed by the intersection of the muscles.

The PIG-NOSED GLASS EEL may be known by the lengthened form of its head and snout, which are far longer in proportion to the dimensions of the fish than in any other member of the family. The generic term Hyoprorus literally signifies swine-beaked, and in former days was applied to a certain kind of galley which had a long and slightly turned-up beak. The sudden height of the body just behind the head is very remarkable, and on close examination, a row of mucous pores will be found along the jaws and on the head. The eyes are not very large, and the general length of the species is between four and five inches. As its specific name imports, it has been taken at Messina.

The HAIR-TAILED GLASS EEL is much longer in proportion than the last-mentioned species, and its body is so extremely compressed that it is hardly thicker than the paper on which this account is printed. This species is also found at Messina. The jaws are short and round, the eye rather small, and the tail tapers away to a hair-like point. The length of this fish is rather more than a foot, and a row of minute points runs along each edge of the body.

The typical genus Leptocephalus is a rather large one, containing more species than the four preceding genera together.

The ROUND-HEADED GLASS EEL derives its specific name of Tænia, or tape-worm, on account of its resemblance to that unpleasant internal parasite. Its head is, as its name denotes, short and much rounded, and the eyes are globular, projecting, and extremely large. The jaws are tolerably well furnished with small teeth. In shape it is long and rather rounded, and the absence of fins renders its resemblance to a tape-worm extremely striking. It seems to be an Asiatic species, having been captured in India and the neighboring islands.

In the ANGLESEY MORRIS (*Leptocephalus morrisi*), another example of this genus, the head is blunt, the eye moderate, the body much compressed, and deepest at the latter third of its length. When living, its polished surface reflects gleams of iridescent light as it winds its graceful way through the sea weeds among which it loves to sojourn, like a ribbon of animated nacre. But when dead and placed in spirits, all the delicate opalescence of its body fades, and soon deteriorates into an opaque dull whiteness like wet parchment.

THE BLIND-FISH.

THE reader will remember that on several occasions it has been deemed expedient to give examples of remarkable deviations from the ordinary system, and to call attention to the wonderful economy of nature, which is most averse to wastefulness, and declines to expend its powers on organs that if existing would be in abeyance. A recent example of such modification has been given in the proteus, on page 186, that curious reptile, or semi-reptile, which inhabits caves wherein penetrates no ray of light, and which, having no need of external eyes, is altogether devoid of such useless organs.

The BLIND-FISH of America affords another instance of similar economy in structure. Living, like the proteus, in a subterranean and perfectly dark grotto, it needs no eyes, and in consequence possesses none, their place being merely indicated by two minute black dots on the sides of the head. The head is naked, but the body is covered with scales and the jaws are furnished with some small but sharp teeth. Its color is whitish-gray, as is, indeed, mostly the case with animals that have been long deprived of the color-giving sunlight. The grotto which contains this very remarkable little fish is in Kentucky.

Of the Blind-fishes, family *Amblyopsidæ*, living in caves, three genera are now known.

Amblyopsis spelæus is found in the caves of Indiana and Kentucky. These Blind-fishes are of small size, living in subterranean streams and ditches of Central and Southern United States.

Four species are now known. F. W. Putnam has given us all the information yet procured about them.

THE HERRING TRIBE; CLUPEIDÆ.

WE now come to that most valuable family of fishes, the Herring tribe, called technically *Clupeidæ*, from the Latin word *clupea*, a herring.

THE well-known ANCHOVY is properly a native of the Mediterranean Sea, though it often occurs on northern coasts. Indeed, one practical writer on fishes thinks that the capture of the Anchovy off northern shores is a task that would be highly remunerative if properly undertaken, and that, with proper pains, the markets in the north might be fully supplied with Anchovies from their own seas.

This little fish has long been famous for the powerful and unique flavor of its flesh, and is in consequence captured in vast quantities for the purpose of being made into Anchovy sauce, Anchovy paste, and other articles of diet in which the heart of an epicure delights. Unfortunately, however, the little fish is so valuable, that in the preparations made from its flesh the dishonest dealers too often adulterate their goods largely, and palm off sprats and other comparatively worthless fish for the real Anchovy. As the head is always removed before the process of potting is commenced, the deception is not easily detected—the long head with its projecting upper jaw and deeply cleft gape affording so clear an evidence of the identity of the fish, that no one would venture to pass off one fish for the other, if the heads were permitted to remain in their natural places. The flavor of the veritable Anchovy is rudely imitated by various admixtures, and its full rich color is simulated by bole armoniac and other abominations.

The very long generic title *Engraulis encrasicholus* was given to it in ancient times, and is still retained, as being quite appropriate. Its literal signification is "gall-tinctured," and the name has been given to it on account of the peculiar bitter taste of the head, in which part the ancients supposed the gall to be placed.

THE COMMON or ALLICE SHAD is extremely plentiful on some of our coasts, but appears to be a rather local fish, and while it abounds in some places, to be wholly absent from others.

The Shad is fond of ascending rivers, especially if the water be clear; and while the Thames was still unstirred by the paddles of multitudinous steamboats, and unpolluted by the contents of countless sewers, this fish would ascend the river for a considerable distance, and has been taken in good condition near Hampton Court. Some person think that the flavor of the fish improves in proportion to its proximity to the river source. Except in size, the Shad bears a very close resemblance to a herring, and in some places is called the King of the Herrings.

The color of the Shad is dark blue on the upper part of the head and back, variegated with glosses or reflections of brown and green, either color predominating according to the angle at which the light falls upon the surface. The remainder of the body is white. There is another species of this genus, the TWAITE SHAD (*Alosa finta*), which is about half the size of the Allice Shad, weighing on an average about two pounds. Both these fish may be at once distinguished by a deep cleft or notch in the centre of the upper jaw.

Shad (*Clupea sapidissima*). This valuable fish is found ranging from Newfoundland to Florida. Mitchell says it is a regular visitor, coming to us from the ocean as in yearly migration; in March ascending towards the head of the Hudson and other rivers, to breed. Its average weight is four pounds. The Shad-fishing of the New England States is considerable—in the Connecticut River especially. Some are taken in the Merrimack River. Unlike most others of its genus that come to us from the north, this fish comes from the south, to deposit its spawn. In Charleston, S. C., it appears in January, proceeding steadily along the coast, at Norfolk in February, and reaching New York in March or April, in accordance with the state of the season. On the coast of Massachusetts it appears in May. The Shad that reach the headquarters of the Hudson attain a distance of one hundred and fifty miles from the mouth. Shad in New England rank low, as the salt-water fishes of the northern coast are superior. East of Boston, Shad are regarded as little better than herring.

THE HERRING (*Clupea harengus*) is undoubtedly the most valuable of fishes, and the one which could least be spared. In Europe it is at once the luxury of the rich and the nourishment of the poor, capable of preservation throughout a long period, easily packed, quick and simply dressed, and equally good whether eaten fresh or salted, smoked or potted.

1. TWAITE SHAD.—*Alausa finta.* 2. SPRAT.—*Clupea sprattus.* 3. HERRING.—*Clupea harengus.* (One-third natural size.)

During the greater part of the year, the Herring lives in deep water, where its habits are entirely unknown. About July or August, the Herring is urged, by the irresistible force of instinct, to approach the shores for the purpose of depositing its spawn in the shallow waters, where the warm rays of the sun may pour their vivifying influence upon the tiny eggs that will hereafter produce creatures of so disproportionate a size, and where the ever-moving tides may fill the water with free oxygen as the waves dash on the shores and fall back in whitened spray, thus giving to the water that sparkling freshness so needful for the development of the future fish.

The Herring is called Alewife in New England, Gaspereau in the British provinces, Spring Herring, Blue-back, Saw-belly, and Cat-thresher in Maine. The Narragansetts called it Aumscrag. Its range is from Newfoundland to Florida. It is thought that Herrings winter in the Arctic Circle and pass southward. They appear off the Shetland Islands in April and May, but the grand shoal is not seen until June. The main body is described as altering the appearance of the ocean miles in extent, divided in columns of six units' length. In America the shad run up the rivers in March, the streams being so full they are trampled on at fording-places.

In Massachusetts the Taunton River is a famous place for Alewives. The erection of dams, however, here as elsewhere, has served to check their increase. The English Herring was once declared to be distinct, but is now regarded as identical with the present species. The celebrated White-bait was once regarded as an English Herring of a peculiar small kind, but now it is definitely known that White-bait is the young of the common Herring. Mr. Blackford informed us that the White-bait, precisely similar to the English, is taken off Coney Island. Young White-bait were kept in aquarium until they had grown to be twelve inches in length.

HARD HEAD, or MENHADEN (*Brevoortia tyrannus*) Moss-bunker, Bony-fish, White-fish, Bug-fish, Fat-back, Yellow-tail, Pogy, Poghagen, Skippaugs, or Bunkers, so called in various places. This is one of the most familiar of native fishes, though it is not a food-fish, but a very valuable one in its services to the fishermen as bait. It is even so numerous at times as to be taken in vast quantities for manuring land. Its oil is used largely in cheap painting.

ANOTHER species of this fish, called LEACH'S HERRING (*Clupea leachii*), is captured during the winter months; the roe being well developed at the end of January, and the spawn deposited in February. It is a small species, between seven and eight inches in length.

The common SPRAT is another very useful fish, though not so extensively valued as the herring.

Like that fish, it swims in vast shoals during the spawning season, which immediately succeeds that of the herring, so that from July to February and March the public can command a continual supply of fresh sea-fish, which can be purchased at so cheap a rate as to be within the reach of all classes, and are, nevertheless, of such excellent flavor that if they were as scarce as they are plentiful, they would be held in high estimation at the tables of the wealthy. To the taste of many persons, however, the Sprat is too rich and too strongly flavored to be in much request.

This fish is captured in nets of various kinds, the nature of the net mostly depending on that of the locality; and as it swims in shoals quite equal in numbers to those of the herring, it is taking in countless multitudes when the boats happen to be fortunate in their selection of a fishing-ground. Now and then the "take" is so enormous that even the European markets, which usually absorb every eatable article which can be brought for sale, and often anticipate the future crops or supplies, are at times so overstocked with Sprats that the fishermen can find no ordinary sale for their perishable goods, and are perforce obliged to dispose of them to the farmers, who spread them over their lands for manure, most unfragrant but exceedingly fertilizing. In color it is very like the herring.

One or two members of this genus demand a brief notice.

The PILCHARD, or GIPSY HERRING (*Clupea pilchardus*), is another of the gregarious fish, and is taken about the month of August by a wonderfully intricate system of boats and nets that seem capable of sweeping every fish out of the sea. Though very like the herring, it may easily be distinguished by the position of the dorsal fin, which is set so far forward that if the fish be held by the first ray of that fin its body slopes upward, whereas in the herring it is nearly balanced and slightly inclines downward.

THE far-famed FLYING-FISH exists in many of the warmer seas, and derives its popular name from its wonderful powers of sustaining itself in the air. Its picture is placed on the next page.

The passage of this fish through the atmosphere can lay no just claim to the title of flight, for the creature does not flap the wing-like pectoral fins on which it is upborne, and is not believed even to possess the power of changing its course. As much of the history of the Flying-fish has been given while treating on the coryphene, the reader is referred to the description of that fish on page 248, where may also be seen an illustration of the attitudes assumed by the Flying-fish as it speeds its course through the air while attempting to avoid its deadly foe beneath.

BEFORE proceeding to our next example of the finny tribes, we must briefly notice a curious fish which seems to be a kind of balance to the sword-fish already mentioned, the "sword" in this instance belonging to the lower instead of the upper jaw, and being formed by a prolongation of its bones. It is known by the scientific name of *Hemiramphus argenteus*, and is found near the surface of the water in the Pacific Ocean. Its color is uniform silvery-white, and its average length is only four inches.

FLYING-FISH.—*Exocœtus volitans*.

THE odd-looking GAR-FISH is known by a vast variety of names, such as SEA PIKE, MACKEREL GUIDE, SEA-NEEDLE, LONG-NOSE, GORE-BILL, HORNFISH, and GREENBONE, the last-mentioned title being given to it because, when it is boiled, its bones are of a bright green hue. The name of Mackerel Guide is owing to the fact that its spawning season exactly precedes that of the mackerel, and the other names explain themselves.

THE fierce and voracious PIKE has well earned its titles of Fresh-water Shark and River Pirate, for though perhaps not one whit more destructive to animal life than the roach, gudgeon, and other harmless fish, the prey which it devours are of a larger size, and its means of destruction are so conspicuous and powerful, that its name has long been a by-word for pitiless rapacity.

SALMON, TROUT, CARP, ETC.

THE SALMON is undoubtedly the king of river-fish; not so much for its dimensions, which are exceeded by one or two giant members of the finny tribe, but for the silvery sheen of its glittering scales, its wonderful dash and activity, affording magnificent sport to the angler, the interesting nature of its life from the egg to full maturity, and last, but not least, for the exquisite flavor and nutritive character of its flesh,

In former days, before civilization had substituted man and his dwellings for the broad meadows and their furred and feathered inmates, the Salmon was found in many rivers. Now, however, there are but few streams where this splendid fish can be seen, for, in the greater number of European rivers, the water has been so defiled by human agency that the fastidious Salmon will not suffer itself to be poisoned by such hateful mixture of evil odors and polluted waters; and in the few streams where the water is still sufficiently pure for the Salmon to venture into them, the array of nets, weirs, and all kinds of Salmon traps is so tremendous, that not one tithe of the normal number are now found in them.

The ingenuity which has been exhibited in the invention of these "infernal machines," as the fixed nets have been justly termed, and the amount of labor which has been expended in

SALMON.—*Salmo salar*.   SALMON TROUT.—*Salmo trutta*.

their manufacture, are worthy of a better cause; for in their arrangement the habits of the fish have been carefully studied, and, in their manufacture, its capabilities have been foreseen. The evil has, of late years, arisen to so great a height, that the Salmon would soon have been extirpated from European rivers, had not the nation wisely interfered to prevent the loss of so much national wealth, and given the fish a fair chance of re-establishing itself in its former plenty.

The short-sighted persons who plant all these obstructions forget that by this wholesale destruction of the Salmon they are acting against their own interests, and that if they destroy the ill-conditioned and young fish, as well as the adult and healthy Salmon, they condemn themselves to the probability of eating bad fish for the present, and the certainty of total deprivation for the future. The fact, however, seems to be, that each petty proprietor of a fishery is jealous of the neighbors above and below him, and indiscriminately slaughters all fish that he can capture in his own waters, simply that they may not pass into those of his neighbor.

The preservation of this noble fish is truly a subject of national importance, and it is to be hoped that, by judicious legislation and active administration of the law, the Salmon may no longer be the rich man's luxury, but again hold its legitimate place as the poor man's cheap subsistence.

The life history of the Salmon is very interesting, and in many parts not a little mysterious. In the short space which is allowable for the subject, I will endeavor to trace the life of a Salmon from its earliest entrance into the world to its exit therefrom; putting forward no particular theories, but merely enumerating the accredited observations that have been made on this curious subject.

We will begin with the cradle that is prepared for the expected brood. This is a groove in the gravelly bed of a river, and is scooped out by one or both of the parents. Even here a discrepancy exists between practical observers, some of whom aver that the groove is made by both parents, by means of rooting with their noses in the ground; others that the male Salmon scoops out the gravel with a hook-like appendage that is developed on his chin during the breeding-season; while others declare that the male never troubles himself about the labor of scooping the groove, his duty being to watch over his mate and to fight any other fish of his own sex and species who may intrude upon their home, and that the whole task devolves upon the female, who executes it by twirling her tail and not by grubbing with her snout.

The whole process of depositing the numerous eggs occupies on the average about ten days, and, after it is accomplished, the parent fish leave the eggs to be hatched by surrounding influences, while they themselves quit the spot and remain in the river for a short period while they recover from the exhaustion caused by the process. During this period they are unusually ravenous, and vast quantities of the young of their own kind, which are about that time abundant in the river, fall victims to their insatiable appetite. After a time, and about the months of March and April, they drop down from pool to pool, in any flood which may seem favorable to them, until they reach the sea, where they are supposed to remain from six weeks to three or four months, when they again seek the river, vastly increased in weight and improved in condition.

The Salmon must be eaten fresh. If it be cooked within an hour or two after being taken from the water, a fatty substance termed the "curd," is found between the flakes of flesh. If, however, more than twelve hours have elapsed from the death of the fish, the curd is not to be seen, and the Salmon is much deteriorated in the judgment of epicures.

The size of this fish is extremely variable, some specimens having been caught that weighed sixty pounds, and Mr. Yarrell mentions one case where a female Salmon was captured and was remarkable for weighing eighty-three pounds. This great weight was owing more to the depth and thickness of the fish than the length.

The Salmon is common to all rivers of the Atlantic coast north of Cape Cod. It is found only in the coldest waters, and is equally distributed in Europe and America. It is not plentiful now south of the St. Lawrence River. The numerous dams have been a potent cause of their scarcity.

A large number of species have come to light in late years over the northern portions of the continent.

The mountain streams of the Great West afford abundance of Salmon trout.

Several Pacific Ocean Salmons are known.

The Grayling (*Thymallus*) is an allied form; two species were found in the rivers and lakes of Michigan.

WHITE FISH (*Correganus clupeiformis*). This is a notable food-fish of the Great Lakes; most highly prized. Numbers of species are also known in the lakes.

The SMELT (*Osmerus mordax*) is a well-known food-fish, highly prized by some. Its range is from Nova Scotia to Hatteras. In the fall, when the frost-fish or tom-cod is appearing, the Smelt come in vast numbers. It is a small fish, yet it is caught by hook.

Angling for it is pleasant sport. Picturesque scenes are often noticed in Boston harbor, when the water is frozen over solid. Holes are cut in the ice, and Smelts taken through them by line and hook. Tents are spread, and the scene becomes exceedingly active and curious. DeKay says this beautiful fish "derives its name from the fact that its smell resembles that of cucumbers."

NEXT to the salmon, the bright-scaled carmine-speckled active TROUT is perhaps the greatest favorite of anglers, and fully deserves the eulogies of all lovers of the rod; its peculiarly delicate flesh, its fastidious voracity, and the mixture of strength, agility, and spirited courage with which it endeavors to free itself from the hook, forming a combination of excellences rarely met with in any individual fish.

The Trout is found in rapid and clear-running streams, but cares not for the open and shallow parts of the river, preferring the shelter of some stone or hole in the bank, whence it may watch for prey. Like the pike, it haunts some especial hiding-place, and, in a similar manner, is sure to take possession of a favorable haunt that has been rendered vacant by the demise of its predecessor or its promotion to superior quarters. Various baits are used in fishing for Trout, such as the worm, the minnow, and the fly, both natural and artificial, the latter being certainly the neatest and most artistic method. The arcana of angling are not within the province of this work; and for information on that subject, the reader is referred to the many valuable works which have been written by accomplished masters of the art.

There is a curious method of catching Trout, much in vogue among the juvenile fishers. This process is called "tickling," and is managed as follows: The tickler gets quietly into the stream, and walks slowly along the banks, feeling carefully for any depression or cavity. One hand is then introduced very gently, while the other is placed over the entrance of the hole, the fingers being spread so as to prevent the exit of any fish that may happen to be resident in that locality. Several such cavities may be tried without success, but at last the smooth side of a fish is felt by the finger-tips.

The startled fish gives a great flounce on being touched, and tries to dash out of the hole, but, being checked by the spread hand, retires to the recesses of its cavern. The finger-tips are then gently brought against the abdomen of the fish, which soon endures the contact, and permits the hand gradually to inclose it. As soon as that is the case, the fish is suddenly grasped, snatched out of the hole, and flung ashore before it can find time to struggle from the captor's hold. Some accomplished ticklers aver themselves to be capable of thrusting the fore-finger into the gill and out at the mouth, and hooking out the fish in this singular manner.

The color of the Trout is yellowish-brown above, speckled with dark reddish-brown, and a number of carmine spots are scattered along, each side of the lateral line. The abdomen is silvery-white, and the lower part of the sides rich golden-yellow. There is, however, considerable variation in the color of the Trout, the locality having considerable influence upon the tints.

One or two other species of this genus require still a passing notice.

The BULL, or GRAY TROUT (*Salmo eriox*) is found plentiful.

It often attains a very large size, but a specimen weighing more than fifteen pounds is not very common.

The SALMON TROUT (*Salmo trutta*) is another species, and in general habits is very like the Salmon, migrating to the sea, and returning to the rivers in a similar fashion. It is illustrated with the Salmon on page 280.

The CHARR (*Salmo salvelinus*), the well-known and delicately flavored SMELT (*Osmerus eperlanus*), called also the SPIRLING or SPARLING, the GRAYLING (*Thymallus vulgaris*), the VENDACE (*Coregonus willoughbii*), and the ARGENTINE (*Scopelus humboldtii*), so useful for bait, all belong to the same family as the salmon and the trout. The accompanying illustration represents two of them.

The PIRAYA, or PIRAI, has been removed from the salmonidæ and placed in another family on account of certain structural differences.

TROUT.

This fish is very plentiful in the rivers of Guiana and Brazil, where it swims in large troops, and is, according to many accounts, a very unpleasant neighbor. It is a most voracious being, with teeth nearly as sharply edged as those of the shark, and a boldness little short of that fish's well-known audacity. It is said, according to Spix, that if even so large an animal as an ox happens to get into one of their shoals, it is immediately assailed, and bitten so severely that it may succumb under its injuries before it can cross a stream thirty or forty feet in width. According to some authors, one of the South American tribes are in the habit of placing their dead in the streams, leaving them to the attacks of the Piraya, which in a single night will clear away the whole of the soft parts, and leave a clean skeleton ready for their peculiar mode of sepulture. Even living human beings seem to enjoy no immunity from this hungry fish, but to be liable to severe bites while bathing.

Be these stories literally true, or only exaggerations of reality, the jaws and teeth of the Piraya are perfectly capable of inflicting such injuries as have been briefly described. The

GRAYLING.—*Thymallus vulgaris.*   CHARR.—*Salmo salvelinus.*

teeth are nearly flat, triangular, and with edges sharp as those of lancets, and are employed by the Macoushi Indians to sharpen the points of those fearful wourali-poisoned arrows so well known to fame since they were brought by Mr. Waterton from Guiana. A part of the jaw containing five or six teeth is carefully cleansed, a hole is bored through the jaw-bone, and a string is passed through the hole and fastened to the edge of the quiver. The arrows are readily sharpened by placing the points between any two teeth and drawing them rapidly through the edges. There are now before me several of these arrows, kindly given me by Mr. Waterton, and which have been sharpened by this process.

IN a neighboring family is placed a very remarkable fish, called the LUMINOUS SCOPELUS (*Scópelus stellátus*).

THE fish which is represented in the illustration on page 285, may fairly take rank as one of the oddities of the finny race.

Flat-headed, round-bodied, and strong-scaled, with projecting eyes of most remarkable formation, the STAR-GAZER has long attracted the attention of naturalists, and given the anatomical investigator much trouble in unravelling the intricate mechanism of its eyes. At a first glance, the fish appears to possess four distinct eyes, each of these organs being divided across the middle, and apparently separated into two distinct portions. In fact, an opaque band runs transversely across the cornea of the eye, and the iris, or colored portion, sends out two processes which meet each other under the transverse band of the cornea, so that the fish appears to possess even a double pupil. Still, on closer investigation, the connection between the divisions of the pupil are apparent, and can readily be seen in the young fish. The lens is shaped something like a jargonelle pear, and is so arranged that its broad extremity is placed under the large segment of the cornea.

The Star-gazer is a native of Surinam, and is one of the viviparous fish. Three species of this genus are now known to naturalists.

THOUGH not so brightly spotted as the trout, nor so desperately active when hooked, and very inferior in flesh, the CARP is yet in much favor with anglers, on account of its extreme

PIRAYA.—*Serrosalmo piraya.*

cunning, which has earned for the fish the name of Fox of the waters. As the number of fish is so great, and our space so small, it will be needful to compress the descriptions as much as possible, and to omit everything that does not bear directly on the subject.

The Carp is found both in rivers and lakes, and in some places, among which the palaces of France may be mentioned, will often grow to an enormous size, and become absurdly tame, crowding to the bank on the least encouragement, and poking their great snouts out of the water in anxious expectation of the desired food. It is most curious to watch these great creatures swimming lazily along, and to see how completely they have lost the inherent dread of man by the exercise of their reasoning powers, which tell them that the once-feared biped on the bank will do them no harm, but, in all probability, will be the means of indulging their appetite with favorite food.

The Carp is one of the fish that retains its life for a lengthened period even when removed from the water; and if carefully packed in wet moss so as to allow a free circulation of air, will survive even for weeks. Anglers never seem sure of the Carp—taking plenty on one day and none at all for a week afterwards, the fish having been aroused to a sense of their danger, and declining to meddle with anything that looks as if it might hide a hook. Even the net, that is so effectual with most fish, is often useless against the ready wiles of the Carp, which will sometimes bury itself in the mud as the ground line approaches, so as to allow the net to pass over it; or, if the ground be too hard for such a manœuvre, will shoot boldly from the bottom of the water, leap over the upper edge of the net, and so escape into the water beyond.

A fine Carp, say of six or seven pounds, is a truly handsome fish, its large shining scales lying on its body in the most beautiful regularity, and gleaming with olive-brown washed with gold. The abdomen is white, with a tinge of yellow.

The beautiful GOLD-FISH (*Cyprínus aurátus*), so familiar as a pet and so elegant as it moves round the glass globe in which it is usually kept, is another member of this large and important genus. It seems to have been brought from China, and has almost acclimatized itself to the cold seasons of some countries. Its habits and splendid clothing are too well known to need description.

Another well-known member of the same genus is the BARBEL, a fine but not brilliant fish, which is common in many of the European rivers.

STAR-GAZER.—*Anableps tetrophthalmus.*

This fish may easily be known, from the four fleshy appendages, called beards or barbules, which hang from the head, two being placed on the nose and the other two at each angle of the mouth. It is one of the mud-loving fish, grubbing with its nose in the soft banks for the purpose of unearthing the aquatic larvæ of various insects which make their home in such places, and being, in all probability, aided by its barbules in its search after food. The Barbel is sometimes so deeply occupied in rooting about the bank, that an accomplished swimmer may dive to the bed of the river, feel for the Barbel along the banks, and bring it to the surface in his bare hand.

THE TENCH prefers the slowest and muddiest rivers, and thriving well in ponds and lakes, or even clay pits. No water, indeed, seems to be too thick, muddy, or even fetid, for the Tench to inhabit, and it is rather curious that in such cases, even where the fishermen could scarcely endure the stench of the mud adhering to their nets, the fish were larger sized and of remarkably sweet flavor.

In the winter months the Tench is said to bury itself in the mud, and there to remain, in a semi-torpid condition, until the succeeding spring calls it again to life and action. The color of the Tench is greenish-olive, darker above than below, and with a fine golden wash.

THE BREAM is mostly found in rather large lakes or in slowly running rivers. Although the flesh of the Bream is not held in any great estimation, being poorly flavored and very full of bones, so that, in spite of the great depth of its body, there is scarcely sufficient flesh to repay the trouble of cooking, still, the fish was formerly in much repute as a delicacy; so that either the fish seems to have deteriorated, or the present generation to have become more fastidious. Spring and autumn furnish the best Bream, and the flesh can be dried something like that of the cod-fish.

The color of the Bream is yellowish-white, except the cheeks and gill-covers, which have a silvery lustre without any tinge of yellow. Sometimes the Bream attains a considerable size, reaching a weight of twelve or fourteen pounds.

LING.—*Lota vulgaris.* SLY SILURUS.—*Silurus glanis.* (One-quarter natural size.)

THE ROACH is a fish especially dear to scientific anglers, on account of its capricious habits, and the delicate skill required to form a successful Roach-fisher.

An angler accomplished in this art will catch Roach where no one without special experience would have a chance of a bite, and will succeed in his beloved sport through almost every season of the year, the winter months being the favorites. So capricious are these fish, and so sensitive to the least change of weather, that a single hour will suffice to put them off their feed, and the angler may be suddenly checked in the midst of his sport by an adverse breeze or change in the temperature.

The Roach is a gregarious fish, swimming in shoals, and keeping tolerably close to each other. It is not a large species, all over a pound being considered as fine specimens, and any that weigh more than two pounds are thought rare. It is a pretty fish, the upper parts of the head and body being grayish-green glossed with blue, the abdomen silvery-white, and the sides passing gradually into white from the darker colors of the back. The pectoral, ventral, and anal fins are bright red, the former having a tinge of yellow, and the dorsal and tail fins are brownish-red.

CLOSELY allied to the roach is the DACE (*Leuciscus vulgaris*), a common and small species that inhabits most English streams. The well-known CHUB (*Leuciscus céphalus*) also belongs to this genus, as does the BLEAK (*Leuciscus alburnus*), in many countries called the TAILOR BLAY by the ignorant, from the idea that whenever any other fish, especially the pike, wounds its skin, it immediately seeks the aid of the Bleak, which, by rubbing its body against the wound, causes the torn skin to close. The beautifully white crystalline deposit beneath the scales was much used in the manufacture of artificial pearls, hollow glass beads being washed in the interior with a thin layer of this substance, and then filled with white wax. The scales of the white-bait were also used for the same purpose. The MINNOW (*Leuciscus phoxinus*) is another member of this large genus, and is too well known to need description.

WE now come to another family, selecting as an example a tolerably well-known species. The SLY SILURUS, sometimes called the SHEAT-FISH, is found in many rivers in different parts of the world.

As may be seen by the engraving, it is a curious-looking fish, and is easily recognizable by the six tentacular appendages of its mouth, the two that are situated on the upper lip being of very great length. The precise object of these tentacles is not quite clear, though some persons believe them to be used as decoys, like the fin rays of the fishing frog, and to be employed in enticing unwary fish within reach of the mouth. Dr. Günther has kindly informed me, that he has often seen these fishes at liberty in their native streams, and that they are capable of directing the points of the tentacles towards any object that they seem anxious to examine. It is, therefore, probable, that these curious appendages are employed as organs of touch. It is one of the mud-loving fishes, and has a custom of hiding itself in holes, or nearly burying itself in the soft alluvium of the river's bed.

The flesh of the Silurus is not held in very high estimation, although its flavor is good ; for it is so fat and gelatinous, that it is difficult of digestion, and not to be eaten by persons of small assimilative powers. A kind of coarse isinglass, or very fine glue, is made from the swimming-bladder of this fish. The eggs of the Silurus are not very numerous, in proportion to the size of the adult fish, and are of a greenish color. They are much eaten by the various fish.

The color is dark green above the lateral line, and of a paler tint below it, and a number of spots are scattered over the body without any apparent arrangement. The abdomen is of a yellowish color, and the fins are tinted with blue and yellow. The Silurus sometimes reaches a considerable size, specimens of seven feet in length and weighing from seventy to eighty pounds having been captured.

PLECTOGNATHI.

A VERY curious order of fishes now comes before our notice. These creatures are called Plectognathi, because their jaws are coalescent.

THE remarkable family of the Trunk-fishes, or Sclerodermi, are known by the curious structure of the external surface, which is composed of a series of hard scales forming a continuous bony armor.

In the genus Ostracion, of which the HORNED TRUNK-FISH, or COFFER-FISH, is a good example, the body is either three or four-sided, and covered with a solid coat of mail formed of six-sided plates or scales, and pierced with holes, through which protrude the mouth, the tail, and the fins. The whole of the interior structure is modified in accordance with this external and inflexible cuirass; and on comparing the general form of this creature with that of certain reptiles, the analogy between the Coffer-fish and the tortoise is too close to escape observation. None of these fishes are in request as articles of food, their flesh being small in quantity, and in some species even thought to have a poisonous effect; but the liver is very large, and yields a tolerable supply of oil. All the Coffer-fishes are natives of the tropical seas, and but few species are known.

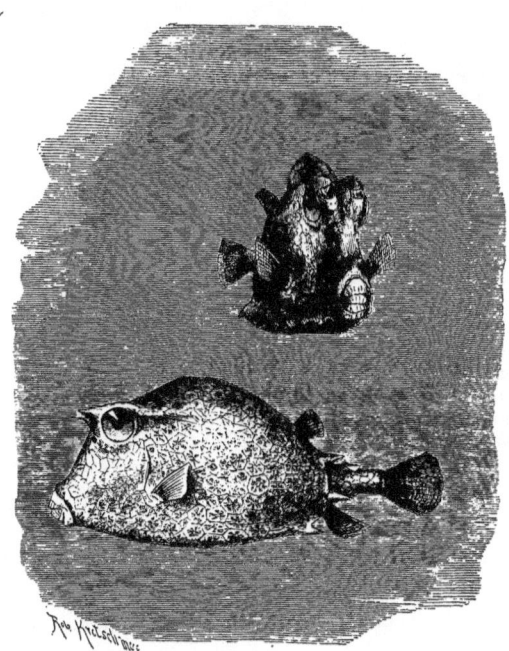

HORNED TRUNK-FISH.—*Ostracion cornutus*. (One-quarter natural size.)

WE come now to a very odd-looking fish, called perforce, for want of a popular title, the OREOSOMA, a name framed from two Greek words, and literally signifying hilly-bodied.

This remarkable little fish was captured in the Atlantic by Peron, and has ever been esteemed as one of the curiosities of the animal kingdom. Upon the body there are no true scales; but their place is supplied by a number of bony or horny protuberances, of a conical shape, and serving no ascertained purpose. These cones may be divided into two distinct sets, the larger set being arranged in two ranks, four on the back and ten on the abdomen, and among them are placed the smaller set. The body of this fish is very deep, in proportion to its length; and the operculum has two ridges, terminating in flattened angles. There are two dorsal fins, the first armed with five spines.

The Trunk-fishes are common objects in the tropical waters of Florida. The Cow-fish is a familiar one. Trigger-fishes, allied forms, are also abundant.

THE very curious TRIGGER-FISH is an example of the moderately large genus *Balistes*, inhabiting the warmer seas of many parts of the world, and which, on account of their rough and fierce exterior, are sometimes called FILE-FISHES, or LEATHER-JACKETS.

The name of Trigger-fish is derived from the peculiar structure of the dorsal fin. When the fin is erected, the first ray, which is very thick and strong, holds its elevated position so firmly, that it cannot be pressed down by any degree of force; but if the second ray be depressed, the first immediately falls down like the hammer of a gun-lock when the trigger is pulled. The mechanical structure of these curious rays is extremely interesting, but the description would occupy too much space to be inserted in this work.

A strong feeling against the flesh of this fish exists among sea-faring men, but, like many other nautical prejudices, is quite without foundation, the flesh being sweet and nutritious, though rather coarse. On the abdomen there is a bony keel, and on each side of the tail there are several rows of horny spines. This species is found in Japanese waters. In color it is one of the most striking of its genus, being tolerably large, and black in color, diversified by some large, pale yellow or white spots upon the sides of the abdomen. Two other species are also given, in order to exhibit the curious variety of form and coloring found in these remarkable fish. The BRISTLY TRIGGER-FISH is notable from the quantity of bristle-like appendages to the tail, while the general appearance of the UNARMED TRIGGER-FISH appears to be smooth all over its body. The name of Balistes is derived from the ancient weapon of war, termed the Balista, which projected a spear or heavy stone with exceeding violence. The curious spine with which the back is armed can be suddenly erected and depressed, as if shot with a spring, and has been compared to the weapon above mentioned. It has been thought, that the flesh of these fishes is poisonous, but the truth of this opinion is very dubious. They are all decorated with bold and sometimes beautiful markings, black, ashen-gray, blue, and yellow being their usual colors.

All the fishes of this genus (which has been divided by some authors into several other genera) are inhabitants of the tropical seas, where they haunt the rocky coasts, and make the ocean radiant with their vivid tintings. To all appearance, they are vegetable-feeders, as nothing but crushed sea-weed has been found in their stomachs.

In the members of the family Gymnodontes, or Naked-toothed fishes, the jaws project from the mouth, and are covered with a kind of ivory or bony substances, composed of very little teeth fused together.

Tile fishes (*Aleutera*) are equally so. Then there are the Puffers (*Tetraodon*), Blowers, and Swell-toad, so called, quite curious to behold as they paddle in the sea, like heavy hulks that have little propelling power. Allied here is the curious Porcupine-fish (*Diodon*), a veritable hystrix in appearance, sometimes reaching three feet in length.

The URCHIN-FISH, or SEA HEDGEHOG, is a good example of the genus Diodon, or Two-toothed fishes; so called because their jaws are not divided, and only exhibit one piece of bony substance above and another below, looking as if the creature only possessed two large teeth.

This curious fish is remarkable for the tremendous array of spiny points which it bears on its skin, and for the power of inflating its body into a globular form, and thus causing the spines to project in every direction, like the quills of an irritated porcupine or a hedgehog that has coiled itself into a ball. From this custom of inflating its prickly body it is sometimes termed the Prickly Globe-fish.

When full-grown, a fine specimen of this fish will measure more than twelve inches in diameter.

The HAIRY URCHIN-FISH is easily recognized by the bristle-like fineness of the spines.

Closely allied to the diodons are the Tetrodons, or Four-toothed fishes, so called because both jaws are divided in the middle, giving them the appearance of possessing four teeth, two above and two below. The spines of these fishes are comparatively small. The Tetrodons are popularly known by the title of BALLOON-FISH, as, like the diodons, they have the power of distending themselves with air, and causing all the spines to erect themselves. When inflated, they necessarily turn on their backs. The STRIPED SPINE-BELLY is a good example of this genus.

BESIDES the tetrodons, this group includes another genus called Triodons, or Three-toothed fishes, the upper jaw being divided into two parts, and the lower remaining entire. The spines are short, and a moderately large sac is seen beneath the body. The POUCHED TRIODON (*Triodon bursárius*) may be selected to represent the genus.

OUR last example of this curious order is the well-known SUN-FISH, which looks just as if the head and shoulders of some very large fish had been abruptly cut off, and a fin supplied to the severed extremity

This odd looking fish can easily be captured. The creature is generally swimming, or rather floating, in so lazy a fashion, that it permits itself to be taken without attempting to escape. In the seas where this fish is generally found, the harpoon is usually employed for its capture, not so much on account of its strength, though a large specimen will sometimes struggle with amazing force and fury, but on account of its great weight, which renders its conveyance into a boat a matter of some little difficulty.

The Mola, or Sun-fish (*Mola rotunda*), Head-fish, so called, as it seems all head. Mola is Latin for mill-stone, and the fish is not unlike a mill-stone in appearance. This is a pelagic, oceanic fish, often seen during summer off our shores. DeKay calls it Short Head-fish. It was known to naturalists of Europe. Its weight reaches five hundred pounds. Its side view presents what would be called a tolerable outline for the head and part of shoulders of a large shark, the posterior part appearing to have been cut perpendicularly through. A thin, narrow fin borders this part, representing a tail, but seemingly of no possible use on such a great, unwieldy creature. Its two great dorsal and ventral fins probably serve it; but curiously, this fish is, as far as we know, never seen swimming upright. It is usually seen lying on its side on the surface of the ocean. When closely approached, it awkwardly sinks out of sight. Rev. Mr. Wood states that the flesh is in good repute among sailors. It certainly is not on this side the Atlantic, as it is like gristle. Boys use it for balls, and it proves to be quite elastic. A large example, captured in Florida and exhibited in New York Aquarium, measured five feet in length.

CREST-GILLED FISHES; LOPHOBRANCHII.

IN the strange-looking fishes, Pegasus and Sea-Dragon, we have further instances of the inexhaustible variations of form and structure with which this world teems, and which seem to be more plentiful, more bizarre, and more incomprehensible in the ocean than on the earth, in the air, or even in the rivers and other fresh waters of the globe.

The order to which these creatures belong is known under the name of Lophobranchii, or Crest-gilled fishes; so called on account of the form of the gills, which are composed of little round tufts, and nearly hidden by the gill-cover. There is but little flesh upon the bodies of these remarkable fishes, which are protected by a hard, bony armor, which, when examined, is found to be most beautifully constructed, so as to protect the animal and to allow of annual increase of dimensions.

IN the family to which belongs the SEA-DRAGON, the breast is developed in a wonderful manner, being always broader than deep, and in some cases the breadth very much exceeding the depth. The mouth is set under the projecting snout in a manner like that of the sturgeon, and the pectoral fins are extremely large and strong, reminding the observer of the same members in the flying gurnards. This seems to be a rather variable species both in form and color.

THE PEGASUS does not possess pectoral fins of such great size as the preceding species, but is yet a very remarkable fish. It is rather long-bodied, and the tail is composed of twelve rings. The much elongated snout is flat and thin, and is furnished on its upper edge with short spines directed backwards. Its color is yellowish-brown. The Pegasus is a Javanese creature.

THE family of the Syngnathidæ is represented by several species, two of which are seen in the accompanying illustration.

The SEA-HORSE is common in many European seas. In all these fishes there is only one dorsal fin, set far back, and capable of being moved in a marvellous fashion, that reminds the observer of a screw-propeller, and evidently answers a similar purpose. The tail of the Sea-Horse, stiff as it appears to be in dried specimens, is, during the life of the creature, almost as flexible as an elephant's proboscis, and is employed as a prehensile organ, whereby its owner may be attached to any fixed object. The specimens represented in the engraving are shown in the attitude which the creatures are fond of assuming. The head of the Sea-Horse is wonderfully like that of the quadruped from which it takes its name, and the resemblance is increased by two apparent ears that project pertly from the sides of the neck. These organs, are, however, fins, and when the fish is in an active mood, are moved with considerable rapidity. It is rather a remarkable fact, that the Sea-Horse, like the chameleon, possesses the power of moving either eye at will, quite independently of the other, and therefore must be gifted with some curious modification in the sense of sight, which enables it to direct its gaze to different objects without confusing its vision.

The color of this interesting little fish is light ashen-brown, relieved with slight dashes of blue on different parts of the body, and in certain lights gleaming with beautiful iridiscent

GREAT PIPE OR BILL-FISH.—*Syngnathus acus.* SEA-HORSE.—*Hippocampus antiquorum.* (One-half natural size.)

hues that play over its body with a changeful lustre. About twenty species of Sea-Horses are known.

In the seas of the Southern Hemisphere, especially in the New Holland waters, there is found sometimes, as a companion of the Sea-Horse, sometimes alone, the HORSE-LIKE PHYLLOPTERYX, a fish which, for its extraordinarily odd aspect, we consider worthy of illustration. This fish, in which are united all the exclusive peculiarities of the family Syngnathidæ, forms a separate genus. As may be seen by reference to the engraving, it is distinguished by many spines, elongated thorns, and tape-like appendages, which float down from all parts of the body. The spines are strong and sharp, the elongated thorns being stiff, while the tape-like streamers are flexible. These three different kinds of appendages take seemingly the function of fins, which, with the exception of the large dorsal fin, and of the small and not clearly visible pectoral fin, are crippled. Its streaming filaments resemble plants, forming a protection to the fish as it floats among algæ; its general appearance like some floating object covered by leaves of seaweeds. Few examples of protective resemblance are so very apparent, for here the beauty of form so often noticed in fins of fishes, is sacrificed to the more practical and useful imitation of straggling weeds. The male of this sea-horse receives its eggs in a pouch on its ventral surface. When they hatch they press the pouch against some hard substance, which forces them out. The Pipe-fish—of this group—also mimics to a certain extent weeds or floating sticks. The male receives the eggs from the female, and carries them in a pouch. In one species found in the Indian Ocean, the female carries its young in a pouch formed by the two ventral fins held together by filaments which extend from its sides. The figure given in the illustration is of natural size.

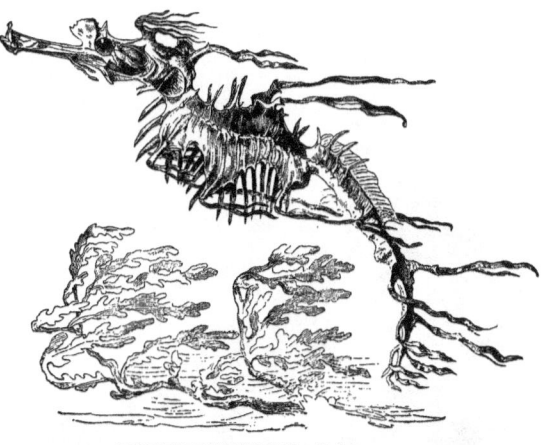

HORSE-LIKE PHYLLOPTERYX.—*Phyllopteryx eques.*

IN the illustration on page 291, is also shown the GREAT PIPE-FISH, which is often called the BILL-FISH and NEEDLE-FISH. It is one of the commonest species of its genus.

This creature is found along the English shores, and can mostly be captured at low water among the seaweed that has been left in the rock pools. To watch these remarkable fishes is an interesting occupation, for they assume such odd attitudes and perform such curious movements, that they never fail to arrest the attention, and never tire the observer. Sometimes they may be seen swimming about with tolerable speed like other fishes, their curious dorsal fins working like an Archimedean screw, and their long snouts being poked into every crevice. Sometimes, assuming a perpendicular attitude, they put their noses to the ground, and hold their tails aloft, while with their beak-like snouts they stir the sand, or, by ejecting water from their mouths, blow little hollows in it, probably for the purpose of disturbing the minute crustaceans and other marine creatures that find refuge in such localities.

The color of the Great Pipe-fish is pale brown, diversified with transverse bars of a dark tint. The average length of a fine specimen is about eighteen inches, but it is said that the fish sometimes attains a length of two or even three feet.

Several species of this genus inhabit European waters, a rather curious example being the SNAKE PIPE-FISH (*Syngnathus anguineus*). This little fish is remarkably slender, and altogether snake-like in form, its length being about fourteen inches, and its thickness scarcely exceeding that of a common goose-quill. The dorsal fin is set very far forward. The tail fin is very tiny, and might easily escape observation altogether.

THE rather quaint-looking species which is represented in the accompanying illustration, is a good example of a remarkable order of fishes, where the body is covered with hard bony scales that do not overlap each other, but are arranged side by side, like the tiles of a pavement, or the cubes of mosaic work. This bony armor is very hard and smooth externally, being covered with a thin layer of a kind of enamel.

Although popularly called BONY PIKE, from the mailed exterior and the lengthened wide-jawed form, which has some resemblance to that of a pike, this fish belongs to a totally different order, and in most points of its construction is formed after a different fashion. The general structure, indeed, of the Bony Pike is very remarkable, and affords another instance of the difficulty with which the fish are classed. The body is elongated, and the jaws are also lengthened and well furnished with teeth, looking very like an exaggerated pike's mouth, or the head of the common gavial of the Ganges. In each jaw there is a single row of sharp and conical teeth, and between them, and on the palate, are numerous other teeth, much smaller in size.

The scales of the Bony Pike are rhombic in form, very like the flat porcelain tiles with which certain ancient chimney-pieces were wont to be decorated, and hardly inferior to those tiles in the polished hardness of their exterior. They are very regularly arranged, being set so as to form a series of oblique rows, extending from the back to the abdomen. As in the sturgeons and sharks, the vertebral column runs along the upper edge of the tail fin. This fish is found in the lakes of America, and sometimes attains a considerable size, being

BONY PIKE.—*Lepidosteus osseus.*

often captured measuring three or four feet in length, and is said sometimes to attain a length of seven feet. Several species are said to inhabit the same waters; but when the remarkable diversity of form and color which often reigns among the fishes is considered, it is highly probable that the supposed species may be nothing more than well-marked varieties. The flesh of the Bony Pike is said to be good.

Bony Pike, Gar Pikes (*Lepidosteus*). Two species of this genus are common in the Great Lakes and rivers of America. Their alliance with forms now extinct renders the species of great interest. Very few are now existing.

THE well-known LAMPREY and its kin are remarkable for the wonderful resemblance which their mouths bear to that of a leech.

They are all long-bodied snake-like fish, and possess a singular apparatus of adhesion, which acts on the same principle as the disc of the sucking-fish, or the ventral fins of the goby, though it is set on a different part of the body. If all had their rights, indeed, the title of sucking-fish ought more correctly to be applied to the Lamprey than to the creature which is at present dignified by that appellation; as the one really applies its mouth to any object to which it desires to adhere, and forms a vacuum by suction, whereas the sucking-fish attains the same object by pressing the edges of the disc against the moving object to which it wishes to attach itself, and forms the needful vacuum by the movement of the bony laminæ.

Several fishes are popularly known by the name of Lamprey, but the only one to which the title ought properly to be given is the larger of the two species in the engraving.

The Lamprey is a sea-going fish, passing most of its time in the ocean, but ascending the rivers for the purpose of spawning. April and May are the months in which this fish is usually seen to enter the rivers; in northern countries the time is postponed according to the climate. In Scotland, for example, the usual month for spawning is June, and, as a general rule, the latter end of spring and the spawning of the Lamprey are synchronous.

The flesh of the Lamprey is peculiarly excellent, though practically unknown to the people. Though it spends so much of its time in the sea, it is seldom captured except during its visit to the rivers, and even in that case is only in good condition during part of its sojourn.

LAMPREY, LAMPERN, AND SAND PRIDE.—*Petromyzon marinus, fluviatilis, and Planeri.*

Practically, therefore, the Lamprey is less persecuted than most of the finny tribe who are unfortunate enough to possess well-flavored flesh, and whose excellencies are publicly known.

Lamprey Eel. The meaning of the generic name refers to a habit of the fish to suck stones and transfer them in such a manner as to protect their spawn, hence Stone-suckers.

In the Merrimack and Connecticut Rivers the Lampreys are taken for food, and much esteemed as such. During the spawning season they ascend the rivers a little before the shad, moving mostly in the night. They are often seen conveying stones, male and female both working. The number and size of these stones are astonishing. Mr. C. F. Holder informs the editor of this edition that he has lately seen in the St. Lawrence River a pile of stones of considerable size, which, collectively, measured nearly four feet in diameter, and about two feet and a half deep, that were transported for the purpose of protecting their spawn, by several of a species of "Stone-toter," the *Semotilus bullaris*, or Chub. In the spring the Lampreys are taken above Albany.

The lesser figure on page 294 represents the LAMPERN, called in some counties of England the LAMPREEN, with that curious faculty of transposition which induces the rustic to speak of thursting instead of thrusting, and to call birds' nests, brids' neesuns.

The Lampern is plentiful in many rivers, and if the generality of residents near the water were only aware of its excellence for the table, would soon be thinned in numbers. The prejudice that exists against the eel and the lamprey is absolutely mild when compared with the horror with which the Lampern is contemplated. Not only do the ignorant people refuse to eat it, but they believe it to be actually poisonous, and would sooner handle an angry viper than a poor harmless Lampern.

The flesh of the Lampern is remarkably good, and is indeed admired by many who have not the least idea of the fish they are eating.

A beautiful adaptation of structure to circumstances is seen in the POUCHED LAMPREY, an inhabitant of the fresh waters of Southern Australia.

This remarkable fish possesses many points of interest, among which the enormous throat-pouch is the most conspicuous. In the common sea-Lamprey of Europe, the throat is dilatable below, but in the present species the skin is distended so greatly as to form a large pouch, such as is represented in the engraving. Taking into consideration the frequent droughts that take place in the country where this creature lives, it is almost certain that the pouch is intended to hold a supply of water, which will enable respiration to be carried on during the days of drought.

The mouth of this species is very large and filled with formidable teeth, the whole interior of the disc being studded with them, and the tongue armed with two long and sharp fangs that seem calculated to do good service to their owner.

A very remarkable species of lampern, termed the SAND PRIDE, or MUD LAMPREY (*Ammocœtes branchiális*), is found in many rivers, and has sometimes been mistaken for the young of the sea-lamprey. It is represented in the lowest figure of the picture on page 294.

It may, however, be distinguished by the form of the mouth, which is of a horse-shoe shape, and incapable of adhesion like that of the fishes belonging to the preceding genus. The Sand Pride, although tolerably common, is not very often seen, owing to its habit of burying itself in the muddy or sandy beds of rivers. Its color is yellowish-brown, the latter tint preponderating on the back and the former beneath. It is a very small species, seldom exceeding six or seven inches in length.

THE MYXINE, or GLUTINOUS HAG-FISH is so remarkably worm-like in its form and general appearance that it was classed with the annelids by several authors, and was only placed in its proper position among the fishes after careful dissection.

The Myxine is seldom taken when at large in the sea, but is captured while engaged in devouring the bodies of other fish, to which it is a fearful enemy in spite of its innocuous appearance. It has a custom of getting inside the cod and similar fishes, and entirely consuming the interior, leaving only the skin and the skeleton remaining. The fishermen have good reason to detest the Myxine, for it takes advantage of the helpless state in which the cod-fish hangs on the hook, makes its way into the interior, and if the fish should happen to be caught at the beginning of a tide, will leave but little flesh on the bones. The cod thus hollowed are technically called "robbed" fish. Six Myxines have been found within the body of a single haddock.

The name of Glutinous Hag-fish is derived from the enormous amount of mucous secretion which the Myxine has the power of pouring, from a double row of apertures, set along the whole of the under surface, from the head to the tail. It is said that the fish is accustomed to envelop itself in a cloud of glutinous matter whenever it is alarmed, and under cover of this substance to escape the view of its enemies; thus presenting a curious parallel to the well-known habit of the sepia and the aplysia, or sea-hare, which, when startled, stain the water with their protective secretion and shoot off under shelter of the sudden darkness.

Around the lips of the Myxine are eight delicate barbules, which are evidently intended as organs of touch; the mouth is furnished with a single hooked tooth upon the palate, serving apparently as an organ of prehension, and the tongue is supplied with a double row of smaller but powerful teeth on each side, acting on the principle of a rasp. The Myxine can scarcely be said to possess any bones, the only indication of a skeleton being the vertebral column, which is nothing more than a cartilaginous tube, through which a probe can be passed in either direction. The structure of the breathing-organs is very remarkable. A double row of branchial cells take the place of gills or lungs, and are supplied with water through a spiracle in the upper part of the head, and two little apertures on the under surface.

The color of the Hag-fish is dark brown above, taking a paler tint on the sides, and grayish-yellow below. Its length is generally about a foot or fifteen inches.

THE last of the fishes is a creature so unfish-like that its real position in the scale of nature was long undecided, and the strange little being has been bandied about between the vertebrate and invertebrate classes. Between these two great armies the LANCELET evidently occupies the neutral ground, its structure partaking with such apparent equality of the characteristics of each class, that it could not be finally referred to its proper rank until it had been submitted to the most careful dissections. In fact, it holds just such a position between the vertebrates and invertebrates as does the lepidosiren between the reptiles and the fishes.

It has no definite brain, at all events it is scarcely better defined than in many of the insect tribe, and is only marked by a rather increased and blunted end of the spinal cord. It has no true heart, the place of that organ being taken by pulsating vessels, and the blood being

LANCELET.—*Amphioxus lanceolatus.*

quite pale. It has no bones, the muscles being merely attached to soft cartilage, and even the spinal cord is not protected by a bony or even horny covering. The body is very transparent, and is covered by a soft delicate skin without any scales. There are no eyes, and no apparent ears, and the mouth is a mere longitudinal fissure under that part of the body which we are compelled, for want of a better term, to call the head, and its orifice is crossed by numerous cirrhi, averaging from twelve to fifteen on each side. Altogether, it really seems to be a less perfect and less developed animal than many of the higher mollusks.

The habits of this remarkable fish are very curious; and it will be better to give the original accounts in the words of the narrators, than to condense or paraphrase them.

The first history of the Lancelet is given by Mr. Couch, who was the first captor of the fish on the North Sea. He saw its transparent tail projecting from beneath a stone on the shore at low tide, and swept it into his hand together with some water. "When alive," writes Mr. Couch, "this fish had a very evident though diaphanous fin, extending from near the snout, round the extremity of the tail, which it encircled in the manner of the same organ in the eel, and terminating at the vent. This specimen was not found in a pool, but lay buried in a small quantity of sand, at about fifty feet from the receding tide; and on turning over a small, flat stone that was on the sand, the tail of the fish appeared exposed.

"When moved, it exhibited signs of great activity, so that the head could not readily be distinguished from the tail; and as there can be no doubt that the fish had sought the shelter of the sand in which it was found, there is little question that mud is its usual habitation; a circumstance still more probable by its want of eyes.

"It was discovered after a heavy storm that had torn it from its native situation, which, from its rarity, we may suppose to be in deep water. Some time later, I obtained two other specimens, which had been thrown up by a tempest. The largest measured two inches and three-tenths in length, which enabled me to discern still more of the internal structure of this fish."

Several other specimens have been obtained, mostly scraped up in the dredge. On the coasts of the Mediterranean this fish is not uncommon; and the following interesting account of some of its habits is given by Mr. Wilde, in his narrative of a voyage to the Madeiras, Teneriffe, and along the shores of the Mediterranean. After describing the general appearance of the fish, he proceeds as follows:—

"These little animals had a power of attaching themselves to each other in a remarkable manner, sometimes clustering together, and at others forming a string six or eight inches long; the whole mass seemed to swim in unison and with great rapidity, going round the vessel in a snake-like form and motion. They adhered to each other by their flat sides, when in line, the head of one coming up about one-third on the body of the one before it; no doubt those sides are of use in forming this attachment.

"The mouth was a circular disc surrounded by cilia that continued in constant motion. When put into a tumbler of water it moved round the glass, and although no eyes were perceptible, it carefully avoided the finger or any substance put in its way, stopping suddenly, or turning aside from it. Both these animals, when taken out of the water, kept up a strong, pulsatory motion for some time. The small one by this means pumped out of its interior a quantity of air and water; and they could be seen coming to the surface to inhale, and a globule of air was observed floating through the internal cavity. In the larger species the internal tube was perfectly distinct and of a blue color. When put into spirits and water it died almost immediately, and turned opaque; a number of circular bands also appeared on it."

The general aspect of the Lancelet is not unlike that of the leptocephalus already described, the delicate, transparent body, and the diagonal arrangement of the muscles causing a considerable resemblance between the two. But the leptocephalus is at once distinguished by its head, which, although very small in proportion to the body, is yet perfect, possessing well-developed eyes, gill-covers, jaws, and teeth, whereas the Lancelet has no particular head, and neither eyes, gill-covers, jaws, nor teeth.

INVERTEBRATE ANIMALS;

INVERTEBRATA.

E now come to the second great division into which all animated beings have been distinguished. All the creatures which we have hitherto examined, however different in form they may be, the ape and the eel being good examples of this external dissimilarity, yet agree in one point, namely, that they possess a spinal cord, protected by vertebræ, and are therefore termed Vertebrated animals.

But with the fishes ends the division of vertebrates, and we now enter upon another vast division in which there is no true brain and no vertebræ. These creatures are classed together under the name of Invertebrated animals; a somewhat insufficient title, as it is based upon a negative and not on a positive principle. Whatever may be its defects, it has been too long received, and is too generally accepted to be disturbed by a new phraseology, and though it be founded on the absence and not the presence of certain structures, it is concise and intelligible.

Numerous as are the species of the vertebrated animals, those of the invertebrates outnumber them as an army outnumbers a company. Although many species of mammals, birds, reptiles, and fishes, are at present known to science, and the yet unrecognized species are necessarily extremely numerous, there is some hope of obtaining an approximate calculation of their respective numbers. But with the invertebrates, any approach to a census even of known forms is well-nigh impracticable; and as it is evident that the ocean alone contains within its fathomless depths myriads of beings as yet hidden from mortal eyes, the reader may conceive the utter impossibility of offering the slightest conjecture respecting their numbers.

SOME EARLY REMINISCENCES BY THE AMERICAN EDITOR.

THE study of invertebrate forms in America is of so recent an occurrence that there are a number now living who remember that, with the exception of the mollusks or "shellfish," forty years since the student had but the merest fragment of recorded knowledge to aid him.

At that time the four great classes of Cuvier were recognized as the legitimate foundations of classification: the two great primary divisions being Vertebrata and Invertebrata—those having an internal bony skeleton, and those having none. At this time, even in the immediate vicinity of the Massachusetts metropolis, he was a wise person, beyond the "general," that had a definite idea of the nature of the very few actinias then known on our coast. The entire amount of knowledge, even with those who recognized them when seen, amounted only to the vague term "animal flowers." That they were animal forms, our few science-reading folks had learned from the science news and gossip that was wafted over from the more scientific centres of England and the continent.

The great branch that embraces the Shell-fish—technically the *Mollusca*—had through various causes become, to a certain extent, familiar. Our clams and oysters were certainly most familiar to us; and the beautiful shells—univalve and others—from their extremely attractive colors and shapes, were sure to be cared for by sailors and visitors of foreign climes, and taken home as interesting mementoes. So with the more common species of our shores. The exterior of these animals, the shells, were favorite objects; and large collections were made by individuals, most of whom were simply collectors who arranged them, more or less in accordance with the simple classification, as Land shells and Fresh-water shells, and as univalves and bivalves.

In the early part of the present century, the French nation associated scientific objects with their expeditions. Napoleon's Egyptian campaign notably resulted in calling forth eminent scientists; and the fine works of these men became important aids to the advancement of natural science.

It was near the middle of this century before the aspiring student in our country could, without difficulty and expense, procure published literature on the invertebrates relating to our own marine forms.

In 1841, Dr. Augustus A. Gould, of Boston, submitted for publication a "Report on the Invertebrata of Massachusetts." This was one of a series of reports of surveys ordered by the State. The matter is contained in an octavo volume, and is very nearly wholly devoted to the Mollusca, or Shell-fish; Dr. Gould being at the time one of the notable American students in that branch. It is now, as it was then, a matter of pride to New Englanders that such an excellent work was produced. The copper-plate figures of each species can never be excelled in fidelity to outline and artistic finish as etchings. This was a gratifying commencement. The author of these introductory lines well remembers the barren field existing at this time. Our school-boy friend, William Stimpson, whose name is enrolled among the pioneers and effective laborers in the various divisions of Invertebrata, was then the enthusiastic learner; joyous to seeming absurdity at sharing our small collection of local marine objects. There lived at this time in Boston an old gentleman, Mr. John Warren, whose occupation was, in the reality, proprietor of a "curiosity shop," but whose tastes and education led him to the study of Mollusca and mineralogy. Primarily, he was a veritable Grandfather Trent, whose commercial requirements were secondary to considerations of taste and science. This was to us a charming resort. The delightful old gentleman was then near eighty years of age. He was of English birth, and impressed us as a courteous gentleman of the old school. He was eminently a handsome man, and, though many years a trader in his wares, he never lost the kindly, friendly manner of dealing with us, that subsists between students embued with the true spirit of science. At eighty, he was our companion—boys, as we were, with few *desiderata* in our minds greater than the speedy approach of every half-holiday. To meet at Mr. Warren's and arrange for a tour of Chelsea Beach, or dredging off Nahant, was the all-absorbing theme during many of these early years. When Mr. Warren could arrange to join us, there was a third party equally joyous, when ranging the beaches was the order, in view of our old friend's infirmities.

While recalling this period, we have before us an old letter-book, in which we have carefully filed away for preservation many choice letters—the volume, now well swollen and embracing the signatures of the most eminent of European and American zoologists—but we recur with sentiments of peculiar nature to the few whose earlier age is suggested by the sealing-wax and the wafer. Such an one bears the signature of William Stimpson, and carries upon its face suggestions of the school-room. These half-holidays of my young friend were coveted periods in the discipline of the Cambridge High School, and most economically were they husbanded. Under date of year 1848, Stimpson writes: "Mr. Warren informs me that you are going to Chelsea Beach on Saturday to collect some of the spoils of the late storm. If you please, I would like to accompany you. . . . I will bring down my dredge and thirty fathoms of line, when we may take a dory and drag around Nahant and Point Shirley."

It is interesting to know that this was some of the first work of the kind done in our country. Some of our Salem friends were also early in the field, most notably Putnam and Dr. Wheatland. In that delightful old neighboring town of seven-gabled houses lived another old gentleman naturalist, Mr. Joseph True, genial and kindly, and possessed of all the virtues and amenities. A visit to his rooms was a treat only next in importance to the Old Curiosity Shop of Mr. Warren. Mr. True was a carver in wood, and his shop stood over a mill-dam, whose fresh and salt waters gave him media close at hand for investigation, but had spent the few minutes daily that he could spare from his work in watching the development and habits of some of our native shell-fish, both marine, fresh water, and terrestrial. Meantime he had collected from our beaches the many species inhabiting the neighborhood.

At this time there were few books to be had on the subjects, and those did not treat of our local objects. Such men as Mr. True, however, observed for themselves, and what they saw was recorded, and known as facts. To the young aspirants these old men, whose knowledge was practical, these valuable associates, "Uncle John" Warren, with his courtly ways and cultured mind, both were delightful companions. Our young friend in an especial manner profited by their teachings and advice. Then came Agassiz, in the homely words of Cuvier, "A pearl from the dirt hills of Neuchatel." Now, a new impetus was given to the study of natural objects. The learned scientist found many collectors in our country, but few investigators. In the glorious school of Cuvier he had been taught to observe. The mere collection arranged and labelled was to him hollowness, meaningless; what to him was the empty shell of the periwinkle, or the impaled carcase of a beetle, so there was no story of its life, nor approximation of its mysteries in death. We were fortunate in residing near his laboratory at Nahant. His frequent presence on the rocky shores and sandy coves and beaches induced a sort of talismanic power, that called up many an unfamiliar form from the vasty deep; and, seemingly, all then

> "Did suffer a sea change
> Into something rich and strange."

The wild rocky promontory is strongly suggestive of the abode of a Prospero, and an Ariel, and the songs of the sea-nymphs,

> Come unto these yellow sands,
> Where the wild waves whyst,"

are readily conjured up. Surprisingly meagre was our knowledge of the indigenous marine invertebrates, but how rapidly did those strange forms come to light at the master's bidding.

A permanent establishment at Cambridge, which afterwards grew into the Museum of Comparative Zoology, offered a place to work up the collections made on the shores during the warmer season.

Stimpson was now an earnest student with Agassiz, one of the first who entered his laboratory as such. His capacity for the study so impressed the master that he selected him as eminently fitted for an investigator, and recommended him, young as he was, as chief naturalist for the Wilkes United States Expedition Around the World. In a letter written at that time Stimpson says, most naively: "I have just been appointed . . . . I shall not be able to go out to see you and bid you good-bye as I would like to do, but I will be back soon, in three years at least." Suggestive of the school-boy yet, he was learned, and eminently capable to enter on the important duties before him. On returning, Stimpson found ample occupation in arranging and classifying the results of his collecting in various portions of the globe.

He now visited Grand Menan, and published a valuable work, embracing the invertebrates of that region. His "Revision of the Synonomy of Gould's Invertebrata of Massachusetts" contained valuable new matter, by pen and pencil. The invertebrates of our coast were now in fair way of being brought to light, and treated with something of the scientific as well as popular care that was accorded such forms in Europe.

An interesting period in the history of our searchings for the invertebrates was when dredging had not been practised in deep water; when the deep sea forms we rescued from the "Maw and gulf of the ravined salt sea-shark;" when the cod and the haddock, by courtesy of the kindly fishermen of Swampscott, yielded each their intestinal contents; when the huge halibut, from the greater depth, responded to the call for pelagic forms; when, too, the diabolic cat-fish, yclept in systematic terms the *Anarrhichas*, came fresh from the rocky beds where some rare mollusk feeds; then were days of enthusiastic working, dampened never a bit, though the odors were never so rank, and the short focus of our near-sighted friend a seeming obstacle. Putnam, of Salem, was now working among the fishes, and Agassiz's museum was rapidly becoming a busy laboratory. Comparative anatomy and some branches of zoology were almost uncultivated in America before this period. Books were not to be had. What little was known was the scattered results of a few foreign expeditions that touched our shores, recorded in European publications. What a commentary on all this do we now witness—the vast amount of published matter of the Smithsonian, as well as that of other institutions. The archives of the Fish Commission now abound in valuable records. Method was now fairly introduced. Agassiz had established his great storehouse and laboratories. Students came to him, each with an especial theme, perhaps, but all to begin labor in the one proper way, with scalpel and lens in hand. The names of those who now would respond with credit to the roll-call of science are so numerous, and the honors are so evenly divided, it would be quite invidious to enumerate any portion.

The Fish Commission, with its admirable appliances, gathered to its organization many promising investigators. Princeton and Johns Hopkins have their schools of biology; and the School of Science at Martha's Vineyard exhibits the increased attention given to natural history. Hyatt, of the Boston Society, has a flourishing school of biology at Cape Ann.

During the pre-Agassiz period, we have seen, little work was accomplished in biology and comparative anatomy. The various State Reports were among the first records of valuable work done in various branches. In Massachusetts the first geological report of Dr. Hitchcock contained catalogues of the birds and mammals then known to inhabit the State. Meagre lists they were, accompanied by no notes of observation. So, also, of the fishes, by Dr. J. V. C. Smith, in the same volume.

A second series, published in 1839-40, was a welcome gift to American science. The large volume on geology, by Hitchcock, was issued separately, and the other subjects were treated in separate volumes.

Dr. Gould's "Report on the Invertebrata of Massachusetts" was immensely creditable. The mollusks occupied the greater part of the volume, as very little research had been bestowed on the other forms by any one in America. The entire matter devoted to "Annelids, Radiates, Tunicates, and Crustacea" was contained within thirty-one pages. The first subject occupied two pages, the second ten pages, and the remainder devoted to a treatise on noxious animals and to crustacea. We have elsewhere spoken of the excellence of the copperplate etching. Binney has since republished this work, with some additional colored plates.

Dr. Storer embodied the results of his investigations in a companion report, issued by the State. This was equally valuable, and good in execution. Afterwards his "Fishes of North America" was issued. Then appeared the fine Reports of the State of New York, Dr. DeKay being prominently identified with some of them. The labors of Mitchell, Binney, Say, and others were recorded here.

The Philadelphia Academy of Sciences was the resort of men of science, Say, Leseur, and Bonaparte, Harlan and Le Conte, Wilson and Audubon, and Sully. Here was a *coterie* of delightful and refined companionship in natural science.

As we unfold the leaves of our old letter-book the honored name of a life-long friend, Spencer F. Baird, occurs. Before the days of the Smithsonian this letter was written, from Carlisle, Penn., to solicit an exchange of a "List of the birds of Carlisle," for a similar "List of the birds of Lynn and vicinity." These are among the earliest local publications. The latter enumerated one hundred and eighty-five species. Putnam had published his observations on the birds noticed in the County of Essex, Mass.

The Lyceum of New York during those days was doing good work. Lawrence, who yet lives, and is honored as one of the most eminent ornithologists, has during his life been identified with the history of our native birds.

Say had published his beautiful work on the insects of North America. Insects and shell-fish were attractive, and greatly, perhaps, on this account there was more interest shown in those divisions. The lower forms "came tardy on." Unfamiliar and plain folks, as it were, naturally less attractive; though after a better acquaintance exhibiting many extremely interesting as well as beautiful members. The sea anemones are among the most exquisite of all Nature's handiwork, both in color and structure, and the numberless microscopic forms now rapidly coming to light challenge our admiration and wonder.

MOLLUSKS.

THE first group of invertebrated animals is called MOLLUSCA. This term is yet retained for this great branch of the animal kingdom, almost meaningless though it be; for it simply expresses one, and a very unimportant, feature of this great group of animal life—that of softness. Probably some naturalist will stop in his work one day, and devote a little time to determining the best and most comprehensive term, one which will express the nature of the division as a whole. The term *Arthropoda*, as applied to the creatures formerly called crustaceans, because they had crusty-like shells, or coverings, seems to be much nearer a natural designation, as the creatures are all jointed or articulated.

Mollusks are bilateral animals, that is, having two equal sides, though in some this is obscured by certain developments, as in the gasteropods. They are either provided with a bivalve or univalve shell, or none, as in the naked Mollusks. The shells are largely composed of carbonate of lime, with more or less animal matter, the whole being secreted by the inclosing mantle, from its outer layer. The shell is entirely without blood vessels. The internal soft parts have a central mouth and digestive tract terminating in an anus, which is primitively at the posterior end, and in the median line. The torsion produced by the growth in the spiral and other irregular shells obscures this, but careful study of the young makes this and many other seemingly difficult points clear.

The nervous system is in the form of ganglia or knots of nerve matter with their connecting nerve lines. The cerebral ganglia, or brain, consists of two knots above the œsophagus, and pairs of knots are distributed around all the important viscera, as presiding centres of nerve power. The heart, situated near what is called the back (dorsally), consists of a ventricle and one or two auricles. It receives the blood from the respiratory organs, and forces it through the body. The whole surface of the body has respiratory functions, but special organs exist in the form of gills, or so-called lungs. As a rule, the sexes are united in the same individual. Numerous marine gasteropods have sexes distinct, and in all of the members of the highest order, the cephalopods, also.

The power of progression in the Mollusks is greatly varied, being in some species almost absent, while in others it is developed to a wonderful degree. Many of these creatures, such as the mussel, the limpet, and the oyster, scarcely stir from the spot where they have once fixed their habitation; the snail and those of a similar form glide slowly along by means of the curiously developed mass of muscular fibres, technically called, from its use, the foot; the scallop drives itself through the water in short jerks or flights, caused by slowly opening and then rapidly shutting its valves; several species are known to jump by a sudden stroke with the foot; the nautilus urges its shell through the waves by the violent expulsion of water from its interior, and is driven along on just the same principle by which a sky-rocket soars into the air; and the flying squid, one of the cuttle-fishes, is able to rival even the flying-fish in its aërial journeys, shooting through the air to considerable heights, and even leaping fairly over both bulwarks of a ship and alighting in the water on the opposite side of the vessel.

The old fable of the nautilus and its sails has long been rejected, but the fabricators of this legend need not have visited the ocean for an example of a molluskan boat. Any one who is in the habit of watching the streamlets that irrigate while they drain our meadow lands, must have seen the common water-snails come floating down the current, lying on their backs, their shells submerged, and the edges of their fleshy foot turned up on all sides so as to convert that organ into a miniature flat boat.

That the Mollusks, or, at all events, some of the species, possess the sense of hearing, is tolerably evident from an examination of the structures. Near the nervous knobs, or ganglia, as they are scientifically termed, of the head, are placed some little vesicles, each filled with a transparent fluid, and containing a tiny knob, or spikelet, of chalky matter, very similar to the well-known ear-bones of fishes, and probably serving a similar purpose. These "bones" appear to be perpetually in motion within their crystalline cell.

The circulation of the Mollusks is tolerably defined, especially in the higher and best-developed species, where the blood is urged on its course by a definite heart, and ramifies through the body by means of well-developed vessels. In the lower forms, however, these vessels can no longer be distinguished, and the blood circulates through a system of little cavities distributed in the body. So completely is this the case, that many Mollusks can be successfully injected, by introducing a fine-nosed syringe at random into the body, and pressing the heated substance very gently into the system.

The movements of the Mollusks are mostly performed by means of the mantle, and through this structure the shell is secreted and molded into form.

Of the secondary services rendered to man by the Mollusks we know but little, owing to the localities in which the greater number of species live, their nocturnal or darkness-loving habits, and their extreme dislike to intrusion. Several species, such as the pholas or burrowing shell, and the teredo, so notorious as the ship-worm, are well known to be actively injurious to man, by destroying the foundations on which his edifices are built, or the vessels in which he trusts his life and property to the waves. Yet even these insidious enemies may have their uses to man, and by destroying the wrecks on which many a noble vessel might be driven and share the same fate, may be the salvation of costly property and invaluable lives.

As to those which are known to be directly useful to mankind, it will be sufficient to give a brief enumeration at present and to mention particulars when we come to the individual species.

Usefulness to man is, when reduced to its essence, the capability of affording him food, and therefore the edible species must take first rank as regards use. The oyster is familiarly known to rich and poor, the latter being often more practically cognizant of its value than the former. It affords at once a refined luxury and a health-giving nutriment; it can be eaten uncooked, or opens a wide field for culinary art; and it has the further advantage of being very plentiful, very cheap, very accessible, and very easily preserved in a living state until needed. The mussel is another largely consumed Mollusk, especially among the lower classes who cannot afford to buy oysters; and in some parts of the world is cultivated and bred in millions, the ever-increasing numbers, together with the peculiar accommodation which they require, threatening to obliterate many a natural harbor, and causing a well-grounded apprehension among ship-owners that their vessels may be deprived of their accustomed refuge by means of this simple Mollusk. The scallop again, with its classically famous shell and coral-red foot, is another of the edible species, as is the cockle, another well-known bivalve. Some of the foreign bivalve Mollusks are considered as very great delicacies, among which the clam takes a very high rank. There is also the huge giant clam, formerly rare, but now perfectly familiar; which has to be cut away from the rock by hatchets, and whose contents are equivalent to a large round of beef, very well flavored, but rather tough and stringy.

The single-valved species furnish many edible examples, such as the whelk and the periwinkle, so largely consumed by the poor, and even the cuttles are capable of affording a tolerably good repast when properly dressed. As a general rule, however, the bivalves are most esteemed, as they are not so fibrous in texture, and therefore not so tough as the univalves.

The shells of the Mollusks are also of much service to mankind. Putting aside the well-known money cowry, perhaps the most infinitesimally divided currency in the world, many species are of exceeding value for the materials furnished by their shelly coverings Some species, where the shell is of that lovely nacreous nature which we popularly term mother-of-pearl, are extensively employed in the manufacture of "pearl" buttons, handles to pocket-knives, ornamental utensils, and in the inlaying of costly furniture; and even pearls themselves, the most precious offspring of the ocean, are composed of the same substance as the nacreous coating of the shell; other shells are largely used in the manufacture of cameos, their alternate coats of creamy white and rich red or warm brown giving beautiful artistic effects when skilfully handled.

In former days, one of the univalve shells, known now as the purpura, and little heeded except by the owners of marine aquaria, was employed for the purpose of producing the celebrated purple of the ancients, which none but the imperial family were permitted to wear.

Shells have in former times been valued at fabulous prices. Collectors were merely such for the pleasure of owning beautiful things, and not for purposes of science; consequently the shells became important objects of commerce.

In 1735 the *Scalaria pretiosa*, which now is sold for about two dollars, then was worth one hundred dollars, and earlier two hundred dollars. Several of the Cypeas were held at one and two hundred dollars. The celebrated Orange Cowrey, in our day, has been held at fifty dollars each. The Cones are proverbially valuable even now. Several species, and the Volutes, have commanded over one hundred dollars each.

The *Argonauta argo* when perfect is a most elegant thing. One in the Boston Natural History Society's Cabinet was purchased for five hundred dollars. Its size is about three inches greater, in diameter, than any other known.

Having now taken a superficial glance at the Mollusks and their uses, we will proceed to the description of individuals, and examine closer into details.

CEPHALOPODA.

The highest of the mollusks are those beings which are classed together under the title of CEPHALOPODA. This is a term derived from two Greek words, the former signifying a head, and the latter a foot, and it is applied to these creatures because the feet, or arms as they might also be called, are arranged in a circular manner around the mouth.

In these animals, which are, as has already been mentioned, thought by many naturalists to be above the mollusks, the organization is highly developed. The nervous system is more like that of the vertebrates than is the case with any other kind of mollusk, the knot of ganglia in the head bearing no small resemblance to a real brain. The Cephalopods breathe by means of a pair of gills or branchiæ, one set on each side of the body, and the circulating system by which the blood is driven through those organs and thence to the remainder of the structures is very complete.

They are all animals of prey, and are furnished with a tremendous apparatus for seizure and destruction. Their long arms are furnished with round, hollow discs, set in rows, each disc being a powerful sucker, and, when applied to any object, retaining its hold with wonderful tenacity. The mode by which the needful vacuum is made is simple in the extreme. The centre of the disc is filled with a soft, fleshy protuberance, which can be withdrawn at the pleasure of the owner. When, therefore, the edges of the disc are applied to an object, and the piston-like centre withdrawn, a partial vacuum is formed, and the disc adheres like a cupping-glass or a boy's leather sucker.

These discs are all under the command of the owner, who can seize any object with an instantaneous grasp, and relax its hold with equal celerity. The arms are almost as movable and as useful to the cuttle-fish as the proboscis to the elephant, for beside answering the purposes which have been mentioned, they are also used as legs and enable the creature to crawl on the ground, the shell being then uppermost.

We will now proceed to a few selected species of Cephalopods, and in the course of describing the several individuals, will examine the curious points of structure which are common to all.

DIBRANCHIATA.

Our first example is the celebrated ARGONAUT, or PAPER NAUTILUS, the latter title being given on account of the extreme thinness and fragility of the shell, which crumbles under a heedless grasp like the shell of an egg, and the former in allusion to the pretty fable which was formerly narrated of its sailing powers. It is rather remarkable, by the way, that the shell of the Argonaut is, during the life of its owner, elastic and yielding, almost as if it were made of thin horn.

Two of the arms of the Argonaut are greatly dilated at their extremities; and it was formerly asserted, and generally believed, that the creature was accustomed to employ these arms as sails, raising them high above the shell, and allowing itself to be driven over the surface by the breeze, while it directed its course by the remaining arms, which were suffered to hang over the edge of the shell into the water and acted like so many oars. In consequence of this belief, the creature was named the Argonaut, in allusion to the old classical fable of the ship *Argo* and her golden freight.

Certainly, the Argo herself could not have carried a more splendid cargo than is borne by the shell of the Argonaut when its inhabitant is living and in its full enjoyment of life and health. The animal, or "poulp" as it is technically called, is indeed a most lovely creature, despite of its unattractive form. "It appeared," writes Mr. Rang, when describing one of these creatures which had been captured alive, "little more than a shapeless mass, but it was a mass of silver with a cloud of spots of the most beautiful rose-color, and a fine dotting of the same, which heightened its beauty. A long semi-circular band of ultramarine-blue, which melted away insensibly, was very decidedly marked at one of its extremities, that is of the keel. A large membrane covered all, and this membrane was the expanded velation of the arms, which so peculiarly characterizes the poulp of the Argonaut.

"The animal was so entirely shut up in its abode, that the head and base of the arms only were a very little raised above the edges of the opening of the shell. On each side of the head a small space was left free, allowing the eyes of the mollusk some scope of vision around, and their sharp and fixed gaze appeared to announce that the animal was watching attentively all that passed around it. The slender arms were folded back from their base, and inserted very deeply round the body of the poulp, in such a manner as to fill in part the empty spaces which the head must naturally leave in the much larger opening of the shell."

Mr. Rang then proceeds to show the real use of the expanded arms, which is to cover the shell on its exterior, and, as has since been definitely proved, to build up its delicate texture and to repair damages, the substance of the shell being secreted by these arms, and by their broad expansions moulded into shape. The expanded extremities of these arms are seen covering nearly the whole shell, and their bases, set with suckers, are bent bridge-like over the rest of the animal. The large eye is seen just protruding out of the shell, the bases of the arms are curved over and behind it, and some clusters of eggs are seen sheltered under the arch of the expanded arms.

The modes of progression employed by the Argonaut are to the full as wondrous as its fabled habits of sailing. Its progression by crawling has already been casually mentioned. While thus engaged, the creature turns itself so as to rest on its head, withdraws its body as far as possible into its shell, and using its arms like legs, creeps slowly but securely along the ground, sometimes affixing its discs to stones or projecting points of rocks for the purpose of hauling itself along.

When, however, it wishes to attain greater speed, and to pass through the wide waters, it makes use of a totally different principle.

As has already been mentioned, the respiration is achieved by the passage of water over the double gills or branchiæ; the water, after it has completed its purpose, being ejected through a moderately long tube, technically called the siphon. The orifice of the siphon is directed towards the head of the animal, and it is by means of this simple apparatus that the act of progression is effected. When the creature desires to dart rapidly through the water, it gathers its six arms into a straight line, so as to afford the slightest possible resistance to the water through which it passes, keeps its velated arms stretched tightly over the shell, and then, by violently ejecting water from the siphon, drives itself, by the reaction, in the opposite direction.

While in the act of swimming, the Argonaut's extremity of the siphon is seen projecting immediately below the eye. If this action forcibly ejects water from the tube, the effect will be to drive the animal rapidly in the contrary direction, *i.e.*, from right to left. An empty shell shows the partially spiral and deeply grooved keel, and an extreme tenuity of the building material.

The animal, or poulp, is very slightly connected with the shell, and, when captured in a net, will sometimes voluntarily leave its home. Many persons have therefore thought that the poulp was not the fabricator of the shell, but only an intruder on the premises of the rightful owner, having taken possession of the shell as a defence for its soft body, just as the well-known soldier or hermit crab arms its soft and unprotected tail with the shell of a whelk, periwinkle, or a trochus. This opinion, ingenious and deserving of examination as it was, has, however, been proved erroneous by a number of experiments, which have shown that the Argonaut is not only the occupier, but the architect of its graceful dwelling, and that the expanded arms are at once the furnishers of the material and the executors of the work.

The precise food of the Argonaut is not ascertained, but Mr. Bennett presumes that, as he always found the globular and translucent, but empty shells of the hyalea, one of the wing-footed marine species, adhering in numbers to the discs on the Argonaut's arms, these creatures must have been captured and devoured by the more powerful mollusk.

As the various cephalopods are so numerous as to preclude all possibility of describing each species, we must content ourselves with a typical form of each family, and a general account of its members.

The species belonging to the family of the Octopodidæ, or Eight-armed Cuttles, possess no external shell like that of the nautilus, its place being taken by two short styles or "pens" in the substance of the mantle. There are eight arms, unequal in length, and furnished with double or single rows of the suckers which have already been described. A good illustration of an Octopus the reader will observe in the right-hand corner of the full-page engraving representing the Red Coral.

They are solitary beings, voracious to a degree, and so active that they find little difficulty in capturing their prey, or in escaping from the attacks of their enemies. Even when pursued into the narrow precincts of a rock pool, the creature is not easily caught. When threatened, or if apprehensive of danger, the Polypus, as the animal was formerly called, darts with arrowy swiftness from one side of the pool to the other, and at the same time so discolors the water with the contents of its ink-bag, that its course is not perceptible, nor, until the water has become clear again, can the precise locality of the creature be discovered. Even if detected, it is not easily captured, as it has a knack of forcing its unprotected body into some crevice, so that no hold can be taken of it, and then affixing itself by its suckers to the rock with such wonderful tenacity that it can hardly be detached as long as life remains.

One example of this family is the WEBBED SEPIA, a very curious animal, found on the coast of Greenland. Its color is violet, and the arms are united by a web almost to their tips. The suckers are set in single rows. Only one species of this genus is known.

In the Octopods the suckers are set directly upon the arms, and the eyes are fixed in their orbits; but in the Decapods, another section of these creatures, the suckers are placed on footstalks, and armed with a bony ring on each. The eyes are movable, and the shell is internal, lying loosely in the mantle. This so-called shell has, however, no real title to the name, being either a spear-shaped body of a horny substance, such as is popularly known by the name of sea-pen, or a curious aggregation of chalky particles, familiar under the title of "cuttle-bone." This "bone" is not attached to the animal by any muscles, but lies loosely in a kind of sac in the mantle, and will drop out if the sac be opened.

Of the order Octopoda, the genus *Stauroteuthis* is a newly discovered one. But a single specimen is known, which was found in the waters about thirty miles east of Cape Sable, Nova Scotia.

The genus *Alloposus* is represented in New England waters by *A. mollis;* a female specimen of which is said to weigh over twenty pounds, and to have a total length of thirty-two inches.

These genera belong to the family which embraces the Argonauts, or Paper Sailors. The latter are familiar to us as tropical species. The United States Fish Commission have dredged about a dozen dead shells a hundred miles off the New England coast.

Octopus vulgaris, of the West Indies, reaches a length of nine feet, and weight of sixty pounds.

Octopus punctatus, of the Pacific coast, reaches a length of radial spread of twenty-eight feet.

There is no evidence of an Octopus having attacked any human being. In habit it is mild and retiring, exhibiting no disposition to lay hold upon anything but its legitimate food, which it finds in abundance on the sea bottom. About fifty species are enumerated, as known throughout the world. The most familiar Octopus to the general reader has been the *O. vulgaris* of the Mediterranean Sea, where it is known as an important edible. During Lent the meat is eaten, and regarded by special Papal indulgence as fish.

Octopus bairdi is an interesting species discovered by the Fish Commission operations off our New England coast. Prof. Verrill, the chief naturalist of the Commission, thus describes its habits, having kept one in confinement: When at rest it remains at the bottom of the vessel, adhering firmly by some of the basal suckers of its arms. While the outer portion of its arms were curled back in different positions, the body was held in nearly a horizontal position, and the eyes were usually half closed and had a sleepy look. When disturbed or in any way excited, the eyes opened more widely, especially at night. It rarely crept about by means of its arms, but would swim briskly. The siphon is used to direct the movements, being bent under the body when it moves forward.

One species of *Eledone*—a genus in which the arms bear a single row of suckers—is found in our waters, but only two specimens have thus far been discovered. In allusion to its warty appearance, Prof. Verrill calls it *E. verrucosa*. There are three species in European waters.

On account of the term Decapoda—ten-footed—being in use also in the class of Crustacea, *Decacera* has been substituted for the next division. The well-known fossil Bellemnites belong to this order. The little Spirula, nautilus-like shell that is abundantly thrown upon the beaches of the Southern States, is another form. These shells have at times been found on Nantucket shores. The animal is not found with the shell. Only three perfect and a few mutilated specimens have ever been found. The United States Coast Survey people dredged one off one of the West India islands in 1878. It came from a depth of nine hundred and fifty fathoms.

The family of the Teuthidæ, popularly known as Calamaries, or Squids, are distinguished by their elongated bodies, their short and broad fins, and the shell or pen which is found in their interior. All the Squids are very active, and some species, called FLYING SQUIDS by sailors, and *Ommastrephes* by systematic naturalists, are able to dash out of the sea and dart to considerable distances. Mr. Beale mentions that he has seen tens of thousands of these animals dart simultaneously out of the water when pursued by dolphins or albacores, and propel themselves through the air for a distance of eighty or a hundred yards. While thus engaged, they have a habit of moving their long tentacles with a rapid, spiral motion, which may possibly help them in their flight, as it undoubtedly does in their propulsion through the water. This peculiar action has been compared by the writer to that of an eight-pronged corkscrew. An interesting account of the Flying Squid may be seen in Bennett's "Whaling Voyage," where it is casually mentioned that these creatures frequently leaped on the deck of the vessel in their daring flight, and sometimes struck themselves violently against the bows, and fell back injured into the sea. This Squid has even been known to fling itself fairly over the ship, and to fall in the water on the other side.

One species of Squid is tolerably common on European coasts, and is often used for bait by European fishermen.

Our present example of this family is the LITTLE SQUID, or SEPIOLA, of which genus six species are known, inhabiting most parts of the world. The specimen from which our illustration was taken was of very large size.

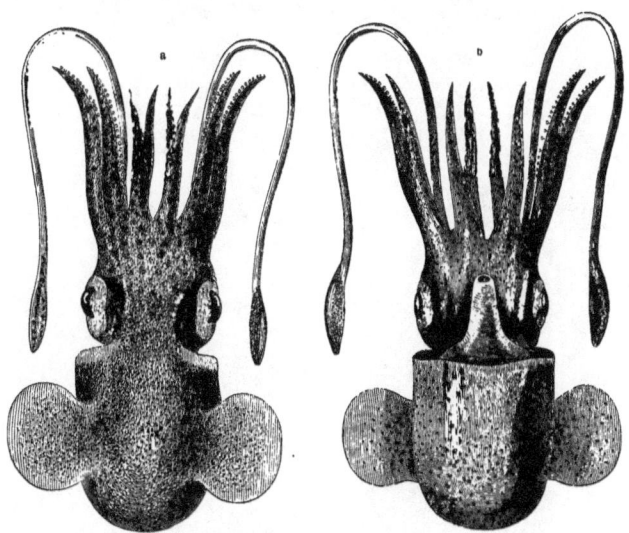

SEPIOLA.—*Sepiola atlantica.* (a, front; b, back.)

One species, the ROCK SQUID, which sometimes attains a large size, may be considered as a formidable antagonist, if irritated.

Squid are not used in America as food, but immense quantities are consumed as fishing-bait.

Family *Sepiolidæ* is represented on our shores by genera *Sepiola*, *Rossia*, and *Heteroteuthis*, the species being rarely seen.

Family *Cranchiidæ* has one genus, *Cranchia*. The body is short and round, with two small fins on the hinder end. The head is small, with large eyes, the corneas of which are perforated, so that the sea-water penetrates to the lenses.

Family *Desmoteuthidæ* is closely allied, having two genera, *Desmoteuthis* and *Taonius*, the bodies of which are longer and pointed posteriorly.

Family *Loligopsidæ*. Forms of this family are longer, and the fins are large; the head very small. Genus *Histioteuthis* is represented by three species only, two of which inhabit the Mediterranean, and the other, *H. collinsii*, the waters off Nova Scotia. One imperfect specimen, and the beaks of two others, are all that have been found. Four other prominent genera are recognized.

The family *Teuthidæ* is characterized by having horny, recurved hooks, in lieu of the suckers on the tentacular arms. These arms have sucking discs by which they are, when necessary, united along their length, leaving the ends to act as forceps in the capture of prey.

ARMED CALAMARY.

A large specimen, called *Moroteuthis robusta*, found in Alaska by Mr. Dall, is allied here. Three mutilated specimens were seen, the largest measuring fourteen feet in length. Certain parts of the structure of the "pen" recalls the Beleminites.

The last family is the *Ommastrephidæ*, in which the body is long and tapers to a point behind. The arms are short, and without hooks, but furnished with two rows of suckers; the tentacular arms are not retractile, but terminate in an expanded club, armed with four rows of suckers. The eyes have lids, and the cornea is perforated so that the salt water penetrates and bathes the lens. The typical genus is *Ommastrephes*, of which one species, *O. illecibrosus*, is the most common Squid north of Cape Cod. Economically, it is an important article; its use for bait by the deep-sea fishermen is very extensive. It swims in large shoals, and is frequently seen following shoals of young mackerel for food. Prof. Verrill once told me of his observing a large specimen while it was making havoc among some fishes. It advances stealthily, says Prof. Verrill, toward the intended victim by undulations of the fins, when it suddenly seizes it by means of the tentacular arms, and kills it by biting the back of the neck with their powerful parrot-bill jaws. As these creatures swim or dart backwards, it is a question how do they so quickly seize upon their prey. In fact, they dart with great swiftness backwards, and then turn obliquely and throw the tentacles to the victim, which close over it like the blades of forceps.

Many quaint old stories are extant concerning these creatures, or imaginary forms called Poulpes. Bishop Pontoppidan, of Norway, is responsible for one notable drawing, which is published in his "History of Norway," and reproduced in numerous later publications. One description of this creature, which stranded in Ireland, was published in 1673, and is as follows: "This monster was taken in Dingle I Cork, in the county of Kerry, being driven up by a great storm; having two heads, one great head out which sprung a little head, two foot or a yard from the great head. With two great eyes, each as big as a pewter dish, the length of it being about nineteen foot, bigger in the body than any horse, having upon the head ten horns, some of six, some of 8 or ten One of eleven foot long, the biggest horns as big as a mans Leg, the least as big as his wrist, which horns it threw from it on both sides, And to it again to defend itself having two of the ten horns plain, and smooth that were the middle and biggest horns. The other eight had one hundred Crowns apeece, placed by two and two on each of them, in all eight hundred Crowns, each Crown having teeth, that tore anything that touched them, by shutting together the sharp teeth, being like the wheels of a watch, the Crowns were as big as a mans thumb, or something bigger. Over this monster's back was a mantle of a bright red color, with a fringe round it, it hung down on both sides like a carpet on a table, falling back on each side, and faced with white—: the Crowns and mantle were glorious to behold. This monster had not one bone about him, nor skin, nor scales, nor feet but had a smooth skin like a man's belly. It swoom by the lappets of the mantle. The little head it could dart forth a yard from the great, and draw it in again at plesure, being like a hawks beak, and having in its little head two tongues, by which it is thought it received all its nourishment. When it was dead and opened the liver wayed 30 pounds."

Any one that has seen the Giant Squids that have recently been discovered, will at once recognize the above as a faithful description of the same.

Whalers have long been in the habit of telling that the sperm whales live on Squids of great size, portions of the latter being often found in the stomachs. Yet it has so chanced that science comes tardily to recognize them.

The first reliable account we have on record is in the year 1873. The jaws of a large Squid were described as taken from the Grand Banks. Since that time, nearly thirty specimens of the species have been seen.

These are referred to three species—*Architeuthis princeps*, *A. harveyi*, and *A. megaptera*. Some five or six species have been described from other parts of the world. Those of our coasts are all from Newfoundland or Grand Banks. The Irish specimen measured thirty-one feet. Measurements of some of the American specimens are as follows: One from the coast of Labrador, which was used for dogs' meat before it could be saved for other purposes, mensured fifty-two feet, the tentacles being thirty-seven, leaving fifteen feet as the length of the body.

Another from the coast of Catalina, in Newfoundland, in 1877, had a head and body nine and a half feet long, and tentacular arms of thirty feet in length. The circumference of the body was seven feet. This was the specimen which was brought to the New York Aquarium, and there exhibited in a large, shallow tank of spirits. We had the pleasure to examine that specimen in company with Prof. Verrill. Such an opportunity had never occurred before, and through the courtesy of Mr. Reiche, the proprietor of the Aquarium, we were allowed all privileges necessary to measure, describe, and sketch the rare creature. Fortunately, it was the best specimen that had been secured, being quite perfect. The body of this creature measured nearly ten feet, as we have seen. What was the astonishment, some years later, to learn of another of twice the dimensions! A body so large, made up of soft flesh, like a gigantic worm, no bones to stiffen it, the only hard part the thin, isinglass "pen," seems to us, as it lies on the shore, surprisingly helpless and out of harmony with its surroundings. Yet, in the vast ocean depths, how well may it not accord,—as the great whale with the same environment. This species of Giant Squid is *Architeuthis princeps*. — *Verrill*.

But few years since, the largest Cephalopod, Cuttle-fish, or Squid, known to naturalists, was scarcely measured by feet. When Victor Hugo wrote "The Toilers of the Sea," his description of the "Devil Fish," a name applied in some countries for a large Squid or Cuttle, was regarded as quite fabulous. The discovery of a portion of an enormous specimen of Squid off the shores of Newfoundland in 1873, revealed the fact that not only were there great species of this form in the deep waters of the North Atlantic, but that the fishermen of Newfoundland have for several years habitually fed their dogs and other animals on fragments of the great creatures that occasionally floated near shore,—always in the shape of dead carcases, no living specimens having been seen until near the present day.

In 1879, the Rev. Mr. Harvey, of Newfoundland, described in the Boston *Traveller* of January 30th, a specimen having the astonishing total length of eighty feet!—the body being twenty feet from the mouth to the point of the tail. He says: "Not far from the locality of the other Devil-fishes (as they are there called), on the second day of November, Stephen Sherring, a fisherman of Thimble Tickle, Notre Dame Bay, observed some bulky object, and as he approached, saw it making desperate efforts to escape. It was aground on the beach, and the tide was ebbing. It was churning the water into foam by the motion of its immense arms and tail. From the funnel in the top of its head it was ejecting large volumes of water, this being its habitual method of moving backward, the force of the stream, by the reaction of the surrounding medium, driving it in the required direction. At times it threw forth its ink, and blackened the sea around it. Its great bulk could not be started by its pumping, and, like a vast hulk, it was hopelessly stranded. At length, as the water receded, and its gills were no longer bathed by the all-needful life-giving medium, it died. Most unfortunately, the fishermen cut the carcase for dogs' meat, but not before reliable measurements were made. As we have seen, it was just twice the size of that hitherto regarded monster of the kind."

The Belemnites, those curious cucumber-like fossils, popularly called Thunderbolts, which are found in various strata, are now known to be the remains of ancient Calamaries, of which the entire animal, with its mantle, fins, ink-bag, siphon, eyes, and tentacles, has been discovered.

Our next example is the common SEPIA, whose wonderful chalky "bone" is so frequently thrown on our shores after the death of the animal in which it was developed.

This so-called bone was formerly in great repute for various purposes, but is now merely employed in the manufacture of pounce and dentifrice, for which latter purpose, however, prepared chalk is quite as effectual, being indeed the same substance, though in the form of powder. It is composed of a vast number of nearly horizontal layers, supported by innumerable little pillars or fibres of the same substance. If one of these shells be snapped across, the structure will be well shown even to the naked eye, while with the help of a common pocket-lens, even the minutest details can be examined. The upper coat will mostly scale off so as to show its smooth surface, while the successive ranges of glittering pillars look like a copy of the Giant's Causeway in minature, as the irregular fracture breaks up their ranked columns into deep caverns and bold projecting rocks. A diagonal cut with a knife will further expose the hard horizontal strata with their myriad pillars; but the method by which the structure exhibits itself in its greatest beauty is to make a very thin transverse section, mount it in Canada balsam on a glass slide for the microscope, and employ polarized light in its examination.

In consequence of its peculiar formation, the cuttle-bone is extremely light when dry, and admits so much air into the interstices that it swims easily in water.

The eggs of the Sepia are dark oval bodies, looking something like a bunch of purple grapes, and from this resemblance termed Sea Grapes by the fishermen. They may often be found on the seashore, flung there by the retiring tide, and left to perish unless rescued by some friendly hand. If these bunches of eggs be placed in a vessel of sea-water, and guarded from danger, they may be seen daily changing in appearance, until at last they burst asunder and let loose the inmates on the world. Nothing can exceed the nonchalant demeanor of the tiny creature not two minutes old. It deliberately makes the tour of its glassy prison, examines every detail with minute attention, and having quite satisfied its curiosity, poises itself for a moment just above the ground, blows out a circular hollow in the sand with a sharp expulsion of water from the siphon, and settles quietly into the bed thus prepared for it.

The family *Sepiadæ* embraces the true Cuttle-fishes. The genus Sepia furnishes the well-known bone and ink of commerce. The flesh is esteemed in European sea-ports. The family *Loliginidæ* includes those forms known to us as Squids. Of the three living genera only one—*Loligo*—is represented on our coast. *L. pealei* is the familiar form, seen on the Cape Cod shores. It reaches a length of about fifteen inches. The species common north of Cape Cod is *Ommastrephes illicebrosa*. A second, *L. brevis*, extends from Virginia to Brazil. *L. galei* inhabits the Gulf of Mexico.

Before proceeding to another large group of cephalopods, it is needful to mention the curious animals called, from the shape of their shell, Spiralidæ. These singular creatures form a distinct though very small family, containing only three species.

In them, the shell is very delicate, and is rolled into a spiral form, something like the proboscis of an elephant when curled up. These shells are very common on the shores of New Zealand, where they are scattered in thousands, and are sometimes thrown on the shores of Europe by the waves of the Gulf Stream. Yet the animal which formed the shell is extremely rare, and is seldom found except in a very fragmentary and battered condition.

TETRABRANCHIATA.

Another order of cephalopods is called by the name of Tetrabranchiata, or Four-gilled animals, because the organs of respiration are composed of four branchiæ. These creatures possess a very strong external shell, which is divided into a series of gradually increasing compartments connected together by a central tube called the siphuncle. As the animal grows, it continues to enlarge its home, so that its age can be inferred from the number of chambers comprising its shell.

In former times these creatures were very abundant, but in our day the only known living representative is the Chambered, or Pearly Nautilus. The spiral home in which the creature resides, and the structure of the chambers, together with their connection by means of the siphuncle, is beautiful.

While the animal still lives, the short tubes that pass through the walls of the chambers are connected by membranous pipes, and even in a specimen that has been long dead, these connecting links hold their places, provided that the shell has not been subjected to severe shocks. In one of these shells now before me, which I have very cautiously opened, the whole series of membranous tubes can be seen in their places, black and shrivelled externally, but perfect tubes nevertheless.

CEPHALOPHORA.

We now take leave of these highly developed mollusks, and pass to other forms where the organization is not nearly so perfect, and where the habits are either so commonplace as to be devoid of general interest, or the animals so shy that they never can be seen performing any act which is likely to attract the attention of an unprofessed naturalist.

It is an enormously large group, containing all the snails, whether terrestrial, aquatic, or marine, the whelks, limpets, and similar animals not so familiarly known. Many species are much used as food, while others are of great service in the arts, furnishing employment to many hundreds of workmen. As the shell of these creatures consists of one piece or valve only, they are sometimes termed univalves, in contradistinction to the oysters, muscles, scallops, and similar shells, which are termed bivalves, in allusion to their double shell.

The larger number of mollusks are divided between the class which embraces the bivalve shells and the present, the Cephalophora, or head-bearers; the former are collectively termed Acephala, or headless. The present class naturally take rank next after the Cephalopods, so-called because the feet are arranged around the head, and both rank higher than the Acephala, the headess, for the reason that they are more like the higher forms of life; that is, they are symmetrical; have heads with a pair of eyes.

The term Gasteropods has heretofore been used to designate this order, from the fact that they crawl upon a flat disc, which was likened to a stomach, hence stomach-footed.

The animals embraced in this order have what is called a lingual ribbon, or tongue, which consists of a band of chitine, a peculiar substance which is characteristic of the skins of insects. This is called an Odontophore, or tooth-bearer. It is attached to the floor of the mouth, and lies free at one end, and bears on its upper surface numbers of hard, tooth-like processes. When in use it is moved by muscles, and drawn over cartilages; a rasping motion brings the hard teeth into contact with any substance taken into the mouth for food.

The mouth of these animals is situated on the under side of the head, and is armed by variously situated jaws or plates of the hard glutinous character. Classification has been greatly aided by the examination of these lingual ribbons. The symmetry of the typical Cephalophoras is lost in the largest number by conforming to the shapes of the external parts—their shells. The cavity of the mouth communicates with an œsophagus, which sometimes dilates and forms a crop, and then the stomach follows, from which the intestine arises.

The circulatory system is well developed; Dentalium being an exception, having no heart. One auricle and one ventricle is usually present. The blood is colorless, the corpuscles of which is nucleated. In both Acephala and the present class, the heart receives the blood from the gills and forces it over the body.

Respiration is by gills or by pulmonary organs, lamellar in form, and by plume-like branchiæ. The nervous system differs in the various groups; ganglia or knots of nervous matter arranged about the anterior parts—around the "swallow," for example—serve the functions of brain. Organs of hearing are present; eyes are generally so, and usually two in number, situated upon the head, or some projecting appendages, called tentacles. The eyes are singularly like those of vertebrates. The sexes are separate in some and in others combined in one individual. Most of these forms lay eggs. In a few the young are produced living, the eggs being hatched within the parent.

Classification of these forms is yet in a most unsettled condition; further study is required for the determination of points of importance. Consequently provisional arrangement is all that the student can look for. This should, of course, be understood, as learners are apt to receive the mischievous idea that classification is fixed.

Our first example of the gasteropods is the BEAKED SPINDLE-SHELL, so called from the rather distant resemblance which its long and pointed form bears to a spindle, and the elongated beak-like process which is seen pointing downwards to the ground as the animal walks along. In the family to which this mollusk belongs, the lip of the shell is always extended and deeply notched.

Another shell belonging to the same family is the THREE-HORNED STROMB.

The Strombs form a large genus, containing about sixty species, and are found in almost every warm sea. They do not appear to be deep-water lovers, being mostly found on the reefs at low water, and seldom extending their range beyond ten fathoms. The operculum of the Strombs is rather curious in its structure, the nucleus being set at one extremity, and the operculum being made up of a succession of horny plates or scales overlapping each other like the tiles of a house, or the successive steel layers of a carriage-spring.

Some species of Strombs attain a considerable size, and are much used in the arts, as, for example, the GIANT STROMB, or FOUNTAIN SHELL (*Strombus gigas*), one of the West Indian species, which sometimes attains the weight of four or five pounds, and is exported to America and Europe by thousands for the use of engravers, who cut the well-known cameos from its beautifully tinted substance. Three hundred thousand of these shells were brought to Liverpool alone in a single year. As the animal increases in age, it gradually fills up the hollow apex and spines with solid matter, and thus materially adds to the weight of the shell without improving its value to the engraver. In some parts of the world, such as Barbadoes, the Giant Stromb is eaten, and sold regularly for that purpose. Pearls of a delicate pink color have sometimes been found in this shell, but their occurrence is not frequent, probably on account of the careless and unobservant habits of the negroes who clean the shells. Pearls are also found in other species belonging to this genus.

The teeth of the Strombs are extremely beautiful and most complicated.

The color of the Three-horned Stromb is brown and yellow of different shades, richly mottled with pale saffron. Its average length is about four inches.

The two species, the COMMON SPIDER-SHELL, and the ORANGE-MOUTHED SPIDER-SHELL, derive their popular names from the generally spider-like contour of their form.

When adult, the outer lip is furnished with several horny appendages, always curved and not precisely of the same shape in the same species, although the general character of their form is sufficiently well marked to distinguish the species. One of these horns is always close to the spine, and is rolled in such a manner as to form a posterior canal. About ten species of these animals are known, and seem to be confined to the Chinese and Indian seas.

The color of the Common Spider-shell is very bright, consisting of boldly mottled chestnut, like the hue of old rosewood, variegated with white, and traversed by lines of orange. The interior of the shell is pale brown, with a dash of yellow. The average length is three or four inches.

The shell of the Orange-mouthed Spider-shell is remarkable for its curious projecting horns, with their sharp points and bold curves. It is worthy of notice that in all the Spider-shells these projections are not developed until the creature has attained adult age, the young Spider-shell resembling that of the stromb. From the peculiar shape of some of the species, these creatures are sometimes called SCORPION-SHELLS.

The color of the Orange-mouthed Spider-shell is creamy-white on the exterior, and rich orange within. The curved spines are white and shining, and bear no small resemblance to the poison-teeth of serpents.

THE shells that are included in the family of the Muricidæ may readily be distinguished by the straight beak or canal in front, and the absence of any such canal behind. The eyes of these animals are not placed on long footstalks, as in the preceding family, but are set directly on the tentacles, without any supporting stalk or projection. All the animals belonging to this family are not only carnivorous, but rapacious, preying on other mollusks, and destroying them with the terrible armature called the tooth-ribbon, and which, when examined with a microscope, proves to be a set of adamantine teeth, sharp-edged and pointed as those of the shark, and cutting their way through the hard shells of their victims as the well-known cordon saw passes through thick blocks of hard wood.

About one hundred and eighty species are known to belong to the typical genus, and there is hardly a portion of the world where a Murex of some kind may not be found.

THE WOODCOCK'S HEAD.

The larger of the two specimens represents the shell which is popularly known under the name of THORNY WOODCOCK, the latter title being given to it, in common with several of its congeners, on account of the long beak, which is thought to bear some resemblance to that of the woodcock, and the former in allusion to the vast number of lengthened spines or thorns which are arranged regularly over its surface. It has also received the equally appropriate and more poetical name of VENUS' COMB.

This shell is found in the Indian Ocean, and varies greatly in dimensions, four or five inches being about the average length. It is evident that as nothing is ever made in vain, or to be wasted, the wonderful array of external spines must play some important part in nature, if not in the economy of the particular species. But what that part may be, and what may be the object of these beautiful structures, is a problem which seems almost insoluble, at all events, with our present means of discovery.

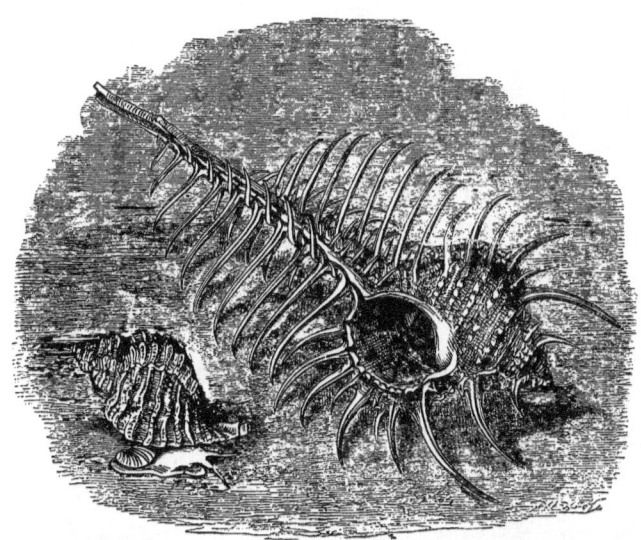

COMMON WOODCOCK-SHELL.—*Murex erinaceus.*   THORNY WOODCOCK.—*Murex tenuispinis.*

The color of the shell is very pale brown, each ridge being slightly tuberculated and edged with white. The spines are uniform drab, or very pale brown, with an almost horny translucence.

Another species is given in the same illustration, in order to show the animal and the position of the eyes, to which reference has already been made. This is the common WOODCOCK, or HEDGEHOG-SHELL. It is very much smaller than the thorny woodcock, and affords a good example of the contrast that can often be effected by different animals which yet belong to the same genus. Its length is hardly more than an inch and a half, and its color is a pale yellowish-brown.

One or two other species belonging to this genus require a passing notice. The WOODCOCK'S HEAD (*Murex haustellum*), remarkable for its long peak and rounded shell, inhabits the same localities as its more beautiful neighbor, being found in the Indian and Chinese seas. It has but few of the spines which decorate the thorny woodcock in such profusion, and even those which are seen upon the surface are comparatively short. The rounded body of the shell, however, together with its long beaked process, does really bear some resemblance to the head and bill of the bird from which it takes its popular name.

The ROYAL MUREX (*Murex regius*) is a very fine example of this genus, and is valued, not only for its rarity, but for the extreme beauty of its form and coloring, which render it an ornament to any cabinet. In color it resembles the thorny woodcock.

THE large empty shell lying in the centre of the engraving represents the SEA TRUMPET, or CONCH-SHELL, so familiar from the use to which it has been put for ages, and which has rendered it a classical appendage to the marine deity whose name it bears.

The Sea Trumpet sometimes attains to a large size, a foot or more in length; and, when it has attained its full dimensions, is employed among the South Sea Islanders and Australians as a trumpet. In order to fit the shell for this purpose, a round hole is bored in the side, at about one-fourth the length from the tip, and the required sound is elicited by laying the shell to the lips, and blowing across the hole as a performer blows the flute. The note—if the noise produced can be called by that name—is hollow and disagreeable; but as it is loud

TWISTED TRITON.—*Triton distortus.*   SEA TRUMPET.—*Triton variegatus.*   WRINKLED TRITON.—*Triton anus.*

and unlike any other sound, it answers the purpose of those who employ it. While blowing the conch, the performer introduces his right hand into the cavity, much in the manner of a player upon the French horn.

Below the Sea Trumpet lies another shell, which would hardly be taken for a Triton until turned over, so as to show the whole of the contour. This is the WRINKLED, or OLD WOMAN TRITON, so called because the corrugated and rudely oval mouth, with its white crumpled folds, is thought to bear some distant resemblance to the face of an old woman surrounded with a close cap. The Wrinkled Triton is comparatively a small species, as may be seen from the proportions preserved in the figure.

Behind the larger figure is seen the TWISTED TRITON, represented in the act of crawling, and given, not so much to exhibit any peculiarity of its shell, which is hidden behind that of the larger species, as to show the form of the animal, its large foot, and eyes placed at the bases of the tentacles. The operculum of this animal is small and leaf-shaped, the nucleus being at one end.

THE FROG-SHELL seems to have been gifted with its popular name on the same principle that caused a well-known dramatic character to detect in a cloud an equal resemblance to a whale and a camel. All the members of this genus possess two rows of ridges, technically called "varices," upon the shell, one row being placed on each side. There are about fifty species of Ranella, spread over all the warm seas. Like the preceding shells, they prefer the shallow to the deep waters, and may be found at almost all depths—from the bare rocks left waterless by the receding tide, to a depth of eighteen or twenty fathoms.

The BULL-FROG SHELL has a roughly tuberculated surface, with deep hollows and bold ridges of thick shelly substance, together with projecting horns on either side. The color of this shell is extremely variable. In the handsomest specimens the ground color is creamy white, largely mottled with bold tints of deepest brown and purest white. But in many instances the entire shell is of a very pale tone, yellow predominating, and the brown entirely subservient, and presenting the same contrast to the full-colored shell as the albino to the negro.

The SPINED FROG-SHELL derives its name from the sharp and rather long spines or projections with which it is furnished. None of these shells are of very great size, their average length being about two inches.

A VERY pretty shell is termed indifferently the LITTLE FIG, or LITTLE PEAR SHELL (*Pyrula ficus*), because its general outline is thought sufficiently pear or fig-like to warrant the application of the name. Both scientific names refer to this far-fetched resemblance, *pyrula* signifying a little pear, and *ficus* meaning a fig.

The foot of the Pyrula is abruptly cut off, or truncated in front, and modified so as to form a short horn or partial crescent at each side.

Nearly forty species of Pear-shell are known to conchologists, and are spread over the warmer seas of the world, living in moderately deep water, varying from sixteen to thirty-five fathoms of depth.

This is a thin and delicate shell, the large expanded lip being especially so, and, in consequence, is very light when the inmate has been removed. The color is very pale yellow, with brown and white arranged in wavy mottlings. Its average length is about four inches.

The delicate thinness of the shell is not, however, a character common to the entire genus, for another species, the BAT-LIKE PEAR-SHELL (*Pyrula carnária*), is quite as remarkable in the opposite direction, its shell being peculiarly large and ponderously constructed. This shell is found in the Indian Ocean, and its general color is dark bay. In all these shells, however, the long canal which projects from the front of the shell is always open, not being filled up with solid matter as the animal increases in age; and the columella, or pillar of shelly substance, which runs up the centre of the whorls, like the solid centre of a screw, is always smooth.

A LARGE and boldly mottled shell, popularly known by the really appropriate name of TULIP WHELK, bears in its rich and variegated coloring some analogy to that of the flower from which it derives its name; while the general shape is sufficiently like that of the whelk to warrant its use, even though the two shells belong to different families. The generic name of this shell is derived from a Latin word signifying a band, and is given to it on account of the boldly banded stripes in which the colors are disposed. As in the last-mentioned genus, the canal, though not so elongated, is always kept open.

Comparatively few living species of Tulip-shells are known to conchologists, sixteen or seventeen being their utmost limit. These shells inhabit the warmer seas, and some of them attain a great size, such as the GREAT TULIP-SHELL, which sometimes reaches a length of nearly two feet.

BEFORE mentioning our last example of the Muricidæ, we have to pay attention to the SPINDLE, or DISTAFF SHELL (*Fusus coius*), so called in allusion to its form. Its scientific names are both given in consequence of its general resemblance to these objects, the former signifying a spindle, and the latter a distaff.

At least a hundred species of Spindle-shells are known, and their range extends over the greater part of the globe. One large species (*Fusus antiquus*), called, from its color, the RED WHELK, is common on European shores, and off some of the coasts of Scotland is extensively captured for sale. When the empty shell is held to the ear, the reverberations of sounds are gathered in its wide lip, and, being returned to the ear in a broken and confused manner, give forth a monotonous sound, rising and falling at intervals, and are thought by the uneducated to be the imprisoned murmurs of the waves. For this reason, the shell is popularly known as ROARING BUCKIE. In some places the empty shell is used as a lamp, the cavity containing the oil and the wick being drawn through the canal, thus producing a charmingly elegant lamp, which even exceeds in beauty the classical forms of the ancients, and quite equals them in efficacy.

Another species, the GIANT SPINDLE (*Fusus colósseus*), is remarkable as being one of the largest living examples of the gasteropods.

The foot of the animal is moderately broad, and the operculum is small, and shaped not unlike a sea mussel-shell. The color of the Spindle-shell is nearly white, and almost uniformly tinted, but darkening slightly towards the point.

WE now arrive at another and rather larger family, of which the common WHELK is a familiar example.

This is one of the most carnivorous of our mollusks, and among the creatures of its own class is as destructive as the lion among the herds of antelopes. Its long tongue, armed with row upon row of curved and sharp-edged teeth, harder than the notches of a file, and keen as the edge of a lancet, is a most irresistible instrument when rightly applied, drilling a circular hole through the thickest shells as easily as a carpenter's centre-bit works its way through a deal board.

The front of the tongue often has its teeth sadly broken, or even wanting altogether, but their place is soon supplied by others, which make their way gradually forward, and are brought successively into use as wanted. As a general rule, there are about a hundred rows of teeth in the Whelk's tongue; each row contains three teeth, and each tooth is deeply cleft into several notches, which practically gives the creature so many additional teeth.

THE sweeping curves, broad swelling lip, and regular ridges, of the next genus of shells, have earned for them the popular title by which they are known.

About nine or ten species belong to this pretty genus, some of which are rare and costly. The IMPERIAL HARP-SHELL is still a valuable shell; but in former days, when the facilities of commerce were far less than at present, it could only be purchased at a most extravagant rate. A small specimen is now valued at from two to five dollars, and a fine one will cost about fifteen dollars; but, in former days, as much as two hundred and fifty dollars have been paid for a specimen. A similar diminution has taken place in the cost of nearly all shells.

The Harp-shells are only found in the hottest seas, and are taken mostly on the shores of the Mauritius, Ceylon, and the Philippine Islands. They frequent the softer and more muddy parts of the coast, and prefer deep to shallow water. None of the Harp-shells possess the operculum.

The color of the Imperial Harp is pale chestnut and white, with a dash of yellow arranged in tolerably regular and slightly spiral bands.

The LITTLE HARP-SHELL is a darker species, and one that seldom attains a greater length than an inch and a half. The peculiar foot is very large, broad, and leaf-shaped, and has a deep fissure just behind the tentacles, nearly cutting the organ asunder. It is said that, when the animal is irritated, the fissure becomes widely expanded. Some writers say that, if the animal is very much terrified, it withdraws itself into its home with such rapidity that the expanded front of the foot is unable to contract sufficiently, so that the fissure is caught against the sharp front edge of the shell, and thus undergoes involuntary amputation.

The general colors are tolerably similar throughout the Harps, but each species always preserves its peculiar individuality. One species, for example, has the spaces between the ridges pencilled in elaborate wavy markings of chocolate on white, and the ribs themselves barred at regular intervals by lines of deep brown; while another, known by the name of VENTRICOSE HARP (*Harpa ventricósa*), has the spaces filled with a succession of arches, one within the other, and of a rich brown color.

A very common shell may often be found on the seashore, looking like a small whelk with a smooth whitish shell, boldly banded with reddish-brown. This is the COMMON PURPLE, or PURPURA (*Púrpura lapillus*), another member of this genus, and worthy of notice as being one of the shells which furnish the celebrated Tyrian purple of the ancients. This color, which, by the way, contains so little blue as to be unlike the tint which we now call by the name of purple, is evidently the analogue of the ink found in the sepia, and is secreted in a little sac by the throat, containing only one small drop.

For the very best dye this material was extracted carefully from the individual shells, but for an inferior kind it was obtained by pounding a quantity of the Purpuræ in a mortar, and straining off the juice, which was thus mixed with the blood and general moisture of the animals, and consequently of less value than the pure dye. So expensive was the dye obtained by this latter process, that a pound of wool stained with it could not be purchased under a sum equalling one hundred and fifty dollars. Any one can try the experiment of dyeing a little strip of linen with the matter obtained from a single shell. After breaking the shell carefully so as not to crush the inhabitant, the cell containing the coloring matter will be seen lying across the head or neck of the animal, and can be removed by opening the sac and taking up the yellowish-white contents with a small camel's-hair brush, or the point of a new quill-pen. When the linen is imbued with this liquid and placed in the rays of the sun, it immediately begins to change its color, and passes through a series of tints with such rapidity that the eye can hardly follow them, unless the slanting rays of the rising or setting sun are chosen for the purpose.

ONE of the strangest, though not the most beautiful, of shells is the MAGILUS, a native of the Red Sea and the Mauritius.

During its stages of development, the Magilus appears once as a small and delicate shell and then as a long, crumpled, and partly spiral tube, with a shell at one end and an opening at the other.

For the purpose, apparently, of carrying out some mysterious object, the Magilus resides wholly in the masses of madrepore, and in its early youth is a thin and delicate shell without anything remarkable about it. As it advances in age, it enlarges in size, as is the case with most creatures; but its growth is confined to one direction, and, instead of enlarging in diameter, it merely increases in length. The cause of the continual addition made to its length is probably to be found in the growth of the madrepore in which it is sheltered, and which would soon inclose the Magilus within its stony walls did not the mollusk provide against such a fate by lengthening its shell and taking up its residence in the mouth.

The most curious point, however, in the economy of the Magilus is, that, as fast as it adds a new shell in front, it fills up the cavity behind with a solid concretion of shelly matter, very hard, and of an almost crystalline structure, so as to leave about the same amount of space as in the original shell. The animal is always to be found in the very front of the shelly tube, and closes the aperture with a strong operculum that effectually shields it against all foes.

The color of the Magilus is whitish. Only one species is known.

In the peculiar formation of the shell there is an evident analogy with the successive chambers formed by the pearly nautilus. In both cases the animal is of small dimensions when compared with the magnitude of its dwelling, and in both cases the creature continually advances forward, taking up its residence in a chamber formed in the front of the shell, and, closing the passage behind in proportion to its advance. The chief difference, however, between the two is, that the Magilus, being a fixed shell and inhabiting a stony tunnel, needs not the delicately structured shell required by the active nautilus, and therefore merely fills up the useless portions of the shell with solid matter, requiring no hollow chambers and no tube of communication.

THE SPOTTED NEEDLE-SHELL, or SPOTTED AUGER, derives its name from the long and sharply pointed form of the shell. More than one hundred species of this genus are known, all inhabitants of the warmer seas, and the greater part resident within the tropics. In all these shells, the aperture is very small and the canal short. The operculum is small and pointed, having the nucleus at the smaller extremity. In many species the animal is entirely blind; and even in those cases where eyes are present, they are very small, and set at the end of the minute tentacles.

The beautiful SPOTTED IVORY-SHELL is also a native of the hotter latitudes.

Few species, not more than eight or nine in number, are known to exist at the present day. They are all very smooth and polished on the exterior, and their substance is so thick

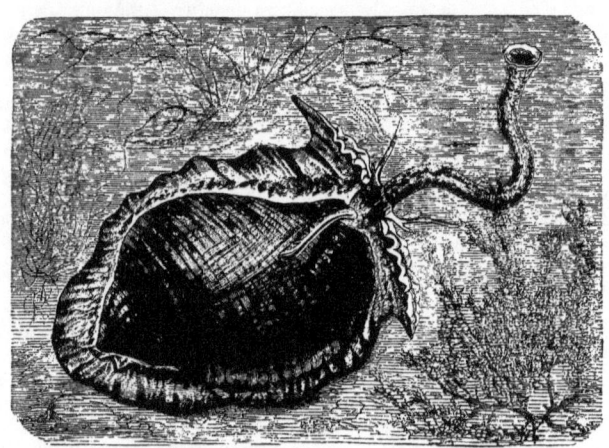

APPLE TUN-SHELL.—*Dolium pomum*.

and solid that they seem almost to be made of earthenware. They reside at a moderate depth, being generally found in twelve or fourteen fathoms of water. It is worthy of notice that the rich spotted markings of the shell are repeated upon the body of the animal. The members of this genus possess tolerably large eyes, set at the base of the long tentacles. As in the preceding genus, the operculum has its nucleus at the pointed end.

The color of the Spotted Ivory-shell is pure porcelain-white, richly spotted with deep brownish-red, something like the tint known to artists as burnt sienna. It is not a very large shell, being about two inches in length.

THE two shells represented in this and in the next illustration belong to the same comprehensive and useful family. The APPLE TUN-SHELL belongs to a moderately strong genus, deriving their popular name from the rounded and barrel-shaped outlines of the shell.

The animal is shown as it appears when crawling, for the purpose of exhibiting the manner in which the siphon is carried bent over the front of the shell, like the uplifted proboscis of an elephant. In these shells the spire is comparatively small and short, and the aperture very large, thus producing a great contrast to the needle-shell. The figure in our illustration is one-eighth the natural size. About fourteen species of Tun-shells are known, all inhabiting the warmer seas.

THE beautiful HELMET-SHELLS are tolerably thick and solid, and their external surface is covered with bold ridges, marking the periodical growth. These ridges are technically called "varices." All the Helmet-shells are natives of the tropical seas, and appear to prefer the shallow waters near the coast. Several of these shells are employed by the engravers in the manufacture of cameos, the differently colored layers producing most exquisite effects when cut by a judicious operator. The colors vary greatly in the different species, and sometimes there is a slight variation even in different individuals belonging to the same species. Cameos, for example, that are cut from the HORNED HELMET-SHELL (*Cassis cornúta*) are white, upon a ground of rich orange; those that are made from the WARTY HELMET-SHELL (*Cassis tuberósa*) are white, on deep dark red; the cameos formed from the shell of the RUDDY HELMET (*Cassis rufa*) are saffron-yellow on warm orange. Another beautiful species, called the QUEEN CONCH (*Cassis madagascariensis*), furnishes a white cameo on a claret-colored ground.

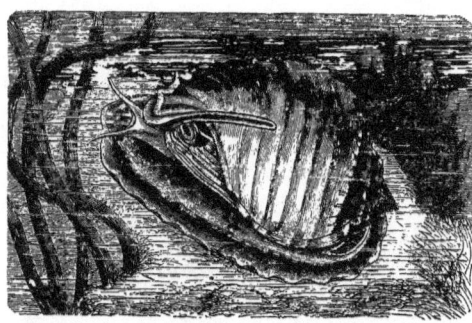

HELMET-SHELL.—*Cassis glauca*. (Small specimen.)

THE next illustration is a dark smooth shell, represented as crawling on the ground, and partially enveloped in the spotted textures of the living creature.

This is the BLACK OLIVE, so called on account of the jetty blackness of its exterior, and the oval, rounded form, which is not unlike that of the fruit whose name it bears. The genus Oliva is a very large one, comprising more than one hundred species, and found in all the warm and tropical seas. As may be seen by the figure, the mantle is furnished with two large lobes, that nearly meet over the back while the animal is moving, and which throw out certain filamentary projections, that look very like tenacles in the wrong place. The foot is very large—so large, indeed, that the shell is partly buried in its soft material—and the eyes are, as may be seen in the figure, placed before the middle of the tenacles.

BLACK OLIVE.—*Oliva mauritana*. (Natural size.)

Owing, probably, to the great development of the foot and mantle, the Olives are active creatures, gliding about with tolerable speed, burying themselves in the sand when the tide leaves the shores on which they are creeping; and if laid upon their backs, they can easily resume their original position by the use of the spreading foot. In spite of their elegant and harmless aspect, the Olives are predaceous and hungry creatures, and can readily be captured by the simple process of tying a piece of meat to a line, lowering it towards the spot where the Olives are creeping, and hauling it up at intervals, carrying with it the various mollusks that have attached themselves to the bait, and do not think of loosening their hold until too late.

The shell of the Black Olive is beautifully polished and of a deep rich black, through which a slight tint of brown can be observed in certain lights. The inside is porcelain-white, and the average length is not quite two inches.

THE next example, the LIGHTNING-COLORED SHELL, or DOVE SHELL, derives its popular name from the peculiar appearance of its markings. This little shell is covered with zigzag white streaks.

ALL the Cones, Cone-shells, or *Conidæ*, a family so called on account of their form, have a similar external outline; the aperture is long and narrow, the head of the living animal is more or less lengthened, the foot is splay and abruptly cut off in front, the tentacles are rather widely separate, and the eyes are placed upon these organs.

The TEXTILE CONE-SHELL is found on the isle of Mauritius. This handsome species is about four or five inches in length, and its markings are curiously disposed, so that it is impossible to say which is the ground color. The dark, narrow, angular lines are dark brown, accompanied by white, and variegated by dashes of yellow umber. The bold triangular spots are pure white, and the inside of the shell is of the same color. The figure shown in the illustration is of natural size.

The ADMIRAL CONE is peculiar for its long and narrow aperture. This species, in common with the other members of the genus, haunts the fissures and holes in rocks, and the warmer pools in coral reefs. They all take a moderate range of depth,

TEXTILE CONE.—*Conus textilis.*

varying from one to forty fathoms. Though slow in their movements, they are extremely voracious, their formidable teeth being well adapted for their predatory habits, and sometimes, it is said, being used effectively upon the hand of their captor. *Conus aulicus* has a bad reputation for such conduct, rather unexpected on the part of a shell-bearing and apparently helpless mollusk.

THE BISHOP'S MITRE, a long, pointed shell with regular spiral markings, belongs to another family, termed the Volutidæ. In these shells the aperture is rather deeply notched in front; the animal has its siphon recurved, and the foot is very large, in some species partly hiding the shell. The eyes are either placed upon the tentacles, or near their base.

The shell of the Bishop's Mitre is spindle-shaped, long-spired, and stout in substance. The proboscis is very long. This mollusk possesses, in common with many others of its class, the capability of protecting itself when alarmed, by the sudden emission of a purplish liquid, having to human nostrils a peculiarly nauseous odor.

The Mitres, etc., are a very numerous genus, about three hundred and fifty living species being known and named. All the large species inhabit the tropics; and although there are some which are found in cooler regions, they are of very small dimensions, and mostly frequent the moderately shallow waters, though a few species are found at a depth of eighty fathoms.

The color of the Bishop's Mitre is very pleasing, being pure, shining white on the background, and the spots being of a rich warm bay, the red predominating.

OF the typical genus of the Volutidæ, which contains about seventy species, and is spread over most of the warm seas, we may describe the Musical and the Bat Volute.

The BAT VOLUTE is remarkable, not only for the bold markings of the shell, but for its own curious form. At each side of the large siphon may be seen a lobe projecting from its base, and the eyes are set on lobes projecting from the base of the tentacles. When the tooth-ribbon of the Volute is examined under the microscope, its armature is seen to consist of a series of three-pointed teeth, forming a very powerful engine of destruction.

The shell called, from its peculiar markings, the MUSICAL VOLUTE, has a series of lines supposed to represent the clefs, the spots doing duty for the notes.

As in the preceding instance, this shell is most variable in the shape and color of its markings, and even the number of lines differs considerably. In this specimen is found the normal number of five lines and four spaces; but in some examples there are only four lines, while in others their number is increased to seven. The color of the shell is a mixture of gray neutral tint and pale brown, the lines being nearly black, and the interior of the shell a very pale drab.

A LARGE, uniformly colored species, called NEPTUNE'S BOAT, is a rather pretty, though simple-looking, shell.

But few species of the genus Cymba are known, nine or ten being their apparent number; and these creatures appear to be found mostly in Western Africa. It has a peculiar form; its oddly-shaped proboscis and recurved siphon giving it a very curious aspect. The foot is of very great size, and deposits a thin enamel on the under side of the shell. When first born, the young animal is of very great size when compared with its shell. The nucleus is large and globular, and in the youth of the animal is sufficiently conspicuous; but, as the inhabitant increases in age, and the home increases in size, the nucleus becomes partly concealed by the growth of the shell, the whorls of which form a flattish ledge around it.

Although not a very large shell, nor remarkable for the variety of its coloring, the Neptune's Boat has yet a pleasing effect to the eye, and, when examined, is really an elegant and delicate shell. Its walls are very thin, in proportion to its dimensions, and the bold, sweeping curves of the surface always call forth admiration. Its color is uniform palish-drab on the exterior, and the inside is beautiful pinky-white, like that of a blush rose.

ON the right hand of the accompanying illustration, and occupying the central portion, the reader will perceive a curious-looking shell represented as crawling upwards, the animal having a very broad and flat foot, and its shell almost covered with the striped mantle. This is the MARGINELLA, our last example of the Volutidæ.

About ninety species of Marginella are known to zoologists, all belonging to the tropical or warm seas. As may be seen by the engraving, the animal is very large in proportion to the size of its home; and the mantle is so formed, that the two lobes almost meet over the back of the shell, nearly concealing it from view. The tentacles are long, and the eyes are placed upon them near their base. The shell is smooth and polished; and when adult, the outer lip has its edge considerably thickened, thus gaining the generic title of Marginella.

The color of the shell is gray, streaked with black lines, and the animal itself is of a pinkish hue, diversified by red rays.

WE now come to the family of the Cowries or Cypræidæ, three representatives of which family are seen in the engraving. As in the last genus, the mantle is expanded into two lobes, which nearly meet over the back of the shell; but in many species these lobes are covered with filaments, like so many tentacles. The eyes are either near the base or middle of the tentacles, and the tooth-ribbon is powerfully armed.

The most familiar example of these shells is the COMMON COWRY, which may be seen on the upper left-hand of the engraving, crawling diagonally upwards, and remarkable for the great length and breadth of the foot, and development of the mantle and tentacles.

The celebrated MONEY COWRY (*Cypræa monéta*) belongs to this genus. These little white shells are well known as being the medium of barter in many parts of Western Africa; and vast multitudes of them are gathered from their home in the Pacific and Eastern seas, and imported into European countries for the purpose of immediate exportation to the African coast.

In the left-hand bottom corner of the engraving may be seen the beautiful PANTHER COWRY, represented as it appears while living, its mantle covered with the curious appendages which look very like the tentacles of the sea anemones. This species derives its name from the rich mottling of the surface. A larger species is called the TIGER COWRY. One of these shells is largely used by the natives of the Sandwich Islands as sinkers for their nets, and a singularly ingenious bait is made from the same shell for the capture of the cuttle-fish.

A number of Cowries are cut into fragments and so fitted together as to form an oval ball of considerable size, with a smooth and mottled surface. Something by way of a tail, or

POACHED EGG.—*Ovulum ovum.*
COMMON COWRY.—*Cypræa europæa.*   WEAVER'S SHUTTLE.—*Ovulum volva.*   MARGINELLA.—*Marginella diádocha.*
PANTHER COWRY.—*Cypræa pantherina.*   WARTY EGG.—*Ovulum verrucôsum.*   DEEP-TOOTHED COWRY.—*Cypræa caurica.*

balance, is fastened to one end of the ball, and the fishing-line tied to the other. The bait is now complete, and is quietly lowered near the spot where the cuttle is known to live, and drawn slowly along the ground. The ever-watchful cuttle is immediately attracted by this novel object, and thinking it to be some hitherto unknown delicacy, darts at it, and arrests its progress by attaching one of its arms to the smooth surface. The fisherman then gives a slight jerk to his line, and the deluded cuttle, fancying that its prey is trying to escape, makes fast another arm. By repeated jerks the cuttle is induced to cling with all its force to the bait, when the fisherman rapidly hauls up the line, and flings the sprawling mollusk on the shore before it is aware of its danger.

Several of these large Cowries can be successfully employed in the manufacture of cameos, especially when human heads form the subject, as the dark mottlings of the shell can be used with singular effect in expressing the deep warm shadows of wavy tresses. The various articles of ornament that are made from these shells are too multitudinous even to be enumerated, much less described. About one hundred and fifty species of this genus are known.

The grooved or wrinkled edges of the lips are well known to every one who has handled a Cowry, and these ridges assume a remarkable development in the DEEP-TOOTHED COWRY, a figure of which may be seen in the right-hand bottom corner of the engraving, the empty shell being laid so as to exhibit the opening and the lips. The color of this shell is extremely variable, but is mostly a mottled wood brown, sometimes diversified with bands, and dark inside. It is not a very large species.

THREE examples of the curious Egg-shells are to be seen in the same engraving. The upper central figure represents the POACHED EGG, a popular and appropriate name, as the peculiar shape and color of the shell bears a singular resemblance to the contour and tints of a well-poached egg as it trembles on the toast. Thirty-six species of the Eggs are known, spread sparingly over the greater part of the world. The under surface and opening of these shells are not unlike those of the cowries, except that in the Eggs the inner lip is without the ridges.

A VERY curious, elongated shell occupies the centre of the engraving. This is the shell probably known by the name of WEAVER'S SHUTTLE on account of its peculiar shape. It is, in fact, one of the Eggs, but has the aperture lengthened into a long canal at either end. The foot of this species is narrower than in the other members of the same genus, but is especially adapted for crawling over the stems of the gorgonia, one of the zoophytes on which the mollusk feeds.

Our third and last example of these shells is the WARTY EGG, remarkable, not so much for the tuberculated exterior of the shell, as for the richly-spotted foot and mantle.

WE now arrive at a vast army of shells called the Sea Snails, and distinguished by having the edges of the aperture without notches, the shell spiral or limpet-shaped, and the operculum either horny or covered with hard, smooth, shelly matter.

Our first example of this family is the NATICA. The mantle of this species is very large, and the front of the foot is developed into a fold, which turns backward over the head and serves as a kind of protection. As the animal is without eyes, this curious structure causes no inconvenience. All the Naticæ, of which about ninety species are known, are found upon the sandy beds of the sea, and sometimes are taken at a depth of nearly six hundred feet from the surface. They are very predaceous in their habits, feeding principally on little bivalves, which they can assault with their short but strongly armed tongue-ribbon. The eggs of these creatures are very remarkable. They are compacted into a kind of spiral roll, broad and rather short, which is suffered to be flung about at the mercy of the waves, and is sometimes found resting on the sands when the tide has retreated.

The colors of the Naticæ are marvellously permanent, and even in the fossil state they are preserved and retain some degree of their original brilliancy. The species which is here represented is yellowish, and marked with gray bands.

The NATICELLA-SHELL is closely allied to the preceding.

AN example of another family, the Neritidæ, is popularly known by the name of the SMOOTH NERITA. The foot of this animal is moderate, the tentacles are exceedingly large, and the eyes are set on footstalks just behind the base of these organs. The Neritas are all inhabitants of the warmer seas, and are found plentifully within the tropics.

The color of the present species is slightly variable, but in the individual specimen the shell is marked with bold, zigzag streaks of white and pale buff, and the interior is pure white at the lip, changing to beautiful canary-yellow in the interior. The operculum is thick, hard, solid, and highly burnished, as if overlaid with glass; its edge is regularly and finely grooved.

Several allied shells are inhabitants of the fresh instead of the salt waters, and are known as Neritines.

One of the most curious of these shells is the SPINED NERITINA. The animal of the Neritina is not unlike that of the preceding genus, but there are one or two minute differences. The operculum is shelly, with a flexible border, and has some small teeth on its straight edge. All the Neritinæ are globular in their general shape, darkly spotted or banded with black and purple, and covered with a polished bone-like epidermis. The color of the Spined Neritina is deep green-black on the exterior, and blackish-white within. The shell is thick and solid at the aperture, but becomes thinner towards the interior.

The CROWN NERITINA. The color of this shell is gray, diversified with dark streaks. One species of this genus, the RIVER NERITINA (*Neritina fluviátilis*), is found in the rivers of Northern Europe.

THE curious shell represented in the accompanying illustration is an example of another family, that of the Clubs, or Cerithiadæ. The shell of the Cerithites is spiral, more or less elongated, and the operculum is horny and spiral. The tentacles are placed rather far apart, and the eyes are set on very short footstalks. These creatures inhabit either marine, brackish, or fresh water.

The PELICAN'S FOOT, sometimes called SPOUT-SHELL, on account of the manner in which the aperture is lengthened into a kind of spout in front, has a rather elongated spire, and is considerably tuberculated on the exterior. As the animal approaches maturity, it adds fresh substance to the lip, until it bears some resemblance to the webbed foot of an aquatic bird. The animal has a short and rather abrupt muzzle, and moderately long, cylindrical tentacles, having the eyes set on protruberances near their base. Only three species of this genus seem to be at present known, but they have a wide range of locality, being spread over the greater part of the world, and found at various depths, sometimes being taken in a hundred fathoms of water.

The color of the Pelican's Foot is white, with a tinge of pink, and white inside. The shell is thickly and strongly made, and heavy in proportion to its weight. As may be seen by reference to our engraving, which is of natural size, it is not a large species, seldom measuring more than two inches in length.

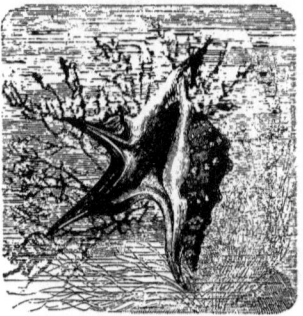

PELICAN'S FOOT.—*Aporrhais pes pelicani.*

The GREAT CLUB-SHELL is considered a species belonging to the typical genus of the family. It is rather a large genus, containing at least one hundred known species, and ranging over the whole world. The largest species are, as is usually the case, to be found within the tropics. The shell is considerably elongated, and with many whorls, and the "varices" or marks of growth are partially visible on the exterior. The aperture is decidedly small when compared with the dimensions of the shell, and has a somewhat twisted canal in front. The outer lip is rather wide, and the inner is much thickened.

One of these shells, the MARSH CERITHIUM (*Cerithium palustre*), is supposed by some persons to produce the strange sub-aquatic musical sounds that exist in several Eastern lakes. A detailed account of these sounds, together with the reason for this conjecture, may be found in Sir J. E. Tennent's "Natural History of Ceylon."

The color of the Great Club-shell is deep chocolate-brown on the exterior, slightly mottled with varying tints, and the interior is brown, but without the chocolate hue.

THE STAIRCASE, OR PRECIOUS WENTLETRAP.

In the family of the Turritellidæ, the shell is either tubular or spiral; the aperture is not waved, notched, or formed into canals; the foot is very small, the muzzle is short, and the eyes sunk rather deeply into the base of the tentacles.

The COMMON TURRITELLA is a species belonging to the typical genus of this family.

In all the Turritellas the shell is long, pointed, and with many whorls; the aperture is rounded and its edge thin; the operculum is horny and with many whorls, and with a slightly fringed edge. About fifty species of these shells are known, spread over the whole world, and inhabiting the moderately deep waters of the shores, ranging from a depth of one to fifty fathoms. They are supposed to be carnivorous. The color of the Common Turritella is whitish.

WORM-SHELL.—*Vermetus lumbricalis.*

The curious WORM-SHELL, which derives its name from its long and twisted form, is a very remarkable shell, and, if carefully examined, affords much instruction as to the mode in which the mollusks build up their wonderful homes. It looks, indeed, much as if it were in the preliminary stage of shell-making, and had completed its arrangements with the exception of pressing the whorls together. When young, the spiral form is tolerably regular, but as it grows in years its regularity decreases, and the shell exhibits the form represented in the engraving, in which the figure is somewhat magnified.

The aperture of the Worm-shell is round, and the operculum is consequently circular, and fits the opening with tolerable closeness. Its external face is concave. When not open, the tube is found to be supplied with many partitions of the same material as its walls. The color of the Worm-shell is grayish-yellow.

A SHELL of somewhat similar construction, but readily distinguishable by the longitudinal slit which extends throughout its entire length, is called the SNAKE-SHELL. About seven species of the Siliquaria are known, all of which are carnivorous in their habits, and are found within sponges. As in the last species, the Snake-shell is regularly spiral at its commencement, where it was constructed by the animal in its youth, but loses its regularity in exact proportion to its age. Its color is whitish. The small head, when just protruding, exhibits the stopper-shaped operculum.

The shell of the STAIRCASE, or PRECIOUS WENTLETRAP, was in former days one of the scarcest and most costly of the specimens of which a conchologist's cabinet could boast. There was hardly any sum which a wealthy connoisseur or virtuoso, as the fashion was then to call those who were fond of natural history, would not give for an especially large and perfect example of this really pretty shell. Now, however, its glory has departed, for a tolerably good specimen may be procured for a very small amount, and a Wentletrap which would twenty or thirty years ago have been sold for two hundred and fifty dollars, can now be purchased for less than one dollar.

Putting aside, however, the question of rarity or cost, this shell is a very interesting one, both for its beauty and the mode of its construction. It is purely white and partly transparent, the elevated ridges being of a more snowy white than the body of the shell, on account of their superior thickness, which does not permit the light to pass through them as in the case of the thinner body. The whorls of this shell are separate from each other, and apparently bound together only by the projecting ridges, so that the general appearance is as if the whorls of a worm-shell had been pressed nearly together, and then kept in their place by a succession of shelly elevations. This beautiful shell is found in the Indian and Chinese seas.

The COMMON, or FALSE WENTLETRAP, is tolerably common upon European coasts.

In this shell the whorls are united together and furnished with a number of circular elevations, which, however, are not nearly so bold as those of the preceding species, but thick in proportion to their height, set obliquely on the shell, and smooth.

The animal has a proboscis-like mouth, which can be retracted at the will of the owner; the tentacles are tolerably long, placed near together; and the eyes are set near the base of the tentacles. The foot is triangular, with the front rather obtuse, and supplied with a fold. When disturbed or alarmed, the creature is capable of exuding a dark purple fluid. Nearly one hundred species of Wentletrap are known, all the largest examples being found in tropical regions. They live at a considerable depth, sometimes being captured in eighty fathoms of water, and little seems to be known of their habits.

The color of the Common Wentletrap is rather varied. Sometimes it is dull white, sometimes it is very pale brown, and in a few specimens the shell is reddish-violet, with the ribs purple.

WE now arrive at another family, termed the Litorinidæ, or Shore Mollusks, because the greater number of them frequent the coasts, and feed upon the various algæ. The shell is always spiral and never pearly, by which latter characteristic it may be distinguished from certain shells belonging to another family, but somewhat similar in external appearance. The aperture is rounded. The animal has its eyes set at the outer bases of the tentacles, and the foot is remarkable for a longitudinal groove along the sole, so that in the act of walking each side advances in its turn. The tongue is rather long, and is armed with a formidable series of sharp teeth, that serve admirably for the purpose of scraping away the vegetable matter on which the animal feeds. The operculum is horny, and rather spiral.

The common PERIWINKLE (*Litorina litórea*) is the most familiar example of this family, and is too well known to need any detailed description. The Periwinkle is found upon our rocks in great profusion, occupying the zone between high and low water, and always being found near the edge of the tide. There is, however, another species (*Litorina rudis*) which occupies a rather higher zone than the previous species, and which, though very plentiful, is not eaten, in consequence of its young obtaining their shells before eggs are laid, and having a gritty and unpleasant effect upon the teeth. Sea birds, however, are not very particular about this drawback, neither is the thrush, which, in winter, when the snails are hidden away in their dark recesses, finds a meal easier to be obtained on the sea-shore than in hunting for its usual prey.

One of the prettiest members of this family is the WINDING STAIRCASE-SHELL, or PERSPECTIVE TROCHUS, so named on account of the peculiar formation of its whorls.

If the shell be held with its top downwards, it looks exactly as if it had been wound around a conical centre which had afterwards been withdrawn, and the projecting edges of the whorls have a wonderful resemblance to the perspective view of a winding staircase seen from below.

Perhaps the most remarkable point about this genus is the singular operculum of some of the species, which differs from that of any other mollusk. Instead of being a nearly flat plate, of horny or shelly substance, it is a conical structure of shelly matter with a riband of membranous substance wound round it, and projecting like the mechanical form so well known as Archimedes' screw. The object of this singular variation is quite unknown.

The color of the shell is rather variable, but consists of mottlings with brown, ochre, and white.

A very curious member of this family is the LOOPING SNAIL (*Truncatella truncátula*), a little species that is remarkable for the habit which has earned for it its popular name. All these creatures inhabit the space between tide marks, and can live for many weeks without water. Their mode of progression is very peculiar, and closely resembles that of the leeches or looping geometric caterpillars with which we are so familar. When they walk they fix the head firmly, then draw up the body in an arch, fix the foot, and then push the head forward. The foot is short and rounded at each end.

The shell is very small, about the size of a split sweet pea, and would escape the eye of ordinary observers. The animal is furnished with short and diverging tentacles, the head is divided into two lobes, and the eyes are placed in the centre behind the tentacles.

THE INDIAN PHORUS, or MINERALOGIST, a name given to the creature in allusion to its extraordinary habit of agglutinating bits of stones and other substances to its shell, has a rather long proboscis, and long tentacles, with the eyes set at their outer bases. The foot is long and narrow behind.

The outer lip is very curious in its structure, being extremely thin, projecting above and receding below. The operculum is horny, and formed by overlapping scales. The color of the Indian Phorus is yellowish-brown above, and pearly-white within. The edges of the lip are ragged and crumpled like those of a withered leaf. Sometimes it prefers other shells, either in fragments or entire, and is then termed the CONCHOLOGIST. In one example shown to me by Mr. Sowerby, the creature had selected a number of shells of a tiny bivalve, and had stuck them round the edges of its own shell in such a manner that they form a spiral line, marking the growth of the shell. One or two little bits of stone accompany them, and they all lie with the hollow upwards.

A MAGNIFICENT species is the SHELL-COLLECTING PHORUS. The long-pointed shells are clubs, or cerithinæ, a Venus-shell is seen at the mouth, and a lucina at the base. The name Phorus is of Greek origin, and signifies a carrier. The movements of the Phorus are said to be very clumsy, the animal staggering and tumbling about like the stromb-shells already described.

IN former days, the PHEASANT-SHELLS were articles of great price and rarity, some specimens almost rivalling the precious wentletrap in the enormous sums asked and obtained for them. Now, however, that their habitations have been discovered, and more frequent voyages are made, they have become comparatively plentiful, although, from the fragility of their structure, a perfect specimen is not at all common, and will still bring a good price in the conchological market.

The Pheasant-shells are now found in great numbers on the sandy beaches of several shores, being especially plentiful on the coast of Port Western, in Bass's Straits. The high tide sweeps them towards the shore, where they are left by the receding waters, and seek for shelter beneath the masses of sea-weed that are always flung on the beach by the tide. On lifting these sheltering weeds, the Pheasant-shells may be found crowded together under their wet fronds. They can move with some speed, the duplicate nature of the foot aiding them greatly in progression.

WE now arrive at the TOP-SHELLS, or TURBINIDÆ, a rather large and important family. In all these creatures the shell is spiral, and beautifully pearly in the interior, the nacre appearing when the outer coating is removed. The animal has a short head, rather long tentacles, with eyes mounted on footstalks at their base, and the head and sides are decorated with fringed lobes. They are all inhabitants of the sea and are vegetarians in their diet, their array of sharp teeth being very useful in rasping away the substances on which they feed.

Order SCUTIBRANCHIA. The Top-shells, so called from their resemblance to a boy's top, and the Neritas, of which the interesting Bleeding-tooth Shell is a representative, are members of this order. An example of the singular distribution of animals was noticed by the editor of this edition on one of the keys or islands of the Florida Reef. The beautiful Bleeding-tooth Nerita was found in considerable numbers on one of the islands, and on no other of the entire reef, along a series of islands one hundred and fifty miles in length. A large Chiton was found on the same island, and in no other locality within the same range.

The COMMON TOP is a little pointed shell.

This shell is a most plentiful species, and may be found by hundreds either crawling among the sea-weeds at low water, or flung upon the sands by the tide. The shell of this creature is beautifully pearly, and when the outer coating is removed the iridescent nacre below has a very lovely appearance. Jewellers and lapidaries employ these shells largely in their art, polishing them carefully and then stringing them together so as to form bracelets and necklaces, or affixing them as ornaments to various head-dresses. Another little shell, called TURBO VERSICOLOR, which is brought from Southern America, is also used for similar purposes. The specimens of Top-shells which are found in the sands are seldom quite perfect, the apex of the spine being usually worn down and rubbed so as to display the sub-lying nacre.

About one hundred and fifty species of Trochus are known, some of them attaining considerable dimensions, and all possessing shells of exceeding beauty. The form of the animal is peculiar. The tentacles are rather long, and the eyes are seen at the extremity of the little footstalks, at their base. The neck-lappets are rather large, and the sides are furnished with lobes and tentacular projections. The operculum is horny, flat, and spiral. Trochi are found all over the world, and have a considerable water range, being captured at all depths, from the shallow waters of the shore to a depth of a hundred fathoms.

Another beautiful species of Trochus is the NILOTIC TOP, a shell which is remarkable for the rich contrast of scarlet flashes on a white ground. One of the rarest species of this genus is the IMPERIAL TOP (*Trochus imperialis*), a shell which has hitherto been found only in New Zealand, and may probably be confined to that strange land. It is a handsome as well as a rare species, and is notable for the bold rounded projections which radiate from the whorls. Its color is violet-brown above and white below. Some authors, however, separate this shell from the Trochi, and place it in a separate genus, on account of the toothed whorls.

THE DOLPHIN-SHELL affords another instance of the entire discrepancy between the shell and the popular name that is given to it, this species bearing no more resemblance to a dolphin than to a roach, a cow, or a peacock.

THE ASS'S EAR is one of the larger species of the genus Haliotis, and is one of the most beautiful among the shells. Even when rough and unpolished, just as it appears after the removal of the animal, the rich iridescence of its interior is almost dazzling in the intense brilliancy of its coloring; and when, by the use of acids, the rough outer coat is removed and the nacreous substance of the shell exposed, there is hardly any marine production that approaches it and none that surpasses it in beauty.

This is a very useful shell to the manufacturer, its thick solid substance, with its lovely iridescence, rendering it well adaptable for the construction of buttons and similar articles, and also for inlaying in the darker woods. Very beautiful sleeve-links are cut out of the muscular impression, its heavy material giving the requisite strength, while the peculiarly corrugated structure produces a very beautiful effect, either when ground and polished or suffered to retain its ordinary contour.

THE GUERNSEY EAR-SHELL is popularly known throughout the Channel Islands by the name of ORMER.

This shell does not attain to so great a size as the preceding, but is, if possible, even more beautiful when polished and the opaque outer coat removed by means of acids and hard labor. The growth of each successive year is marked by a bold ridge, sweeping in a curve from the spine to the edge, and rapidly enlarging towards the margin. These ridges are caused by a regular series of furrows, in reality very shallow, but, on account of the peculiar manner in which they reflect the light, appearing to possess considerable depth. The effect presented by these ridges is really marvellous, the rich iridescence of delicate pink, green, and blue, with the slightest imaginable lines of golden light marking them, being quite beyond the powers of description, or even of artificial colors. Each ridge is perforated by a single hole near its extremity, and their course is marked even on the interior of the shell.

The animal of the Guernsey Ear-shell is largely eaten, but requires careful management in the cookery, as it is liable to be tough and stringy if badly handled. Before being subjected to the culinary art, it is well beaten, like a beef-steak, and is then cooked in various ways.

A similarly shaped shell, but without any perforations on the edge, is the STOMATIA, or FURROWED EAR-SHELL, so called because the place of the holes is supplied by a single groove or furrow. This shell is a native of the hotter seas. Its color is pale reddish-gray on the exterior, and pearly within.

A VERY curious snail-like shell is the VIOLET SNAIL (*Janthina communis*), so called from the beautiful violet-blue of the shell.

The Violet Snail inhabits several seas, and is most common in the Atlantic Ocean, though it is also found in the Mediterranean. Though in the look of the shell there is nothing sufficiently remarkable to attract notice, the habits and structure of the animal are most curious and interesting. The Janthina is essentially a surface species, always floating about, incapable of directing its course, and not even able to sink when threatened with danger. Being quite at the mercy of the winds and waves, it is often seen floating in great numbers, thus denoting the existence of some aërial or marine current, and may in such cases be swept up by thousands.

The food of the Janthina is said to consist mostly of the small blue velellæ; but, as the animal is without eyes, and is incapable of directing its course, it cannot be very rapacious. Some minute brown shells have been found in the stomach of several specimens.

The Janthina secretes a rather richly-colored fluid, respecting which many conflicting opinions have been given. Mr. F. D. Bennett, who has made some valuable observations on this curious mollusk, has the following remarks upon the fluid :—

"The body of this mollusk contains a very blue liquid, which, when the animal is punctured, exudes to the amount of three or four large drops. It is readily diffused through water or colorless spirit—to the former it communicates a faint tinge of its own peculiar hue, and to the latter a pink color, with a purple shade. It communicates its color to paper, and may be conveniently used as a blue ink ; several memoranda and pages of my journal, written with this fluid, have, after a lapse of more than five years, retained their original appearance both in color and intenseness. For this use, however, it must be employed from the recent animal, as it will not keep in any quantity, but becomes thin and discolored.

"It is believed that this fluid is analogous in use to the black secretion which the cuttle-fish pours forth to obscure the water and elude the pursuit of its enemies ; but this opinion must be received with some qualification. The living examples of Janthina which I have irritated when they have been confined in a vessel containing sea-water have not emitted any of the colored fluid ; when taken in hand, they would sometimes allow a little to exude ; but the entire quantity obtained from one animal by artificial means was never sufficient to cloud or obscure, although it would stain about half a pint of pure water."

Order ZYGOBRANCHIA. The Ear-shells (*Haliotis*), and the Patellas, Little Knee-pans, etc., are of this order.

The order CTENOBRANCHIA includes four sub-orders, and embraces some of the handsomest and best-known of shell-fish. The Janthinas are not familiar in the temperate regions, but the beaches of the tropics are strewn with their cast-off shells at times. They are essentially oceanic in habit ; resting, if at all, on the rafts of sea-weed, their long floats of bubbles supporting them safely. These creatures have a rich blue and purple coloring, and seem to have no feature of protective resemblance ; consequently, they would present to hungry fishes a tempting morsel. Their eggs are supported under the raft or float. They have, however, a means of protection that may prove all sufficient : that of throwing out a thick colored liquid when approached, which stains the surrounding medium, and thus affords a certain means of escape from enemies. An excellent figure of Janthina with its float attached is seen in the group of Ear-shells.

The well-known Volutes, Olivas, Murices (Rose-buds), Pyrulas (Pear-shells), Buccinum (Trumpets), Purpuras (Purple Shells, that throw out purple liquid for protection), Cones, Naticas, Ovules, Cypreas, Strombi (Conch), the curious Pteroscerods, and the very beautiful

Cassis, or Queen Conch, the Doliums, and the exquisite Ranellas—all are of this order, under their respective sub-orders. In this enumeration are some of the most valued of shells as well as most beautiful. Many of the Cones are of great value as rarities, and the Strombi and Cassids are of considerable value commercially.

In our waters there are not many of especial interest excepting to the student. Our semi-tropical waters, however, bear some of marked beauty. The most beautiful Queen Conch (*Cassis madagascarensis*) has been found frequently on the Florida Reef, and along the coast as far as Charleston. The Bahamas are good localities for these and the large Strombi that are used in cameo-cutting. The great Horse Conch (*Strombus gigas*) is abundant in Florida waters, and it is very beautiful when first removed from the water; its rich colors fade quickly when the animal dies. Several species of Volutes are common on the Reef and along the west coast of Florida. Olivas are also found. The Murex, in many species, is found, though sparsely, but of exceeding beauty and singularity of structure or ornament. On the New England coast and on the Grand and Georges banks are many interesting forms, though none of great beauty. The smaller Buccinums are common, and many interesting shells are found in the stomachs of fishes.

THE well-known univalves, so familiar under the name of Limpets, are divided into several families, on account of certain variations in the structure of the shell. The first family is termed Fissurellidæ, on account of the fissure which appears either at the apex or in the front edge of the shell.

All the Limpets are strongly adhesive to rocks, as is well known by every one who has tried to remove one of these mollusks from the stony surface to which they cling. The means by which the animal is able to attach itself with such firmness is analogous to the mode in which the suckers of the cuttle-fish adhere to the objects which they seize, the formation of a vacuum, and the consequent pressure of the atmosphere, being the means employed. The foot of the Limpet is rounded, broad, thick, and powerful; and when the animal wishes to cling tightly to any substance, it presses the foot firmly upon the surface and retracts its centre, while its edges remain affixed to the rock. A partial vacuum is therefore formed, and the creature becomes as strongly attached to the rock as a boy's leathern sucker to the stone on which he has pressed it.

The KEY-HOLE LIMPET is so called on account of the aperture at the top of the shell, which serves as a passage through which is expelled the water that has passed over the gills. This aperture is found in all the species of the genus Fissurella, but varies greatly in form and comparative dimensions, being, in some cases, a mere rounded hole in the shell, while in others it is a long and curiously-shaped aperture, very like the key-hole of a lock. The aperture increases with the shell, being hardly perceptible when the animal is young, but encroaching rapidly until it removes the whole of the sharp apex. These animals are mostly found at the same depth with the great tang sea-weeds, but are sometimes to be taken in fifty fathoms of water. The genus Fissurella is a large one, comprising about one hundred and twenty species.

The curious DUCK-BILL LIMPET inhabits the hotter seas, and is found on the shores of New Zealand, the Red Sea, and the Cape. It belongs to a small genus, containing about ten species.

This shell derives its name from its peculiar shape, which certainly does bear some distant resemblance to the beak of a duck. The animal is of very great comparative dimensions, and while living covers the shell with its mantle. Its color is black, and its sides are edged with short fringes. The eyes are set on the outer bases of the tentacles. The color of the shell is very pale yellow.

A number of nearly allied shells, belonging to the same family as the preceding species, are called CUP-AND-SAUCER LIMPETS, from the peculiar cup-shaped process on the interior, the shell itself taking the place of the saucer. This process forms the base, to which are attached the muscles which draw the animal to the rock. None of these Limpets appear to be active, seldom quitting the spot on which they have settled themselves in their infancy.

The form of the shell is extremely variable, depending greatly on the substances to which it adheres, and the color seems to be quite as mutable as the form. A specimen in my possession has an exceedingly thick shell, with very deep ridges, and a boldly waved edge. Its color is brown, of various shades, diversified with a little ochreous yellow. The "cup" is very much lighter than the interior of the shell, and is of a grayish-white with a slight yellow tinge, and marked with wavy streaks that give it a singular resemblance to chalcedony. The substance of the cup is very delicate, hardly thicker than the paper on which this account is printed.

The species called LADY's BONNET (*Calyptræa equestris*) belongs to the same genus. The generic name is derived from the word *calyptra*, which signifies a lady's cap. The food of these mollusks seems to be rather varied, as they are known to eat the minute algæ, and one specimen has been observed in the act of devouring a little sea-slug which we placed in the same vessel.

The HUNGARIAN BONNET LIMPET is almost invariably found adhering to oysters in a moderate depth of water, varying from five to fifteen fathoms, though it sometimes prefers a greater depth. The finest specimens are, however, taken in the shallower waters. The popular name is sufficiently appropriate in this instance, as the shell is exceedingly like the celebrated Phrygian bonnet of the ancients, or the republican cap of a later period.

The COMMON LIMPET is so familiar that it need not be figured nor described. One species of its genus attains to an enormous size, measuring a foot in diameter, and having a shell of very great thickness.

THE next family, called appropriately Dentalidæ, or the Tooth-shells, have long puzzled zoologists to assign their right position in the scale of nature, and even baffled the wide experience and penetrative acuteness of Cuvier himself. The general opinion of the systematic naturalists of his time referred the Dentalidæ to the annelids or worms; but Cuvier always expressed his doubts as to the accuracy of their views, and remarked that the solution of the problem would be found in the nervous and respiratory systems.

Sub-class Scaphopoda is one of the late divisions, embracing the Tooth-shells, so called from their resemblance to long teeth. There are not many, but certain characters render them of especial interest. They are the lowest in rank, being the most closely allied to the *Acephala*. The Dentalium was a favorite object with the aborigines of the west coast of America, its value as wampum, or money, being very great.

SUPER-ORDER ISOPLEURA, meaning equal-sided. It embraces quite singular appearing forms, which are included in three orders. The Chitons are the more familiar of them; once placed in a group as multivale shells. The Chitonidæ, or Mail-shells, are appropriately so called, because their shells are jointed like the pieces of plate armor. When separated from each other, the plates bear a strong resemblance to the joint of a steel gauntlet, and overlap each other in a similar fashion, a thick and strong mantle taking the place of the leather. There are eight of these plates, and all of them have a somewhat saddle-like shape. A similar arrangement may be observed in the lower abdominal plates of many beetles. Each of these plates is fixed to the mantle by certain rounded processes from their front edge, and when the plates are examined separately the processes will be plainly seen, white and pearly as the interior of the shell.

The genus Chiton is an extremely large one, containing more than two hundred species. Some of them are found at a depth of ten or fifteen fathoms, while a few of the smaller species are found in eighty or a hundred fathoms of water.

The PRICKLY CHITON is remarkable for the array of rather long spines with which the movable plates are armed, and which, when the creature contracts itself, give it a curious resemblance to the hedgehog. Its color is reddish-brown on the exterior, and pinky-white within. Although this shell attains a very great size, a large specimen measuring about five or six inches in length, it is not as valuable as in its youth, the curious spines being gradually lost as it approaches old age, just as human beings lose their hair, and the shell being by degrees rubbed tolerably smooth in some places and encrusted in others with corallines, calcareous matter, and the shelly coatings of various marine zoophytes. Sometimes the sea-weeds find a lodgment on the shell, as is often the case with other comparatively stationary mollusks, such as the common limpet; and in that case the algæ not only find a home, but conceal their protector by their waving fronds.

The accompanying illustration represents the MARBLED CHITON, a rather prettily colored shell, its exterior being rusty-red mixed with brown and yellow, and edged with brown. The SHORT-SPINED CHITON is covered with short spines. Its color is sooty-black, but this dull uniformity of a sombre hue is more than redeemed by the beautiful and minute pencilling with which its surface is engraved. The BANDED CHITON, or CHITONELLA, has been removed by modern naturalists into a separate genus, on account of the formation of the armor. The plates do not cover the entire surface as in the preceding genus, as only a portion is seen above the mantle. The defence is, however, nearly as perfect as in the previous genus, as the projections approach each other beneath the surface of the mantle, and would act as effectually in shielding the internal organs as if the plates had met on the surface. These creatures are generally found in the clefts of canal rocks.

MARBLED CHITON.—*Chiton marmoreus.*

The animal is more active than the limpet, but does not appear to be very locomotive in its habits. Its broad creeping disc adheres very strongly to the rocks, and holds the animal so firmly that, if it should happen to have taken up its abode within a crevice, to extract it without tools would be an impracticable task. Like the dentalium, this creature possesses neither eyes nor tentacles. The figures in our illustration are of natural size.

INOPERCULATE AND OPERCULATE GASTEROPODS.

PASSING from the sea to the land, we come to those gasteropods which breathe atmospheric air, and are furnished with respiratory organs suited to the lower element in which they live. These creatures fall naturally into two large sections, the one being destitute of an operculum and the other possessing that remarkable appendage. They are respectively called inoperculate and operculate gasteropods, and it is with the former that we have now to deal. The inoperculate are generally furnished with large shells; but in some, such as the slugs, the shell is either very small or wholly absent. The shell of these animals, when present, is not nearly so hard and porcelain-like as that of the sea-snails, and contains a much larger proportionate amount of animal matter. It is worthy of notice, that in order to prevent the waste of moisture in those species which live on land, and the entrance of water in those which inhabit the ponds and rivers, the respiratory passage is small, and closed with a kind of valve.

This group embraces the largest number of species of mollusks, including snails, slugs, whilks, limpets, couriés, etc. The head is well developed, and one or two pairs of tentacles are present. The sexes are usually separate.

THE first family is that of the Snails, or Helicidæ, containing a vast number of species. Most of the Snails have a shell large enough to permit the animal to withdraw itself wholly into the protecting domicile. During the time when they are active these creatures require no closure of their shells, and accordingly have no vestige of an operculum, as may be seen by looking at a common Snail. In the winter, however, when they retire from active life, and need that the aperture of their domicile shall be closed, the place of the operculum is supplied by a layer of hardened mucus, sometimes strengthened with the same substance of which the shell is composed, and always being perforated with a little hole to permit the inhabitant to respire. Any one may see this structure, called technically the epiphagus, by examining a Snail drawn from the crevice in which it ensconces itself during the winter months.

The animal has a rather short head, furnished with four tentacles, the upper pair being the largest, and bearing at their tips the little black specks which are supposed to act as eyes. These tentacles are retractile; and it is very interesting to watch them drawn back or pushed out like the finger of a glove, and to see the curious manner in which the eye speck is shot, as as it were, through the tentacle attached to the slender black thread which runs up its centre.

THE genus Helix, which is universally accepted as the type of this family, is of enormous extent, both in numbers and range of locality, containing more than fourteen hundred species, and spread nearly over the whole earth. The common garden Snail is a too familiar instance of this genus to need a description. I may, however, mention, that its depredations can, in a great measure, be checked by searching for it in the winter months, and taking it from the crevices in which it hides itself, or even by destroying the eggs which it lays just under the surface of the soil, and which look like pellucid peas. The much maligned thrush, too, is a mighty hunter of Snails, and, in spite of its autumnal raids on the fruit, does such good service in Snail-killing before the world is astir, that it ought to be encouraged by the gardener, and the fruit which it eats considered as the wages paid for killing the Snails.

THE great EDIBLE SNAIL is largely consumed in many parts of the world, and is regularly fed and fattened for that purpose. It is a remarkable fact, that in many spots where the Romans—great connoisseurs in Snails—had fixed their establishments, the Edible Snail is still to be found. Regular houses were built for the purpose of fattening the Snails, which were bred to an enormous size by constant feeding with a mixture of meal and new wine. There are even now on the European continent several snaileries, where the inmates are abundantly supplied with food, though they are not fattened with the elaborate precaution of the Roman times.

Even the common Snail is thought a delicacy by those who are sufficiently strong-minded to eat it; and it is quite common to see, even in Paris, the poorer orders dressing their dinner of Snails on an iron plate, heated over burning charcoal. I once knew an old woman, one of the few surviving wearers of scarlet cloaks, who used daily to search the hedges for Snails, for the purpose of converting her milk into cream. This cheap luxury was obtained by crushing the Snails in a piece of linen, and squeezing their juice into the milk. She showed me the whole process, which I afterwards imitated as far as the mixture with the milk, but could not bring myself to test the result by taste.

THE LEMON BULIMUS is an example of another large genus, containing more than six hundred species. These shells can be distinguished from those of the snails, to which they are closely allied, by the greater comparative length, the oval shape of the aperture, and the thickness of the outer lip. The last whorl is always very large. Some of the exotic species, such as *Bulimus ovatus*, attain a large size, and lay eggs even larger than those of the chaffinch, the young animal having a shell measuring an inch in length when hatched. Several species are very beautifully colored. Many species of Bulimi are excellent food, and are sold in the markets.

IN the illustration on opposite page will be seen a shell as if climbing up a tree. It is the largest of all the land snails, and is known as the great AGATE-SHELL (*Achatina*) of Africa. This Snail will attain a length of eight inches, and lay eggs larger than those of the bulimus, and with hard calcareous shells. The figure is drawn of the natural size.

We now come to the CHRYSALIS-SHELL. This shell belongs to a large genus, containing about one hundred and sixty species, and has received its popular name from its shape, which bears some resemblance to that of a chrysalis. This animal has always a short foot, pointed behind, and very short lower tentacles.

We now arrive at the great family of Limacidæ, or Slugs, a race of beings which many a gardener doubtlessly wishes extinct.

In these creatures the foot and body are indistinguishable from each other; the head is retractile; and the whole creature can be gathered into a short rounded mass, looking so like a pebble that it would escape a casual glance. At the first view, the Slugs appear to be destitute of shell, but on a closer examination, the shell is found upon the fore part of the body, and either entirely or partially buried beneath the integuments. When removed, it is not unlike the operculum of many mollusks, being small, flattish, and with an evident nucleus. They have four tentacles, like those of the snails, the eye-dots appearing, as in those mollusks, on the tips of the upper and longer pair. The respiratory orifice is placed on the right side of the body.

The GREAT GRAY SLUG is the largest of the European species, and when furnished with abundant food, on which it can fatten itself during the night, and a secure hiding-place,

AGATE-SHELL.—*Achatina mauritiana.*

whither it can retreat during the day, often attains an enormous size. The careless gardener, who has suffered heaps of old rubbish to collect in his dominions, is often horrified, when he at last removes the stones or sticks, to find under them a number of huge Gray Slugs, that have been silently consuming his flowers and vegetables, and lie slimy and obese at his mercy, bewildered with the unaccustomed light, and unable to escape their impending and deserved fate. It is true that Slugs, snails, and all similar creatures, must have been created for some useful purpose, and, in their proper place, discharge the duty for which their forms were designed and their instincts implanted; but it is clear that a garden is not the proper place for Slugs, and that if they make their appearance within its precincts, they must be extirpated; just as rats, which are useful in a sewer, are noxious in a house, and must pay with their lives the penalty of their intrusion.

The well-known BLACK SLUG (*Limax ater*) belongs to the same genus as the preceding species, and is very common during the summer, coming out of its hiding-place during the evening, and making its appearance along the sides of roads, in hedgerows, and similar situations. It is nearly, but not quite so large as the gray species.

The common RED SLUG, or LAND SOLE (*Arion rufus*), is another member of this family. It may be known by the deep red-brown of its body, which sometimes approaches to black. It is very plentiful in gardens, and as, on account of its color, it is not readily seen in the dark, it escapes observation, and does much damage without being discovered. Those who desire to rid their gardens of these pests will find that a very effectual plan is to search the grounds after dark, by the aid of a "bull's-eye" lantern.

THE semi-spiral shell, called TESTACELLA, is one of the very few carnivorous land mollusks. The Testacella, although plentiful, is seldom seen, on account of its peculiar habits. It feeds almost wholly on earth-worms, which it pursues through all the windings of their retreats, its long lithesome body enabling it to insinuate itself wherever the worm can burrow, and its hard little shell securing it from danger by stopping up the tunnel behind its progress. This curious Slug can be obtained in gardens by digging up the loose soil, but, on account of its services to the gardener, should be released, and permitted to resume its destructive avocations.

The tooth-ribbon of this creature is most formidably armed, having about two thousand teeth arranged in fifty rows. The teeth are needle-shaped, barbed, sharply pointed, slightly curved, and converge towards the centre of the ribbon, thus forming a weapon which no worm is capable of resisting. Only three species of Testacella are known; the English species is supposed to have been introduced from Southern Europe.

WE will now pay attention to the Water-snails, several of which can be found in every large pond or stream, and at first we may regard two species of APPLE-SNAILS, belonging to a genus remarkable for several peculiarities of formation. Although the Apple-snails belong more properly to the gill-bearing mollusks, and follow in the systematic arrangement the phorus, described on page 328, we placed them with the pond-snail and planorbis, for the reader's convenience of having combined on a few pages the various water-snails.

The Apple-snails are found throughout the warmer parts of the world, inhabiting the lakes and rivers, and, in case of drought, burrowing deeply into the mud and remaining buried for a lengthened period, sometimes for a term of years, until a fresh supply of water arouses them from their strange torpor, and urges them again to seek the upper regions.

In his "Natural History of Ceylon," Sir J. Emerson Tennent mentions this curious habit. "The *Ampullaria glauca* is found in still water in all parts of the island, not alone in tanks, but in rice-fields and the water-courses by which they are irrigated. When, during the dry season, the water is about to evaporate, it burrows and conceals itself till the returning rains restore it to activity and reproduce its accustomed food. There, at a considerable depth in the soft mud, it deposits a bundle of eggs with a white calcareous shell, to the number of one hundred or more in each group.

"The *Melania paludina*, in the same way, retires during the droughts into the muddy soil of the rice-lands, and it can only be by such an instinct that this and other mollusks are preserved when the tanks evaporate, to reappear in full growth and vigor immediately on the return of the rains.

"A knowledge of this fact was turned to prompt account by Mr. Edgar S. Layard, when holding a judicial office at Point Pedro.

"A native who had been defrauded of his land complained before him of his neighbor, who, during his absence, had removed their common landmark, diverting the original water-course and obliterating its traces by filling it up to a level with the rest of the field. Mr. Layard directed a trench to be sunk at the contested spot, and discovering numbers of the Ampullaria, the remains of the eggs, and the living animal which had been buried for months, the evidence was so resistless as to confound the wrong-doer and terminate the suit." After a few hours of rain, the Apple-snails may be observed emerging from their muddy retreat as if to welcome the newly found moisture.

The animal of the Apple-snail is very curiously formed. The long siphon, formed by a development of the neck-lappet, is seen on the left. Projecting just without the shell are seen the eyes, set at the extremities of short and stout footstalks, and the enormously long tentacles are placed just in front of the eyes. At the first glance the creature appears to have four tentacles, but on a closer examination, the front pair are seen to be merely developments of the muzzle. In one respect, the Ampullaria seems to be a connecting link between the gill-bearing and lung-bearing mollusks, being said by high authorities to possess a pulmonic or lung sac, in addition to its gills.

POND-SNAIL.—*Limnæa stagnalis.* (Natural size.)

In the accompanying illustration the common POND-SNAIL, or LIMNÆA, is shown in the act of climbing up the stem of a water-plant. In all the members of this family the shell is thin, and sufficiently capacious to contain the entire animal when it desires to withdraw itself into its home. The aperture is simply rounded, without notches or ridges, and the lip is sharp.

In the water-glass of the illustration will be observed the pond-snail, an equally common European shell, called from its flattened whorls the PLANORBIS. In this animal, the foot is short and round; the tentacles are long, slender, and leave the edges at their inner bases. Both this and the preceding species are in the habit of burying themselves in the mud during a drought, and there passing a semi-torpid existence.

A very remarkable species, called, from its peculiar shape, the FRESH-WATER or RIVER LIMPET (*Ancylus lacustris*), is found in various parts of America, Madeira, and some portions of Europe. It inhabits swiftly running streams, and is mostly seen attached to stony and aquatic plants. Although the shell is so limpet-like, the animal does not partake of the resemblance, being very like that of the pond-snail, and having triangular tentacles with the eyes at their bases. The generic term, *Ancylus*, is of Greek origin, and signifies a small round shield or target.

PLANORBIS.—*Planorbis corneus.* (Natural size.)

The little elongated POUCH-SHELL, a species of a rather small genus, extending over the greater part of the globe, is thin, spiral, polished, and the aperture is rounded in front. In the greater number of species, the mantle is fringed with long filamentary appendages, but in the present example the edges are quite plain. It is, however, always flat and much expanded. The tentacles are long and slender, and the eyes placed at their bases.

The Physas and Planorbis, Lymneas and the delicate Pond-shells are of the order PULMONATA. The Helices are included; quite familiar to us as Garden-snails. The common Garden-snail of Europe, *Helix hortensis*, was introduced into Essex County, Mass., many years since, and now is quite frequently found in the vicinity of Salem and Lynn. *Helix alternata* is the most abundant species in Eastern New England. A small area on Bass Point, Nahant, is a favorite locality, and in no other place is it found within several miles. *H. albolabris*, or White-lipped Snail, is also common. Numerous species of *Pupas* are abundant in the same region. A familiar form of this group is the Naked Slug, which has but a thin film of shell on its back, and is called Garden Slug (*Limax*).

OPISTHOBRANCHIATA.

WE now come to some of the strange and almost grotesque forms which are assumed by many of the mollusks. These belong to a fresh order, in which the shell is sometimes altogether wanting, and even when present is of very small dimensions, and is almost, if not wholly, concealed by the soft parts. In fact, they may be considered as the marine analogues of the common land slugs. The gills of these animals are rather curiously formed, not being placed in a definite cavity, as is the case with the previous species, but projecting boldly from the surface of the animal, and set towards the rear of the body. On account of this position of the gills, the animals are termed Opistho-branchiæ, or Rearward-gilled mollusks. The whole internal structure of these creatures is fully as curious as their external form, and well repays dissection, the organs of digestion especially being rather complicated, and possessing many points of interest.

The BUBBLE-SHELL, spotted on the exterior, is one of a moderately large genus of mollusks, that are found in almost all tropical and temperate seas, and may generally be captured where the bed of the ocean is of a sandy nature. In all the family to which the Bubble-shells belong, the shell is very thin, globular, and cylindrical, and the aperture is long and rounded. The large side lobes are said to be often used as fins for swimming.

The APLUSTRUM is a closely allied species, formerly placed in the same genus, but now separate on account of several structural differences. It has a shell rounded and slightly colored, but small in proportion to the size of the animal. The foot of the Aplustrum is exceedingly large, capable of concealing the shell entirely in its folds. Behind the tentacles are four large and flattened lobes, and the eyes are very small and set at the inner bases of the tentacles.

A VERY curious creature is the Bulla Ampulla. This is an example of a genus termed CYLICHNA; it is remarkable for one or two peculiarities. The animal is smaller in proportion than is generally the case with this family, being short and broad, and not able to wrap its lobes over the shell. The head is flat, blunt in front, and the eyes are deeply sunk in the tissues. The shell is cylindrical and mostly smooth, and the aperture is narrow.

A curious animal is the BOATMAN'S SHELL, a specimen of a very small genus, containing only five species. It is a sand-loving creature, mostly remaining in a rather considerable depth, and being found in fifty fathoms of water. This animal has no eyes; and although the side lobes are turned up, and are very large, they do not envelop the shell.

Sub-order TECTIBRANCHIATA. In some of its families the animal has an ovoid shell, as in the Bullas, or Bubble-shells; the curious Aplysias, or "Sea Pigeons," great masses of flesh, with no shell, and wide flaps for swimming. These are common forms in the tropical waters. Sea Hare is another name, from fancied resemblance. We have seen these creatures throw out their purple ink of great density, clouding the surrounding water, evidently for the purpose of defeating its enemies and securely changing its locality.

Several species of the *Aplysia* inhabit the waters of the Florida Reef. The ornamentation is variable. The specimens were about the size of a large cowrey.

The shell is never visible, being either very small and covered by the mantle, or wholly wanting, and in some instances being translucent and flexible as horn. The common SEA HARE (*Aplysia depilans*) is a well-known example of this curious genus, which has given its name to the entire family.

The Sea Hares possess the power of throwing out at will a rather large amount of a peculiar colored fluid, mostly of a deep violet, which is secreted by part of the mantle. This habit formerly caused the Sea Hare to be held in great dread, the popular opinion attributing to the violet fluid the most virulent properties, such as staining the skin indeli-

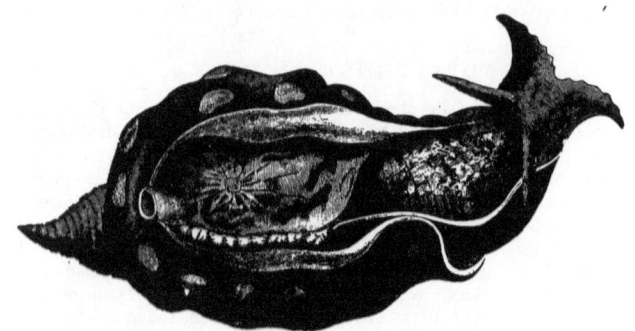

SEA HARE.—*Aplysia depilans.*

bly with the purple dye, injuring its texture like strong caustic, and causing the hair to fall off. In allusion to the last-mentioned idea, the animal has received its specific title of *depilans*. The illustration is remarkably true to nature, the figure being of natural size.

We will now briefly examine a few of the more remarkable species.

An oddly-shaped creature, looking as if it were made of some rough membrane covered with little projections, is the DOLABELLA, or HATCHET-SHELL, so called on account of the form assumed by the shell, which is, however, so entirely hidden under the softer parts that it is not visible until the lobes have been put aside.

When separated from the animal, the shell appears small in proportion to the size of the creature which formed it, and as it is attached only to the hinder part of the body, it is of little use in protecting the soft parts from injury. In color the shell is pure, shining white, and pearly; its substance is very thick, and it is covered with a tough membrane, technically called the epidermis. The color of the animal is dull olive-green.

Our last example of this family is the LOBIGER. While swimming, it uses the side lobes as fins. This creature is also called Lobe-bearer, in consequence of the rounded and flattened lobes that project from each side of the body, much like the four wings of a butterfly. The tentacles are also flattened and rather oval, and the eyes are very small and set on the sides of the head without any footstalks. The foot is small, and the hinder part of the body is lengthened and pointed so as to resemble a tail. The shell is small, oval, transparent, flexible, and set on the body so as to act as a shield to the plume-like gills. This species is found on the coasts of Sicily.

THERE is a curious animal belonging to the next family of mollusks, called, from the mode in which the shell is carried, the INDIAN UMBRELLA. In this creature the body is large, roundish, and covered with tubercles, and in shape something resembling a great limpet;

while the flat, white, pearly shell is perched horizontally on the very middle of the back, just like an Eastern umbrella held over the palanquin of some great potentate. The color of the animal is dull ochreous-yellow.

WE now arrive at a very remarkable series of mollusks which have been separated by systematic naturalists into a distinct section, appropriately called Nudibranchidæ, or Naked-gilled Mollusks, because their gills are always external and placed on the back or sides of the animals. Many of these strange creatures are to be found on the European coasts; and if the reader should wish to gain a further insight into their habits, and to examine the marvellous forms which the different genera assume, as well as their exquisitely delicate and varied coloring, he is referred to the magnificent work of Messrs. Alder and Hancock. The entire structure of the Nudibranchs is most curious and well worthy of examination, but is too purely anatomical for admission into these pages. A few, however, of the more notable structures will be mentioned in the course of our description.

DORIS.—*Doris pilosa.*

Our first example is the COMMON DORIS, a slug-like animal, which is represented in our illustration in the act of swimming. The figure is much magnified. All the members of the family to which this creature belongs may be known by the plume-like gills set in a circle on the middle of the back, like the feathery coronet with which the Blackfoot Indian adorns the head of his horse, and the two tentacles placed more towards the front. In the skin are imbedded a vast number of little spiculæ.

Of the family *Doridæ*, the *Polycera lessoni* is a familiar form. It may be found on the algæ in still pools left by the tide. In the bath-houses at Cragie's Bridge, in Boston, it is common. It is the same as the European species. It is a pleasing form, and proves an agreeable addition to the aquarium.

DORIS (*Doris bilamellata*). About an inch in length. This is also a cosmopolitan form. It inhabits similar localities as the former, at Beverly and Nahant. Stimpson dredged it in Boston harbor. *Doris tenella*, about half the size, is found in same places. Several other species are found on our New England coast.

THE next family is represented by two species, each of which will be briefly described. In this family, called Tritonidæ, the gills are arranged in lines along the sides of the back, and the tentacles can be withdrawn into their sheaths.

DENDRONOTUS.—*Dendronotus arborescens.*

The DENDRONOTUS, which is represented in the accompanying illustration, derives its very appropriate name from two Greek words, the former signifying a tree, and the latter the back.

The beautiful branched gills are set in a very shrub-like fashion upon the back, and even the tentacles and appendages of the head are branched so as to correspond with the gills.

The Dendronotus is common in the waters along the New England coast. It is one of the finest, most showy of the race in this region, and forms a pleasing inmate of the aquarium. Our illustration gives the size of the American form.

Our next example is the beautiful Doto. It may be here remarked that the word Doto is the name of one of the sea nymphs of mythology, and that in consequence of their exquisite coloring and beautiful forms, the names of nearly all the nymphs have been given to different species of nudibranchs. The tentacles of this animal are slender, and can be retracted into certain trumpet-like sheaths, which are seen projecting from the body. In this creature the processes of the digestive system pass into the large appendages on the back; and it is a curious fact that, although they fall off when the animal is handled, they are soon reproduced, and the creature seems to suffer little inconvenience from their loss. Examples of this creature can be found on the European coasts.

Of the family *Dotonidæ*, the *Doto coronata* is common, inhabiting the same localities as the preceding forms—Nahant, Back-bay, and Beverly Beach. It is equally common in Europe.

Another family is formed by the Eolidæ. In these creatures the theory of phlebenterism finds its best proofs, as the processes of the digestive organs extend throughout the beautiful projections on the back, even though, as in one genus, they are placed on footstalks.

The beautiful Eolis is common on the coasts of Europe, and has often been seen moving over the plants and stones with tolerable activity, and always keeping its tentacles and papilla in motion, sometimes contracting and sometimes extending them, while the movement of the water causes it to wave in a very graceful manner. These papilla possess the property of discharging a milky kind of fluid when the animal is irritated. The fluid, however, is quite harmless, at all events to the human skin. As in the previous case, the papillæ are liable to fall off at a touch. While using the dredge, the naturalist is sure to bring plenty of nudibranchs to the surface; but owing to their habit of contracting themselves into a shapeless mass, an uninitiated observer will probably fail to notice them, and fling them overboard again, together with the sea-weeds, stones, and other refuse substances. The Eolis is a voracious being in spite of its delicate beauty, and if several of them are kept in a vessel and not supplied with the sertularia and other zoophytes on which they feed, they will attack and devour each other.

Family *Eolidæ* is represented in American waters by *Eolis papillosa;* found in same localities as the last. It is one of the most common species in northern seas. Several beautiful species are found in the usual places—Boston, Back-bay, Beverly, Nahant, and Lynn. *Eolis bostoniensis* is a notable one, and the salmon-colored species.

The last and most remarkable example of the nudibranchs is the Glaucus, or Sea Lizard, a strange-looking creature. In this animal the gills are slender, cylindrical, and supported on three pairs of lobes or footstalks.

The Sea Lizard is very common in many parts of the Atlantic, where it is found in vast numbers during a calm and when the sea is smooth. Mr. F. D. Bennett writes as follows about this strange and eccentrically formed being: "These creatures obtain in greatest number where currents most prevail; they are active and very predatory in their habits, and would appear, from the observations of my brother, which I have already confirmed, to subsist chiefly upon the soft parts of the defenceless genera Velella and Porpita. The specimens we captured and kept in sea-water contracted their bodies into many convulsive attitudes, but seldom employed their branchial fins, and floated buoyantly while passive. When immersed in fresh water they contracted themselves into a very small compass, assumed a globular form, cast the tentacles from off their branchial fins, lost their color, and expired in a very few moments."

Super-order Anisopleura now (1885) embraces the largest division of the Gasteropods. The naked mollusks are of them. Nudibranchs, so called because their respiratory organs are external, and derive the oxygen from the surrounding medium. Most beautiful forms are seen in this group. A few forms are found on our New England coast. The *Dendronotus arborescens* is an elegant example—the one seen in the above cut is identical with ours. Protective resemblances are common in these forms; the latter appears like a bunch of pretty algæ, with its numerous branching frond-like respiratory organs floating gracefully in the water. *Eolis* and *Doris* are also represented by pleasing species. *Elysia* is much like the common slugs or naked snail, but somewhat more decorated.

NUCLEOBRANCHIATA AND PTEROPODA.

WE now arrive at a new order of mollusks, if possible stranger than that which has just been briefly described. The animals of this order are inhabitants of the sea, but differ from their kind in living almost wholly on the surface of the waters instead of crawling upon the stones or plants of the ocean bed.

It will be seen that a division is here made of more importance than the ordinary one of Orders, because of striking differences that can only be thus expressed: The Pteropods are all rather unfamiliar, being pelagic in habits, their delicate forms being borne upon the surface of the great deep. They derive their name, PTEROPODA, or WING-FOOTED MOLLUSKS, from the fin-like lobes that project from the sides, and are evidently analogous to the similar organs in some of the sea-snails. A fine specimen of this group of mollusks is seen in the illustration on page 343. The appendages are used almost like wings, the creature flapping its way vigorously through the water, just as a butterfly urges its devious course through the air. They are mostly found in the hotter seas, swimming boldly in vast multitudes amid the wide waters.

The first family of these creatures is represented by the CARINARIA. In this genus, the gills are protected by a small and very delicate shell of glassy translucence, bearing but little proportion to the size of the animal. The creature itself averages two inches in length, and is very transparent, permitting the vital functions to be watched by the help of a microscope. When swimming, the Carinaria reverses its attitude, and keeps the tiny shell downwards. The curiously modified foot of the animal is formed into a fin wherewith the creature can propel itself through the water, or a rudder by which it can guide its course.

In the ORDER HETEROPODA the delicate glass-like shells found on the broad ocean, called *Carinarias, Atlantas, Pterotracheas,* etc., are embraced. The curious eel-like *Leptocephalus* and the *Sagittas* are seen in our waters.

THE curious figure HYALEA is remarkable not only for the two wide fins which are found in all the family to which it belongs, but for the long appendages which pass through certain apertures in the shell, and trail behind as the creature proceeds on its course. The wings are united by a nearly semicircular lobe. The empty shell is placed below in order to show its curious structure.

A CREATURE, smaller than the Hyalea, and, with an odd-looking three-pointed shell, is the CLEODORA. It is a very beautiful and interesting animal, of which Mr. F. D. Bennett writes as follows: "On that part of the body which is lodged in the apex of the shell, there is a small, globular, pellucid body, resembling a vesicle, and which at night emits a luminous gleam, sufficiently vivid to be visible even when it is opposed to the strong light of a lamp. It is the only example of a luminous shell-fish I have ever met with; nor would the luminosity of this species be of any avail, did not the shell possess a structure so vitreous and transparent. Examples were chiefly captured at night or in the evening.

AN example of an allied genus, notable for the straight-pointed shell, is the SPIKE-SHELL. The fins of this little animal are rather narrow, and the apex of the shell soon loses its sharpness, being by degrees divided into compartments and gradually broken off. The Spike-shell is mostly found near floating sea-weed.

A GOOD example of the Pteropoda is the large-winged CYMBULIA. Though greatly resembling the carinaria in general appearance, it is divided from that creature by many

WING-FOOTED MOLLUSK.—*Pteropoda*.

important structural differences. Its shell is flexible, and in shape and translucency somewhat like the glass-slipper of fairy mythology, the point, or toe, being set forward. Only three species of this genus are known.

The ORDER THECOSOMATA is a division embracing the *Hyaleas* and *Cleodoras*. The family *Cymbulidæ* includes some comparatively large species, which secrete peculiar looking shells. They are slipper-shaped, and very much like a mass of jelly, thick, transparent, and flexible. Species are *Cymbulia* and *Tiedemania*.

The ORDER GYMNOSTOMA embraces the Naked Pteropods. *Clione* is a more common genus. *C. borealis* is the familiar Arctic form which is seen in vast patches on the ocean. This, with the *Limacina*, a member of the preceding order, forms the principal food of the whalebone whale of the North Atlantic. *Clione papilionacea* is found in our waters as far south as New York. Its resemblance to a butterfly gives it the specific name.

BRACHIOPODA AND CONCHIFERA.

As group after group of mollusks passes before our notice, each seems to be more extraordinary than its predecessor, and to present us with stranger and more unexpected forms.

The mollusks of the next group are the first of the bivalves, but stand alone in many particulars, and evidently form a transition between the gasteropoda and the ordinary bivalves. They are all inhabitants of the sea, and, when adult, are found attached to rocks, coral branches, and even other shells; but in their earlier stages are apparently able to swim freely through the water, as is the case with many other mollusks.

In the ordinary bivalves, the two shells correspond with the right and left side of the animal; but in the Brachiopoda, as these creatures are called, the one covers the upper and the other the lower portion, and are called accordingly the dorsal and ventral valves. Of these, the former is smaller than its companion, to which it is jointed by means of certain interior sockets, which receive corresponding hooks in the ventral valve, and lock them together so tightly, that they cannot be separated without something being broken. The ventral valve is large, and is marked by a decided beak, not unlike the bill of a parrot. In most instances the beak is perforated with a round hole, through which passes the peculiar organ by which the animal attaches itself to the substance on which it rests; and when this is not the case, the hooked beak itself answers that purpose.

In the interior is a rather complicated internal skeleton. The food is obtained in a singular manner. The animal is furnished with a pair of rather long arms, covered with vibrating fibres or cilia, and by means of the constant action of the cilia a current is caused, which drives a continual stream over the mouth, and enables the animal to seize the minute animals that dwell in the sea and are distributed throughout the waters.

WE will now proceed to the examination of our selected examples of these curious mollusks.

The genus Terebratula is the first to mention. This name is derive from a Latin word signifying a wimble, and is given to the animal in allusion to the round hole which perforates the beak. The popular name of LAMP-SHELL also refers to the same aperture, because it looks like the round hole through which the wick of an ancient lamp is drawn. The structure of the shell itself is very curious, being made up of innumerable flattened prisms laid side by side and arranged in a slightly oblique position, so that their ends project over each other, something like the slates in a house-roof. The substance of the shell is also perforated by multitudes of very minute circular apertures.

Next comes the PARROT-BILL LAMP-SHELL, so-called from the shape of the beak, which is long and hooked in a manner which much resembles the beak of the bird whose name it bears. The color of this species is black.

Our last example of these remarkable mollusks is the GOOSE-BILL LAMP-SHELL. All the members of the family to which this animal belongs are known by the long and comparatively narrow valves, and the footstalk which attaches them to the rocks, and which passes from between the valves. The substance of the shell is rather soft and perforated. The valves are slightly open at each end, and blunted in front. Very little is known of its habits in the living state, but it is worthy of notice that the Goose-bill Lamp-shell is the oldest known form of organic life.

ACEPHALA.

The Headless Mollusks are the lowest in rank in the scale of life. The common Clam is a good example of the class. The long fleshy process of the Clam is popularly called the head, but it is the foot, in one sense, being opposite the place where the head would naturally be. The true foot is midway, and is the tougher triangular part which is the locomotive organ. The long fleshy part which is called the head consists of two tubes, one cavity absorbing water and the other throwing it out after it has served its purpose.

In some of the bivalves, the mussel, for example, there is a gland near the foot which secretes the byssus, a bundle of threads by which the animal fastens itself to any object.

Bivalves are hinged, and re-enforced as it were by a stout ligament on the outer side. One or two adductor muscles are placed within, attached to each shell within the depressed portion that shows on the inside of the bare shell. The ligaments, by contraction, tend to force open the shell's valves; the muscles on the inside draw them tightly together.

Classification of *Mollusca* is yet in a very unsatisfactory state. As in the case of some other divisions of natural history, as long as there is no satisfactory guide to classifying, a consideration of the forms under family heads is most convenient and useful.

Though not possessing so many species as the gasteropoda, this group surpasses it in point of numbers, the bivalves being produced in countless myriads, and, perhaps, less exposed to the attacks of foes than most of the race. They are extremely useful in both salt and fresh water, feeding on the particles that would otherwise pollute the element in which they live. Their mode of feeding is somewhat similar to that of the last-mentioned group, the water being driven over the mouth by the continual action of certain appendages, and there cleared of all its solid portions. So completely does a bivalve effect this purpose, that it even intercepts the microscopic plants and animals which are invisible to the naked eye, and conveys them to the stomach with marvelous certainty.

In the first family, of which the common Oyster is a very familiar instance, the two valves are unequal in size, and the animal inhabits the sea. The Oyster is too well known to need description; but it may be mentioned, that the practical naturalists have for some years been carefully studying its habits, for the purpose of breeding the valuable mollusk artificially, and so of securing a constant supply throughout the four months of the year during which the creature is out of condition. In this country the system is being gradually carried out, but in France it is developed to a very large extent, and with great success.

The details of the process are too elaborate to be here described, but the general idea may be given in a few words. The very young spawn, or "spat," as it is technically named, is removed from the natural beds, and is dispersed in shallow "banks," so that each tiny Oyster has plenty of room, and can affix itself to the bed of the bank without being injured by the pressure of its fellows. Fascines, made of slender branches, and sunk into banks paved with stone, birch, and broken earthenware, are found to be most useful for this purpose. In the banks near Dieppe, the Oysters are seen lying in regular rows like the tiles of houses, and are at all times ready to be taken from the bed and sent to market.

This process possesses a double value, inasmuch as an oyster-bed, if left to itself, would increase to such an extent as to endanger navigation; and these inland banks are always accessible, whatever may be the weather. In some cases, when confervoid growths are rife, the Oysters attain a decided green hue, and are thought very valuable by connoisseurs in such matters. In all improvements, however, there is always some drawback. The Oysters produced by artificial culture are acknowledged to be fatter and finer than those which are suffered to grow in the open seas; but their artificial size is said to be a poor compensation for their comparative want of flavor, the artificially bred Oyster being to the marine mollusk what the capon is to the pheasant.

In the sea, thousands of Oysters perish by the attacks of a strange enemy. The reader has doubtlessly remarked that the shells of many Oysters are partially perforated by little round holes. These are the marks left by a kind of sponge, called Cliona, which burrows into and gradually destroys the shells of this mollusk, causing them to fall to pieces by its ravages.

As a matter of economic value, the Virginia Oyster and the common Clam are all important.

The VIRGINIA OYSTER (*Ostrea virginiana*) extends along our coast from the St. Lawrence River to the Gulf of Mexico. Many years since Oysters were natural to the shores of New England. Old extinct beds are found around Cape Cod, and along the coast to Mount Desert. Huge heaps of shells are seen at various places—notably in Casco Bay, Maine. They are of enormous size. Professor Verrill inclines to the opinion that climatic changes have conspired to produce a scarcity, and, in some places, extinction of Oysters in New England. South of Cape Cod they flourish abundantly. In the Oyster of our shores the sexes are separate; the eggs are fertilized after they are deposited, and develop in the water. At first the young Oyster swims freely, and after the shell begins to develop it settles permanently. Our Oysters are all of one species, notwithstanding several specific names have from time to time been used to designate them. Two species of Oysters are edible on the Pacific coast, *O. conchophila* and *O. lurida*. The European Oysters are insufficient to supply the market, and the American Oysters are imported there in great quantities. Edible Oysters are found in Japan, Cape of Good Hope, and in Australia. Some of the Asiatic ones measure three feet in length.

The Oyster industry in America exhibits the following statistics: The Census of 1880 gives the number of persons employed in the business as over fifty thousand, and over four thousand vessels; involving an investment of ten million dollars. The number of bushels of Oysters sold is over twenty millions.

A CURIOUS and rather valuable shell is the CHINESE WINDOW-SHELL. It is found in the country from which it takes its name.

This shell is extremely flat, and of a beautiful translucence, and in many parts of China is employed for windows, just as is ground-glass among ourselves, the nacreous substance permitting the light to pass through, but effectually preventing an inquisitive eye from distinguishing objects within the apartment. Very small pearls are found in this shell, too minute and too opaque to be employed by jewellers. They are, however, collected and exported to India, where they are calcined and formed into lime for the use of wealthy betel-chewers. They are also burned in the mouths of the dead.

The shell is of great use in commerce, affording the substance from which is cut those large flat "pearl" buttons that were formerly so fashionable, but seem now to have descended to the denizens of the stable. The button "moulds" are cut from the shell by an instrument that somewhat resembles the trephine, by which portions of the skull are removed in case of severe injuries, and in their rough state look like gun-wads. They then pass through a series of processes in which they are polished and pierced, and made ready for sale.

THE SADDLE-SHELL is remarkable for the way in which the shell is attached to other substances. The contrivance by which it is attached is most remarkable. The animal deposits a plug or peg of shelly matter on the oyster, and in the right valve there is a hole or notch into which the peg fits, much after the fashion of a button. When the left valve is in its place, this contrivance is hidden. The shell of this creature is beautifully thin—hardly thicker, indeed, than the paper on which this account is printed—and elegantly waved. It inhabits the European seas.

THE LIMA, or FILE-SHELL, of the same species, is worthy of notice on account of the curious refuge which it constructs by binding together a large mass of shells, corals, sand, and other materials, by means of the silken threads or "byssus" which it is capable of secreting. It often appears almost entirely buried in the mass of nullipores which it has gathered around its shell. The long tentacular appendages are kept in constant movement, possibly without the will of the animal, keeping up their writhing contortions just as our hearts continue to beat without our knowledge. Even after the death of the animal, and when they have been separated, the filaments continue to move, twining and twisting like so many worms.

The File-shell can pass through the water with some rapidity, urging itself along by the sharp closing of its valves. Its color is crimson, with the exception of the mantle, which is orange. The shell is pure white, so that a living and healthy specimen is a most beautiful creature.

A very curious example of this family is found in the THORNY OYSTER, a species that is remarkable for the singularly long projections from the shell. The object of these spines is rather obscure, but is said to answer a double purpose; the one being to act as a *cheveaux-de-frise*, whereby the attacks of marauding fish or other foes may be repelled, and the other to aid in fixing the animal to the spot on which it has established itself. Any fish, however, that would be strong-jawed enough to crush the shell, even without the spikes, would care no more for them than does a donkey for the prickles of a thistle; and the smaller and more insiduous enemies would receive no check from the hedgehog-like array of bristling points. The animal of the Thorny Oyster is eatable, and in many places is looked upon as a delicacy.

The group including the curious *Spondylus*, with its numerous projecting processes, also embraces the *Malleus*, or Hammer-shell, and the *Lithodomus*, or stone-borer, and the *Modiolus*, a large mussel-like shell, of our shores.

The *Unionidæ* rank next, the family embracing the large number of fresh-water shell-fish, ranging from the small unios of our creeks and rivers to the great bivalves of the western waters, lakes and rivers.

Family *Lucinidæ* embraces some small circular shells, prettily ornamented by concentric ridges. Two species are found on our coast. It is a singular fact, that certain shells are so confined to special localities, and that some are so exceedingly scarce, unless indeed some cause has been actively at work to decimate them. The *Lucinia radula* is an example of both these conditions. On Chelsea Beach, in Massachusetts, broken valves of this shell are occasionally seen; but only one perfect shell, with the animal in it, was found up to the time William Stimpson published his work on the marine shells of Massachusetts. This example of Lucinia we now have before us. It was figured by Stimpson, and recorded standing lonely as the only perfect example found on our coast. The Lucinia resembles the Cytherea above figured.

The *Cyprinidæ* are represented on our shores by the large bivalve called Quahog, or Round Clam, very much resembling the Venus, but larger, and having an epidermis covering of greenish-brown, the *Cyprina islandica*, although the *Venus mercenaria*, is the proper Quahog of the Indians.

The pretty little chestnut *Astarte* is one of the most attractive of our bivalves; about the size and exactly the color of a chestnut, and not very unlike it in shape.

THE last example of this family is the HINNITES, a shell remarkable for its exceeding variability of form. When young, it wanders freely through the ocean; but when it finally settles down in life, it acts like weak-minded men, and molds itself to the locality in which it happens to reside. If it gets among scoriæ, as is not unfrequently the case, the shell follows all the irregularities of its resting-place; and in one instance, where one of these shells had settled upon a group of serpulæ, it had accumulated itself to them in the most curious manner, actually overlapping the shell, so as to form its edge into the half of a hollow cylinder. The colors are red, brown, and white, but their relative amount and the manner of their disposal are as variable as the form.

THE next family are termed Wing-shells, or Avicularidæ, because the apices, or "umbones," as they are called, are flattened and spread on either side, something like the wing of a bird. The interior of the valves is pearly, and the exterior layer is composed of a kind of mosaic work of five or six sided particles. This structure is easily to be seen by means of a moderately powerful simple lens, merely by holding up a scallop or other shell before the window, so as to allow the light to pass through it.

A dark, whitish species is the curious HAMMER-SHELL. Only for the oddity of its form, which somewhat resembles that of the hammer-headed sharks, it attracts some attention. As it lies on the ground, it would hardly be taken for a shell by one who was not acquainted with it, the enormously expanded ears and strangely crumpled valves giving it a most unshell-like aspect. This strange form is, however, only to be seen in the adult specimens, or when young. The shape of the Hammer Oyster is very like that of the pearl Oyster, presently to be described.

The VARIEGATED SCALLOP is, in common with many other mollusks, able to move with considerable swiftness by means of repeated strokes of its valves, a single stroke carrying it for several yards.

The animal is very beautiful, its color being orange or fine scarlet, and the mantle marbled with brown of different hues. A series of round black dots, called ocelli, and thought to answer the purpose of eyes, are ranged around its edge, and surrounded by long, tentacular filaments. Like the Oyster and mussel, the Scallop is considered as a delicacy, and eaten dressed in various ways. The shell is of little value, its chief use in these days being as a vessel in which Oysters are "scalloped;" but in the ancient times it was in great request, as the sign of one who had made a pilgrimage to the shrine of St. James. When at rest, the Scallop lies on the right valve.

The family *Pectinidæ*, or Scallops, follows in order. The *Pecten irradians*, or Common Scallop, is much used in New York and southern part of New England as an edible. The adductor muscle alone is used. It is prized by many, but is not uniform in its effect on others; while some dislike its sweetish taste.

THE well-known PEARL OYSTER is one of the most valuable of the shell-bearing mollusks, furnishing the greater part of the pearls which are set by jewellers and worn by ladies. The specimens represented in the engraving are half as large as in their natural size. These creatures are found in Ceylon, Madagascar, Swan River, Panama, etc. Not only the pearls themselves are valuable, but the shells are of great importance in the commercial world, furnishing the best "mother-of-pearl," as the nacreous lining of the valves is called.

The pearls are secreted by the animal in precisely the same manner as the nacre of the shell, and are, indeed, the same substance, formed into a globular shape, and disposed in concentric layers, so as to give that peculiar translucency which is quite indescribable, but is known among jewellers by the name of "water." As to the precise method and object of their formation opinion differs, the general impression being that they are morbid secretions, often stimulated originally by a grain of sand or some such substance finding admission into the shell. These objects may be obtained by introducing into the shell certain extraneous bodies, around which the nacre is secreted so as to form very good imitations of the pearls formed after the usual manner. Examples of such artificial pearls will be mentioned in the course of the following pages.

The Pearl Oyster does not produce its costly harvest under six or seven years of age, and it is, therefore, a matter of importance that the bed should be so managed that the young Oysters may be suffered to remain in peace until they have attained an age which renders them capable of repaying the expense of procuring them, and that no part of the bed should be harried where the Oysters are too small to produce pearls. It is hoped that the increasing knowledge of the mollusk and its habits will enable proprietors to sow the sea with pearls just as they sow a field with grain, and that the harvest may be equally certain in either case.

The Oysters are now obtained by means of men who are trained to the business, and who can remain under water for a considerable time without being drowned. Each diver takes with him a net bag for the purpose of holding the Oysters, puts his foot into a stirrup, to which hangs a stone weighing about thirty pounds, and after taking a long breath is swiftly carried to the bottom. He then flings himself on his face, fills his bag as fast as he can, and when his breath begins to fail, shakes his rope as a signal, and is drawn up together with the bag.

Very exaggerated accounts have been given of the time passed under water by the divers, from two to seven minutes being mentioned as the usual periods. The real fact, however, is, that one minute is the ordinary average; a few men being able to endure an immersion of a minute and a half. This is a long period, as any one will confess who has attempted to repeat the feat. Yet, with a little practice, it can be achieved, even by those who can lay no claim to extraordinarily capacious lungs; and I have more than once performed it with tolerable

PEARL OYSTER.—*Meleagrina margaritifera.*

ease. If the lungs be thoroughly filled four or five times in succession, and emptied to the last gasp, so as to expel all foul air that may be lingering in the tiny vesicles, the blood becomes so well oxygenized, that a further supply of breath will not be needed for some time, and a deep inspiration will serve to keep the blood in a healthy state for a marvellously long time. All swimmers who are fond of diving will find that they can remain under water nearly twice their usual time by taking this simple precaution.

The best plan for procuring the Pearl Oyster is evidently the employment of the diving bell, so that the best shells might be leisurely selected, the spot left undisturbed, and the sharks outwitted. In the illustration are given specimens of the shell in various stages, as well as the interior, showing the pearls as they appear when the animal is removed.

A LARGE, flattish, wedge-shaped shell, generally moored to a stone, or fastened to the bottom by a number of short threads, is the PINNA, so called from the Latin word, signifying a wing.

The aggregate mass of these threads is termed the byssus, and is, indeed, a very curious object. The threads are spun by the foot, and are attached to the centre of each valve, thus forming a powerful cable by which the shell is moored to the rock. The threads are wonderfully strong, silken in their texture, and, had the mollusk been sufficiently plentiful, might have been employed in various manufactures. I have seen a pair of gloves that have been woven from the byssus of the giant Pinna, a species which sometimes attains the length of two feet, and has a most singular appearance when old, owing to the mass of parasitic creatures, such as serpulæ, balani, and sundry zoophytes, that always congregate on such substances.

It is remarkable that a little crab, called, from its habits, Pinnotheres, is often found within the shell of this mollusk, and was formerly thought to have entered into a tacit agreement with its host to act as sentinel and to bring in food as a return for the hospitality afforded to it. This, however, is not a solitary instance of such strange alliance, several other mollusks being known to shelter their particular crustacean guest. When at rest, the Pinna is mostly buried in the sand, with the exception of the upper edges of the shell, which are permitted to protrude just above the substances in which the rest of the creature is immersed.

WE now come to the large, useful, and even beautiful family of the Mussels, although, in most cases, their beauty is not perceptible until the shell has been polished and the rich tints thereby brought out. Rough and polished mahogany are not more unlike each other than the Mussel-shell before and after the polishing process. Some species are marine, while others inhabit the fresh water, and all may be known by the peculiar shape of the shells.

The EDIBLE MUSSEL, so common in the fishmonger's shop and the costermonger's barrow, is found in vast profusion on European coasts, where it may be seen moored to rocks, stones, and fibres, alternately covered with water or left dry, according to the flowing and ebbing of the tide. The heedless bather is sometimes apt to come unexpectedly upon a collection of these mollusks, and if he once meets with that misfortune, his lacerated limbs, cut in all directions by the knife-like edges of the shells, will serve as effectual warnings not to repeat the same imprudence.

At some periods of the year the Mussel is extremely injurious as an article of food, though the effects seem, like those produced by eating the bonita, to depend greatly on the constitution of the partaker, some being able to eat it with impunity, while others who have shared the same meal are visited with asthma, violent rash, nausea, and many other symptoms which, though not absolutely dangerous, are peculiarly annoying. The Mussel is largely used for bait as well as for human consumption, more than thirty millions being collected annually in one locality for that purpose. Little, ill-shapen and badly-colored pearls are often found in this mollusk, but are quite useless for the market. Attempts have been successfully made to propagate the breed of Mussels; and the vast plantations, as they may be called, of these creatures have increased to such an extent, that they threaten to obliterate several useful bays for all maritime purposes.

An allied species, the DREISSENA, inhabiting the fresh waters, has of late years rapidly overrun England, having been originally imported into the Surrey Docks, whence it has spread with astonishing fertility, passing from one river to another, getting into all the little rivulets that trickle between meadows, and even obtaining entrance into artificial basins by means of the water that feeds them through pipes. The shell is like that of the edible Mussel, but shorter, and without the beautiful nacreous lining.

THE FORK-TAILED DATE-SHELL is a little, ochre-colored shell, without any peculiar beauty of form or color, but yet as remarkable a creature as any that has been or will be mentioned.

This little being has the power of burrowing deeply into the hardest stone. I know an instance where the substance in which the Lithodomi were imbedded was a shell of the gigantic limpet from Madagascar, measuring about six inches in diameter and half an inch or so in thickness. This specimen, which I have carefully examined, was a really wonderful one, the thick, hard, and solid substance of the shell being literally riddled with the holes of the Lithodomus, whose forked processes projected from the circular aperture much like the eggs of the common scatophagus from the substance in which they are sunk.

The method by which this little mollusk contrives to excavate its chamber is a complete mystery. It is known that in its earlier stages it spins a byssus, and attaches itself to substances like the common Mussel, but that in process of time it begins to bore its way into the object to which it is moored. As the shell increases in size, the chamber is enlarged in dimensions; but the original aperture remains of the same diameter as when first bored, and therefore effectually prevents the animal from making its exit.

Some persons have suggested that the animal employs an acid for the purpose of dissolving the rocks; but if such were the mode of operation, the shell would suffer equally with the stone. A continual current of water forms the basis of another theory; and provided that the animal were sufficiently long-lived, there is no doubt but that the constant action of water would in process of time wear away the stone, however hard it might be. But as yet no theory has sufficiently accounted for the fact that the creature excavates these chambers with wonderful rapidity, and that, in all cases, the chamber corresponds with the shape and size of the shell. It is evident, also, that the shell itself is not the means by which the chamber is bored, as the peculiar shape of the hole prevents the shell from rotating.

The Lithodomus seems to drive its curious tunnels through everything that comes in its way, for, in one case, a specimen has bored through the upper part of the limpet-shell, broken into a chamber already excavated by another individual, and forced its way fairly through the inhabitant as well as the habitation.

The animal is slightly luminous, as is the case with most of the burrowing mollusks. The color of the shell is uniformly pale brown.

The FINGER DATE-SHELL is a rock burrower, and so beautifully decorated that it seems a sad pity to bury so lovely a shell in so dark a recess. At a little distance it is quite ordinary in appearance, being apparently a plain, mahogany-colored shell; but when examined closely, it is found to be elegantly formed, colored with a peculiarly rich ruddy brown hue, and sculptured with myriads of minute waved ridges and channels drawn crosswise over the shell, which give wonderful effects of light and shade, and heighten the tints materially. The animal is edible, and is eaten like that of the common Mussel.

A RATHER curious-looking-shell, which, from its rude resemblance to the familiar toy of childhood, is called the NOAH'S ARK.

The Ark-shells are found all over the world, hidden under stones, in the crevices of rocks, or even within the forsaken burrows of the pholas or the date-shell. Owing to their retiring habits, and the nature of the localities in which they live, they are mostly distorted or damaged. They can move themselves very fairly by means of a curious conical byssus, composed of a series of thin plates, which can be cast off or re-formed at the will of the animal.

EXAMPLES of pearl-bearing mollusks which inhabit the fresh waters, are the European and the Chinese Pearl-mussel.

The EUROPEAN PEARL-MUSSEL was once a valuable inhabitant of English rivers, on account of its contents. It is now, however, seldom sought except for bait, and in the latter capacity is more useful than in the former, as it is estimated that not more than one per cent. contain any pearls, and not more than one per cent. of the pearls is of any commercial value. The older and more irregular the shell, the better chance is there of finding a pearl; and a diligent collector may soon obtain a tolerable series of these objects for his cabinet. Now and then, however, a really fine pearl is found; and one, that was obtained from the Conway, now holds a place in the crown of England. This Mussel is tolerably active, and, if laid on the sandy or muddy floor of an aquarium, will soon assume its usual attitude.

The genus to which the CHINESE PEARL-MUSSEL belongs is distinguished by the thin elastic wings into which the valves are produced.

From this species the Chinese, those incorrigible tricksters, are in the habit of producing imitation-pearls by a very simple process. A string of small shot is introduced between the valves, and the animal restored to its native element. The irritation caused by the presence of the foreign body forces the mollusk to deposit the nacreous secretion upon the intruding substances, and after a while the shot are covered with layer upon layer of pearly substance, the thickness of the coating depending upon the length of time occupied in the construction.

The same ingenious people are also accustomed to make curious little pearl-covered josses, by stamping them out in thin bell-metal, slipping them into the shell, and leaving them between the valves until they are sufficiently coated with pearl.

In the THORNY CLAM, a curious member of another family, the shell is covered with long and branching projections, something like the horns of a young roebuck. All the Clams are natives of the warmer and tropical seas, especially among coral reefs, and their color and shape are extremely variable. Mr. Broderip writes of them as follows: "The shells are attached by their external surface to submarine bodies, such as coral rocks, and shells have been observed at depths varying from points near the surface to seventeen fathoms. These shells appear to be subject to every change of shape, and often of color, that the accidents of their position may bring upon them. Their shape is usually determined by the body to which they are fixed; and the development of the foliated laminæ which form their general characteristic is effected by their situation; and their color most probably by their food, and their greater or less exposure to light. The Chama that has lived in deep and placid waters will generally be found with its foliations in the highest state of luxuriance, while those of an individual that has borne the buffeting of a comparatively shallow and turbulent sea will be poor and stunted." The Clams are generally attached by the upper valve. The animal is edible, and is considered a great delicacy. About fifty species of Clams are known.

The Tridacnidæ are easily known by their deeply-waved shells, with the indented edges fitting into each other, and the overlapping foliations of the surface. Although separated from the true Clams, they are popularly called by the same name. The YELLOW CLAM is often buried in a mass of white madrepores. A well-known species, called from its enormous dimensions the GIANT CLAM (*Tridacna gigas*), was formerly rare, but is now tolerably plentiful. It attains to a gigantic size, sometimes weighing more than five hundred pounds, and containing an animal which weighs twenty pounds, and can furnish a good dinner to nearly as many persons.

The natives of the coasts on which it is found—namely, those of the Indian seas—are extremely fond of this creature, and eat it without any cooking, just as we eat oysters. The substance of the shell is extremely thick and solid, and enables it to be used for many ornamental purposes.

In former days, when this species was very rare, a magnificent specimen was presented to the church of St. Sulpice, in Paris, where it may now be seen, the valves being set up as *bénitiers* for containing the holy water. This shell dates from the time of Francis I. It is evident, that the byssus by which so enormous a shell is moored to the rocks must be of great size and strength, and, indeed, is so strong as to require an axe for its severance. The muscles, too, by which the animal contracts its shell are enormously powerful; and it has been remarked by Mr. Darwin, that, if a man were to put his hand into one of these shells, he would not be able to withdraw it as long as the animal lived.

The SPOTTED BEAR'S-PAW CLAM has been placed in a separate genus, on account of a difference in the number of projections on the hinge, technically called hinge teeth. The mouth is marked by a coronet-like circlet around it, and the foot is seen below just projecting from its groove.

This animal also spins a byssus, which is, however, weak and slight compared with that of the gigantic species just described.

THE family of the Cockles, or Cardiadæ, so called from their heart-like shape, is well represented by the common COCKLE (*Cárdium edúle*). Generally, the Cockle is a marine animal; but it sometimes prefers brackish water to the salt waves of the ocean, and a small variety is found in the Thames nearly as high as Greenwich, when the water is sensibly flavored with salt at each high tide. Another species, the PRICKLY COCKLE (*Cárdium aculeátum*), is found on the southern coast, and regularly brought to market.

The Cockle is gathered in great numbers for the purpose of being eaten, although, as the greater number are consumed in the open air, they can hardly be said to be procured for the table. According to Mr, Maxwell, "a crowd of the more youthful description of the peasantry are collected every spring tide to gather Cockles on the sands by daylight when the tide overruns. The quantities of these shell-fish thus procured would almost exceed belief ; and I have frequently seen more than would load a donkey collected in one tide by the children of a single cabin. They form a valuable and wholesome addition to the limited variety that the Irish peasant boasts at his humble board ; and afford children, too young for other tasks, a safe and useful employment."

This mollusk frequents sandy bays, and remains about low-water mark, burying itself in the sand by means of the powerful foot, which also enables it to leap to a surprising height.

The common HEART-COCKLE and the remarkable SPIRAL HEART-COCKLE differ in their form, according to their name. The latter is notable for the boldly spiral umbones.

The Spiral Heart-Cockle is in the habit of burrowing in the sand, leaving only the openings of the siphon above the surface. In the TUBERCULATED COCKLE these organs are at once recognizable by their fringed edges ; and the large foot is seen below, carrying the superstructure along. Even when taken out of the water, the Cockles are very lively; and if placed in a pan or basin they tumble about with great energy, knocking their shells against each other and the sides of the vessel with remarkable activity.

WE now come to a group of these shells where the siphons are extremely long. The first family is represented by the BANDED VENUS-SHELL, so called on account of its beautiful colors and elegant form, and the bands which traverse its surface. All the Venus-shells are handsome, and have well deserved their name. The shells are extremely hard in texture, thick, and smooth, and are mostly found in the warmer seas.

About one hundred and seventy species of Venus-shells are known, spread throughout all parts of the world, and ranging from low-water mark to a depth of one hundred and forty fathoms.

The *Venus mercenaria* of America, or QUAHOG, is an important bivalve commercially, ranking in this respect after the oyster and common clam (*Mya arenaria*). This shell-fish is, fortunately for the inhabitants of our coast, who depend on some kind for food and profit, distributed where the mya is not found. This shell was a very important article among the North American Indians. Beside depending greatly upon it as food, their money and ornaments were made from the shells. The blue of the interior of the shells was esteemed, and bits of certain shape were used as media of trading transactions.

The purple *wampum* was called by the New England Indians *Suckauhock*. This was valued at twice that made from the white shells.

The beautiful CYTHEREA is closely allied to the genus Venus, and is therefore appropriately named Cytherea, that being one of the classical epithets applied to Venus in consequence of her predilection for the island of Cytherea in the Ægean Sea. In this animal the two portions of the siphon do not diverge.

IN the family of the Mactridæ, or Trough-shells, the valves are of equal dimensions, and rather triangular in shape. The animal has the two channels of the siphon united as far as the extremity, and the foot is ample and strong.

The common Trough-shell is found on many coasts, always preferring those of a sandy nature, where it can hide itself by sinking just below the surface. The foot is capable of considerable motion, and can be extended to some length ; and when the movements are rapidly performed, it enables the creature to jump about nearly as actively as the cockle. The Trough-shells are found in all parts of the world, and in some coasts of the British islands are so plentiful that they are gathered for the purpose of feeding pigs. The species which is usually employed for this purpose is *Mactra subtruncáta*, and, like the cockle, it is taken at low water. Although so usually inhabiting the zones just below and above low-water mark, these shells are sometimes found as low as thirty fathoms beneath the surface.

The largest bivalve on our shores is the *Mactra solidissima*, or Great Clam, so-called in

New England. Hen Clam, Sea Clam, and Surf Clam are other local names. It is distributed between Carolinas and Labrador. This shell is prized by some, but it is extremely tough; the eatable part being the stout foot which composes the largest portion of the animal. By this foot the Clam is enabled to plough its way through the mud, and to leap considerably.

At extreme low tides on the Nahant beaches these clams are found imbedded just below the water-mark. Being so near shore, heavy storms throw up great numbers. The common Clam (*Mya arenaria*) is vastly more important, and is prized as an edible in all parts of the New England coast. South of Connecticut, along the coast, although this Clam is very large, it seems to be less palatable, and is consequently less esteemed. The cold waters farther north seem to add a certain excellence to this shell-fish; and this is noticeable in the case of most fishes.

SCROBICULARIA.—*Scrobicularia piperita.*
SWORD-BLADE RAZOR-SHELL.—*Solen ensis.*
COMMON TROUGH-SHELL.—*Mactra stultōrum.*
COMMON RAZOR-SHELL.—*Solen vagina.*

There are several clam-like shells, having dark colored epidermis, that are found in more northern waters; and some on the Grand Banks, and "Georges." The *Glycimeris* and *Panopœa*, we frequently took from the stomachs of fishes caught in those localities.

Besides these were many beautifully shaped small species, which the cod and haddock in browsing along the shelving rocks of the "Banks" have fed upon.

Among the most interesting of the bivalves that find a home in the waters on our New England coast are the *Solemyas*, *Solens*, *Ensis*, and *Siliquas*. Their beautiful shapes and glassy veiled coverings ranging from the pleasing shades of olive to dark chestnut of exquisite polish.

Thracia conradi is a notable shell, looking at first like the venus, but having a singular inequality of valves. One valve is quite convex, the other somewhat flattened. Chelsea Beach, near Boston, was the only locality known on the New England coast during many years. During heavy north-east storms numerous valves, more or less damaged, were thrown upon the beach, but in only one instance, during many years of frequent visiting at the locality, was a living specimen found. Dredging in all directions failed to discover them.

The very remarkable shell from which protrude two enormously long siphon tubes is the SCROBICULARIA, an example of the family Tellinidæ, all the members of which are notable for the length and divergence of these tubes.

WE now come to the well-known Solenidæ, or Razor-shells, so called on account of their shape.

These curious mollusks always live buried in the sand in an upright position, leaving only an opening shaped like a key-hole, which corresponds with the two siphon tubes. Those who are fond of examining the sand and rocks at low water will doubtlessly have been startled and amused by little jets of water which spirt some few inches in height, but never reappear.

These are caused by the RAZOR-SHELL; and if the locality whence the jet started be watched, the little keyhole-like orifice will be seen. To catch the mollusks that emitted the water is no easy task, but may be managed in two ways. The simplest but roughest method is to take an iron rod hooked at the end, plunge it into the sand like a harpoon, and pull it out smartly in an oblique direction, bringing with it the shell. This method, however it may answer for those who only want the creature for the purpose of eating the animal, or using it as bait, is by no means suited to those who wish to capture the inhabitant uninjured and to experimentalize upon it. These, therefore, must employ a different plan.

IN the next family, called Gaper Shells, because the valves when closed do not unite completely, but leave a moderately wide aperture at the hinder part, the shell is strong, thick, and opaque; the foot is comparatively small, and the siphons are united and retractile.

The GAPER SHELL inhabits sandy and muddy shores, and is especially fond of frequenting the brackish waters of river-mouths, where the streams are sure to bring with them a soft deposit of mud and sand. The species which is represented in the engraving burrows nearly a foot in depth into the sand, and is able to breathe and gain subsistence by the long siphons, which just protrude above the surface. In looking at this animal, and observing its habits, the entomologist is forcibly reminded of the manner in which the rat-tailed maggot, *i. e.*, the larva of *Eristalis tenax*, the great bee-like fly, with enormous eyes, is in the habit of hovering for a moment over a flower or leaf, settling for a moment, and then darting off again with lightning speed. Like the Gaper Shell, this larva spends its life deeply buried in the mud, carrying on the business of respiration by means of a long tube which, like the siphon of the mollusk, can be retracted or extended at will.

The Gaper Shell is much sought after in many places as an article of food, not only by man, but by birds and beasts, such as the walrus and the blue fox.

GAPER SHELL.—*Mya arenaria*.

THE nearly cylindrical WATERING-POT SHELL is a curious creature found in some of the hotter seas.

This species is a good example of a family termed the Gastrochænidæ, in which the valves are thin, gaping, and when adult, often connected with a rather long calcareous tube, as in the present instance.

The Watering-pot Shell derives its name from the curious perforated disc which closes its lower extremity, and bears no small resemblance to the rose of a watering-pot. In allusion to the same peculiarity, the French writers call the animal by the name of Arrosoir. All the species are burrowers, some into coral, some into stone, some into shells, and others into sand, as is the case with the creature which we are now examining. From the other end of the tube the siphons can be protruded to some extent, and withdrawn when the animal is alarmed.

One species belonging to this family, the *Gastrochæna modiolina*, has been known to drive its burrow fairly through some oyster-shells into the ground below, and then to make a permanent home by cementing all kinds of materials into a flask-like case and fixing its neck into the perforated oyster-shell.

THE very curious and common shells, popularly called PIDDOCKS, are found in profusion along the sea-coast.

The common Piddock may be found in vast numbers in every sea-covered chalk rock, into which it has the gift of penetrating, so as to protect itself from almost every foe.

Every one is familiar with the beautiful white shell of the Piddock, crossed by a series of elegantly curved projections, something like the teeth of a file. According to some writers, it is by means of these projections that the creature is able to burrow into the rock; and the possibility of such a feat has been proved by the simplest possible means, namely, by taking a Piddock into the hand and boring a similar hole with it. Mr. Robertson, who kept these creatures alive in their chalky burrows, devoted much time to watching them, and finds that during the process of burrowing they make a half turn to the right, and then back to the left, never turning completely round, and, in fact, employing much the same kind of movement as is used by a carpenter when boring a hole with a bradawl.

Mr. Woodward remarks very justly, that "the condition of the Pholades is always related to the nature of the material in which they are found burrowing; in soft sea-beds they attain the largest size and greatest perfection, whilst in hard and especially gritty rock, they are dwarfed in size, and all prominent points and ridges appear worn by friction. No notice is taken of the hypothesis which ascribes the perforation of rocks, etc., to ciliary action, because, in fact, there is no current between the shell, or siphon, and the wall of the tube." As soon as the animal has completely buried itself it ceases to burrow, and only projects the ends of the siphon from the aperture of the tunnel.

Some species of Piddock are eaten, *Pholas costáta*, one of the South American species being a good example. In Europe it is seldom used, except for bait, its fine white foot, which looks, when fresh, as if cut out of ice, answering that purpose admirably, its glittering whiteness serving to attract the attention of the fish, and its toughness causing it to adhere strongly to the hook.

Several other genera are worthy of notice, among which the Martesia is, perhaps, the most curious, shells belonging to this genus having been found in cakes of wax floating on the waves off the Cuban coast, and others in masses of resin on the shores of Australia. The PAPER PHOLAS is another species of this interesting genus.

Family *Petricolidæ* embraces certain clam-like shells that, as the name imports, live in stone. Our American species bores into wood, or more commonly it is found in the hard bottom, exposed between tide-waters. The shell is a chalky-white, and is long and considerably ridged. The celebrated Date-clam is much the same shape, but has a beautiful chestnut epidermis.

IN the Ship-worm we have an example of a creature, which, though useful enough in many ways, and doing good service in transmuting dead and decaying substances into living forms, is yet the dread of mariners and the terror of pier-builders.

The SHIP-WORM derives its name from its depredations on the bottoms of ships and all submerged wooden structures. It is found in most seas, and works fearful damage by eating into piles, planks, or even loose wood that lies tossing about in the ocean. I have now before me a portion of a pier which is so honeycombed by this terrible creature that it can be crushed between the hands as if it were paper, and in many places the wood is not thicker than ordinary foolscap. This piece was broken off by a steamer which accidentally ran against it; and so completely is it tunnelled, that although it measures seven inches in length and about eleven in circumference, its weight is under four ounces, a considerable portion of even that weight being due to the shelly tubes of the destroyers.

I have also a block of oak, where the Ship-worm has been nearly, though not quite so destructive as in the former instance. This specimen is notable, as giving an example of a principle on which many piers, etc., have been protected from this mollusk. A large iron bolt passes through the midst of the block, and the rust of the projecting head has spread itself for some distance over the wood. Multitudes of holes, large and small, surround the bolt, but not one has pierced that portion over which the rust extends. Knowing the objection entertained by the Ship-worm to rust, engineers have been in the habit of driving a number of short iron nails, with very wide heads, into the timber, arranging them in regular rows, with their heads at no great distance from each other. The action of the salt water soon causes the rust to spread over the spaces between the heads, and upon these spots the Ship-worm refuses to settle.

Another plan, and a very effective, though rather expensive one, consists in forcing a solution of corrosive sublimate into the pores of the wood. This salt of mercury is very destructive to animal life, and M. Quatrefages asserts that one twenty-millionth part of corrosive sublimate is enough to destroy all the young Ship-worms in two hours, and the ten-millionth part would have the same effect in forty minutes. He therefore proposes that ships should be cleared of this terrible pest by being taken into a closed dock, into which a few handfuls of corrosive sublimate should be thrown and well mixed with the water. The salts of copper and lead have a similar effect, but are not so rapid in their operation. The wooden piles on which jetties and piers are supported can be preserved in the same manner. Iron, however, is now rapidly superseding wood for such structures, and is quite impervious to the attacks of any mollusk, no matter how sharp its teeth.

When removed from the tube, the Ship-worm is seen to be a long grayish-white animal, about one foot in length and half an inch in thickness. At one end there is a rounded head, and at the other a forked tail. The burrow which the creature forms is either wholly or partially lined with shell, and it is worthy of notice that the Ship-worm and its mode of burrowing was the object that gave Sir I. Brunel the idea of the Thames Tunnel.

The Teredo did not always lead this fixed and darkling life, but at one time of its existence it swam freely through the ocean, having organs of sight and hearing for the purpose of guarding itself against the dangers of the deep.

Of all shell-fish, the Teredo is the most important in its relations to commerce. Its ravages are such that nothing short of an entire coating of copper plates on the hulls of vessels will suffice to prevent the serious injury sure to come to them when exposed in warm and temperate seas. On our coast, south of Cape Cod, spars and buoys are coated with verdigris paint.

It is an interesting fact that in the tropical regions, where the waters swarm with the eggs of the Teredo, there flourishes the palmetto-tree, the wood of which is a perfect resistant to the attack of the dreaded shell-fish. Piers are constructed of the palmetto logs, and prove of immense importance in our Southern harbors. In the warm waters of the Gulf of Mexico, the Ship-worms work with great rapidity. A pier of ordinary wood may seem to the eye wholly sound. On close inspection of it there will be observed on the surface minute holes, which, to the uninformed, are little suggestive of imperfection. Make a section of that wood, and we will see the interior of the log wholly replaced by the white, hard shells of the creatures, which entered in the young state, just before hatched upon the outer surface. These minute holes show where each young shell-worm penetrated. From these points they progress, eating the interior wood, and leaving nothing behind but the lime-shell tubes. Thus, when the pier seems to the eye intact, its integrity is wholly destroyed; the least jar or movement suffices to throw the structure down.

An enormous species of this genus, called from its dimensions the GIANT TEREDO (*Teredo gigantea*), has been found at Sumatra. This huge mollusk sometimes attains the length of six feet, and a diameter of about three inches, but, fortunately for timber, does not make its habitation in that substance, contenting itself with boring into the hardened mud of the sea-bed. The color of the shelly tube is pure white externally and yellow within. On account of its mud or sand burrowing habits, the specific title of *arenária* has been applied to this species.

TUNICATA.

The strange-looking creatures, as the Plonæa, the Sea-Squirt, the Clavellina, etc., have long perplexed systematic naturalists, and even now, although they have been the subject of careful examination by accomplished zoologists, many parts of their economy are enigmatical in the extreme. The order to which they belong is called by the name of Tunicata, because the animals possess no shell, but are covered with an elastic tunic. Some of them are transparent and really beautiful, while others are apparently little more than shapeless masses of gelatinous substance, studded with minute stones, fragments of shells, and coarse sand, overgrown with sea-weeds, and perforated by certain bivalve mollusks.

The simple or solitary tunicates are classed together under the name of Ascidiadæ. The common SEA-SQUIRT is a good example of the typical genus.

This animal, in common with all its kin, feeds mostly, if not wholly, upon the minute vegetable organisms, such as the desmids, diatoms, etc., which abound throughout the water, and the manner in which these substances are brought to the digestive organs is equally simple and beautiful. "The mouth," writes Mr. Rymer Jones, "is quite destitute of lips or other extensile parts, and situated, not at the exterior of the body, but at the very bottom of a capacious bag inclosed in the interior of the creature.

"It is obvious, then, that whatever materials are used as aliment, must be brought into the body with the water required for respiration ; but even when thus introduced, the process by which they are conveyed to the mouth still requires explanation.

"A truly miraculous apparatus is provided for this purpose. The whole surface of the respiratory chamber is covered over with multitudes of vibratile and closely-set cilia, arranged in millions, which by their united action cause currents in the water, all of which flow in continuous streams directly towards the mouth. It is sometimes possible, in very young and transparent specimens, by the aid of a good microscope, to witness the magnificent scene afforded by these cilia when in vigorous action.

—— ' salientia viscera possis
Et perlucentes numerare in pectore fibras.'

The effect upon the eye is that of delicately-toothed oval wheels revolving continually from left to right, but the cilia themselves are very much closer than the apparent teeth, the illusion being caused by a fanning motion transmitted along the ciliary lines, producing the appearance of waves, each wave representing a tooth of the supposed wheel."

Another tunicate is the CYNTHIA, one of a rather numerous genus, not uncommon on European coasts. The AGGREGATED CYNTHIA (*Cynthia aggregáta*) is to be found on almost any substance that has remained for any length of time below low-water mark, and stones, rocks, wooden piles, or even the larger sea-weeds, are frequently covered with these curious creatures, sometimes set in solitary state, and sometimes gathered together in groups by means of the interlacing of the fibres by which they attach themselves. Some species are eaten, *Cynthia microcosmus* being the most in favor, and regularly brought to market for sale. This animal derives its specific title from the multitude of animal and vegetable parasites that grow upon it, and so transform it into a little world.

Our next example is the PELONÆA, so called from two Greek words, the former signifying mud, and the latter to inhabit. This animal, as its name imports, is in the habit of burying itself in the mud, where it remains fixed and nearly motionless, respiring and obtaining nutrition by means of two open tubes seen at the smaller end. Only two, or perhaps three, species of this genus are known, and the animal is found in northern Scotland and Norway.

THE curious BOLTENIA, so called after Dr. Bolten, a naturalist, of Hamburg, is found in rather deep waters, being sometimes drawn up by fishermen's lines from a depth of seventy fathoms. The animals of this genus are attached to long footstalks, at the end of which the creature sways like a fritillary on its slender stalk. The two orifices by which water is admitted into and ejected from the system are seen, and their remarkable four-cleft openings are well displayed. When very young, the Boltenia is often found affixed to the stem of its parent.

The Boltenia, several species, is an unfamiliar animal, unless the observer is interested enough to go to the beaches after storms, when it will be found cast ashore, with great quantities of sea-weeds, kelp, etc. It is always an attractive creature, looking more like a rich peach or damson, with its beautiful pink and lemon coloration.

Many of the Ascidians are very uninviting in appearance.

The *Cynthia pyriformis* is one of the most beautiful of the race. It is called Sea-Peach, from its rich velvety surface and bright pink blush, precisely the aspect of a blood-peach.

WE now arrive at the Social Ascidians. Our first example of them is the CLAVELLINA. Its blood circulates through channels of communication, passing to and fro through separate tubes. It is a small creature, and extremely transparent, the latter characteristic making it a valuable species to the physiologist, who is enabled to watch its structure, and the methods in which the different organs perform their duties, without needing to dissect it. The Clavellina may be found on the European shores at low water, adherent to rocks, stones, or sea-weed, to which it attaches itself by means of the tiny root-like projections which are developed from the outer tunic, something like the little rootlets by which ivy clings to a wall.

Our second example is the SYNTETHYS, another European species. When full-grown, a group of these creatures forms a largish mass, nearly six inches in diameter, and as many in height, each member of the group being about two inches long. They are rather transparent and of a greenish color, and, when touched, they will contract themselves violently, and vanish into the common mass on which they are seated. These animals are propagated both by eggs and buds, the buds being produced on offshoots of the creeping tube. Sometimes the young one severs its connection with the parent, and fixes upon some fresh locality, there to form the basis of a new colony, but it frequently remains on the same spot, and only serves to increase the general mass.

OF the Botryllidæ, or Compound Ascidians, we may mention the common STAR-SHAPED BOTRYLLUS. The "tests," or equivalents of the shell of these animals, are fused into a common mass in which these individuals are imbedded. In the present genus the animals are arranged in a star-like form, each group consisting of a number of individuals, not less than six, and not more than twenty, in number. Many of these groups, or systems as they are technically called, are found upon the common test. The branchial orifices are simple, and the other orifice is common to all the members of the group, and forms, as it were, the centre of the radiating star. Six European species are known, which may be found on stones and sea-weed at low-water mark.

A VERY beautiful and curious mollusk, called from its luminous appearance the PYROSOMA, *i.e.*, Fire-body, is an example of the next family. This is one of the compound tunicates, and looks like a gelatinous cylinder, open at one end, and closed at the other, and having its body covered with numerous zoïds grouped in whorls. A large Italian-iron tube, studded with daisies, will give a good idea of its general shape.

The ejecting orifices of the aggregated animals all open into the hollow interior of the cylinder, and the consequence of this structure is, that by the constant flow of the rejected water, the whole mass is driven slowly and regularly through the waves. When seen at night they look just as if they were made of glowing white-hot iron, and they are at times so numerous as to choke up the nets of the fishermen, and diffuse so strong a light around them that even the fishes are rendered visible when they happen to swim within the sphere of its radiance. There is generally a greenish hue about the light.

Of the appearance presented by these animals when existing in great numbers, Mr. F. D. Bennett gives the following vivid and valuable account: " When assembled in the sea, and, as is usually the case, near the surface, these creatures present a gorgeous spectacle; their vivid phosphoric light being sufficient to illuminate, not only the extent of ocean they occupy, but also the air above, rendering all surrounding objects visible during the darkest night, and permitting a book to be read on the deck, or near the stern cabin-windows of a ship. They are occasionally collected together in incredible numbers. On two occasions, at midnight (in lat. 20° and 40° N. Atlantic Ocean), the ship sailed over many miles of water which they had illuminated, and in which they were so densely crowded as to be taken in any amount by buckets or nets.

"When captured, they exhibited no signs of animation, and emitted a peculiar half-fishy odor. When left in a vessel of sea-water, and allowed to be tranquil, their light was withheld, or only sparingly displayed; but when they were handled, or the water in which they were contained was agitated, their body instantly became one blaze of phosphoric light, which, upon close examination, could be observed to proceed from myriads of luminous dots, occupying the situations of the small brown specks, noticeable in the fleshy structure of the mollusk. Upon the irritating cause being removed, the phosphoric light gradually expired, and the Pyrosoma remained in darkness until again disturbed, when it once more illuminated objects with its vivid gleam; and this was repeated until after the death of the animal, when no luminous effect could be produced.

"When living specimens were immersed in fresh water, they not only existed for some hours, but emitted a constant light. Even after they had been so much enfeebled as to cease to give light in sea-water, or after they had been seriously mutilated, their phosphorescence invariably reappeared when they were put into fresh water, which appears to act as a peculiar stimulus in reproducing the phosphoric light of these, as well as of most other marine luminous animals.

"The Pyrosoma does not communicate its luminosity to water, nor to any object in contact with it (like many luminous Medusæ), its body being enveloped in a membrane that has no luminous secretion. But when the mollusk is cut open in water, some of the brown specks before mentioned will escape, and, diffusing themselves through the fluid, shine independent of the animal; in this respect, as well as in their structure and color, bearing some resemblance to the luminous scale on the abdomen of the small fire-fly of Bengal."

Our last example of these remarkable mollusks is the SALPA, which is mentioned on account of the curious phenomenon called "alternate generation," which is exhibited by this creature.

The Salpa takes two distinct forms, so entirely unlike each other that no one who was unacquainted with the circumstance would imagine that they could possibly belong to the same species. Sometimes the Salpæ are seen united in long chains, and swimming through the ocean with a beautifully graceful movement that greatly resembles the undulations of a swimming serpent. Sailors often call these chains of Salpæ by the name of Sea-Snakes.

The remarkable characteristic in this creature is, however, that the solitary Salpa produces a chain of united individuals, and that each of the united Salpæ becomes the parent of a solitary one. So that, as Mr. Rymer Jones happily remarks, "a Salpa mother is not like its daughter or its own mother, but resembles its sister, its granddaughter, and its grandmother." When swimming at ease through the water, the Salpa, like many other inhabitants of the ocean, is hardly perceptible, on account of the extreme transparency of its structure, the only indication of its presence being a kind of iridescence as the light plays upon the delicate membranes. The motive power is obtained by regular contractions of the body, by which the refuse water is rejected with some force, and thus drives the creature along by direct action, just as a rocket is propelled through the air. It is a remarkable fact, that in the chain of united Salpæ, each individual expands and contracts in exact unison, so that the force is applied to the water in the strongest possible manner. Sometimes the chains become broken up, but the fragmentary portions do not seem to be at all inconvenienced by the change in their condition, swimming about as actively as before. The creature is very slightly luminous, giving forth its phosphorescent light when touched, and especially when pressed.

POLYZOA.

INFUNDIBULATA.

THE very remarkable beings which now come before our notice are appropriately termed POLYZOA, from two Greek words, signifying "many animals;" because a large number of individuals are massed together in groups of various forms and textures. Some naturalists mostly designate them by the term of Bryozoa, or "moss animals," on account of their frequent resemblance to the various mosses; but, as this term has been employed in far too wide a sense, grouping under one common designation a number of beings belonging to different classes, the more recent observers have decided on the more appropriate title of Polyzoa.

For very many years—indeed, from the earliest days of natural history until comparatively modern times—the Polyzoa were ranked among the vegetables; and a learned Italian observer who ventured to express his opinion that they partially, at least, partook of the nature of animals, was persecuted by the professors of the day with the usual acrimony excited by a discoverer who is in advance of his time. Even the acute and experienced Linnæus could not receive the new doctrine, which was for a while "exploded" by the researches of another naturalist, who announced that he had seen corals in flower, thus setting the question at rest in the minds of those who desired to be so convinced.

Truth, however, stood its ground, and though for a time suppressed by those who had a personal interest in maintaining the theories which they had so long promulgated, in the due course of events became triumphant.

The true animal nature of these and many other beings, which had been formerly classed among the vegetables, was at length fairly proved by the researches of two eminent men, Trembley and Ellis, the latter of whom may lay claim to the honor of having produced the best and most comprehensive work of his time; a work, indeed, which is valuable even at the present day, owing to the invariable clearness and occasional brilliancy of the descriptions, and the number and accuracy of the engravings.

Ellis called all these creatures by the name of Corallines, a title now given to one of the true vegetables, but discovered many anatomical and physiological details, and set their animal nature beyond a doubt. All his researches were conducted with the aid of instruments which in our day would be thought almost useless, the microscope employed being only a simple lens mounted on a stand, and devoid of the complicated apparatus for magnifying and illuminating that now afford such aid to the observer.

After the animal nature of the Polyzoa had been fairly established, they were confounded with many other marine and aquatic inhabitants, such as the corals and the various zoophytes, in consequence of the superficial resemblance between their external forms. Lately, however, their true place in the animal kingdom has been discovered, and their affinity with the lower mollusks clearly proved, the tunicates forming the connecting link between the mollusks proper and the molluscoids, as these animals are sometimes called.

Having glanced at the general history of these curious and really beautiful animals, we will proceed to examine the form and characteristics of the individual species.

Should the reader obtain from the sea or fresh water a being which is evidently either a zoophyte or one of the Polyzoa, he may set his doubts at rest by examining the tentacles, and if he finds that they are furnished with cilia, or minute filaments, he may assure himself that they belong to the group of animals on which we are now engaged.

The forms assumed by the general mass of the various species of Polyzoa are extremely different, some resembling twigs or mosses; others looking like lumps of spongy substance adhering to sticks, stones, or leaves, or even lying freely in the water; others being flat and ramified, like broad-leaved sea-weeds; others spreading film-like over leaves, stones, shells, or similar objects; while a few are able to crawl at liberty, the entire organism being animated by some wonderful instinct, which urges all the myriad individuals of which it is composed to employ their force in the same direction.

The number of these creatures is so vast, that it is impossible to give more than a brief description of them; but in the following pages it will be found that a careful selection has been made of the typical forms, and that sufficient details of their structure will be given to enable the reader to form a general idea of the subject, and in most cases to refer any specimens which he may find to their genera or families. Those who desire further information on the subject will find it in Busk's elaborate catalogue of the Marine Polyzoa, and the large work by Allman on the Fresh-water Polyzoa.

Putting aside the classification of the polyzoa until the termination of the work, we will proceed at once to the description of the many species of this class.

The first family of the polyzoa is known by the manner in which the cells are arranged around an imaginary axis, and connected with each other by flexible stalks. The general shape of the whole group, or "polyzoary," as it is termed by some authors, is very shrub-like, standing boldly erect, and giving out branches by two and two, after the fashion called by botanists "dichotomous."

An example of one of these beings, the LITTLE CHAIN, or BREAST-PLATE, is plentifully found in the sea, and is properly classified among the zoophytes. The *Catenicella hastata* is somewhat remarkable for the shape of the cells, the form of their mouths, the method in which they give out their branches, and the peculiar organs called technically "avicularia" and "vibracula;" the former being processes that in many species bear an almost absurdly close resemblance to the heads of birds; and the latter, curious hair-like projections, which move regularly backward and forward as if impelled by machinery. These remarkable organs will be presently described more fully.

The members of the present genus are found most commonly in the Australian seas, seldom in the southern hemisphere, while in the northern hemisphere they are almost entirely unknown. Many specimens have been taken from Bass's Straits, at a depth of forty-five fathoms. As a general rule, however, the polyzoa prefer the shallower waters, and are most commonly found a little below low-water mark.

Another species belonging to the same genus is remarkable for the long pointed spines that project from the margin, like a pair of cow's horns. In allusion to this peculiarity it is called *Catenicella cornúta*.

Another curious polyzoon, termed *Calpidium ornatum*, is also found in Bass's Straits, at the same depth as the preceding species. It shows a singular method of construction.

Each cell is extremely wide in proportion to its depth, and instead of possessing but one mouth, it is pierced with three apertures shaped something like keyholes. It is conjectured that each cell is inhabited by three separate individuals, a supposition which is strengthened by the great comparative dimensions of the cell and the thickness of its walls. Still, no sign of internal partitions have been discovered, although some remains, apparently of the inhabitants, have been seen at the bottom of the cell. In some cases there are only two apertures to each shell.

An example of the typical genus of this family is the *Salicornaria farciminoides*. The strange specific name of this creature is given to it on account of its external resemblance to the Farciminaria, another genus of polyzoa which will be presently described. In this genus the cells assume a kind of honeycombed aspect, being almost hexagonal in their shape and pressed closely together.

In this place it may be as well to mention that in all the species belonging to the first subdivision of the polyzoa the mouth is not quite at the extremity of the cell, is of a somewhat crescentic form, and furnished with a movable lip or door, which closes the aperture when the animal retreats. In many cases this lip is membranous. All the marine polyzoa are termed Infundibulata, or Funnel-shaped animals, on account of their form; their fresh-water relations being called Hippocrepia, or Horseshoe animals, because the tentacles are arranged in a shape resembling that of a horseshoe. It will be, perhaps, hardly necessary to apologize for the introduction of so many technical terms, the fact being that the minute dimensions of the objects have caused them to escape popular observation, and to depend for their nomenclature upon the learned and scientific. Still, the technical phraseology is never employed where its use can be avoided, and when circumstances render its introduction inevitable, its meaning and the reasons for its employment are always given.

We now arrive at another family, the Cellularidæ, where the general shape resembles that of the preceding family, but the cells, instead of being arranged round an imaginary axis, and so forming cylindrical branches, are arranged on the same plane. A magnified example of this family is the *Cellularia peachii*, so called in honor of the eminent naturalist, Mr. Peach.

In a creature belonging to the genus Menipea, found in Tierra del Fuego, and termed from its *habitat*, *Menipea fugueris*, the curious "operculum" closes or rather guards the mouth of the cell. In this genus it is in the form of a simple spike. This species is found at low water.

The avicularium is an object which is set somewhere about the middle of a cell, and always upon its outside, and assumes various shapes in the different species of polyzoa. What may be the precise nature of the avicularia is at present rather a mystery, and no one can definitely pronounce them to be actual portions of the cell, or merely parasites that remain affixed to the same spot. In all cases there is a decided resemblance to the head of a bird, though in some species the similitude is closer than in others. Only one avicularium is to be found on a single cell, though many cells do not possess these strange appendages.

By close examination, it will be seen that the avicularium can be roughly distinguished into three portions; namely, a base by means of which it is attached to the cell, a rather large head, and a movable spine like the lower mandible of a bird's beak. In those examples where the avicularium is seated directly upon the cell, the only movement is that of the lower mandible, which opens and shuts with a continual motion, as if it were a veritable head of a hungry bird snapping at its food. In those cases, however, where the base is lengthened into a neck, the entire head is endowed with motion, nodding up and down in the most lively manner, very like those wooden birds sold in the toy-shops, whose head and tail are alternately raised and depressed by means of strings and a weight. But whether the head moves, or is still, the jaws continually open and shut, and will often inclose between their parts any small worm that may happen to come across their path, and have even been known to seize each other in their grasp.

When the beak has seized a victim, and the mandibles closed upon it, they retain their grasp with astonishing tenacity, and when, as sometimes happens, two avicularia have seized the same worm, the unfortunate victim is rendered entirely helpless by the grasp of its foes.

The purpose of these objects seems to be rather dubious, but two conjectures have been offered, which at all events are worthy of notice.

According to the opinions of some observers, the avicularia answer the purpose of police, and force intruders to leave the spot where their presence might do harm to the creature on which they are placed. This duty seems, however, to be performed by the vibracula, and we must search for another theory for the true object of the avicularia. Mr. Gosse has put forward a conjecture which is not only highly ingenious, but bears with it the elements of probability.

"More than one observer," he remarks, "has noticed the seizure of small roving animals by these pincer-like beaks, and hence the conclusion is pretty general, that they are in some way connected with the procuring of food. But it seems to have been forgotten, not only that these organs have no power of passing the prey thus seized to the mouth, but also that this latter is situated at the bottom of a funnel of ciliated tentacles, and is calculated to receive only such minute prey as is drawn within the ciliary vortex. I have ventured to suggest a new explanation.

"The seizure of a passing animal, and the holding it in a tenacious grasp until it dies, may be a means of attracting the proper prey to the vicinity of the mouth. The presence of decomposing animal substance in water invariably attracts crowds of infusory animalcules, which then breed with amazing rapidity, so as to form a cloud of living atoms around the decaying body, quite invisible in the aggregate to the unassisted eye; and these remain in the vicinity, playing round and round until the organic matter is quite consumed. Now, a tiny annelid or other animal caught by the bird's head of a polyzoon and tightly held, would presently die; and though in its own substance it would not yield any nutriment to the capturer, yet by becoming the centre of a crowd of busy infusoria, multitudes of which would constantly be drawn into the tentaculean vortex and swallowed, it would be ancillary to its support, and the organ in question would thus play no unimportant part in the economy of the animal."

We now proceed to the vibraculum. It is hollow, the interior being filled, during the life of the animal, by a fibrous contractile substance, which enables the organ to perform its curious movements. These movements are very irregular as regards time, but very regular in their directions, each vibraculum sweeping slowly over the whole surface within its reach, first moving in one direction and then in the other, and it is sufficiently notable that these movements will continue for several days after the death of the polype to which it is attached.

The mouth of a cell belonging to another polyzoon shows a curious operculum, with a branched form, like the horn of a fallow deer, and may be contrasted with the simple spiny operculum.

I AM now going to describe several curious and bizarre forms of Marine Polyzoa. One of them is the Bull's-horn Coralline of Ellis—the Ladies' Slipper, as it is more elegantly and equally appropriately named at the present day. The cells of this species bear a considerable resemblance to a series of delicate, slender-toed slippers, adherent to each other, while from the opening protrudes the beautiful bell-shaped circle of tentacles. Sometimes a rudimentary cell may be found, but always below the aperture.

A common creature is the Snake-head Coralline, so called from the extraordinary similitude with a reptile.

In another species, which is called *Beánia mirábilis*, the mouth is surrounded with a series of thorns or spines. It is found mostly on shells. Each cell is united to its predecessor and successor by a slender tube.

The curious Farciminaria, remarkable for the array of short and stout spines with which its surface is thickly studded, is a New Zealand species, and appears to be the sole representative of its family. It grows in slender branches, which are dichotomous.

In the family to which the *Gemellária loricáta* belongs, the cells are arranged in pairs and opposite each other, the orifices of the pairs looking in the same direction. This species is the Coat-of-Mail Coralline of Ellis, deriving its name from the shape of the cells, which bear no slight resemblance to steel corslets.

The succeeding family, of which the *Dimetópia spicáta* is an example, may be known by the arrangement of the cells, which are in pairs, but with their mouths placed at right angles to each other. When growing, it is a very pretty species, being white, nearly transparent, and attaining a height of about three inches. It grows in thick tufts, and is found in Bass's Straits.

The Shepherd's-purse Coralline of Ellis (*Notámia bursária*) is a common European species, and its peculiar avicularium shows a tobacco-pipe-like head.

A most curious vibraculum, which is toothed like a saw, belongs to a creature called *Caberéa patagónica*, living in the country from which it takes its specific name.

On a very remarkable species, the *Bicellária ciliáta*, the cells are surrounded by long processes. An avicularium belonging to another species of the same genus is conspicuous for the enormously long stalk of the head, and the three finger-like appendages at the base.

A tolerably common European species is the Bird's-head Coralline (*Búgula avicularia*), popularly so called on account of the number, shape, and activity of the avicularia. Our attention is now called to a well-known polyzoa, which may be found lining the sides of rock-pools, or affixed to shells, and even to living crustaceans, the spider-crab being often enveloped in its soft, plumy branches to such an extent that it marches beneath their shade, like Macduff's army under its leafy disguise. I know scarcely a more wonderful sight than is presented by a living specimen of the Bugula, with its wonderful appendages in full action. As if moved by machinery, they nod up and down like automata, sometimes throwing themselves back like the head of a fan-tail pigeon; the mouth opens slowly, with a wearied kind of air, that almost forces the observer to yawn in sympathy with the deliberate movement, while ever and anon the jaw suddenly closes with a snap so sharp that the ear instinctively watches for the sound.

EVERY one who has walked along the sea-shore must have observed the pretty, leaf-like Sea Mats strewn on the beach, and admired the wonderful regularity of their structure, perceptible to the naked eye; but when magnified even by a pocket lens, their beauty increases in proportion to the power employed, and the marvellous arrangement of the cells, and the orderly system in which they are placed, are almost beyond belief. Beautiful, however, as they are in this state, they are but the dead and lifeless habitations of the creatures who built the wondrous cells, and the only method of showing the Sea Mat in its full glory, is to take a living specimen from the stone or shell to which it is affixed, and watch it under the microscope while the creatures are still in full activity. The common Sea Mat is sometimes called the Hornwrack.

The peculiar manner in which the polypes of the Sea Mat protrude themselves is quaintly and accurately described by Mr. Gosse in his "Evenings at the Microscope." After pointing out the cradle-like shape of the cells, he proceeds as follows: "Suppose that a coverlid of transparent skin were stretched over each cradle from a little within the margin all round, leaving a transverse opening just in the right place, viz., over the pillow, and you would have exactly what exists here. There is a crescent-form slit in the membrane of the upper part of the cell, from which the semicircular edge and lip can recede if pushed from within.

"Suppose, yet again, that in every cradle there lies a baby with its little knees bent up to its chin, in that zigzag posture that children, little and big, often like to be in. But stay, here is a child moving! Softly and slowly pushes open the semicircular slit in the coverlid, and we see him gradually protruding his head and shoulders in an erect position, strengthening his knees at the same time. He is raised half out of bed, when lo! his head falls open, and becomes a bell of tentacles. The baby is the tenant polype."

The Toothed Sea Mat is a variation with curious tooth-like appendages from which it derives its name.

A curious polyzoon, bearing the name of *Carbasea episcopalis*, is found in Bass's Straits at a depth of forty-five fathoms. This species is found in two forms, either parasitic on sertularia and various polyzoa, and then of small size, or leading an independent existence, and reaching considerable dimensions. It is chiefly remarkable for the singular form of the ovicells, which bear a wonderfully close resemblance to bishops' mitres, and have earned for the species the title of *episcopalis*.

Two specimens of another genus are called *Diachoris magellanicus* and *Diachoris crotali*. The latter shows remarkable appendages which guard the mouth; and the former, *Diachoris magellanicus*, exhibits the method in which each cell, except at the margins of the fronds, is connected with six others, something like the stellate cells in pith. In fact, the Diachoris is a flustrum dissected, the cells being drawn away from each other and connected by stalks. The central cell is by connecting stalks united to the six that surround it.

Any one who picks up a piece of a dark sea-weed, will find that many parts of its structure are covered with a peculiar growth, that looks as if a portion of Sea Mat had been cemented upon it. This substance is indeed closely allied to the Sea Mat, and is chiefly to be distinguished by the membranous nature of the polyzoary, which will not permit it to stand boldy erect after the manner of the true Sea Mat. This species is called *Membranipora pilosa*.

The feathery plume of tentacles is extremely graceful, and, when the creature is living, has a remarkably elegant effect. In a specimen now before me, viewed by a power of only thirty diameters attached to the binocular microscope, the polypes of the Membranipora are beautifully exhibited, some shut up closely in their homes, some just putting forth their heads from the cells, others half protruded, and a few with the plumy tufts displayed in all their beauty. It is as well to view this and other polyzoa with different kinds of illumination, both as opaque and transparent bodies; artificial light is, however, to be avoided.

An allied species is called *Lepralia landsborovii*. In the Lepralia, as well as the Membranipora, the process of development is very interesting, especially as it can be readily watched under the microscope.

Towards the end of May, specimens of Lepralia, Flustra, and Membranipora should be procured and placed in shallow glass vessels containing sea-water. After a little time, minute beings, much resembling the ordinary infusoria, are seen swimming about. Presently, the "gemmules," as these creatures are technically named, become stationary, affix themselves to some definite spot, and develop a feather-covered polype, being now similar in shape to a single cell of the species from which it was produced. Buds, or projections, are soon formed at the sides, which are rapidly developed into new cells, and in their turn are the means of putting forth new cells. Thus it will be seen that each polyzoary spreads from a centre; and that, although a free gemmule is capable of producing stationary cells, the greater number of cells have never passed through the state of their original progenitor. When a polyzoary has attained a considerable size, it is not unfrequent to find the margins of the group filled with vigorous and lively polypes, being those last produced, while the centre is composed of empty cells, the original inhabitants having died out from old age.

NET-PORED ANIMAL.—*Retepora cellulosa*.

The *Lepralia spinifera* differs through the short sharp thorns with which the edges of the cell are guarded, and the curious ovicell, notable for its beautiful sculpturing of ridges radiating around a centre. The *Lepralia trispinosa* is distinguished by three long spines from which it derives its name.

A LARGE and interesting genus with about forty or fifty known species, comprises the species of Lepralia, Eschara, Lunulites, Cupularia, and Selenaria. Among these especial attention must be drawn to *Lepralia monoceros*, or the Unicorn Lepralia, so called on account of the single horn, or club, with which it is furnished; the *Lepralia alata*, or Winged Lepralia, remarkable for the classically elegant sculpturing and the projecting "wings;" and the *Lepralia variolosa*, so called on account of the peculiar mottlings which are thought to bear a resemblance to the face of a person seamed with small-pox.

The spoor-like avicularium of the *Cellepora fusca* is a creature notable for its urn-shaped and chalky, stiff cells, arranged either irregularly or in the form of a quincunx, *i. e.*, like the cinque spots on a die, ∶·∶ It is a native of Bass's Straits. Two tolerably common species are the *Eschara foliacea*, so called from its superficial resemblance to the scar left by a wound, and the *Eschara flabellaris*, a very curious polyzoon of the same genus. The latter is remarkable for the hood or helmet-like ovicells. About eleven species of this genus are known.

In the illustration on the opposite page is shown one of the most curious of the polyzoa, named, from the external resemblance which its apertures present to the meshes of a net, the *Retepora*, or Net-pored animal. The polyzoary of this species is hard, chalky, is only pierced on one surface, and has so much the appearance of the true stony corals, that it might easily be mistaken for one of these objects. It is a European species, and the specimen represented of the natural size.

The next interesting family of polyzoa is called Selenariadæ, because they are round as the full moon, or Norval's famous shield. In the circular form of the margin they all agree, but differ considerably in their curvatures, some being with one side plane and the other convex, while others are convex-concave, like a watch-glass, or, to speak more accurately, like the lens technically called a meniscus. They are all remarkable for their very large vibracula.

One of these creatures is called *Cupularia lowei* on account of its resemblance to a dome.

A good example of a very shield-like genus of this family is the *Selenaria maculata*. The reader will not fail to remark its exact resemblance to the target-shield used by many nations, ancient and modern. It may be here mentioned that in this last-mentioned family the vibracula are thought by some naturalists to act as locomotive organs.

Our space is so rapidly drawing to a close, that it will be hardly possible to give much more than a brief account of a few more interesting examples.

IN the group of polyzoa I am going to describe, there is a great external similarity between their forms and those of the true stony corals so familiar to us. It will, moreover, be found, that in many details of their structure, there is a decided analogy between them and the true zoophytes which will be described in the latter part of the work.

In the *Crisia eburna* the arrangement of the cells is simple and elegant, and the various branches are connected with each other by means of certain horny joints. The use of these joints is, in all probability, to enable the polyzoary to resist the action of the waves, and so to avoid the fractures which would probably result if the joints were as stiff and inflexible as the cells. A similar provision will be seen in the *Gorgonia*, a zoophyte which will be described on a future page.

A remarkably constructed species, the *Idmonea atlantica*, looks something like a many-legged spider, with its branches protruding from a rounded centre which represents the body of that animal.

A polyzoon remarkable for the profusion and great comparative length of the cells, is termed *Pustulopora delicatula*, the generic name being on account of the minute dots with which the surface of the cells and polyzoary is studded, and the specific title in reference to the delicate structure and soft brown hue with which it is colored.

A stoutly-built polyzoon named *Hornera*, is notable for its resemblance to several corals.

IN the accompanying illustration, the resemblance to the corals, the madrepores, and even the stony habitations of certain marine worms, is very close and striking.

Fig. A represents a polyzoon fancifully entitled *Alecto*, in honor of one of the Furies of that name. It is seen of its natural size as it appears while spreading itself over the inner surface of a shell. At Fig. B, a portion of the same species is shown as it appears when magnified, and is given to exhibit not only the method in which the cells are sunk into the polyzoary, but the mode in which the branches are developed from each other.

At Fig. C is delineated a portion of the appropriately named *Tubulipora serpens*, a being which has the cells even more elongated than in the *Pustulopora* which has been lately described. The singular resemblance between the lengthened cells of this species and the hard shelly tubes of the well-known *Serpula*, so familiar on account of its scarlet and white plumes and marvellously engraved stopper, must be evident to every one who has seen the little creature, or even noticed its empty habitation.

Fig. D represents a curious species, called from its shape *Discopora patina*, the former word being of Greek origin and given in allusion to its disc-like form, and the latter being a Latin word signifying a flat dish like our present champagne glasses. The numerous pores or orifices through which the animal protrudes, are seen upon the surface. A magnified example of the same species is seen at Fig. E, having been broken asunder in order to show the manner in which the cells are massed together, so as to produce a honeycomb-like aspect at their mouths. If the reader will compare this with the last-mentioned species, the evident connection between the two will be readily perceptible.

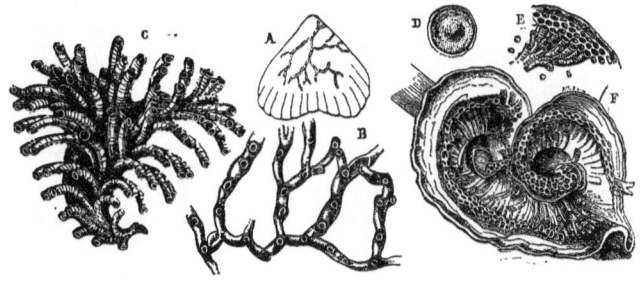

POLYZOA.

A. *Alecto dichotoma*.   B. *Alecto dichótoma*. ×   C. *Tubulipora serpens*.   D. *Discópora pátina*.
E. *Discópora pátina*. ×   F. *Discópora pátina*. × (Contorted.)
(The sign × signifies that the object is magnified.)

A very remarkable modification of the same species may be seen, rather magnified, at Fig. F. The original gemmule from which the whole mass sprang had made a mistake in its settlement, having fixed itself upon a slender stem where it could find no space for its expansion into the normal circular form. Being fixed, it could not move, but philosophically made the best of the situation, and finding itself unable to spread into a single disc-like body, and equally unable to extend beyond the supporting substance to which it was affixed, adopted a compromise, and coiled itself into the singular form here represented.

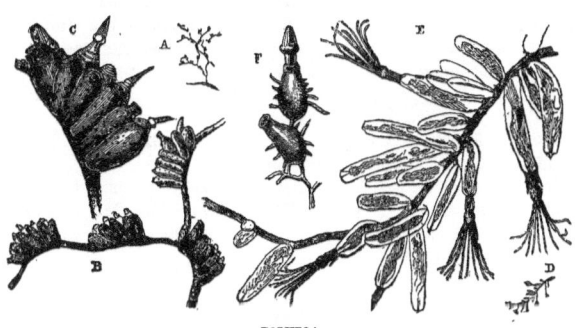

POLYZOA.

A. *Serialária lendigeri*.   B. *Serialária lendigeri*. ×   C. *Serialária lendigeri*. × ×
D. *Bowerbánkia imbricáta*.   E. *Bowerbánkia imbricáta*. ×   F. *Buskia nitens*.
The sign × signifies that the object is magnified.

Our next group of polyzoa exhibits some very remarkable forms. At Fig. A is seen a specimen of the *Serialaria lendigeri*, a species which without the aid of a lens presents no particular points of interest, but, when magnified, is seen to be a really curious being. Two branches of this creature are represented at Fig. B, as seen when moderately magnified, and the further details of its structure are given at Fig. C, where the polypes are shown protruding from their bases, and the peculiar dottings of the cells are seen. It will be noticed that the cells are gathered into groups, connected with each other by the stalk-like processes of the polyzoary.

Fig. D represents a sprig of the *Bowerbankia imbricata* of the natural size, and the same species is shown much magnified at E. In this species the cells are also placed upon the footstalk formed by the polyzoary, but they are not grouped together as in the last-mentioned species. The polypes are long and slender, and the walls of the cells are delicately transparent, thus allowing the observer to examine the structure of the polype through the walls.

A part of the digestive organs of the *Bowerbankia* is deserving of a passing notice. Like all the other species, it possesses a feathery crown of tentacles sprouting around the mouth, and directing the minute objects which serve it for food, from the mouth into the œsophagus, popularly called the gullet. In this genus, however, a further provision is made, for immediately below the œsophagus comes a kind of contractile gizzard, lined with a marvellous pavement of teeth arranged in a tesselated formation, and capable of bruising and crushing the food before it is passed into the stomach and thence to the intestine. One species of this genus, *Bowerbankia densa*, is common on the English coasts, being found parasitic on *Flustra foliacea* in patches of about an inch or so in diameter, and may be readily obtained by those who are desirous of studying its habits and structure.

At Fig. F is given a small portion of a polyzoon named *Buskia nitens*, the former title being in honor of the eminent naturalist, and the latter alluding to the shining appearance of the species.

WITH the next descriptions our examples of the Marine Polyzoa are terminated.

Of the *Alcyonidium gelatinosum*, popularly called the Sea Ragged Staff, Mermaid's Glove, or more commonly, Dead Man's Fingers, in allusion to the cold dampness of its surface, great numbers can be found on the sea-shore, especially after a storm, when it may be seen lying among the masses of sea-weed and other débris that are flung on the beach by the angry waves. In its natural state it is affixed by its base to stones, shells, and other supports, and is always extremely irregular and variable in its form, no two specimens being alike. When picked up, its aspect is anything but attractive, but when placed in sea-water and suffered to remain at rest for a while, it becomes a most beautiful object. From each of the tiny pits with which its surface is thickly studded, projects a polype, with a beautiful crown of waving tentacles, and so numerous are these polypes, that they densely cover the surface and render microscopic observation rather difficult.

As in other species, fresh colonies of the Alcyonidium are formed by gemmules, which are given forth from the general mass, swim about freely for a time, by means of the cilia with which their surface is thickly studded, and when they have attained a proper age, settle down and at once begin to develop fresh cells on all sides. The little vesicles wherein the gemmules are originally formed, may be seen in the spring scattered through the transparent substance of the polyzoary, and looking like little white points. Each vesicle contains about five or six gemmules, and as it can be easily isolated, its rupture and the consequent escape of the gemmules can be easily seen in a moderately powerful microscope.

Our next ample is the *Pedicellina echinata*. These little creatures look wonderfully like the common moss that grows so plentifully on walls. The cilia by which the necessary currents are formed in the water for the purpose of obtaining food, is similar to a tulip in its form.

PHYLACTOLÆMATA.

ONE of the most remarkable polyzoa that at present are known to exist, is the *Cristatella mucedo*. The entire polyzoary is not only free and unattached to any object, but even possesses the power of locomotion. It is frequently seen to crawl over the stem of some aquatic plant. In order to qualify it for this process, the lower surface of the polyzoary is modified into a flattened disc, which thus becomes analogous to the foot of the gasteropodous mollusks already described. The substance of the disc is contractile.

To an ordinary eye, that any creature should crawl, would not appear a very surprising fact, but to the mind of a naturalist, the whole phenomenon is full of wonder. It is easy enough for a single being to advance in a given direction, and even though it has a very army of legs, like a centipede or a julus, the limbs are all directed by the same mind. But in the present case, there is no common centre to which the wills of the myriad polypes that compose the group can be referred ; and the locomotive capacities of the Cristatella remain one of the many unsolved mysteries with which nature abounds.

In all respects, this is a remarkable species. Instead of hiding in darkness and coveting the shade, as is the case with nearly all the polyzoa, the Cristatella exults in light, and loves to crawl in shallow waters where it is exposed to the full blaze of the meridian sun. The ordinary length of the Cristatella is from one to two inches, and its general aspect reminds the observer of a yellowish-brown hairy caterpillar, softer than the well-known woolly bear, or larva of the tiger-moth, and indeed looking much as if it were made from the soft silken substance denominated "chenille."

It is one of the fresh-water species, and the plume of tentacles is not funnel-shaped, but formed as if set in a horseshoe.

A most marvellous production, which requires some explanation, is the so-called "statoblast." It is a rather formidable object. The statoblasts are developed within the cavity of the parent, where they may be seen of different sizes, and in most cases arranged like beads on a string. They consist of two nearly hemispherical or oval discs, which are united at their edges, and the line of junction strengthened by a more or less deep ring, so that the general aspect of the statoblast is not unlike that of the planet Saturn.

In the Cristatella and another genus, the Pectinatella, the statoblast is armed with a double series of hooks, starting from the edges of the discs, those of the opposite discs alternating with each other and extending well beyond the ring. The spines are gradually developed, and force their way through an enveloping substance which surrounds the statoblast. After a while it often happens that the soft gelatinous envelope is washed away.

These wonderful objects contain the future young ; and the process of development is briefly as follows : Within the walls of the parent they attain their full size, and when the parent dies at the end of the season, they are liberated and pass from its body. They then attach themselves to subaquatic substances, such as vegetables, sticks, stones, etc., and at the beginning of the next season the two discs separate, and out comes the young Cristatella, ready to take upon itself the tasks for which it was created. It often happens that the two discs of the statoblast cling to the young for some time after it has given up its contents, and the little creature carries about the separate halves in a manner that reminds the observer of a bean newly sprouted from the ground and bearing the two halves of the seed which was planted. The Cristatella also produces buds, and in fact, the statoblast is a kind of bud of rather peculiar construction. The disc of the statoblast is brown.

An example of an interesting polyzoon found in ponds and streamlets, and often adhering to the rootlets of duck-weed, is the *Lophopus crystallinus*. It deserves peculiar interest as being the first species of polyzoa that was detected. The honor of its discovery rests with Trembley, who named it appropriately "Polype à panache," the plume-like group of tentacles being sufficiently large to be seen with the naked eye. In this creature, the place of the external wall or ectocyst, is taken by a soft gelatinous envelope.

Mr. Allman remarks that in the interior of the Lophopus are often to be seen a vast number of little glittering particles of a pear-like shape, which move about through a series of tubes connected with each other like the capillary vessels of the vertebrates. After much investigation of the subject, he came to the conclusion that they were merely parasitical.

Our history of the polyzoa is soon concluded. A fresh-water polyzoon called scientifically the *Alcyonella fungosa*, and, popularly, the Fresh-water Sponge, because when dry it has a very sponge-like aspect, is found in masses of considerable size, sometimes weighing as much as a pound, adherent to various substances which are constantly beneath the waters in which it lives. It frequently develops itself round the pendent twigs of the weeping-willow and other trees which dip the extremities of their branches into the water. I have seen the timbers of locks quite encrusted with the Alcyonella in many places.

When carefully removed and placed in fresh water, it gradually develops a kind of white downy appearance over its entire surface, which disappears with the rapidity of magic if a hand is moved quickly over the vessel in which it is lying. This downy appearance is caused by the tentacles which protrude themselves in vast numbers, and instantly retract when the creatures are alarmed.

Even in swallowing its food, the Plumatella displays considerable powers of discrimination, accepting some particles as they pass over the mouth, and rejecting others as unworthy of reception. Its usual places of abode are under stones, submerged branches, floating leaves, and similar substances. Sometimes it attains a considerable size, spreading over a square foot of surface, and having some branches more than three inches in length. It is in best condition towards the end of summer.

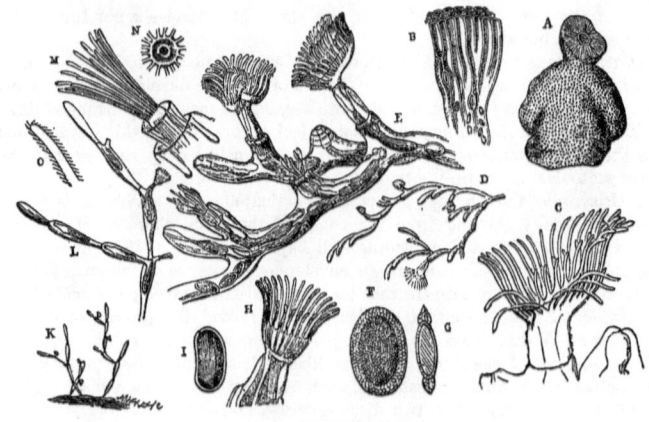

POLYZOA.

A. *Alcyonella fungosa.*   B. *Alcyonella fungosa.* ×   C. *Alcyonella fungosa.* (Tentacles of a polype.)   D. *Plumatella repens.*
E. *Plumatella repens.* ×   F. *Plumatella repens.* (Statoblast. ×)   G. *Plumatella repens.* (Statoblast, side view. ×)
H. *Fredericella sultana.* ×   I. *Fredericella sultana.* (Statoblast. ×)   K. *Paludicella.*   L. *Paludicella.* ×
M. *Paludicella.* (Tentacles. ×)   N. *Paludicella.* (Tentacles. ×)   O. *Paludicella.* (Tentacle. × ×)
The sign × signifies that the object is magnified.

A lovely polyzoa is the *Fredericella sultána*; the former title being given to it in honor of M. Fr. Cuvier, the celebrated naturalist, and the latter being earned by its graceful and queenly beauty. This is a common species, and is found plentifully in tufts on submerged stones, plants, sticks, and similar objects. It also inhabits rivulets, but seems to prefer tolerably still waters. As it is tolerably hardy, it is useful to microscopists, who can keep it alive in a common vial of water and place it under the microscope whenever they choose.

Our last example of the Polyzoa is the *Paludicella ehrenbergii*. Its peculiar form is not a horseshoe outline, but a funnel-like shape of the marine polyzoa. This arrangement of the tentacles seems to be unique among the fresh-water species; for, although the tentacles of Fredericella appear at first to assume the circular form, a more careful examination will show that this is not really the case.

The mechanism by which the floating particles contained in the water are inevitably driven towards the mouth is of a knife-blade shape, on which the cilia is arranged in such a manner that all those of one side point upwards and those of the opposite side downwards. The tentacular plume viewed from the front shows that the arrangement of these organs is really circular. Only the bases of the tentacles are delineated.

ARTHROPODA.

REATURES that compose this great BRANCH OF THE ANIMAL KINGDOM were regarded by Cuvier as articulated animals having a symmetrical body, in that both sides were equal. One of his four great Divisions embraced these forms, under the title Articulata. The bodies of these animals are characterized by a peculiar feature, the series of rings, of which the earth-worm is a simple example.

The circulating system is represented just under the back by a long vessel, the heart, connecting with vessels that propel the blood over the system, and return it to the gills, or lungs. The stomach and intestine lie in the median line of the body. The nervous system has ganglia, or enlargements of the nervous cords.

The Arthropoda have certain features in common: bi-lateral symmetry, one side being like the other; rings, or articulating parts, segments, arranged one upon another, each ring—theoretically—bearing a pair of limbs, which are also jointed. The blood is usually colorless, yet in some instances yellowish, or red, or purple. The globules of the blood, however, are not colored, the coloring matter being held in the fluid itself. The alimentary canal is usually nearly straight. The eyes are usually confined in the head.

Authors have formerly divided the Arthropoda into two classes, the Insects and Crustaceans; but the places of some examples are so obscure they are held somewhat in reserve. The Horseshoe Crabs and Trilobites, Water-Bears, Sea-Spiders, and *Linguatulina* are now resting between the two classes, Insecta and Crustacea, where authors are inclined to believe they will ultimately find a permanent place.

INSECTS; INSECTA.

THE INSECTS afford the first examples of the Articulata, *i. e.*, the jointed animals without vertebræ. Their bodies are composed of a series of rings, and they are separated into at least two and mostly three portions, the head being distinct from the body. They pass through a series of changes before attaining the perfect form; and when they have reached adult age they always possess six jointed legs, neither more nor less, and two antennæ, popularly called horns or feelers.

In most instances their preliminary forms, technically called the larva and pupa, are extremely unlike the perfect Insect; but there are some in which, at all events externally, they retain the same shape throughout their entire life. The whole of the growth takes place in the preliminary stages, so that the perfect Insect never grows, and the popular idea that a little Insect is necessarily a young one is quite incorrect. It is true that smaller and larger specimens occur in every species, but this difference in size is due to some external influences that have acted on the individual; and we find large and small examples of an Insect, say a wasp, or a beetle, just as we find giants and dwarfs among mankind.

Insects breathe in a very curious manner. They have no lungs nor gills, but their whole body is permeated with a net-work of tubes through which the air is conveyed, and by means of which the blood is brought in contact with the vivifying influence of the atmosphere. These breathing tubes, technically called tracheæ, ramify to every portion of the creature, and even penetrate to the extremities of the limbs, the antennæ, and even the wings, when those organs exist. Their external orifices are called spiracles, and are set along the sides.

They have very little internal skeleton, the hard materials which protect the soft vital organs being placed on the exterior, and forming a beautiful coat of mail, so constructed as to defend the tender portions within, and yet to permit perfectly free motion on the part of the owner. Certain projections of this substance are often found in the interior, especially in the thorax, a central portion of the creature, and are used for the attachment of muscles where considerable power is needed.

This external skeleton is quite unique in its chemical composition, being made almost entirely of a substance called chitine, to which are added several other materials, such as animal matter, albumen, and the oil which gives the bright colors so prevalent in most of the species.

There are many other interesting points in the structure of the Insects, such as the eyes, the wings, the tracheæ, etc., which will be described in the course of the following pages.

The systems on which the Insects have been arranged are as perplexing as numerous, differing according to the characteristics chosen by their authors. In this work the system employed is that of Mr. Westwood, which seems to combine many advantages to be found in the different arrangements of various authors, and is sufficiently intelligible to be understood without any painful exercise of the memory.

BEETLES; COLEOPTERA.

THE first order, according to this author, is called the Coleoptera, a word of Greek origin, signifying sheathed-winged animals, and includes all those insects which are more popularly known under the title of Beetles. In these insects the front pair of wings are modified into stout horny or leathery cases, under which the second pair of wings are folded when not in use. The hinder pair of wings are transparent and membranous in their structure, and when not employed are arranged under the upper pair, technically called the elytra, by folds, in two directions, one being longitudinal and the other transverse. On examining these wings carefully, it will be seen that their supporting nervures are furnished with hinge-like joints, which permit them to be folded in the right direction and no other. One of the best examples of a folded wing among the Beetles is to be found in the common Cocktail Beetle (*Staphylinus*), where the large and beautiful wings are packed away under two little square elytra, just as a folded map is packed into its covers. In other instances where the elytra are very long, as in the common Musk Beetle (*Cerambyx*), the wings are first folded longitudinally and then a little piece doubled over at the tip, so as to fit within the cover.

The mouth is furnished with jaws, often of considerable power, which move horizontally.

The last character that must be considered in the Beetles is the mode of the metamorphosis or change of form which is undergone by them before they attain their perfect state. After being hatched from the egg, they take the form that is popularly known under the title of grub, and is quite unlike the shape of the perfect insect. In this state they remain for various periods, according to the species and the climate, and then pass into the second, or pupal state, when they look much like the perfect insect, but are unable to move about. This characteristic seems to separate them from the earwigs, cockroaches, and grasshoppers, which would otherwise have been included in the same order with the Beetles, but are now placed in separate orders on account of the character of their preliminary stages, where the shape of larva, pupa, and perfect insect are very similar, and the pupa is active.

Passing over, for the present, the details of classification, we come to the first family of insects, scientifically called the Cicindélidæ, and popularly known by the name of Tiger Beetles, or Sparklers. These Beetles are represented by several species, among which the common TIGER BEETLE (*Cicindéla campestris*) is the most common and perhaps the most beautiful. Well does this little creature deserve its popular name, for what the dragon-fly is to the air, what the shark is to the sea, the Tiger Beetle is to the earth; running with such rapidity that the eye can hardly follow its course; armed with jaws like two reaper's sickles crossing each other at the points; furnished with eyes that project from the sides of the head and permit the creature to see in every direction without turning itself; and, lastly, gifted with agile wings that enable it to rise in the air as readily as a fly or a wasp. Moreover, it is covered with a suit of mail, gold embossed, gem studded, and burnished with more than steely brightness, light yet strong, and though freely yielding to every movement, yet so marvellously jointed as to leave no vulnerable points even when in full action, and, in fine, such a suit of armor as no monarch ever possessed and no artist ever conceived. True, to the naked or unobservant eye it seems to be but a dully green Beetle with a blue abdomen, but if placed under the microscope, and a powerful light directed upon it, it blazes out with such gorgeous brilliancy that the eye can scarcely endure the glory of its raiment.

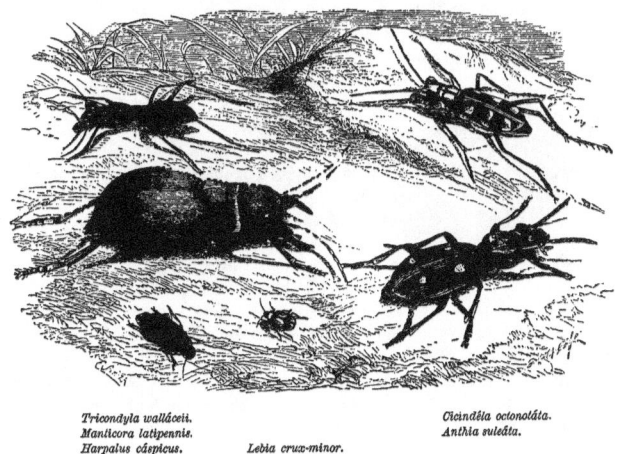

Tricondyla wallâceii.
Manticora latipennis.
Harpalus câspicus.
Lebia crux-minor.
Cicindéla octonotáta.
Anthia suleâta.

The typical species which is represented in the illustration is the EIGHT-SPOT TIGER BEETLE of India.

The European Tiger Beetle is remarkable for exuding a powerful scent, much resembling the odor produced by a crushed verbena leaf.

The family *Cicindelidæ*, which embraces the group of Tiger Beetles, so called, probably, from their singular markings and stripes, is represented in North America by a number of species. Their habits are terrestrial. During hot mid-summer days they are met with in dirty roadsides, or roadbeds, running or flying so swiftly they are difficult to capture. In the tropics the species are fond of trees.

An example of a very large genus belonging to this family is given in the engraving under the title of *Manticora latipennis*, the generic title being given to it because its great dimensions and ferocious habits are thought to bear some analogy with those of the fabled Manticora, a beast which the older naturalists were accustomed to describe with great zest, and in an engraving now before me had figured with the face of a human being, with hair carefully parted, six rows of shark's teeth, and a tail armed with a very arsenal of projectile spikes.

A VERY large and important family of Beetles, the Carabidæ, now comes before us, which is represented by very many species, the common Ground Beetles being familiar examples. The accompanying illustration represents the celebrated BOMBARDIER BEETLE (*Brachinus crépitans*), which belongs to this family. This little beetle is plentifully found in many places. When this beetle is handled, a sharpish explosion is heard, and on looking at the creature, a tiny wreath of bluish vapor is seen to issue from the body. This vapor has a very pungent odor, and when discharged against the skin, leaves a yellow mark like that produced by nitric acid. Originally, it is a liquid, secreted by a certain gland, but as

BOMBARDIER BEETLE.—*Brachinus crépitans*. (Magnified.)

soon as it comes in contact with the atmosphere it becomes suddenly volatilized, thereby producing the explosion and causing the smoke-like vapor to arise. The insect can fire off its miniature artillery seven or eight times in succession. Even after the death of the insect, the explosion can be produced by pressure.

One species of this genus, *Lebia crux-minor*, is given in the former illustration. It is notable for the cross-like mark from which it derives its name. The largest species belonging to this family are to be found in the exotic genus Anthia, an example of which is given in the same illustration. The males are remarkable for the enormous size of the mandibles, and the thorax, or chest, seems to be divided into two parts. Most of the species are found in Southern Africa. *Anthia sulcata* is a native of Senegal.

It may be here remarked that the very largest of the Carabidæ is a Javanese beetle, named MORMOLYCE, which is mostly found under the branches of trees. Mr. Westwood mentions that one of these insects in his possession has attained the extraordinary length of three inches and a half. As may be seen from the engraving, it is a very odd-looking insect, hardly recognizable as a beetle, and more resembling the mantis than the beetles. It will be noticed that this creature has a very long neck, a very flat body, elytra wide and flattened like those of the leaf-insect, and a thorax also flat and deeply toothed at the edges.

MORMOLYCE.—*Mormolyce phyllodes*.

The *Carabidæ* are represented in North America by numerous species, one of the most familiar of which is the one prettily shaped, black, and with gold spots, *Calosoma calidum*, very common in fields. Its habits are somewhat voracious, the Junebug sometimes being assailed by it, and torn to pieces. A species of *Anophthalmus* is found in the Mammoth Cave, in Kentucky, which is blind, no eyes being visible.

PASSING by the group of Carabidæ known as the Scaritidæ, a sub-family of beetles which are mostly found under stones or in holes near the sea-shore, we come to the Harpalides, of which the *Harpalus caspicus* is our present example. They are mostly rather small, and seldom bright colored, with the exception of a few species, such as the well-known SUN BEETLES, which are so familiar to us as they run actively over gravel walks or roads as if enjoying the blazing sunbeams.

WE now come to the large group of WATER BEETLES, which are divided into several families. The fresh waters of many rivers, ponds, and lakes are very populous with the Water Beetles, which may be seen by thousands on a summer day, swimming, diving, rising to the surface, and evidently enjoying life to the utmost.

In order to enable them to perform the various movements which are necessary for their aquatic existence, their hind legs are developed into oars with flattened blades and stiff hairy fringe, and the mode of respiration is slightly altered in order to accommodate itself to the surrounding conditions. It has been already mentioned that in all insects the respiration is conducted through a series of apertures set along the sides, and technically called spiracles,

In the Water Beetles, the spiracles are set rather high, so as to be covered by the hollowed elytra, and to be capable of breathing the air under those organs. When, therefore, the Beetle dives, it is in noways distressed for want of air, as it carries a tolerable supply beneath the elytra. When, however, that supply is exhausted, the beetle rises to the surface, just pushes the ends of the elytra out of the water, takes in a fresh supply of air and again seeks its subaquatic haunts. Any one may see in almost any ditch the Great Water Beetle (*Dyticus dimidiatus*), ever and anon rising to the surface, poking its tail out of the water, and then diving to the bed of the stream.

Towards evening, this, in common with many other Water Beetles, is accustomed to leave the streams, to spread its wide wings, and to soar into the air. In the early morning it again seeks its watery home, and is accustomed to save time and exertion by closing its wings and dropping like a stone as soon as it perceives the water below. The larva of the Dyticus is a

(Larva.)　　　　(Male.)　　　　(Female with egg-sac.)
GREAT WATER BEETLE.—*Dyticus dimidiatus*.

terribly ferocious creature, both in aspect and character. It inhabits the waters, and is a very hyena in the terrible grasp and power of its jaws. The perfect insect is quite as voracious, and when a number are kept in a single vessel, they are sure to attack and kill each other. No one who cares for the animated inhabitants of his aquarium should permit a Dyticus to be placed among them, as a fox makes no more havoc in a chicken-roost than a Dyticus in an aquarium.

A smaller species is called *Ilybius ater*.

WHIRLWIG BEETLE.—*Gyrinus mergus*.

To this group belong the WHIRLWIG BEETLES, or GYRINIDÆ, so plentiful on the surface of many rivers and ponds, but always choosing a still spot, where they are overshadowed by the bank or an overhanging tree, for the locality wherein they perform their mazy dance. These insects are very hardy, and even on a winter's day the Whirlwigs may be seen taking advantage of the last gleam of sunshine, and wheeling around their complicated maze as merrily as if the warm winds of summer were breathing on them. The reader will see a magnified specimen in the engraving; its natural length is signified by the line aside.

The Whirligigs of North America, the country boy will tell, "give milk." For certain they emit a milky liquid when caught, which latter is not easily accomplished, as they dive with exceeding celerity, when they adhere for a time by their claws to the bottom. They

BURYING BEETLES, HORNET, WATCHMAN BEETLE, ETC.

carry down a bubble of air on the tip of the abdomen, and when the supply is exhausted rise for more.

Passing by several large and interesting families, we come to curious creatures, popularly known by the name of Rove Beetles, or Cocktails, the latter name being given to them on account of their habit of curling up the abdomen when they are alarmed or irritated. The common Black Cocktail has, when it assumes this attitude, standing its ground defiantly with open jaws and elevated tail, so diabolical an aspect that the rustics generally call it the devil's coach-horse. It has, moreover, the power of throwing out a most disgusting odor, which is penetrating and persistent to a degree, refusing to be driven off even with many washings.

These beetles are termed Staphylinidæ, or Brachelytra, the latter term signifying short elytra.

Two species, scientifically termed *Ocypus olens* and *Creophilus maxillosus*, are common throughout Europe. The latter is plentiful in and about drains or dead animal matter, and may be known by the gray hairy look of the elytra. There is a smaller species (*Staphylinus erythrópterus*) which has the elytra of a dusky red, and is not so common as the preceding insect. I have often remarked that the red-backed shrike is very fond of this insect, and used to find the nests of the shrike by means of the beetles that the bird had stuck upon the thorns near its home.

The Staphylinidæ include a vast number of species that may be found in almost every imaginable locality, and live on almost every imaginable kind of food.

The *Staphylinidæ*, or Rove Beetles, are extremely common in the United States, and useful as scavengers. The *Historidæ* and several other families include the common Dung or Carrion Beetles. *Necrophorus* is a very common form.

Next to the Staphylinidæ are placed some insects that have become quite famous for their curious and valuable habits. These are the Necrophagæ, popularly and appropriately termed Burying Beetles.

It is owing to the exertions of these little scavengers that the carcases of birds, small mammals, and reptiles are never seen to cumber the ground, being buried at a depth of several inches, where they serve to increase the fertility of the earth instead of tainting the purity of the atmosphere. These beetles may easily be captured by laying a dead mouse, mole, bird, frog, or even a piece of meat on the ground, and marking the spot so as to be able to find the place where it had been laid. It will hardly have remained there for a couple of hours before some Burying Beetle will have found it out, and straightway set to work at its interment. The plan adopted is by burrowing underneath the corpse and scratching away the earth so as to form a hollow, into which the body sinks. When the beetles have worked for some time they are quite hidden, and the dead animal seems to subside into the ground as if by magic.

The object of burying dead animals is to gain a proper spot wherein to deposit their eggs, as the larvæ when hatched feed wholly on decaying animal substance.

In the accompanying full-page illustration many figures are given of the Burying Beetles, showing them while in the act of interring a dead bird.

We now come to the Lamellicorn beetles, so called from the beautiful plates, or lamellæ, which decorate the antennæ. This family includes a vast number of species, many of which, as, for example, the Common Cockchaffer, are extremely hurtful to vegetation both in the larval and adult form. In this family are found the most gigantic specimens of the Coleoptera, some of which look more like crabs than beetles, so huge are they and so bizarre are their shapes. In all these creatures the lamellæ are larger and more beautiful in the female than in the male insect.

The Common Cockchaffer is too familiar to need any description of its personal appearance, but the history of its life is not so widely known as its aspect. The mother beetle commences operations by depositing the eggs in the ground, where in good time the young are hatched. The grubs are unsightly-looking objects, having the end of the body so curved that the creatures cannot crawl in the ordinary fashion, but are obliged to lie on their sides. They are furnished with two terribly trenchant jaws like curved shears, and immediately set to work at their destructive labors.

They feed mostly upon the roots of grasses and other plants, and when in great numbers have been known to ruin an entire harvest.

Of the STAG BEETLE, the largest of the genus Coleoptera, we present a beautiful colored illustration. When it has attained its full dimensions it is an extremely powerful and rather formidable insect, its enormous mandibles being able to inflict a very painful bite, not only on account of the powerful muscles by which they are moved, but in consequence of the antler-like projections with which their tips are armed. These horn-like jaws only belong to the male, those of the female being simply sharp and curved mandibles, in no way conspicuous.

The larvæ of the Stag Beetle reside in trees into which it burrows with marvellous facility, and as after they have emerged from their holes they appear to cling to the familiar neighborhood, they may be found upon or near the trees in which they have been bred.

From the formidable shape of the mandibles it might be supposed that the Stag Beetle was one of the predaceous species. This, however, is not the case, the food of this fine insect consisting mostly, if not wholly, of the juices of vegetables, which it wounds with the jaws so as to cause the sap to flow. It is true that specimens have been detected in the act of assaulting other insects, but they never seem to have been observed in the act of feeding upon their victim. Whether the food be of animal or vegetable nature, it is always liquid, and is lapped, or swept up, by a kind of brush which forms part of the mouth, and looks like a double pencil of shining orange-colored hairs.

It seems that during the winter the Stag Beetle hibernates, as there is in the Ashmolean Museum, at Oxford, an earthen cell, or cocoon, in which was found a Stag Beetle very neatly packed, with its horns bent over its thorax. A popular name for this beetle is Hornbug.

The Stag Beetle is equally well known in the United States as in Europe; its large size and stag-like horns giving it an attractive appearance.

IN the accompanying, as well as in the next illustration, beetles are represented that have been rendered forever famous by the honors which the ancient Egyptians paid to them, and the frequency with which they are represented upon their hieroglyphs, and even sculptured on a gigantic scale in the hard granite which that wondrous race could work so easily. The present is the SACRED SCARABÆUS of the Egyptians, an insect which deserves a passing notice on account of its curious habits.

SACRED EGYPTIAN SCARABÆUS.—*Scarabæus sacer.*
(Natural size.)

The reader will remember that the burying or sexton beetles are in the habit of interring the dead bodies of various small animals in order to form a convenient nidus in which to deposit their eggs, and insure for their young a bountiful supply of food as soon as they enter the world. The Scarabæus is urged by a similar instinct, but exercises it upon different materials. Every one who has walked in the field must have noticed the singular rapidity with which patches of cow-dung disappear, and many may have observed that this phenomenon is caused by the efforts of sundry beetles, which burrow beneath the mass and convey the substance deep into the ground. The common watchman beetle, so well known from its habit of flying on droning wings in the evening, is one of the best known of these valuable beetles; and it is worthy of notice that, despite of the nature of the substance in which they work, not a speck adheres to their bright and polished armor.

The Egyptian Beetle employs similar substances for the cradle of its future young, but not in the same manner, kneading into irregular balls in which it deposits its eggs, and then rolling it away by means of its odd-looking hind legs. After it has made the ball, which is often larger than itself, the beetle sets to work to roll it to a convenient spot where the earth is soft, and performs this curious operation by a retrograde motion, the

STAG-BEETLE AND LONGICORN BEETLE.

hind legs directing the ball, while the four other legs are employed in locomotion. During this operation the beetle seems to be standing on its head, the hind legs being necessarily much elevated in order to guide the ball, which by dint of much rolling becomes nearly spherical. A tolerably deep hole is then excavated in a suitable spot, the ball rolled into it, and the earth filled in.

Many beetles perform this useful operation; and even in several European countries where the beauty of the climate is only equalled by the uncleanliness of the inhabitants, these beetles are of inestimable service, and are, perhaps, the only means whereby the towns and villages are rendered endurable, at all events to unaccustomed eyes and nostrils. Fortunately these insects fly by day as well as by night, and being gifted with extraordinary powers of scent, are sure to be on the spot as soon as their labors are required.

There are few parts of the globe where beetles possessing similar habits are not to be found, and although they do not display equal skill in the construction of egg-containing

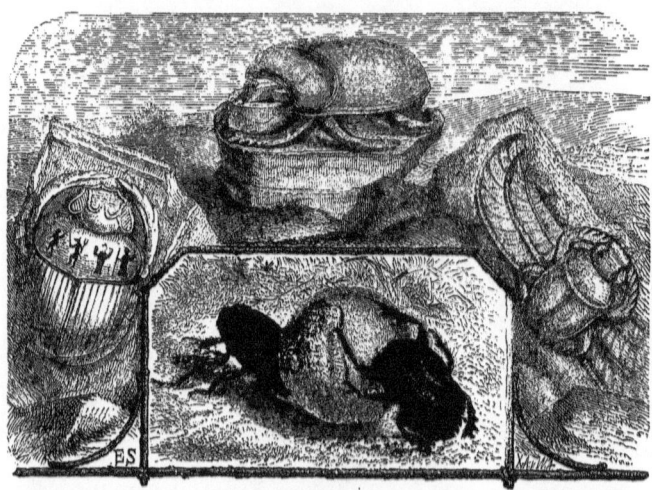

SPOTTED SCARABÆUS.—*Ateuchus variolosus.* (In natural size, surrounded by Scarabees in diminished form.)

balls, they are equally efficacious in the results. It may be here mentioned, that the watchman beetle (*Geotrúpes stercorárius*) is the "shard-borne" beetle mentioned by the poets, the title being due to the shelly elytra which are held aloft during its flight. They are marvellously tenacious of life; and, as an example of this property, I may mention that I once caught a Geotrupes in the air which had been mulcted of one elytron, lost several of its limbs, and the whole of its abdomen, the contents having been evidently scooped out by some bird. Yet it was quite strong on the wing, and seemed little the worse for its injuries. This beetle is represented in the full-page illustration on page 376.

Several species of this kind of beetle, called Dung-beetles, or Tumble-bug, are found in the northern United States. The Bronze Dung-beetle (*Copris carnifex*) is the most attractive of these scavengers. It is a more southern species than the Common Dung-beetle, or Pellet Beetle (*Ateuchus volvens*), which, however, is found in all the States. The latter is closely related to the Sacred Scarabæus, which, by some authorities, is of the same genus as the SPOTTED SCARABÆUS (*Ateuchus variolosus*). The ancient Egyptians, being so impressed by what they regarded as a benefit conferred by these scavengers, they looked up and treated their beetle as sacred, too, representing them, as we see above, in sculpture of their tombs, houses, temples, etc. The latter species is found in the south of France.

A VERY fine Lamellicorn is the ATLAS BEETLE, a native of the Philippines and part of India. Its colors are as follows :—The male is of a brilliant metallic olive-green, brightly polished and shining; but the female is of a much duller hue, having the thorax and the base of the elytra rough, and the green of a blackish cast. The length of the male is about three inches.

A VERY odd-looking beetle is the CHRYSOPHORA. It belongs to the family Rutelidæ, the members of which belong entirely to the hot countries of the globe, and are most plentiful in the tropics. They do not seem to attain the gigantic dimensions which are found among the allied families, such as the Dynastidæ, but are all very beautiful insects, on account of the extreme brilliancy of their coloring. The Chrysophora is quite remarkable for its curious form and glowing colors. The hind legs are extraordinarily developed, and seem disproportionately long and stout when compared with the moderately sized body. Another point of interest in this beetle is the structure of the "tibia" of the hind leg, *i. e.*, the joint immediately preceding the jointed foot. The lower part of this joint is prolonged into a stout and sharp spur, not unlike that on the leg of most gallinaceous birds. The object of this curious modification is not known.

THE HERCULES BEETLE, which is represented in the accompanying full-page illustration, is an example of the family termed Dynastidæ, or powerful beetles, on account of their enormous size and strength. They are the giants among insects; for, although many others exceed them in length or width, these creatures are so stoutly made, that any other insect becomes dwarfed when placed by their side.

In this family, the males are remarkable for the strange and often grotesque horny processes which are developed from the head and thorax, the females being destitute of these ornaments. Most of the Dynastidæ inhabit tropical regions, only a very few species being found in the moderate climates. They are generally night-fliers, ascending to considerable elevations, and during the day they hide themselves in holes in the earth, in hollow trees, or similar situations. Their food seems to be nearly, if not wholly, of a vegetable nature.

We have one example of the *Dynastidæ*, the family that embraces the giant *Dynastes hercules*, a beetle about six inches in length. Our species is found in the Southern States, and measures about two inches.

PASSING by one or two families of more or less importance, we arrive at the Buprestidæ, a family of beetles remarkable for the extraordinary gorgeousness of their tints, almost every imaginable hue being found upon these brilliant insects.

They are found in many portions of the globe, but, as is generally the case with insects, their colors take the greatest intensity within the tropics. They fly well, and seem to exult in the hottest sunshine, where the bright beams cause their burnished raiment to flash forth its most dazzling hues. They are, however, slow of foot, and, when alarmed, have a habit of falling to the ground with folded limbs, as if they were dead.

The CHRYSOCHROA is one of the finest of this splendid family. The sides of the thorax are covered with little round pits, something like the depressions on the head of a thimble, and are of a fiery copper hue. The head and middle of the thorax are light burnished blue, like that of a well-tempered watch-spring, and the elytra are warm cream-colored, diversified with a patch of deep purple-blue at each side, and another at the tip. The Chrysochroa is a native of India.

WE now come to the celebrated CUCUJO, or FIREFLY OF BRAZIL. Each side of the base of the thorax shows two light patches, which in the living insect are of a pale yellow, and at night burn with a lustre far surpassing that of the common glow-worm. When the insect expands its wings for flight two more fire-spots are seen beneath the elytra; and when the creature approaches near the observer, the whole interior of its body seems to be incandescent. These insects are nocturnal in their habits, and at night in the forests, when the air is filled

HERCULES-BEETLE.

with myriads of blazing stars, crossing and recrossing in every direction, making the deepest glades luminous with their flaming lamps, and appearing and vanishing as if suddenly brought into existence and as suddenly annihilated, they present a sight almost too magnificent for description. So splendid are these beetles, that the ladies are often in the habit of catching them and trimming their dresses with these living diamonds, taking care to fasten them in such a way as not to injure them.

When in full glow, the light is so intense that a letter or book may be read by its aid, provided that the insect be slightly squeezed so as to excite it to throw out the luminous element. There are very many species of Fire-flies, but this is the best known, and one of the most luminous of its kind. Mr. Westwood mentions that one of these insects was brought in a living state to Europe, and was kept alive by continually moistening the woodwork of its cage.

The Elateridæ, or Spring Beetles, so well known from their habit of jumping with a slight clicking sound when laid on their backs, are allied to the Buprestis beetles.

The *Elateridæ* comprise several prominent beetles. They are well known in America as Snap-beetles. *Pyrophorus* is the genus that embraces several species of Fire-flies of Central and South America. The genus *Photinus* has several species, most of which have phosphorescent glands. Our common New England Fire-fly is a familiar example.

The celebrated GLOW-WORM belongs to the typical genus of its family.

Contrary to the usual rule among insects, where the male absorbs the whole of the beauty, and the female is comparatively dull and sombre in color and form, the female carries off the palm for beauty, at all events after dusk, the male regaining the natural ascendancy by the light of day. Either through books, or by actual observation, almost every one is familiar with the Glow-worm, and would recognize its pale blue light on a summer's evening. Many, however, if they came across the insect by day, would fail to detect the brilliant star of the night in the dull, brown, grub-like insect crawling slowly among the leaves, and still fewer would be able to distinguish the male, so unlike are the two sexes.

THE family *Dermestidæ* embraces the pests of our museums, *Dermestes*. They are also very destructive to small fruit shrubs when in leaf. The *Anthrenus* is equally destructive, and is the most common pest in museums.

NEXT to this family is another, called the Telephoridæ, which is represented in Europe by the well-known beetles, popularly called, from their red or bluish colors, SOLDIERS and SAILORS. They are found in great quantities in the spring, and upon the umbelliferous flowers they assemble plentifully. They are carnivorous, voracious, and combative to a degree, and in my school-days the fashionable spring amusement consisted in setting Soldiers and Sailors to fight with each other. They fly readily, but slowly, and only to short distances, and may be known while in the air by their peculiar attitude, the long body hanging nearly vertically from the wings.

A VERY destructive family, termed Ptinidæ, must now be briefly noticed. To this family belong the insects which are so well known by their labors, though themselves are mostly hidden from sight. Among the Ptinidæ are placed the little beetles that eat holes in our furniture, books, etc., and do such irremediable damage in so short a time. Mr. Westwood mentions one instance where a new bedpost was wholly destroyed by one species of these beetles (*Ptilinus pectinicornis*) in a space of three years.

The celebrated DEATH WATCH, represented in the accompanying engraving, belongs to this family. That peculiar name is popularly given to several species, such as *Anobium striatum* and *tesselatum*, on account of the ticking sound which is made by knocking their heads against the woodwork, and which is used as a signal to their mates. The exact natural length of the beetle is indicated by the line next to the illustration.

DEATH WATCH.—*Anobium tessellatum.*

THE OIL BEETLE.

Towards the middle of spring and for the next month, may be found certain very handsome looking beetles of a deep, rich, red color, and remarkable for the beautifully-toothed antennæ. This insect is to be seen mostly upon flowers, and is popularly known by the name of CARDINAL BEETLE. The scientific title is *Pyrochróa rubens*. This is the only European genus of the family to which it belongs, and which is called Pyrochroidæ, in allusion to the typical genus. The word Pyrochroa, or Flame-colored, is given to this beetle on account of its bright red exterior.

A succeeding family, the Mordellidæ, is chiefly remarkable for the curious fact that the larvæ of several of its genera, those of the *Ripiphorus*, for example, inhabit the nest of the common wasp, undeterred by the poisoned stings of their involuntary hosts from taking possession of their home. It seems that each specimen of this beetle monopolizes a single cell, and entomologists are of opinion that the nurse wasps feed the intruders, together with the rightful owners of the cells, not being able to distinguish between them.

The insect represented in the accompanying illustration is found in Europe, and is here given as an example of the family Cantharidæ, of which the BLISTER FLY, sometimes called

BLISTERS, OR SPANISH FLIES, WITH LARVA.—*Lytta vesicatoria*.

the SPANISH FLY, is the typical species. In the illustration, both insects and the larva are magnified.

In the whole of this family, certain noxious elements are strongly developed, which, like all other noxious things, can be transmuted and modified into benefits by those who know how to use them. There is a certain substance secreted within these creatures technically called Cantharidin, and looking, when separated from extraneous matter, like minute crystalline flakes of snowy whiteness. It can be dissolved in spirit, but not in water.

Spain is famous for the multitudes of Blister Flies which are found within its limits, and the whole of South-western Europe is prolific in this remarkable beetle. Whenever it may be present, its vicinity is known by the powerful odor which it exhales, just as the musk and tiger beetles may be detected by the nostril, though unperceived by the eye. On account of its peculiar properties, it is not easily prepared, the dust which flies from the dried and drying insects being light, searching, pungent, and inflammatory to the last degree.

The larva or grub of this beetle is said to reside under ground, and to feed upon the roots of vegetables.

The Spanish Fly is a handsome insect, nearly an inch in length, and of a rich silken green, with a gold gloss in certain lights. It is a very remarkable fact that fish will eat the Cantharis without injury, and anglers have found, rather to their surprise, that if they could fix a Cantharis on their hook, it proved to be a very effectual bait for fish, the chub seeming particularly fond of this very stimulating food. The common hedgehog has been known to eat these insects with impunity.

BELONGING to the same family, and very common in Europe, is an insect which popularly goes by the appropriate name of OIL BEETLE, because, when handled, it has the property of pouring a yellowish, oily fluid from the joints of its legs.

The abdomen is extremely large in proportion to the rest of the body, and the short, diverging elytra descend but a very little way below the thorax. Insects of this genus—especially the males, where the elytra are longer than in the other sex—are used by unprincipled druggists for the purpose of mixing with the true blister fly, which they resemble sufficiently to deceive an inexperienced eye. In some parts of the world, however, they are always employed in connection with the blister beetle, or even used instead of that insect. The oily matter that is poured from the joints is considered in some countries to be a specific for rheumatism, and is expressed from the insect for medicinal purposes.

The Oil Beetle's color is dull indigo-blue, and its natural length is not much more than one inch and a quarter.

A few other insects of this family are rather remarkable in their habits. One of these is the SITARIS, the larva of which is found in the nests of several of the mason bees (*Anthóphora* and *Osmia*), and the general opinion of naturalists is that they feed upon the larvæ of those insects. Some, however, think that their only object in this intrusion is to eat the provision of pollen that has been laid up for the young bee.

The MEAL-WORM, so well known to bird-fanciers as a wholesome diet for nightingales and other birds; to millers, for its ravages among the grain; and to sailors, for its depredations among the biscuit, is the larva of a beetle named *Tenebrio molitor*, the former word being given to it in allusion to its love of darkness, and the latter to the damage which it occasions to the miller. This is one of the maggots which have caused sailors to knock the edge of a biscuit upon the table before eating it, an action which in many old voyagers has become so deeply rooted a habit, that they are actually unable to resist the movement. These larvæ are terribly sharp-toothed, eating their way through the sides of casks while in search of food. Some species of the same genus have the power of ejecting an acrid fluid to the distance of more than a foot; the one most remarkable in this respect being a Brazilian insect, *Tenebrio grandis*.

WE now arrive at a vast group of beetles, embracing several thousand species, which are popularly classed under the name of Weevils, and may all be known by the peculiar shape and the very elongated snouts. Many of these creatures have their elytra covered with minute but most brilliant scales, arranged in rows, and presenting, when placed under the microscope, a spectacle almost unapproached in splendor. They are mostly slow in their movements, not quick of foot, and many being wholly wingless.

Many of these creatures are extremely injurious to vegetables, both while growing and when stored up in barns or granaries. Most persons are too familiar with the little maggots that infest peas, and frequently ruin whole pods at a time, each pea containing a single white grub. These are the larvæ of the PEA WEEVIL (*Bruchus pisi*), which feed upon the soft substance of the pea, and make their escape just about the time when the vegetable is sufficiently ripe for gathering. One of the CORN WEEVILS (*Bruchus granarius*), so destructive to grain, also commits great ravages among the peas. One species of this genus inhabits the cocoa-nut, and the creatures are infinitely more abundant in tropical than in temperate climates. It is thought, indeed, that several species of these destructive insects have been imported into Europe in cargoes of grain, and finding the country suitable to their habits, have thriven there.

Another species of Weevil, the GRASS WEEVIL, or LISETTE (*Rhynchites bacchus*) commits terrible devastations among the growing vines, sometimes stripping the bushes of their leaves, which it rolls up and lines with silk.

The most brilliant of the Weevils are to be found in the typical family Curculionidæ, to which belong the well-known Diamond Beetles, in such request as objects for the microscope. Magnificent, however, as are these insects, some of the common little field Weevils, which may be found abundantly on peas, nettles, and other vegetables, yield to them not a single jot, when properly magnified and illuminated, the successive rows of glittering scales with their numerous facets being quite as splendid as the scale-lined pits which cover the elytra of the Diamond Beetle.

The maggots that are so frequently found in nuts, and which leave so black and bitter a deposit behind them that the person who has unfortunately tasted a maggot-eaten nut is forcibly reminded of the Dead Sea apple, with its inviting exterior and bitter dusty contents, also belong to the Weevils, and are the larvæ of the NUT WEEVIL (*Balaninus nucum*). All the members of this genus are remarkable for the extraordinary length of the snout, at the extremity of which are placed the powerful jaws. Fig. a shows the beetle in its natural size, while in the other figures it is magnified. A foreign species, a native of Cayenne, is termed *Balaninus proboscideus*, and is also remarkable for the inordinate length of the curved snout.

While the nut is yet soft and undeveloped, the female Weevil bores a hole at the base of the fruit, deposits an egg therein, and makes the best of her way to another nut, which she treats in a similar manner. As the nut increases, the young grub feeds on the interior of the nut, which is at first soft and milky, so as to suit its infant needs, and by degrees hardens into a fruity substance more fit for it after its jaws and digestive organs have acquired strength. After it has attained its full growth, it gnaws a round hole through the shell of the nut, allows itself to drop to the ground, buries itself below the surface, and in the ensuing autumn emerges in the perfect form.

The common CORN WEEVIL (*Calandra granaria*) is perhaps the most destructive of its tribe, its depredations far exceeding those of the insects that destroy nuts, acorns, apples, cherries, flowers, and other vegetables. This pest of corn-dealers is of very small size, not larger than the capital letter at the beginning of this sentence, and is therefore able to make its way through very small crevices. Like the preceding species, it passes its larval existence within the grain on which it feeds, devours the whole of the interior, and then, gnawing its way through the shell, becomes transformed in process of time into its perfect shape, which is that of a little long-beaked Weevil of dull red color, which, however, under the microscope, is singularly beautiful.

Many species belonging to this destructive genus are equally plentiful all over the world, and equally injurious. There seems, indeed, to be no vegetable substance that is not eaten by the Weevils, which appear to have a peculiar liking for those that are used for human food. Almost every article has its peculiar Weevil. There is the RICE WEEVIL, for example (*Calandra oryzæ*), known from the previous species by the four red spots on the elytra, which is nearly as destructive towards rice and Indian corn as the Corn Weevil towards wheat.

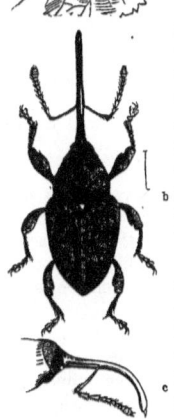

NUT WEEVIL.—*Balaninus nucum. a.* Laying eggs. *b.* Back. *c.* Head. (The line indicates the natural length.)

One of the largest species is a native of the West Indies, and is known by the name of the PALM WEEVIL (*Calandra palmarum*). This huge Weevil sometimes attains the length of two inches, and its color is a dull, velvet-like black. The larva of this large beetle is a great fat white grub, called gru-gru by the negroes, and considered by them to be a great dainty. The more educated inhabitants know this grub by the name of *Ver palmiste*. This grub is especially fond of the newly planted canes, and is sometimes so terribly destructive among them that a fresh planting becomes necessary. When this creature is about to attain its pupal condition, it weaves for itself a kind of cocoon formed from the fibres of the plant in which it lives.

Before noticing the long-horned insects, we must briefly mention a terribly destructive family of beetles, that are certainly allied to the Weevils, but whose precise degree of relationship does not seem to be very accurately understood.

To this family belongs the far-famed *Scolytus destructor*, a little dull colored insect, insignificant in appearance, but able to lay low the loftiest elm that ever reared its leafy head. Hundreds of the finest trees have fallen victims to the devouring teeth of this tiny beetle, a creature hardly the sixth of an inch in length. These insects not only burrow into the trees for the purpose of obtaining food, but therein they deposit their eggs, and therein are the young larvæ hatched.

The mother beetle deposits the eggs in a row, and the young, immediately upon entering the world, begin to eat their way through the wood, all diverging at right angles from the burrow in which they were laid, and all increasing the diameter of the burrow in exact proportion to their own growth. Hundreds of these quondam dwellings may be seen on roadside fences and railings, and so numerous are they on many trees that the bark falls off in flakes, the course of the sap becomes arrested, and at last the tree dies from the injuries to which it has been subjected by these minute but terrible foes, who work in darkness, unseen and secure. The grubs or larvæ may often be found in these tunnels. They are thick, round, and fat, without feet, and of a whitish color, except the horny head with its powerful jaws.

THE destructive beetles that are embraced in the Curculio family have been more notable than almost any group of insects, as the small fruit trees have suffered in all parts of North America where such fruits are grown. The term Weevil is applied to these insects. One species attacks the roses, both wild and cultivated. Another is found feeding on pine trees. The WHITE PINE WEEVIL (*Pissodes strobi*) is especially destructive to the white pines. The PLUM GOUGER (*Anthonomus prunicida*) resembles the Plum Curculio very much. The latter is named *Conotrachelus*. When the fruit is set the beetles sting them. Apples and peaches are also subject to the same pest. *Sitophilus* is the grain Weevil.

Other species are, *Centorhynchus*, the European turnip Weevil, introduced into Maine, where it stings the radish.

A common pest in the Western States is the Potato-stalk Weevil (*Barideus binotatus*), and *B. vestitus* eats the tobacco plants in the Southern States.

The Colorado potato beetle (*Doryphora decemlineata*) is a pest sufficiently well known at this time, having reached as far as it can go eastward, on the farms of Maine.

WE now come to the Longicorn beetles, so called on account of the extraordinary length of the antennæ in many of the species, an example of which will be found in the colored illustration on page 378. These insects are well represented by many species, though none have the antennæ of such wonderful length as the *Xenocerus semiluctuosus*. While the length of its body is only seven-eighths of an inch, its antennæ measure four inches, and often still more.

As in the preceding family, the Longicorn beetles pass their larval state in wood, sometimes boring to a considerable depth, and sometimes restricting themselves to the space between the bark and the wood. The grubs practically possess no limbs, the minute scaly legs being entirely useless for locomotion, and the movements of the grub being performed by alternate contraction and extension of its ringed body. In order to aid locomotion the segments are furnished with projecting tubercles, which are pressed against the sides of the burrow. Those of the common wasp beetle (*Clytus arietis*) may be found at the beginning of summer in fir trees, or in palings and posts of that wood.

The just-mentioned XENOCERUS, so remarkable for the inordinate length of the antennæ, is one of the finest examples of the Anthribidæ, not only for the long and thread-like antennæ, but for the beauty of its coloring. It is a native of Amboina, where it was found by Mr. Wallace. The male is jetty-black diversified with small white stripes on the head and thorax. The elytra are boldly decorated with the same contrasting hues. The female is also white and black, but the former color greatly predominates, the black being reduced to marks on the sides of the head and thorax, the tips of the elytra, and four black spots, two on the middle of the elytra and the other two on the thorax.

The well-known MUSK BEETLE (*Cerambyx moschátus*) belongs to this group. The scent, which more resembles attar of roses than musk, is extremely powerful, and is often the means of betraying the presence of the insect as it lies hidden among the leaves. The larva is a wood-borer, and I have taken numbers out of old willow trees, which I split with wedges for the express purpose.

A beetle with a large tuft of hair on each of the antennæ is termed *Disaulax cayennensis*. It is a native of the country whence it takes its name. The stout bases of the elytra are yellowish-orange. The whole of the body is boldly marked with deep black and snowy-white of a silvery lustre. The *Plectodera scalator*, a much larger species, belongs also to the Longicorns, and, like the preceding species, is marked with black and white, though the arrangement of the tints is different.

THE largest of the Tortoise Beetles, or Cassididæ, is the *Aspidomorpha amplissima*. This broad and flat insect is found in the Philippines. These insects derive their popular name from the tortoise-like shape of the body, which is so expanded that the whole of the limbs are concealed under its shelter. Many of these beetles are a light green, or greenish brown, and when they are stationary upon a leaf they can with difficulty be distinguished. The larva is remarkable for possessing a large forked appendage upon the end of the tail, which turns over the back and is loaded with excrementitious substances, so that the creature can hardly be seen under the load which it bears.

In the present species the body is chestnut-brown, and the elytra are furnished with wide, thin, and semi-transparent margins. Their centre is spotted with black.

PASSING by several families, we come to our last example of the Coleoptera, the *Chrysomela cerealis*, a member of a very large family.

All the Chrysomelidæ are round-bodied, and in most cases are very brilliantly colored with shining green, purple, blue and gold, of a peculiar but indescribable lustre. They are slow

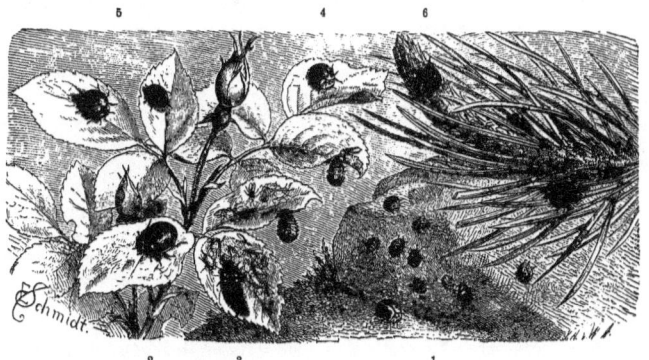

LADYBIRDS.—1. *Micraspis duodecimpunctata*. (In natural size.) 2. *Coccinella septempunctata and two larvæ*. (In natural size.) 3. Its magnified larva among aphides. 4. *Coccinella impustulata*. (In natural size.) 5. Two different specimens of *Coccinella dispar*. 6. *Chilocorus bipustulatus*. (In natural size.) The line indicates the average length of these beetles.

walkers, but grasp the leaves with a wonderfully firm hold. One of the genera belonging to this family contains the largest European specimen of these beetles, commonly known by the name of the BLOODY-NOSE BEETLE (*Timarcha tenebricosa*), on account of the bright red fluid which it ejects from its mouth and the joints of its legs when it is alarmed. This fluid is held by many persons to be a specific in case of toothache. It is applied by means of permitting the insect to emit the fluid on the finger and then rubbing it on the gum, and the effects are said to endure for several days. The larva of this beetle is a fat-bodied, shining, dark-green grub which may be found clinging to grass, moss, or hedgerows in the early summer. They are so like the perfect insect that their identity cannot be doubted.

THE family of the Coccinellidæ, or Ladybirds, is allied to the Chrysomelidæ, and is well known on account of the pretty little spotted insects with which we have been familiar from our childhood, and of which our illustration gives an interesting collection. Though the LADYBIRD is too well known to need description, it may be mentioned that it is an extremely useful insect, feeding while in the larval state on the aphides that swarm on so many of our favorite plants and shrubs. The mother Ladybird always takes care to deposit the eggs in spots where the aphides most swarm, and so secure an abundant supply of food for the future offspring.

EARWIGS.

TAKING leave of the beetles, we now proceed to a fresh order, distinguished by several simple characteristics, among which may be mentioned the soft and leathery elytra, or forewings, the wide and membranous hind-wings, and the forceps with which the tail is armed. The insects belonging to this order are popularly known by the name of Earwigs.

Before proceeding to the description of individual species, it is necessary to remark that the word Earwig is slightly incorrect, and owing to a popular notion that the insects crawl into the human ear, thence into the brain, and complete their work by causing madness in the minds of those who are afflicted by their presence.

The fact is, as all must know who have the least smattering of anatomy, that the insect never could gain admission to the brain by means of the ear. In the first place, the cerumen which is secreted in the ear and serves to preserve the natural moisture of the tissues, is so inexpressibly bitter, and so entirely opposed to the habits of the Earwig, that if one of these insects should by chance happen to crawl into the ear, its first impulse would be to retreat. In the second place, the drum of the ear would present an impassable obstacle, and in the third place, supposing the drum to have been ruptured, and the Earwig to have passed the spot where it existed, the complicated bony passages through which the auditory nerve passes would be too small to admit of its passage, even if the nerves which fill the channels were removed.

In point of fact, the correct name of this insect is the Earwing, so called because its spread wings have an outline somewhat resembling that of the human ear.

The membranous wings of the Earwig are truly beautiful. They are thin and delicate to a degree, very large and rounded, and during the day-time packed in the most admirable manner under the little square elytra. The process of packing is very beautiful, being greatly assisted by the forceps on the tail, which are directed by the creature with wonderful precision, and used as deftly as if they were fingers and directed by eyes. The Earwigs seldom fly except by night, and it is not very easy to see them pack up their wings. Some of the smaller species, however, are day-fliers, and in spite of their tiny dimensions, may be watched without much difficulty.

Earwigs feed on vegetable matter, especially preferring the corollas of flowers. Pinks, carnations, and dahlias are often damaged greatly by these insects, which sometimes occur in vast quantities, and ruin the appearance of a well-tended flower-bed. Fortunately for the florist, the habits of these insects are constant, and they can be destroyed in great numbers by those who desire to kill them. Being intolerant of light, they avoid the sunshine by every means in their power, and creep into every crevice that may hide them from the unwelcome light.

In consequence, they are fond of crawling among the thick and shadowy petals of the dahlia, and are frequently found in the slender spur of the nasturtium, so that any one who is about to eat one of these flowers will do well to examine the spur before he makes the attempt. Knowing these habits, horticulturists catch them by hundreds by the simple plan of placing lobsters' claws, bits of hollow reed, and similar objects on the tops of sticks, knowing that the Earwigs will crawl into them at the dawn of day, and may be shaken out and killed when the gardener goes his rounds.

The Earwig is remarkable for a parental affection quite exceptional in the insect race, the mother watching over her eggs until they are hatched, and after the young have entered the world, taking as much care of them as a hen takes of her chicken.

There are about seven or eight European species, some of them being of very small size. I have often seen them flying about at midday, when they might easily be mistaken for beetles. They have several times alighted on the sleeve of my coat, and afforded good opportunities of watching the curious manner in which the wings are tucked under their cases. The largest species is the Giant Earwig. It is of very rare occurrence, and seldom seen, as it only inhabits the sea-shore, and never shows itself until dusk.

GRASSHOPPERS, LOCUSTS, CRICKETS, ETC., ORTHOPTERA.

A LARGE and important order succeeds the Earwigs, containing some of the finest and, at the same time, the most grotesquely formed members of the insect tribe. In this order we include the grasshoppers, locusts, crickets, cockroaches, and leaf and stick insects; and its members are known by the thick, parchment-like upper wings, with their stout veinings and their overlapping tips. As in all the orders, there are exceptional species, wherein one or more of these attributes are wanting. But the characters are in themselves constant, and in most cases the indications of the missing member can be found. For example, many species never obtain wings at all, in many others the males only are furnished with these organs, and in others they are so small as to escape a casual notice.

THE first family of Orthoptera is the Blattidæ, a group of insects familiar under the title of Cockroaches.

In these insects the body is flattened, the antennæ are long and thread-like, and the perfect wings are only to be found in the adult male. The common COCKROACH, so plentiful in our kitchens, and so well known under the erroneous name of black-beetle—its color being dirty-red, and its rank not that of a beetle—is supposed to have been brought originally from India, and to have found itself in such good quarters that it has overspread the land in all directions.

The Cockroaches are particularly fond of heat, and are found in greatest abundance in kitchens, bake-houses, and other places where the temperature is always high. They are nocturnal in their habits, very seldom making their appearance by daylight, but leaving their hiding-places in swarms as soon as darkness brings their day. On board ship they become an almost intolerable nuisance, pouring out of the many hiding-places afforded to them by a ship's timbers as soon as the lights are put out, and drive sleep far away by their pestilent odor and their continual crawling over the face and limbs of those who are vainly endeavoring to seek repose.

Together with the rats and mice, these insects sometimes increase to such an unbearable extent, that, when the vessel comes to a port, the crew are sent on shore, pots of lighted sulphur are placed in the hold, and the hatches battened down for four-and-twenty hours. This severe treatment kills all the rats and mice and all the existing generation of Cockroaches, and is so far a temporary relief. But the eggs, which are laid in great profusion, retain the elements of life, in spite of the sulphureous fumes; and in a few months the ship will be nearly as much overrun as before with these pests.

There are several means of destroying the Cockroaches in houses, and if they are perseveringly carried out, a dwelling may be kept comparatively free from them. The common red wafers, if scattered over the floor, are rapid and effectual poison to these insects, and meal mixed with plaster of Paris has the same effect. Traps, too, can be readily made by twisting a funnel of paper, putting it into the neck of a jar with a little sugar and water at the bottom, and laying slips of wood or pasteboard as ladders by which the Cockroaches can reach the treacherous banquet. Those that enter will never escape with life, and the quickest way of killing them is to pour boiling water into the jar.

A hedgehog is also a good remedy against Cockroaches, and, if allowed the run of the kitchen during the night, will be wonderfully efficacious in keeping down their numbers.

The eggs of the Cockroach are not laid separately, but inclosed in a hard membranous case, exactly resembling an apple puff, and containing about sixteen eggs. Plenty of these cases may be found under planks or behind the skirting boards where these insects love to conceal themselves. Along one of the edges of the capsule there is a slit which corresponds with the opening of the puff, and which is strengthened, like that part of the pastry, by a thickened margin. The edges of the slit are toothed, and it is said that each tooth corresponds with an egg. When the young are hatched, they pour out a fluid which has the effect of dissolving the cement which holds the edges together, the newly-hatched Cockroaches push themselves through the aperture, which opens like a valve, and closes again after their exit, so that the empty capsule appears to be perfectly entire.

The shape of the young much resembles that of the perfect insect, except that in neither sex are the wings in existence. In the pupal stage the resemblance is preserved, the creature is active, and exhibits the rudimentary wings. The reader may often have seen white, brown, and mottled Cockroaches. These are the insects that have lately changed their skins; and if one of these creatures be taken, it will be found that in a day or two it will attain the same reddish-brown color as its companions.

The Cockroach is a very active insect, running both backwards and forwards with astonishing speed, and is furnished at the extremity of the abdomen with two short projections resembling miniature antennæ, and popularly regarded as such.

The accompanying illustration gives a figure of a short, stumpy insect with large hind legs. This is the FIELD CRICKET, a noisy creature, inhabiting the sides of hedges and old walls, and making country lanes vocal with its curious cry, if such a word can be applied to a sound produced by friction. The Field Cricket lives in burrows, made at the foot of hedges or walls, and sits at their mouth

FIELD CRICKET.—*Gryllus campestris.*

to sing. Our illustration shows both male and female in their natural size, the former just coming out of its burrow. It is, however, a very timid creature, and on hearing, or perchance feeling, an approaching footstep, it immediately retreats to the deepest recesses of the burrow, where it waits until it imagines the danger to have gone by. Despite of its timidity, however, it seems to be combative in no slight degree, and if a blade of grass or straw be pushed into its hole, it will seize the intruding substance so firmly that it can be drawn out of the burrow before it will loosen its hold. The males are especially warlike, and if two specimens be confined in the same box, they will fight until one is killed. The vanquished foe is then eaten by the victor. In White's "Natural History of Selborne" there is a careful and interesting description of the Field Cricket and its habits.

MOLE CRICKET.—*Gryllotalpa vulgaris.*

The well-known HOUSE CRICKET (*Acheta doméstica*) is a near relation of the above-mentioned species, and is so familiar as to need no description.

One of the oddest-looking of the insects is the MOLE CRICKET, so called on account of its burrowing habits and altogether mole-like aspect. This insect is illustrated in the natural size, and, as may be seen, attains considerable dimensions. The right-hand figure represents the Mole Cricket while in its larval stage. Those who like to give the needful time and trouble will find the internal anatomy of the Mole Cricket to be highly developed, remarkably interesting, and easily dissected.

Like those of the mole, the fore-limbs of the Mole Cricket are of enormous comparative size, and turned outwards at just the same angle from the body. All the legs are strong, but the middle and hinder pair appear quite weak and insignificant when compared with the gigantic developments of the front pair. This insect is rather local, but is found in many parts of Europe, where it is known by sundry popular titles, Croaker being the name most in vogue.

The wings of the Mole Cricket are large and handsome; and when folded, their hardened outer edges project along the back like two curved spines. Some persons have thought that this insect is the cause of the well-known phenomenon called the Will of the Wisp, or Jack o' Lantern, because in a locality where one of these deceptive lights was fluttering after its uncertain wont, a Mole Cricket was captured on the wing.

The food of the Mole Cricket is chiefly of a vegetable nature; but the insect will eat animal food when offered, having been known to feed upon raw beef with great zest. Like the field cricket, it is very combative, and when it has vanquished its foe is sure to eat him. As may be imagined from the tasks which it performs in driving burrows through the earth, the muscular strength of the Mole Cricket is exceedingly great; and when the insect is held in the hand, its struggles for escape are apt to inflict rather sharp scratches on the skin of the captor.

The color of the Mole Cricket is brown of different tints, darker upon the thorax than on the wing-covers, both of which organs are covered with a very fine and short down.

As might be surmised from the extraordinary muscular power of the fore-legs, the Mole Cricket can burrow with great rapidity. The excavation is of a rather complicated form, consisting of a moderately large chamber with neatly smoothed walls, and many winding passages communicating with this central apartment. In the chamber are placed from one to four hundred eggs of a dusky yellow color; and the roof of the apartment is so near the surface of the ground that the warmth of the sunbeams penetrates through the shallow layer of earth, and causes the eggs to be hatched.

The Mole Cricket (*Gryllotalpa*) is very common, and destructive to vegetation in the warmer portions of the United States. Its ravages on the sugar-cane is of a serious nature. The *G. borealis* is found in New England, in moist earth near ponds.

There is a singular species, called *Schizodáctylus monstrósus*, now common in the insect cases sent from India, which is notable for the manner in which the enormously long wings and their covers are rolled at their tips into spiral coils. This belongs to the same family as the mole cricket, and, like that insect, is a burrower, making holes nearly a yard in depth.

THE MIGRATORY LOCUST, represented in the accompanying colored illustration, is a well-known instance of a very large family of insects represented in our own land by many examples. All the Locusts and Grasshoppers are vegetable feeders; and in many cases their voracity is so insatiable, their jaws so powerful, and their numbers so countless, that they destroy every vestige of vegetation wherever they may pass, and devastate the country as if a fire had swept over it.

Such is the case with the Migratory Locust, so called from its habit of congregating in vast armies, which fly like winged clouds over the earth, and, wherever they alight, strip every living plant of its verdure. So assiduously do they ply their busy jaws, that the peculiar sound produced by the champing of the leaves, twigs, and grass blades can be heard at a considerable distance. When they take to flight, the rushing of their wings is like the roaring of the sea; and as their armies pass through the air, the sky is darkened as if by black thunder-clouds.

The family *Cicadariæ* includes an interesting group of insects, called in New England, incorrectly, locusts. *Cicada* is the generic name of the common "locust." Another species, called seventeen-year locusts, is notable for the great length of time the grubs live. During seventeen years the grubs live under ground, feeding on the roots of trees. The oak-tree is a favorite. At the termination of the period the grubs have attained their adult condition, when

SWARM OF MIGRATORY LOCUSTS.

they, being in the pupa state, come to the surface and the perfect Cicada appears, leaving the empty cases behind. Myriads of these creatures infest the oak forests, making the stridulous concerts so characteristic of them during the entire day. Whittier says of them, they

> "Stab the noon silence
> With their shrill alarm."

Or, in popular language, *zeeing* expresses their note, if it can be called a note, for it is the result of a mechanical rasping.

Now and then is found in the fields a very large, locust-like insect, of a beautiful grass-green hue, and having at the end of its tail a long, flat-bladed instrument called an ovipositor, and used for the purpose of boring holes in the earth and placing its eggs below the surface. This is the GREAT GREEN GRASSHOPPER (*Phasgonura*, or *Acrida viridissima*), which unfortunately loses its soft, light green color soon after death, and as it dries becomes a dirty yellowish-brown. It is a very fine insect, often measuring two inches in length, and three inches and a half over the expanded wings. It seems to be rather capricious in its appearances, in some years being quite plentiful, and in others hardly to be seen. The jaws of this insect are wonderfully powerful, and its captor will act wisely to keep his finger out of their reach. The internal structure of this grasshopper is extremely interesting, and on account of its large dimensions are easily studied. The gizzard is especially worthy of notice.

A SINGULAR insect is the Eyed Pterochroza. It is one of those beings in which are found a strong resemblance to other parts of creation. In this insect, we have an example of a member of the animal kingdom reproducing with startling fidelity the forms, colors, and even the accidental variations of leaves and flowers, thus exhibiting another phase of that wonderful adaptive power, which gives to many flowers, such as the orchids, a striking resemblance to bees, butterflies, and other insects. In this instance, the resemblance to leaves is not only due to the peculiar outline and the leaf-like nervures, but to the presence of certain spots which look exactly like the tracks of leaf-mining or leaf-devouring caterpillars. These creatures belong to the same family as the locusts, and their habitation is Brazil.

The locusts (*Locustariæ*) of North America include some very interesting forms. The Katydid, whose notes so invade night's attribute, stillness, during autumn, and some smaller ones, *Æcanthus*, are notable for their characteristic notes.

The grasshoppers (*Acrydæ*) are familiar enough in America, particularly in view of their monstrous destructive habits in the grain-fields. A species in Florida, called the Lubber Grasshopper, feeds on the orange-trees.

A strange-looking insect, with an attenuated body and long, slender limbs, is the WALKING-STICK INSECT. It is one of a most remarkable family of Orthoptera, none of which are found excepting in the hottest parts of the earth. That the Walking-stick Insect fully deserves its name, will at once be recognized by reference to the engraving. This insect belongs to the family of Phasmidæ, an appropriate title, derived from a Greek word signifying a spectre, many of these creatures being, as it were, the mere unsubstantial visions or shadowy outlines of insects.

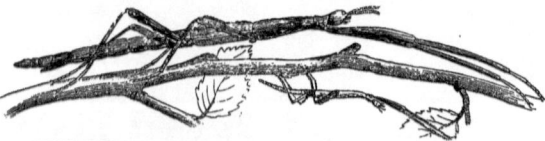

WALKING-STICK INSECT, grown and as larva.—*Bacteria trophina*. (In natural size.)

The chief point of interest in these creatures is their marvellous external resemblance to certain portions of the vegetable kingdom; some assuming the forms of a broken branch and twigs with such extraordinary fidelity that the most practised eye is often deceived, and others taking not only the flat outline and half curl of fallen leaves, but even reproducing their peculiar nervures and soft vegetable green with such marvellous exactness, that those who see them for the first time can hardly be made to believe that they are not the objects which they so faithfully represent. As if to add to the singularity of these creatures and to keep up the illusion, the eggs of several species are ribbed and colored precisely like the seeds of certain plants.

The *Phasmidæ* embrace some very extraordinary creatures, the Walking Sticks and Spectres being prominent and familiar members of this family. Our North American species of Walking Stick (*Diapheromera femorata*) is not over two inches in length of body, resembling the larger species of the East Indies.

One of the singular species which have such a wonderful resemblance to fallen leaves is the LEAF INSECT. The peculiar, leaf-like elytra, and also the singular manner in which the limbs are furnished with wide, flattened appendages, in order to carry out the leafy aspect, have often astonished people. Only the females possess the wide, veined wing-covers, those of the male being comparatively short. The wings, however, are entirely absent in the female, while in the opposite sex they are very wide and reach to the extremity of the body. One of these has lived for a considerable time in a greenhouse.

THE *Mantidæ*, or Praying Insects, also belong to the Orthoptera. These creatures derive their name from their habit of sitting with their long and flattened fore-legs held up and joined as if in the attitude of prayer. The form of this insect can be best seen from the drawing. So remarkable an insect could not fail to be the subject of many wild fables, some of which

PRAYING INSECT.—*Mantis religiosa*. (Female, and a cluster of eggs from which some larvæ are making their exit. Natural size.)

may take rank as popular superstitions. For example, it was long thought that if any one lost his way in a forest and met with a MANTIS, he had only to ask the insect to direct him on his road, when the obliging creature would stretch out one of its arms and point out the proper direction. According to old legends, one of these insects, being met by St. Francis Xavier and commanded to chant a prayer as well as to act it, responded to the request of the saint by singing a canticle—we presume in the Latin language.

Unfortunately for the character of the Mantis, the real reason for holding up its feet is, to be in readiness for seizing its prey or to defend itself from an enemy, the creature being voracious as a wolf and combative as a game-cock. It feeds chiefly upon other insects, stealing upon them quietly and catching them in its claws by a rapid movement, just as the loris takes its winged prey; and should it meet with another of the same sex and species, the two begin to fight with dauntless courage, cutting at each other with their fore-legs with the skill of practised swordsmen, and making their strokes so truly and with such force, that they have been known to sever the body of their antagonist with a single blow. The winner, that is to say the survivor, generally consummates his victory by devouring the body of his slaughtered foe.

The Chinese are fond of keeping these insects in cages and matching them against each other like game-cocks or bull-dogs. These creatures are said by some authors to be cowardly, because, if ants are put into their cages, they endeavor to escape in all directions. True as the fact may be, the inference is quite unwarrantable, the Mantis being entirely justified in trying to escape from such direful foes as the ants of its own country. During the last war of the English in India, a picket of soldiers contrived to disturb a large wasps' nest, and were forced to scatter in all directions in order to avoid the attacks of their small but formidable antagonists, for whose assaults they, being Highlanders, were very ill prepared. Yet no one would impugn the courage of the soldiers (the officer in command, an old pupil of my own, having won the Victoria Cross); and the ants are even more terrible foes to the Mantis than the wasps to human beings, their dimensions being quite disproportionate, and their usual prey being insects whom they overpower by numbers and united action, so that the size and courage of the Mantis are impotent when opposed to such foes.

Our *Mantidæ* are also small compared with those of the tropics.

FRINGE-WINGED INSECTS; THYSANOPTERA.

THE next order, according to Mr. Westwood's arrangement, is that called the Thysanoptera, or Fringe-winged Insects, on account of the manner in which the wings are edged with long and delicate cilia. They are all little insects, seldom exceeding the tenth or twelfth of an inch in length, but, although small, are capable of doing considerable damage. They are mostly to be found on plants and flowers, especially those blossoms where the petals are wide and deep and afford a good shelter. The convolvulus is always a great favorite with them. Greenhouses are sadly liable to their inroads, and owing to their numbers they are very injurious to melons, cucumbers, and similar plants, covering their leaves with a profusion of decayed patches, that look as if some powerful acid had been sprinkled over them. Only one family of these insects is acknowledged by entomologists.

TERMITES, DRAGON-FLIES, ETC.; NEUROPTERA.

WE now come to an order of insects containing some of the most beautiful and a few of the most interesting members of the class. They are known by the possession of four equal-sized membranous wings divided into a great number of little cells technically called areolets. The mouth is furnished with transversely movable jaws, and the females do not possess a sting or valved ovipositor. In this order are comprised the ant-lions, the dragon-flies, the termites, the lace-wings, and the May-flies.

THE first family in Mr. Westwood's arrangement is that of the Termites, popularly but erroneously known by the name of White Ants, because they live in vast colonies, and in many of their habits display a resemblance to the insect from which they take their name. All the Termites are miners, and many of them erect edifices of vast dimensions when compared with the size of their architects. For example, the buildings erected by the common White Ant (*Termes bellicôsus*) will often reach the astonishing height of sixteen or seventeen feet, which in proportion to the size of the insect would be equivalent to an edifice a mile in height if built by man. The dwelling is made of clay, worked in some marvellous manner by the jaws of the insect-architects; and is of such astonishing hardness, that although hollow, and pierced by numerous galleries and chambers, they will sustain the weight of cattle, which are in the habit of ascending these wonderful monuments of insect labor for the purpose of keeping a watch on the surrounding country. A full-sized habitation of the warlike Termite resembles a large irregular cone, having a diameter about equal to its height, and covered with turrets and smaller cones. Nor is this all, for the subterranean excavations are every whit as marvellous as the building, consisting of galleries, chambers, and wells some fourteen inches in width, and penetrating about five feet into the earth. These excavations serve for homes, for nurseries, and for roads of communication between the several portions of the vast establishment.

To give a complete history of the Termites would be a task demanding so much time and space, that it cannot be attempted in these pages; and we must, therefore, content ourselves with a slight sketch of their general history, premising that many parts of their economy, and especially those which relate to their development, are still buried in mystery.

The most recent investigations give the following results:—

Each Termite colony is founded by a fruitful pair, called the king and queen, who are placed in a chamber devoted to their sole use, and from which they never stir when once enclosed. These insects produce a vast quantity of eggs, from which are hatched the remaining members of the colony, consisting of neuters of both sexes, the females being termed workers and the males soldiers, the latter being distinguished by their enormous heads and powerful jaws; of larvæ of two forms, some of which will be fully developed, and others pass all their lives in the worker or soldier condition; of pupæ of two forms; and, lastly, of male and female perfect insects, which are destined to found fresh colonies. The neuters of either sex are without wings.

PASSING by, for the present, several families of the Neuroptera, we come to the Libellulidæ, or Dragon-flies. These insects are very familiar to us by means of the numerous Dragon-flies which haunt our river sides, and which are known to the rustics by the very inappropriate name of Horse-stingers, they possessing no sting and never meddling with horses or any other vertebrate animal. The name of DRAGON-FLY, on the contrary, is perfectly appropriate, as these insects are, indeed, the dragons of the air, far more voracious and active than even the fabled dragons of antiquity.

Even in their preliminary stages the Dragon-flies preserve their predatory habits, and for that purpose are armed in a most remarkable manner. During the larval and pupal states, the Dragon-fly is an inhabitant of the water, and may be found in most of our streams, usually haunting the muddy banks, and propelling itself along by an apparatus as efficacious as it is simple, and exactly analogous to the mode by which the nautilus forces itself through the water. The respiration is carried on by means of the oxygen which is extracted from the water; and the needful supply of liquid is allowed to pass into and out of the body through a large aperture at the end of the tail. On taking one of these creatures from the water, the extremity of the tail seems to be pyramidal, but on examination will be seen to consist of several pointed flakes which can be separated and then disclose the aperture above mentioned.

By means of this apparatus, water is admitted into the body, and, after giving up its oxygen, is violently expelled, thereby forcing the insect forward with a velocity proportioned to the power of the stroke. If one of these creatures be put into a glass vessel, it appears at first to move by simple volition; but if a little sand be allowed to settle at the bottom, the disturbance caused among the grains by the ejected water will show the mode of progression. If the larva be allowed to take in the water, and then suddenly moved into the air, the force with which it expels the contained water will drive it to a distance of three or four inches.

Such are its means of locomotion; those of attack are not less remarkable or less efficacious.

The lower lip, instead of being a simple cover to the mouth, is developed into a strange-jointed organ, which can be shot out to the distance of nearly an inch; or, when at rest, can be folded flat over the face, much as a carpenter's rule can be shut up so as to fit into his pocket, and can be rapidly protruded or withdrawn, very like the instrument called a "lazy-tongs." Like that instrument, it is furnished at its extremity with a pair of forceps, and is able to grasp at passing objects with the swiftness and certainty of a serpent's stroke.

The creature remains for some ten or eleven months in the preliminary stages of existence, and when the insect is about to make its final change, the undeveloped wings become visible

DRAGON-FLIES, LAYING EGGS.

on the back. When its time has come, the pupa leaves the water, and crawls up the stem of some aquatic plant until it has reached a suitable elevation; it clings firmly with its claws, and remains apparently quiet. On approaching it, however, a violent internal agitation is perceptible, and presently the skin of the back splits along the middle, and the Dragon-fly protrudes its head and part of the thorax. By degrees, it withdraws itself from the empty skin, and sits for a few hours drying itself, and shaking out the innumerable folds into which the wide gauzy wings have been gathered. After a series of deep respirations of the unwonted air, and much waving of the wings, the glittering membranes gain strength and elasticity, and the enfranchized insect launches forth into the air, in search of prey and a mate.

There are very many species of Dragon-flies, all very similar in their habits, being fiercely predaceous, strong of wing, and gifted with glittering colors. Unfortunately, the rich azure, deep green, soft carnation, or fiery scarlet of these insects fade with their life, and in a few hours after death the most brilliant Dragon-fly will have faded to a blackish-brown. The only mode of preserving the colors is to remove all the interior of the body, and to introduce paint of the proper colors. This, however, is but an empirical and unsatisfactory sort of proceeding; and no matter how skilfully it may be achieved, will never be worth the time bestowed upon it.

In many species, the sexes are of different colors, as, for example, in the beautiful DEMOISELLE DRAGON-FLIES, where the male is deep purple, with dark spots on the wings, and the female a rich green, with the wings uncolored. The wings are of an exceedingly fine quality, and the structure of the whole body can be best learned by reference to our full-page engraving. Another form of Dragon-fly is the *Libellula trimaculata*. It is an example of the restricted genus Libellula.

THE singular group of insects termed SCORPION-FLIES also belong to this family. These insects derive their popular name from the curious appendage with which the abdomen of some of the species is armed. The male of the common Scorpion-fly has the sixth and seventh rings of the abdomen rather slender, and capable of movement in every direction; while the last ring is modified into a stout, thick, rounded form, furnished with a pair of forceps not unlike those of the earwig. While at rest, the creature sits with the abdomen curled quietly over the back like a pug-dog's tail; but, when irritated or alarmed, it brandishes the tail about in a very alarming manner, snapping at the same time with the forcipated extremity, and, if it seizes the finger, can inflict a very perceptible nip. Few persons, indeed, who are not accustomed to the insect can summon up sufficient moral courage to hold it while its scorpion-looking tail is being flourished in so menacing a manner.

THE beautiful LACE-WING FLIES, or Hemerobiidæ, are also members of this order. Several species of the Lace-wings are also called by the appropriate name of Golden Eyes, on account of the extreme brilliancy of the large and projecting eyes, which glow as if with internal fires, and give forth flashes of gold and ruby light. Unfortunately, there is a sad drawback to their beauty, for, when handled, they exhale a most powerful and indescribably odious stench, unlike any imaginable combination of evil savors, but quite unique, and never to be forgotten after a single experience. The Lace-wings may be taken in the evening as they fly from tree to tree, and in the daytime may be found clinging to the under side of leaves.

THE far-famed ANT-LION is one of the insects that are more celebrated in their preliminary than in the perfect stage of existence. As may be seen by reference to the illustration, their perfect form is very light and elegant, and closely resembling that of the dragon-flies, save that the wings are lighter, softer, and broader.

In their larval condition, however, as will be noticed, they are by no means attractive-looking creatures, somewhat resembling flattened maggots with their rather long legs and their very large jaws, the legs being apparently useless as organs of progression, all movements being made by means of the abdomen.

Slow of movement as is this creature, and yet predaceous, feeding wholly on living insects, the mode of obtaining its food seems to be rather a problem. The solution, however, is simple enough, the creature digging a pitfall, and lying ensconced therein while the expected prey approaches.

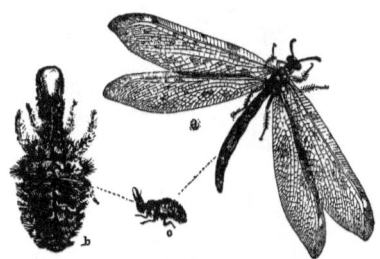

ANT-LION.—*Myrmeleon formicarius.* a, Ant-Lion; b, c, larvæ (Figs. a and c are of natural size, while Fig. b is magnified.)

The beautiful NEMOPTERA COA belongs to an allied family of this order, and is remarkable through the curious development of the hinder pair of wings, a peculiarity which is repeated, though not on so extensive a scale, in many of the butterflies.

Our accompanying full page engraving represents the MAY-FLY, or EPHEMERA, the best and most familiar type of the family to which it belongs, and which is scientifically called the Ephemeridæ.

This insect has long been celebrated for its short space of life, a single day sometimes witnessing its entrance into the perfect state and its final departure from the world. The popular idea concerning these insects is, that the whole of their life is restricted to a single day. This, however, is an error, as they have already passed at least two years in their preliminary stages of existence. In the larval and pupal states, they are inhabitants of the water, and are fond of hiding themselves under stones, or burrowing into the muddy banks. Under the latter circumstances they make a very curious tunnel, something like a double-barrelled gun. It is said that the larva feeds upon mud, and, as a proof of this assertion, it may be mentioned that Swammerdam always found mud within those specimens which he dissected. I can personally vouch for the accuracy of his remarks, but would not like to assert that, although mud was always found in the stomach and intestines of those larvæ which I have dissected, it might not have been swallowed with the food rather than composed it.

The May-fly is peculiarly notable for a stage of development which seems to be quite unique among insects. When it has passed through its larval and pupal state, it leaves the water, creeps out of its pupa case, and takes to its wings. After a period, varying from one to twenty hours, it flies to some object, such as the trunk of a tree or the stems of water-plants, and casts off a thin membranous pellicle, which has enveloped the body and wings, the dry pellicle remaining in the same spot, and looking at first like a dead insect. After this operation, the wings become brighter, and the three filaments of the tail increase to twice their length. Some authors call the state between the leaving the water and the casting the pellicle the "pseudimago" state.

Some of these insects are well known to fishermen under the names of green and gray drake, the former being the pseudimago, and the latter the perfect form of the insect, which is represented in the illustration. Sometimes these insects occur in countless myriads, looking like a heavy fall of snow as they are blown by the breeze, and having on some occasions been so plentiful, that they have been gathered into heaps and carted off to the fields for manure.

The Perlidæ, known to anglers by the name of STONE-FLIES, belong to the Neuroptera. Several species of the same family are popularly called Yellow Sally and Willow-fly. They may be known by the large folded front pair of wings, and the two bristle-like appendages at the tail.

MAY-FLY.

CADDIS-FLIES; TRICHOPTERA.

QUITTING the Neuroptera, we must give a few lines to another order of insects, the TRICHOPTERA, popularly known by the name of CADDIS-FLIES.

These insects, of which there are many species, are chiefly remarkable in their larval state, on account of the curious portable habitations which they construct. All anglers are familiar with the Caddis, and the singular variety of form and material employed in the construction of its home. Being a soft, white grub, totally unarmed, and presenting a most delicate morsel to every river-fish, the Caddis is forced to conceal itself in some way from its innumerable foes. For this purpose, it builds around itself a nearly cylindrical tube, open at each end, and composed of substances varying according to the locality and the species. Sometimes these tubes are made wholly of short pieces of stick, laid sometimes side by side, and sometimes in a partly spiral form, something like the wires of the submarine telegraph. Sometimes the tubes are made of sand or little stones, while the deserted shells of the planorbis, and other freshwater shells, are very common materials.

FLIES AND BEES; HYMENOPTERA.

WE now come to a vast order of insects, technically called the HYMENOPTERA. In these insects the wings are four in number, transparent, membranous, the veins comparatively few, and the hinder pair smaller than the others. Their mouth is furnished with powerful horny jaws, and with a tongue guarded by the modified maxillæ. The females are armed with a many-valved sting or ovipositor. In this enormous order are included all the bees, wasps, and their kin, the great family of saw-flies, the ichneumons, the gall-flies, and the ants, each single family being so large, and presenting so many points of interest, that an entire volume could be devoted to them with great profit. Our space, however, prohibits us from attempting more than a slight sketch of each family, together with descriptions of a few typical species. Without, therefore, enumerating the various arrangements of this large order, or the characteristics on which they are founded, we will proceed at once to the family of the Tenthredinidæ, or Saw-flies, the first in Mr. Westwood's system.

In this and the next family, the females are furnished with a peculiar ovipositor, composed of several pieces, and which, though connected with a gland secreting an irritant fluid, are not envenomed as in the bees, wasps, and their kin. All these insects are comprised under the general term of Terebrantia, or borers, and fall easily into two large groups, in one of which the abdomen proceeds directly from the thorax, and in the other is connected with the thorax by means of a footstalk. Each of these groups is further subdivided, as will be seen in the course of the following pages.

The true Saw-flies are known by the curious piece of animal mechanism from which they derive their name. The females of this family are supplied with a pair of horny saws, placed side by side on the lower extremity of the abdomen.

These saws are of various forms, according to the particular species to which they belong, and may be seen even in the dried specimens, the top of their sheath slightly projecting, and their shapes plainly visible after the removal of a portion of the abdomen. When taken from the insect and placed under the microscope, they present a very pretty appearance, owing to the gently-curved ribs with which their sides are strengthened and decorated. The saws act alternately, one being pushed forward as the other is being retracted. Their object is to form a groove in some plant, in which the eggs of the mother insect can be deposited, and wherein they shall find a supply of nourishment in order to enable them to complete their development; for it is a most remarkable fact that, after the egg is deposited in the groove, it rapidly increases in size, obtaining twice its former dimensions.

In the genus Cimbex, of which an example is given in the illustration, the larvæ possess twenty-two feet, and have the power of discharging a translucent greenish fluid from certain

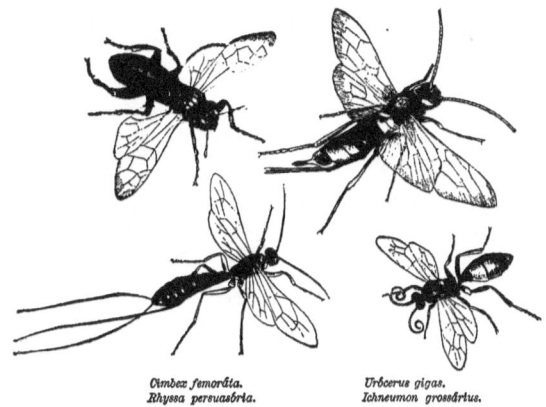

Cimbex femoráta.
Rhyssa persuasória.
Urócerus gigas.
Ichneumon grossárius.

pores placed on the sides of the body just above the spiracles. This feat they can repeat six or seven times in succession. When they have eaten their way to the next stage of existence, they spin a cocoon of a brownish color and of a stringy, tough consistency, and either suspend it to the branches of a tree on which they have been feeding, or hide it under fallen leaves. In this cocoon they remain for a comparatively short time, and then emerge as perfect insects.

The terrible TURNIP-FLY (*Athalia centifóliæ*) belongs to this family. The larva of this species is popularly called the Nigger, on account of its black color. Our engraving shows the insect in both its stages of development. A very small species of Athalia is called *Athalia spinarum*. Its larva feeds upon the various cabbages, eating away the whole of the soft green parts of the leaves, and only rejecting the thick nervures. It makes no cocoon, but retires into the ground, excavates a kind of oval cell, which it lines with a slimy substance, and there awaits its final change.

The well-known black GOOSEBERRY-FLY (*Nématus grossuláriæ*) is another of the Sawflies. Its larva, so destructive to the fruit, is blackish-gray. These tiresome creatures are often seen in great numbers, more than a thousand having been taken on a single goose-

TURNIP-FLY.—*Athalia centifolia.* (Natural size.)

berry-bush, and there are two broods in the course of a year. Without going into further details, it is sufficient to say that there is hardly a plant without its especial Saw fly, and that any one who can discover a really effectual mode of checking their ravages, will confer no slight benefit on mankind.

THE fine insect in the illustration at top of this page, which is known by the name of the GIANT ICHNEUMON (*Ichneumon grossárius*), is an example of the next family, in which the ovipositor is converted into a gimlet instead of a double saw. With this powerful instrument, the female is enabled to drill holes into living timber for the purpose of depositing the eggs. When they are hatched, the young grubs immediately begin to gnaw their way through the wood, boring it in every direction, and making burrows of no mean size. Those of the present species prefer fir and pine, and I have had specimens of the wood sent to me which have been riddled by the grubs until they looked as if they had harbored a colony of the ship-worm. The perfect insects often make their appearance in houses, the larvæ having been concealed in the timbers and rafters; and I know of one case where a gentleman who had built a wooden garden-house, was sadly annoyed by the multitudes of the Sirex which emerged from the timber. In such cases the insects do not seem to attain their full dimensions, but appear dwarfed and stunted. All wood-boring insects are, however, extremely variable in size.

The next group of the Terebrantia is called Entomophaga, or Insect-eaters, because the greater number of them are parasitic upon other insects, just as the Saw-flies are parasitic upon vegetables. In these insects the ovipositor is furnished with two delicate spiculæ, and the last segments of the abdomen are not formed into a telescope-like tube.

The first family is that of the Cynipidæ, or Gall insects, the creatures by whose means are produced the well-known galls upon various trees, the so-called oak-apple being perhaps the best known, and the Ink-gall (also found on the oak) the most valuable. These Galls are formed by the deposition of an egg in the leaf, branch, stem, twig, or even root of the plant, and its consequent growth. The well-known Bedeguar of the rose, with its soft mossy envelope and delicate green color, relieved by bright pink, is caused by one of these insects (*Cynips rosæ*); and the celebrated Dead Sea-apples are nothing but galls formed by the *Cynips insâna*. The spherical oak-galls, which contain a single insect, and are about the size of a large marble, are closely allied to the true Ink-galls; and if one of these objects be cut with a knife, the action of the astringent juice upon the iron of the blade will produce a kind of ink. The best galls are those which are gathered before the insect makes its escape, as the astringent quality is then more powerful.

The true Ichneumons, of which a specimen is given in the illustration, form a vast group of insects, the Ichneumonidæ alone numbering many more than a thousand described and acknowledged species. In them the ovipositor is straight, and is employed in inserting the eggs into the bodies of other insects, mostly in their larval state. In some cases, this slender and apparently feeble instrument is able to pierce through solid wood, and is insinuated by a movement exactly like that which is employed by a carpenter when using a bradawl. When not engaged in this work, the ovipositor is protected by two slender sheaths that enclose it on either side.

Were it not for the Ichneumons, our fields and gardens would be hopelessly ravaged by caterpillars and grubs of all kinds, for practical entomologists always find that when they attempt to rear insects from the egg or the larval state, they must count upon losing a very large percentage by the Ichneumons.

Take, for examples, three or four caterpillars of the common white cabbage butterfly, place them under water, and open the body from end to end. It will be found that, in almost every case, the caterpillar bears the seeds of death within its body in the shape of tiny white grubs, like very minute grains of rice. These creatures are the young of an Ichneumon-fly (*Microgaster glomerátus*), and retain their place within the caterpillar until the time for it to change into the perfect form. They then simultaneously eat their way out of the skin, spin a number of bright yellow silken cocoons, and in process of time change into tiny flies and set out on their destructive mission. The caterpillar never survives their attacks, and is seldom able to move away from the spot whereon it happened to be when the Ichneumons make their escape, the body being enveloped in their yellow cocoons.

All the Ichneumon-flies may be distinguished by their fussy restless movements, as they run up and down any object on which they may settle, and the continual quivering of their antennæ. The two lower figures in the illustration belong to this family, that on the left showing an example of the long ovipositor with which several species are furnished, and the other being given in order to show the wasp-like abdomen and the curled antennæ.

The *Rhyssa persuasoria* is the largest European Ichneumon, and is parasitic on *Urocerus juvencus*, another species of the same genus as that to which the giant Ichneumon belongs. The larva on which it preys bores deep holes in fir-trees, and, in consequence, the Rhyssa may be seen running up and down the trunks in search of some spot where the ovipositor may be introduced so as to lodge in the hidden larva. So deeply does the insect contrive to force its weapon into the wood, that it sometimes is unable to withdraw it, and may be seen hanging dead and dry to the tree in which it has buried the ovipositor too firmly.

Crabro cribrárius. *Philanthus tridngulum.*

PASSING by several families belonging to this group, we must briefly mention the beautiful RUBY-TAIL FLIES, or CUCKOO FLIES, so plentiful in old walls and similar localities. These are distinguished by the fact that, in the females, the last segments of the abdomen are formed into a telescopic tube, which can be projected or retracted at pleasure, and is furnished with a minute sting. These are, perhaps, the most brilliant in color of any European insect, and are veritable humming-birds of the insect tribes, their bodies literally flashing with ruby, sapphire, and emerald, as they flit restlessly in the sunbeams. They are parasitic insects, and haunt the walls for the purpose of depositing their eggs in the larva of sundry solitary bees and wasps.

IN the next great division of Hymenopterous insects, the ovipositor of the female is changed into a sharply pointed weapon, popularly called a sting, and connected with a gland in which is secreted a poison closely analogous to that which envenoms a serpent's tooth. These are again divided into the Insectívora, or those which have fore-wings not folded, and the larvæ solitary and feeding on other insects; the Sodáles, where the fore-wings are not folded, and the larvæ are social; and the Diplóptera, where the fore-wings are folded, and the larvæ (in the social species) inclosed in separate cells.

The first of these sections comprises all those curious and interesting insects known popularly by the names of Sand Wasps and Wood Wasps. These creatures are in the habit of making burrows into the ground or in posts, and placing therein their eggs, together with the bodies of other insects which are destined to serve as food for the future progeny. Spiders are sometimes captured and immured for this purpose. In many instances the captured insects are stung to death before they are placed in the burrow, but it is often found that they only receive a wound sufficient to paralyze them, so that they lead a semi-torpid life until they are killed and eaten by the young grub. Two of these Sand Wasps are given in the illustration. That on the left is one of the wood-borers, drilling its burrow into posts, palings, and similar substances, and feeds its young with the larvæ of one of the leaf-rolling caterpillars that lives in the oak, and is scientifically known by the name of *Tortrix chloránna*. It also employs for this purpose several two-winged insects. One species of these burrowing wasps prefers the well-known cuckoo-spit insect for this purpose (*Aphróphora spumária*), pulling it out of its frothy bed by means of its long legs.

The right-hand figure represents a species that is in the habit of provisioning its burrow with the hive-bee, which it contrives to master in spite of the formidable weapon possessed by its victim, and then murders or paralyzes by means of its sting. M. Latreille mentioned that he saw from fifty to sixty of these insects busily engaged in burrowing into a sandbank not more than forty yards long; and as each female lays five or six eggs, and deposits a bee with each egg, the havoc made among the hives is by no means inconsiderable.

IN the accompanying illustration is shown a Brazilian species, belonging to a genus which is represented in Europe by more than twenty species. In these insects the legs are very long and spider-like, enabling their owners to run about among grass with great vivacity, their wings quivering all the while with violent agitation. Some of the species are in the habit of catching spiders, and provisioning the burrows with them. It is worthy of notice, that the largest specimens of Hymenoptera are to be found in exotic insects belonging to this family, the genus Pepsis being most remarkable for the great dimensions of its members.

The right-hand figure represents an insect which is common in Southern Europe. Judging by the habits of those species which have been studied, the whole of the family to which it belongs are sand-burrowers, and seem to be cruelly predaceous, mastering insects of considerable size, and dragging them into their burrows. One of these insects (*Scólia bicincta*) has been known to capture and inter a large locust, the tunnel being some eighteen inches in depth and very wide at the mouth.

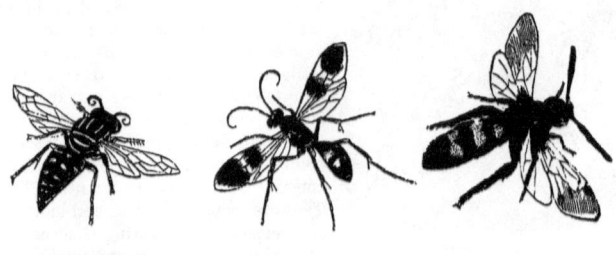

Monédula signata.   *Pómpilus nóbilus.*   *Scólia pratórum.*

A FORMIDABLE but useful insect is the *Chlorion lobatum*, which wages fierce war against cockroaches, those pests of American and Oriental houses, and its services are fully appreciated by the natives, none of whom would kill one of these insects on any account, or permit any one to injure it. With the slaughtered cockroaches it stocks its nest as a provision for the young when they escape from the egg. These insects are tolerably numerous, and are all remarkable for the bright and yet deep purple and green of their bodies, and sometimes of their wings.

OUR next subject is the LARGE-HEADED MUTILLA. It is a curious, wingless insect, with head disproportionately large, when the size of its body is taken into consideration. This is an example of a family where the females, although armed with a powerful sting, are quite destitute of wings. Most of the Mutillidæ are exotic, requiring a large amount of heat to preserve them in health, only a very few being natives of Northern America and Europe. In some of the larger species the sting is fearfully poisonous, a single insect having been known to make a man so seriously ill that he lost his senses a few minutes after being stung, and his life was despaired of for some time. A child has been known to die from the effects of the sting inflicted by the Scarlet Mutilla of North America, an insect whose weapon is as long as the abdomen. All these insects appear to be sand-borers.

WE now come to the Wasps, in which the wings are folded throughout their entire length when at rest. A wasp distinguished through the slenderness of the middle part of its body is a native of Australia. It belongs to the Solitary Wasps, many of which are found in Europe. The curious nest of this insect is formed like a globe. The creature makes a separate nest for each egg, the material being clay well worked. The nest is stocked with the larvæ of moths or butterflies.

To this family belongs that wonderful Burrowing Wasp, which is a builder as well as an excavator, and which erects a tubular entrance, often more than an inch in height, with the fragments of sand which it has dug from the tunnel. It is thought, and probably with correctness, that the object of the insect in making this edifice is to deter its parasitic foes from entering so long and dark a channel. The tube is always curved. When the burrow is completed, the Wasp lays its egg in the tunnel, and packs in it a series of little green caterpillars, which serve as food for the larva. When the arrangements are completed, the Wasp takes down her tube, and employs the materials in closing the mouth of the tunnel. The technical name of this insect is *Odynerus muraria*. Another species is also known to possess this curious faculty.

The true Wasps, or Vespidæ, come next in order. These insects are gregarious in their habits, building nests in which a large, but uncertain number of young are reared. The common Wasp makes its nest within the ground, sometimes taking advantage of the deserted hole of a rat or mouse, and sometimes working for itself. The substance of which the nest is made is a paper-like material, obtained by nibbling woody fibres from decayed trees or bark, and kneading it to a paste between the jaws. The general shape of the nest is globular, and the walls are of considerable thickness, in order to guard the cells from falling earth, a circular aperture being left, through which the inhabitants can enter or leave their home.

Many species of Wasp inhabit Europe, the HORNET (*Vespa crabro*) being the largest, and, indeed, being nearly equal in dimensions to any tropical species. This formidable insect makes a nest very similar to that of the wasp, but the cells are necessarily much larger. The nest is generally placed in hollow trees, but I have known a colony of these insects to establish themselves in an outhouse, and to cause great annoyance before they could be expelled.

A very pretty nest is also found in Europe, the work of the *Vespa britannica*. It is suspended to branches, is nearly globular in shape, and extremely variable in size, some specimens being nearly a foot in diameter, while others are comparatively small. A very pretty specimen in my possession is about the size of a tennis ball. Some exotic species make nests, the covering or outer case whereof is thick and tough as pasteboard, and nearly white in color. One of these nests, which is found in the Brazils, is popularly called the Dutchman's pipe, its shape somewhat resembling an exaggerated pipe-bowl, the aperture for ingress and egress doing duty for the mouth, and the branch on which it is suspended taking the place of the stem. I believe that the insect which forms this curious structure belongs to the genus Chartergus. The central orifice penetrates through all the layers of combs.

The left-hand figure on the engraving at page 401 represents a fine insect, a native of Brazil, belonging to the Bembecidæ. This species is in the habit of catching grasshoppers of considerable size, carrying them off, and stocking with these insects the habitation made for its young. A very fine species of Chrysis is parasitic upon it.

THERE are, perhaps, few insects so important to mankind as those which procure the sweet substance so well known by the name of honey. Nearly all the honey-making Hymenoptera are furnished with stings, and in many species the poison is fearfully intense. Some of these insects, such as the HIVE BEE, which is represented in the accompanying engraving, make waxen cells of mathematical accuracy, the larvæ being placed in separate cells, and fed by the neuters. In some cases, such as the common HUMBLE BEE, the cells are egg-shaped, each cell being either occupied by a larva, or filled with honey; while in some species the eggs are placed parasitically in the nests of other bees, so that the larvæ feed either upon the stores of food gathered for the involuntary host, or upon the body of the deluded insect itself.

The Hive Bee is the typical example of the honey-gatherers, but its general economy is too well known to need much description. Suffice it to say that, as in the ants, the community consists of males, females, and neuters, but that in the Bees, all the members of the establishment are winged, and the wings are permanent. In each hive there is one fully-developed female, called the queen, several others in process of development, and intended to be the heads of future establishments, a limited number of males, and a vast band of neuters, *i. e.*, undeveloped females. The males have no sting, but both the females and neuters are armed with this tiny, but formidable weapon. Since in civilized countries the Hive Bees are kept in habitations of limited size, their numbers soon outgrow their home, and a large number accordingly quit the hive under the government of the old queen, the rule of the hive being taken up with one of the young queens, which has burst from its cell in the meanwhile. A fresh colony is founded as soon as the Bees can meet with shelter, and their new residence is speedily filled with honey and young. The cells of the Bee-comb are set back to back, and each comb hangs like a thick curtain from the top and sides of the hive, so that the cells lie nearly horizontally.

In gathering honey, the Bees lick the sweet juices from flowers, swallow them, and store them for the time in a membranous cup, popularly called the honey-bag. When this cup is filled, the Bee returns to the hive, and discharges the honey into cells, closing its mouth with

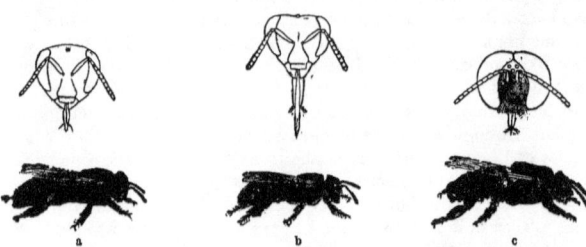

HIVE BEE.—*Apis mellifica*. (a, Queen; b, Laborer; c, Drone; with front of the heads. The latter are magnified, and each belongs to figure beneath.)

wax when it is filled. The structure of the Bee-cell, its marvellous adaptation to the several purposes for which it is intended, its mathematic accuracy of construction, whereby the best amount of material is found to afford the greatest amount of space and strength, are subjects too complicated to be here described, but may be found in many works which have been written upon the Hive Bee.

THE members of the genus NOMADA are very wasp-like in their general aspect, are not hairy, and are, indeed, often taken for small wasps by inexperienced observers. They are, however, true bees. Their habits are rather obscure, but they are thought to be parasitic insects.

The CARPENTER BEE of Southern Africa is one of those curious insects which construct a series of cells in wood. After completing their burrow, which is open at each end, they close the bottom with a flooring of agglutinated sawdust, formed of the morsels bitten off during the operation of burrowing, lay an egg upon this floor, insert a quantity of "bee-bread," made of the pollen of flowers and their juices, and then cover the whole with a layer of the same substance that was used for the floor. Upon this is laid another egg, another supply of bee-bread is inserted, and a fresh layer of sawdust superimposed. Each layer is, therefore, the floor of one cell and the ceiling of another; and the insect makes, on the average, about ten or twelve of these cells.

THE numerous HUMBLE BEES are a group of insects readily recognized by their thick, hairy bodies and general shape. Their nests are placed underground, often in banks, and contain a variable number of cells, sometimes not more than twenty in number, and sometimes exceeding two or even three hundred. The cells are loosely connected together, and are of an oval shape, their texture being tougher and more paper-like than those of the hive bee. In these, as in the ordinary bees, there are the three kinds of inhabitants; but with the Humble Bees, both the females and neuters take part in the labors of the establishment, while the number of perfect females is comparatively large.

The honey made by these insects is peculiarly sweet and fragrant, but to many persons, myself included, is rather injurious, always causing a severe headache. Some of the Humble Bees (*Bombus muscôrum*) employ moss in the construction of their nests, and pass it, fibre by fibre, through their legs, in a manner that reminds the observer of carding cotton. They are, on that account, popularly called CARDER BEES. Others, again (*Bombus lapidarius*), prefer to make their nests in heaps of stones, or similar localities, and these are the fiercest of their kind. Generally, the Humble Bees are quiet and inoffensive, even permitting their nest to be laid open and the cells extracted, without offering to molest the invader. The ORANGE-TAILED HUMBLE BEE, however, is large and fierce; and possessing a powerful sting, with a very large poison-gland, becomes no despicable foe to those who offend it, or whom it chooses to consider as foes.

The nests of the Humble Bees are not permanent like those of the hive bee, but perish during the winter, the only survivors being a few females, who are destined to found fresh colonies in the succeeding year.

Another species, the BANDED BEE, is so greatly in use in Egypt, and is fed by being placed on board of barges, and transported down the Nile, so as to insure a bountiful supply of honey. The owners of the hives pay a small sum to the owners of the boats, and, in return, their bees are carried along the fertile stream during the honey season, and afterwards returned with full combs. Payment is mostly in kind, thus insuring the proper fulfilment of the compact.

For want of space, we are compelled to pass by many interesting Hymenoptera, such as the Leaf-cutter Bees, the Wood-borers, and the Mason Bees, each of which creatures would demand more space than can be given to the whole of the insects.

STREPSIPTERA.

A VERY small, but very remarkable order now comes before our notice—the STREPSIPTERA, comprising insects of very minute proportions, all of which are parasitic upon the bodies of different bees and wasps, five, and even six, having been discovered within a single wasp. Their presence may generally be discovered by the peculiar swollen aspect of the abdomen; and, in many cases, the heads of the parasites may be seen protruding from between the segments.

The name Strepsiptera signifies, literally, twisted wings, and is given to these creatures because the front pair of wings are transformed into short and twisted appendages, quite useless for flight or for defending the second pair of wings. These are almost disproportionately large, membranous, and with a kind of milky look as the insect flies through the air. The eye is composed of a very few lenses, in some species only fifteen on each side, two or three thousand being the ordinary average among insects. The antennæ are of a remarkable form, branched and forked like the horns of a stag. The thorax is enormously large, and the abdomen of very small size; but, as the creature does not appear to take food during its life in the perfect state, this is of little moment. Curiously enough, the larvæ of these insects are themselves subject to internal parasites; and it is very possible, that they, in their turn, may be infested by other creatures less than itself, and equally disagreeable.

BUTTERFLIES AND MOTHS; LEPIDOPTERA.

WE now come to an order in which are included the most beautiful of all insects, namely, the Butterflies and Moths. On account of the feather-like scales with which their wings are covered, and to which the exquisite coloring is due, they are technically called Lepidoptera, or scale-winged insects.

The wings are four in number, and it is occasionally found that the two pairs are connected together by a strong bristle in one, and a hook-like appendage in the other, so that the two wings of each side practically become one member, in a manner similar to the formation of many hymenopterous insects. Those species which take any nourishment subsist entirely upon liquid food, which is drawn into the system by suction, and not by means of a brush, as

Mechanitis lysimnia.
Thecla.
Heliobpis cupido.
Mesosemia misipsa.

Gynæcia dirce.
Papilio thoas.
Epicalia ansœa.

Catagramma marchalii.

Papilio protesilaus.

is the case with the liquid-feeding beetles and bees. The wings are strengthened by nervures, which are of great use in determining the position of the insects.

IN the system which is adopted in this work, the Lepidoptera are divided into two sections—the Butterflies and Moths, technically called Rhopalocera and Heterocera—which may generally be distinguished from each other by the form of the antennæ, those of the Butterflies having knobs at their tips, while those of the Moths are pointed. The first family is that of the Papilionidæ, in which are included the largest and most magnificent specimens of this order. The fanciful names with which so many of these insects have been honored are chiefly due to Linnæus, who was so struck with the splendid dimensions and gorgeous coloring of these insects, that he deemed them worthy of Homeric titles, called them Equites, or knights, and, separating them into two divisions, gave to all those which had red spots on the sides of the breast the names of the Trojan heroes, and to those which were without the red spots, and had an eye-shaped mark on the lower wings, the names of the Greek warriors who fought against Troy. Unfortunately, for this division, it happens that the two sexes of many species are very diverse, and cause great confusion, so that Polycaon, one of the Greeks, and Laodocus, a Trojan hero, have been found to be the two sexes of the same insect, the latter being the female.

The splendid insect which is shown in the illustration is one of a genus which, by common consent, takes the first rank among the Lepidoptera, in consequence of their great size, the elegant boldness of their shapes, and the richness of their coloring. In them, the two fore-legs are not stunted, as is the case with so many butterflies, but are large, strong, and can be used in walking; and the tips of the antennæ have a very slight bend upwards. The caterpillar of this insect is furnished with two retractile tentacles, placed on the neck in a fork-like shape, and is able to protrude these organs at will, at the same time emitting a very unpleasant odor. The chrysalis is hung up by means of silken threads from the tail, and the body is prevented from swinging about too rudely by a pair of very stout silken cords, one of which is affixed to each side, and moors the pupa firmly to the substance against which it hangs. The shape of the pupa is rather peculiar, being angular in its outline, and having also an irregular curve.

AMPHRISIUS.—*Ornithóptera Amphrisius.*

The perfect AMPHRISIUS is a boldly marked insect, though without much variety of coloring. The upper wings are rich blackish-brown, and the lower are fine king's-yellow, edged with jetty-black, and having a fringe of long hair-like scales upon their inner edges. The under side is nearly of the same colors, except that a few dashes of chalky-white are seen upon the upper wings, as if dashed in with a quick sweep of a dry brush, and a similar dash of yellow is seen upon the lower wings. The abdomen is bright yellow below and dark brown above, and round the neck is a narrow collar of fiery crimson, rich and silken as "chenille."

Another species of this genus, the POSEIDON, is a great favorite with the natives of the Darnley Islands, who are accustomed to catch several of these fine insects, to tie one end of a fine thread to the butterfly, and the other to their hair, so as to permit the insects to flutter about their heads. This style of head-dress is much admired, and, indeed, is equally poetical and artistic. Many of these insects differ greatly, according to sex, the upper wings of the male PRIAM, for example, being velvety-black, striped with silky-green, and the hinder wings entirely silky-green, spotted with black and orange; while the female is dark brown, spotted with white.

IN the genus Papilio we find the insects to be of nearly as magnificent proportions as in the former genus, though none of them reach the enormous size of the Priam, which will sometimes measure nearly eight inches across the spread wings. The colors are, however, more varied, and quite as brilliant, while a curious feature is often added by the prolongation of the hinder wings into two long tail-like appendages. The larva is of varied form, sometimes smooth, sometimes covered with fleshy protuberances, sometimes long, and able to throw out or to withdraw at pleasure the two first segments of the body, sometimes short, thick, and grub-like, and in one or two instances marvellously resembling snails in the general form. The genus is a very comprehensive one, including between two and three hundred known species, among which may be found almost every imaginable tint in every gradation, and exhibiting bold contrasts of color which scarcely any human artist would dare to place together, and which yet produce a result equally striking and harmonious.

Our first example is the SARPEDON, one of the most common of the genus, being found plentifully throughout many parts of Asia, Australia, and the Sandwich Islands. Its flight is rather swift, and easily recognizable, and, in common with many allied species, it has regular beats, traversing the same ground time after time with almost mechanical regularity. Entomologists take advantage of this habit, and if they see one of these butterflies pass over a certain spot, they just go and sit down where they saw the insect, and catch it as it comes round on its next circuit.

THE HECTOR forms a fine contrast to the preceding insect, its colors being almost wholly black and flaming crimson. On its upper surface, the front pair of wings are sooty-black, with a broad dash of gray-brown over the centre, and a little pencilling

SARPEDON.—*Papilio sarpedon.*   HECTOR.—*Papilio hector.*

of the same color near the tips; and the lower wings are deep velvety-black, diversified with spots of intense crimson. The wings have a very narrow edging of white. The chest and part of the abdomen are black, and the head and rest of the abdomen of the same rich crimson as the spots on the wings. The under surface is colored much in the same way, except that the crimson spots are larger.

On the large engraving at page 405, and in the right-hand lower corner, may be seen a butterfly, with two very long straight tails to its under wings. This insect belongs to the genus Papilio, and is known by the name of Protesilaus. The colors of this species are comparatively dull, but when examined, their soft contrasts are so pleasing to the eye, that any change would only be for the worse. Above, the upper wings are partly transparent, a large patch of white scales being set near their base. The under wings are mostly white, with the exception of a little dash of scarlet on their inner edge, a few half moons of dull yellow near their tips, and streaks of blackish-brown on their edges and along the centre of the tails. On the under side is a little more variety, the shining transparent membrane of the upper wings being crossed with dark bars, and the central stripe of the lower wings being edged with scarlet. It is a native of Demerara.

Our last example of this genus is the THOAS, a very striking insect, whose colors are almost wholly black and yellow. This insect is to be seen in the same illustration as the last, and its colors can be well imagined from the fact that, excepting a very small spot of orange-red on the inner edge of the lower wings, all the dark parts are black, and all the light are rich yellow. Below, it is almost wholly yellow, but of an ochreous and duller cast. In Northern Europe there exists but one acknowledged example of the genus Papilio. This is the beautiful SWALLOW-TAILED BUTTERFLY (*Papilio machaon*), a rare and brilliant creature. The flight of this insect is rather high, swift, and straight.

WHITE BUTTERFLY (*Pieris cratægi*), WITH EGGS, CATERPILLAR AND LARVA.

Two examples yet remain of the Papilionidæ. The first is the very remarkable insect which is known by the name of LEPTOCIRCUS, and which, until comparatively late years, was as rare as it is singular. This insect is not of great size, the expanded wings seldom exceeding an inch and a half, and being usually rather less in their measurement. The general color of this butterfly is brown, with the exception of a moderately broad greenish band along the centre of the wings. In the female the band is nearly colorless, and the light patch on the upper wings is transparent. The under parts are nearly of the same colors, except that the outer edges of the tails are fringed with a narrow line of glittering white, like burnished silver. The insect is a native of Siam and Java.

Lastly, we come to the prettily-marked Thais, one of a genus of Papilionidæ, which can always be known by the peculiar markings of their wings. The colors are, in all the species, yellow, black, and red, and the wings are edged with a series of bold festooned marks. The inner edges of the hinder wings are deeply scooped, as if to permit free motion of the abdomen.

WE now come to another family, called the Pieridæ, which may be known at once by the manner in which the inner edges of the hinder wings are folded, so as to form a kind of gutter in which the abdomen rests. In all these insects, the colors are comparatively sober, the upper surface being generally white and black, and the under surface sparingly colored with red and yellow. Our accompanying illlustration represents the COMMON WHITE BUTTERFLY. It is a true representative of the family Pieridæ, as well as the Brimstone Butterfly, the harbinger of spring; all the Marbled Butterflies, the Orange-tip, and the now scarce Veined-white, which last-mentioned insect belongs to the typical genus. The EPICHARIS is almost wholly white and black above, a slight tinge of rose-color appearing on the lower edge of the hinder wings, and being due to the rich orange-red spots on the under surface. All the color is concentrated upon the under surface of the lower wings, the groundwork of which is bright yellow traversed by black nervures, and which are adorned by six large oval spots of orange-red. Our well-known Orange-tip Butterfly is a familiar example of a similar gathering of the color upon the under surface of the lower wings.

There is a pretty butterfly, called the SPIO, which also belongs to this large family, and may be distinguished from the succeeding group by the angulated front wings. It is a native of tropical America, Java, and India, in all of which countries it is tolerably plentiful. The colors of the upper surface are deep black, largely mottled with yellow and orange. The under surface is washed with pale yellow, purple, and brown of various depths.

ON the accompanying illustration are seen some specimens of a beautiful group of butterflies placed in the family Heliconia, because their graceful forms and elegantly disposed tints are presumed to render them worthy of the companionship of Apollo and the Muses.

EPICHARIS.—*Pieris epicharis*.

The uppermost figure represents the Phono, a native of Jamaica, Brazil, and the neighboring parts. The wings of this curious insect are almost wholly transparent, the opaque and colored portions being confined to a narrow band round the edge, and a few spots and streaks upon the wings. All these markings are blackish-brown, except on the under side, where the edge of the hinder pair of wings is tinged with yellow, and sometimes marked with a series of little white spots. An allied butterfly, the Transparent Heliconia (*Heliçónia diáphana*), so closely resembles this species, that the two are often confounded together.

PHONO.—*Ithónia phono*.
MARSÆUS.—*Mechanitis marsæus*.
ERATO.—*Heliçónia erato*.
SPIO.—*Léptalis spio*.

The lowermost figure at the right hand is the ERATO, a native of Surinam. In this insect there is always some variation in color, and the sexes are so different that they might easily be supposed to belong to separate species. In the male the upper wings are rich brownish-black with large spots of yellow, and the lower wings are also blackish-brown, streaked in a radiating manner with blue, and edged with little oval spots of pure white. The female has the ground color of the same hues as her mate, except that the base of the upper wings is boldly striped with rusty red, and the radiating streaks on the lower wings are of the same warm tint. In both sexes the under surface is brown, with pale yellow spots on the upper wings, and narrow streaks of pale red on the lower wings. The spread of wings is about three inches.

One species of this genus (*Helicónia charitónia*) is very gregarious in its habits, great numbers gathering in some particular spot, and playing about like the gnat assemblies that are so common in the summer time. So plentiful are they, that when tired they can hardly find a place to rest upon, as crowds are continually settling upon the neighboring trees, and as continually driving off the crowds which have just sat down to rest.

MIDAMUS.—*Euplœa midamus.*

The last figure represents the MARSÆUS, a very elegantly shaped butterfly, a native of tropical America. The ground color of the wings is black, diversified with many bold stripes and patches of orange, and a large golden-yellow mark across the extremity of each upper wing. In the illustration, the white patch on the upper wings represents the golden-yellow of the insect.

In the upper left-hand corner of the engraving on page 405 is another example of this genus. The upper wings of the LYSIMNIA are chestnut at the base, and thence black to the tip, with the exception of two bold patches of nearly transparent membrane. The under wings are chestnut, edged with black, and having a jagged black streak across them, above which is a transparent stripe. The under surface is colored in nearly the same manner, except that a row of white spots runs around the edge.

To the same elegant family belong the butterflies of the genus Euplœa, a good example of which is the MIDAMUS. This insect strongly reminds the British entomologist of the purple emperor, the sober brown of the wings changing to rich shining purple when the light falls at a particular angle. At first sight, the butterfly appears to be quite a dull and inconspicuous insect, its colors being hardly more attractive than the simple black and white of the engraving. But if it be moved so that the light falls diagonally on its wings, the dull brown suddenly changes as if by magic into imperial purple of a richness exceeding the power of man to imitate, and more than realizes the metamorphosis achieved by the fairy god-mother's wand. This transformation is confined to the upper wings, the lower retaining their simple brown hue. The upper wings are sprinkled with some pale spots. The under side is grayish-brown, marked with spots similar to those on the upper surface.

PERHAPS the most interesting of these butterflies is the now celebrated BUGONG (*Euplœa hamáta*), the so-called "moth," on which the aborigines of New South Wales are in the habit of feeding.

The Bugong is found chiefly upon a range of granite hills called the Bugong Mountains, and it is rather remarkable that the insects congregate upon the outcropping granite masses in preference to the wooded sides of the mountains, and are found in greatest plenty at a considerable elevation.

The color of the Bugong is dark brown, with two black eye-like spots on the upper wings. The body is rather stout, filled with a yellow, oily substance, and covered with down. It is not a large insect, the spread of wing averaging an inch and a half.

WE now arrive at another family, of which the ARCHIPPUS affords a good example of the typical genus. This fine insect measures about four inches and a quarter between the points of the outspread wings, of which the entire contour is bold and sweeping. There is but little diversity of coloring in this butterfly; rich chestnut striped and streaked with black being the ground tint, and relieved round the edges with white spots, arranged in a rather irregular double series. The under surface presents similar hues, but of a paler cast. The head, thorax, and abdomen are deep, velvety-black, decorated with small spots of snowy-white.

ARCHIPPUS.—*Dánais archippus.*

THE large and important family of the Nymphalidæ contains a vast number of species, most of which are notable for their brilliant coloring, and many of which are well-known natives of Europe. These insects are, indeed, so numerous, that only a very slight sketch can be given of them.

The large and boldly-marked insect in the lower left-hand corner of the engraving at top of next page is the DIDO, a native of Brazil and Guiana, and is here represented of the natural size. The ground color of its wings is blackish-brown, and all the lighter parts are soft, leafy-green, with a slight pearly gloss. On the under surface, the ground color is chocolate, the green marks are much paler, and rather more opalescent than on the upper surface, and are edged with silvery-white. There are, besides, several bands of the same delicate hue on various parts of the wings. The caterpillar of this insect is green, diversified with a red and white stripe on each side of the body, and covered with several rows of short spines, besides two rather long appendages to the tail.

The uppermost figure in the same engraving on next page represents the THYODAMAS, an insect marked in a very unique fashion. Having a ground color of grayish-white, the whole surface is scribbled over with lines and streaks of brown, differing greatly in width, some being fine, as if traced with a crow-quill, and others broad and decided, as if drawn with a brush. Along the edges of the wings are a few double lines of rusty-brown. The under side of both wings is much paler, and the markings are finer and farther apart.

The right-hand upper figure is an example of the genus Marpesia, and is remarkable for the bold contour of wing, and the elongated tail with which it is decorated. The color of the THETIS is by no means various, but has, nevertheless, a decided and pleasing effect. The upper surface is uniform ruddy chestnut, over which are drawn several narrow stripes that traverse nearly the entire wings, passing from the edge of the upper pair to the extremity of the lower. From the lower margin of each under-wing start two projections, or tails, one being rather short, and the other very long, narrow, and slightly enlarged at the tips. The under side is pale rusty-red, with a very slight gloss of blue when seen in certain lights.

The last figure in this illustration represents the *Agraulis moneta*, an insect that closely resembles the well-known Adippe Fritillary of England, save that the color is deeper, and the metallic spots of the under surface larger and brighter. The upper surface of this handsome insect is rich ruddy chestnut, and on the under side of the wings are a number of large spots which shine as if they had been plated with silver, and then carefully burnished. It is necessarily impossible to represent this peculiar metallic lustre in a simple engraving, but a good idea of its real beauty may be formed by imagining the ground color of the upper wings to be pale chestnut, that of the under wings wood-brown, and all the spots to be composed of highly-burnished silver leaf.

THYODAMAS.—*Cyrestis thyodamas.*
DIDO.—*Colhôsia dido.*

THETIS.—*Marpesia thetis.*
Agraulis moneta.

To this family belongs the brightly-colored genus Vanessa, of which the common PEACOCK BUTTERFLY is a familiar British example. This insect, which is one of the finest butterflies, may be seen very plentifully in fields, roads, or woods, when the beauty of its coloring never fails to attract admiration.

One of the most notable peculiarities in this butterfly is the uniform dark hues of the under side, which present a great contrast to the varied shades of blue and red which decorate the upper side. The object of this arrangement seems to be that the insect may be able to conceal itself from its foe at will, a purpose which is readily attained by a very simple manœuvre. When the Peacock Butterfly thinks itself in danger, it flies straightway to some shaded spot, such as a tree-trunk or old palings, closes its wings over its back, and remains motionless. The effect of this proceeding is, that the wide expanse of bright colors is suddenly replaced by a flat, dark, leaf-like object, which looks more like a piece of bark torn from the tree than an insect. The apparent vanishing of the butterfly has always a rather startling effect, even to those who are accustomed to it, the large, brilliant creature disappearing as mysteriously as if annihilated, or covered with the cap of darkness.

PEACOCK BUTTERFLY.—*Vanessa Io.*

The beautiful SCARLET ADMIRAL, so well known by the broad, scarlet stripes that are drawn over the wings; the LARGE and SMALL TORTOISE-SHELL BUTTERFLIES; the COMMA

BUTTERFLY, so called from a comma-shaped white mark on the under wings, and the rare and beautiful CAMBERWELL BEAUTY, are all members of this genus.

WE now come to the genus Catagramma, which is remarkable for the manner in which the under surface of the lower wings is colored. There is in all a somewhat circular arrangement of lines, which in many species take the form of a figure of 8, more or less distinctly outlined. The generic name Catagramma refers to this peculiarity, and is derived from a Greek word signifying a delineation. They are all inhabitants of the warmer portions of the New World.

The Catagramma Peristera (or the Pigeon Catagramma) derives its name from the resemblance which the changing shades of the wings bear to the opaline hues of a pigeon's neck. The ground color of the upper surface is black, with two large patches of scarlet in the centre of each wing, the scarlet changing to violet when the light falls obliquely on the wings. The under surface of the upper wings is of paler tints, but colored in a similar manner, except a slight streak of blue on the edge, and a stripe of buff across the tip. The under wings are yellowish-buff, variegated with two black patches in the centre, each of which is garnished with a pair of azure spots. Just above these marks are two black streaks, and a curved blue stripe edged with black runs round the lower margin.

If the reader will turn to the engraving on page 405 he will find a figure in the upper right-hand corner, that represents the *Catagramma marchalii*, an insect that is marked more boldly than the last-mentioned species. The upper surface is black, with a short azure band on the upper wings, and a very narrow gray-blue streak round the lower edge of the second pair. The under surface of the first pair of wings is scarlet from the base nearly to the edge, where a broad band of black streaked with white completes the wing. The markings of the under wings are blackish-brown or very pale wood-brown, except one tiny patch of scarlet on the upper edge.

THE POLLUX, a large and boldly colored insect, is a native of Ashantee and Guinea. As is evident by the enormous dimensions of the thorax, which contain the muscles that work the wings, so wide and strongly made, the butterfly is swift and enduring of flight. The upper surface of both wings is deep rich black-brown, and the body is of a similar, but rather paler hue. The somewhat indistinct markings on the upper wings are ochreous-yellow, and those at the base of the lower wings are likewise yellow, which fades into white towards the base. The slight edging of the lower wings is blue, except the little streak at the angle, which is yellow. The under surface is very richly mottled, though without any brilliant colors. The basal half of the wings is jetty-black, with streaks and rings of white; then follows a broad white belt changing gradually into buff, and on the upper wings the remainder is brown, marked indistinctly with shades of gray. In the lower wings the white belt is followed by a broad stripe of chocolate, then by festoons of gray upon brown, then of a row of deep blue spots, then by a waved band of yellow, and lastly by a border of black. The legs are black and white like the base of the wings, from which they can hardly be distinguished when folded.

A BUTTERFLY which is known by the appropriate name of ACONTHEA (which word is of Greek origin, signifying thorny), is a native of Java and India.

Although not remarkable for any brightness of hue, its tint being peculiarly sober, the regular shape of the larva and pupa render it worthy of observation. The caterpillar is mostly found on some species of Bryonia, and is remarkable for the wonderfully long projections from its body, which are evidently analogous to, though far surpassing in size, those upon the caterpillar of the peacock-butterfly, which is represented on page 412. When it has cast its skin for the last time, and is about to change into the pupa state, it prepares for the coming event by spinning a large web of stout and shining silken threads, which often nearly cover the under surface of the leaf to which it is afterwards suspended. It then bursts through the caterpillar-skin, hitches itself to the silken web, and hangs there until its final change into the perfect form. As may be seen from the illustration, the shape of the pupa is very remarkable, reminding the observer of an ancient jousting-helmet with the visor down.

The two beautiful insects, known under the terms *Hetæra piera* and *Hetæra dracontis*, are examples of the family Satyridæ. Both these creatures, unlike as they appear to be, belong to the same genus. The *Hetæra piera* bears a wonderful resemblance to the transparent heliconia. Its wings are delicately transparent, and with the slightest imaginable tinge of yellow. On the lower wings there is a blush of orange-red, and the marks are darkish brown.

The *Hetæra dracontis* is a delicately marked, though not brilliant insect. The upper wings are very soft brown, traversed by a band of a grayish hue, and with a very slight tinge of chocolate. The lower wings are also brown, but with a faint wash of blue, and the light marks are azure. On the under side it is wholly brown, with two round spots of black edged with buff, and two or three whitish blotches.

THE family of the Erycinidæ comes next in order, and, as may be seen from the specimens upon the colored illustration, embraces insects of very differing forms and colors. The strange-looking insect, Zeonia Batesii, derives its name from Mr. Bates, who discovered it.

The white portions of the wings are membranous and transparent, and the dark portions are nearly all black, except that the base of the projecting portions of the lower wings is deep blue. The light-colored bar is rich scarlet. This specimen represents a male; the hinder wings of the female are closer together, and the tails are nearly straight.

A SMALL but elegant butterfly is the *Calydna calamita*. The upper surface of this insect is black, diversified with numerous blue and white spots. Below, the ground color is brown, spotted profusely with black and white, and having some short transverse lines of yellow. This insect inhabits the regions about the Amazon.

A butterfly called *Eurygone opalina* is of simple but extremely beautiful coloring. Unless held in a favorable light, the insect seems to be of a simple orange color, but if held with its head towards the observer's eyes, and the sun being behind his back, its wings glow with a golden effulgence that surpasses all power of description. As the insect is gently turned or held so as to communicate a quivering motion, all the tints of the rainbow play over the trembling wings, and the glory reflected from its surface is almost intolerable to the eye. As is the case with all the butterflies, this insect is represented of its natural size.

ON the illustration at page 405, two more examples of this family may be seen. The first is placed in the centre of the left-hand side, and immediately under the tip of the left wing of the great Thoas butterfly. This is the *Helicopis cupido*, an insect which, if only viewed on its upper surface, seems, except for the long and slender projections of the hinder wings, to be hardly worthy of much observation, the color being pale and dull brown, changing to pale rusty-red towards the base of the wings, and having a rather large whitish spot in the centre of the upper wings. But on turning it over, so as to bring the under surface into view, it proves to be a really wonderful insect. The upper wings have little remarkable about them, their color being brown, becoming paler towards the edge, and having a sharply defined whitish-yellow mark in the centre. But it is on the lower wings that the chief interest is concentrated. On a ground of ochreous-yellow are a number of large spots which look exactly as if they were made of gold-leaf artificially affixed to the wings, the resemblance being so close, that without the aid of a magnifier which shows their real structure, a person who had seen them for the first time might well imagine that they had been veritable pieces of gold-leaf, and fastened to the wing by cement. This butterfly is a native of Demerara, while the *Misipsa* inhabits the regions about the Amazon. It may be seen in the left-hand lower corner of the same engraving. The color of this pretty little insect is silvery blue, over which are drawn a number of black bands, thus producing a very bold effect. The under surface is simply light brown, with some bands of a darker hue.

ERYCINIDS.

THE accompanying fine engraving represents the magnificent insect called the NEOPTOLE-
MUS. It belongs to the genus Morpho, in which are contained some of the most resplendent
beings to be found in the world, all being
beautiful, and some endowed with a gorgeousness of coloring that is almost inconceivable.
In the present species the upper wings are of
the richest azure, glittering like burnished
metal, and iridescent as the opal, but with
far greater intensity of hue. In some lights
the colors are sombre enough, being only pale
gray and darkish brown; but when the light
falls favorably upon the wings, their colors are
truly magnificent. Around the edges of the
wings is a broad belt of black, very deep
towards the tips, and narrowing towards the
angle. The under side is soft brown, decorated
with many irregular stripes of yellowish gray,
and besprinkled with a number of eye-like
spots arranged in a tolerably regular row, three
on each of the upper wings, and of nearly
equal size, and four on each of the lower wings,
one being very large and separate from the
rest, and the remaining three small and close
together. In the centre of each eye there is a
little white spot, round which is a broad ring
of black, then a narrower ring of buff, then a
line of black, and lastly a gray line.

NEOPTOLEMUS.—*Morpho neoptolemus.* (Natural size.)

JUST above the left-hand corner of the Thoas' wing in the illustration on page 405, may
be seen a little butterfly of simple coloring. This is one of the HAIR-STREAK butterflies,
belonging to another family called the Lycænidæ. In this family are contained the beautiful
blue butterflies so common in the fields, and whose exquisitely spotted under surface never
fails to attract admiration. All the Copper Butterflies belong to the same family.

The present species is a native of Demerara, and is very scarce, not yet having received a
name in the scientific catalogue. The color of the upper wings is brown, with light streaks of
blue radiating from the bases, and that of the lower wings is blue, edged with brown. Below
it is brownish-gray, with a single narrow line of rusty-red crossing both pairs of wings, and a
dash of the same color on the hinder edges.

BEFORE taking a final leave of the butterflies, it is necessary to mention a family of
Lepidoptera, which possess so many of the characteristics belonging to the butterflies, and so
many of those belonging to the moths, that entomologists find some difficulty in placing them,
in their proper position, some considering them as members of the Rhopalocera, and others as
belonging to the Heterocera. These insects are popularly known by the name of SKIPPERS,
on account of their short and irregular flight. Several of these insects may be found mostly
along hedge-banks towards the end of the day. They do not seem to fly very high, but
pass in their peculiar jerking fashion along the banks, flitting in and out of the herbage
with restless, eager movements which can never be mistaken for the flight of any other
insect. All these creatures have rather large heads, their antennæ have a slight hook at the
tip, and their wings are small when compared with the dimensions of the body, thus producing
the peculiar flight.

The second great division of the Lepidoptera is that of the Moths, distinguishable by
means of the pointed tips of their antennæ, which are often furnished with a row of projections
on either side, like the teeth of a comb; and in the males are sometimes supplied with branching

appendages. In most instances the wings are conjoined by means of the bristle and loop which have already been mentioned.

The first family of the Moths is the Sphingidæ, a group which contains a great number of swift-winged insects, popularly and appropriately called Hawk-moths, from the strength and speed of their flight. In many instances the proboscis is of great length, sometimes equalling the length of the entire body, and in such instances it is found that the insect is able to feed while on the wing, balancing itself before a flower, hovering on tremulous wing, and extracting the sweets by suction. In some cases, however, such as the well-known death's-head moth, the proboscis is very short, barely exceeding the length of the head. In the long-tongued Hawk-moths the chrysalis is furnished with a distinct horny case, in which the elongated proboscis can be packed during the period occupied in development. In the genus Smerinthus the wings are sharp and angulated, and the tongue is short.

ONE of the commonest species of this genus is the LIME HAWK-MOTH, so called because the larva feeds on the leaves of the lime-tree. It is a green caterpillar, thick-bodied, covered with little protuberances, and upon each side are some whitish streaks edged with red or yellow. Just at the end of the tail there is a short knobby protuberance, and the fore part of the body is rather narrow. When the larva has completed its time of feeding, it descends to the ground, and buries itself about eighteen inches deep in the earth, whence the chrysalis may be extracted in the winter by the help of a pickaxe and trowel. Beside the lime, the elm and birch are favored residences of this insect.

Although very common in some places, it seems to be rather local, being scarcely, if ever, found in many spots where the trees which it loves are abundant. The color is very variable, but the general tints are leaf-brown and green, with a few blackish spots and stripes, the brown being towards the base and the olive-green towards the tips of the wing.

An allied species, termed *Smerinthus ocellatus*, is seen in the engraving on page 419.

The splendid insect, appropriately named the DEATH'S-HEAD MOTH, is tolerably common throughout Europe, though, from its natural habits and the instinct of concealment with which the caterpillar is endowed, it is not so frequently seen as many rarer insects. Owing to the remarkably faithful delineation of a skull and bones upon the back of the thorax, the insect is often an object of great terror to the illiterate, and has more than once thrown a whole province into consternation, the popular idea being that it was some infra-natural being that was sent upon the earth as a messenger of pestilence and woe, if not indeed the shape assumed by some witch residing in the neighborhood.

I once saw a whole congregation checked while coming out of church, and assembled in a wide and terrified circle around a poor Death's-head Moth that was quietly making its way across the churchyard-walk. No one dared to approach the terrible being, until at last the village blacksmith took heart of grace, and with a long jump leaped upon the moth and crushed it beneath his hobnailed shoes. I keep the flattened insect in my cabinet, as an example of popular ignorance and the destructive nature with which such ignorance is always accompanied.

Although in itself a perfectly harmless creature, it yet has one unpleasant habit, and is said to make its way into bee-hives, for the purpose of feeding on the honey. Still, its numbers are so inconsiderable, that it could do but little harm in an apiary, and need not be dreaded by the owner.

The caterpillar of this moth is enormously large, sometimes measuring five inches in length, and being very stoutly made. It feeds on various plants, the jessamine and potato being its favorites, and may be best found by traversing potato-grounds in the night, and directing the light of a bull's-eye lantern among the leaves. It can be readily kept and bred, but requires some careful tending, and it must be remembered that it will only eat the particular food to which it has been accustomed, and if bred among the potato will refuse the jessamine leaf, and *vice versâ*. When the caterpillar is about to change into its chrysalis state, it should be placed in a vessel containing seven or eight inches of earth, which should be kept moderately damp by means of a moist sponge or wet piece of moss laid on the top. If this precaution be not taken, the shell of the chrysalis is apt to become so hard that the moth is unable to break its way out, and perishes in the shell. I have several specimens where the moth has thus perished. The caterpillars are also much infested by ichneumon-flies, so that the collector often finds his hopes of a fine insect destroyed by these small and fatal flies. It is worthy of remark, that, when this moth first emerges from the chrysalis shell, its wings, legs, and antennæ are enveloped in a fine and delicate membrane, which soon dies when exposed to the air, and falls off in pieces, permitting the limbs to unfold themselves. Mr. Westwood regards this membrane as analogous to the pellicle upon the pseudimago of the may-fly, described at page 396.

One of the most curious points in the history of the Death's-head Moth is its power of producing a sound—a faculty which is truly remarkable among the Lepidoptera. The noise is something like the grating, squeaking cry of the field-cricket, but not nearly so loud. The mode of producing the sound is rather doubtful; but modern investigations seem to confirm the opinion of Huber and Rösel, who thought that the sound was produced by friction of the abdomen against the thorax just at the junction. At all events, it is certain that the moth always bends its abdomen downwards whenever this squeak or cry is heard, and a circular tuft of orange-colored hairs below the wings is seen to expand at the same time.

The color of the caterpillar is bright yellow, and the body is covered with many small tubercles. Along each side run seven oblique bands of a fine green. At the end of the tail is

PINE HAWK-MOTH.—*Sphinx pinastri*. With eggs and caterpillar. (Natural size.)

a granulated kind of horn, and upon the back are many spots of black and blue. The color of the moth is briefly as follows:—On the upper surface, the front pair of wings are blackish-brown covered with waved stripes and dashes of deep black and powdered with white. There are also some stripes of rusty-red on the edges. The lower wings are ochre-yellow, and marked with two bands of deep bluish-gray, the upper band about half the width of the lower. The thorax is blackish-brown, and has on its surface a marvellously accurate semblance of a human skull and collar-bone. The plumes, or lengthened scales, of which this is composed are beautifully soft, with a rich deep pile, and feel like velvet under the fingers. A fine specimen of the Death's-head Moth is almost the largest insect found in Europe, the spread of wing sometimes reaching nearly six inches. The antennæ are remarkable for their stiff and sturdy make and the curious hook with which they are terminated.

We now arrive at the typical genus of the family, of which the CONVOLVULUS HAWK-MOTH affords a good example. It may be mentioned that the term Sphingidæ is derived from the peculiar attitude sometimes assumed by the caterpillars, which have a custom of raising the fore part of the body so as to bear a fanciful resemblance to the well-known attitude in which the Egyptians were accustomed to represent the mysterious Sphinx.

The fine insect seems to be found sparingly in most parts of Europe, especially towards the south. As is the case with many of the nocturnal moths, its eyes shine brightly at night, and on account of their great size are very conspicuous in this respect. The specific name of the moth has been given to it because the caterpillar is known to feed on the common field convolvulus or bindweed, and it is sometimes known by the title of Convolvulus or Bindweed Hawk-moth. The caterpillar is mostly green, spotted and splashed with black and brown, and having a row of oblique stripes on each side. Generally the stripes are yellow, and edged with black, but they are sometimes wholly of the bolder color, while the entire caterpillar sometimes assumes a brownish hue. Upon the end of the tail there is a sharp curved horn, quite harmless, and whose use is at present unknown. The color of the wings is mostly wood-brown, checkered with ash, gray, and white, and the abdomen is ringed with broad bands of rose-color and narrow stripes of black, while down its centre runs a broad streak of gray.

OLEANDER HAWK MOTH. *Sphinx nerii*. With caterpillar and larva.

Of several other fine insects belonging to this genus, we mention the PRIVET HAWK-MOTH (*Sphinx ligustri*), and the PINE HAWK-MOTH (*Sphinx pinastri*). The latter has been chosen for an illustration on account of the nice pattern with which the caterpillar is inscribed. (See page 417.)

The beautiful OLEANDER HAWK-MOTH, which is here represented of the natural size, belongs to another genus, in which the caterpillar has the power of prolonging or withdrawing the head and neck like the proboscis of an elephant, a faculty which has earned for another insect the name of elephant hawk-moth.

OUR next illustrated example is the HUMMING-BIRD MOTH. Although not gifted with the brilliant hues which decorate so many of the Hawk-moths, it is a more interesting creature than many an insect which can boast of treble its dimensions and dazzling richness of color. This insect may be readily known by its very long proboscis, the tufts at the end of the abdomen, and the peculiar flight, which so exactly resembles that of the humming-bird, that persons accustomed to those feathered genus have often been deluded into the idea that Europe actually possesses a true humming-bird.

In the curious moths of which the HYLAS is a good example, the wings are as transparent as those of the bee tribe, and, indeed, the hymenopterous idea seems to run through the whole of these creatures so thoroughly, that the shapes of their bodies, the mode of flight, and even the manner in which they move the abdomen, are so bee and wasp-like, that an inexperienced observer would certainly mistake them for some species of the hymenoptera. Others there are which bear an equal resemblance to the gnats, and are of correspondingly small dimensions.

IN the next family, the Anthroceridæ, we find a number of moths of no great dimensions, but possessing great brilliancy of coloring, and flying by day. A very familiar example of this group is found in the GREEN FORESTER, a pretty little insect, not exceeding an inch and

SMERINTHUS.—*Smerinthus ocellatus.*   HUMMING-BIRD MOTH.—*Macroglossa stellatarum.* (Natural size.)

a quarter in the spread of wing, but colored with extremely pure hues. It may be found plentifully in the month of June, and is most common on the outskirts of woods. The caterpillar of this insect feeds on the common dock and several allied plants, and, like the perfect insect, is of a green color, but diversified with two rows of black dots along the back, and a row of red dots on either side. The color of the moth is very simple, the upper wings being of a soft golden-green, with a peculiar silken gloss, and the under wings brown. The body is green, but with reflections of blue.

The well-known BURNET-MOTH, so familiar on account of the rich velvety-green, spotted with scarlet, which decorates its wings, also belongs to this family. The caterpillar feeds on many plants, and is notable for making a spindle-shaped cocoon in which it passes through its pupal state. This cocoon is of a light brown color, and is usually fastened to an upright stem of grass.

In the Ægeriidæ, the wings are as transparently clear as in the Sesiadæ, and the general aspect is equally unlike that of a moth. A species called CURRANT CLEAR-WING (*Ægeria tipuliformis*) is very common, and is fond of haunting currant-bushes, where it may be captured without much difficulty, being rather dull and sluggish in taking to flight, though when once on the wing it is quick and agile in its movements. On account of its resemblance to the large gnats, it is popularly called the GNAT CLEAR-WING. The caterpillar of this insect feeds upon the pith of the currant-trees.

A LARGE insect, of tolerably, but not very frequent occurrence, is the LUNAR HORNET CLEAR-WING. Its popular name is given to it in allusion to its singular resemblance to a hornet, the similitude being so close as to deceive a casual glance, especially when the insect is on the wing. In common with all the members of this genus, the Hornet Clear-wing is a rather sluggish insect, being oftener seen at rest than on the wing, and being mostly found while clinging to the trunks or leaves of the trees on which they lived in the larval state. Their flight is rather slow and heavy, and as their tongues are comparatively short, they are not able to poise themselves on the wing, and sip the sweets of flowers while balancing themselves in the air.

The larva of the present species feeds upon the willow, boring into the young wood and sometimes damaging it to a serious extent. All these insects inhabit, while in the larval state, the interior of branches or roots, and make a kind of cocoon from the nibbled fragments of the wood. Just before undergoing the transformation, the larva turns round so as to get its head towards the entrance of the burrow, and after it has changed into the pupal form, is able, by means of certain projections on the segments, to push itself along until the upper half of the body protrudes through the orifice, and permits the perfect moth to make its escape into the open air.

The wings of this insect are transparent, with orange-red nervures and dusky fringes. The head and thorax are shining brown-black, with a yellow collar, and the abdomen is ringed with orange and dark brown.

THE Uraniidæ form a curious and somewhat doubtful family, some authors having considered them to belong to the butterflies rather than the moths. Many of these insects are of most gorgeous coloring; their form, including the tailed wings, is very like that of a butterfly, and they are diurnal in their habits. Still, the preliminary stages of the caterpillar and pupa are such that they prove the insects really to belong to the moth tribe. All these insects are inhabitants of the hotter parts of the earth, and are most plentiful within the tropics.

The *Urania sloanus* is a native of Jamaica.

The *Castnia licus* comes from Brazil and Central America. Its coloring is bold and yet simple. The upper surface of the first pair of wings is dark blackish-brown shot with green, the latter color being best seen by looking along the wing from point to base. Near the outside edge of the hinder wings is a row of azure spots, and the narrow fringe is white and brown. A bold white band runs through the centre of both pairs of wings.

A VERY curious moth is the NEW ZEALAND SWIFT (*Hepialus virescens*). It is a foreign example of a genus well known in Europe by some curious though common insects belonging to a family called the Hepialidæ. From the head of the larva rises, in a nearly perpendicular line, a horn as long as the body of the insect. In the typical genus the larva is entirely subterranean, feeding on the roots of plants, and, as in some of the preceding insects, the chrysalis is able to ascend its burrow when near the time of assuming the perfect form. All these moths are very quick of wing, darting in a nearly straight line with such swiftness that they look like mere light or dark streaks drawn through the air. Yet they are captured with comparative ease, as they are not so agile as swift, and can be taken by quickly striking a net athwart their course. From their great speed, they are known by the popular name of Swifts.

The New Zealand Swift is a truly curious insect, not so much for its form or colors, but for the strange mischance which often befalls the larva, a vegetable taking the place of the ichneumon-fly, and nourishing itself on the substance of the being which gives it support. A kind of fungus affixes itself to the larva, and becomes developed on its strange bed, taking up gradually the fatty parts and tissues of the caterpillar, until at last the creature dies under the parasitic growth, and is converted almost wholly into vegetable matter.

The well-known GOAT-MOTH is, next to the death's-head moth, one of the largest of the British Lepidoptera, its body being thick, stout, and massive, and its wings wide and spreading.

Some readers may perhaps have observed certain large, round holes in the trunks of trees into which a finger can be readily thrust, and out of which an empty chrysalis case often projects. These are the burrows made by the caterpillar of the Goat-moth, and often are very destructive to the trees. The larva itself is but little smaller than that of the death's-head moth, and is by no means an attractive-looking creature. Its body is smooth and shining, mostly of dull mahogany-red tinged with ochreous-yellow, and having a large oval patch of chestnut on the back of each segment. It is gifted with a curiously wedge-shaped head, and its muscular power is enormous, as may be proved by actual experiment during the life of the creature, or inferred from the marvellous arrangement of muscles which are made visible upon dissection.

It exudes a liquid of powerful and fetid odor, thought by some to resemble the unpleasant effluvium exhaled by the he-goat. Its influence extends to a considerable distance, and a practised entomologist will often detect the presence of a Goat-moth caterpillar simply by the aid of the nostrils. In spite, however, of the repulsive aspect and unpleasant odor, this creature is thought to be the celebrated Cossus of the ancients, a grub which was found on trees, and, when dressed after some particular fashion, was looked upon as a very great dainty.

A much smaller moth, the WOOD LEOPARD, is a very prettily-marked insect, though without the least brilliancy of color. The caterpillar of this insect feeds upon the interior of many trees, seeming to prefer the wood of the apple, pear, and other fruit-trees. It is a naked, fleshy-looking larva, of a light yellow color, and having a double row of black spots upon each segment. Like the goat-moth, it prepares a cocoon-like cell when it is about to take the pupal form, but the lining is of stronger materials, cemented firmly together with a glutinous substance secreted by the insect. The moth is seldom seen until July, and is tolerably plentiful in some places, appearing to be decidedly local and rather intermittent in its visits.

THE family of the Bombycidæ includes several insects of inestimable value to mankind, the various silk-producing moths being included in its ranks. The common silk-worm is the most useful of all of them. The accompanying oleograph is a true illustration of this familiar insect. The valuable results of its habits are too well known to need any description. But as it is not generally known that upwards of forty silk-producing moths exist in different parts of the world, a short history will be given of some of them, together with a brief description of one of the finest species.

All these insects secrete the silk in two large intestine-like vessels in the interior, which contain a gelatinous kind of substance, and become enormously large just before the caterpillar is about to change into a pupa. Both the silk organs unite in a common tube at the mouth, technically called the spinneret, and through this tube the semi-liquid is ejected. As soon as it comes into contact with the air it hardens into that soft, shining fibre with which we are so familiar. If a single fibre of silk be examined through a good microscope, it will be seen to consist of two smaller fibres laid parallel to each other, like the barrels of a double gun, this structure being due to the double secreting vessels. The goodness of silk chiefly consists in the manner in which these semi-fibres are placed together. Silk-worm "gut," as it is called by anglers, is made by steeping the caterpillars in strong vinegar for a time, and then pulling them suddenly until they elongate into the well-known threads to which hooks are attached.

The caterpillar employs the silk for the purpose of constructing a cocoon in which it can lie until it has assumed the perfect form; and proceeds with wonderful regularity and dispatch in its work, its head passing from side to side, always carrying with it a thread, and the cocoon being gradually formed into the oval shape which it finally assumes. The few outermost layers are always rough and of poor quality; these are stripped off, and the end of the thread being found, it is fastened to a wheel, and spun off into a hank of soft yellow fibre. The coloring matter is very variable, sometimes being hardly visible, and at others giving the silk a bright orange tint. It fades much on exposure to light.

Among the many silk-worm moths may be mentioned the DASEE-WORM of Bengal (*Bombyx fortunátus*), an insect that makes an inferior silk, with which the bales are often adulterated unless the owner or purchaser is very careful in examining them. The silk is yellow, and there are several crops annually. A much more valuable insect is also cultivated in Bengal, by the name of BORO POLOO (*Bombex textor*). The caterpillar is small, and the cocoon of proportionate dimensions. The silk is very good, and of a pure white. One of the commonest insects reared by the same nation is the TUSSER or TUSSEH of the Bengalese (*Anthérea páphia*), called by different names by the various tribes which cultivate it. It is very abundant, and as it is hardy and feeds on many kinds of food, is a truly valuable insect. It supplies the natives with great part of their clothing, and is even imported into Europe. There are several large manufactories of this silk, the most important of which is at Bhagulpore. The habits of this insect seem to vary much according to the locality.

The AILANTHUS SILK-WORM has lately attracted great attention, and appears likely to supersede the ordinary silk-worm in many respects. It is a native of China, and has been largely used for the purpose of supplying clothes for the people. As the name implies, the caterpillar feeds upon the Ailanthus tree (*Ailanthus glandulosus*), which, although imported from China into the moderate climates of Europe and America, grows well and fast in these countries, and has been firmly acclimatized. Rearing the Ailanthus-moth is one of the easiest of processes, the caterpillars remaining quietly on the trees and spinning their cocoons amid the branches. The eggs are hatched in a similar manner to those of the common silk-worm, and after being fed through their first moult with picked leaves, are transferred to the trees, and there left. It is of course necessary to cover the trees with netting in order to prevent the birds from feeding on such delicate morsels.

The color of the caterpillar is green, marked with black, except the head and the last segment, which are yellow. The general color of this moth is grayish-yellow above, with splashes and markings of dull violet, black, and white. The transparent crescent is worthy of notice. The silk is strong, and takes dye easily, but does not possess the peculiar gloss which has long been proverbial. It is a truly fortunate circumstance that this insect has been so opportunely brought into notice, as it is wonderfully hardy, not subject to many diseases to which the common and delicately constituted silk-worm is liable, and being apparently free from that strange fungoid parasite which occasionally commits such fearful ravages, and has been known to depopulate a whole district in a single night.

An allied species, the ERIA SILK-WORM (*Attacus ricini*), has long been in use in many parts of Asia, where it is cultivated by the peasants, and affords them raiment of a marvellously enduring character, and yet sightly. Although the cloth that is woven from the silk of this insect is loose and seemingly flimsy of texture, it is so wonderfully durable, that a garment is said to last during nearly an entire lifetime.

THE family of the Arctiidæ, so called because some of the hair-covered larvæ have a bear-like look, is represented in Europe by many examples, some being really handsome insects, and others remarkable for some peculiarity in themselves or the larvæ.

Perhaps the most curious example of this family is the HOUSE-BUILDER MOTH, which derives this name from its habits. It is common in many parts of the West Indies, and is in some places so plentiful as to do considerable damage to the fruit-trees. As soon as the larva is hatched from the egg, it sets to work in building its habitation; and even before it begins to feed, this industrious insect begins to work. The house is made of bits of wood and leaves, bound together with silken threads secreted in the interior. When the creature is small, and the house of no great weight, it is carried nearly upright; but when it attains size and consequent weight, it lies flat and is dragged along in that attitude. The entrance of this curious habitation is so made that the sides can be drawn together, and whenever the creature feels alarmed, it pulls its cords and so secures itself from foes.

THE LOBSTER-MOTH derives its name from the grotesque exterior of the caterpillar. This larva is one of the oddest imaginable forms, hardly to be taken for a caterpillar by one who

SILK-WORM AND MOTHS.

was not acquainted with it. The apparently forced and strange attitude in which this caterpillar is represented is that which it assumes when at rest. The second and third pair of legs are much elongated. The moth itself displays no very notable points of structure except the raised tufts on the disc of the fore wings.

The well-known TIGER-MOTH (*Arctia caja*), with its red and brown coloring, is a well-known example of this family, and its caterpillar is no less familiar under the name of Woolly Bear. This is a very harmless creature, feeding almost wholly on the dead nettle, but some of its allies are terrible plagues to the agriculturist, or even to the country at large, having been known to inflict serious damage to crops, and in some parts of Germany even to strip whole forests of their foliage.

One of these insects, called the VAPORER-MOTH (*Orgyia antiqua*), is especially remarkable for the strange contrast between the sexes, the male being a wide-winged moth of the ordinary kind, and the female a fat grub-like creature with hardly a vestige of wing, and scarcely

GIPSY-MOTH.—*Hypogymna dispar*.

stirring from the spot on which it is placed. The well-known PUSS-MOTH (*Cerura vinula*), so called because its markings bear some resemblance to those of a tabby cat, belongs to this family. The caterpillar of this moth is a handsomely colored creature, remarkable for the odd, sphinx-like attitude which it assumes when at rest, the pink St. Andrew's cross which is drawn over the back, and the forked appendage at the extremity of the body, from which a pair of long and delicate filaments can be thrust or withdrawn at pleasure. This caterpillar constructs a cocoon of wonderful strength, composed of bits of wood cemented together, and of such hardness that a penknife cannot penetrate it without risk of being snapped in the attempt.

As may be seen by reference to the engraving, the GIPSY-MOTH differs much in its coloring, according to the sex, the male being blackish-brown and the female grayish-white. The upper wings of both sexes are marked with four waved transverse bands of moderately light brown, and a dark brown mark near the middle of the front edge like the letter V, inside of which is a blackish spot. On the European Continent this moth is very abundant, and the caterpillar is often extremely injurious to the trees.

Another moderately winged moth, called the PALE TUSSOCK-MOTH, was also chosen for an illustration. This name the insect derives from its color and the tufts of hair that decorate the body of the caterpillar like tussocks of grass upon a field. The caterpillar goes by the popular name of the Hop-dog. The color of the Pale Tussock-moth is light brownish-gray, the fore wings being diversified with several marks of blackish-brown, the shape and dimensions of which may be seen by reference to the engraving. The hinder wings are much paler, and the band is dark brown.

PALE TUSSOCK-MOTH.—*Dasychira pudibunda*. Male, cocoon, caterpillar.
(Everything of natural size.)

The HERA, our next example, belongs to a genus which is known in Europe by the beautiful DOMINULA, or SCARLET TIGER-MOTH, with its rich green and scarlet wings. In the present instance, the fore wings are cream-colored with broad markings that look at first sight as if they were black, but when viewed in a good light are seen to be of the deepest imaginable green with a velvety lustre. The hinder wings are rich crimson scarlet, decorated with three or four black spots. This species is found in several parts of Europe.

THE STING-MOTH is a native of New South Wales, and the caterpillar feeds on the leaf of the stringy bark-tree. About the month of February it changes into the pupal state, and resides for some time in a curious kind of habitation. Just before it throws off the last larval skin, the caterpillar weaves a small and close cocoon or case, of an egg-like shape, which it suspends to the stem of a leaf, and therein awaits its final change.

The color of the moth is simple, but rather pretty. The fore wings are chestnut, edged with green and white, and the hinder wings are bluish-gray, edged with yellow and marked with green, yellow and brown.

The family of the Lithosiidæ is represented in Europe by several moths, of which the CINNABAR-MOTH (*Callimorpha jacobæa*) is perhaps the best known, on account of its vermilion and scarlet wings of precisely the same color on both sides. The ISSE, which, like the Heliconia, is a native of Brazil and the neighboring countries, has the upper wings black, beautifully diversified by some red longitudinal stripes at the base, succeeded by two broad yellow patches. Near the edge there are some white spots. The hinder wings are red, veined with black and bordered with a broad black band on which are some red spots.

WE now come to the large family of the Noctuidæ, containing a very great number of species, many of which so closely resemble each other that to distinguish them is not a very easy matter.

The delicately colored PEACH-BLOSSOM MOTH derives its name from the colors upon the wings, which closely resemble the soft pink upon the peach blossom. Although spread over the northern parts of Europe, it does not appear to be very plentiful, and does not assemble near one spot, as is the case with many rare and local moths. The caterpillar is easily known on account of a large hump that projects just behind the head, the summit of which is cleft into two bands, and also by the series of triangular elevations along the back upon which runs a pale zigzag line. The color of this larva is originally dark brown, but it sometimes assumes a paler hue. It may be found sparingly upon the common bramble about May. The fore wings of the Peach-Blossom Moth are soft brown, with a few waved lines running nearly parallel with the edges, and having five spots of delicate pink. The hinder wings are simple grayish-yellow, with a single waved line running across the middle.

The LARGE SWORD-GRASS MOTH derives its generic name from the handsome appearance of the caterpillar, the word calocampa being derived from two Greek words, the former signifying beautiful, and the latter a grub. The larva may be found in summer and autumn upon many plants, but especially on spinach, lettuce, and asparagus, and is not very common. Its color is rich green; a double row of white spots runs along the back, the rows being divided by a yellow line, then a row of white spots arranged in groups, and lastly a line of scarlet. The moth itself, although of pleasing tints, is not nearly so handsome as the caterpillar. The general color is brown, in some individuals marked with yellow and in others with chestnut. The curiously shaped marks upon the wings are brown-black. The hinder wings are gray, and the fringe is yellow.

When this moth is alarmed it has a habit of falling to the ground, with the upper wings drawn closely round the body and the antennæ and legs folded. In this attitude it looks more like a stray piece of stick than a moth, and would escape any one who was not searching carefully for it and was not acquainted with its habits.

The insect in the illustration is the CLIFDEN NONPAREIL, a fine and rare example of the Underwing-moths, so called because the hinder pair of wings are mostly of some bright color, while the upper pair are of comparatively sober tints. All these insects have a habit of settling on trunks of trees, or objects of similar dark hues, and drawing their upper wings so closely over each other as to conceal the brilliantly colored lower wings entirely beneath their shelter. When so seated, or rather suspended, as they always hang in a vertical attitude, it is almost impossible to discover them, even though they be marked down to the very tree on which they

CLIFDEN NONPAREIL.—*Catocala fraxina.* (Natural size.)

alight. They require some little care on the part of the pursuer; for although they depend much on their dull coloring for concealment, they are very alert on the wing, and the moment that they take alarm they speed away with wonderful alacrity.

The SWALLOW-TAILED MOTH is a well-known European species, very common in woods, and being mostly found among the underwood, whence it may be dislodged by beating the branches. The caterpillar feeds on many shrubs, but prefers the willow, the lime, and elder trees, the elder being its chief favorite. The cocoon is made of withered leaves.

The PEPPERED MOTH derives its name from the color of the wings, which are white, covered with little black dots, that look as if they had been shaken out of a pepper-castor. The stripes on the fore wings are black.

The V-MOTH, another of a very common species of this family, is so called on account of the dark brown mark upon the fore wings, which much resembles the letter after which it is named.

THERE are several other families of moths, many of which contain numerous species, but our space does not allow to treat them all. Some of them are very small and apparently insignificant, though their vast numbers often give them powers of destruction which are unequalled by the larger but scarcer insects.

The PEBBLE HOOK-TIP MOTH is one of these insects, and one that has greatly perplexed systematic entomologists to place it in its proper position. The Geometridæ, as a rule, have the antennæ perfectly simple and thread-like; but the male of this insect has those organs in a feathery form, like those of other families. The larva, again, is of rather eccentric shape, with projections along its back, with tufts of stiff hairs, and assuming an attitude very like that which is characteristic of the puss-moth larva already described.

The popular name of this moth is derived from the hook-like tips of the wings. Its color is reddish-buff, over which are drawn a number of waved dusky streaks. In the centre of the wing there is a dusky spot, and an orange-brown stripe is drawn from the inner margin to the extremity.

A VERY pretty and well-known moth is the OAK-LEAF ROLLER (*Tortrix viridana*), a moth of a beautiful apple-green upon the upper wings. In the illustration the insect is represented in its natural size. In some places, these moths swarm to a fearful extent, stripping whole trees of their leaves. I have known the oaks to be surrounded with whole clouds of these moths, fluttering about like gnats, and forming an exhaustless banquet to the empis-flies, which were catching them by thousands, embracing them in their long legs, and flying about with their prey, sucking their juices like so many winged vampires.

OAK-LEAF ROLLER AND CATERPILLAR.—*Tortrix viridana.*

Other species live beneath the bark of trees, or even burrow into the wood, while others are hatched in the interior of fruits, and live unsuspected in their retreats until they are on the point of changing to the pupa form, when they eat their way out, and leave a round hole as a memento of their presence. The CODLING-MOTH is one of the commonest of these tiresome insects, living in the middle of the fruit from which it takes its name, and giving rise to the condition which is termed "maggoty." The larva is a round, fat, white grub, which may too often be found in the interior of an apparently sound and ripe apple, and which gives to every part which it has touched a very bitter and nauseous flavor, like that of a worm-eaten nut. None of the Tortricidæ are of very brilliant colors, the Oak-leaf Roller being one of the most conspicuous. The fore wings are dark grayish-brown, striped transversely with a darker tint. On the outer part of each wing there is a dark brown space streaked with golden bars. The hind wings are simple dusky-brown.

The rose suffers sadly from the ravages of several of these moths, some of which feed within the bud, and others tie the young leaves together and feed upon the interior.

THE Tineidæ form a very large family of moths, all of which are of small dimensions, and some exceedingly minute. From several points in their structure, Mr. Westwood seems doubtful whether they ought not to be united to the Yponomeutidæ; the general narrowness of their wings, and the rare occurrence of labial palpi, being the points by which they have been separated. The larvæ of most of the species form portable cases of various materials, in which they reside, some feeding upon animal, and others upon vegetable substances. The too well-known CLOTHES-MOTH (*Tinea tapetzella*) belongs to this family. There is another species of the same genus, popularly called the WOLF-MOTH (*Tinea gravella*), which haunts granaries and malt-houses, and does great damage by feeding on the grains and fastening them together with its silken web.

The pretty little LILAC-LEAF ROLLER MOTH belongs to this family. Those who possess gardens have doubtlessly noticed that many leaves of the common lilac are rolled into a cylindrical form, bound together by silken threads, and that, if this little case be opened, out tumbles a small whitish caterpillar with a black head, who loses no time in letting itself to the ground by means of a silken fibre spun from its mouth. How the larva rolls the leaf is quite a mystery, and though it has been watched by many careful observers and seen to fasten its threads, the precise force which makes the leaf assume its cylindrical form is as yet undiscovered. The caterpillar lives within, and feeds upon the rolled portion of the leaf, thus surpassing the feat of Ascanius recorded in the "Æneid." The fore wings are golden-brown, with pale brown transverse markings, and the hinder pair are edged with long grayish fringes. The larva feeds on the ash and privet as well as on the lilac.

Our last example of the Lepidoptera is the beautiful WHITE-PLUME MOTH, an insect which never fails to attract attention, on account of the singular elegance and beauty of its form.

This insect belongs to a small family which is remarkable for the fact, that, except in one genus, the wings, instead of being broad membranous structures, are cleft into narrow rays, feathered in a most soft and delicate manner. The White-plume Moth is to be seen in the evenings, flying in a curious, uncertain manner, and looking not unlike a snow-flake blown casually by the wind. It seems never to fly to any great distance, settling quite openly on leaves or plants, without taking the precaution of clinging to the under side, as is the custom with so many of the smaller moths. When it rests, it folds the wings so that they only look like a single broad ray. The legs of this moth are very long and slender. The color of this insect is pure white.

An allied insect, the TWENTY-PLUME MOTH (*Alúcita hexadactyla*), has its wings cleft into a great number of plumes, thus giving rise to its popular name. In reality, there are twenty-four plumes, each of the fore wings being cleft into eight divisions, and the hinder wings into four. It is much smaller than the White Plume, and is fond of haunting houses, where it may be seen moving up and down the window panes with much agility. Autumn is the best time for finding this little moth. Its general color is ashen-gray, with two darkish bands and a white fringe.

HOMOPTERA.

In the next order are comprised some very grotesque insects, some of which have been thought to belong to other orders, and a few not being known to be insects at all until comparatively late years. They have rounded bodies, not more than three joints in the tarsi, and their wings are four in number, wholly membranous, the fore pair being larger than the hinder, but not overlapping in repose. The mouth forms a kind of tube, sometimes nearly as long as the body, and often sufficiently hard and stiff to pierce the skin.

In this curious order are placed the Aphides, those little green insects that swarm upon roses and other plants, and are termed "blights" by gardeners, who employ that term in a strangely wide sense; the Cicadæ, with their beautiful membranous wings, their large heads, and their loud voices; the tribe of Hoppers, of which the Cuckoo Spit insect, known in its perfect state under the name of Frog-hopper, and the beautiful Scarlet Hopper, are familiar examples; the wonderful Lantern-flies, also leapers, which are found only in hot climates; the Wax Insects of China; and lastly, the Scale Insects, or Coccidæ, from which the "lac," so important in commerce, is obtained.

The Cicadæ, which are represented in the accompanying oleograph, have three joints to their feet, these members affording useful characteristics in settling the precise position of the various species. They are very large insects, sometimes measuring more than six inches between the tips of the expanded wings. Their mouth or beak is three-jointed and very long, being tucked under the body when not required. The females are furnished with a curious apparatus, by which they are enabled to cut grooves in the branches of trees for the purpose of depositing their eggs therein, and which is clearly analogous to the instrument possessed by the saw-flies. On the under side of the body, and nearly at the extremity, are seen a pair of jointed valves, which form the scabbard to the boring instrument. At first sight, the borer appears like a spear-head deeply notched along both edges; but on a closer examination it is seen that this apparently single instrument is composed of three pieces, namely, two saw-edged blades, set back to back, and a central support in which they both slide. There seems little doubt that these instruments work alternately, like the saws of the tenthredo.

The slits made by these curious saws are wonderfully deep, considering the instruments with which they are cut, and look as if little splinters of wood had been partially detached by a pen-knife, but left adherent at one end. Each of the burrows under these elevations is about a third of an inch in length, and contains from four to ten eggs. Altogether, each female deposits between six and seven hundred eggs. As soon as the young are hatched, they emerge from the cell, and make their way to the ground. At this period of their existence they are not unlike the common flea, both in size and shape. They grow rapidly, and when they are changed to the pupal form exhibit but little alteration in form, except that the rudimentary wings are visible externally. They live for some time in the preliminary stages, and guard themselves against the frosts of winter by burrowing into the ground to a depth of nearly a yard. When the perfect insect makes its escape, it leaves the empty pupal shell nearly entire, except a slit along the back through which the creature has passed.

The male Cicada has the power of producing a shrill and ear-piercing sound, so loud in many species that it can be heard at a considerable distance, and becomes a positive annoyance, like the same tune played for several hours without intermission. The organ by which the sound is produced is internal, but its position may be seen externally by looking at the under side of the body, just behind the last pair of legs, where a pair of horny plates may be seen. These plates are the protecting covers of the sound-producing apparatus, which consists of two drum-like membranes and a set of powerful muscles. The color of the perfect insect is mostly of a yellowish cast, and the wings are firm, shining, and membranous, somewhat resembling those of the dragon-fly in texture, but having larger cells or spaces between the nervures.

One species of Cicada is a native of Europe (*Cicada anglica*). Generally, however, the Cicadæ are tropical insects, or, at all events, inhabit the warm countries, those in the cooler parts of the world being comparatively small. Several species of Cicada are eaten like the locusts.

THE wonderful LANTERN-FLIES are known by the three-jointed antennæ and the two ocelli beneath the eyes.

It may here be remarked that the eyes of insects are of two-fold character, namely, the compound and the simple, the former being constructed of a variable number of facets, so arranged, that each, though a separate eye, with its own optic nerve, is made to coincide with the others, and to produce a single image in the sensorium. Many insects, especially those which fly or run rapidly, have a vast number of facets in the compound eye, the common peacock butterfly possessing about thirty-four thousand of these lenses, seventeen thousand on each side. The average number, however, is about six or eight thousand. The ocelli, or simple eyes, are round, lens-like objects, mostly set in front of the head; and it is imagined that the two sets of eyes perform distinct offices, the compound eyes for the purpose of observing distant objects, and the ocelli in order to examine the food or any substance within close proximity.

In many of the Fulgoridæ, the head is formed into the oddest imaginable shapes, sometimes lengthened into a curved horn, like that of the Lantern-fly, sometimes broad, with a deep keel above, and sometimes with a raised edge of knife-like sharpness. The head is said to emit a phosphorescent light, similar to that of the fireflies.

The Wax Insects belong to this family. These creatures are plentiful in China, where the waxen secretion is manufactured into many useful articles, and is equal, if not superior, to that obtained from the bee. That this creature should produce wax is thought to be very marvellous, but there is no reason to consider the fact more wonderful than that the bee should secrete a similar substance. There is this difference, that the bee produces the wax from six little pockets arranged along the abdomen, whereas the Fulgora pours it from various parts of the body, just as the oil is emitted by the meloë-beetle already described.

CICADÆ, LANTERN FLY, ETC.

THE Cercopidæ, or Hoppers, are well known in Europe, mostly from the habits of the larva, and the saltatorial powers of the perfect insect. The CUCKOO-SPIT, or FROG-HOPPER, is very plentiful in this country, and is often a great annoyance to amateur gardeners, who dislike to find their hands or faces suddenly wetted with the frothy exudations in which the creature lives enshrined. The larva fixes itself upon various plants, and sucks their juices through its long beak, which it plunges into the soft substance. When the accumulation of froth is very great, which usually happens in the heat of the day, a drop of clear water begins to form at the lowest part, into which the froth drains itself, and is presently relieved by the falling of the drop. The scientific name of this insect is *Aphrophora spumaria*. Another species of Frog-hopper (*Aphrophora goudotii*), a native of Madagascar, pours out clear water without the preliminary process of forming the liquid into froth. In its perfect state it can leap to an extraordinary distance, the spring being so smartly made as to cause a sharp tap on the object from which it leaps. As it alights it often tumbles over, and loses some little time in kicking about before it can again get on its short legs.

A SMALL but very remarkable insect is the Coccus, popularly known as the SCALE INSECT, or MEALY BUG, the former title being applied to the exterior of the female, and the latter given on account of the white mealy substance that is found within her body. The male of this insect is winged. To gardeners the Cocci are sad pests, infesting various fruit-trees, and increasing with such rapidity that their progress can scarcely be checked. The young, too, are of such minute size that they can hardly be seen or destroyed. It appears, however, that the most effectual way of checking their depredations is to make a kind of semi-liquid paste of fine clay and water, and with a brush to wash it well into the bark of the affected trees, so as to cover the insects, deprive them of air, and debar them from removing. Three or four coats are necessary, in order to stop up the minute cracks which are sure to take place in the drying clay, and which would afford ample opportunities of egress to these tiny creatures.

Within the shell-like body the young Cocci are hatched, amid an abundant supply of white substance, something like flour. The mother by this time has died, but her shelly skin still remains, and forms a house wherein her children live until they are strong enough to enter the world. They are usually hatched towards the end of June, and the young escape at the end of July.

The COCHINEAL INSECT (*Coccus cacti*), of which we give a very fine illustration, belongs to the same genus. This species is a native of Mexico, and lives upon a kind of cactus, called, from its insect guest, the *Cactus cochinellifer*. The wonderful amount of rich coloring matter which these insects contain is well known. The beautiful colors, carmine and lake, are obtained

COCHINEAL INSECT.—*Coccus cacti*. (a, The insect alive upon the Opuntia, covered with its waxen sweat; b, male; c, female. The objects are magnified, and the lines indicate the natural size.)

from this insect, and the best scarlets are likewise produced from the Cochineal, the difference of hue being due to a mixture of chloride of tin. The trade in the Cochineal is very great; and as the substance is very costly, and permits a parcel of great value to be compressed into a small bulk, it is often used in lieu of cash in mercantile transactions, and a package will go travelling backwards and forwards for a long time before it reaches its final destination.

Several other species, such as the POLISH SCARLET GRAIN (*Coccus polonicus*), and the KERMES (*Coccus ilicis*), are also valuable to colorists, and impart a very fine scarlet to substances treated with them, although the hues are not equal to those obtained from the cochineal. The latter of these insects was known both to the Greeks and Romans, and was used by them for the purpose of obtaining the purple dyes which were so much worn by the higher classes.

The LAC INSECT (*Coccus lacca*) is another member of this most useful genus. This species resides in India and the hotter parts of Asia. It is found attached to the twigs of trees, and is then called stick-lac, the shell-lac being the waxen secretion purified and shaped into thin, shell-like plates.

THE Grape Phylloxera (*P. vitifoliæ*) is the most destructive of the Aphides. Though first characterized in Europe, North America seems to be the home of the genus, for, while there are but two well-defined species so far known in Europe, sixteen distinct species are found in various parts of the United States. They are gall-inhabiting creatures. For a long period the Phylloxera was only an object of interest to the naturalist; but, five or six years since, the Grape Phylloxera came suddenly to be a creature of great concern to the public. Indeed, this species has become so prominent that it is entitled *the* Phylloxera, though fifteen other species are known. It is found from Canada to the Gulf States, and east of the Rocky Mountains.

Early in the history of the grape culture in the United States, the gall-making type was observed on the leaves of certain varieties, particularly on the Clinton, and in 1856 this was briefly described by Dr. Fitch, State Entomologist of New York, by the name *Pemphegus vitifoliæ*. The more normal root-inhabiting type was not suspected, however, until it was discovered by Prof. Riley, of Washington, in 1871. A kind of grape-root disease began to show itself in France, where the grape interest is of vast importance. Large sums were offered by the government. It was found eventually that it was the identical species that is indigenous on American vines, and that it was imported into France from America, probably during our civil war, on our vines sent to French nurserymen. It presents two forms or types, the root-eating and the gall-inhabiting. The insect is about a sixteenth of an inch in length. The appearance of the vine-roots, after being infested, is that of a collection of young potatoes just formed. There seems no practical remedy for the vines already affected. The only serviceable practice is to substitute new vines; and though America furnished the enemy, yet it is found that the American vines resist the pest much better than those of France. Consequently, the vineyards of the Old World are being replanted by vines shipped from our American vineyards. The Cunningham, Norton's Virginia, Clinton, Concord, Taylor, are preferred. Those especially preferred are of the species *æstivalis*, as Herbmont, Cynthiania. The French have now learned to appreciate our vines, when formerly they scorned the idea of importing them. American grape culture, meantime, is advancing favorably from this communication of the culturists of both countries.

HETEROPTERA.

THE insects belonging to the large and important order which is placed next to the Homoloptera, are readily known by several conspicuous characteristics. The wings are four in number, and the front pair are very peculiar in their structure, the basal portion being horny, like the elytra of beetles, and the remaining portion membranous, like the hinder wings of the same insects. In some species, however, the wings are wanting, as in the common Bed-bug (*Cimex lectularius*). The body is always much flattened, the mouth is beak-like, and in the pupal stage the creature is active and resembling the perfect insect, except in its want of wings.

As the space which can be devoted to the remaining insects is extremely limited, it will be merely possible to give a brief sketch of the different families.

WATER-BOATMAN AND WATER-SCORPION.

In the first section of the Heteroptera, the species are aquatic and predaceous, some being very quick and active in the water, and others slow, but yet fierce and voracious, and obtaining by craft the prey which the others win by fair chase.

The first family of the Notonectidæ, or Back-swimmers, is represented in Europe by many species, of which the common Water Boatman (*Notonecta glaucus*) is the commonest. As may be seen by reference to our full-page illustration, these insects are in the habit of lying on the back and propelling themselves by means of the hind legs, which are very long, and with the extremities expanded and fringed with stiff hairs, so as to resemble and do the duty of oars. When lying on the back, the insect is wonderfully boat-like, the general shape much resembling an Indian canoe. These creatures breathe atmospheric air, for which they are forced occasionally to come to the surface, receiving it under their wings, just as is the case with the water-beetles already described. When the air has been taken in, it is prevented from escaping by means of the stiff hairs with which the segments are furnished. Any one who has watched a pond in the summer-time must have noticed the Water Boatmen coming to the surface, poking their tails out of the water, and then descending. The beak of all this family is very sharp and strong, especially in the genus Corixa; and a heedless captor will often find a sudden pang shoot through his fingers, caused by the determined thrust of the insect's sharp beak, which is armed with a pair of spear-like points, the edges of which are deeply jagged. The wings of the Water Boatman are large and handsome, and the insects are powerful on the wing. They fly by night.

THE family of the Nepidæ is represented in Europe by the common WATER SCORPION, a very flat and leaf-like insect, which is found abundantly in slow running streams, ditches, and ponds. It is figured in the same engraving. This insect derives its popular name from its scorpion-like aspect, the two slender filaments appended to the abdomen representing the sting-tipped tail, and the raptorial fore-legs resembling the claws. It is with these legs that the Water Scorpion catches its prey, which, when once grasped in that hooked extremity, is never able to make its escape. The beak is short, but very strong and sharp, and is not bent under the thorax, as is the case with that of the water boatman.

The Reduviidæ comprise a great number of terrestrial insects, mostly exotic, but a few being natives of Europe. Some of them are very large, and one species, the WHEEL-BUG (*Arilus serratus*), is said to possess electric powers. Its popular name is derived from the curious shape of the prothorax, which is elevated and notched, so as to resemble a portion of a cog-wheel. One species of the Wheel-bug (*Reduvius personatus*), inhabits houses, and is said to feed upon the bed-bug. The larva and pupa of this insect are difficult to discover, on account of their habit of enveloping themselves in a coating of dust. The HAMMATOCERUS belongs to this family. The insect is remarkable for the curious structure of the second joint of the antennæ, which consists of numerous small articulations. The generic title is derived from two Greek words, signifying Link-horned, and is given to the insect in allusion to this peculiarity.

WHEEL-BUG WITH LARVA.—
Reduvius personatus. (Natural size.)

THE remarkable insect, *Dalader acuticosta*, belongs to the Mictidæ, and is notable for two peculiarities, namely, the flattened expansion of the third joint of the antennæ, and the singular width and flatness of the abdomen, which is so very leaf-like as to remind the observer of the leaf-insect The abdomen is greatly expanded, and extends on either side far beyond the wings, which, when opened, exhibit the curious formation of the parts below. The thorax is also expanded and developed into a semi-lunar shape, the points being directed forward. The third joint of the antennæ is seen also to be expanded, flattened, and formed into a kind of battledore shape. In color, this insect is very inconspicuous, being altogether of a dull brown, like a withered oak-leaf. It is a native of Northern India.

ANOTHER insect, termed *Diactor bilineatus*, belongs to the same family, and is remarkable for the curious formation of the hinder pair of legs. In the genus of which this species is an example, the tibiæ are expanded like the blade of a South Sea Islander's paddle, being very flat, and not so thick as the paper on which this account is printed. In the present species, the coloring is very splendid, the whole of the thorax being rich emerald-green, with a peculiar lustre, as if incrusted with the minutest of gems, and diversified with two longitudinal streaks of light red. The latter color, however, always fades in process of time, as is generally the case with all the shades of red in insects. The elytra are also green, but not so sparkling in effect. It is a native of Brazil, and the family in which it is placed is the Anisoscelidæ. The broad leafy expansions of the legs are chestnut-brown, spotted with a paler hue.

APHANIPTERA.

WE are now come to another order, deriving its name from the invariable absence of wings, the name being derived from two Greek words, the former signifying invisible, and the latter a wing. There are not many species belonging to this order, and they are all known by the popular name of Fleas. A magnified representation of the common FLEA is given in our illustration.

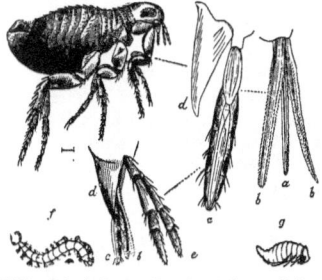

FLEA.—*Pulex irritans*. a, Upper lip. b, Jaw. c, Feeler of the under lip. d, Under lip. e, Feeler of the short, not visible jaws. f and g, Larvæ. (All the objects are magnified; the line underneath the large figure indicates the natural size of the Flea.)

These insects are notable for their extreme agility and the hard shelly substance of their integuments, two characteristics which are very useful in defending them from foes, for in the first place they leap about so quickly that they are not easily caught, and in the second place they are so hard and polished, that even when seized they are apt to slip through the fingers before they can be immolated to the just wrath of the captor. As may be seen by reference to the engraving, the mouth of these insects is very complex in its structure, and is a veritable surgeon's case of lancets, saws, and probes. Although eager for blood to a proverbial extent, Fleas can endure a very long fast without much inconvenience. I have known a room to be unused for years, and yet, when I became its unfortunate first occupant, being rendered helpless by a broken leg and dislocated ankle, the Fleas came swarming in positive armies to their long-delayed feast, like the locust hosts descending upon a cornfield, and caused unspeakable miseries until they were routed by continual slaughter. What food these insects may have found in an empty room is not easy to say, as, though the larvæ might, perhaps, have continued to subsist on the feathers of the pillows, the perfect insects could not eat such juiceless substances, and must either have gone altogether without food, or drawn their subsistence from some unknown source.

Another species of Flea, the CHIGOE (*Pulex penetrans*), sometimes corrupted into JIGGER, is a terrible pest in tropical countries, attacking human beings, and by its peculiar habits causing severe injuries, unless they are checked at once. They mostly attack the feet, generally preferring the bare spot just between the toe and the nail. When they have made their way fairly under the skin, they swell to a very great size, the body becoming about the size and shape of a sweet pea, and being filled with a vast number of eggs. Generally, those who live in the Chigoe-infected regions are careful to have their feet examined every day, and the offending insects dislodged with the point of a needle. Sometimes, however, one may escape observation until it has obtained its full development, when its only external sign is a slight swelling, with a bluish color. To extract one of these swollen insects is a matter of no small difficulty, for if the body be burst, and a single egg suffered to remain, the creature will be hatched in the wound, and the result will be a painful festering sore. If such an event should take place, the best plan is to pour a drop of spirit of turpentine into the wound, a process sufficiently painful, but yet preferable to the risk of the future sores.

The young negroes are very subject to the Chigoe, and every evening a chorus of outcries is usually heard, being sounds of lamentation from the children, whose toes are undergoing maternal inspection. The little creatures, with the short-sighted cunning of childhood, always try to hide the Chigoe bite, in hopes of escaping the resulting needle. But their cunning only meets its due reward, as when the Chigoe has made her burrow, the sharp eye of the negress is sure to discover it, and then the whole nest has to be excavated, and rendered untenable by red pepper, rubbed well into the hollow. Indeed, if it were not for the terror inspired by the red pepper, the children would hardly have a sound foot among them.

It may seem curious that the insect should be able to burrow under the skin without being discovered, but the fact is, that it sets about its work so quietly, and insinuates itself so gently, that the only perceptible sensation is a slight but not unpleasant irritation.

DIPTERA.

WE now pass to the DIPTERA, or Two-winged Insects, which may be known not only by the single pair of wings, but by the little appendages at their base, called halteres or balancers, and which are the only vestiges of the hinder pair of wings. Moreover, the wings are not capable of being folded. This order is of vast extent, and includes a whole host of species, many being extremely minute, and many others displaying so many uncertainties of form and habit, that the arrangement of this order is one of the greatest difficulties with which systematic entomologists have to contend. In the following engravings a few examples are given of this order, for the purpose of illustrating some of the principal families.

THE COMMON GNAT is an example of the family Culicidæ. The mouth of this pretty and graceful but very annoying insect, is fully as complicated as that of the flea, and under the microscope is a truly beautiful object. The male Gnat, which is easily known by the plumed antennæ, is not to be feared, not being a bloodsucker, that characteristic belonging solely to the female.

The eggs of the Gnat are laid in, or rather upon, water, and are built, as fast as laid, into a boat-like shape, which possesses such powers of flotation, that, even if water be poured upon it, the mimic vessel turns out the water, and rights itself as well as any life-boat. When hatched, the larvæ fall into the water, and begin at once to make themselves very conspicuous by their continual twisting and jerking themselves about. They are long-tailed, large-headed insects; and when they are at rest, they hang with their heads downwards, the whorl of hairs at the tip of the branched tail serving as a float. Through this tail the respiration is carried on, the little creature requiring to breathe atmospheric air. In process of time, the larva changes into an active pupa, and, lastly, when the perfect insect is about to make its appearance, it rises to the surface, the pupal skin splits along the back, and forms a kind of raft, on which the Gnat stands until its wings have attained sufficient strength for flight.

The Tipulidæ are very familiar to us through the well-known insects called DADDY LONG-LEGS, or CRANE-FLIES. In their perfect state, these insects are perfectly harmless, although ignorant people are afraid to touch them. But, in their larval condition, they are fearful pests, living just below the surface of the ground, and feeding on the roots of grasses. Whole acres of grass have been destroyed by these larvæ; and, two or three years ago, Blackheath Park was so infested with them, that the turf was much injured, and in the beginning of autumn the ground was covered thickly with the empty pupa cases of the escaped insects.

ONE of the wonders of natural science is the ARMY-WORM (Heerwurm). At first sight, it appears to be a single being, but by closer observation it will be seen to consist of a vast multitude of larvæ, or caterpillars. There is a European and an American Army-worm, distinguished from each other by the nature of the tiny creatures which collect in a body for procession or migration; the one being a collection of the larvæ of the small light yellowish gnat (*Sciara militaris*), belonging to the family *Mycetophilidæ* of this class, while the other consists of the caterpillars of one of the moths of the family *Noctuidæ*.

When in such a large collection, the larvæ of the *Sciara militaris* move forward in a snake-like manner. They look pallid, and are kept closely together by their mucous surfaces, so that they really appear as one body. So strongly do they stick together, that the tail-piece of the worm may be lifted with help of a stick for a moment without becoming refracted. The faculty of moving on consists in the uniform motions of all the larvæ. Every one shoves forward with the back of the body, and then stretches out the fore-part as if feeling. The whole appears like a little stream slowly gliding along.

Sometimes this larvæ-procession has to overcome obstacles in its way which often cause a dissection. Small hindrances the Army-worm surmounts, but larger ones cause a temporary disunion. Sometimes one part of the mass of bodies disappears under leaves, but generally a reunion takes place. A break by force, caused, for instance, by the hoofs of a horse, or by the wheels of a wagon, soon becomes joined, just as is the case with the so-called procession-caterpillar of the moth *Cnethocampa processionea*, belonging to the family *Bombycidæ*, of the class *Lepidoptera*.

For centuries, the mysterious movements of the Army-worm have given rise to all kinds of superstitious beliefs among the people of Europe, where from time to time it was seen in Silesia, Thuringia, Hanover, Denmark, Norway, and Sweden. They predicted luck or misfortune from its appearance, some prophesying war, others the result of the harvest. The inhabitants of the Silesian mountains predicted luck, whenever the Army-worm took its way down the valley, but unfruitfulness whenever it crawled up. The people in the Thuringian woods predicted peace when it took the former direction, and war when it took the latter. Some even believed their own destiny to be connected with the worm. They threw clothes and ribbons in its way, and felt happy, especially hopeful women, when it crawled over the things; but they regarded one as a dead man whose things it avoided.

These funny beliefs are now all overthrown by the studies and close observations which Mr. Beling, an inspector of the German forests, has made of the Army-worm. He detected its exact nature and origin, and he has delivered many treatises on the subject since the year 1868. His observations fully convinced him that the only cause for the appearance of the Army-worm is the longing for food of larvæ of the *Sciara militaris*. The Army-worm generally measures fifteen feet in length, and three to four inches in width.

Though really not belonging to the *Diptera*, but, like the *Cnethocampa processionea*, to the *Lepidoptera*, we may undertake here, for sake of conformity, to treat the AMERICAN ARMY-WORM. The caterpillars forming it are termed *Leucania extranea*. They collect, like the European Army-worm, in vast numbers, and devastate whole meadows in a short time. When they can no longer find grass, they emigrate to other fields, and attack even rye and wheat. Large clusters of these caterpillars have been observed in the Western States and on Long Island. In 1861, such a gathering of caterpillars proceeded fully sixty yards in two hours.

The Texas and Mexican collections of the same or of a similar species of caterpillars are known by the name of WIRE-WORM.

GAD-FLY (Female).—*Tabanus bovinus*. With the head. (Natural size.)

THE accompanying illustration represents the common BREEZE-FLY, a well-known example of the Tabanidæ. It is also known by the popular names of GAD-FLY and CLEG. As in the gnats, the females are the only bloodsuckers, but they exert their sanguinary ability with terrible force. While staying in forests and suffering greatly from the bites of the Gad-flies, I used to keep a little naphtha in a bottle, and rub it occasionally over my face and hands, for the purpose of repelling these blood-thirsty insects which selected me for their victim, leaving my companions untouched. I have found the whole of the unprotected space round the neck covered with their bites, and my ears thickly stained with blood from the effects of their weapons.

To this family belongs the terrible TSETSE, the curse of Southern Africa, which destroys horses, dogs, and cows by thousands, though it causes no harm to man or to any wild animal. Fortunately, it is a very local insect, its boundaries being as sharply defined as if drawn on a map, one side of a stream being infested with this active insect, while the other is perfectly free. The figure is drawn most accurately.

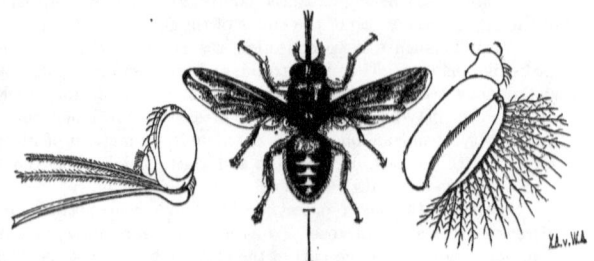

TSETSE.—*Glossina morsitans.* a, Head, with parts of the mouth. b, Antenna, or feeler. (The objects are differently magnified. The line underneath the figure of the Fly indicates the natural size of the Tsetse.)

The following account of its habits and the effects of its bite are given by Dr. Livingstone:—"In the ox the bite produces no more immediate effects than in man. It does not startle him as the gad-fly does; but a few days afterwards the following symptoms supervene: the eyes and the nose begin to run; the coat stares as if the animal were cold; a swelling appears under the jaw, and sometimes at the navel; and, though the animal continues to graze, emaciation commences, accompanied with a peculiar flaccidity of the muscles; and this continues unchecked until, perhaps months afterwards, purging comes on, and the animal, no longer able to graze, perishes in a state of extreme exhaustion. Those which are in good condition often perish soon after the bite is inflicted, with staggering and blindness, as if the brain were affected by it. Sudden changes of the temperature, produced by falls of rain, seem to hasten the progress of the complaint; but, in general, the emaciation goes on uninterruptedly for months, and do what we will, the poor animals perish miserably.

When opened, the cellular tissue on the surface of the body beneath the skin is seen to be injected with air, as if a quantity of soap-bubbles were scattered over it, or a dishonest, awkward butcher had been trying to make it look fat. The fat is of a greenish-yellow color, and of an oily consistence. All the muscles are so flabby, and the heart often so soft, that the fingers may be made to meet through it, and the lungs and liver partake of the disease. The stomach and bowels are pale and empty, and the gall-bladder is distended with bile."

The insect which occasions these terrible results is hardly larger than a house-fly. It is curious that, although horses perish under its bite, mules, asses, and goats escape injury, and it seems that the bite of a single fly is sufficient to cause death. Another curious symptom is, that the blood loses its redness, and hardly stains the hands of the person who dissects the smitten animal. The source of all this mischief is to be found in a little poison-gland at the base of the mouth, not larger than a mustard-seed, and yet infinitely more deadly than the venom of the rattlesnake. The color of the Tsetse is brown, with a few yellow bars across the abdomen. When it bites a man, the pain which it causes is very slight, and the worst results are a trifling irritation not more severe than that caused by the bite of a gnat.

A large insect is the BANDED HORNET-FLY. It is an example of the Asilidæ, among which are found the most gigantic specimens of the order. The body of these insects is long, and clothed with stiff hairs. They are fierce and voracious, feeding mostly upon other insects which they catch on the wing, and out of which they suck the vital fluids through their powerful proboscis. One species of this family has been known to capture and carry off a hive-bee, a remarkable instance of a stingless insect attacking and overcoming a creature so formidably armed as the bee. Some of them are said to attack cattle after the manner of the Tabanidæ. As with the preceding family, the larvæ of the Asili reside under ground, and feed upon the roots of plants.

The family of the Syrphidæ, or Hoverer-flies, is rather large, and contains many interesting insects. Among them may be mentioned the Volucella flies, which feed, while in the larval state, on the larvæ of bees and wasps, and, as if to aid them in gaining admission into the nests of those formidable creatures, are shaped and colored so like the insects which they invade, that at a little distance it is almost impossible to distinguish between them.

The DRONE-FLY (*Eristalis tenax*) belongs to this family. This insect bears a wonderful resemblance to the hive-bee, and has a habit of moving the abdomen in a manner that leads an unaccustomed observer to fancy that it possesses a sting. The larva of this insect is popularly known by the name of Rat-tail maggot, on account of its peculiar construction. This larva resides in mud, with the head downwards. In order to enable it to breathe, the respiratory tubes are carried into a long and telescopic appendage attached to the tail, the end of which is furnished with a brush of hairs something like that on the tail of the gnat larva. The extremity of this curious organ is always held out of the muddy water, and it is most curious to see the grubs elongate their tails as the depth of water is increased.

All the vast family of Muscidæ, or Flies, are members of this order, and as at least eight hundred species of this one family are known, it may be imagined that no description of them can be attempted.

THE large and bold looking fly, represented in our illustration, belongs to the family of the Œstridæ, and is popularly known by the name of BOT-FLY. All these insects are parasitic in or upon animals. The larva of this Bot-fly resides in the interior of horses, and is conveyed there in a very curious manner. The parent fly deposits her eggs upon the hairs near the shoulders of the horse, where the animal is sure to lick them in order to rid itself of the unpleasant feeling caused by agglutinated hairs. The eggs are thus conveyed to the stomach, to the coats of which organ the larvæ cling, and there remain until they have attained their full growth. They then loosen their hold, are carried, together with the food, through the interior of the animal, fall to the ground, and immediately begin to burrow. They remain underground until they have undergone their metamorphoses, and then emerge in the shape of the perfect insect. They do not seem to inflict any damage upon the animal from whose bodies they have drawn their nourishment, and some veterinary surgeons believe that they are rather beneficial than injurious.

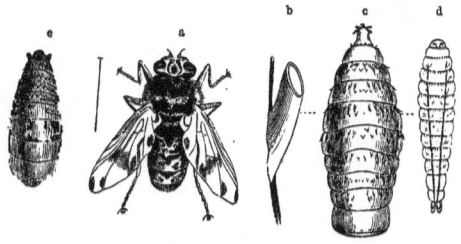

HORSE BOT-FLY.—*Gastrophilus equi*. a, Fly. b, Egg on a hair. c, d, e, Larva in its transformations.

Another kind of Bot-fly (*Œstrus bovis*) resides in the cow, but instead of being taken into the stomach, it burrows into the skin, and there forms large tubercles, that are popularly called worbles or wurbles. An aperture is always left on the top of the tubercle, and the larva breathes by means of keeping the two principal spiracles opposite to the orifice. When full grown, they push themselves out of the aperture, fall upon the ground, and there burrow and undergo their transformations.

The spiracles, to which allusion has been often made, are the apertures through which air is admitted to the system. Insects breathe in a very remarkable manner, the air being conducted through curiously-constructed vessels to every part of the body, even to the extremities of the feet and antennæ. It will be seen that the structure of these vessels must be very remarkable, on account of the opposite duties they have to perform. As they penetrate the entire insect, it is needful that they should be flexible, in order to permit the creature to move about at will, as, if they were stiff-walled, the joints would be rendered useless, and the insect would be unable to move a limb. Another characteristic, however, is required. They must be always kept sufficiently open for the free passage of air, and it is not easy to see how these qualities should be united, as a flexible tube will mostly, if abruptly bent, as is continually the case with the air-tubes of the limbs, lose its roundness at the angle, and shut off the communication. An India-rubber gas-tube is a familiar instance of this property of flexible tubes.

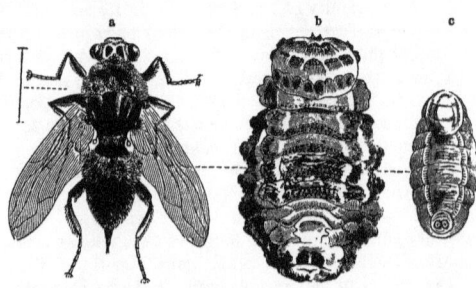

CATTLE BOT-FLY.—*Hypoderma, or Œstrus bovis.* a, Fly. b, Larva. c, Pupa. b and c, As seen from the under side. (All magnified; line at left-hand side natural size of Fly.)

The difficulty is, however, surmounted by a simple and yet most effectual plan. The tubes are double, one within another, and in the interspace a fine but very strong hair-like thread is closely wound in a spiral. It will be seen that, by means of this structure, the tube can be bent in any direction without losing its roundness. The long flexible tubes of Turkish pipes are made in a similar manner, a spiral wire forming the basis, upon which is sewn the leather and silken outer tube—one of the many instances where the art of man has been anticipated in the animal creation. A third species (*Œstrus ovis*), of which we give also an exact illustration, is parasitic in the sheep, inhabiting the frontal sinus, *i. e.*, the open space between the bones on the forehead and between the eyes.

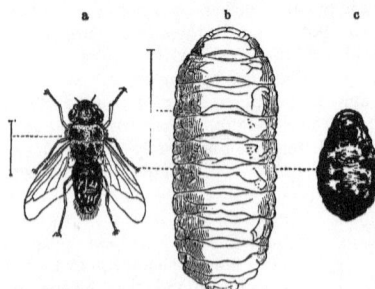

SHEEP BOT-FLY.—*Œstrus ovis.* a, Fly. b, Larva. c, Pupa, as seen from the under side. (Everything magnified.)

The Hippoboscidæ, popularly known under the name of Forest-flies, deserve a short notice. These are round-bodied insects with legs that can cling with wonderful force, and are capable of moving backwards, forwards, and sideways with equal swiftness, so that they are not easily captured, even when they do not take to wing. These insects are mostly found in or near forests, and are very annoying to horses and cattle. As may be seen by reference to the engraving, their integuments are covered with hair, and are remarkably tough and leathery. The various species of Ticks belong to this family, and are closely allied to the Forest-fly.

HORSE, OR FOREST-FLY. *Hippobosca equina.* (Magnified.)

Although not included in Mr Westwood's list of insects, the Lice are mostly considered as belonging to this class, under the name of APTERA. There are very many species of these obnoxious creatures infesting different animals and tribes, and at least three species are found upon the human subject.

LOBSTERS, CRABS, SHRIMPS, ETC.;

CRUSTACEA.

HAVING now completed our brief survey of the insects, we proceed to the CRUSTACEA, a very large class, in which are included the lobsters, crabs, shrimps, water-fleas, and a host of other familiar beings. Even the Cirrhipeds, popularly known under the name of Barnacles, are members of this large class, and a number of curious animals, which have until lately been classed with the spiders, are now ascertained to belong to the Crustacea.

These beings can be easily separated from the insects on account of their general structure, the head and throat being fused into one mass, called technically the cephalo-thorax; the number of limbs exceeding the six legs of the insects; and the mode of breathing, which is by gills, and not by air-tubes. As a necessary consequence of the last-mentioned structure, the Crustaceans possess no spiracles, such as are found in all the stages of insect life, from the larva to the imago. They undergo a well-marked metamorphosis, and in those creatures whose development is best known, the change of shape is so entire as to have led the earlier zoologists to consider the undeveloped Crustacea as separate species. They may be also distinguished from the spiders by the presence of a series of feet, or rather of locomotive organs arranged under the abdomen, as well as by the metamorphosis of their earlier stages, a phenomenon which is not known to take place among the spiders.

The name of Crustacea is sufficiently appropriate, and is given to these creatures on account of the hard shelly crust with which their bodies and limbs are covered, a covering which, in some cases, is of such flinty hardness as to be used for the purpose of sharpening knives, and in others, attains a glossy polish which reminds the observer of glazed porcelain.

As our space is rapidly diminishing, we must proceed at once to the different families and genera, simply noting the more important characteristics as we proceed through the class.

TEN-LEGGED CRUSTACEANS; DECAPODA.

THE first section of these creatures are called the Podopthalmata, or Stalk-eyed Crustaceans, because their eyes are set upon footstalks. The first order is that of the Ten-legged Crustaceans, so called on account of the five pairs of legs that are set in each side. These are exclusive of the complicated apparatus of the mouth, and the jaw-feet which guard its entrance. The Crabs are placed first in the list of Crustaceans, and are technically called Brachyura, or Short-tailed Crustaceans, because their tails are of comparatively small size, and are tucked under the large shielded body. In the preliminary stages, however, the Crabs have tails as proportionately long as those of a lobster or a cray fish.

In the accompanying illustration of the STENORHYNCHUS may be seen one example of the first family of these animals, being a group of Crustaceans distinguished chiefly by the singular form of the carapax or upper shell, which is wide and abrupt at the base, but is prolonged in front, so as to form a long and pointed beak. In all these creatures the legs are long in comparison to the body, but in the LEPTOPODIA they are of such inordinate length, as to remind the observer of the round-bodied, long-legged harvest spider, which scuttles over the ground so rapidly when disturbed. On account of this great length of limb and small size of body, these crabs are often called Sea-spiders. The eyes of the Leptopodia are rather large, and not retractile. It may here be mentioned that the eyes of Crustacea bear some resemblance to those of insects, being compound organs, with a large number of facets, some square and some hexagonal. The eyes of the common shore crab or the shrimp afford excellent examples of this structure. It is a native of the West Indies.

SEA-SPIDER.—*Stenorhynchus longirostris.*

In the Stenorhynchus, the projecting beak is proportionately shorter than in the preceding genus, is cleft at the tips, and very sharp. The fore-limbs, which are furnished with large claws, are stout and strong.

As the shelly armor of the Crustaceans is, in most cases, so hard, strong, and unyielding, the mode of growth might be considered a problem not very easy of solution. For with the Crustaceans the growth continues during nearly the whole of life, or at all events for several years after they have passed through the various changes to which they are subjected in their imperfect stages of existence. Their increase of size and weight is marvellously rapid, and how it can be accomplished without subjecting the Crustaceans to the lot of the starveling mouse, who crawled into a jar of corn, but could not crawl out again after feasting on its contents, seems to partake of the character of an animated puzzle.

The answer to the problem is simply that the creature sheds its armor annually, expands rapidly while yet covered only by a soft skin, and is soon protected by a freshly-deposited coat of shelly substance. Even this answer contains a second problem little less difficult than that which it solves. How can a Crustacean, say a crab or a lobster, shed its skin? It is true that the cast shells are found, showing that the creature has escaped from its old and contracted tenement by a slit in some part of the body, such as the top of the carapace, and has left its shell in so perfect a state that it might easily be mistaken for the living animal.

But how did it manage about the claws? We all know what large muscular masses they are, how very small is the aperture in which the joint works, and how stiff and firm is the broad tendinous plate which is found in their interior. Examination shows that there is no opening on the claws through which the creature might have drawn the imprisoned limb, and it is also evident that the only method by which these members can be extricated, is by pulling them fairly through the joints. As a preliminary step, the hard, firm, muscular fibres which fill the claw and give it the well-known pinching power, become soft, flaccid, and watery, and can thus be drawn through the comparatively small openings through which the tendons pass from one joint to another. The sharp and knife-like edges of the plates cut deeply through the muscle, which, however, is little injured, on account of its soft consistency, and heals with great rapidity as soon as the animal recovers its strength, and is gifted with a new shell. In the common edible crab, the flesh is quite unfit for consumption during this process, as any one can attest who has attempted to dress and eat a "watery" crab. Yet in some of the exotic crustaceans, these conditions are exactly reversed, and the crabs are never so fit for the table as while they are soft and shell-less, after the old suit of armor has been thrown off, and before the new integument has received its hardening. It may here be observed, that the bases of the crustaceous armor is composed of chitine, the remarkable substance of which the elytra and other portions of the insect skeleton are composed.

The name of Leptopodia signifies slender feet. When full grown, the limbs of this species will mostly attain a length of one foot and more. The Stenorhynchus is shown of its natural size.

The Leptopodia of the West Indies resembles that of the Florida Reef. Besides the singular slenderness of the entire body and limbs, there are pretty brilliant blue markings on its tentacles and claws. Quite appropriately it is called Spider Crab. Its resemblance is much like the Daddy Long-legs. The American species we have seen inhabits shallow water. Around artificial works it chooses to crawl over the piers just under the surface. It measures about ten inches in spread of limbs, and is much more slender even than that figured above. Also, it has smooth limbs, and no hairy appendages.

The largest crab, in point of spread of limbs, is the GREAT SPIDER CRAB OF JAPAN (*Microcheira*), belonging to this group. The largest specimen known in any collection is that of the British Museum, which measures between the tips of the first pair of legs eighteen feet. Reliable information is recorded of measurements reaching twenty-two feet. The body of one of the latter measurements is about the size of a "Derby" hat. A photograph of one in our possession is taken with such a hat hanging near it. But the limbs are so long that as a man stands holding them upright, they tower above him a long distance.

Our next example is the *Camposcia*. This creature is quite different in its appearance. When its exterior is free from extraneous substances, it looks slender and small; while burdened with sponges and other marine growths, its form is clumsy and twice as large.

The hairy limbs, as well as the whole of the body, are encrusted so thickly that their true shape is quite undistinguishable, and the animal seems to masquerade under a domino of living disguises. Even the joints can barely be ascertained, and, but for the continual movements to which they are subjected, it is very probable that the sponges would increase with such rapidity, that in a short time the limbs would be rendered immovable. These growths are so constant and rapid that the creature can only free itself at the time when it changes its skin; and it is likely that the crab may feel as comparatively light and disburdened after throwing off all this encumbrance of heavy voluminous substances, as does a thick-wooled sheep after the shears have removed the heavy fleece, and enabled the lightened animal to skip about the field astonished at its own activity and the sudden coolness shed over its body.

The Camposcia possesses all the characteristics already mentioned as appropriate to the family in which it is classed, and that the snout—if we may be allowed to employ the word—

is elongated and very deeply cleft, so as to form a forked protuberance. The body is rounded at the base, and small in proportion to the limbs, though the apparent disproportion is not so marked as in the Leptopodia. This species is a native of the Philippines. The genus to which this creature belongs does not seem to be very large, only three, or perhaps four species, being known.

Still keeping to the same family, we come to a curious genus containing some very remarkable creatures, among which the *Doclea calcitrapa* is one of the most notable.

In this genus the beak is comparatively small, but still contains the cleft tip, although the notch is not nearly so deep as in other members of this family. The claws are of considerable size and power, and the legs are long and furnished with an array of stiff bristles. The chief peculiarity, however, which most strikes the sight, not to say the touch, is the formidable display of long and pointed spikes, which radiate from the body like the spines of a hedgehog. The eyes are not very prominent, being set on rather short footstalks, and nearly concealed by the projection from the shelly covering. This genus seems to be widely spread over the hotter portions of the globe, specimens having been taken off the Mauritius, in China, India, and the Philippines, of which latter locality the present species is a native.

All the crabs of this family are marine, and prefer the deeper parts of the sea, where they lurk among the waving masses of sea-weeds, or crawl upon the oyster banks. As might be imagined from the length and slenderness of their limbs, they are but slow of progress, and seem to tumble over the ground in a very unsteady manner. Still, their long limbs are admirably calculated for the peculiar substances on which they pass their lives, and they are able to stride, as it were, over obstacles which would seriously encumber a creature with shorter legs. Their food consists almost wholly of small mollusks and other marine animals.

WE now come to another family wherein many of the same characteristics are preserved, but the legs are of moderate size. These creatures are popularly known by the name of Spider-crabs, and scientifically are termed Maiadæ.

We will describe a few examples of this genus. The GOUTY CRAB has been gifted with its very appropriate name on account of the knobby and swollen limbs, which give it an aspect as if it were suffering from the painful but unpitied disease from which it derives its name. The specific title of Chiragra is of Greek origin, and bears a similar signification. The Gouty Crab is known to be an inhabitant of the Mediterranean, and is thought also to be a native of the West Indian seas.

A creature which looks as if it had been made almost at random out of a thistle-bud and a handful of thorns, is known under the name of THORNBACK CHORINUS (*Chorinus acanthonotus*). This species can hardly be mistaken for any other, on account of its altogether eccentric shape, and the branching spines which spring on every side from its body and the joints of its limbs. It is but feebly provided with claws, these members being little larger than the ordinary limbs; and the eyes stand out on tolerably long footstalks.

Nearly related to these species, we find two moderately common European crustaceans, which are interesting in their habits, though not particularly pleasing in their aspect. One of these is the FOUR-HORNED SPIDER-CRAB (*Arctopsis tetraodon*), a rather long-legged creature that seems to be very local in its habits, being rarely or never seen in some localities, while in others it is found in vast numbers. This crab generally hides itself under the overhanging masses of fuci which cover the submerged rocks, and thence is fond of descending into the lobster and crab pots, and so is made captive by the fishermen. The color of this species is yellow, and the body and greater part of the limbs are densely covered with thick hair. The male is larger than the female.

The second species is the HARPER-CRAB, or GREAT SPIDER-CRAB, or SEA-TOAD (*Hyas araneus*), as it is sometimes called. This is commonly found on nearly all the coasts of Northern Europe, and prefers to range among the weeds just about the zone beneath low-water mark. It is one of the day-feeders, and will often leave the waves for the purpose of feeding upon the fish and other animal substances that have been flung upon the shore by the tide.

In some places it haunts the stake-nets, and there makes a regal feast before it is disturbed by the proprietors.

An example of a very common and a very useful European species is the common THORN-BACK SPIDER-CRAB, or SQUINADO. It is plentiful upon European coasts, and is not a very prepossessing creature in external appearance, its body being one mass of sharp and not very short spines, and its whole frame possessing a weird-like and uncomely aspect.

Another curious creature is the THREE-SPINED SPIDER-CRAB, so called from the peculiar shape of the body, which, on account of the projecting beak and the strange modification of the carapace, has a kind of three-cornered aspect. Several species of this genus are known, and are found in the West Indian seas and off the Philippines. The present species is one of the most common, and is found in the West Indies.

Two remarkable examples of this family are the Ram's-horn Crab (*Criocarcinus superciliosus*), a species which is distinguished by the two long, horn-like projections from the snout, and the Thorn-claw Crab (*Acanthonyx zebrida*). The generic name Criocarcinus, which is of Greek origin, and signifies Ram-crab, is given to the animal on account of this structure. The body is thorny, though not so wholly beset with spikes as in the spider-crab, and the eyes are placed on moderately long footstalks. The specific term, superciliosus, refers to a Latin word signifying an eyebrow, and is given to this crab on account of the overhanging plates under which the eyes are hidden when the footstalks are laid close to the head, as is the custom of the creature when alarmed. The present species is found in the New Hebrides.

The THORN-CLAW CRAB is a curious-looking little creature, especially notable for the large and boldly hooked extremities of the limbs. The name of ACANTHONYX, or THORN-CLAWED, is given to the genus on account of this structure. At first sight, the Acanthonyx hardly seems to belong to the same family as the preceding species, the shape of the body being apparently the reverse to that which is characteristic of the Maiadæ. But on a closer examination, it is found that this difference is more apparent than real, and that though the body seems to be wider across the head, or rather, the cephalo-thorax, to speak accurately, the width is owing to mere projections and not to any increase of the actual body. The Thorn-claw Crab is found in many European seas, and is tolerably common in the Mediterranean.

Our last examples of the Maiadæ are the Heraldic Crab (*Huenia heraldica*), the Long-snouted Crab (*Huenia elongata*), and the Micippa (*Micippa philyra*).

In these three species can be observed a curious variation of form that takes place in animals that belong to the same family, and even to the same genus. The body of the MICIPPA is very large in proportion to the limbs, rounded, and covered with numerous protuberances of various sizes, mostly small tubercles, but sometimes being developed into bold spikes. The claw legs are remarkably small in proportion, and the claws themselves are even more feeble than might be inferred from the dimensions of the entire limb. Several species of Micippa are known, all of which are obtained from one or other of the Philippine Islands.

The Heraldic Crab and the Long-snouted Crab are very dissimilar in external appearance, and yet belong to the same genus. The HERALDIC CRAB derives its name from the shape of its carapace, which presents a fanciful resemblance to the shield and mantle employed by heraldic painters in depicting coat armor. The sides of the carapace are developed into four singular projections, flat, and looking very much as if pinched out of the shell while its material was plastic. The snout is tolerably long and very sharply pointed, and the eyes only just project from under the protecting shell.

The LONG-SNOUTED CRAB is a creature in which the carapace, instead of being wide, flattened, and formed with ring-like projections at the side, is drawn out to a wonderful length, and possesses two angular projections towards the base. Both these crabs are natives of Japan.

IN the family of crabs which is known by the name of Parthenopidæ, we have a very different form, the carapace being more or less triangular, the beak or snout small and not notched, and the eyes very retractile. The claw-legs are generally large in proportion to the other limbs, which are often very short.

The Domed Crab is a very remarkable example of this family, and in addition to certain generic peculiarities, well displays the characteristics of the family. The claw-legs are very large throughout their entire structure, and are furnished at their extremities with short but powerful nippers. The carapace of this creature is extremely wide, but the width is due, not so much to the body as to the shell, which is expanded in such a manner as to conceal the legs under its shelter.

The generic name Cryptopodia is derived from two Greek words signifying Hidden-legs, and is an extremely appropriate title. Even the large claw-legs can be folded up and tucked away so neatly under the carapace, that, when the creature lies still on the ground, no vestige of limbs can be seen, and it might easily be mistaken for a stone thrown casually on the shore. In fact, the whole contour of this crab, whether when moving or quiescent, irresistibly reminds the observer of the tortoise tribe, and bears a special analogy to the box-tortoise, which has already been described and figured. The eyes of this genus are very small, and, like the limbs, can be wholly retracted and hidden under the shell. The Domed Crab is a native of Japan.

A very singular and unprepossessing crab, called SPINOSE PARTHENOPE (*Parthenope horrida*), belongs to the typical genus of the family. At present, this genus seems to be very small, the number of known species being decidedly limited. Owing to the marine residence of these creatures, and the extreme difficulty, not to say impossibility, of watching them in their watery homes, the habits of these Parthenopidæ are but little known, and in most cases can only be conjectured from the bodily form, just as the fossil animals are known to be carnivorous or herbivorous by the structure of their teeth and jaws, to be swimming creatures because they possess fins and paddles, or to be capable of flight because they are furnished with wings.

In the Spinose Parthenope, the carapace approaches to a five-sided figure, rather wider than its length, moulded into a series of the oddest imaginable protuberances, and covered with knobs, tubercles, and spines. The beak is sharp, short, pointed, and has a strong tooth just between the antennæ. The claw-legs are very large, armed with powerful forceps at their extremities, and covered thickly with such a multitude of knobs, spikes, and protuberances, that they really seem as if they were subject to disease and had thrown out a crop of unhealthy growths. The hinder limbs are comparatively small, but yet are strongly made, and armed with a whole array of thorny spines, so that, what with the claws and what with the spines, the creature is a truly formidable being, and one that may not be grasped with impunity by a careless hand.

This species inhabits some of the hotter parts of the world, and specimens were procured from the Mauritius.

The little STRAWBERRY-CRAB is very appropriately named, as its color is of a pleasing red, and its surface studded with numerous tubercles, so as to bear some resemblance to the fruit whence it derives its popular name. It is a European species, and is generally found in deep water, so that the dredge is the instrument usually employed in its capture.

The SPINE-ARMED LAMBRUS is a member of a moderately large genus, inhabiting the Mediterranean and the warmer seas of the world in general. In many respects the genus Lambrus resembles the parthenope, but is distinguishable by having one plate fewer in the abdomen, and by the manner in which the antennæ are jointed. The eyes of this Spine-armed Lambrus are retractile and placed on footstalks of an elaborate and curious construction.

THE large family of the Canceridæ now comes before us, and is familiarly known through the medium of the common EDIBLE CRAB, which is represented in the accompanying illustration, the figure being drawn from a young specimen.

This is a very common species, being plentiful around rocky coasts, and generally remaining just under low-water mark. The fishermen catch it in various ways; but the most usual method, and that by which the greatest number of these crustaceans are captured, is by means of certain baskets, called crab-pots, cruives, or creels, according to the locality. These baskets are round, and in shape something like a flattened apple, and have an aperture at the top through which the crab gains access to the interior. When once within the basket, it cannot escape, because the opening is guarded by an inverted cone of osiers, like the entrance to a common wire-mouse-trap, so that the elastic sticks yield to the expected prey while passing downwards, but effectually prevent all upward movement.

The Edible Crab of Europe resembles greatly the *Cancer sayi* of New England shores, north of Cape Cod.

In many external points the ÆTHRA resembles the domed crab, which has already been described on page 443. Like that being, the carapace is very wide, flat, and expanded at the edges. The limbs, too, are comparatively short, and can be concealed under the shell, which, from its hilly surface, covered with tubercles, and the irregular, notched, and ridged carapace, has but little of the cancerine aspect. Zoologists of the present day, however, have placed it in the same family with the edible crab. The claw-feet, with their forceps, are very like those of the parthenope, but are not so proportionately large, and their surfaces are concave, so as to fit into the trunk. The eyes are very small, and their orbits nearly circular.

EDIBLE CRAB.—*Cancer pagurus.*

All the species of this genus inhabit the East Indian and African seas. Large specimens attain a length of three, and a width of four and a half inches.

We still have to describe three more curious examples of this large family, each being notable for some peculiarity of form or habit.

MONTAGU'S CRAB belongs to a genus which finds several European representatives. It is a flat-bodied and strongly-made creature, very restless in disposition, and with a curious fondness for getting under stones, and turning them over; probably for the sake of obtaining a meal from the smaller marine animals that are accustomed to shelter themselves in such localities.

The shelly covering of this crab is remarkably strong and flinty, and the muscular power of the claws is gigantic, when the small size of the creature is taken into consideration. It is a tolerably common species on several European coasts, appearing to be peculiarly plentiful on the southern side of England.

The RED-SPOTTED ÆGLE is a curiously marked crab, the carapace being divided into a number of partitions, in which is a certain, though not very definite regularity. It inhabits the warmer seas. The Mauritius and the Philippines are favored haunts of the Ægle. The color is red and whitish spotted.

The TOOTHED PERIMELA is our last example of this family.

The name of Toothed Perimela is given to this species in allusion to the shape of the carapace, which has the front edge rather flattened, and cut into a series of four or five strong teeth, like those of a saw. The surface of the carapace is smooth, and is swollen into several decided projections, something like those softly rounded hills called by the French "*mamelons.*" Over the region of the liver, the carapace is concave.

Nearly allied to the preceding species is the HAIRY CRAB (*Pilumnus hirtellus*), a creature which derives its popular name from the curious hairy covering with which it is decorated. The convex carapace is studded more or less thickly with longish hairs, and the four hinder pairs of legs are also protected in the same manner. This crab is not a very common one, and is mostly found on the northern coasts of Europe. It seems to prefer moderately deep water, fifteen fathoms being the usual depth at which it is captured.

It may be easily known by the following characteristics: On the front edge of the carapace are arranged four spines set in the same line, and the front is divided by a deep notch down the middle. The claw-legs of this species are always unequal in size, and the first joint of the outer pair of antennæ is short. The ground-color of the Hairy Crab is chestnut-brown, with a reddish tinge of greater or less intensity, according to the individual. The legs are dusky-red, relieved by a series of bands of a grayish-yellow color.

In many characteristics, the genus Zozymus resembles the genus Ægle so closely, that the two genera have been blended together by several systematic zoologists.

As a general fact, it is much to be wished that the modern fashion of breaking up the old and established genera into a host of new ones, many of which contain but a single species, had not proved so fascinating to the authors. In many cases, the characteristics employed as generic differences are so very trifling, that they are barely of sufficient importance for the establishment of a species. New families also have been invented with reckless profusion, and in many instances, known to every naturalist, the characteristics on which the family is founded serve equally for the genus and the species. Needful as is some definite system of nomenclature, and admirable as is the system which Linnæus founded, and which has since formed the basis of all arrangement, it can be pushed too far, and, as is well known to be the case, is so widely abused, that merely to learn the multitude of sesquipedalian titles with which the study of zoology is now encumbered, requires a greater exercise of memory than to study the habits and peculiarities of structures which alone form the true objects of zoological science.

The BRASSY CRAB is remarkable for the curious protuberances into which the carapace is moulded, and which cover the claws and legs. Only three or four species are ranked under this genus, and all of them are natives of the Mauritius or the Philippines.

The genus to which the SPOTTED CRAB is assigned is rather more comprehensive than the preceding, and contains six or eight species, all being remarkable for their round and smooth bodies, the peculiar notches and projections on the edge of the carapace, and the huge claws that terminate the first pair of legs.

The members of this genus belong to the tropical regions. Some species are found about Jamaica and the West Indies in general. Others inhabit the Philippines, and others, again, are natives of the Mauritius. Except from the peculiar spots with which the carapace is decorated, this crab might easily escape detection while lying with its limbs withdrawn, and its frame in a state of quietude, for it is so round and so smooth that it looks very like a large pebble that has received a partial polish from the action of the waves. Many specimens are covered more or less with vegetable and animal growths, such as corallines, algæ, barnacles and zoophytes, and are therefore almost undistinguishable while they are quiet.

The spots upon the carapace are bright red.

Examples of two dissimilar species of the same genus are the TUBERCLED GALENE and the SMOOTH GALENE. The Tubercled Galene derives its name from the profuse warty excrescences which grow upon the claw-feet and the pincers. In the Smooth Galene the claws are very much smaller in proportion, and destitute of the tubercles which are so characteristic in the former species. Both these crabs are natives of the East Indies.

WE now arrive at the family of the Portunidæ, or Swimming Crabs, in which the last pair of feet are flattened sideways, and have the last joint dilated into a thin oblique plate, which answers as an oar or a fin, and enables the creature to propel itself through the water. The first example of this family is the GREEN, or SHORE-CRAB, so familiar to every one who has passed even an hour on the coast between the time of high and low water. Although one of the commonest of the crustaceans, it is at the same time one of the most interesting, and, owing to its diurnal habits, its fearless nature, and its love for the shallow waters, it is very easily observed. I have spent many a pleasant hour in watching the habits of this little creature, and could hardly have imagined the activity, the piercing sight, and the cleverness with which this crab is endowed.

It is a fierce and even voracious animal, chasing and fairly running down living prey, and actually leaping upon its victim with a spring like that of the hunting spider. I have seen the Green Crabs run after and catch even the active sandhoppers, calculating with nice precision the spot on which they alighted, and pouncing on them before they could get themselves into position for a second leap. If the prey should be of tolerable size, the crab does not leap at it, but darts out one of its claws with a stroke so sharp and quick, that the eye can scarcely follow it, and as true of aim as the serpent's dart, draws back the victim, seizes it immediately with the other claw, and begins to pull it to pieces before it can recover from the shock.

The Green Crab has a most extended distribution, the North American species being the same as the above. It is also found in European seas, South America, and the Sandwich Islands. Its range here is from Cape Cod to Maryland. Martha's Vineyard is a good locality for this species.

The little crustacean which is called by the name of the VELVET FIDDLER-CRAB, derives its popular and appropriate title from the movements which it makes while swimming through the water.

The last joints of the hinder feet are extremely flat, and it is by their movements that the crab is enabled to swim. Their motions are very like those of an oar when used in "sculling" a boat, and are popularly thought to resemble the movements of a fiddler's arm while playing a lively tune. The word "velvet" is affixed to the name, because the entire shell of a perfect specimen is thickly covered with shining hairs, short, silken, and soft, something like the pile of velvet or fine plush. It is seldom, however, that a really perfect specimen is seen, as the soft velvety pile is easily rubbed off, and in almost every instance has sustained some damage, so that the blackish shell is seen, with its polished surface. The edges are very seldom clothed with their normal coating of hair. When tolerably perfect, a full-grown specimen is a really handsome creature, with its coat of velvet pile, its striped feet and legs, its scarlet and blue claws, and its vermilion eyes set in their jetty sockets.

This species is not one whit less voracious or cruel than the edible or the green crab, and as it enjoys all their activity, with the additional privilege of swimming through the water, it is even a more formidable animal, chasing and killing every creature that it can overcome. Even the hermit-crab, that lies so snugly in its shelly cell, with the large fighting-claw guarding the entrance, and its body withdrawn into the inmost recesses of the shell, is frequently captured and killed by this doughty warrior. Every one who has tried to pull a hermit-crab out of its house, knows the difficulty of the task. The creature has the art of retreating into its dwelling so far, and pressing its claws and legs so firmly against the inner mouth of the shell, that there is nothing by which the animal can be grasped, except, perhaps, the antennæ; and the crab will allow itself to be pulled to pieces rather than loosen its hold. Yet the Fiddler-crab makes little account of the hermit, but pokes his claw into the shell, pinches the poor hermit across the thorax, and drags him out of his cell. It then pulls off and eats the soft abdomen, tears up the body and limbs, and flings them away in fragments, as if for sheer wantonness of destruction.

There are many species belonging to this genus, which are scattered all over the world, especially where the seas are warm or temperate. The Velvet Fiddler is tolerably common. Of these the MARBLED FIDDLER (*Portunus marmoreus*) is perhaps the handsomest, on account of the regular patterns of buff, brown of various shades, and red, which are seen upon the body. The shape of the patterns is variable, but their arrangement is always symmetrical. These colors are, however, very fugitive, and can only be preserved by removing the whole of the soft parts, and dyeing the carapace with great care. All the species seem to be decidedly local, so that in the space of two or three miles of coast as many species of Portunus may be found, each in its own particular locality.

Perhaps the very best swimmer in the family is the OCEANIC SWIMMING CRAB, a creature to which the generic name of Neptunus has been given on account of its wonderful mastery over the waves.

This crab is apparently made for speed, its flattened limbs and body being calculated to offer the least possible resistance to the dense fluid through which it has to pass. The Oceanic Swimming Crab is among crustaceans what the albatross is among birds, being able to sustain itself for days together without needing rest, and whenever it does seek a brief repose, needing nothing but the floating algæ as a temporary resting-place. The movements of this species are achieved with an easy grace and freedom that remind the observer of the swallow's flight, as the crab flies swiftly through the water, its claws ready to seize their prey, and its limbs held in such an attitude that they offer scarcely any resistance to the element in which the creature lives.

This species has a very wide range, and is found throughout the warmer seas. It is common around India, Australia, and the Philippines, and from its bird-like fleetness and activity, has never failed to attract attention. Like others of its family, it feeds upon living prey, and chases its victims through the water with a speed as rapid, an aim as certain, and a voracity as unfailing as are exhibited by the shark itself. The species is notable for the shape of the carapace, and the sharp spine into which each side is developed.

The EDIBLE CRAB of America (*Neptunus hastatus*) forms a somewhat important item of commerce in certain portions of our coast. In New England, north of Cape Cod, it is practically unknown as an edible. When in the soft state, after moulting, it is highly prized, and ranks even higher than the best oysters. They are not produced in sufficient numbers to render them common in markets, excepting those of the Middle States. The region of Hampton Roads is the central point of this luxury. This species is also consumed largely in its hard-shell state. It extends southward to the Gulf of Mexico, where it is abundant on the reef.

Another strange-looking creature is nearly as good a swimmer as the oceanic crab, and has many of the same habits. Like that crustacean, the FORCEPS-CRAB roams the ocean as freely as the bird roams the air, shooting through the waves with arrowy swiftness in chase of prey, gliding easily along just below the surface, hanging suspended in the water while reposing, or occasionally lying across some floating sea-weed. The chief peculiarity of the Forceps-crab is the structure from which its name is derived, the wonderful length of the first pair of limbs, and the attenuated forceps with which they are armed. Though not possessing the formidable power with which some crabs are armed, the Forceps-crab is yet as terrible an enemy to the inhabitants of the sea, for it can dart out these long claws with a quick rapidity that almost eludes the eye, and grasp its prey with unerring aim.

No one who has not watched the crabs in their full vigor and while enjoying their freedom, can form any conception of the many uses to which the claws are put and the wonderful address with which they are used. Their bony armor, with its powerful joints, appears to preclude all delicacy of touch or range of distinction, and yet, the claws are to the crab what the proboscis is to the elephant. With these apparently inadequate members the crab can pick up the smallest object with perfect precision, can tear in pieces the toughest animal substances, or crack the shell of other crustaceans as a parrot cracks a nut in its beak. It can direct them to almost every part of its body, can snap with them like the quick, sharp bite of a wolf, or can strike with their edges as a boxer strikes with his fists.

The paddle-legs are broad and well developed, so as to ensure speed, the front of the carapace is sharply and deeply serrated, and the sides are drawn out into long pointed spines. It is a native of the West Indian seas, and is represented about the dimensions of an ordinary specimen.

The NIPPER-CRAB (*Polybius henslowii*) is a better swimmer than the fiddler-crab, being able, according to Mr. Couch's account, to ascend to the surface of the sea, and to pursue its prey through the waters. So well does this creature swim, and so voracious is its appetite, that it captures and eats even the swiftest sea-fish, having been known to pounce upon the mackerel and the pollack. Its method of proceeding seems to be to dart upon its prey, grasp it firmly with its sharply-pointed and powerful claws, and retain its hold until the unfortunate victim is quite fatigued and falls an easy prey. It is not so handsome as the velvet fiddler, having none of the beautiful scarlet and azure tints which decorate that species, and being mostly colored with different shades of brown.

Our last example of this interesting family is the SENTINEL-CRAB, so called from its extreme watchfulness, and the wonderful manner in which its eyes are arranged so as to explore objects in every direction, without needing to move, or even to raise itself from its flat and crouching attitude. The generic name of this creature is of Greek origin, being composed of two words, the former signifying a foot, and the latter an eye, and is given to it on account of the strangely long footstalks on which the eyes are set. When the creature is at rest, the footstalks lie horizontally upon the body, and are received into two channels or grooves, where they lie hidden and safe from danger.

A somewhat similar disposition is found in some of the land-crabs, but differing in the arrangement of the footstalks. Each of these curious organs consists of two pieces, and in the Sentinel-crab the first is long and the second very short, while in the land-crab exactly the reverse takes place, the length of the footstalk depending on the second joint. Only one species of Sentinel-crab is at present known, and is a native of the Indian Ocean. It never attains very great size, its length varying from two to four inches.

WE now leave the swimming and marine crabs, and turn to those which are able to spend a great part of their existence out of the water. The FLATTENED MUD-CRAB belongs to a tolerably numerous genus of crabs, which live along the banks of rivers or in damp forests, and are evidently a link between the aquatic and the true land-crabs. The THELPHUSA lives in burrows, which it excavates in the mud to a considerable depth, and gives the fisherman no small trouble before it can be dug out.

One species of this genus, the GRANCIO of the Italians, is very common around Rome, and is largely captured for sale in the markets, as its flesh is very delicate, and in great request on the fast days of the church. It is dug out of the mud and kept alive for sale, as it can endure removal from the water for a very long time, sometimes living a month upon dry land, the only precaution needful being that it should be kept in a damp spot, such as a cellar. It is a most useful species, as it can be eaten throughout the entire year, but is thought to be in best condition during and immediately after the moult. There are many ways of dressing this delicacy, some persons killing it by long immersion in milk, and others asserting that its flesh has more flavor if eaten raw, like that of the oyster. In the market these crabs are tied to strings, but always at such lengths that they cannot reach each other, or if they should do so they would of a certainty attack and maim their nearest neighbors.

The Lake of Albano is a very favorite resort of these crabs, which absolutely swarm in its soft muddy bed. On the first view, the Mud-crab looks very like the common green crab of the sea-shore, but can be distinguished by its color, which is of a whitish or livid hue. It runs about with great speed, and when it fears the approach of an enemy, hurries into the water, burrows under the mud, or hides itself beneath a friendly stone. Should, however, its retreat be cut off, it proves that it can fight as well as run, and grips with such force, that it makes the blood flow before it can be shaken off. During the winter it dives deeply into the mud, and there remains hidden, until the warmth of spring induces it to leave its retreat.

ANOTHER family of land-crabs is well represented by the TOULOUROU BLACK-CRAB, or VIOLET-CRAB of Jamaica (*Gecarcinus ruricola*).

This singular creature is found in vast numbers, and for the most part lives in burrows at least a mile from the shore, and sometimes at a distance of two or even three miles, seldom, indeed, visiting the sea but for the purpose of depositing its eggs. About the months of December and January the eggs begin to form, and the crab is then fat, delicate, and in good condition for the table. In May, however, it is quite poor and without flavor, and does not recover its proper condition until it has visited the sea, deposited the eggs, and returned to its home. About July or August the Violet Crab is again fat and in full flesh, having, in fact, laid in a stock of fat which will afford it sufficient nourishment through the time in which it remains in a torpid state. It retires to the bottom of its burrow, into which it has previously conveyed a large amount of grass, leaves, and similar materials, closes the entrance, and there remains until the next year.

It is a very quick and active creature, scuttling off to its hole with astonishing rapidity, and is not to be captured without the exercise of considerable skill and quickness. Nor must it be handled without caution, for as it runs, it holds up its claws ready to bite, and if it succeeds in grasping its foe, it quickly throws off the limb—which continues to gripe and pinch as sharply as if still attached to its former owner—and makes good its escape while the claw is being detached. For the table, this crab is esteemed as one of the greatest delicacies, and is treated in various modes, sometimes stewed, but mostly cooked in its own shell.

The PEA-CRAB, a curious little crustacea, is found within the shells of the horse-mussel, and one or two other bivalves. That this crab was a frequent inhabitant of the pinna was a fact well known to the ancient naturalists, who put forward a number of ingenious but rather fabulous theories to account for the singular alliance. By some writers it was said that the Pea-crab supplied the place of eyes to the blind pinna, and that its especial task was to warn it of the approach of the polypus or cuttle-fish, receiving board and lodging as a reward of its labors. Some thought that the Pea-crab performed the office which ancient tradition attributed to the jackal, and was sent out by the mollusk for the purpose of obtaining food, the host and guest dividing the spoil.

What may be the real reason for this strange habit is not quite clear, for though the Pea-crab will live in the same shell without inflicting any apparent injury to its host, it is yet very fond of mussel-flesh, and will eat it with much eagerness. Indeed, several specimens have been kept alive for more than a year by being fed upon that diet. Perhaps it may feed upon the juice and less important parts of the mollusk, just as the ichneumon larva feeds on the juices of the caterpillar. Sometimes two and even three specimens are found within a single shell, and on examining the mussels taken from an old bank where they have been permitted to rest quietly, almost every shell will contain one specimen of the Pea-crab.

The color of the Pea-crab is reddish cream-color, and the dimensions are small. The average diameter is half an inch. It is a very timid creature, as might be inferred from the remarkably retired spot in which it passes its life; and when it is alarmed, it contracts its limbs and pretends to be dead, remaining motionless for a very long space of time, and not moving until it feels sure that its enemy is out of the way.

The little Pinnotheres, so commonly found ensconced in the American oysters, is designated specifically *P. ostrea*, on account of its habitual sojourn there. The female only is seen, the male maintaining an independent existence.

THE LONG-ARMED MYCTIRIS is an example of a moderately large family of crustaceans, all of which inhabit the warm seas, and are most plentiful under the tropics. In this genus the carapace is very delicate, convex, and somewhat circular in form; the limbs are long and slender. In the present species the carapace is curiously divided by two longitudinal furrows into three convex protuberances, and projects slightly in front. The claw-feet are long and armed with pincers that are very powerful in proportion to the dimensions of the animal. It is a native of the Australian seas.

WE now arrive at another family, called the Ocypodidæ, or Swift-footed Crabs, from their extraordinary speed, which equals or even exceeds that of a man.

The accompanying engraving represents the FIGHTING CRAB, a creature whose name is well deserved. As the reader may observe, one of its claws is enormously large in proportion to the body, being indeed, nearly equal in dimensions to the whole carapace, while the other claw is quite small and feeble. It is remarkable that sometimes the right and sometimes the left claw is thus developed. This animal is a most determined fighter, and has the art of disposing its limbs like the arms of a boxer, so as to be equally ready for attack or defence. The figure shows the crab in its natural size.

FIGHTING CRAB.—*Gelasimus bellator.*

The Fighting Crab lives on the sea-shore or on the border of salt marshes, and burrows deeply in the earth, the holes being tolerably cylindrical and rather oblique in direction. In some places these holes are so close together that the earth is quite honeycombed with them, and the place looks like a rabbit-warren. Each burrow is tenanted by a pair of crabs, the male always remaining in the post of danger at the mouth of the tunnel, and keeping guard with his great claw at the entrance.

While running, it has a habit of holding the large claw aloft, and moving it as if beckoning to some one, a habit which has caused one of the species to be named the Calling Crab. This action has in it something very ludicrous, and those who have watched the proceedings of a crab-warren say that there are few scenes more ridiculous than that which is presented by the crustaceans when they are alarmed and go scuttling over the ground to their homes, holding up their claws and beckoning in all directions. The generic name is derived from a Greek word signifying laughter, and is given to the crabs because no one can look at them without laughing. These crustaceans possess very long footstalks, on which their eyes are placed, but, as has already been mentioned, the second joint of the footstalk is long and the first is short.

The FIDDLER CRABS, or, as they are called also, Fighting Crabs, are represented in America by the *Gelasimus pugillator*. They are characterized by the singular difference between the two fore-arms. The above description and figure apply very closely to the American form. We have seen thousands of these crabs, of the same uniform size, throughout the army, which they simulated, covering an area of many yards on a smooth beach.

The ludicrous uplifting of the great arm—though sometimes earning for them the name of fiddlers, the arms looking like bass viols—when these creatures were moving together, suggested most readily an army on the march, and manœuvring meantime. An interesting feature was observed, in that being crowded closely, each touching the next, their movements were the result of simultaneous impulse. The whole army would be seen approaching you, steadily as a heavy column of troops; anon the entire mass wheeled, or changed instantly, and with the greatest precision, to oblique march or in echelon. We observed this at Cedar Keys, in West Florida.

This crab is not uniformly distributed in New England, being found in scattered localities. We never saw it in the vicinity of Boston, Mass.

A beautiful species called LADY-CRAB, or SAND-CRAB (*Platyoniculus ocellatus*), was once found in the harbor-side waters near Boston, but is now nearly if not quite extinct in the eastern portions of New England.

Closely allied to these creatures is the RACING CRAB (*Ocypode cursor*), sometimes called the Sand-crab, from its habit of burrowing in the sand. In our illustration it is represented of the natural size. Sir J. Emerson Tennent, in his "Natural History of Ceylon," writes as follows of this crab: "In the same localities, or a little inland, the Ocypode burrows in the dry soil, making deep excavations, bringing up literally armfuls of sand, which, with a spring in the air, and employing its other limbs, it jerks far from its burrows, distributing it in a circle to the distance of many feet. So inconvenient are the operations of these industrious pests, that men are kept regularly employed at Colombo in filling up the holes formed by them on the surface of the Galle Pace. This, the only equestrian promenade of the capital, is so infested by these active little creatures, that accidents often occur through horses stumbling in their troublesome excavations."

RACING CRAB.—*Ocypode cursor*.

These crabs run with surprising swiftness, and it is by no means easy to catch them before they escape into their burrows. Sometimes they are made to afford a few hours' amusement to military officers and other persons who have too much time on their hands, the struggle between man and crab being as exciting as the battle between an eagle and a salmon. One device is ingenious, simple, and often successful. Long strings are attached to flat pieces of slate or stone, which are carefully laid near a burrow, and some tempting food laid outside. The crabs crawl out to feed on the bait, and while they are engaged, the slates are quietly drawn over the entrance of the burrows. A sharp rush is then made, the crabs scuttle away to their homes, and one or two are generally captured before they have recovered their presence of mind sufficiently to leave their barricaded doors and ask for admission into another habitation. Another amusement is to chase the crabs on horseback, trying to ride them down by main speed, and to kill them with a gun. They mostly take an oblique line when running, so that a pursuer who is acquainted with their habits is more likely to succeed in his endeavors than one who employs nothing but main speed in the chase.

None of these crabs care much about the water, being quite satisfied if they can obtain sufficient moisture to keep their gills in working order. As is the case with most of their kindred, they seek the ocean when the time for laying their eggs has arrived. Even then, they remain but a very short time in the water. It is, however, conjectured that the first stages of existence must be passed either in the water or underground, as a very small Racing Crab never seems to be found. On account of the great speed of these creatures, the Greeks were accustomed to designate them by a name which signifies a horseman or knight. Opinion appears to be divided with respect to the value of their flesh, some species being highly esteemed, while others are totally rejected, and even decried as poisonous. It may be, however, that locality has some influence in these opposite opinions, and that in some places the crabs may feed on wholesome food and therefore be eaten with impunity, while in others they may perforce mix with their diet certain substances injurious to human health, and so become in some degree poisonous. The reader will doubtlessly remember that the common edible mussel is at one time perfectly harmless, and at another is so injurious as to cause serious effects upon the health of those who eat it, life itself having been threatened by the mysterious influence.

The Racing Crab alluded to above has a representative species considerably larger than this on the beaches of the sub-tropical portions of America. It is called the SPIRIT CRAB. On the Florida Keys it abounds; its colors so accord with the yellowish-white sand of the beaches, one is quite deceived at first glance. Were nothing stirring, a few moments after your advent, you would notice nothing of animal life; move never so lightly, and the light-colored ghosts flit in great numbers to their holes in the sand.

The ANGULAR CRAB is one of the European species, and in many respects bears some resemblance to the preceding species. The eyestalks of this crab are also long and movable, the carapace is wider than long, and the legs of the male are nearly five times the length of the carapace; in the female they are only twice the length. The Angular Crab is taken off the southern coasts, and is either dredged out of rather deep water or found within the stomachs of fishes. It is a burrower, forming excavations in hardened mud, and always having each extremity of the habitation open. A Mediterranean variety of the same species prefers to live among rocks, and is a good swimmer, frequently coming to the surface of the water, but not being known to frequent the land. The claw-legs are of great length, and the claws themselves are large and powerful. Its name of Angular Crab is given to it in allusion to the shape of the carapace.

IN the next family, of which the PAINTED CRAB (*Grapsus pictus*) is a good example, the eyestalks are very short, and the carapace is squared. The members of this family are found in nearly all warm parts of the globe, not, however, being natives of the European coasts. Now and then a FLOATING CRAB (*Planes linneana*) is swept into the seas together with masses of the well-known Gulf-weed; but its presence is purely accidental, and cannot entitle it to rank among the European species.

The Painted Crab is a native of the Antilles, and is a very active as well as beautiful species, haunting the sea-shore and running about nimbly in the spray. It is a good climber, and can ascend or descend nearly perpendicular rocks, provided that they are washed by the waves. Some species of this genus prefer the mouths of tidal rivers, and remain mostly at the edge of the water. They seem to rejoice in the hottest rays of the tropical sun, and run about nimbly hither and thither, with the sunbeams flashing on their wet bodies. They are all wary and timid beings, betaking themselves to the water on the least alarm, and flinging themselves into the waves with such force that their flat bodies skim for some little distance over the surface, much as the flying squirrel skims through the air between two trees. While running along, they strike their claws against each other as if for the purpose of menacing their pursuer, and when a number of these crabs are startled in one locality, the clatter which they make is surprising. The color of the Painted Crab is reddish, covered with spots and variegations of yellow. It is not at all a large species, the carapace being seldom more than two inches in length.

The Painted Crab resembles closely a species that inhabits the waters of the Gulf. At Fort Jefferson, Tortugas Islands, we were continually amused by the actions of these Crabs. They were called Spider Crabs, not from their slenderness, but from a resemblance to the Hunting or Zebra Spiders (*Salticus*), and particularly from the singularly furtive movements that characterize that group of spiders. Being amphibious, they would be found usually upon the brick walls or piers of the fort. When approached they suddenly flattened themselves closely against the surface, and their antennæ or eyestalks moved quickly, as we have seen those of the above named spiders.

Two remarkable species of Crabs are the CRESTED and the ARMED CRAB.

The former, a curious animal, inhabits Japan. Most of the species of this genus are found in the hot parts of the world, such as the Sandwich Islands, the Mauritius, and the West Indies. The whole shape of this creature is strange in the extreme, its carapace being covered with all kinds of tubercles and spines, and edged with saw-like teeth. Even the claws are covered with unexpected spikes and tubercles, and when folded in front of the body, assume a very crest-like aspect. The creature instinctively makes use of its extraordinary shape for the purpose of concealment, and when it is alarmed, it tucks its legs away under the broad carapace, folds its claws over its front, and remains perfectly motionless in spite of all annoyances. A sailor has been known to find one of these crabs on the sea-shore, to take it for a curious stone, and so to put it in his pocket. Some time afterwards, when he had laid down the supposed stone, he was not a little surprised to see it put forth a number of legs, and run away at best speed.

The name of Crested Crab is given to this species in allusion to the form of the closed claws.

The Armed Crab is also a native of Japan and China, and belongs to the same family as the preceding animal. The chief peculiarity in this creature are the four sharp spines with which the carapace is armed, those at the side being of very great dimensions, each measuring half the length of the body. The claws are sharp and powerful, and are formed in a manner somewhat resembling the same members in the crested crab.

Allusion has more than once been made to the power of voluntarily throwing off a limb, a faculty which is inherent in all the crustacea, but in some species is prevalent to a wonderful degree. The land-crabs, for example, will always sacrifice their best claw as a means of purchasing safety, and seem able to part with almost any number of legs without feeling the loss. If, for example, a land-crab or mud-crab be taken up by the legs, it suddenly shakes itself loose, leaving in the captor's hands the limbs which he has grasped, and making off with the remainder. The animal always throws off its limbs at one of the joints, seeming to achieve the feat by a sudden muscular contraction, like the movement which shakes off a blind-worm's tail, snaps away the wings from a flying ant, or breaks up the whole anatomy of a brittle-star into fragments. If the limbs be cut or severed between two of the joints, there is a flow of blood, and the creature seems to feel the injury acutely. It soon, however, heals itself by shaking off the injured portion at the joint immediately below the wound, and then seems to recover itself from the shock. This faculty is very needful to creatures who depend upon their claws for obtaining food, and who are so quarrelsome in disposition. As has already been mentioned, the crustaceans fight terribly, and in those cases where the combat is not *à l'outrance*, both parties have usually to deplore a limb or two crushed in the nippers of the opponent. Were no means provided for replacing the injured members, the poor creatures would die of starvation, as would an elephant if deprived of his proboscis, or a lion whose feet had been cut off and teeth drawn.

Every injured limb, therefore, is at once discarded at some joint, no bleeding takes place, and the stump heals almost immediately. After a short time, a little button seems to be protruding from the joint, and before many days have passed, a very small but perfect claw is seen to protrude. This new member grows regularly though slowly, and so in process of time the creature is re-supplied with its full complement of limbs. Every one has noticed the frequent inequality in the size of lobsters' claws, how one side is armed with a huge weapon nearly as large as a man's hand, while the other can only boast of a puny, soft-shelled claw an inch or so in length. This inequality is the result of some injury that has been inflicted on the limb from which the little claw has sprouted, and in almost every instance the original claw has been lost in battle. After the moult, and the induing of a fresh suit of armor, the growth of the new claw proceeds more rapidly.

It must be noticed that this power of reproduction of a lost or injured members always denotes that the creature possessing this capability is not very highly organized. Very few of the vertebrates, and those mostly belonging to the reptiles, are able to reproduce a lost member, and even in these few instances, the restorative power is very limited. A very few examples have been recorded where a limb has been lost and replaced, but such phenomena are extremely rare, and can only be looked upon as variations from the usual system.

The faculty of avoiding danger by closing all the joints of the limbs and merging them as far as possible under the carapace, is carried to a wonderful extent in the TORTOISE-CRAB, a crustacean that derives its popular title from its general similitude to the reptile from which it derives its name. There are, indeed, many of the tortoise tribe which are not able to enclose themselves nearly so perfectly as does this crab, and excepting the box-tortoise, there is perhaps none that exceeds it in the very perfect concealment of all vestige of their limbs. The carapace is wide, flattened at the edges, and dome-like in shape, so as to afford a perfect cover to the limbs. Owing to the manner in which these crabs conceal their limbs, Cuvier called them by the appropriate name of Cryptopods, or Hidden Feet. In all of these creatures the carapace is domed or vaulted, so as to form a shelter for the legs, while in the typical genus, the claw-feet are very large and compressed, with a decided upper edge which is notched or toothed so as to form a crest. The French know these crabs by various names, such as Migranes, Coqs de mer, and Crabes honteux.

The claws are broad, flattish, notched at the edge, and scooped in a peculiar fashion, so that when folded over the body they exactly fit to the shell, as if they were part of the same piece. Two sides of this species are given in order to show the crab in its upper and under aspects. The Tortoise-crab is a native of the Mauritius.

Crested and Armed Crabs, and the Tortoise-crab (*Oamara*), are found on the Florida Reef. The latter reminds one strongly of the box-tortoise, its parts shut so admirably together. The creature seen from above, appears when at rest or alarmed, as if there were no limbs.

IN the family of the Leucosiidæ, the carapace is more or less rounded, and projects somewhat in front.

The URANIA-CRAB is an example of the typical genus. It has a smooth carapace with rounded edges, and the claw limbs are very large and powerful in proportion to the size of the body. On their edges they are covered with rounded tubercles, and one or two of these projections are scattered upon the surface of several joints. In all these crabs, the apertures through which water passes to cover the gills are in the form of canals. One curious characteristic of these crustaceans is that the external antennæ are very small indeed, and are inserted in a narrow but deep notch near the eyes. In common with nearly all this genus, it is a native of the Philippines.

A much smaller species of the same genus is called SPOTTED LEUCOSIA (*Leucosia hermatostica*). In it are observable the same characteristics which have already been mentioned, namely, the very small and narrow snout, the round and flask-like body, the strong claw-feet, and the very small size of the external antennæ, which cannot be seen from the upper surface. A very strange looking creature is the KEELED CRAB. It derives its name from the form into which its carapace is moulded, being pinched, as it were, into a kind of keel throughout its length.

In this crab the chief points of interest are the long arms and the apparently shapeless carapace, which is moulded as if squeezed out of clay by a single grasp of the hand, and the very long claw-feet. In consequence of this latter structure, the Japanese call one of the species of this genus, the Tenkô-gani, or Long-handed Crab. In this creature, as in the last, there is no appearance of external antennæ when viewed from the upper surface, the claws are feeble in proportion to the long and somewhat powerful limbs to which they are attached, and the end of the carapace is drawn out into a long and sharp point. This species inhabits the Philippines.

THE Nut-crabs are members of the same family, and are found off the British shores. PENNANT'S NUT-CRAB is generally to be caught in about fifteen fathoms of water. It is rather a sluggish and inactive species, burying itself in the sand or mud at the bed of the sea, much after the fashion of the toad, and only leaving its eyes and claws at liberty to act. Thus it sits and waits for prey, behaving much like the ant-lion in its pitfall of sand. It seldom moves about except at night, and even when it does travel, its motions are very slow and deliberate. It is rather a pretty little crab, being of a tolerably bright yellow, with a red patch on the snout.

This, and other species of the same genus, are often found in the stomachs of marine fish, and as the shell is very hard, specimens are discovered in a good state of preservation.

In the course of the preceding pages we have seen many instances of curious structures which seem to be wholly supplementary and of no use whatever to the creature. The very fact of their existence is a proof that there is some use for them, although their office is so obscure as to elude all present researches. Such a crab is the IXA CYLINDRUS. On each side of the body is a large cylindrical projection, so that the extreme measurement from side to side is nearly equal to three times the length of the body; the claws are long and feeble, all the force seeming to be thrown into the two projections. This crab is a native of several parts of Asia, and is found both in India and off the Philippines.

Another odd-looking crab, having its carapace scooped and grooved in a wonderful manner, is the *Nursia plicata*. It is found in the Indian Ocean.

But we will not forget to mention the SEVEN-SPINED CRAB, so called on account of the seven sharp points that project from the carapace. There are several large species of this genus, one of which has nine spines instead of seven. In this curious creature the arms are longer and more slender than in the preceding species, and the claws at their extremities are exceedingly delicate and feeble. This crab is a native of the Eastern seas, and is generally captured off the coasts of India.

Our next example is the MASK CRAB. It buries itself in the sand or muddy bed of the sea, and only permits its snout to project, with the long antennæ, so as to feel (or, as some suppose, to listen) for approaching prey, and the eyes to look in all directions for any eatable creature that may haplessly wander within reach, and the claws, in order to seize the prey when it passes within the grasp of their long and formidable hands.

The antennæ are apt to become clogged with mud, and the crab is thereof gifted with an apparatus whereby they can be perfectly cleansed. In order to effect this object, the crab bends each antennæ sideways, until it rests on the hairy base of its companion; it then draws it completely through the stiff bristly hairs, until every particle of extraneous matter is brushed away. Sometimes the Mask-crab buries itself deeply, that it only leaves the tip of the antennæ above the sand. The name of Mask-crab is given to this crustacean, because the carapace is so formed that its two waved grooves mould the surface into an obscure likeness of the human face.

In the genus of the POLISHED CRAB (so called from the smooth, shining surface of the carapace), the carapace is somewhat heart-shaped, and very narrow behind; the claw-feet are small and short. Although this is one of the European species, it is not very commonly found, probably on account of its habit of burying itself rather deeply in the sand, so that the eye cannot perceive it, and the dredge passes over its sunken body without sweeping it into the net. It is rather a pretty little crab, though unfortunately its beauty is only skin-deep, and perishes after death. When living and in good health, the carapace is of a soft rose color, and has a very pleasing appearance; but when the shell is emptied of its contents, or even after the death of the inhabitant, the pink hue rapidly fades into the dull grays so prevalent among dead crustaceans. In the Mediterranean the Polished Crab is very plentiful.

WE now come to another family, of which the WOOLLY CRAB is an excellent type.

This creature derives its name from the coating of thick short hair with which its body is covered. All the species of this genus possess several peculiarities; at each side of the shell,

HAIRY CRAB.—*Dromia lator.* (See page 456.)

and just at the base of the claw-legs, is an aperture that looks as if it had been cut for a button-hole, and partly closed with a membrane. These apertures are in fact the openings through which the water passes for the purpose of supplying the gills with the needful moisture, and allowing it to escape when it has performed that office.

The limbs are very remarkable, both as to their shape and their disposition; they are unequal in size, and the two last pairs are elevated on the back in a very curious fashion. At their extremities is a large hooked nail, which is jointed to the limb, and can be folded back so as to take a firmer grasp. The Woolly Crab seldom approaches the shore, but prefers the deeper waters, and is only to be caught by a dredge with a very long line. On account of its locality very little is known of its habits, though much is conjectured. It is mostly found in the Mediterranean.

The SCALLOP-CRAB has derived its name from its habits. The general shapes of this crab are not at all unlike those of the pea-crab, which has already been described; and the curious analogy that exists between form and habits, cannot but strike every one who has an opportunity to see the two creatures.

IN the sub-order which now comes before our notice, is seen a modification of structure which evidently forms one of the connecting links between the crabs and the lobsters, or, to speak more accurately, between the short-tailed and long-tailed crustacea. The two large divisions of the body bear scarcely any ordinary proportion to each other, the abdomen being exceedingly small, and the "cephalo-thorax" enormously large. Some of these creatures extend the abdomen from the body like the lobsters, while others bend it under them like the crabs. In some species, of which the common hermit-crab is a familiar example, the last pair of legs are totally useless for walking, and are modified into a pair of appendages, by means of which the animal is enabled to grasp with a hold so firm, that it may often be torn asunder rather than be forced to loosen its gripe.

The next descriptions refer to a curious species belonging to the typical genus of the first family. In all these crustaceans the body is rather globular, and the carapace is bent downwards in front. The eyes are short. One of these crustaceans, called the HAIRY CRAB, is found in the hotter seas, and has been captured off the Cape of Good Hope. The two hinder pairs of legs are very small in proportion to those limbs which are evidently intended for progression. They are furnished at their tips with a hooked claw. These modified and apparently stunted limbs are, however, extremely useful, their office being ascertained by studying the economy of the animal. With the claws at the end of these limbs the crab seizes pieces of sponge, shells, and other marine substances, so as to conceal its form under their shelter, thereby exhibiting a curious analogy to the well-known habits of the tortoise beetle while in its larval state. Some crabs are admirable examples of this peculiarity, as, for instance, the *Dromia lator*, which has been chosen for an illustration. It is drawn as being nearly hidden under the mass of sponges under which it lies concealed, the sponge being nearly as large as a man's fist, while the crab is about the size of half an ordinary walnut.

One species of this genus, called, from the shape and mouldings of the carapace, the DEATH'S-HEAD CRAB, is found among the Channel Islands of Europe. The scientific name of this crab is *Dromia vulgaris*. Its color is deep brown, changing to pink upon the claws, the carapace is strongly knobbed above, and the edge is notched so as to form four broad teeth. Some species of this genus are thought to be poisonous, but without any apparent reason. The hairy covering is not so extensive in other species, for the carapace of the common Death's-head Crab is quite smooth and polished, the hairs being restricted to the limbs, where they afford an excellent basis for sponges, corallines, and zoophytes.

THE BEARDED CRAB is an example of another family, in which the carapace is formed into a kind of beak, and is almost always covered with sharp spines. The fifth pair of legs are comparatively short, and are not employed in walking. In the Bearded Crab the eyes are very large and round, and the carapace is covered with short but sharp spines. The antennæ are long and the claws powerful, and are well suited for detecting and securing prey. The Bearded Crab is found in the Mediterranean.

A strange and weird-like creature, which is called by the appropriate name of the PORCUPINE-CRAB, is a native of Japan. In this species the characteristics of the family seem to be carried to the very utmost. The last pair of legs are extremely small; so diminutive, in fact, that they are folded under the body and not visible when the creature is viewed from its upper surface. The carapace is triangular and thickly covered with spines; and even the limbs bristle with thorny points set as closely as the horny bayonets of the hedgehog. It appears to be rather dull and sluggish in its movements, crawling along the bed of the sea with slow, monotonous action.

One species of this genus, the NORTHERN STONE-CRAB (*Lithodes maia*), is found off the European coasts, and is plentiful on many of the Scottish shores. It is covered with short, thick spines which extend over its legs and claws, and in its general shape bears some resemblance to the spider-crabs already described. In spite, however, of its thorn-studded surface, it is much eaten by fishes, and is not unfrequently found entire in the stomachs of the fish that are taken off European coasts.

The Northern Stone-crab may be known by its very long beak, furnished at the end with two short and rather diverging teeth, and by its bright scarlet color when it is first taken from the water. The Porcupine-crab, which has the characteristics of its genus almost exaggerated, is found in Japan.

A species, *Lithodes arctica*, found in the northern Atlantic waters, is surprisingly armed with spines. It is large, measuring ten or twelve inches across. Specimens are brought up from the waters on Grand Banks, from which source we have received specimens.

The last of the Homolidæ is an uncouth-looking creature which is called the NODULED CRAB, on account of its singular conformation.

This crustacean, instead of being covered with thorny points, as in the stone-crabs, has its entire carapace, limbs, and claws so studded with tubercles, that it can scarcely be recognized as a living creature, and looks more like a rough stone encrusted with marine growths. The carapace is rather triangular in form, but its sides are so scooped into hollows and projections, its surface so moulded into elevations and depressions, and its shell so covered with tubercles of various shapes and sizes, that its true proportions are not easily distinguished.

The claw-limbs are large and powerful, and are even more obscure in shape than the body, for the substance of the shell is thrown out into such a forest of tubercles that at first sight it seems to be covered with a very fertile crop of fungi, algæ, or the thick and fleshy molluscoids which spread so rapidly when once they have obtained a resting-place. Even the antennæ of this strange animal are furnished with long projecting points, and look something like the beautiful comb-shaped antennæ of the larger moths. The Noduled Crab is found in the Columbia River.

THE next family is a very small one, and is called Raninidæ, from the fancied resemblance which its members bear to the shape of a frog. In these crabs the carapace is something like the half of a jargonel pear, from which about half an inch has been cut at each end. The broader end is towards the front, and is scooped so as to form a number of tooth-like projections. The abdomen of these creatures is extremely small, and may be represented by about three-quarters of an inch of a French bean stuck on the small end of the pear which answers to the carapace.

The limbs are moderately large, and the crab is said to leave the water and travel on land. Some persons say that it climbs to the tops of houses, but without mentioning the height of the houses or the materials of which they are composed. The claws are rather large, flattened, something of a triangular shape, deeply toothed, and with the pincers bent inwards at almost a right angle. All the legs are very close together at their bases, and the last two pairs ascend upon the back. All the Raninidæ inhabit hot countries, and are found chiefly in the Mauritius, the Philippines, and India. The TOOTHED FROG-CRAB (*Ranina serrata*) is a good and tolerably common example of this family. It is a native of the Mauritius and Japan.

It is altogether a curious-looking creature, with a broad, flattish carapace, edged in front with the most formidable-looking teeth, that hardly seem to belong to the shell, but to have been taken from the mouth of a shark and fastened artificially upon the front edge of the carapace. The legs of this creature seem quite insufficient to carry the great, broad carapace, and the abdomen is almost absurdly small. The color of the shell is very pale pink, and the spines which cover its surface are of a whiter hue, looking almost as if they had been pricked into the carapace by human means. All the points are directed forward, and have a very rough effect when the hand is drawn from front to rear. When full grown, the Toothed Frog-crab is about as large as a man's fist.

The family *Raninidæ* is represented in hot climates by singular creatures—large, frog-like bodies, with limbs reduced to short appendages that, in many instances, would seem to disappear beneath the shells.

A RATHER pretty-looking and decidedly curious crab, which is an example of another family, are termed the Hippidæ. In this family the carapace is long, rounded, and rather thimble-shaped, in most cases slightly flattened above. The abdomen is very small, and from the upper view of the body is hardly visible. In one species of the typical genus, the ASIATIC HIPPA (*Hippa asiatica*), the carapace is very round, elongated, and altogether egg-shaped, so that it would hardly be taken for a part of a crustacean. Even its color is a hue rarely seen among the shelly race, being a soft, pale yellow, very like the chrysalis case of the oak egger-moth, which, indeed, it also resembles in shape.

The color of the SYMNISTA is very pale yellow. Its claws are suddenly broad, rather sharp, and bent over at a right angle. The antennæ are long and beautifully fringed with hairs. It is a small species, only two or three inches in length.

The general shape of the OAR-FOOT CRAB points it out as allied to the preceding species, although the antennæ are not so long, and their fringe not so conspicuous.

The name of Oar-foot is given to this species on account of the curious modification by which the false feet are developed into oar-like appendages with flat blades, which serve for swimming like the hinder feet of the swimming-crabs. The last ring of the abdomen is changed into a flattened and pointed paddle. The carapace is convex and of a tolerably regular oval. This species is a native of New Holland, and never attains to any great size.

WE now come to a singular group of crabs which are remarkable for their soft and shell less tails, and the mode employed to protect them. From their solitary habits they are called Hermit-crabs, and from their extreme combativeness they have earned the title of Soldier-crabs.

The best known of these crustacea is the common HERMIT-CRAB (*Pagurus bernhardus*), which we have chosen for the accompanying colored illustration. Like all its race, the Hermit-crab inhabits the shell of some mollusk, in which it can bury its unprotected tail, and into which it can retreat when threatened with danger. The Hermit-crab usurps the deserted home of various mollusks, according to its size, so that, when young and small, it is found in the shells of the tops, periwinkles, and other small mollusks; and when it reaches full age, it takes possession of the whelk-shell and entirely fills its cavity.

Any one may find these odd crabs by watching a rock pool after the tide has gone down. There are always plenty of shells in such places, and if the observer will remain very quiet, he will see one of the apparently empty shells suddenly turn over, and begin to run along at a great pace, much faster than if it had been inhabited by its usual occupant. On the least movement of the spectator, the shell stops as abruptly as it had started into action, and rolls over as before, seemingly dead and empty.

On picking it up, the mystery will be revealed, for within the shell will be seen an odd little crab, with a body curved so as to fit exactly round the shell, with one claw small and one very large. If it be touched it retreats still farther into the shell, and defies any attempt to pull it out. Even if a claw be grasped, the creature cannot easily be withdrawn, and clings so tightly to its home that in most cases it may be torn asunder rather than loosen its hold. It is enabled to hold thus firmly by means of a pair of pincers situated at the end of the tail, and which are indeed the last pair of legs modified for that purpose. Sometimes the creature can be coaxed, as it were, out of its shell by a long and steady pull, but, as a general rule, to get a Hermit-crab uninjured out of its abode is a very difficult task.

I have often accomplished it by putting the shell upon an open actinia. The crab, feeling the tentacles of the actinia gradually surrounding its limbs, and not liking the aspect of the living gulf into which it is rapidly descending, makes the best of its way out of the shell, and can be snatched up before it has found time to recover its presence of mind. Sometimes a Hermit-crab may be captured while the inhabitant is three parts outside its shelly house, but, on the least alarm, the creature flies back to the farthest recesses of its home as if worked by a spring.

This crustacean is wonderfully combative, and will fight on scant provocation. Anything will serve for a cause of war, such as a piece of meat, a smaller crab, or a shell to which another individual happens to take a fancy. If two Hermits be removed from their houses, and put into a rock pool with only one shell in it, the combats which take place for the possession of that solitary shell are as fierce and determined as any that have taken place in the tourney or the field of battle. As with most of the crustaceans, the victor always eats his fallen foe; and even though he be bereft of a few legs, he seems to care nothing for the loss, but eats away with perfect appetite.

HERMIT-CRABS.

Even when the crabs are suited with homes, their combats are fierce, deadly, and active, in spite of the heavy shell which they drag behind them, and which seems to incommode them no more than the hundredweight of steel inconvenienced an ancient knight. They spar with great address, guarding the only vulnerable point with the large claw, and threatening the adversary at the same time with that weapon. At last one of them makes a dash, the pair grapple, the weaker is gradually overcome, the stronger pushes his claw into the failing adversary's shell, crushes his unprotected breast, draws him dying out of his shell, picks him to pieces and eats him.

To see a Hermit-crab fitting itself with a new shell is a very ludicrous sight. The creature takes the shell among its feet, twirls it about with wonderful rapidity, balances it as if to try its weight, probes it with the long antennæ, and perhaps throws it away. Sometimes, however, when the preliminary investigations have proved satisfactory, it twists the shell round until the tail falls into the opening, and then parades up and down for a little while. Perhaps it may be satisfied, and after twirling the shell about several times, whisks into it with such speed that the eye can scarcely follow its movements. Indeed, it seems rather to be shot into the shell from some engine of propulsion than to move voluntarily into the new habitation. When the number of empty shells is great, the Hermit is very fastidious, and will spend many hours in settling into a new house.

A Hermit-crab when deprived of its shell presents a most absurd appearance. It is dreadfully frightened, crawls about with a terrified kind of air like that which is put on by a beaten dog, and will put up with anything by way of a house. I have seen a very large whelk-shell inhabited by a very little crab, so small and weak that it could not drag its huge home about, and was tumbled backwards and forwards as the waves washed over the shell. It was much too small to fix itself in the mouth of the whelk-shell, as is the usual custom of Hermit-crabs, and had been forced to content itself with a hole that had been broken near the point.

This crab may be kept in an aquarium, as it is hardy, and can be fed with perfect ease. It is, moreover, less liable to fight with and kill its companions than the other crabs, probably on account of the shell, which protects the body, and renders a battle a very laborious undertaking. So that if two or three Hermits of similar dimensions are put into an aquarium, they will live on terms of armed neutrality, and if care be taken to feed them separately, they will survive for a long time. It is rather remarkable that when they become sickly, they are sure to leave their shells and lie listlessly on the stones or sea-weed. As soon as one of them is seen to act in this manner, it should be at once removed.

There is a curious notion prevalent respecting the Hermit-crabs. All the fishermen, and the sea-side population in general, firmly believe that the Hermit-crab is the young of the lobster, and that when it becomes large enough to protect itself, it leaves the shell, gets a hard tail, and changes into a real lobster. Any one who wishes to study the structure of the Hermit-crab can do so by visiting a fishmonger's shop, and looking over the stock of periwinkles, many of which are sure to be occupied by a Hermit-crab instead of a mollusk.

There are very many species of Hermit-crabs, those of the tropics being the largest and handsomest. Some of these larger species inhabit the trumpet-shell, some are found in the large turbos, and are handsome creatures, richly mottled with black and brown, and there are one or two species which live in the cone shells. These curious Hermits are shaped so as to suit the shell in which they reside, their bodies being quite flat and almost leaf-like, so as to enable them to pass freely into and out of the long narrow mouth of the shell. The CRAFTY HERMIT-CRAB is found in the Mediterranean, and among other shells which it inhabits, the variegeted triton is known to be a favorite.

The Hermit-crabs are among the most interesting of the crustaceans. The singular habit of adopting cast-off domiciles of other creatures is quite unique. The structure of the Crafty Hermit is very closely like, if not the same as that of the great Hermit of the Florida waters. These crabs select the shells found most suitable around them. The Horse Conch (*Strombus gigas*) is the favorite in the above locality; although we have seen them in Triton, which is not so abundant. One is often met with so large as to fill the largest Horse Conch, nearly one foot in length. This is an interesting feature of this animal;

its exposed claws and head being reinforced, as it were, by handsome scaled armor. These portions are extremely hard, and of a bright brick color. Once ensconced in the shell, which in its stoutness is like a castle, the armored front that Hermit presents may defy any enemy. But strategy sometimes succeeds. Once the creature unconsciously peeps too far away from his sally-port, he is outflanked, and forced to give battle outside his portcullis. Few objects are more entertaining. The great Land Hermits are especially so. They never go to the sea, but live in dry places, where they burrow under stones or logs.

The DIOGENES HERMIT-CRAB is a handsome and rather large species that inhabits Brazil and the West India Islands.

It occupies the shells of various mollusks, mostly, however, giving the preference to some large species of turbo ; and Mr. Bennett mentions that he possesses an unique shell which he found on the branch of a tree, having been taken from the sea by one of these crabs. While living, the Diogenes gives out a very unpleasant odor ; and as the crabs are in the habit of assembling in great numbers, the aggregate effect is rather overpowering. They gather together at the foot of trees or under bushes or brushwood, and even contrive to clamber up the branches or the trunks of trees, drawing themselves up by their powerful claws and limbs, and caring little for the heavy burden which they bear on their backs. They have the power of producing a curious noise, something like the croak of a frog alternating with sounds as if of drawing water through the lips.

They are active and voracious beings, and feed with equal avidity on animal and vegetable substances. They prefer animal food, such as fish, crabs, and, indeed, almost any kind of flesh, but they will greedily eat yams, cocoa-nut, and other fruits. They are timid creatures, croaking when disturbed, and seldom attempting to fight, but draw themselves smartly within their homes as soon as they feel alarmed. They are mostly nocturnal in their habits ; and as they bring a large supply of shells to land, and are very fastidious about their accommodation, a great heap of empty shells is to be seen upon the shore, and there is a continual rattle during the night as the creatures knock the shells about in their movements.

In all these creatures the larger claw is very much developed; so that when the crab has withdrawn into the shell, the claw lies over the entrance and closes it like a living door, which has the further advantage of being used as an offensive weapon. The footstalks on which the eyes are set, are moderately long, stout, and jointed, and enable their possessor to see in all directions. The color of this species is reddish-brown, spotted thickly with black.

A common species at Tortugas is the Diogenes. In an old wooden building attached to the Post, we had a room for the examination and care of natural objects. Under this building the Diogenes lived, several of them. One exhibited symptoms of restlessness, and after several days of fruitless wandering in and out of our room, he ultimately came to a halt, and a determination to scale the corner of a bookcase, where a saucer of fresh water chanced to be. With his heavy Trochas shell hanging like a soldier's equipments from his back, he reached, with tolerable celerity, the shelf, or top, of the bookcase. Here he sipped, and moistened his gills, and cautiously commenced the return. This he accomplished much as any climbing mammal would, hand over hand, with the body hanging behind. This crab repeated the feat often, apparently for the purpose of moistening its gills. It became quite tame ; would take food from the hand. It was sent north to Massachusetts, where a land Hermit is a strange thing. There it moulted twice, but though it was protected under glass, it died after a year's captivity. Several others were placed among loose packages in a box and sent north, but the strongest had battled with the others and destroyed them, leaving nothing but dried shells. A small specimen of the same had picked up a cast-off clay pipe, and this we cherished for a time as a comical example. The bodies of these Hermits are soft and worm-like.

Closely allied to the hermit-crabs, we find a very remarkable creature, called from its habits the ROBBER-CRAB. It is also known by the name of PURSE-CRAB. The habits of this creature, which will presently be described, are most remarkable, and there are several singular peculiarities of structure. The abdomen, for example, is no longer soft as in the hermits, but is covered above with strong plates, which overlap each other like those of the lobster's tail. The under surface of the abdomen is soft and membranous. This is one of the crustacea that can endure a long absence from water, and is fitted with a peculiar addition to the breathing apparatus. There are twenty-eight gills, fourteen at each side of the body, and enclosed in a large hollow, which they do not nearly fill. Even when the footstalks are considered, on which the gills rest, they hardly occupy the tenth part of the hollow.

The Robber-crab is found in several parts of the Indian Ocean, is very common in Amboyna, and has been taken off the Mauritius. Mr. Darwin gives the following interesting account of this crab:—"It would at first be thought impossible for a crab to open a strong cocoa-nut covered with the husk, but Mr. Liesk assures me he has repeatedly seen the operation effected. The crab begins by tearing the husk, fibre by fibre, and always at that end under which the three eye-holes are situated. When this is accomplished, the animal commences hammering with its heavy claws on one of these holes till an opening is made; then, turning round its body, by the aid of its posterior and narrow pair of pincers, it extracts the white albuminous substance of the nut.

"I think this is as curious a case of instinct as ever was heard of, and likewise of adaptation of structure between objects apparently so remote from each other in the scheme of nature as a crab and a cocoa-nut tree. This crab is diurnal in its habits, but every night it is said to pay a visit to the sea, no doubt for the purpose of moistening its branchiæ. The young are likewise hatched and live for some time on the coast. These crabs inhabit deep burrows, which they excavate beneath the roots of trees, and here they accumulate surprising quantities of the picked fibres of the cocoa-nut husk, on which they rest as on a bed. The Malays sometimes take advantage of their labor by collecting the coarse fibrous substance, and using it as junk."

In the missionary voyage of Messrs. Tyerman and Bennett, a very spirited account is given of these crabs, and one or two interesting details are mentioned. For example, when the crab walks it raises itself well off the ground, standing nearly a foot in height, and gets along quickly, though with a clumsy and stiff gait. The antennæ are very sensitive, and it is said that if they are touched with oil, the creature immediately dies. Another mode of opening the shell is employed by these crabs besides that which is mentioned by Mr. Darwin, for, according to Messrs. Tyerman and Bennett, the crab, after tearing off the husk, insinuates the smaller joint of the claw into one of the holes at the end of the nut, and then beats the fruit against a stone until the shell is broken.

This crab is by no means handsome, but is a very large and remarkably shaped creature. A fine specimen, when stretched out at length, will measure between two and three feet in length, and as it is stout in proportion to its length, it may rank with some of the largest of the crustaceans. The abdomen is of a curious form, and is evidently one of the structures intermediate between the crabs and the lobsters. Its general color is pale yellowish-brown, and its limbs are covered with little projections of a nearly black hue.

During the day the Robber-crab mostly hides in the fissures of rocks, or in holes at the foot of the trees, and in the evening issues from its concealment to prey upon the cocoa-nut. Its wonderful skill and power in opening this huge fruit have already been mentioned, but some writers give it credit for more extensive qualities, and say that it is in the habit of climbing up the palm trees for the purpose of obtaining the fruit. The particular palm which it is said to climb is the *Pandanus odoratissimus*.

It appears to be fierce in proportion to its strength, and Mr. Cuming found that if intercepted in its passage, it at first tried to intimidate its disturber by holding up the claws and clattering them loudly; and that even when it found itself obliged to give ground, it retreated with its face to the enemy, still maintaining a threatening attitude. The eyes of the Robber-crab stand on rather long but stout footstalks.

WE now come to the Porcelain-crabs, so called because their shells are smooth and polished as if made of porcelain, and have much of the peculiar semi-transparent gloss of that manufacture. The specimen shown in the engraving is of natural size.

Several of these crabs are natives of the European seas, among which we may mention two species. The first of these is the common BROAD-CLAW PORCELAIN-CRAB, so called from the singular width and flatness of the claws, each of which is nearly as large as the whole body. Altogether this is a flat crab, and, like all flattened beings, is formed for a life under stones or in narrow crevices. It may be found easily enough by going to the very verge of low-water-mark, and quickly turning over the loose stones which lie piled upon each other by the waves. Under these stones lies the Broad-claw, flat and quite at its ease, its great claws fitting beautifully into its shell, much like the same members in the domed crabs.

PORCELAIN CRAB.—*Porcellana platycheles.*

The food of the Broad-claw consists mostly of animalcules, which it catches by making regular casts with its hair-covered jaw feet, and sweeping its prey into its mouth by the action. The mode in which the hairs or bristles are set upon the foot is very beautiful, and is exactly calculated to act as a net, which will sweep up every object that crosses its path. Several of the terminal joints of these jaw-legs are edged with long and slightly curved hairs set nearly at right angles with the joint. It follows, then, that when the limb is flung out nearly in a straight line, these hairs diverge; but that as the limb is bent while being withdrawn, the hairs become nearly parallel to each other, some of them cross, and form a very complete net-work of stiff bristles that sweep everything before it. Moreover, each separate bristle has a double row of still smaller hairs, projecting from each side, something like the vanes of a feather, and nearly touching those of the next hair on either side. This structure is evidently intended to ensure the capture of the very minute animalcules, which might be able to escape through the comparatively large meshes formed by the bristles.

Though the Broad-claw loves to hide in this manner, and remains so quiescent, it is fully able to move about, and can dart through the water with astonishing celerity, flinging out the abdomen, and giving a series of sharp flaps that urge it along just in the manner adopted by the lobster and its kin. Still, it does not attempt to swim, but merely darts towards some spot where it can find a hiding-place, and whence it will not stir for weeks together, finding in its narrow home all that it needs in the way of food.

Another European species is the LONG-HORNED PORCELAIN-CRAB, a little creature that is common in similar localities. It is much eaten by various fishes, and the codfish makes great havoc among its ranks. Both these crabs can pinch smartly with their flat but powerful claws, and, in spite of their insignificant appearance, cannot be handled with impunity. In them the last pair of legs undergo a curious modification, being very small, nearly hidden in the abdomen when not in use, and apparently objectless. They are, however, very useful limbs, being employed as brushes, and used for the purpose of cleaning the abdomen and part of the carapace from adhering substances.

WE now come to the second great division of the Crustacea, namely, those which have long and powerful tails. The lobsters and shrimps are examples of these creatures. In swimming rapidly through the water, the tail is the organ of propulsion which is employed, and a glance at its form will soon explain its use. This powerful mass of solid muscle is first stretched out to its utmost, and the fan-like appendage at the extremity is spread to its widest; the creature then closes its tail smartly under the body, so as to assume the attitude in which lobsters and shrimps are mostly brought to table.

The effect of this sudden contraction is, that the creature shoots swiftly through the water. Of course, the animal darts backwards, but so sharp are its eyes, and so true is its aim, that it can fling itself into a crevice barely large enough to contain it. Any one who wishes to see this manœuvre practised in all its force, may do so by watching the little seaside pools wherein the shrimps and prawns are accustomed to disport themselves as long as the water lasts, and where, when it dries up, they bury themselves in the sand to await the coming tide.

This shooting mode of progression is not their only means of movement. By the ordinary use of their legs, nearly all the species can crawl among the sea-weed, or upon the rocks and bed of the sea, just as an insect crawls on the ground. And, when they are balancing themselves in mid-water, and are only desirous of moving gently about, they can do so by means of the numerous false legs under the body, which may be seen moving with great rapidity. Those who are fortunate enough to possess a marine aquarium, and can keep a prawn or a shrimp in the miniature ocean, will have many opportunities of watching the easy and graceful movements of these elegant crustaceans.

THE first family is called the Galatheidæ, in honor of the beautiful and unfortunate nymph vainly beloved by Polyphemus. Several species of this family are found on the European coasts, one of which is the common PLATED LOBSTER. This is a handsome little creature, the general ground color being red, upon which are drawn a number of blue spots and streaks. Its activity does not correspond with its beauty, for, according to all accounts, it is a dull, sluggish creature, and, from Mr. Couch's observations, is "incapable of any motion but backward, and rarely rises above the bottom, where, by a laborious motion of its tail, it contrives to retreat from its enemies; but its usual progress is by creeping, and by the legs only." Yet, although it is thus tardy while crawling, it can dart backward with all the agility of its race; and if alarmed, flashes through the water with arrowy speed, and can hardly be captured or its exact direction ascertained.

The beak of this species is triangular, and armed with seven strong teeth. By these characteristics it is distinguished from another species, MONTAGUE'S PLATED LOBSTER (*Galathea squamifera*), which has a short and wide beak, cut into nine spine-like teeth. The color of this creature is greenish-brown, tinged with red. It is to be found under stones at low-water mark.

THE small but important family of the Scyllaridæ is easily recognized by the wide, flat carapace, the large and leaf-like outer antennæ, and the partly flexible tail-fan, by which the creatures drive themselves through the water. In consequence of their shape, they go by the popular name of FLAT, or BROAD LOBSTERS. The habits of these crustaceans seem to be much alike. They live in moderately shallow water, where the bed of the sea is soft and muddy. Into this substance they burrow rather deeply, so as to be entirely concealed, and only issue from their retreat for the purpose of seeking food. In all the members of this genus, the carapace is longer than wide, and the sides parallel to each other. The common BROAD LOBSTER is exceedingly plentiful in Greenland, where it forms the chief food of the Arctic auk (*Alca arctica*). The beak-like projection of its carapace is very wide, but does not project. The carapace is covered with little tubercles, and along the central line runs a series of spines. The outer antennæ are large and deeply toothed. The color of this species is brownish, covered with red marks, dispersed in a simple but very pretty pattern, which would serve as a model for embroidery, and would be particularly suitable for the heavy metallic ornamentation upon uniform coats. It is but a small species, measuring only three inches in length.

Some species of this family are eatable, and in Japan are considered as delicacies.

The Broad Lobsters are represented in the sub-tropical waters of our Southern States. In the moat at Fort Jefferson were numerous smooth, round holes of three inches diameter. Much watching failed usually to discover any living thing in them; but a vigorous spading underneath sufficed to unearth a scylla of about eight inches in length. Another species, smaller, is found in northern waters.

One species of these creatures, known by the name of the SPOTTED IBACUS, is a great favorite with the Japanese. In this genus the carapace is extremely wide, and is expanded in such a manner as to hide the feet, so as to remind the spectator of the domed crab, already described on page 443. The color of the Japanese Scyllarus is red, covered with blue points.

THE accompanying full-page illustration represents the well-known SPINY LOBSTER (*Palinurus vulgáris*), which belongs to the next family of crustaceans. In all this family the outer antennæ are very long and stout, and their basal joint is large.

The Spiny Lobster is also called the SEA CRAY-FISH, or the RED CRAB. Its claws are very small, and by no means formidable. It is mostly found on the western and southern coasts, and is caught in crab-pots, like the common lobster. Its flesh is good and well-flavored, though rather tougher and coarser than that of the lobster; moreover, the want of the claws is a drawback to its excellence, so that it is not esteemed nearly so much as the true lobster. Sometimes it is found entangled in the nets, and even upon the fishermen's lines.

The average length of this species is eighteen inches, and its weight about five pounds, when adult. Its color is purple-brown, with some irregular white spots, and its legs are reddish-white, banded longitudinally with brown. One species of this genus, *Palinurus ornatus*, sometimes attains to an enormous size, measuring from the end of the antennæ to the tail rather more than four feet.

The Spiny Lobster is abundant on the Florida Reef, and there serves as a tolerable substitute for the Lobster, which does not inhabit south of New York. It is called Craw-fish at Key West. As an edible it lacks the pleasant flavor of the former, being more like the common edible crab.

THE LOBSTER OF AMERICA inhabits from St. Lawrence River to New York State. Formerly, the specimens obtained for the markets were of good average size of eighteen inches. They are now reduced to smaller numbers, and one a foot in length of body is rare. Legislation has become necessary for the protection of this most useful and highly-prized food crustacean.

THE two next examples belong to a family called the Thalassinidæ, in which the abdomen is long, its integuments rather soft, and the carapace small and compressed on the sides.

The first one, the MUD-BURROWER, is not very often seen, as it lives in a burrow some two feet under the surface of the mud. It forces itself beneath the mud by means of the third pair of legs, and there passes the greater portion of its time. The shell of this species is very thin, and but for the enormous claw with which it is furnished it would seem quite a helpless creature.

One species of this genus, the GREAT BURROWING CRAB (*Callianassa major*), inhabiting Florida and other parts of America, forms a very remarkable burrow. Mr. T. Say, who found this creature by digging in the sand, gives the following account of its habits: "It had formed a tubular domicile, which penetrated the sand in a perpendicular direction to a considerable depth; the sides were of a more compact consistence than the surrounding sand, projecting above the surface about half an inch or more, resembling a small chimney, and rather suddenly contracted at top into a small orifice. The deserted tubes of the Callianassa are in many places very numerous, particularly where the sand is indurated by iron into the incipient state of sandstone; they are always filled up, but may be readily distinguished by the indurated walls and summit often projecting a little above the general surface."

The Mud-burrower is rather a pretty little creature, being of a soft pink hue, sometimes changing to yellow on the sides. Very soon after death these colors fade, and change into dull gray. The haddock seems to feed largely on this species, as fragments are mostly found in the stomach of the fish.

The second species is the SLOW SHRIMP, a rather curious looking creature, much resembling the common shrimp, except that it possesses a pair of large and stout claws. Its popular name is derived from the sluggishness of its movements, as it has scarcely any idea of running or swimming away if alarmed, but only attempts to escape by burrowing in the mud. If, therefore, it should be intercepted upon some harder ground, where it is not able to burrow, it exhausts all its strength in unavailing efforts, and is easily taken prisoner. The best way of obtaining this creature is to dig it out of the sand. It is but a small species, measuring about three inches in length. Several other burrowers are inhabitants of the European shores. One of them is the MUD-BORER (*Gébia stelláta*), a small species, measuring about two inches in length. This creature often takes possession of the burrows which have been made and forsaken by the razor-shell, but it is doubtlessly able to bore holes for itself. It is rather a pretty little crustacean, being of a pale yellowish-white, covered with very little

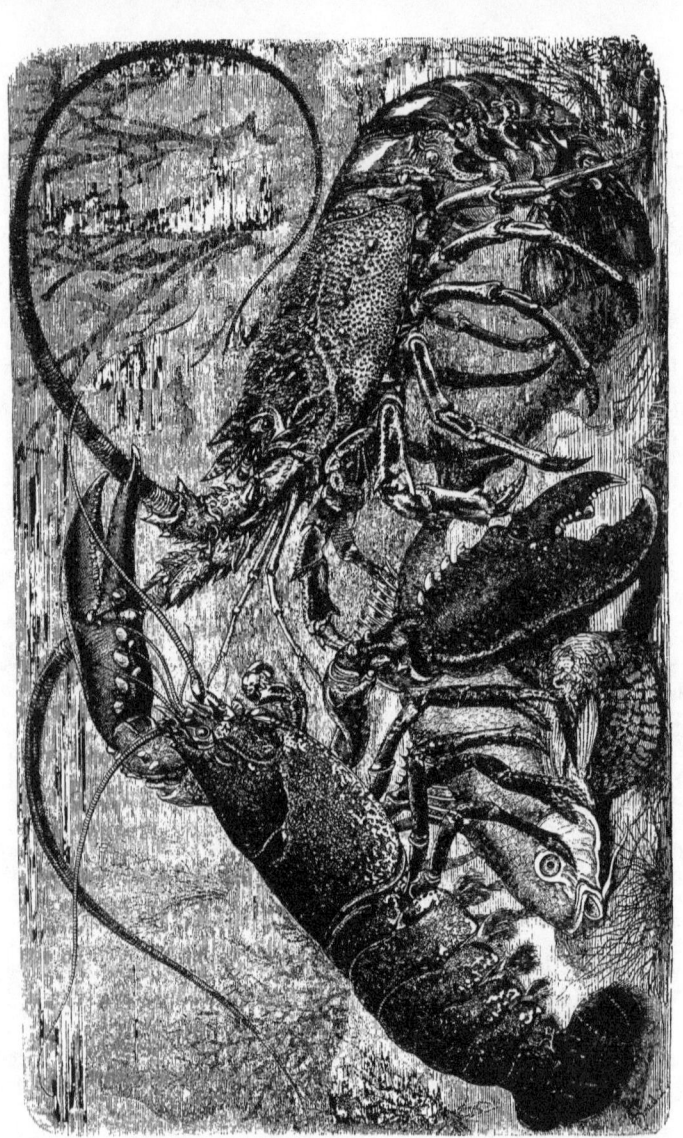

LOBSTER AND SPRING LOBSTER.

star-shaped orange spots. On the front of the carapace are multitudes of little spines, arranged in longitudinal rows.

Another species, the DELTURA (*Gebia deltura*), inhabits the same and similar localities as the mud-borer. In many points it much resembles that crustacean, and has been thought by some persons to be the female of the same species. It furnishes abundance of food to various fishes, especially those belonging to the ray family, and its remains are found abundantly in their stomachs. It is much larger than the mud-borer. All the members of this genus have the carapace formed into a triangular beak, and the outer pair of fore-feet formed for walking. One more remarkable species of burrowing crustacea is the CALOCARIS (*Calocaris macandrei*), which resides at a very great depth, having been ascertained to live at the bottom of the sea, more than a thousand feet from the surface. Here, like the rest of these creatures, it burrows in the mud, passing a kind of sub-marine mole-like existence. As, at this great depth, and under the mud, the ordinary visual powers would be of no avail, the creature has but the rudiments of eyes, which are small and quite without coloring. The Calocaris is mostly to be obtained from the stomachs of haddocks, rays, and flat-fishes.

The color of this curious species is delicate pale rose while living, but, as is usual with this fleeting tint, it soon fades after death. The shell of the Calocaris is very delicate and thin, and the whole of the feet are covered with hairs.

WE now come to the family of the Astacidæ, which includes two well-known and very similar creatures, the fresh-water cray-fish, and the salt-water LOBSTER. The latter is illustrated on the preceding full-page illustration, together with the Spiny Lobster. The Lobster is not much of a rover, seldom straying far from the spot on which it was hatched. It is rather remarkable that Lobsters are liable to permanent varieties, according to the locality in which they reside, and a good judge will be able to determine at a glance from what part of the country any given Lobster has been taken.

Sometimes a green specimen is brought to market, and the salesmen have a theory that it has obtained this change of color by living in some spot where the ores of copper impregnate the earth. They consequently believe it to be poisonous. Both ideas, however, seem to be groundless.

Lobsters are always sold by number and not by weight, and their value is necessarily dependent on the accurate eye of the dealer. The Lobsters are caught in creels or pots, like the crabs, but with greater ease and economy, as they are very fond of meat, be it fresh or tainted, and even if it should be putrefying will be attracted to it. Bright and shining objects seem quite to fascinate the Lobster, which will enter a "pot" even though the bait be nothing more than a number of empty oyster-shells placed so as to exhibit the shining white of the interior. A few years ago a curious bait was employed with great success. It was very simple, consisting of nothing more than a common phial bottle, silvered on the inside. This was hung in the lobster-pots, and served to attract the creatures to the bait. It has been suggested that the potency of this strange allurement may be attributed to its resemblance to the phosphorescent shining of putrid animal substances. But it is quite as probable that the glittering object may serve simply to attract the Lobster's attention, and that when it has approached in order to satisfy its curiosity, it perceives the bait, and immediately enters the trap. It is found that both bait and bottle are required, as if the latter is used alone, the Lobsters discover their mistake and quit a spot where they find no food.

Like many other crustaceans, the Lobster is a most combative animal, quarrelling on the slightest pretext, and fighting most furiously. In these combats it mostly loses a claw or a leg, being obliged to discard entirely a wounded member. A fresh leg or claw sprouts from the scar, and it is to this circumstance that the frequently unequal size of Lobster-claws is owing. Lobsters, indeed, part with these valuable members with strange indifference, and will sometimes shake them off on hearing a sudden noise. It is said that the commanders of certain preventive sloops were accustomed to levy a tax upon the Lobster-fishermen, threatening that unless a certain number of Lobsters were furnished to them they would fire cannon over the Lobster-grounds and make the creatures shake off their claws.

If the fishermen find that they have wounded a Lobster, they have recourse to a very strange but perfectly efficacious remedy. Supposing one of the claws to be wounded, the creature would soon bleed to death unless some means were taken whereby the flow of blood may be stopped. The method adopted by the fishermen consists in twisting off the entire claw. A membrane immediately forms over the wound, and the bleeding is stopped. The new limb that is to supply the place of that which was lost, always sprouts from the centre of the scar.

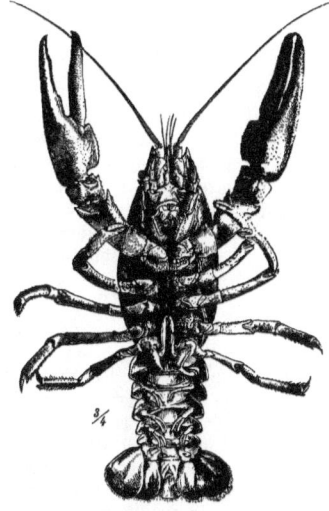

CRAY-FISH, OR CRAW-FISH.—*Astacus fluviatilis*.

The accompanying illustration shows the common Cray-fish, or Craw-fish (*Astacus fluviatilis*). This species has an almost exact resemblance to the marine lobster, which it resembles in many of its habits and qualities. Like that creature, it hides itself in some crevice, and does not issue from its concealment except for the purpose of obtaining food. It is equally quarrelsome, and also displays many tokens of its combats in the shape of lost or minute members. It is quite a rare thing to find a large Cray-fish with both its claws of the same size. The illustration is three-quarters of the natural size.

This creature mostly hides under stones or holes in the bank, sometimes partially scooped out by the inhabitant, but mostly being the deserted tenement of a water-vole. Herein the creature sits, with its head towards the orifice, and its claws thoroughly protecting its home. Even the sharp spikes of the head form no inconsiderable protection, for, if the hand be thrust into a hole tenanted by a Cray-fish, a sensation is perceived as if the fingers had been pushed against a quantity of needle-points. From these dens it issues in search of prey, which consists of dead fish and any similar substances.

Cray-fish can be caught in various ways. There are large "pots" or "creels," made of wickerwork, into which the creature is enticed by a bait, but out of which it cannot escape. There are Cray-fish nets, by which many hundreds can be caught in an afternoon. These are simple circular nets fastened inside an iron hoop and having a piece of meat tied in the centre by way of bait. A long string is attached to each net, and a forked stick, something like a clothes-prop, used for laying or taking them up. The fisherman always has several dozen of these nets, which he disposes along the river-bank in the spots which he thinks best suited to Cray-fish. By the time he has laid his last net, he must visit the first, which he pulls up quickly, and in which he mostly finds three or four Cray-fish eagerly eating the bait. The net is then replaced, and he proceeds to the second. On an average, each net produces three Cray-fish every round.

The flesh of the Cray-fish, is something like that of the lobster, but far more delicate and without the indigestible qualities of the larger crustacean. It is only in season for a comparatively short time, and in the other months of the year the flesh is soft, watery, and flavorless.

THE next family includes the true Shrimps, and contains but one genus. The Shrimp, which is so familiar on our tables, and which, until the marine aquaria became so common, was equally unknown in its living state, inhabits the shores of England, where it is produced in countless myriads. In every little pool that is left by the retiring tide, the Shrimps may be seen in profusion, betraying their presence by their quick, darting movements as they dash about in the water and ever and anon settle upon some spot, flinging up a cloud of sand as they scuffle below its surface, their backs being just level with the surrounding sand. In consequence of this manœuvre, the fishermen call them "sand-raisers." The small prawns are often confounded with the Shrimps and popularly called by the same title. They can, however, be easily distinguished from each other, the beak of the prawn being long, and deeply saw-edged, while that of the Shrimp is quite short.

While living, the Shrimp wears tints so exactly like those of the sand, that when it is lying motionless, it harmonizes exactly with the tawny bed of the sea, and cannot be discerned except by a practised eye. When boiled, it does not change to so bright a red as is usually the case with eatable crustacea, but assumes a duller and more opaque hue. During life the Shrimp is a most beautiful creature, nearly translucent in many points, and when seen against the light seeming to possess some inward illumination. Its habits are interesting, and can be successfully watched by means of an aquarium, though it is necessary to bestow some care on the creature, and keep it properly supplied with food, as it is, though so delicate, a very voracious animal and requires much feeding.

Shrimps are caught for sale in a peculiar wide and purse-like net set crosswise upon a pole, and pushed along the sand at the depth of about two feet or a little more. By this method of procedure great numbers of Shrimps are gathered into the net as they dash along the sand, and together with them are various other inhabitants of the sea, quite useless to the shrimper, but very valuable to the seaside naturalist. Any one who is studying the habits of the marine animals will do well to pay a shrimper for the right of examining the net and retaining whatever is useful or interesting. The method of burying itself in the sand is by using the hinder legs as scoops, settling into the small hollow made by them, and then flinging the sand over its back with its antennæ.

There are several species of true Shrimps, all good for food, and, although comparatively scarce, taken together with the common Shrimp. There is the BANDED SHRIMP (*Crángon fasciátus*), known by the narrow and rounded abdomen and the brown band that crosses the fourth ring. It is about an inch in length. It seems to be rather a rare species. Another Shrimp is called the SPINY SHRIMP, on account of five rows of teeth-like points upon the carapace. It is of a rather light brown color, banded and striped above with grayish white, and spotted below with crimson. Another species, BELL'S SHRIMP (*Crángon sculptus*,) is very small, being little more than three-quarters of an inch in length. There are several raised lines on the carapace, each with a few small teeth. In color it is extremely variable, but is mostly drab, with little black spots and chestnut specks, and is adorned with patterns of pale brown edged here and there with blue.

THE SHORT-BEAKED RED SHRIMP belongs to the family of the Alpheidæ. This is a Japanese species. Its carapace projects over the eyes in a hood-like shape, and the beak is very small, sometimes indeed being altogether absent. The first pair of legs are always very stout and strong, and one claw is much larger and more powerful than the other. The greater number of the species belonging to the genus Alpheus live in the tropical seas, and those that have been found within the waters of moderate climates have clearly resided at a considerable distance from land. One species, for example, EDWARDS' RED SHRIMP (*Alphéus ruber*), has been found in the stomach of cod-fishes, mostly in fragments, but very rarely entire. Another species, the SCARLET SHRIMP (*Alphéus affinis*), is of a deep scarlet except the claws, which are marked with yellow. This Shrimp has been taken in the Channel Islands.

Another species, MONTAGUE'S SHRIMP (*Athanas nitescens*), is popularly thought by the fishermen to be the young of the lobster, its deep green color and large pincers giving it a great resemblance to that crustacean. It is a sociable little being, congregating in some favored spot and assembling in considerable numbers. In fact, it is seldom found alone; and in clearing out a little sand pool, six or seven may often be found in close companionship.

WE now arrive at the PRAWNS, a family which is easily known by the long and saw-edged beak that projects from the carapace. This family is very rich in species, many of which are most lovely creatures, resplendent in scarlet, azure, green, purple, and orange, and of a beautiful transparency, which gives double effect to the colors with which they are adorned.

In the RING-HORNED PRAWN the beak is extremely long, and slightly turned upwards.

While it lives at some distance from the shore, it cannot be captured in the ordinary shrimp nets. The fishermen call it the Red Shrimp. The spines, or teeth in the upper edge of the long beak, do not spring at once from the substance of the beak, but are simply jointed to it, so that they can be moved slightly by pressure. A large number of species belong to the genus Hippolyte. In these creatures the beak is very large and strong. Several of the Æsop Prawns belong to this genus. They derive their popular name from the hump-like manner in which the abdomen is raised towards the centre and then bent downwards. COUCH's ÆSOP (*Hippolyte couchii*) is perhaps the most common of these beautiful little creatures, and has the characteristic hump strongly defined. It may be found plentifully in the shore-pools, flitting about the water with a movement much like the flirting and fluttering of a robin in a garden, and displaying its beautiful colors to the best advantage. It is a lovely little being, very variable in color, but always marked with bright and peculiarly pure hues, mostly white, purple, and scarlet. Many of these Æsop Prawns are charming inhabitants of an

EDIBLE PRAWN.—*Palæmon serratus.*

aquarium, their pellucid bodies and beautiful colors making them fit inhabitants of the drawing-room or the conservatory. One species, WHITE's ÆSOP PRAWN (*Hippolyte whitei*), is an especially beautiful creature, being green with a white streak running along the back, and having a number of azure specks scattered over the body.

Even the large EDIBLE PRAWN (*Palæmon serratus*), the figure of which is drawn of natural size, is a beautiful inhabitant of an aquarium. No one who has only seen Prawns on the table, red, opaque, and with their tails folded under them, can form the least conception of their wonderful beauty while living. As they swim gracefully through the water, the light passes through their translucent bodies and their beautifully streaked integuments, rich with transparent browns, pinks, and grays of various depths. Their delicate and slender limbs are ringed with orange and purple, and stained with pale blue.

At night, when a lamp is brought into the room, the effect produced by the Prawn is really surprising. The large globular eyes glow as if illuminated by some powerful light within; and as the creature comes out of the darkness its eyes alone are visible, as they shine like two globes of living fire.

It is very interesting to watch the habits of this beautiful creature. It is extremely voracious, and seems always to be ready for food. I used to feed my own Prawns with the bodies of shrimps, hermit-crabs, and other marine crustacea that had died in the aquaria. All that was needful was to drop the dead animal into the water so that it should pass the spot where the Prawn had made its home. As soon as it approached, the Prawn used to dart out like a tiger from its den, its long antennæ waving in great excitement, and its forceps open and extended so as to be in readiness. The claws appear to be very feeble, but they are stronger than they seem, and are perfectly adequate to the task which they are called upon to perform. The creature would quickly grasp its prey with one claw, carry it off to its home, and there leisurely pick it to pieces, displaying considerable discrimination in choosing the most delicate morsels, and abandoning the remainder to its smaller companions who still lived in the same tank, and preserved their lives by hiding themselves in little nooks and crevices, wherein they were safe from their giant kinsman. The air of utter contempt with which the Prawn would twist off and fling aside the legs and antennæ of a shrimp or a hermit-crab was very amusing. Its greatest dainty, for which it would leave almost every other kind of food, was the soft abdomen of the hermit-crab.

The forceps employed for this purpose are those at the extremity of the second pair of feet, those of the first pair being used for a different purpose. Mr. Gosse has given the following account of those limbs and their use. After mentioning that they are covered with hairs set at right angles to the limb, like the bristles of a bottle brush, he proceeds as follows:—
"These are the Prawn's washing brushes, especially applied to the cleansing of the under surface of the thorax and abdomen. When engaged in this operation, the animal commonly throws in the tail under the body, in that manner which we see assumed in the finest specimens that are brought to table, which is not, however, the ordinary position of life, the body being nearly straight. Then he brings his fore-feet to bear on the belly, thrusting the bottle brushes to and fro into every angle and hollow with zealous industry, withdrawing them now and then, and clearing them of dirt by passing them between the foot-jaws.

"The reason of the inbending of the tail is manifest. The brushes could not else reach the hinder joints of the body, and still less the swimming-plates, but by this means every part is brought within easy reach. Sometimes the brushes are inserted between the edge of the carapace and the body, and are thrust to and fro, penetrating to an astonishing distance, as may be distinctly seen through the transparent integument. Ever and anon the tiny forceps of the hand are employed to seize and pull off any fragment of extraneous matter which clings to the skin too firmly to be removed by brushing; it is plucked off and thrown away clear of the body and limbs. The long antennæ and all the other limbs are cleaned by means of the foot-jaws principally."

THE SWORD-SHRIMP, a native of Japan, belongs to another family, termed the Penæidæ. All the members of this family have a very long and much compressed abdomen, and the beak very small or absent. One of them is the GROOVED SHRIMP (*Penæus sulcátus*), a common species in the Mediterranean. It has three grooves on the carapace, two long and one shorter in the middle. It is a large species, sometimes attaining the length of seven inches.

Another species is the SIVADO, sometimes called the SWORD-SHRIMP, or the WHITE SHRIMP, the last-mentioned term, however, being applied very loosely by the fishermen. It is a very beautiful little creature, being of a translucent white color, dashed and spotted with rich crimson. It is said that this species cannot endure exposure to the air, and that it dies immediately on being removed from the water.

MOUTH-FOOTED CRUSTACEANS; STOMAPODA.

ANOTHER order of crustaceans now comes before us, called the Stomapoda, or Mouth-footed Crustaceans, so called because their legs mostly issue from the neighborhood of the mouth. The gills are external, and are formed in a most curious manner of a series of tiny cylinders. The greater number of Stomapods live in the hotter seas, but a few are inhabitants of the English coasts.

Our first example of these odd-looking creatures is the CHAMELEON-SHRIMP, perhaps the most common of its kind. This species is abundant on European coasts, and derives its popular name from the extreme variability of its coloring. It seems to alter according to the locality in which it resides. Those, for example, which live upon a sandy coast are of a gray hue, those which are found among the large dark sea-weeds are brown, and those that prefer the ulva and zostera beds are green, like the vegetation among which they live. These creatures are sometimes called Opossom-shrimps, from a curious modification of their structure. The last two feet are furnished with an appendage that forms a sort of pouch. In the male this pouch is small, but in the female it is large, and capable of containing a large number of eggs, which are carried about by the crustacean just as the opossum carries its young.

In the Northern seas these Opossum-shrimps exist in vast multitudes, and form much of the food on which the great whale of those seas depends for its subsistence. Several species are thus eaten, and one of them, *Mysis flexuosus*, is largely eaten by the enormous shoals of salmon that visit these regions in the months of July and August, thereby aiding in giving to the fish that fineness of condition and fulness of flesh which ought to be possessed by a well-nurtured salmon. These creatures are fond of congregating at the mouths of rivers, probably because they find plenty of food in such localities, and during the winter, haunt the whole line of coast.

Many species of Opossum-shrimps are found upon European shores, and can be captured by the simple plan of hauling up masses of sea-weed, and seizing the little crustaceans before they can escape.

Another example of these beings is the CLUB-HORNED PHYLLOSOME, a member of another and a very remarkable family. These crustaceans are in the habit of floating on the surface of the water, extending their legs, and there lying quite at their ease. The body is beautifully transparent, and it would be almost impossible to see the Phyllosome were it not that the eyes are of a most beautiful blue, and serve as indications of their owner's presence. This species is a native of the Atlantic Ocean. The name Phyllosoma is derived from the Greek, and signifies Leaf-bodied. One or two examples of this creature have been found floating near the Channel Islands. All the members of this family have the body exceedingly flat and leaf-like, formed by the carapace and part of the thorax. The abdomen is extremely small in proportion to the enormous size of the cuirass, and the limbs are so formed that they can be spread from the body so as to present a large radiating outline. Our illustration is a true representation of a rare species of this family. It is drawn in natural size.

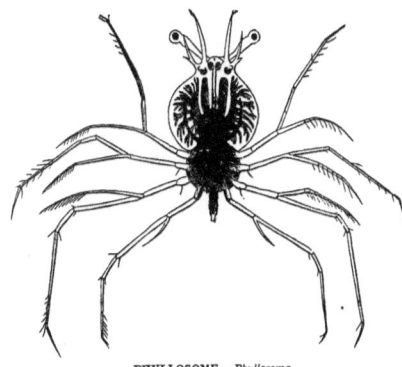

PHYLLOSOME.—*Phyllosoma*.

The two next examples belong to the remarkable genus of the Stomapod Crustaceans. In these creatures the upper part of the body is defended by a single and large cuirass, covering much of the head, being wide and free behind. The members of the genus Ericthus have the cuirass enormously developed, prolonged in front into a kind of beak, which projects over the head, and having behind several strong and rather long spines. These creatures have smaller claws than is found to be the case with the generality of the family, and all the limbs are of only moderate dimensions. The last segment of the abdomen is developed into a wide and flat fan-like blade. The eyes are large, round, and set on stout footstalks.

The GLASSY ERICTHUS derives its name from the translucency of its integuments, and the ARMED ERICTHUS is so called in consequence of the sharp spines that defend its shield. Both these species are inhabitants of the Atlantic.

BEFORE passing to the next family, we must cast a brief glance at a very strange-looking crustacean, called the TRANSPARENT ALIMA (*Alima hyalina*). This remarkable animal looks much as if an Ericthus had been drawn out like wire to a considerable extent, retaining all the characteristics of the family, and some which belong to the genus. The abdomen is extremely long, something like the tail of a scorpion, and terminated by a flat paddle. The cuirass is so large and so loose that it hardly seems to belong to the creature, but to have been taken from some larger crustacean, and dropped upon its back. The eyes are large and globular, and stand on slender curved footstalks, bearing no small resemblance to a dumb-bell with a long and rather curved handle, each eye answering for the heads of the bell, and their united footstalks for its handle. The claw-feet are long, slender, and can be used with much quickness.

These creatures are natives of the warmer seas, such as the tropical portions of the Atlantic, the South Seas, and New Guinea. They all live at some distance from the shore.

MANTIS-SHRIMP.—*Squilla mantis*.
(Somewhat diminished.)

WE now come to a curious family, called the Squillidæ. In these creatures the body is long and mostly flattened, and the first pair of legs are very large, and used for seizing prey; the last joint folding over serves to answer the purpose of a claw. The carapace is divided into three lobes. The best known of these crustaceans is the MANTIS-SHRIMP, so called from its great resemblance to the insect from which it takes its title. As will be seen by reference to the accompanying illustration, the carapace of the genus Squilla is small but long, and shields the mouth, the antennæ, and their appendages. The abdomen is very long and boldly jointed, and the appendages at its extremity are made in a manner that much resembles the fan-like tail of the lobster.

All the Squillæ are voracious, fierce, and active beings, and can strike as sharply with their long claw-feet as can the mantis with the corresponding limbs. From all appearance it seems as if the creatures were in the habit of hiding themselves in dark crevices, and from their dens striking quickly at passing prey.

This theory is much strengthened by the observations of Dr. Lukis, who kept a Mantis-shrimp alive for a short time. "It sported about, and after a first approach exhibited a boldness rather unexpected. When first alarmed, it sprang backwards with great velocity, after which it placed itself in a menacing attitude which would rather have excited the fear of exposing the hand to it. The prominent appearance of the eyes, their brilliancy and attentive watching, the feeling power of the long antennæ, evinced quick apprehension and instinct. I brought a silver teaspoon near them, which was struck out of my hand with a suddenness and force comparable to an electric shock. This blow was effected by the large arms, which were closed and projected in an instant with the quickness of lightning."

The Squillæ are seldom seen near land, specimens being mostly taken nearly six miles at sea, where the bed of the ocean is known to be of a sandy nature. They are good swimmers, darting quickly through the water by the action of the paddle at the end of the tail. The GOUTY SQUILLA derives its name from the largely-tuberculated limbs, which look as if the animal were badly attacked with the gout. It is taken off the Mauritius.

SESSILE-EYED CRUSTACEA.

Our attention is now drawn to the second great group of crustaceans, called the Sessile-eyed Crustacea, because their eyes, instead of being placed on footstalks, are seated directly upon the shell. The body is divided with tolerable distinctness into three parts, for which the ordinary titles of head, thorax, and abdomen are retained, as being more convenient and intelligible than the ingenious and more correct, though rather repulsive, titles that have lately been affixed to these divisions of the body.

They have no carapace, like the stalk-eyed crustaceans, nor do they breathe with gills, but by means of a curious adaptation of some of their limbs. None of the Sessile-eyed Crustacea obtain any large size, an inch and a half being nearly their utmost limit in point of length. Most of these animals reside along the sea-shores, where they are of very great use in clearing away the mass of dead animal and vegetable matter which is constantly found in the sea.

AMPHIPODA.

The first order of the Sessile-eyed Crustaceans is termed the Amphipoda, a word derived from the Greek, and signifying "both kinds of feet," because they are furnished with limbs for walking and swimming; whereas, in the Isopoda, or similar-footed crustaceans, the feet are all of the same character. The females are in the habit of carrying their eggs under the thorax, mostly between certain flattened appendages attached to the base of the legs.

The next family is called by the name of Orchestidæ, or Jumpers, because they possess the power of leaping upon dry ground. The most familiar of these little crustaceans is the well-known SAND-HOPPER, or SAND-SKIPPER, seen in such myriads along sandy shores, leaping about vigorously just before the advancing or behind the retiring tide, and looking like a low mist edging the sea, so countless are their numbers. Paley has a well-known passage respecting this phenomenon, too familiar for quotation.

The leap of the Sand-hopper is produced by bending the body and then flinging it open with a sudden jerk—in fact, the exact converse of the mode of progression adopted by the lobster and shrimp. The Sand-hopper feeds on almost anything that is soft and capable of decay, and seems to care little whether the food be of an animal or vegetable nature. Decaying sea-weed is a favorite article of food, and wherever a bunch of blackened and rotting sea-weed lies on the sand, there may be found the Sand-hoppers congregated beneath it, and literally boiling out when the sea-weed is plucked up.

Wherever there is sand, the Sand-hopper is to be found, even though no traces may be perceptible; and an experienced shore-hunter will seldom fail in obtaining as many as he wishes in the space of a few minutes. Even where the sand is extremely dry and level, and seems unfit to nourish Sand-hoppers, these little creatures are often snugly ensconced beneath, having burrowed deeper and deeper as the sand became dry. If a smart stamp of the foot be given, a vast number of little holes will make their appearance, as if by magic. These are the burrows of the Sand-hoppers, which have been made while the sand was still wet, and over which a film of moist sand had formed itself. The shock caused by the stamp of the foot breaks the dried films, and the hole is at once made apparent.

To catch the Sand-hopper in fair chase is no easy task, but it can be captured without any difficulty by simply digging up the sand and throwing it aside. The Sand-hoppers seem so bewildered with their sudden change, that they merely sprawl about listlessly, and can be picked up at leisure.

The teeth of this creature are strong and sharp, as indeed is needful for the tasks imposed upon them. The Sand-hopper will eat anything; and on one occasion, when a lady had allowed a swarm of these little crustaceans to settle on her handkerchief, it was bitten to rags when she took it up. It is very fond of worms, will eat any kind of carrion, and sometimes, when pressed by hunger, has no scruple in eating its own kind. It has many enemies, as is sure to be the case when a little creature is produced in absolute clouds, when it is quite harmless, easily obtained, and excellent food. Sea-birds feed largely upon the Sand-hoppers, and many land-birds are in the habit of passing much of their time upon the shore, and eating their fill of these crustacea. The green crab is a terrible enemy to the Sand-hopper, even running it down in fair chase, as I have witnessed, and displaying wonderful ingenuity in pouncing upon the active little creature just as it descends from its leap. Even a little beetle, not a quarter its size, feeds upon the Sand-hopper, instinctively attacking it from below, where it is comparatively undefended by its shelly coat. Sometimes three or four beetles will unite in attacking upon a single Sand-hopper. The technical name of this beetle is *Cillenum laterale*.

The SHORE-HOPPER (*Orchestia littorea*) is also plentiful on sandy coasts, preferring those where the sand is sprinkled with rocks. It may be known from the sand-hopper by its more compressed body, the partly-clawed character of the two first pairs of legs, and the comparatively small size of the first pair. Though it hops on the sand, like the preceding species, and has many similar habits, it is seldom found occupying the same locality, the sand-hopper taking to one part of the coast and the Shore-hopper to another.

Another strange-looking creature is the common SAND-SCREW, an example of the next family. In these creatures the antennæ end in a lash-like point, called appropriately the flagellum, or little whip. The Sand-screw is so called from the odd movements which it makes when laid upon dry sand, wriggling along while lying on its side, and displaying an awkwardness, in this respect, which contrasts greatly with the wonderful power and freedom with which it can force its way through wet sand. In the course of its burrowings, it makes many tortuous tracks in the sand, that are generally taken for the trace of some worm's passage.

There are many fossil remains said to be the relics of certain worms, but which are now thought by Mr. Albany Hancock to have been produced by some crustacean of similar habits to the Sand-screw. He has given a most interesting account of this discovery, and the following passages are extracted from his account:—"I went down to the beach, just as the tide was leaving the spot where the broad tracks were usually in great profusion. The sand was quite smooth, all irregularities having been obliterated by the action of the water. Here and there, however, the tracks had already made their appearance, but were as yet of very limited extent, and there was no longer any difficulty in taking the whole in in one view, and, moreover, the extremities were perfectly distinct. It was only necessary to watch attentively, to note the formation of the numerous and labyrinthine windings that had been so long a puzzle.

"I had not long to wait before the sand at one of the extremities was observed to be gently agitated, and, on this agitation ceasing, the track was found to have added nearly half an inch to its length. In the course of two or three minutes, the sand was again put in motion, and the track once more a little prolonged. These movements were repeated over and over again, until it was quite clear, that the track was formed by slow, intermitting steps, and not, as might have been supposed, by one continuous gliding motion. Having satisfied myself of this, I took up the morsel of sand at the end of the track, just as it was again becoming agitated, and found that I had captured a small crustacean, the species of which was unknown to me, though in general appearance it was not altogether unlike the common sand-hopper, but not quite so long. I soon took in this way five or six specimens, all of the same species, and all forming tracks of precisely the same character, namely, broad, slightly elevated, flattened, and grooved.

"While forming its track, the animal is never seen; it moves along a little beneath the surface of the sand, which it pushes upwards with its back, and the arch or tunnel thus formed partially subsides as the creature presses forward, and, breaking along the centre, the median groove is produced."

A more slender and delicate-looking crustacean is KROYER'S SAND-SCREW, a creature which possesses some of the same habits as the last-mentioned species. It burrows horizontally beneath the sand, like the common Sand-screw, but differs in its mode of action, the back always appearing above the sand.

The LONG-HORNED COROPHIUM, a curious-looking and very interesting species, inhabits the muddy parts of the sea-shore. This creature is common in the summer and early autumn, at which times it walks boldly upon the wet shore. During the later part of autumn and the winter, it resides in holes which it burrows into the mud and clay, and in some places is so plentiful, that the mud is quite honeycombed by its tunnels. This species is very common on the French coasts, especially in the great mussel preserves near Rochelle. M. D'Orbigny, who observed their habits closely, has given a very animated account of their manner of feeding.

The whole of the muddy deposit along the shores is inhabited by myriads of marine worms, such as the nereis and lug-worm, and upon these the Corophium feeds. As the tide rises, the worms ascend to the mouths of their burrows, for the purpose of eating the little animalcules that swarm on the shore. The Corophium wages continual war against these worms, darts at them with surprising speed, fastens on them, and eats them. Sometimes a great lug-worm will be surrounded by thirty or forty of these curious crustacea, all attacking it simultaneously, and forming a strange group as the worm writhes in its endeavors to escape, and carries with it the small but pertinacious foes under whose attack it is sinking.

Hundreds of the Corophium may be seen beating the mud rapidly with their enormous antennæ, for the purpose of discovering their prey, and the energy of the movement and the evident excitement under which the creatures labor partake largely of the ludicrous. They do not restrict themselves to the worms, being equally ready to prey upon fishes, oysters, or indeed any animal substance that comes in their way. The fishermen, who know it by the name of Pernys, are very angry with this little creature, and declare that it robs them of their mussel harvest. They even assert that it climbs the posts of the complicated wood-work to which the mussels cling, cuts the silken threads by which these mollusks are attached, and, having thus let them fall into the sea, eats them at leisure. As is the case with the sand-hopper, the Corophium is greatly persecuted by larger creatures, and is eaten in vast numbers by birds and many fishes. All the members of this genus can be recognized by the enormous dimensions of their antennæ, which are extremely thick at the base, and look much more like a very large pair of legs than true antennæ.

We now come to some very curiously shaped crustacea, whose habits are fully as remarkable as their forms. Their scientific name is Phronima, and their best known species is FLEMING'S HERMIT-SCREW. This creature incloses itself in a nearly oval and transparent sac, which is found to be the body of one of the medusæ. M. Risso tells us that, like the argonauts and carinariæ, these creatures may be seen in calm weather voyaging along in their glassy boats, and rising to the surface or sinking through the water at will. They live on animaculæ, and for the greater part of the year remain in the muddy depths of the ocean, ascending to the surface in the spring. How they enter their habitations, and their general economy, are subjects at present obscure.

There are several species of Phronima, all inhabiting similar dwellings. *Phronima sentinella*, for example, chooses the bodies of the æquoriæ and geroniæ for its home. These creatures are called by the name of Hermit-screws on account of the solitary life which they lead, each shut up in its cell or cocoon, as it may possibly be called. In all the Hermit-screws, the head is large and vertical, with two little antennæ, and the body is soft, nearly transparent, and ends in a number of bristle-like appendages. All the legs are long, slender, and apparently weak, except the fifth pair, both of which legs possess a large and powerful claw, and are directed backward.

A little crustacean belonging to an allied genus is not uncommon on European coasts. It has habits of a somewhat similar nature, dwelling in the chambers within several common medusæ. It will occasionally leave this curious residence, and return to it at will. It is about half an inch in length, has the two first pairs of feet shortest, tipped with a claw, and has the three last pairs of legs longer than the others. The name of this crustacean is *Metœcus medusarum*. Mr. Spence Bates separates all these parasitic animals into a distinct family, under the name of Phronimadæ. All the members of this family have the mandibles very large, some of the legs prehensile and oddly formed, and the head of enormous comparative size. Some of them attach themselves to fishes, and others to medusæ.

Another strangely formed and closely allied crustacean is the *Dactylocera nicæensis*, whose habits are, however, very imperfectly known, though it is presumed that they resemble those of the hermit-screw and its kin. In this genus, the head, though large, is not of such enormous comparative dimensions as in Phronima, and is rather squared in form. Some of the strange and grasping legs possess great muscular development, and are armed at their extremities with formidable claws, the movable joint bending over at right angles.

A small, but very remarkable crustacean, one of the few which really construct a home for themselves, is the CADDIS-SHRIMP, scientifically called *Cerapus tabularis*. The close resemblance between this creature and the well-known caddis-worm cannot but strike an observer. All the animals belonging to this genus inhabit a case which they are able to carry about with them. In spite of the awkwardness of such an appendage, the Caddis-shrimp passes along at a brisk pace, moving by means of the two pairs of long antennæ, which not only look like feet, but are used for locomotion. The real feet are kept within the tube, with the exception of the two front pairs, which are almost wholly used for catching prey and feeding itself.

Some persons imagine that the tube of this creature is not of home manufacture, but is the deserted residence of some annelid. There is, however, no reason why a crustacean, which is much higher in the scale of creation, should not make as good a tube. The material of which these tubes are made resembles rough leather or papier-maché, and grayish-brown in color, and very tough. They are very small, in some species being not more than the sixteenth of an inch in length, and proportionately small in diameter. Sometimes the tubes are set so thickly upon the plant as to conceal its surface from view. They are set without the least order, and look as if they had been simply flung upon the sea-weed to which they adhere. The common carrageen (*Chondus crispus*), from which the well-known Irish moss is made, is the plant that is most favored by their presence. When taken out of its cell, the little animal is not unlike a sand-hopper, except that the two pairs of antennæ are enormously developed, and the first few pairs of legs are furnished with small claws.

The generic name Cerapus is taken from the Greek, and is very appropriate, signifying "horn-footed." These strange antennæ are continually flung forward, grasping at everything that comes within their reach, and reminding the observer most forcibly of the peculiar actions of the cirripedes or barnacles. The Caddis-shrimp does not love the very shallow waters, and, except by use of the dredge, cannot be obtained but at the very low tides of March and September, those precious days so invaluable to the practical naturalist, where he finds laid out before him large tracts of the ocean-bed that, except for a few days, at intervals of six months, remain covered with water, and hide their treasures from all eyes.

The accompanying illustration represents the common FRESH-WATER SHRIMP, or FRESH-WATER SCREW.

FRESH-WATER SHRIMP.—*Gammarus pulex*. (Twice natural size.)

In common with the other Screws, this creature derives its name from its movements when taken from the water and laid upon the ground. Not being able to stand upright upon its feeble legs, it is forced to lie on its side, so that the perpetual kicking of its legs only forces it round in a screw-like fashion, similar to the conduct of the marine screw-shrimp when laid on the sand.

The Fresh-water Shrimp is extremely plentiful in every stream, and may be seen in great numbers even in the little rivulets that conduct the water from the fields. They lurk in recesses in the bank or under stones that form the bed of the stream, occasionally darting out to seize some prey, and then making their way back again. Occasionally they push themselves a yard or two up the rivulet, but are sure to come floating back again before very long, allowing themselves to be passively swept along by the force of the water as if they were dead, but starting suddenly into active exertion as soon as they reach their former haunts.

In the water this crustacean moves by a series of jerks, and mostly lies on its side, though it often swims with its back uppermost, and frequently rotates as it passes along. It is a voracious creature, feeding upon dead fishes or any similar carrion. It is fond of the muddy parts of the stream, liking to conceal itself in the soft alluvium when fearful of danger. The eggs of the female are kept for some time under the abdomen, and the young remain in that situation until they have attained sufficient strength to shift for themselves.

Three other species are marine. These are the Wood-boring Shrimp, the Skeleton-screw, and the Whale-louse. The WOOD-BORING SHRIMP is a crustacean that nearly rivals the shipworm itself in its destructive powers. It makes burrows into the wood, wherein it can conceal itself, and at the same time feast upon the fragments, as is proved by the presence of woody dust within its interior. Its tunnels are made in an oblique direction, not very deeply sunk below the surface, so that after a while the action of the waves washes away the thin shell and leaves a number of grooves on the surface. Below these, again, the creature bores a fresh set of tunnels, which in their turn are washed away, so that the timber is soon destroyed in successive grooved flakes.

According to Mr. Allman, its habits can be very easily watched, as if it is merely placed in a tumbler of sea-water, together with a piece of wood, it will forthwith proceed to work and gnaw its way into the wood.

In this creature the jaw-feet are furnished with imperfect claws, and the tenth segment from the head is curiously prolonged into a large and long spine. The great flattened appendages near the tail seem to be merely used for the purpose of cleaning its burrow of wood dust which is not required for food. The creature always swims on its back, and when commencing its work of destruction, clings to the wood with the legs that proceed from the thorax. The Wood-boring Shrimp is one of the jumpers, and, like the sand-hopper, can leap to a considerable height when placed on dry land.

Another wood-boring shrimp will be described in a succeeding page.

In the illustration is seen the marine crustacea called appropriately the SKELETON-SCREW, or MANTIS-SHRIMP. The bodies of the Skeleton-screws are indeed skeleton-like in their bony lankness, but their appetites are by no means small in proportion to their size. They are furnished with terrible instruments of prehension, their first and second pairs of legs being devoted wholly to this purpose. The last joint but one is enormously large, and the last joint is thin, and shuts down like the blade of a claspknife into its haft, the groove being represented by a double row of spines between which the blade is received. The blade itself is finely notched along the edge. These claw-like terminations to the legs are used not only for seizing prey, but for grasping the branches and drawing the long attenuated body from one part to another.

MANTIS-SHRIMP.—*Caprella linearis.*

Mr. Gosse, who has paid much attention to these curious beings, remarks that their movements among the marine vegetation are wonderfully like those of the spider monkeys among the branches, their long thin bodies adding to the resemblance. They run about with great agility, and are always to be found in the branches of the *Plumatella cristata*. The same writer has given a very interesting history of the Mantis-shrimp :—

"Their manners are excessively amusing. The middle part of their long body is destitute of limbs, having instead of legs two pairs of oval clear vesicles, but the hinder extremity is furnished with three pairs of legs armed with spines, and a terminal hooked blade like that already described. With these hindermost legs the animal takes a firm grasp of the twigs of the polypidom, and rears up into the free water its gaunt skeleton of a body, stretching wide its scythe-like arms, with which it keeps up a see-saw motion, swaying its whole body to and fro. Ever and anon the blade is shut forcibly upon the grooved haft, and woe be to the unfortunate infusorium, or mite, or rotifer that comes within that grasp! The whole action, the posture, figure of the animal, and the structure of the limb, are so closely like those of the tropical genus *Mantis* among insects, which I have watched thus taking its prey in the Southern United States and the West Indies, that I have no doubt passing animals are caught by the crustacean also in this way, though I have not seen any actually secured.

"The antennæ, too, at least the inferior pair, are certainly, I should think, accessory weapons of the animal's predatory warfare. They consist of four or five stout joints, each of which is armed on its inferior edge with two rows of long, stiff, curved spines, set as regularly as the teeth of a comb, the rows divaricating at a rather wide angle. From the sudden clutching of these organs, I have no doubt that they too are seizing prey; and very effective implements they must be, for the joints bend down towards each other, and the long rows of spines interlacing must form a secure prison, like a wire cage, out of which the jáws probably take the victim, when the bending in of the antennæ has delivered it to the mouth.

"But these well-furnished animals are not satisfied with fishing merely at one station. As I have said above, they climb nimbly and eagerly to and fro, insinuating themselves among the branches, and dragging themselves hither and thither by the twigs. On a straight surface, as when marching (the motion is too free and rapid to call it *crawling*) along the stem of the zoophyte, the creature proceeds by loops, catching hold with the fore limbs, and then bringing up the hinder ones close, the intermediate segments of the thin body forming an arch, exactly as the caterpillars of metric moths, such as those, for example, that we see on gooseberry bushes do. But the action of the crustacean is much more energetic than that of the caterpillar. Indeed, all its motions strike one as peculiarly full of vigor and energy.

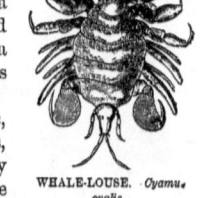

WHALE-LOUSE. *Cyamus ovalis.*

"I have seen the large red species swim, throwing its body into a double curve like the letter S, with the head bent down, and the hind limbs turned back, the body being in an upright position. It was a most awkward attempt, and though there was much effort, there was little effect." In our illustration the creature is enlarged.

The WHALE-LOUSE is, like all the species of this genus, parasitic, residing on the whale and dolphin. Their hooked and diverging legs, armed with their sharply-curved claws, enable them to cling so tightly that not even the swift movement through the water, or the active exertions of the creature on which they reside, are sufficient to shake them from their hold. The different species of Whale-louse seem to prefer various parts of the body, one species clinging to the head, another to the side, and another to the fin. They all burrow rather deeply into the rough and thick skin of these marine mammalia.

Their bodies are flattened and rather oval; they have five pairs of legs, all prehensile; and on the second or third joint of the thorax, instead of legs there are long appendages for respiration, which usually are bent over the back. The illustration is of natural size.

ISOPODA.

IN the Isopod crustacea, the signification of which word has already been given, there is a great resemblance to the common wood-louse, and many of them might easily be mistaken for those common and destructive beings. The females have large horny plates on their legs, so formed as to produce a large pouch under the thorax, wherein the eggs are contained. In many species some of the rings of the abdomen are connected so as to resemble a single joint.

The BAFFIN'S BAY ARCTURUS is one of the best developed of the whole order. In all the species belonging to this genus the body is long, and the first four pairs of legs are beautifully feathered at the ends. These cannot be used for walking, the three last pairs of legs being devoted to this purpose. The long antennæ are used as organs of prehension, and with them the creature captures its prey. The young are said to cling by their legs to the antennæ of the parent.

Several of these species take possession of the corallines, each selecting a particular branch, and not permitting any other to intrude upon its premises, fighting with great valor against any assailant.

They resemble the fly-catchers in some of their habits, sitting patiently on their branch until they see some little creature passing within reach. They then dart at their prey, seize it, return with it to their resting-place and there eat it leisurely. They sit in a curious erect attitude, swaying the body about and occasionally cleaning the antennæ by drawing them through the tufted feet.

The common FISH-LOUSE is parasitic upon many species of fish, clinging tightly by means of their hooked legs. It is thought by many fishermen that the creature is by no means hurtful to the fish, but that it is absolutely beneficial, causing death if removed.

A rather curious and tolerably plentiful species of Isopod crustacean is the SHRIMP-FIXER, so called from its habit of affixing itself to shrimps and prawns, concealing itself under the side of the carapace. Any number of these curious parasites may be obtained from a fishmonger's shop, by the simple process of looking over his stock of prawns, and picking out those which have a swelling at the side of the carapace. The fishermen, who have the oddest ideas about marine objects, and know as little about shrimps as a ploughman about worms, generally fancy that these parasites are young soles! probably on account of the general shape of the male.

The female of this crustacean is generally found with a mass of eggs which are congregated beneath the body, and are kept in their places by the pouch formed by the plates attached to the legs. Owing to the pressure caused by the carapace of the prawn, the sides of the Shrimp-fixer are dissimilar, and distorted individuals are very common.

In all the members of this genus the male is much smaller than the female, being barely one-sixth the size of his mate, and is narrow and elongated, whereas she is wide, pear-shaped, and ending in a point. The false legs are ten in number, five on each side, and modified into triangular membranous plates, forming a pouch for the reception of the eggs.

The color of this species is greenish, with a slight lustre above, and dark at the edges of the plates.

The members of the genus Ione may be known by the appendages of the abdomen, which are thread-like and arranged round the body. The female is also larger than the male.

This creature is also a parasite like the preceding, but makes its home within the thoracic plate of the burrowing crab (*Callianassa subterranea*), which has already been described on page 464. It forms a tumor on the side, and can be removed in a living state. It seems that both sexes are to be found under the same shell, the tiny male holding firmly to the appendages of his mate like a little child holding to its mother's dress. The color of this species is orange-yellow, and the appendages are white.

Mr. Tuffen West has favored me with the following remarks upon an allied crustacean:—
"Some years ago, I assisted in the dissection and made drawings of the male and female of a remarkable crustacean taken from the gills of a hermit-crab. It was thought that a new genus would have to be constituted for it. In the female of this species there was a distinct space left between the plates covering the ova, for the accommodation of the male, and it is thought doubtful whether he ever takes any food. The males are model husbands; having once selected a mate, they never leave her."

ANOTHER wood-boring crustacean is called the TIMBER-BORING SHRIMP, or GRIBBLE.

Though belonging to another family, this creature is as destructive as that which has already been described, but makes its tunnels in a different manner, burrowing deeply into the wood instead of driving oblique passages. It proceeds in a very methodical manner, the tunnels being quite straight unless they happen to meet a knot, when they pass round the obstacle and resume their former direction. Small as is this crustacean, hardly larger, indeed, than a grain of rice, it is a sad pest wherever submarine timber is employed, for it works with great energy, and its vast numbers quite compensate for the small size of each individual. It appears to attack equally any kind of wood, though its progress is slower in oak and other harder woods than in deal. Sometimes it is found attacking the same timber as the chelura.

As with most of these creatures, the male is smaller than the female, being about one-third her size. The female may be distinguished by the pouch in which the eggs and afterwards the young are carried. About six or seven young are generally found in the pouch.

The Gribble is ashen-gray in color, with darker eyes. The timber into which these creatures have been boring looks very like old worm-eaten furniture. The creature is able to roll itself into a nearly spherical form, like the well-known pill-woodlouse. The tail is composed of many segments, and the antennæ are in pairs, set above each other.

A creature much resembling the common woodlouse, is the GREAT SEA-SLATER, or SEA-WOODLOUSE, a species which, though extremely plentiful, is not seen as often as it might be imagined, owing to its extremely retiring habits and hatred of light. The Sea-slater lives on the stone and rocks of the sea-shore, and hides itself carefully during the day in the crevices, its flattened body enabling it to crawl into very small chinks. At early morning, however, and in the evening, these creatures may be found by thousands, and any one who will take the trouble to search the rocks by the aid of a "bull's-eye" lantern will find himself repaid by the vast number of nocturnal animals that have ventured out of their dens.

The female carries her young in a kind of pouch formed by the development of a number of horizontal plates along the abdomen. They remain in this natural cradle for some time, and even after they are able to run about, may be seen clinging to their parent. Mr. Tuffen West tells me that on one occasion he picked up a very large Sea-slater, but nearly let it fall again, startled by seeing four or five little ones run from the body. More and more followed, until twenty had made their appearance. Thinking that he had taken up a dead specimen, he put it down again, and was hardly less surprised to see it run off quite briskly.

The substance of the Sea-slater is rather softer than that of the common woodlouse. It appears to feed either on animal or vegetable substances, and is itself much preyed upon by birds and other enemies. The fish are very fond of these creatures, and some species have been known to hover about rocks during a storm for the purpose of preying on the Sea-slaters that are washed into the water. The color is very variable, but is mostly some shade of brown or gray. This, as well as the succeeding species, belongs to the family of Oniscidæ.

The WATER HOG-LOUSE is the aquatic representative of the sea-slater just described. This species is plentiful in fresh water, whether still or running, and in general walks very leisurely, though when alarmed it can run swiftly. In this genus the proportion of the sexes is reversed, the male being larger than the female. Its average length is about half an inch.

The common WOODLOUSE shows an equal development of the legs. This creature is very plentiful in all damp places, and especially exults in getting under logs of wood or decaying timber. In cellars and outhouses they are common, and are generally to be found in dark and damp localities. Fowls are very fond of them, and there is no surer way of extirpating these sharp-toothed creatures than by allowing some fowls to scrape and peck about in the places where they have taken up their residence. Under the bark of dead and decaying trees is a very favorite residence with the Woodlouse, and in such localities their dead skeletons may often be found, bleached to a porcelain-like whiteness.

The color of the Woodlouse is a darkish leaden hue, sometimes spotted with white.

An allied species, the LAND-SLATER (*Oniscus asellus*), is equally plentiful. This species may be distinguished by the two rows of yellow spots and the same number of white spots that run along the back. There are also eight joints in the outer antennæ, whereas there are only seven in the same members of the woodlouse.

The well-known PILL-WOODLOUSE, or PILL-ARMADILLO, when rolled up into a globular shape, bears a strong analogy to the common hedgehog, and a still stronger to the manis, as in the latter case the creature is defended by horny scales that protect it just as the external skeleton protects the armadillo. While rolled up this creature has been often mistaken for a bead or a berry from some tree, and in one instance a girl, new to the country, actually threaded a number of these unfortunate crustaceans before she discovered that they were not beads.

As they bear such a resemblance to pills, they have often had to pay the penalty of their likeness; for in the earlier days of medicine, and even up to the present time, they have been employed in the pharmacopæia. Even now, though no modern physician would prescribe them for the cure of any disease, the Pill-woodlice may be seen in the recesses of druggists' shops. I have often seen a drawer half-filled with these creatures, and used to convert them into marbles, bullets for a toy cross-bow, and various other purposes, in which they were quite as useful as if they had been employed according to the original design. The color of the Pill-woodlouse is a dark grayish-brown, with a slight polish.

ENTOMOSTRACA.

WE now enter upon a subdivision of the crustacea, called scientifically the Entomostraca, a term derived from two Greek words, the former signifying an insect, and the latter a shell. All these strange creatures are aquatic, and their bodies are protected by a shell of horny or leathery consistence, sometimes in one single piece and sometimes formed of several portions. The gills are attached to the feet, or the jaws and the feet are jointed and fringed with hairs.

This sub-class embraces a numerous group of small creatures, important as being food for fishes. As parasitic forms they prove considerably injurious to our food-fishes, fastening about the gills, and eventually destroying them. Many inhabit fresh water.

The well-known *Cyclops* is an example of one of the principal orders called the *Copepoda*.

Many of the species of orders *Siphonostomata* and *Ostracoda* are familiar as parasitic on the sharks, and especially the sluggish molar, or sun-fish. *Penella* of the latter is large, and has a length of several inches.

In the first section of these creatures the gills are attached to the feet, and they are therefore termed Branchiopoda, or gill-footed. They all swim freely in the water. The first order, the Phyllopoda, or Leaf-footed Entomostraca, have the joints of the feet flat, leafy, and gill-like, and are fitted for respiration. Sometimes the body is naked, and at best, only the head and thorax are covered with the carapace. The first family of these creatures is the Apodidæ, or Footless Entomostraca, so called because all the feet are formed into breathing organs. There are no less than sixty pairs of these feet, all with many joints, and, indeed, the number of joints which are required to form one of these apparently insignificant creatures, is almost incredible. With the wonderful patience of the German nation, Schæffer counted the joints, and found that they fell very little short of two millions.

One species, the CRAB SHIELD-SHRIMP (*Apus cancriformis*), is found in Europe. It swims either on its back or in the usual attitude, and uses the branchial feet in its progression. Its food seems to consist of the smaller Entomostraca. The mandibles of this creature are very powerful, and capable of breaking up the shells of the creatures on which it feeds. Its color is brownish-yellow clouded with brown. One species was discovered by Mr. Tuffen West, and named after him.

A very remarkable being is the BRINE-SHRIMP. It loves to reside in water so strongly charged with salt that every other creature dies in so saturated a solution, about a quarter of a pound of salt being contained in one pint of water. These animals may be seen by thousands in the salt-pans at Lymington, Hants, where the workmen call them Brine-worms.

They congregate thickly in the strongest brine, while in the ordinary sea-water they do not trouble themselves to venture. The workmen believe that the continual movements of these creatures have the effect of clearing the brine, and if they find that their own salt-pan is without the Brine-shrimp, they always fetch some from another pan.

The movements of this little creature are most graceful. It mostly swims on its back, its feet being in constant motion, and its course directed by means of its long tail. It revolves in the water, bends itself into varied curves, turns fairly over, wheels to the right or left, and seems thoroughly to enjoy the very fact of existence. Its color is mostly red, and in some of the pans the Brine-shrimps congregate in such multitudes near the surface that the water looks quite pink with their bodies.

The FAIRY-SHRIMP is appropriately named, as a more fairy-like creature can hardly be conceived. It is to be found in several parts of Europe. In spite of its comparatively large size, measuring more than an inch in length, it may easily escape observation, as its body is of glassy transparency, and scarcely visible in the water, except by the red and blue tints of its tail, branchiæ, and feet. It always inhabits stagnant water, and may even be found in the half-putrid mass of mud and water that lies at the bottom of casting-nets.

In his valuable work on the Entomostraca, Dr. Baird gives the following account of this beautiful creature:—"They swim upon their back, and in fine, warm weather, when the sun is not too strong, they may be seen balancing themselves, as it were, near the surface, by means of their branchial feet, which are in constant motion. On the least disturbance, however, they strike the water rapidly with their tail from right to left, and dart away like a fish, and hasten to conceal themselves by diving into the soft mud, or amongst the weeds at the bottom of the pool.

"It is certainly the most beautiful and elegant of all the Entomostraca. The male is especially beautiful. The uninterrupted undulatory waving motion of its graceful branchial feet, slightly tinged as they are with a light reddish hue; the brilliant mixture of transparent bluish-green and bright red of its prehensile antennæ, and its bright red tail, with the beautiful plumose setæ springing from it, render it exceedingly attractive to the view.

"The undulatory motion of its branchial feet serves another purpose in addition to that of keeping the animal suspended in the water. The thorax or body of the animal has been described, when floating on its back, as like the cavity of a little boat, the feet representing the oars. When these are in motion, they cause the water contained in this cavity to be compressed, and to mount up as along a canal, carrying in the current the particles destined for its food towards the mouth. It seems to be constantly, when in this position, employed in swallowing and digesting its food, its masticatory organs being in perpetual motion."

BRANCH-HORNS; CLADOCERA.

WE now come to some of the minute species of these curious animals. The creatures belonging to this order are termed the Cladócera, or Branch-horns, because their antennæ are forked and branched. With the exception of the head, the body is wholly enclosed within two shells, like the half shells of a walnut. There is only one eye, of very large comparative size; there are two pairs of antennæ, and the lower pair are used in swimming. In the first family, the Daphniadæ, the lower antennæ are very large and upper very small. There are five or six pairs of feet enclosed within the shell, and the intestine is straight.

The common WATER-FLEA is a good example of the typical genus. Eight or nine species of this genus are now known, and the habits are much the same in all. They are to be found in fresh water, whether still or running, but seeming to prefer the former. Even the horse-ponds, which are so trodden about by cattle that they seem to consist of equal parts of bad pea-soup and duck-weed, are favored residences of the Water-fleas, and several species may sometimes be found in one little pond.

The common Water-flea has five pairs of feet, the lower antennæ branched and branchleted, the lower pair being very large and powerful. The head is lengthened downwards into a prominent beak. These little creatures exist in vast numbers, and sometimes assemble in such heavy masses that they quite change the color of the water. Dr. Baird has remarked that they will sometimes assemble so as to form a belt of a foot or so in breadth, and ten or twelve yards in length, and that the whole belt will pass round the pond, thus obeying some strange instinctive command that, though inaudible to human ears, is conveyed to the myriads whom it concerns, and marshals their hosts with the same wonderful discipline that governs a flock of starlings. If a shadow should fall upon this belt of Daphnias, they all suddenly vanish, and appear again as soon as the darkness has passed away.

The eggs of the Daphnia are placed within the space between the shell and the body, and are there kept until they are hatched, and the mother permits them to escape from under her shelly covering. The young are not in the least like their parent, having no shell, no abdomen, and not nearly all their limbs. Each of the antennæ is divided into two branches, and each of these branches is subdivided at its extremity into three branchlets.

Another of these wonderful little beings is called the MOINA, a genus of which very few species are at present known. The color of this species is olive-green, its head is round and blunt, and the lower antennæ are peculiarly large and muscular at the base. The carapace is much smaller behind, and at the end of the abdomen are eight very short spines and two long and stout claws. This species is to be found in Europe, especially in ponds of woods, in which localities the European microscopic naturalists have detected vast numbers of beautiful and even rare creatures that are seldom seen in other places.

On the back of the Moina is seen a dark mass, which, on account of its shape, is called the ephippium, or saddle. This remarkable appendage appears upon the back of the female, particularly in summer, and is divided into two capsules, each portion containing one egg. The eggs which are placed in this singular resting-place are found to remain unhatched through the winter, and are probably defended by this living saddle from the effects of cold, just as pistols are sheltered from the inclemency of the weather in their holsters. Probably, also, the saddle may serve to protect the eggs from the effects of drought, when the pools are dried up by a hot and rainless autumn. When the mother moults, the saddle and its contents are thrown off, together with the shell, and it is not uncommon to find specimens of the young swimming about with a portion of the saddle adhering to them, and looking like young beans just sprouting from the earth, and carrying with them the two lobes of the seed from which they sprang.

A very pretty little Entomostracan, belonging to a small sub-family called the Sidinæ, is called *Sida crystallina*. In all these beings there are six pairs of feet, the lower antennæ have two branches, and a row of sharp and rather strong filaments springs from the edge of the larger branch. In this genus, one branch of the lower antennæ has three and the other two joints. Though it is occasionally very active, passing through the water with great rapidity, it is mostly dull and stationary, having a curious habit of pressing the back of its head against some object, and there remaining for a considerable period without moving. It derives its specific name of crystalline from its beautifully transparent aspect.

IN the Lynceidæ there are two pairs of antennæ, the upper being very short, and the branches of the lower having three joints. They have five pairs of legs, and one eye, with a black spot in front of it. The abdomen is jointed. All the species are rapid swimmers, and their food consists of both vegetable and animal matter.

In the Chydorus, of which one or two European species are known, the body is nearly spherical, the lower antennæ are very short, and the beak is very long, sharp, and curved downwards. The color is olive in the present species, and has a smooth, shining exterior. It may be found in ponds and ditches throughout the year.

THE *Chydorus sphæricus*, a curious globular-looking creature, is an example of another family, called the Polyphemidæ, having only four pairs of feet, which are not included in the shield. Their single eye is very large, and has given rise to the name of Polyphemus, which belonged to the one-eyed giant overcome by Ulysses and his companions. The lower antennæ have two branches—one with four joints and the other with three. In the lower part of the carapace there is a large, empty space for the accommodation of the eggs and young.

An example of the typical genus is the common POLYPHEMUS (*Polyphemus pediculus*), found in ditches and ponds. In this creature the abdomen is long and projects from the shell, and in the adult the eye is enormously large, seeming to occupy the whole head. There is a deep notch or groove in the Polyphemus, seeming to separate the body from the head. It appears always to swim upon its back, and uses both the antennæ and legs to drive it through the water.

OSTRACODA.

In the order called Ostracoda—a term derived from a Greek word, signifying a shell—the cuirass is in two parts, and incloses the animal like a bivalve shell. The hind jaws are furnished with gills. In the family of the Cypridæ, the upper pair of antennæ are long, have numerous joints and a pencil of long filaments; the lower pair are short, thick, and used as feet. There are two pairs of real feet. One of these creatures is called CYPRIS. It belongs to a genus which has many European species, and it may be found in almost every pond or ditch. The body is inclosed thoroughly in its valved cuirass, something like a walnut in its shell, the fringed antennæ and legs protruding from between the valves and permitting the creature to move. It is a most elegant little being, the shell being gracefully curved, and the antennæ being fine and transparent as if they were threads of glass. Dr. Baird tells us that the valves are very brittle, and that on their exterior they are washed with a kind of varnish which protects them from the action of the water. Owing to this varnish, these creatures cannot venture even to rise to the surface; for as soon as the shell is exposed to the air, it becomes quite dry, and so buoyant, that no exertion of the Cypris can sink it again.

These tiny animals will often live through a hot summer which dries up the pond in which they reside, and at the first rain will make their appearance again, swimming merrily about as if nothing had happened. As soon as they feel themselves being deserted by the water, they bury themselves deeply in the mud, and even their eggs retain their vitality, though the mud should be baked quite hard. When the Cypris changes its skin, it throws off the whole shell, the internal parts of the body, the beautiful comb-like gills, and the tiny hairs which clothe the bristles of the antennæ.

Two other examples of this pretty genus are named *Cypris claváta* and *Cypris vídua*.

In the family of Cytheridæ, the upper pair of antennæ have no long filaments.

The members of the typical genus Cythere are mostly marine, and may be found in the little rock-pools at the sea-side, darting about among the branches of sea-weeds and zoophytes that live so plentifully in such situations. Safe in these sheltered spots, they care nothing for wind and waves, and the storm which flings the huge whale on the shore will fail to injure these tiny beings, whose very minuteness is their safety. One species, *Cythere minna*, is remarkable for being the largest one seen by Dr. Baird. Its valves are white. It was found in deep water and taken in a dredge. *Cythere inopinata* derives its specific name of *inopinata* or unexpected, from the fact that the creature was found where no one would have expected its presence, namely, in small ponds. It is a very small species, and always remains at the bottom. Its color is white, and there is a little orange-colored mark on the upper edge. An oblique view of this species has been chosen, in order to show the curious rounded projections upon the middle of each valve.

The *Cythere impressa* was found in sand at Torquay. The shell is dull black in color, and is covered with little punctures impressed upon its surface, whence is derived its specific name.

A closely allied genus is remarkable for the manner in which the valves are ridged, irregular, covered with tubercles, and having their edges boldly toothed. This species was taken in the Isle of Skye.

In the family of the Cypridinadæ there are two eyes, set as footstalks, and two pairs of feet, one pair being always within the shell. There is only one genus of these creatures, and all the species are marine. The shell is oval, sharply pointed at each end, and the front edge is deeply notched. The pair of feet that are retained within the shell are modified into one organ, which seems to be intended for the purpose of supporting the eggs. Some other species are luminous.

OAR-FOOTED ENTOMOSTRACANS; COPEPODÆ.

THE above term is chosen for this order of crustaceans because their five pairs of feet are mostly used for swimming. The body is divided into several rings, the cuirass covers both the head and thorax, and the mouth is furnished with foot-jaws.

In the family of the Cyclopidæ the head and body are merged together with the first ring of the thorax. There are two pairs of foot-jaws, and the fifth pair of legs are very minute.

A species called *Cyclops quadricornis* is very common in every pond and ditch, and the female may at once be recognized by the little egg-bags which she bears on the sides of the abdomen, like John Gilpin's wine-bottles at his belt. The color of this species is exceedingly variable, differing according to the locality where the creature happens to reside. It is mostly white, but some individuals are brown, others greenish, while a few are red. Both salt and fresh water are inhabited by the CYCLOPS, and some of the marine species are so highly luminous, that they add in no slight degree to the phosphorescence of the ocean.

Canthocamptus minutus is the name given to a very little species. It is a creature with a long abdomen, which it is able to turn over its back, something after the fashion of the earwig or the cocktail beetles. In this Canthocamptus the thorax and abdomen are merged into each other, and gradually diminish in size to the extremity. All the species belonging to this genus have very small and simple foot-jaws. It inhabits ponds and ditches of fresh water. Mr. Tuffen West tells me that a short time ago he was examining some of the slime that had gathered upon the roof of the Cramlington Pit, at a vast depth from the surface, and that he found in the slime some of these minute crustaceans quite brisk and lively, whisking their tails up and down smartly. These creatures must have been washed down the pit while still unhatched, and have been thus carried down from the open air into the bowels of the earth.

Another creature of the same genus is termed *Cetochilus septentrionalis*. Though very small, not more than the sixth or seventh of an inch in length, it is of exceeding importance to commerce, as it affords food to the herring, several whales, and other valuable beings. In the seas where this little creature lives, whole tracts are reddened with the multitude of their hosts, which swarm near the surface and congregate in such vast numbers, that the wind has been known to catch up a whole bank of them, like a wave, and fling it into the vessel, covering the deck and the sailors with their bodies. The codfish feeds largely and luxuriously upon these abundant creatures, needing not to take any pains about them, but swimming lazily through their masses and opening its mouth, into which they pass without the least trouble.

The long antennæ are used as oars, being thrown backward at every stroke until their tips touch each other. This attitude, however, is only assumed while the creature is in haste, as it is often seen to pass gently through the water, with its antennæ at right angles to the body.

Dr. Sutherland, in his "Voyage to Baffin's Bay," writes of these elegant little beings: "They are always on the alert to elude and escape from their pursuers. When the water is but slightly agitated, they dive from the surface, and in a few minutes, when it becomes still, they can be seen ascending slowly, but rarely using the antennæ. I could only obtain specimens by including them in a large quantity of water taken up suddenly, from which they could be separated subsequently by straining through a calico bag. A bucketful (two gallons) of water often produced twenty to thirty individuals, and sometimes twice that number. They never survived a single night, even though kept in their native element in a vessel. From their constant darting from side to side of the vessel, perhaps it is a safe inference that the fear of danger in their new situation may be one of the chief causes of the early extinction of life."

The color of this species is light red, and the body is nearly translucent.

Another curious species deserves a word of mention. This is the *Notodelphys ascidicola*, which is found swimming in the bronchial sac of the ascidia.

VARIOUS-FOOTED ENTOMOSTRACA; PŒCILOPODA.

TUBE-MOUTHED ENTOMOSTRACA; SIPHONOSTOMA.

WE now come to another group of Entomostraca which are parasitic upon fish and other inhabitants of the waters. They belong to Dr. Baird's third legion, called the Pœcilopoda, a term derived from two Greek words, signifying various-footed. They are so named because they are partly formed for walking or seizing prey, and partly for swimming and breathing. In the first order, the SIPHONOSTOMA, or tube-mouthed Entomostraca, the mouth is furnished with a tube containing sharp, spike-like mandibles. The foot-jaws are well formed. The object of the tube and its sharp mandibles is evidently for the purpose of piercing the skin and sucking the juices of the beings upon which they cling; and the strong foot-jaws enable them to hold so firmly, that they cannot be shaken off. The first tribe is called Peltocephala, or buckler-headed, because the head is shaped something like an ancient buckler; the head is also furnished with plates in front, and small antennæ of two joints. The first family of these creatures is called Argulidæ, and may be known by the circular-shaped head-shield, and the manner in which the second pair of foot-jaws are modified into a pair of powerful suckers.

The FISH-ARGULUS may be seen upon many of the ordinary river-fishes, the stickleback being its favorite. I have seen it on the roach, and even upon the golden carp. It is not very small, being about the diameter of a small sweet pea, and may easily be watched if placed in an aquarium in which any fish are swimming. The little creature at once makes for the fish, darting along with considerable speed, and fixes itself to the side just under the pectoral fins. It does not, however, remain fixed to the fish, but occasionally leaves it, and starts off on little voyages of discovery, always, however, returning at short intervals, as if for the purpose of assuring itself of a meal. It is wonderfully flat, looking very like the shed seed-vessel of some plant, and the resemblance is increased by its pale green color.

The female is considerably larger than the male, and may at once be known by the black spot on each side of the abdomen.

The CALIGUS is referred to another family.

This creature is mostly found upon the codfish and brill, and clings with great firmness. Mr. Tuffen West tells me that he has examined the Caligus carefully with the microscope, and assured himself that the suckers are present. "They are hemispherical, shallow in front, where their margin thins off to a translucent membrane; and deep behind, where their concavity is bounded by a strong, transversely striated membrane."

A remarkable parasite, adherent to the gills of the lobster, is called *Nicothoë astaci*. This creature belongs to a different tribe, which may be known by the small and mostly blunt head and the long and well-jointed antennæ. The family Ergasilidæ have the head rounded, the body oval, the abdomen well developed, and the feet small and branched.

The LOBSTER-LOUSE is sometimes found in considerable numbers fixed to the gills of the lobster, from which the female never moves after she has once taken a firm hold, though the male is more erratic in his habits, and swims about as he chooses. During her early youth, the female is not much larger than the male; but, as soon as she attaches herself to her new home, a pair of strange projections are seen to grow from the side, and by degrees become so large, that they seem to constitute the entire creature. Below these projections the egg-sacs are developed.

A curious parasite that infests the sturgeon is rather more than half an inch in length and the twelfth of an inch in breadth. It is termed *Dichelertium sturionis*. This creature insinuates itself deeply into the skin, making its way to the bony arches upon which the gills are supported, but not appearing to touch the membranous gills themselves. Sometimes as many as ten or twelve are taken from a single fish. They can grasp very firmly by means of their forceps, and are able to turn round whenever they please. This curious creature belongs to the order of the Lerneadæ, in which the mouth is formed for suction, and the limbs scarcely visible. All these beings are parasitic upon fishes, and are often so deeply buried in the tissues, that the whole body is concealed and only the egg-bearing tubes suffered to appear. As is the case with many creatures, especially those that occupy a low place in the scale of creation, the young enjoy a wider range than the parent, being able to roam about at will, and not settling down to a motionless existence until they have attained maturity.

LERNEADA.

THERE seems to be no bound to the wondrous forms which these parasites assume, as may be learned from the following example: the *Chondracanthus zei*. It is called so because its body is covered with cartilaginous spines or tubercles. The name is derived from two Greek words, the former signifying cartilage and the second a thorn. This strange being is found upon the gills of the John Dory.

The two most extraordinary beings, which are called *Lernæodiscus* and *Jacculina*, were discovered under the abdomen of a lobster. In both these creatures (which certainly seem to belong to the Lerneans), the whole of the head becomes modified into a set of branching fibres, much resembling the roots of a tree. There is no mouth whatever, all nourishment being transmitted through these fibres. They are quite recent discoveries.

Though our space is rapidly diminishing, we may still mention a few more of these creatures. One of common occurrence is the PERCH-SUCKER, in which exists a great dissimilarity between the female and her small mate. Another species is termed *Anchorella uncinata*. In this parasite the arm-like appendages are very short, and united from the base so as to look like a single organ. The body of the female Anchorella is white, and the short arms end in a rounded knob. This creature is rather more than half an inch in length. The male of the same species would hardly be recognized as having any connection with the long-bodied creature that has just been described. The length of the male is about the forty-eighth of an inch. Another species of the same genus is the *Anchorella rugosa*, so called because the body is notched at the side. This creature is about the seventh of an inch in length. All these creatures infest the cod, haddock, and similar fishes.

A wonderful example of a parasitic crustacean is the *Tracheliastes*, with its long egg-bags and strangely-developed upper extremity.

IN the next tribe of Entomostraca the head is kept buried in the tissues of the animal to which the parasites cling, and are there held firmly by some horn-like processes that spring from the back part of the head. They are, in fact, living spears, the barbed heads being sunk into their prey. The two best-known members of this tribe are the *Lamproglena pulchella* and the *Lernentoma asellina*.

Not the least strange-looking among them is termed *Chalimus scombri*. It is, like many others of its class, parasitical upon a parasite, and it is found adhering to the caligus. With its long tube and sucker it adheres to its prey, and it may often be seen hanging to the lower part of the caligus like a fish at the end of a line. This is one of the many instances that prove the truth of that quaint and far-seeing old saying, namely—

> "Big fleas and little fleas
> Have lesser fleas to bite 'em;
> The lesser fleas have smaller fleas,
> And so, *ad infinitum*."

A creature that is found upon the sun-fish, and adheres to the gills, is called *Cecrops*. It is not always fixed to this habitation, but floats about by thousands in the Mediterranean, where it is preyed upon by many fishes.

Our next example is the SHARK-SUCKER, a species that is found adherent to the eyes of the Arctic shark, and appears to blind it. The sharks to which this unpleasant appendage was attached seemed to be quite destitute of sight, and did not flinch in the least when a blow with a lance was aimed at them. The arm-like appendages of this creature are inserted into the corner of the eye for nearly one-fourth of their length. This parasite attains to the length of three inches. An allied species, called *Lernæopoda galei*, is found on one of the common dog-fishes known by the name of tope, and described on page 199 of this volume. A strange, elongated creature is the *Penella filosa*, so called from its extreme length. This species is found to penetrate into the flesh of the sword-fish, the tunny, and the mole-fish, all of which have been described in this volume. It is said to cause them considerable pain. A parasite of even stranger form, but belonging to the same genus, is *Penella sagittata*. One of these parasites, called the SPRAT-SUCKER, is sometimes tolerably common, many specimens being obtainable at a single fishmonger's shop, while for several years hardly one will be seen. The color of this parasite is pale sea-green, with a slight bluish cast. The eggs are very green.

A strange and seemingly shapeless parasite, that is found to affix itself to the carp tribe, is the *Lernæocera cyprinacea*. The *Lernæa*, a creature of somewhat similar form, is notable for being found upon the gills of the codfish. This creature belongs to the typical genus.

PYCNOGONIDES.

It is hardly possible to imagine any forms that are so strange, any habits so astonishing as those which are found in the crustaceans described in the following lines. Although they have been known for some time, their proper place in the scale of creation has long been a disputed point among systematic naturalists, some considering them to belong to the crustaceans and others to the spiders. As, however, they undergo a true metamorphosis, which is not the case with any spider, they are now admitted to be real, but unique crustacea. Even such naturalists as Siebold and Milne-Edwards differed about them, the former placing them among the spiders, and the latter ranking them with the crustacea.

Such strange creatures as these are not easily described, especially when the space that can be granted to them is so limited, for their whole economy is so thoroughly unique that they require a volume rather than a page. They are found upon the European coast, and their history is briefly as follows.

Two of these strange-looking creatures with wonderfully small bodies and enormous legs, jointed and arranged in such a manner as almost to preclude the idea of their real character, are called *Pycnogonum littorale* and *Phoxichilidium coccinium*. Indeed, it seems passing strange how the tiny abdomen can absorb sufficient nutriment for the supply of those marvellous limbs. Their economy is as strange as their form.

Some specimens of a well-known zoophyte (*Coryne eximium*) are often seen attached to the rocks or sea-bed. The *Phoxichilidium* is frequently found as a nodule. In spite of the long limbs, it appears packed away in a very complete manner, the limbs being rolled round the body so as to form the creature into a kind of ball. During its growth the young Phoxichilidium has to pass different stages. Sometimes it possesses the rudiments of limbs, with long filamentous appendages; sometimes it throws them off, and contents itself with a pair of stout claws, and then again grows a fresh set of limbs and a pair of small and feeble claws.

Strange as are these habits, there is still a kind of analogy with other modes of animal life. On page 474 is mentioned the curious little crustacean which resides within the body of a beroë, and in the present instance there is an evident analogy with the various galls and their inhabitants, the cells of the Phoxichilidium being in fact the galls of the coryne.

SWORD-TAILED CRUSTACEA; XIPHOSURA.

The crustacea abound in strange forms. The LONG-TAILED MOLUCCA CRAB belongs to a separate order, called by the name of Xiphosúra, or Sword-tailed Crustacea, in allusion to the long and sharp spine which projects from the shell. These creatures, of which several species are known, can easily be recognized by their general shape. The body and limbs are covered by a curious shield, composed of two parts, the junction taking place across the centre of the body. Though perfectly harmless, these creatures can be made very offensive, for the natives of Molucca are accustomed to use the long sharp tail spine as the head for an arrow or lance, and thus make a most formidable weapon. Many of these crustacea attain the length of two feet, so that the spike is nearly a foot in length, and is capable of inflicting a deadly wound.

The edges of the hinder portion of the shield are deeply toothed, and the space between the teeth is occupied by a rather long and sharply-pointed spine, which is not fixed, but is movable on its basis. The feet are mostly furnished with tolerably strong claws.

The Molucca Crabs often leave the sea and crawl upon the sand, where they may be taken without much difficulty. They cannot endure the heat of the sun's rays, and are in the habit of burrowing into the sand when the sunbeams beat too fiercely on their shells. Sometimes they do not bury themselves very deeply, and then they are discovered by the projecting tail-spike, which shows itself above the level of the sand, and betrays the position of the animal. As they pass over the sand they present a very curious appearance, as their large shield-like shell entirely covers the limbs, and the creatures seem to be carried along by some external agency rather than to be propelled on their own limbs. Owing to the shortness of the legs, and the large rounded shell, the Molucca Crabs are almost helpless if laid on their backs, being obliged to wait until some friendly wave may strike them and enable them to resume their proper attitude. These crustaceans occur largely in certain strata, and are found in a fossil state, many species attaining to a very great size. One living species (*Limulus cyclops*) is a native of the East Indies, and goes by the popular name of PAN-FISH, or SAUCEPAN-CRAB, because the shell, when the limbs and body have been removed and the tail spine permitted to retain its place, has some resemblance to the useful culinary article from which it derives its name. It is often used as a ladle for dipping water out of a vessel.

BARNACLES; CIRRIPEDIA.

WE now come to the last members of the crustacea, creatures which were for a long time placed among the mollusks, and whose true position has only been discovered in comparatively later years. Popularly they are called Barnacles, but are known to naturalists under the general term cirripedes, on account of the cirri, or bristles, with which their strangely transformed feet are fringed.

When adult, all the cirripedes are affixed to some substance, being either set directly upon it, as the common acorn-barnacle, so plentiful on European coasts; placed upon a footstalk of variable length, as in the ordinary goose-mussel; or even sunk into the supporting substance, as is the case with the whale barnacles. When young, the cirripedes are free and able to swim about, and are of a shape so totally different to that which they afterwards assume, that they would not be recognized except by a practised eye. More will be said on this subject.

Along the under surface are set six pairs of limbs not furnished with claws, but being developed at their extremities into two long filaments, jointed and covered with hairs. By means of these modified limbs the cirripedes obtain their food. The common acorn-barnacle affords a familiar and beautiful example of the mode by which this structure is made subservient to procuring a supply of food. The closed valves at the upper part of the shell are seen to open slightly, a kind of fairy-like hand is thrust out, the fingers expanded, a grasp made at the water, and the closed member then withdrawn into the shell.

This hand-like object is in fact the aggregated mass of legs with their filaments. As the limbs are thrust forward, they spread so as to form a kind of casting net; and as they return to the shell, they bring with them all the minute organisms which were swimming in the water. This movement continues without cessation, as long as the Barnacles are covered with water, and appears to be as mechanically performed as the action of breathing is performed by the higher animals.

We will now cast a hasty glance at the transformations through which these creatures pass before attaining their perfect state. It has already been mentioned that the young cirripedes are free and able to wander about at will; and as is generally the case in such instances, they are apparently of a higher organization when young than when adult. For example, the young Barnacle can swim freely with certain limbs. When adult, it loses those limbs. When it is young, it possesses eyes; but when it attains maturity, it loses those valuable organs, which, although indispensable to a wanderer, are needless for a being which is fixed to one spot and needs not to move in order to obtain subsistence.

When first set free from the parent, the Barnacle is extremely minute, and has a striking resemblance to the young of one of the Entomostraca already described. It has three pairs of legs, with imperfect joints and ending in bristle-like appendages. By the vigorous flapping of these limbs the young Barnacle is driven quickly through the water, with a sharp but uncer-

GOOSE-MUSSEL.—*Lepas anatifera.* (On pumice-stone.)

tain movement. In fact, a microscope of low power, when applied to the water wherein a number of these tiny creatures are swimming, discloses a swarm of merry little beings playing about just like the clouds of gnats over water, or the dancing motes in the sunbeam.

Just in the middle of the part of the body which by courtesy we will call the forehead, a single eye is placed, black, round, and shining as if it were a little jet bead inserted into the body. There are also two very large antennæ, which serve two useful purposes, for they aid the free and imperfect Barnacles to proceed through the water, while they are the means whereby the creature fixes itself to the rock when about to undergo its last change.

In the accompanying illustration is seen a group of the common GOOSE-MUSSEL, or DUCK-BARNACLE,, so called on account of the absurd idea that was once so widely entertained, that this species of barnacle was the preliminary state of the barnacle-goose, the cirri representing the plumage, and the valves doing duty for the wings.

This Barnacle is tolerably universal in its tastes. It clings to anything, whether still or moving, and is the pest of ships on account of the pertinacity with which it adheres to their planks. Its growth is marvellously rapid, and in a very short time a vessel will have the whole of the submerged surface coated so thickly with these cirripedes that her rate of speed is sadly diminished by the friction of their loose bodies against the water.

When once the Goose-mussel has affixed itself to any object, the rapidity of its growth is positively startling. The minute young are poured from its shells in such multitudes that they look like cloudy currents in the water; and after they have enjoyed their brief period of freedom, they settle down, attain maturity, and in their turn become the origin of a countless posterity.

I have seen a large log of timber, about fourteen feet in length by one foot square, so thickly covered with these Barnacles that the wood on which they rested was not visible. The same log, which had evidently formed part of the cargo of a timber ship, had been attacked by the ship-worm as well as the Barnacle, and had been tunnelled from end to end by that insatiable devourer. The log was so entirely covered by the Barnacle and the ship-worm, that the wood of which the beam was composed was quite invisible, and could not be seen until the heavy masses of Barnacles were lifted up by the hand.

The old boatman who had picked up the log while fishing, and had ingeniously built a trough to receive the log, a tank of sea-water to supply the trough, and a kind of tent composed of sails to hold the trough and the log together, was very full of a discovery that he had made. He was fully persuaded that the ship-worm and the Barnacle were identical, and that when the ship-worm was tired of boring into wood, it came to the surface, and was immediately changed into a Barnacle. He was quite impervious to reason, and always went into a passion whenever the facts seemed to contradict his theory.

If the objects were enumerated to which the Barnacle will cling, a volume would hardly be sufficient for the mere catalogue. It has been found on ships, boats, floating timber, shells, turtles, whales, and marine snakes. A moment is sufficient to give them a firm hold of any object, and when once they have fixed their antennæ, the fiercest storm cannot shake them off. Even after death, the force with which they cling is as great as during life, and they seem almost to form part of the substance to which they adhere. The length of the footstalk is extremely variable, in some measuring three or four times the length that it does in others. This species is found in nearly all temperate and warm seas.

A second, but smaller Stalked Barnacle, is the FASCINE-BARNACLE, a larger and finer species, which can be distinguished by the number and shape of its shelly valves. These valves afford most important indications of the genus to which any species belongs, and in the arrangements of some zoologists they play the principal part in the formation of the system.

The Fascine-barnacle is found in the Indian Ocean.

A rather singular form of Barnacle, resting on short, stoutly-shaped footstalks, and having somewhat triangular valves, is the MITELLA-BARNACLE. This species comes from China, the Philippines, etc.

Our next example, the EARED BARNACLE, derives its appropriate name from the curious tubular projections which stand out boldly from either side, like the ears of a quadruped from the head. This species lives in the warmer seas.

A group of Eared Barnacles have been found attached to another genus of Barnacle, which lives on, or rather in, the skins of cetacea, and to which we shall presently allude. Indeed, these beings seem to care little about the substance to which they adhere, one species of Stalked Barnacle having actually been taken upon the delicate surface of a living Medusa.

WE now leave the stalked barnacles and proceed to some other species. One of them, the BELL-BARNACLE, which is found off the coast of Madeira, Africa, and other hot parts of the ocean, forms generally a small group of upright shells, surrounded by buttress-like and pointed projections. It sometimes attains a very considerable size, and is eaten by the Chinese, who think that it resembles the lobster in flavor.

Other species are also eaten, such as the PARROT'S-BEAK BARNACLE, a creature deriving its name from a curved projection something like the bill of a parrot. This enormous Barnacle is sometimes found measuring between five and six inches in height, and between three and four inches in diameter. It is found in large bunches, sometimes consisting of a hundred individuals, some adhering to the rocks and others to the shells of their companions. The bunches of Parrot-beaked Barnacles bear a decided resemblance to the strange cacti whose leaves are set so oddly upon each other.

This Barnacle is gathered, or rather hewn from the rocks in large quantities, and exported to Valparaiso and other places, where it is held in high estimation as a delicacy for the table. It is generally boiled, and eaten cold, like the common crab, and is said to resemble that crustacean in general flavor. It is a South American species, and is found most plentifully and of the largest size at Concepcion de Chile, and the best specimens are taken from a little island called Quiquirina, which lies across the mouth of the bay.

A creature which is found plentifully on some coasts is called the common ACORN-BARNACLE. On many coasts the surface of every stone and rock that is washed by the sea, the exterior of every pile of masonry that is lashed by the waves, is covered with the shells of this curious little creature, which is extremely valuable to the naturalist, as its habits are easily studied, and from its exceeding plenty any number of specimens can be obtained. They are very pretty inhabitants of an aquarium, but they require peculiar conditions to keep them in health, and if they die, are sure to corrupt the surrounding water to such an extent, that nearly every other inhabitant of the aquarium will share their fate. Spots over which the tide only runs for a few hours are thickly studded with these Barnacles, and it is interesting to see how quickly they open their valves and fling out their arms as soon as the water covers them at each returning tide. When the sea withdraws, they close their shells firmly, and retain within their interior a sufficiency of water wherewith to carry on the business of respiration until the next tide brings a fresh supply. Total submersion seems to be hurtful to them.

They are very awkward to the shore bather who does not know the coast, as the edges of their shells are exceedingly sharp and knife-like, and inflict very painful scratches when brought into collision with the unprotected skin. Even to those who are searching on the rocks for marine curiosities the Barnacles are very annoying, as they are constantly scratching the hands when an incautious searcher happens to stumble and tries to save himself by grasping at the rocks.

A rather curious cirripeda is the CORAL-BARNACLE, which, as is evident from the material on which it is supported, will only be found in those seas which are warm enough to produce corals. Sometimes the growth of the coral is too rapid for the Barnacle, which is gradually covered by the increasing stony deposit, and at last is actually buried deeply in the mass, where it dies from starvation. The reader may remark that one of the mollusks described on page 318, is also in the habit of making its residence upon coral, and were it not for a peculiar adaptation of structure, would perish for the same reason. But the Magilus is able to extend its shell as fast as the coral deposits fresh substance, and therefore always contrives to keep itself within reach of the water. In the Pyrgoma the cone is composed of a single piece, very thick, rather compressed, and open above.

Some very strange forms of cirripedes are now presented to us. One of them is figured in the illustration. The CORONET-BARNACLE, so called on account of the coronet-like shape of the body, is always found upon the skins of the cetacea which inhabit the Arctic Seas, such as the Greenland whale and the long-armed whale (*Balænoptera longimana*).

CORONET-BARNACLE.—
Balanus crenatus.

The specimen exhibited in the illustration is represented of its natural size, but Mr. Sowerby informs me that in a piece of whale skin only four inches in length, no less than six specimens of this creature are attached, all larger than that shown in the engraving. The cirripeda does not merely adhere to the skin, but in process of time actually buries itself deeply into the tissues, and would seem to cause much annoyance to the creature on which it was parasitic.

A still stranger example of these curious parasites is the BURROWING BARNACLE, which sometimes are found deeply sunk into the skin of a whale. This species plants itself in the skin of the whales belonging to the Southern seas. This pest of the cetaceans is nearly cylindrical in shape, and remarkable for a series of raised rings, which surround it like the hoops upon a barrel. As the creature increases in age, it also increases in length, and adds ring after ring, in proportion to the depth of its imbedment in the skin. The Burrower-barnacle is found in great numbers, and actually studs the whale's skin with its shells. Not only does the skin suffer from their presence, but the blubber is also infested by them, as they often pass completely through the skin, and sink deeply into the fatty tissues beneath. I have seen several fine examples of these sunken cirripedes, and could not but admire the wonderful adaptation of their structure to their mode of living.

Barnacles were collected and arranged as multivalve shells formerly. They are subdivided and embraced under several orders, among which the *Protolepas*, *Cryptophialus*, and *Alcippe* are known.

Members of the family *Lepadidæ* are numerous on our coast. The *Conchoderma virgata* is a curious form, often found on floating stuff in our waters. They have fleshy stalks by which they fasten to floating *débris*. *Lepas* is a familiar genus.

Family *Balanidæ* includes more species than others. The Acorn-barnacles are numerous. Species are found attached to sea-turtles and sluggish fishes. *Coronulas* are found on whales.

Xenobalanus is found on turtles and the black-fish dolphin.

CRAB-SPIDER, OR MATOUDOU.

SPIDERS, SCORPIONS, AND MITES;

(ARACHNIDA).

TRUE SPIDERS; ARANEIDEA.

ANOTHER class of animated beings now comes before us, which, under the general term of Arachnida, comprises the Spiders, Scorpions, and Mites.

These beings breathe atmospheric air, they have no antennæ, and they have four pairs of legs attached to the fore parts of the body.

In some of the higher Arachnida, there is a bold division into thorax and abdomen, and the former portion of the body is clearly divided into separate segments. By the earlier naturalists, the Arachnida were placed among the insects, but may readily be distinguished by several peculiarities. In the first place, they have more than the normal number of six legs, which alone would be sufficient to separate them from insects. They have no separate head; the head and thorax being fused, as it were, into one mass, called the cephalo-thorax. In many of the lower species there is not even a division between the thorax and abdomen; and the body, thorax, and abdomen are merged into one uniform mass, without even a mark to show their several boundaries. They undergo no metamorphosis, like that of the insects, for, although the young Spiders change their skin several times, there is no change of form.

Beginning with the true Spiders, we find that their palpi (*i.e.* the jointed antennæ-like organs that project from the cephalo-thorax) are more or less thread-like, and in the males are swollen at the extremity into a remarkable structure, as indicative of the sex as is the beard of man, the curled tail-feathers of the drake, and the gorgeous train of the peacock. In the different genera, these palpi are differently formed, and afford valuable indications for systematic zoologists.

Several examples of these Spiders will be described in the course of the following pages. They are remarkable through their exceeding diversity of form, and they can be readily distinguished from each other. They are very small, and the largest specimen is hardly equal to the head of a minikin pin. Still, their structure is not very difficult to be comprehended, and a moderately good magnifying-glass will mostly be sufficiently powerful to answer the purpose. The Spiders all breathe by means of certain lung-like organs, called the pulmonary sacs, though some species are also furnished with air-tubes. These sacs communicate with the external air by means of small apertures called "stigmata," which are analogous to the spiracles of insects. There are seldom more than two of these stigmata, and never more than four.

In these strange creatures, the mandibles are furnished with a curved claw, perforated at the extremity, something like the poison-fang of a venomous snake, and used for a similar purpose. A gland furnishes a secretion which is forced through these organs, and is injected into any object that may be wounded by the sharp claw. The fluid which is secreted for the service of the fangs is nearly colorless, and is found to possess most of the properties that exist in the venom of the rattlesnake or viper. The very existence of this fluid is denied by some writers, and its poisonous nature by others. I can, however, state from personal experience, that the bite of an angry Spider inflicts a really painful injury, not very dissimilar to the sting of a wasp. I have seen a lady's hand and arm swollen so as to be hardly recognizable as belonging to the human figure, in consequence of a bite inflicted by a large Spider on the back of her hand.

They all spin those remarkable nets which we popularly call "webs," and which differ wonderfully in the various species. These webs are, in very many instances, employed as traps, wherein may be caught the prey on which the Spider feeds, but in other cases are only used as houses wherein the creature can reside. Some of the uses to which these wonderful productions are put, as well as some details of their structure, will presently be mentioned.

We now pass to the typical species of these curious animals.

The Spiders belonging to the family Mygalidæ may at once be known by the shape of their mandibles and the terrible claws which proceed from them. In the greater number of Spiders, the claws are set horizontally, but in the Mygalidæ they are bent downwards, and strike the prey much as a lion clutches at his victim with his curved talons. Several species of these Spiders are known, most of which attain to considerable dimensions, and some are so enormously large as to become really formidable creatures, which man himself does not like to attack except with a weapon of some kind, or, at all events, with a shod foot.

The GREAT CRAB-SPIDER, which is represented in the fine colored illustration, belongs to the typical genus of this family, and is one of the formidable Arachnida that are said to prey upon young birds and other small vertebrates, instead of limiting themselves to the insects, and similar beings, which constitute the food of the generality of the Spider race. All Spiders are carnivorous, the dimensions of their prey varying with those of the destroyer, and it is by no means an illogical supposition that a Spider whose spread of limb equals that of a human hand, might suck the juices of some of the smaller and more helpless vertebrates.

In Madame Merian's well-known work on the insects of Surinam, there is a careful and forcible sketch of one of these great Spiders (*Mygale avicularia*) engaged in preying upon a humming-bird, which it seems to have taken out of its nest. She gives also a description of this Spider, mentioning that it chiefly feeds upon ants, but that when they fail, it climbs the trees and catches the humming-birds. For a time this account was believed, and the Spider received the specific name of *avicularia* in consequence of its bird-catching propensities. After a while, however, several persons ventured to discredit the story, and at last both the account and the illustration were set down as simple fabrications of the imagination. Experiments were also tried, dead humming-birds being put into the dens of these Spiders, without any result, and the whole of Madame Merian's account was boldly denounced as fabulous.

Yet there were many observers of nature who continued to think that so painstaking a naturalist as Madame Merian, who had spent many years of her life in constant investigations, was not likely to have given so circumstantial an account without some grounds for it. That she was quite correct in saying that the Spider fed generally on ants, was conceded even by her opponents, and it was just possible that she might not be wholly incorrect in the latter part of her statement.

Moreover, they thought that the experiments were by no means conclusive, and that the natural conditions were not fulfilled. It was true enough that when a dead humming-bird was pushed into the nest of a Mygale, the creature did not attempt to eat it, but retreated to the back of its den, or tried to get away. They thought that the Mygale could not be expected to act otherwise, and that there was a vast difference between a dead humming-bird pushed into a burrow in the daytime by a huge heavy-footed biped, and a living humming-bird, asleep at night in its nest upon a tree. An animal of any kind must be left undisturbed, if the observer wishes to gain an insight into its habits; and if he deliberately violates all the conditions, he can hardly expect favorable results. If a practical naturalist wishes to learn whether the Mygale, a nocturnal being, is in the habit of visiting the trees at night and robbing the nests of the humming-birds when it could not obtain its proper supply of ants, he would hardly set to work in so clumsy a manner as to poke a dead humming-bird into the creature's burrow by day.

Surely, the only method would be to ascertain, in the first place, that the Spiders could not obtain the ants on which they usually fed, and then to watch the nests of the humming-birds at night, to see if the Mygale paid them a visit. The experiments were simply futile. Humming-birds never think of getting into subterranean burrows, and if a Mygale saw such a bird making its way into his domicile, he would be justified in running away as fast as he could from so strange a phenomenon. Lately, however, the Mygale has been seen repeatedly to kill the young, not only of the humming-bird, but of other vertebrates, and thus Madame Merian's reputation for veracity remains intact. It is true that, in one or two places, she narrates circumstances which are not true; but then she always takes care to mention that such events were related to her by a third person; and whenever she speaks of any circumstance as having been witnessed by herself, her statements may be implicitly relied upon.

As a proof of her perfect veracity on this habit of the Mygale, I will quote a passage from M. Moreau de Jonnès, who spent many years in Martinique, and watched carefully the habits of these enormous Spiders:—

"It spins no web to serve it as a dwelling. It burrows and lies in ambush in the clefts of hollow ravines, in volcanic tufas, or in decomposed lava. It often travels to a considerable distance, and conceals itself under leaves to surprise its prey, or it climbs on the branches of trees to surprise the colibris (*i. e.* humming-birds) and the *certhia flaveola* (a bird allied to our common tree-creeper). It usually takes advantage of the night to attack enemies, and it is commonly on its return towards its burrow that one may meet it in the morning and catch it, when the dew, with which the plants are charged, slackens its walk.

"The muscular force of the Mygale is very great, and it is particularly difficult to make it let go the objects which it has seized, even when their surface affords no purchase, either to the hooks with which its tarsi are armed, or to the claws which it employs to kill the birds and the anolis (a kind of tree-lizard). The obstinacy and bitterness which it exhibits in combat ceases only with its life. I have seen some which, though pierced twenty times through and through the corslet, still continued to assail their adversaries without showing the least desire of escaping them by flight.

"In the moment of danger, this Spider usually seeks a support against which it can raise itself and mark its opportunity of casting itself upon its enemies. Its four posterior feet are then fixed upon the ground; but the others, half extended, are ready to seize the animal which it is about to attack. When it darts upon it, it fastens itself upon the body with all the double hooks that terminate its feet, and stretches to attain the superior base of the head, that it may sink its talons between the cranium and the first vertebra. In some of the American insects I have recognized the same instinct of destruction.

". . . . . The Mygale carries its eggs inclosed in a cocoon of white silk of a very close tissue, forming two rounded pieces, united at their body. It supports this cocoon under its corslet by means of its antennulæ, and transports it along with itself. When very much pressed by its enemies, it abandons it for an instant, but returns to take it up as soon as the combat is concluded.

"The little ones are disclosed in rapid succession. They are entirely white; the first change which they undergo is the appearance of a triangular and hairy spot which forms on the centre of the upper part of the abdomen.

"I had preserved from 1,800 to 2,000 of these, all of which proceeded from the same cocoon. They were all devoured in the same night by some red ants, which, guided by an instinct that set at defiance all my cares, discovered the box in which I had inclosed the Spiders, and insinuated themselves into it by means of an almost imperceptible aperture, through which myriads of them passed, one by one, in the space of a few hours. It is owing, in all probability, to the destructive war waged upon the avicularia by these insects that the number of these Arachnida is confined within such narrow limits, which by no means correspond with their prodigious capability of reproduction."

The talons of the Spiders are scientifically called by the appropriate name of "falces," the word being Latin, and signifying a reaping-hook. By this name they will be called in the course of the following pages. The falces of the great Crab-spiders are of enormous size, and when removed from the creature and set in gold, they are used as tooth-picks, being thought to possess some occult virtue, which drives away the tooth-ache.

A much smaller example of this family is SULZER'S ATYPUS (*Atypus sulzeri*). This creature is of a peculiar structure. The eyes are mounted on a kind of pedestal or watch-tower, so as to allow the creature to see objects in its front, which would otherwise be hidden by the enormous and elevated mandibles.

This Spider is rare, but has been found in several places. It frequents damp situations, and makes a rather curiously shaped burrow, nearly horizontal at its commencement, but afterwards sloping downwards. The tunnel is lined with a kind of web of white silk, very strongly compacted, which serves to prevent the earth from falling into the burrow. Part of the tube projects outside the entrance, and acts as a protection. The female places her eggs in a little cocoon also composed of white silk, and keeps them at the bottom of the burrow until the young are hatched.

The length of this creature is nearly half an inch, and its color is reddish-brown, becoming paler and more ruddy on the limbs. The male is smaller and darker than his mate.

A CURIOUS spider is the TRAP-DOOR SPIDER of Jamaica, erroneously called the Tarantula.

Like the preceding species, this Spider digs a burrow in the earth, and lines it with a silken web; but, instead of merely protecting the entrance by a portion of the silken tube, it proves itself a more complete architect by making a trap-door with a hinge that permits it to be opened and closed with admirable accuracy. The door is beautifully circular, and is made of alternate layers of earth and web, and is hinged to the lining of the tube by a band of the same silken secretion. It exactly fits the entrance of the burrow, and, when closed, so precisely corresponds with the surrounding earth, that it can hardly be distinguished, even when its position is pointed out. It is a strange sight to see the earth open, a little lid raised, some hairy legs protrude, and gradually the whole form of the spider show itself.

The strength of the membrane is very considerable. One of the nests in my own collection has endured a large amount of rough handling, and has yet suffered but little damage. It will permit a man's finger to be slipped into its interior, and has a very soft and silken feel to the touch. The trap-door has, however, been somewhat injured, as most of the dry earth has been shaken off, and only the layers of web left in their places. I have also several of the Spiders themselves preserved in spirits, and though they have lost their colors, as is the wont of most preserved Arachnida, their falces are very perfect, and the peculiar barbed mandibles are clearly perceptible.

The mode in which these Spiders procure food seems to be by hunting at night, and, in some cases, by catching the insects that are entangled in the threads that the creature spins by the side of its house. There are several species of Trap-door Spiders, and all seem to possess similar habits. In the daytime they are very chary of opening the door of their domicile, and if the trap be raised from the outside, they run to the spot, hitch the claws of their fore feet in the silken webbing of the door, and those of the hind feet in the lining of the burrow, and so resist with all their might. The strength of the Spider is wonderfully great, in proportion to its size, and few persons would anticipate the force of its resistance.

Small Spiders which construct trap-door domiciles are not uncommon in North America. But the discovery of Dr. Holmes, of Charleston, S. C., of a gigantic species on his plantation caused some surprise.

This specimen, with the young just hatched, is in the American Museum, Central Park. The body of the Spider is larger than an ordinary mouse. The legs are short and stout, and, with the body, are covered with coarse hairs. The nest, as it is now seen, is a cylinder of about one and three-quarters of an inch in diameter, and seven inches long. It is like an adobe tube built in the earth. Unlike other Spiders, this does not weave a web, but depends upon his subterranean castle, which it defends with astonishing power. The ground is not only excavated for his purpose, but it is opened as a stone-mason works. Instead of plasters on the sides of the excavation, he digs a large hole, and then commences at the bottom to build a wall of mud, which, on completion, forms a tube. This is closely woven with delicate silvery silk, satin-like in appearance. At the lower end is a small hole left to let out the water or any moisture that may accumulate. In the West Indian species the door is constructed entirely of silken threads. In the present species the door is a flat disc fitting as tightly within the tube, at its entrance, as a stopper to a bottle. This is made of mud, and lined compactly with the silken satin threads. At one point it is hinged and so hung, that, while the creature may pass out by pressing against it, it closes of its own weight when left. Nests are observed to be placed on sloping ground, thus rendering it natural for the doors to close readily.

Of all the many surprising natural objects, of all instances of marvellous beauty of adaptation exercised by the lower forms, this seems especially notable.

Some specimens of Trap-door Spiders and their nests are in the Museum, sent from Mentone, France. They are in blocks of earth removed for the purpose. When these little blocks of earth teemed with the grasses that once grew upon them, the Spiders were pretty safe from intrusion, so far as seeing their nests would lead to it. Now, when nearly all trace of green is vanished, the nests are nearly invisible. There are circular lines, each indicating the periphery of a nest; and these appear as if a metallic punch had been used to punch a sharp outline in the earth. This outline is seemingly perfectly circular. Some nests are supplied with a second, or an offshoot, which is underground. A tightly fitting valve is introduced at the entrance of this. The external hinged door is precisely like a valve. It is fitted exactly to a depressed shoulder, and, in this respect, is like a wide-mouthed glass jar used by druggists.

THE Lycosidæ, or Wolf-spiders, are all ground-livers, and take their prey in fair chase instead of catching it in nets. They are mostly found among herbage, low bushes, fallen leaves, and similar localities; and if they should happen to feel alarmed, they run for safety under stones, mosses, rocks, and into any accidental crevice in the earth. The family includes an immense number of species, which are found in almost every part of the world. They are fierce and determined hunters, chasing their prey wherever it may seek shelter. Some of them are semi-aquatic in their habits, and are not only able to run fearlessly upon the surface of water, but can descend along the aquatic plants until they are deeply immersed, breathing by means of the air which is entangled among the hairy clothing of their bodies.

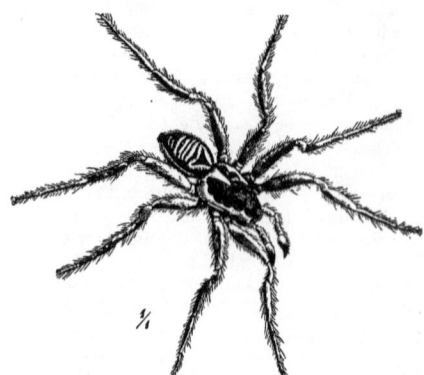

TARANTULA-SPIDER.—*Tarantula apuliæ.* (Natural size.)

The accompanying engraving represents the celebrated TARANTULA-SPIDER, so called from the town of Tarentum, in Italy, where this Arachnid is very plentiful.

There was a deeply-rooted belief among the inhabitants of that town and its neighborhood, that if any one were bitten by the Tarantula he would be instantly afflicted with a singular disease called tarantismus, which exhibited itself in one of two extremes, the one being a profound and silent melancholy, and the other a continual convulsive movement of the whole body. It was also thought that this disease could only be cured by music, and that a certain tune was needful in each particular case.

The disease undoubtedly existed, and might, not improbably, be cured by music; but its source was entirely unconnected with the Tarantula. It ran through towns and villages like wildfire, drawing into its vortex hundreds of persons of both sexes who came within the sphere of its influence. The patients would leap, and dance, and wave their arms, and shriek, and sing, as if the ancient Dionysia were being re-enacted in Christian times; and, indeed, it is by no means unlikely, that the frenzied gesticulations of the ancient bacchanals were attributable to a similar cause. As soon as the music ceased to play, the patients ceased to dance, and fell back into the profound stupor from which the brisk sounds had aroused them. The disease was evidently a nervous affection, tending to propagate itself, like chorea and hysteria at the present day, and, in fact, seems to be little more or less than a rather aggravated form of the former of these maladies—if, indeed, they are not different developments of the same ailments.

That the tarantismus should be cured by music and consequent dancing, is a natural result. The patient indulged in long and continuous exercise, fell into a violent perspiration, fell exhausted, slept calmly, and awoke cured. The Spider, upon whom the odium of this strange disease rested, is perfectly innocent, being as harmless to man as any other Spider, and only formidable to the insects on which it preys.

Another species of Lycosa, inhabiting the south of France, has sometimes been confounded with the true Tarantula of Italy. The habits of this species have been carefully studied by M. Olivier, and have afforded some interesting details respecting the economy of the creature. It frequents dry and uncultivated soils, and sinks therein a little pit, of a depth varying with its size and the length of its residence. The interior of this cell is strengthened with a web. At the entrance of this burrow it sits watching for its prey, and as soon as an unfortunate insect passes within range, it darts forward, seizes it in its talons, and bears the victim away to its den, where it feasts in peace and solitude.

The female is a kind parent, and extremely fond of her eggs and young. She envelopes the eggs in silk, and forms them into a globular ball, which she always carries about with her until the young are hatched. When the time comes for the little spiders to make their appearance in the world, the mother tears open the envelope, and so aids her young to escape. As soon as they are fairly out of the egg, they transfer themselves to the body of their parent, where they cling in such numbers, that she is hardly visible under her swarming brood. They remain with their mother through the winter, and in the following spring the bonds of mutual affection are loosened, and the young disperse to seek their own living.

A Wolf-spider, termed *Lycosa andrenivora*, derives its name from its habit of killing the smaller bees, such as the andrena and its kin. It is mostly found in old pastures and commons, and its color is extremely variable, though brown and yellow are the prevailing tints. A very common species (*Lycosa campestris*) may be seen plentifully in all meadows and pasture-lands, or even on the lawns in our gardens. It is a brisk and nimble creature, running quickly along, as, indeed, is needful for a being that depends on its agility for its living.

About June, the female has made up her little packet of eggs, inclosed in a snowy-white silken envelope, and carries this burden about wherever she goes. Nothing will separate her from her eggs. If the packet is forcibly removed, she remains on the spot, hunting in every direction, and evidently suffering great distress; and if the white ball be laid near her, she soon spies it, darts at it almost fiercely, and carries it off. Her affection for her egg-ball is, however, quite instinctive; like the feeling which induces a hen to sit upon a piece of white chalk, which she takes for an egg. If a little bit of white cotton-wool be rolled up so as to resemble the lost egg-packet, the spider will seize it and make off with it, not at all suspecting the imposition.

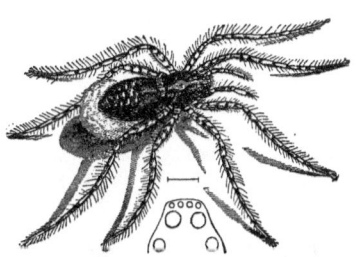

WOLF-SPIDER.—*Lycosa saccata*. Female with the egg-sac. Magnified representation of the eyes as seen from behind, beneath.

There are, on an average, about one hundred eggs in each packet. They are quite round, and very tiny, like the palest yellow translucent dust shot; and their silken covering is drawn so tight, that their globular forms give it an appearance of being embossed. The color of this species is greenish-brown, with a few little dark spots, and the body is also banded and spotted with yellowish-brown of various shades.

Another species of Wolf-spider is shown in our illustration. This is rather a prettier species than that which has just been mentioned, being of a rich chestnut-brown, with a longitudinal bar of yellow along the body, and a number of yellow spots on each side of the bar, where it runs over the abdomen. The colors of the male are rather duller than in the female. The cocoon of this species is yellowish-brown in color, and contains about fifty eggs. A band of slighter texture and lighter hue surrounds the cocoon.

One species of this genus, the PIRATE-SPIDER (*Lycósa pirática*), deserves also a brief notice. This creature is mostly found near water, or on marshy land. It is very quick and active, and can run on the surface of the water without sinking. If alarmed, it immediately takes refuge below the surface, crawling down the stems of aquatic plants, and can remain in that position for a long time. The egg-packet contains about one hundred eggs, generally rather less, and seems to be no impediment to the activity of the mother, who can run over the water even when thus encumbered.

The color of the Pirate-spider is rather complicated. The cephalo-thorax is brownish-black, edged on either side by a white band, and having a dull yellow streak along its centre. Along the upper part of the abdomen runs a chestnut patch, edged with white spots, and having an arch-like mark of pure white, the point of the arch being directed towards the tail of the spider. The rest of the abdomen is simple gray-brown. The male is smaller and duller colored than his mate.

I may here remark, that a full account of these, and many other European Arachnida, may be found in Mr. Blackwall's splendid work on this subject.

A handsome spider, termed *Dolomedes mirabilis*, is found in well-wooded districts. We learn from Mr. Blackwall's researches, that the cocoon of this species is of a dull yellow color, smooth within and rough without, and containing more than two hundred yellow eggs, loosely tied up in the cocoon. She carries her yellow burden under the thorax, and supports it, not only by her limbs, but by some silken threads which serve to bind it to the body. When the young are about to leave the cocoon, the mother spins a rather large silken nest among grass or low bushes. This nest is of a dome-like shape, and under its shelter the young spiders are first set free. They immediately cluster upon the silken lines spun by themselves, and remain under the dome until they are strong enough to go out into the world on their own account.

The color of this fine spider is yellowish-brown, and at each side of the body runs a deep black band, having a narrow white line along its centre. When the female has laid her eggs, she loses these fine tints, and resumes a sober gray color.

Another pretty species belonging to the same genus is termed *Dolomedes fimbriatus*. Its body is nearly an inch in length. Like the Pirate-spider, it is found in the vicinity of water. It will often descend voluntarily below the surface of the water, its respiration being conducted by means of the air-globules which cling to the mass of hairs with which its body is covered.

The cocoon of the female is brown in color, and of considerable size, containing more than two hundred eggs. It is carried, like that of the preceding species, under the thorax. The color of this spider is rich dark brown, with a broad band of yellowish-buff down each side, and a double row of little white spots on the abdomen. The legs are paler, and of a more ruddy hue.

WE now come to the beautiful Hunting Spiders, a family which is spread over the world. They are the very chetahs, or hunting leopards, of the Spider race, and have the mottled beauty as well as the active limbs of the mammalian leopards. They can all run fast, and have also the power of leaping upon their prey to a considerable distance. They are mostly found upon walls, among stones, or upon leaves.

The handsome little spider that is called by the name of *Eresus cinnabarinus* is by no means common, and may indeed be considered as one of the rarest species. It is not very large, being only one-third of an inch in length, but its color is extremely beautiful, the cephalo-thorax being deep velvety-black, edged towards its hinder margin with vermilion, and the whole upper part of the abdomen being colored with the same brilliant hue. On the upper part of the abdomen are six square black spots, the first four being large and the last two small. Each of these spots is edged with pure white, and their effect against the rich scarlet of the abdomen is very fine.

The common HUNTING SPIDER, sometimes known by the name of Zebra-spider, from its boldly-striped markings, is very frequently found, and in the summer time may be seen on almost every wall and tree-trunk, busily hunting for prey. Even upon the window-sills the Hunting Spider pursues its chase ; and as it is very bold and allows itself to be approached quite closely, its proceedings are easily watched. When it sees a fly or other insect which it thinks suitable for food, it sidles quietly in the direction of its intended victim, keeping a most careful watch, and ever drawing nearer to its prey. As the fly moves, so moves the Spider, until the two beings almost seem to be urged by a common instinct. Surely and gradually it makes its way towards the unsuspecting fly, and then, with a leap so quick that the eye can scarcely follow its movements, it springs upon its prey, rolls perhaps over and over in a short struggle, and in a few moments emerges victorious from the contest, its former antagonist dead or dying in its grasp. I have witnessed such a scene hundreds of times, as the garden in which I passed many years was furnished with long ranges of old walls full of crevices that were exactly suited to the purposes of the Hunting Spider.

Even on a perpendicular wall the Spider will make these leaps. It is sure not to fall to the ground, because it always draws a silken cord behind as it moves, and so, whenever it leaps upon its prey, it is saved by its self-woven ladder, and reascends, bearing its dead victim in its grasp.

While engaged in its search, the Hunting Spider is all full of animation. It traverses the wall with great speed and in a very jerky manner, first darting this way, then running that way, then diving into a crevice, then running out and looking around. Sometimes, when it wishes to extend its sphere of vision, it raises the whole front part of the body by simply straightening the fore-legs, and it is surprising what a knowing look it assumes when in that position.

This is a handsome species to examine when under a low power of the microscope, say about twenty-five diameters. Its color is brown, banded obliquely with white. The female does not carry her eggs with her, but wraps them in either one or two cocoons, and hides them in some secure spot, such as the crevices in rocks, and under the bark of trees. Only fifteen or sixteen eggs are placed in each cocoon.

Other species of this interesting genus are termed *Salticus blackwallii* and *Salticus formicarius*. The former is a really large species, measuring one-third of an inch in length. Its color is grayish-black, spotted with a darker hue, and sundry short bands of the same color are drawn diagonally over the cephalo-thorax and the edges of the abdomen. A band of dull yellow is drawn along either side of the abdomen. The latter species is extremely rare, and is remarkable for its ant-like shape. The great mandibles are rather dark brown, and the front half of the cephalo-thorax is nearly of the same hue, but with more black. The entire centre of the body is buff, and the latter half of the abdomen is black-brown, divided from the buff by a white band.

The last species we will mention is the very remarkable *Myrmarachna melanocephala*. It is even more ant-like than the preceding species. Its mandibles are of very great size, and its attenuated abdomen is acorn-like in form. It is a native of Bengal, and is wonderfully like the mutilla, that terrible ant which has already been described on page 401. It is notable for several reasons, among which may be the fact that its head seems to be nearly distinct from the thorax, a structure quite unlike that of the arachnida, from the mygale to the cheesemite. It is thought to eat ants as well as to resemble them. The head, if it may be so called, of this curious Spider is black, and the remainder is red. It is about half an inch in length.

I may mention here, that Spiders, like the crustacea, are apt to be terribly quarrelsome ; and the strangest part of their nature is, that they are most combative during the season of love. In many species, especially those where the male is of insignificant dimensions compared with those of the female, all courtship is conducted under the most unexpected difficulties. A male in love is equally a male in a fright, for if his addresses are not received favorably, he runs a great chance of being eaten on the spot. And even when he has not been repulsed, he still stands in great danger; for many of the Arachnidan beauties are as cruelly deceitful as the enchantress of the "Arabian Nights," and kill their lovers ruthlessly as soon as they have granted their prayers. So, as Alphonse Karr well remarks, the stereotyped exclamation of "Love me, or I die!" is by no means a metaphor, but a simple enunciation of a fact.

When Spiders of nearly equal powers fight with each other, the battle rages vehemently, and if the weaker can escape with life, it is sure to have lost several of its limbs. As with the crustaceans, however, the deprivation is only temporary, for the severed members are reproduced; and though they hardly seem to attain the same dimensions as the original limbs, are yet to a degree serviceable.

The Spiders belonging to the genus Thomisus are, like the hunting-spiders, dependent for their subsistence on their bodily powers and activity. Some, which are rather slow of limb, are in the habit of concealing themselves under leaves or in crevices, and thence pouncing suddenly on the insects that venture too near the treacherous precincts, but the generality are active creatures, running about swiftly, and much resembling the saltici in their movements. Sometimes these creatures are popularly called Crab-spiders, because they can move in any direction without needing to turn their bodies.

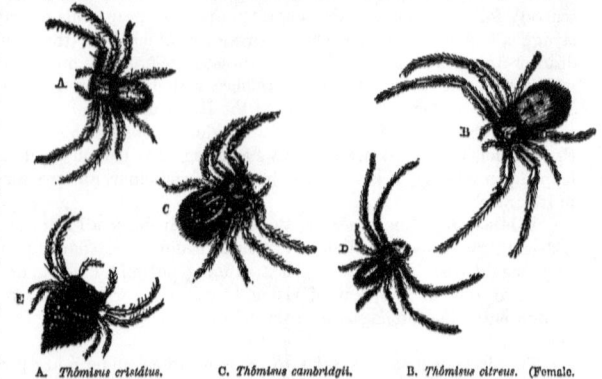

A. *Thômisus cristátus.* C. *Thômisus cambridgii.* B. *Thômisus citreus.* (Female.)
E. *Arkys lancier.* D. *Thômisus citreus.* (Male.)

Fig. A of the accompanying illustration shows one of the common species of Thomisus. It is mostly found on the ground, or lurking among the foliage of old pasture-land. In its color, and indeed in its whole appearance, it is singularly variable, and exhibits so many differences that the simple varieties have been treated by several zoologists as distinct species.

This is one of the many species which, when young, is accustomed to take aërial excursions, and to form that delicate substance popularly known as "gossamer." There is no gossamer spider, as is generally supposed, but many species are in the habit of spinning long loose threads and allowing themselves to be wafted into the air. Lycosæ are very fond of the same curious habit. Sometimes these gossamer webs, each with its minute aëronaut, may be seen floating by thousands in the air, glittering with iridescent light as the morning sunbeams fall on them, and covering the fields with their pearl-strung threads as far as the eye can reach.

The whole question of the spider's web is very curious and interesting; and although our fast waning space will not permit of a full description, a few lines must still be granted to these beautiful structures.

The web is produced primarily from a fluid contained within the body of the spider, and secreted within certain glands, varying in number and dimensions according to the species. Like the thread of the silkworm, this substance becomes hard on exposure to the atmosphere, and is drawn out through tubes of exceeding minuteness. In the silkworm, these spinnerets, as they are called, are two in number, but in the spider they are almost innumerable, so that the apparently single thread of the tiniest spider, minute as it may seem, and really is in fact, is composed of many hundred finer threads all collected into one strand, like the fibres of hemp in a rope. The strength obtained by this form of structure is very great, and the line is not only strong, but elastic, capable of being drawn out like an India-rubber thread and resuming its original length when the extending force is removed.

As regarding the gossamer web, Mr. Blackwall makes the following observations:—
"Although spiders are not provided with wings, and consequently are incapable of flying, in the strict sense of the word, yet, by the aid of their silken filaments, numerous species, belonging to various genera, are enabled to accomplish distant journeys through the atmosphere. These aërial excursions, which appear to result from an instinctive desire to migrate, are undertaken when the weather is bright and serene, particularly in autumn, both by adult and immature individuals, and are effected in the following manner.

"After climbing to the summits of different objects, they raise themselves still higher by straightening the limbs; then, elevating the abdomen, by bringing it from the usual horizontal position into one almost perpendicular, they emit from their spinners a small quantity of viscid fluid, which is drawn out into fine lines by the ascending current occasioned by the rarefaction of the air contiguous to the heated ground. Against these lines the current of rarefied air impinges, till the animals, feeling themselves acted upon with sufficient force, quit their hold of the objects on which they stand, and mount aloft.

"The webs named gossamer are composed of lines spun by spiders, which, on being brought into contact by the mechanical action of gentle airs, adhere together till, by continual additions, they are accumulated into irregular white flakes and masses of considerable magnitude. Occasionally, spiders may be found on gossamer webs after an ascending current of rarefied air has separated them from the objects to which they were attached, and has raised them into the atmosphere; but as they never make use of them intentionally in the performance of their aëronautic expeditions, it must always be regarded as a fortuitous circumstance."

The same writer also remarks that the various directions in which these gossamers are known to sail is in no way attributable to the will of the spider, but merely to the currents of air through which the webs float. He also reviews the opposite opinions regarding the production of the first lines of the web. Some writers say that the spider has the power of projecting its threads in any direction which it may choose, while others assert that it has no such power, and that the creature is forced to wait for a current of air which can bear the slender thread on its breath. After noticing the arguments and experiments on both sides of the question, he comes to the conclusion that the spider is indebted to the air and not to its own projectile capabilities.

Strong and elastic as these webs may be, they have never yet been put to any useful purpose, save to check the bleeding of a cut finger, or to form the cross-wires of an astronomer's telescope. The thread of many species is suitable enough for manufacture, but it cannot be supplied in sufficient quantities. Spiders cannot be kept in any number, as they would be always fighting and eating each other; and they are so voracious that they could not be properly furnished with food, flies being difficult to catch in many parts of the year, and in the cold months quite unattainable. As a proof that if the web could only be obtained in sufficient quantity it might be woven into various articles of apparel, there are now in existence several pairs of gloves, stockings, and other fabrics that have been made, though with very great difficulty, from this substance.

The odd-looking spider called *Arkys lancier* is seen at Fig. E. It is a native of Brazil and the surrounding countries. The cephalo-thorax of the spider is orange-yellow, with a line drawn transversely over it, and changing to a brilliant red at the point on each side. The round spots on the abdomen are bright yellow; the hinder feet are covered only with short down, but those in front are furnished with strong spines.

At Figs. B and D are represented the two sexes of *Thomisus citreus*, for the purpose of showing the great difference in their dimensions and general shape, the female being twice as long as the male, and, as a necessary consequence, very much larger in cubic dimensions. This species is tolerably common, and is usually found on flowers, whether growing in gardens or in the field.

The female is a light citron-yellow, with some dark streaks on the cephalo-thorax, and a double row of round dark spots on the upper part of the abdomen. The yellow color extends over all the limbs. The male, on the contrary, is light leafy-green, with two black bands running down the abdomen, and a darker streak on each side of the cephalo-thorax approaching to brown. The first and second pairs of legs are dark chestnut-brown, while the others are green like the body, so that it is a very pretty-colored creature, and so unlike the female that few persons would believe it to belong to the same species.

A VERY active spider is the *Philodromus dispar*. It can run swiftly even upon polished substances. It is found in well-wooded districts, and is remarkable for the speed with which it runs. The cocoon made by the female is rather large, being nearly a quarter of an inch in diameter, and containing about seventy pale yellow eggs laid loosely in a white cell. This cocoon is not carried about by the female, but is lodged in a larger cell of dull white silk; and this cell is generally placed within a leaf, the edges of which are drawn together by stout lines of the same silken fabric. A dead and already withered leaf is chosen for this purpose.

The color is quite different in the two sexes. The female is rather prettily marked with brownish-chocolate upon a ground color of reddish-yellow, while the male is deep black-brown, with a curious scribbled pattern of a paler hue along the back. The specific name of "dispar," or unlike, is given to the spider on account of this dissimilarity. It is worthy of notice, however, that in the immature state the colors are alike in both sexes. The reader will doubtlessly remember that this is the case with many birds, and that even when the adult male glows with all the hues of the rainbow and the adult female wears a mere dress of sober brown, black, and gray, the young birds are so similar in their plumage that it is hardly possible to distinguish one sex from another.

In a species termed *Philodromus oblongus*, the two sexes are colored in nearly the same manner, and the male is chiefly to be distinguished from his mate by the smaller extremities of the palpi.

Our last example of this genus is the *Philodromus pallidus*. It is a small but rather pretty species, in which the male is rather smaller and slightly darker than the female. The cocoon of this species is slightly made, and white in color, and contains a large number of little spherical eggs, not adhering to each other. The color is pale grayish-brown, profusely speckled with tiny black dots, and marked in a very peculiar manner with dark chocolate-brown. On the upper part of the cephalo-thorax there is a large and nearly triangular patch of this color, with a point directed towards the tail, and around it are arranged several short streaks all converging towards its point. At the end of the abdomen a number of similar stripes are drawn, but without the triangular patch.

A certainly remarkable spider which belongs to another genus, is termed *Sparassus smaragdulus*. The sexes are wonderfully dissimilar, but instead of one sex being brilliantly colored, and the other only tinted with dull hues, as is mostly the case, both sexes are equally beautiful, though with boldly-contrasting colors. This difference of hue is only in the adult spider, as, when immature, the male and female are colored alike.

This spider is more than half an inch in length, and is found in tolerable plenty in northern Europe, its beautiful colors rendering it very conspicuous. The adult female is pale green, with some darker stripes painted, as it were, upon the upper surface of the cephalothorax, and all drawn from the sides towards the centre; while along the middle of the abdomen runs a deep green streak, edged with greenish-white. The male, which is smaller than his mate, has the whole front of the body colored like that of the female. But the abdomen is totally different. The ground color is pinky cream, speckled with brown, and three broad crimson bands are drawn longitudinally throughout its entire length, the central band having several protuberances at intervals.

THE family of the Drassidæ is spread over the greater part of the world. They all have a rather remarkable habit of concealing themselves, not in holes or crevices, but in silken cells, spun by themselves among leaves, under stones, in chinks of walls, and, in fine, wherever their instinct leads them. They are active creatures, and catch their prey by fair chase, in one instance even pursuing the victim beneath the surface of the water.

The species shown at Fig. B never attains to any great size, two-fifths of an inch seeming to be the ordinary length of a female, the male, of course, measuring rather less. It generally resides under stones, and but for that habit would be seen oftener than is now the case. The cocoon of the female is pure white, and contains rather more than one hundred eggs of a very pale yellow color. The cocoon is then placed in a larger sac, also made of white silk, and placed in a hole in the ground. The mother spider generally includes herself in this second sac. The color is alike in both sexes, being of a pale reddish-brown.

At Fig. F. is shown another species of the same genus, and at Fig. E is drawn a portion of a twig, showing the manner in which the female deposits her eggs. The reader will probably have seen these curious little egg pyramids on the branches of various trees. This species is very small, the male measuring barely the eighth of an inch in length. It is rather prettily colored. The limbs are buff, with a large patch of chocolate-brown on the first two

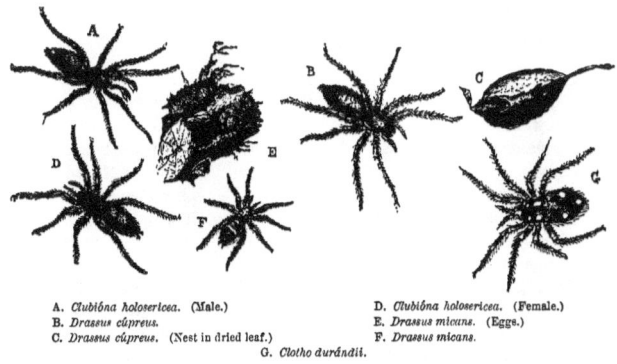

A. *Clubióna holosericea.* (Male.)
B. *Drassus cúpreus.*
C. *Drassus cúpreus.* (Nest in dried leaf.)
D. *Clubióna holosericea.* (Female.)
E. *Drassus micans.* (Eggs.)
F. *Drassus micans.*
G. *Clotho durándii.*

pairs. The cephalo-thorax is ruddy leaf-brown, with six white streaks, their points converging to a spot in the central line. The abdomen is deep black in the shade, but when the light shines upon it, various iridescent hues of purple, green, and copper are given out, rendering the creature a really beautiful species. The name of "micans," or glittering, is applied to the creature on account of its changing colors.

At Figs. A and D are shown the two sexes of a curious and prettily marked spider of moderate dimensions, the female measuring nearly half an inch in length.

This species is mostly found in well-wooded districts, living in a pretty white silken house, which it spins under the shelter of rough bark or shady leaves. The cocoon containing the eggs is placed in this cell, and affectionately tended by the parent. The cocoon is also made of white silk, and generally contains rather more than one hundred spherical eggs. These are very pale yellow in color, and laid loosely in the cocoon. The cephalo-thorax of this species is pale dull green, and the abdomen is soft silken gray, with a peculiar velvety lustre, produced by the dense clothing of hair with which it is covered. The specific name "holosericea," signifies silken, and is therefore very appropriately given to the species.

Another species of this genus is termed *Clubiona nutrix.* It is rather larger than the generality of the Clubionas, being nearly three-quarters of an inch in length. The reader must understand that the length is exclusive of the limbs, and is measured from the front of the cephalo-thorax to the end of the abdomen. This is a very rare creature.

The last example upon this illustration (Fig. G) is a really remarkable creature, whose habits have been studied by M. Dufour. That careful naturalist found it in the Pyrenees, in Catalonia, and in the mountains of Narbonne. It has also been discovered in Egypt and Dalmatia.

Of this Arachnidan, M. Dufour gives a most interesting description, from which the following passage is extracted:—

"It makes at the inferior surface of large stones, and in the clefts of rocks, a cocoon, in the form of a cap, or little dish, a good inch in diameter. Its contour presents seven or eight emarginations, of which the angles alone are fixed upon the stone, by means of bundles of thread, while the edges are free. This singular tent is of an admirable texture; the exterior resembles the finest taffetas, composed, according to the age of the worker, of a greater or less number of doublings.

"Thus, when the Uroctea (another name for the Clotho), as yet young, commences to establish its retreat, it only fabricates two webs, between which it remains in shelter. Subsequently, and, I believe, at each moulting, it adds a certain number of doubles. Finally, when the period marked for reproduction arrives, it weaves a cell for this very purpose, more downy and soft, where the sacs of eggs, and the young ones newly disclosed, are to be shut up. Although the external cap or pavilion is designedly, without doubt, more or less soiled by foreign bodies, which serve to conceal its presence, the apartments of the industrious fabrication are always scrupulously clean."

THE curious and interesting WATER SPIDER is now far better known than was formerly the case, as the aquaria that have been established have tended to familiarize many people with this as well as with many other inhabitants of the water.

This creature leads a strange life. Though a really terrestrial being, and needing to respire atmospheric air, it passes nearly the whole of its life in the water, and, for the greater part of its time, is submerged below the surface. To a lesser degree, several other spiders lead a somewhat similar life, sustaining existence by means of the air which is entangled in the hairs which clothe the body. Their submerged existence is, however, only accidental, while in the Water Spider it forms the constant habit of its life.

Like the pirate-spider, this creature is purposely covered with hairs, which serve to entangle a large comparative amount of atmospheric air, but it has other means which are not possessed by the species already described. It has the power of diving below the surface, and carrying with it a very large bubble of air, that is held in its place by the hind-legs; and in spite of this obstacle to its progress, it can pass through the water with tolerable speed.

The strangest part in the economy of this creature is, that it is actually hatched under water, and lies submerged for a considerable time before it ever sees the land. At some little depth the mother spider spins a kind of egg or dome-shaped cell, with the opening downwards. Having made this chamber, she ascends to the surface, and there charges her whole body with air, arranging her hind-legs in such a manner that the great bubble cannot escape. She then dives into the water, proceeds to her nest, and discharges the bubble into it. A quantity of water is thus displaced, and the upper part of the cell is filled with air. She then returns for a second supply, and so proceeds until the nest is full of air.

In this curious domicile the spider lives, and is thus able to deposit and to hatch her eggs under the water without even wetting them. The reader will have noticed the exact analogy between this sub-aquatic residence and the diving-bell, now so generally employed. As to the spider itself, it is never wet; and though it may be seen swimming rapidly about in the water, yet the moment it emerges from the surface, its hairy body will be found as dry as that of any land spider. The reason for this phenomenon is, that the minute bubbles of air which always cling to the furred body repel the water and prevent it from moistening the skin.

The eggs of this spider are inclosed in a kind of cap-shaped cocoon, not unlike the cover of a circular vegetable dish. This cocoon usually contains about a hundred little spherical eggs, which are not glued together.

The Water Spider is a truly active creature, and its rapid movements can be watched by means of placing one of these Arachnida in a vessel nearly filled with water. If possible, some water plant, such as the vallisneria, or anacharis, should be also placed in the vessel. Here the spider will soon construct its web, and exhibit its curious habits. It must be well supplied with flies and other insects thrown into the water. It will pounce on them, carry them to its house, and there eat them.

It is a tolerably common species, being especially fond of inhabiting quiet and rather deep ditches, where it is well sheltered, and the stream is not rapid enough to endanger the security of its domicile. It is necessary that the water plants to which the nest is fixed should be sufficiently firm to prevent the nest from being swayed on one side, as, in that case, the air would escape, and the water make its entrance. I have often watched its active movements through the water. Whenever it swims, it always keeps its head downwards, just as is the case with a human diver, and it urges itself through the water with quick smart strides of its hairy legs.

The limbs and cephalo-thorax of this species are brown, with a slight tinge of red; and the abdomen is brown, but washed with green. It is densely covered with hairs. On the middle of the upper surface of the abdomen are found round spots arranged in a square. The male is rather larger than the female, and his legs are larger in proportion. He may, however, be distinguished by the large mandibles and longer palpi.

We now come to the family of the Ciniflonidæ.

All these spiders are fond of residing in crevices in rocks, walls, and stones, or under leaves, or sheltered by old projecting bark; and near their hiding-place they weave nets of a most elaborate structure, not flat, like those of the common garden-spider, but inclosing spaces of considerable size in comparison with the small dimensions of their architects. These webs are woven chiefly by means of a peculiar apparatus on the hinder legs, consisting of two rows of parallel and movable spines. The web is most intricate in its arrangements, and connected with the hiding-place of the spider by means of a silken tunnel of variable length, through which the creature darts when it feels the vibration of an insect in its web, and to the bottom of which it retreats if it apprehends danger. Sometimes the spider makes more than one of these tubes.

Several species of Ciniflo are very plentiful, and may be found hidden in their dark silken caverns even in houses. Cellars often contain them, and they frequently swarm in the belfries of old churches. They are extremely ferocious, and mostly seem to be hungry, killing fly after fly with untiring assiduity.

The *Ciniflo ferox* is moderately plentiful, and may be found in old buildings, especially in the dark crevices behind the windows, and under stones. Its length is a little under half an inch. The cephalo-thorax is heart-shaped, of a pale yellowish-brown, and clothed thinly with long black hairs. The abdomen is dark brown, and is variegated with buff markings.

A small, but interesting spider, termed *Ergatis benigna*, is not unfrequent upon heaths and commons, and makes an irregular web at the tips of the gorse and heather. This web passes from one twig to several others, and is studded with the bodies of the captured prey. Within the web the female spider places her cocoons, which are two or three in number, dish-shaped, and are fastened to the stems of the plants upon which the web is built. There are about thirty eggs in each cocoon, and they may mostly be found about June.

The color of the female is very dark brown, upon which is described a bold pattern of buff. The male is smaller, darker, and the markings on the body are of a duller hue. Fierce as is this little creature in its own way, it often falls a victim to the voracious asilidæ, or hornet-flies, which completely reverse the usual order of things, and instead of being devoured by the spider, act the part of its destroyers. The soft skin of this spider is easily pierced by the jaw-lancets of the harvest-fly, and, owing to this structure, the poor little spider learns practically the discomfort of being eaten.

A PRETTY species of another family is the *Agelena labyrinthica*. It is found plentifully on heaths and commons, and derives its specific name from the complicated nature of its web. This is a very large structure when compared with the dimensions of its architect, and is spread almost horizontally over the tops of furze, heath, and the other plants which are found on commons. It is a tolerably massive web, and well calculated for catching prey. Unlike the garden-spider, which boldly sits in the middle of the web, trusting to the delicate meshes escaping the eyes of flying insects, the Agelena does not trust itself in sight, but sits in its dark cavern, which communicates with the web by means of a silken tunnel.

The *Tegenaria domestica*, belonging to another genus, is a fine spider which is mostly found in old houses, chiefly haunting the corners, and spinning a thick, horizontal sheet of web, and forming a rather stout, silken tube as a communication between the web and the den where the spider sits and watches for its prey. Both figures of our illustration are of natural size. The well-known Cardinal-spider, which frightens ignorant persons by its large size, is probably a variety of this species. The web is always very completely supported by guy ropes, which are laid with the greatest care, and disposed as artistically as if arranged by a professional architect.

Like the preceding species, this spider makes several dish-shaped cocoons, surrounds them with a coarse envelope, and covers the white silken cases with bits of old mortar, fragments of wood, particles of whitewash, or any other substance that can be easily obtained. These cocoons are to be seen in June and July.

The color of this species is ruddy brown, dark brown, and buff, the first tint being drawn in a broad band along the whole body, the second being the ground tint, and the third being formed with rows of spots on each side of the central line. The limbs are banded with reddish-brown and black. The male is smaller than the female, darker colored, and has his legs proportionately longer. The average length of the body is nearly three-quarters of an inch, so that the spread of limb is, in a fine specimen, very considerable. This species occasions dire tumults in the minds of housemaids, who sweep away the webs with ruthless broom, and give the spider no hope of a peaceful home.

There are several species of Tegenaria, all having very similar habits. In one species, it has been found that the spider changes its skin, or moults, nine times before they reach adult age, the first moult being achieved while in the cocoon, and the remaining eight after the

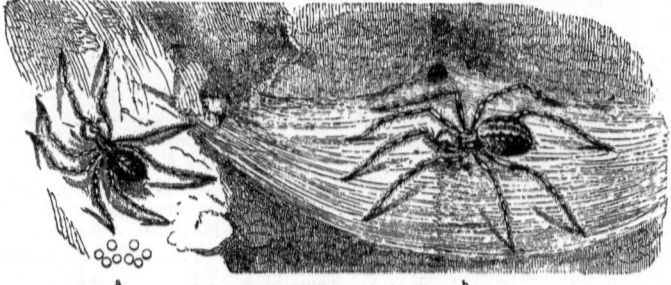

HOUSE SPIDER.—*Tegenaria domestica.* (a, male; and beneath, the position of the eyes; b, female.)

young have left their silken cradle. It has, moreover, been proved by actual experiment, that a limb may be removed at the joint and renewed many times in succession, the new limb appearing after the next moult. The life of this species averages four years.

A PRETTY spider, which is termed *Cœlotes saxatilis*, is rather more than half an inch in length ; but, owing to the shortness of the limbs, it does not present so formidable an aspect as many of less real dimensions. The female of this species makes a curious and rather large cocoon of a dish-like shape, measuring nearly half an inch in diameter, and attaches it firmly to the under surface of stones, by means of a series of strong silken threads. The cocoon is found about May and June. The outer surface of the cocoon is rather profusely studded with patches of mud, in all probability to take off their too great brilliancy.

The color of this species is simple, but pleasing. The cephalo-thorax and limbs are reddish-brown, and the abdomen is yellow, over which is drawn a broad black streak, narrowing as it approaches the extremity of the abdomen.

A very large genus, Theridion, is spread over the greater part of the world. These spiders are mostly of small dimensions, some being extremely minute. Several of the tiny spiders, popularly called Money-spinners, belong to this genus; and, fortunately for themselves, they are protected from destruction by the prevailing notion that it is unlucky to kill a money-spinner.

A rather large species, inhabiting Corsica and known by the name of MARMIGNATTO, or MARMAGNATTO (*Theridion tredecim-guttatum*), seems to be rather a formidable creature, its bite causing much pain, even to man, and, according to Rossi, inducing most serious symptoms, which are only removable by sharp treatment and copious perspiration. It lives in the open fields, and preys mostly upon insects of the grasshopper kind, stretching long threads across the furrows, which serve to entangle the feet of the active insect, and enable the slower Arachnida to make sure of its victims. When the spider finds a locust thus entangled, it further secures the struggling insect by fresh threads spun over its feet and legs; and when it has fairly bound all its limbs, it mounts upon its victim and inflicts a fatal wound at the junction of the head with the neck. As soon as the locust has received the bite, it is attacked with a violent convulsion through its whole frame, and dies almost instantaneously.

This action seems to be universal throughout the Theridia, wherever a spider attacks a large and powerful insect. In Webber's "Song Birds of America," there is an animated account of a battle between a large cockroach and a spider, which seems to belong to this genus. In this case, the cockroach struggled furiously, and was nearly escaping, had not the little spider bethought itself of a new manœuvre. "We had noticed him frequently attempting to bite through the sheath armor of the cockroach, but he seemed to have failed in piercing it. He now seemed determined to catch the two fore-legs that were free. After twenty trials at least, he noosed one of them, and soon had it under his control. This pair of legs was much more delicate than the others; he instantly bit through the captured one.

"The poison was not sufficient to affect the large mass of the cockroach a great deal, but the leg seemed to give it much pain, and it bent its head forward to caress the wound with its jaws; and now the object of the cunning spider was apparent. He ran instantly to the old position he had been routed from on the back of the neck, and, while the cockroach was employed in sóothing the smart of the bite, he succeeded in enveloping the head from the back in such a way as to prevent the cockroach from straightening it out again, and, in a little while more, had him bound in that position, and entirely surrounded by the web. A few more last agonies, and the cockroach was dead, for the neck, bent forward in this way, exposed a vital part beneath the sheath; and we left the spider quietly luxuriating upon the fruit of his weary contest. This battle between brute force and subtle sagacity lasted one hour and a half."

The color of the Marmignatto is deep black, with thirteen round spots on the abdomen, one spot being blood-red.

Another Theridion has been seen to catch its prey in a somewhat similar manner, netting the insect in its silken toils, spinning thread after thread, and binding it tighter and tighter to the spot, and at last killing it when fairly tied down, and then carrying it off to its domicile.

The genus Linyphia. As in the preceding genus, the generality of these spiders are of very small dimensions. One species (*Linyphia triangularis*) is very plentiful, and towards the end of summer or the beginning of autumn, its webs may be seen stretching across the branches. Though but a very little spider, not so large as a grain of rice, it makes webs of wide spread, laid horizontally, and carefully sustained by guy ropes attached to different objects around. Sometimes the guy ropes are so strong, and their elasticity so great, that they actually draw the net out of its flat horizontal direction, and make it swell into a very shallow dome.

The structure of the web is rather loose, and the fibres are necessarily very slender, but is yet strong enough to arrest and detain tolerably large insects. The spider generally remains near the middle of and below the web, and, as soon as a passing insect becomes entangled in the treacherous meshes, the spider runs nimbly to the spot, wounds the insect through the web, and so kills it. The next move is to bite a hole in the web, pull the dead insect through, and then to suck the juices from its body.

The curious spider seen in the illustration is called the *Tetragnatha*. In this spider the jaws are very large, long, widened towards their tips, and diverging from each other. The eyes are nearly of the same size, and are arranged in two regular lines, nearly parallel to each other. The web which this creature spins is vertical, like that of the garden-spider.

WE now arrive at the Epeiridæ, a family containing some of the strangest members of the spider race. The best known of this family is the common GARDEN-SPIDER, sometimes called the CROSS-SPIDER, from the marks upon its abdomen. It is illustrated in the accompanying illustration. This is thought to be the best typical example of all the Arachnidæ. It is found in great numbers in gardens, stretching its beautiful webs perpendicularly from branch to branch, and remaining in the centre with its head downwards, waiting for its prey. This attitude is tolerably universal among spiders; and it is rather curious that the Arachnidæ should reverse the usual order of things, and assume an inverted position when they desire to repose.

MALE OF THE TETRAGNATHON.—*Tetragnathon extensa*. Above the position of the eyes as seen from behind. (Magnified.)

The web of this spider is composed of two different kinds of threads, the radiating and supporting threads being strong and of simple texture. But the fine spiral thread which divides the web into a series of steps, decreasing in breadth towards the centre, is studded with a vast amount of little globules, which give to the web its peculiar adhesiveness. These globules are too small to be perceptible to the unassisted eye, but by the aid of a microscope they may be examined without difficulty. In an ordinary web, such as is usually seen in gardens, there will be about eighty-seven thousand of these globules, and yet the web can be completed in less than three-quarters of an hour. The globules are loosely strung upon the lines, and when they are rubbed off, the thread is no longer adhesive.

Many interesting circumstances can be narrated of this spider, but our space will not permit of more than a brief description. Several species of Epeira are inhabitants of England, and have different habits. The following account of an Epeira and its web is given by the Rev. D. Landsborough, in his "Excursions to Arran":—

a, FEMALE OF THE CROSS-SPIDER.—*Epeira diadema*. b, The eyes as seen from the front. (Magnified.)

"As he was rather a gigantic spider, his tent, instead of being on the ground, was elevated, like the house of a giant of whom in early life we have all read. It was built on the tops of the common grass, *Holcus lanatus*, more than a foot above the ground. Had he built his house on the top of one stalk of grass, the house and its inhabitant might have borne down a single slender stalk. But he had contrived to bring together several heads whose roots stood apart, and, with cordage which he could furnish at will, had bound them firmly together, so that his elevated habitation was anchored on all sides. From whatever *airt* the wind blew it had at once halser and stay. Not only did he bind the heads together, but he bent, doubled, and fastened them down as a thatch roof, under which his habitation was suspended.

"As he was a larger spider than usual, his house was large; the more capacious apartment, which I believe was the nursery, being below; and the smaller one, which was his observatory or watch-tower, being above, from which he could pounce upon his prey, or, in case of hostile attack, could make his escape by a postern gate, so as to conceal himself among the grass.

"During my visit in June last, I was anxious, as we returned from Whiting Bay, to ascertain whether this interesting colony of tent-makers was still in a thriving state, and not seeing any at first, I began to fear that a Highland clearance had taken place. When I at last discovered a few of them, I saw that, as there are times of low trade among our industrious two-footed artisans in town, so are there occasionally hard times among our six-footed operatives in the country. The field in which they encamped had, I suppose, been overstocked. The stately *Holcus* had been eaten down; but these shifty children of the mist had availed themselves of the heather, doubling down the tops of some of the heath-sprigs, and under this thatched canopy forming their suspension-tabernacles. As yet, however, it was too early in the season. The house had only one apartment; the web of which it was formed was as yet thin, so that through it I could see the spider, which, being but half-grown, had not yet got in perfection its fine tiger-like markings. 'Go to the ant, thou sluggard;' go also to the spider. He who taught the one taught the other; and learning humility, let both teach thee."

SEVERAL strange-looking creatures, having their bodies covered with points, knobs, and spines, in a most formidable array, belong to the families termed *Acrosoma*, *Eripus*, *Garteracantha*, and *Iteniza*. These curious spiders inhabit several of the hot parts of the earth, and are remarkable for the extreme hardness of their skin and the brilliancy of their coloring. The skin of these arachnids is as hard and firm as the shelly armor of the crustaceans, and really startling to the touch. There is, however, one spider, the Sclerarachne, which even surpasses them in the hardness of its skin. This is a very small species, with six eyes, a native of Cuba, and evidently forms one of the links between the true spiders and the mites. The name Sclerarachne is of Greek origin, and literally signifies "hard-spider."

During their life-time these spiders literally glitter with resplendent hues, and gleam like living gems set in the deep verdure of the forests. Crimson, azure, emerald, and purple adorn these remarkable Arachnida, and in several species the skin looks exactly as if it were made of burnished gold and silver. After death these glaring colors vanish and change into dull browns and blacks, but in many cases a few relics of the former beauty are still discernible, especially in those specimens where the surface once glittered with metallic radiance.

A collection of spiders belonging to the genera Acrosoma and Gasteracantha presents a most extraordinary appearance. There seems to be no bound to the variety of spines and spikes with which the bodies of these creatures are armed; and had it not been for the lack of space, a few illustrations would have been wholly filled with their strange and weird-like forms. The object of these appendages is quite unknown. Some writers have suggested that they may be intended as defensive armor, and given for the purpose of deterring birds from eating them. But this opinion is quite untenable, as there is no reason why they should be thus guarded more than any other spiders. Indeed, this is another of the many mysteries of zoology, which will never be unveiled until we learn to look beneath the surface and to inquire not only the object of a color or formation, but its meaning.

In the illustration of the GARDEN-SPIDER, only the female is given, which is one of the fiercest Amazons of the spider race; and in case she should object to the attentions of her intended spouse, he must needs flee for his life, a feat which he generally performs by flinging himself out of the web, and lowering himself quickly to earth with his silken ladder. This creature derives its name from a triple yellow cross upon a dark brown band that runs along the central line of the back of its abdomen.

A SPIDER which presents a very strange appearance is called the Nops. It is an arachnid which has only two eyes instead of eight or six, but in which these organs are so enormously large that their dimensions compensate for their paucity of number. On the front portion of the cephalo-thorax there is a black spot, and on this spot are seated the two eyes, round, globular, black, and brilliant. It is one of the hard-skinned species, and appears to be allied both to Gasteracantha and Epeira, in spite of its two eyes.

It is mostly found under stones in woods, and in such localities is tolerably plentiful, but is very rare in houses, though it does sometimes make its appearance in the dwellings. The coloring of the Nops is very simple, the cephalo-thorax being ruddy brown and the abdomen dark brown. It is not a large species considering that it lives in a hot country, measuring rather less than half an inch in length. It is believed to be the only known spider that possesses only two eyes. This species is an inhabitant of Cuba.

Another remarkable arachnid, termed OTIOTHOPS, is especially notable from the fact that its two hinder eyes are united together. This spider is a native of Cuba, and is generally found under stones in well-wooded places, and, like the last-mentioned species, has a hard and shelly skin. In length it is rather under half an inch.

SIX-EYED ARACHNIDA; SENOCULATA.

WITH the exception of the curious spiders just mentioned, the species which have been described bear eight eyes, or rather ocelli, very like the organs of the same name in insects, and arranged upon the cephalo-thorax in various patterns. One well-known writer on the Arachnida has based his system entirely upon the number and arrangement of these ocelli; but the zoologists of the present day seem to think that such a system is insufficient for such a purpose, though very useful—and, indeed, palpably so—as a subordinate means of arrangement. The next group of spiders are in reality separated by the fact that they possess only six eyes, and are therefore called Senoculata, or Six-eyed Arachnida, the preceding belonging to the group of Octonoculina, or Eight-eyed Arachnida.

This species of Dysdera has lately attracted much attention, for, although it is properly a native of Southern Germany, it has lately been discovered in other European countries too.

It can easily be identified by its straight jaws, its powerful falces, and its six eyes arranged in a form something like that of a horseshoe, two small ocelli in front and four larger behind. It has altogether a reddish cast; and its length is more than half an inch.

A pretty spider, which is known under the name of SCYTODES, is found both in Europe and Africa, but only in the hotter parts of the former continent. It may be identified by its six eyes arranged in pairs, and its elegant coloring, which is pinky-white, with two rows of black spots on the abdomen and black rings on the legs. Its eyes are brilliant yellow. The female always uses her jaws in carrying the cocoon, which is about the size of an ordinary pea.

In the illustration of the SEGESTRIUM, both sexes are given, in order to show their different shape and comparative dimensions. The three forms are magnified, and the lines underneath the male and the female indicate the natural size of both.

The Segestrium also has six eyes, and is found in Europe. It lives mostly in hollows of walls and rocks, spinning a silken tube in which it conceals itself, and holding in its feet the lines which communicate with the exterior. The tube is open at both ends, so that when the spider feels either of the lines shaken, it can dart out at once upon its prey.

The common HARVEST-SPIDER, or HARVEST-MAN (*Thalangium longipes*), is a very common and well-known inhabitant of Europe, and, whether in gardens or in the open field, is to be found in very great numbers.

Sometimes the Harvest-spider is seen scrambling over the grass with wonderful speed, its little round body hardly discernible as it moves along, and its long straggling legs looking like animate hairs. Sometimes it prefers to cling to a wall or fence, and there remains perfectly quiet, with its legs stretched out to their full extent, and occupying a wonderful spread of surface. Sometimes again, especially on windy days, it seeks sheltered spots, such as crevices in old walls, or the rough bark on the leeward side of tree-trunks.

One summer day, as I was bathing in the river, just below a lasher, I happened to look under the cross-beam of the wood-work, and there saw something which I took for a mass of black horsehair. Wondering how such a substance could get into such a situation, I went to examine it, and then found that the supposed horsehair was nothing more or less than a legion of Harvest-spiders, all gathered together, their little bodies nearly hidden by their bent legs. There must have been some thousands of the creatures under the beam, all perfectly motionless. An intelligent countryman, to whom I pointed out this curious assemblage, was quite as surprised as myself, never having seen anything of the kind before.

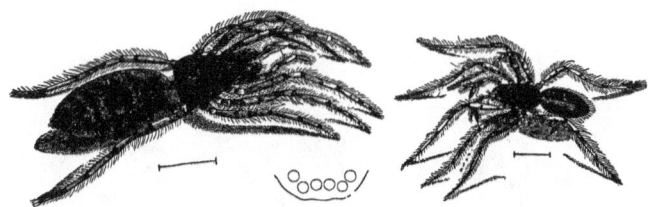

SEGESTRIUM.—*Segestria senoculata*. Male and female; beneath the position of the eyes. (See page 511.)

Like many other very long-limbed creatures, the Harvest-spider seems to set little store by its legs, and will throw off one or two of them on the slightest provocation. Indeed, it is not very easy to find a Harvest-spider with all his limbs complete; and if such a being should be captured, it is nearly certain to shed a leg or two during the process. It appears to be totally indifferent to legs, and will walk off quite briskly with only half its usual complement of limbs. I have even known this arachnid to be deprived of all its legs save one, and to edge itself along by this solitary member, in a manner sufficiently ludicrous. The cast legs contain much irritability, and even after they have been severed from the body continue to bend and straighten themselves for some little time.

A strange genus, termed Gonoleptes, is closely allied to the Phalangium. These curious spiders have the palpi very broad, very flat, and armed with thorns; and the body is flat, expanded behind, and covered with a hard shelly skin. The legs are extremely long, and the hinder pair are longer than the others. All the members of this genus are exotic.

PSEUDOSCORPIONES.

THE formidable-looking arachnid GALEODES, which is represented in the accompanying illustration, by no means belies its appearance, but, from many accounts, seems to be a really dangerous creature. It is drawn of its natural size.

The bite of the Galeodes is much dreaded in the countries where the creature lives, and is said to produce very painful and even dangerous effects. Still, we may leave an ample margin for exaggeration; and when we consider the black catalogue of crimes that are attributed to the newt, the blind-worm, and various other harmless creatures of our own land, we may well imagine that the popular opinion of the Galeodes is not likely to be very favorable.

The Galeodes is fond of warm, sandy situations, and, like many of the Arachnida, is seldom seen except by night, when it comes from its hiding-place in search of prey. Under such circumstances, it is very likely to retaliate if injured by a bare hand or foot, and to inflict a wound causing considerable pain. There are several species belonging to this genus.

Even the scorpion itself is hardly more formidable in aspect than the Galeodes, and to the generality of the insect tribe it is even a more dreadful foe. Armed with two pairs of powerful mandibles placed side by side, like the claws of lobsters without their jointed footstalks, the Galeodes sets off at night in search of prey. It runs with wonderful rapidity, more like a mouse than a spider, and, from the large size to which it sometimes attains, covers much more space than a mouse as it darts over the floor. One specimen I saw was about two inches in length, exclusive of the limbs, and measured exactly ten inches in total length. With straightened limbs the length would have been very greatly increased.

A large specimen of the Galeodes will attack any insect and almost any creature of small size. It has been known to leap upon a lizard, to cling to its back as the combatants rolled about on the ground, to kill it by driving its fangs into the spinal cord at the junction at the head with the neck, and, finally, to eat it entirely with the exception of some of the larger bones. The lizard measured three inches in length, exclusive of the tail.

GALEODES.—*Solpuga, or Galeodes araneoides.*

Much attention has been given to the Galeodes and its habits by Lieutenant-General J. Hearsey, who has kindly communicated to me the following observations:—

When the Galeodes approaches any creature that it desires to attack, it thrusts out its long palpi, touches the body with the rounded tips of those members, and immediately raises them aloft, as if fearful lest they should be injured. The whole action is wonderfully like the manner in which an elephant flings its proboscis in the air after touching anything of which it is not quite sure. The tips of the palpi are rounded and soft, and when they are applied to any object a sort of phosphorescent flame seems to be emitted from them. Having satisfied itself by the touch, the creature rushes in at once to the attack.

In order to ascertain whether the Galeodes would really attack and eat vertebrated animals, an ordinarily-sized specimen was captured and placed under a bell-glass. A very young musk-rat was then inserted under the glass, the Galeodes being on the opposite side. As the creature traversed its transparent prison, it came suddenly on the young musk-rat, which was quite a baby and could not open its eyes. Without hesitation it sprang on the little animal, killed it, and in a very short time had eaten it.

The manner in which the Galeodes kills its prey is really remarkable. The double set of pincers are sharply hooked, like the beak of an eagle, and are capable of being separately opened and shut like lobsters' claws, and of being used conjointly to secure prey between them; and, moreover, the upper joint of each claw can be pushed far over the lower. When the creature seizes a large animal, such as the lizard above mentioned, it buries the pincers in the flesh, and deliberately shears its way onwards, each pair of pincers working alternately, one pair being engaged in holding the prey and the other in cutting.

The same Galeodes was then pitted against a little bat, about three or four inches across the wings. Though small, it was full-grown and lively. When placed under the glass shade, it fluttered about, but was speedily arrested by the spider, which leaped upon it, proceeded to drive its fangs into the neck, and clung so tightly that it could not be shaken off. In vain did the bat try to beat off the enemy with its wings, or to rid itself of the foe by flying in the air. Nothing could shake off the Galeodes; the long legs clung tightly to the victim, the cruel fangs were buried deeper and deeper in its flesh, the struggles gradually became weaker, until the point of a fang touched a vital spot, and the poor bat fell lifeless from the grasp of its destroyer.

The next antagonist of this redoubtable warrior was a scorpion, about four inches in length. The Galeodes seemed nothing daunted, seized the scorpion by the root of the tail, just where it could not be touched by the sting, sawed its way through the tail, severed that deadly weapon from the body, and then killed and ate the scorpion, together with its tail. There was, however, much uncertainty as to its mode of attack in this instance, for no one could exactly ascertain whether it was directed to the one point of safety by chance or instinct. Another similar scorpion was then procured and placed in the glass bell. The Galeodes darted as usual to the attack, but unfortunately seized its foe by the front. The scorpion immediately grasped the Galeodes in its nippers, quickly brought its tail over its back, and by a well-directed stroke succeeded in stinging its enemy. At the moment of receiving the stroke, the Galeodes started back, opened all its limbs, began to quiver throughout its whole frame, and rolled over quite dead.

The color of the Galeodes is palish-yellow, and the tips of the fangs are black. Their surface is very hard and polished; and when the light falls upon them, they gleam as if covered with burnished gold. In a specimen now before me, the array of hairs with which the fangs are fringed glitter as if tinged with the rainbow.

One species of Galeodes inhabits the New World, being found in Havana, but the greater number of them are inhabitants of the hotter portions of the Old World. In India the present species is plentiful, and is apt to be rather annoying, especially to a new-comer.

BOOK-SCORPION.—*Chelifer cancroides.*

In this engraving is seen a much magnified representation of the curious BOOK-SCORPION, or CHELIFER, a little arachnid very much resembling a tiny scorpion without a tail. The body is flattened, and the palpi are much elongated and furnished with a regular claw at the end, like that of a true scorpion. The Chelifer is an active little being, running with much speed, and directing its course backward, forward, or sidewise, with equal ease. It lives in dark places in houses, between books in libraries, and similar localities, preferring, however, those that are rather damp. It does no harm, however, to the books, but rather confers a favor on their owner, feeding on woodlice, mites, and other beings that work sad mischief in a library.

Its general color is brownish-red, and it is remarkable that the palpi are twice as long as the whole body. This, as well as an allied genus called Obisium, is found in Europe. The two genera can be easily distinguished by the cephalo-thorax, that of Chelifer being parted by a cross grove, and that of Obisium being entire.

WE are now approaching the true Scorpions, and pause on the way to describe the remarkable arachnid which is called PHRYNUS (*Phrynus palmatus*). In this, as well as the

THE ROCK-SCORPION.

Scorpions, the abdomen is divided into segments, the palpi are very large and foot-like, and are furnished at their tips with claws like those of the crustacea. The cephalo-thorax is broad, semicircular, and very slightly separated from the abdomen.

Of all the spider race, the Scorpions are most dreaded; and justly so. These strange beings are at once recognized by their large claws and the armed tail. This member is composed of six joints, the last being modified into an arched point, very sharp, and communicating with two poison glands in the base of the joint. With this weapon the Scorpion wounds its foes, striking smartly at them, and by the same movement driving some of the poison into the wound.

The effect of the poison varies much, according to the constitution of the person who is stung, and the size and health of the Scorpion. Should the creature be a large one, the sting is productive of serious consequences, and in some cases has been known to destroy life. Generally, however, there is little danger to life, though the pain is most severe and the health much injured for the time, the whole limb throbbing with shooting pangs, and the stomach oppressed with overpowering nausea. The poison seems to be of an acrid nature; and the pain can be relieved by the application of alkaline remedies, such as liquid ammonia, tobacco ashes, etc. Melted fat is also thought to do good service, and the nausea is relieved by small doses of ipecacuanha. Some of the poison can mostly be brought to the surface by means of pressing a tube, such as a tolerably large key or the barrel of a small pistol, upon the spot; and the duration, if not the severity of the pain, is thereby mitigated. The great ROCK-SCORPION of Africa is much dreaded by the natives, whose only idea of cure is to tie a bandage firmly above the wound, and then make the patient lie down until the effects have gone off.

The Scorpions inhabit most warm countries, and everywhere are held in the greatest detestation. All kinds of precautions must be taken to guard against a sudden wound, for these creatures are very fond of warmth and afraid of light, and therefore crawl into houses, and conceal themselves in the warmest and darkest spots that can be found. They get into beds, creep under pillows, make their way into the toes of boots, crawl into clothes, hide themselves under cushions, and are, withal, so plentiful, that no careful person thinks of thrusting his hand under a pillow or his foot into a shoe without ascertaining that no Scorpion has taken up its abode there.

They are fierce and rapid creatures, perfectly aware of the terrible weapons with which they are armed, and not unfrequently routing a foe only by the ferocity of their aspect. When threatened or alarmed, the Scorpion curls its tail over its body, flourishes the venomed weapon about in a most menacing style, and if it thinks that it cannot conveniently escape, it takes up the offensive, and boldly rushes to the attack, its claws and tail ready for the assault.

It is a rather remarkable fact, that the poison of the Scorpion gradually loses its effect upon a human being, and that a man suffers less and less each time that he is stung. One bold philosopher had the courage to follow out this principle to the furthest extent, and made Scorpions sting him repeatedly until he had become poison-proof, and suffered but little inconvenience beyond the transient pain of the puncture.

The Scorpion, however repulsive in appearance and venomous in action, yet may excite some admiration for its attachment to its young. While they are yet small and feeble, they congregate upon the person of the mother, swarming over her back, her forceps, her limbs, and even clinging to her tail, and exist in such numbers that they quite conceal the outline of their parent. The little Scorpions remain upon the body of the mother until they are about a month old, when they separate, and are able to shift for themselves. It will be remembered that the young of several spiders behave in a similar manner.

In all these creatures the tail is composed of the six last joints of the abdomen, and the powerful limbs, with the lobster-like claws at the tips, are the modified palpi. The eyes of the Scorpions differ in number, some species having twelve, others eight, and others only six; these last constitute the genus Scorpio. The lower surface of the Scorpions has two remarkable appendages, called the combs, the number of teeth differing in the various species. In the Rock-scorpion the teeth are thirteen in number, while in the red Scorpion there are never less than twenty-eight teeth. The Rock-scorpion is a large creature, measuring about six inches in length when fully grown.

Like the other Arachnida, the Scorpion is carnivorous, and feeds upon various living creatures, such as insects and the smaller crustacea. They mostly seize their prey in their claws, and then wound it with the sting, before attempting to eat it. Even the hard-mailed coleoptera, such as the ground beetles, the weevils, etc., fall victims to this dread weapon, while the grasshoppers and locusts fall an easy prey before so terrible a foe.

MITES; ACARINA.

WE will now turn our attention to the little, but annoying, creatures called Mites.

None of the Mites attain large dimensions, and the greater number of them are almost microscopic in their minuteness. Everywhere the Mites are found, in the earth, in trees, in houses, beneath the water, and parasitic upon animals. They haunt our cellars and swarm upon our provisions—cheese, ham, bacon, and biscuits are equally covered with these minute but potent destroyers; and even our flour stores are ravaged by the countless millions of Mites that assail the white treasures. Whether the cause or the effect of the malady, Mites are found in many forms of disease, both in man and beast, and will certainly propagate the infection if they are removed from the patient and transferred to a healthy person. They are even found deep within the structures of the vital organs, and Mites have been discovered in the very brain and eye of man.

A very common and most annoying species is the well-known HARVEST-BUG.

This little pest of our fields and gardens is very small, and of a dull red color, looking exactly like a grain of cayenne pepper as it glides across a leaf. It is seldom seen until June or July, and is most common in the autumn, in some places swarming to such an extent that the leaves are actually reddened by their numbers. They are especially plentiful on the French bean; and I well remember that when I was a little boy I was horribly tortured by the Harvest-bugs, which came from the leaves of the French beans among which I was employed, and, crawling over my shoes, left a scarlet ring of intolerable irritation round my ankles.

While we are walking through the stubble-fields, the Harvest-bug is terribly apt to make successful attacks upon our ankles; and in the case of persons endowed with a very tender skin almost drives the sufferer to the verge of madness. Gilbert White, in his "Natural History," tells us that warreners are "so much infested by them on chalky downs, where these insects swarm sometimes to so infinite a degree, as to discolor their nets and to give them a reddish cast, while the men are so bitten as to be thrown into fevers."

The Harvest-bug does not confine its attacks to human beings, but equally infests horses, dogs, sheep, and rabbits. It burrows under the skin in a very short space of time, and after a little while a red pustule arises, sometimes as large as a pea, occasioning great irritation at the time, and much pain if it be broken or wounded. On account of its red color, the French call the Harvest-bug the ROUGET.

A RATHER pretty species is called *Ixodes venustus*. It derives the name of "venustum," or beautiful, in consequence of the pretty coloring of its surface. The ground color of this creature is deep black, upon which are set some patches of rich orange-red, edged with yellow. The little lines arranged round the body are also yellow, and its legs are red. It is moderately large, being about one-sixth of an inch in length.

Two species are parasitic upon the rhinoceros and the hippopotamus, and derive their name from the creatures which they infest. The HIPPOPOTAMUS-MITE, or TICK, as it is sometimes wrongly called, is of pale straw color above, and deep liver-red below, the limbs being of the same color as the upper surface, but rather paler. The lines and streaks upon the body are black. Its body is decidedly convex, and there is a very slight indication of a thorax. Its length is about a quarter of an inch.

The RHINOCEROS-MITE has also a convex body, the head and palpi are orange, and the blotches upon the body and the limbs are of the same rich hue. This creature is slightly larger than the preceding. It also belongs to Africa, being found on the Borele, sometimes called the Rhinaster (*Rhinoceros bicornis*).

Another species of Ixodes is termed *thorácicus*. All these creatures are furnished with suckers, through which they can draw the juices of the animals on which they are parasitic, and with a peculiar barbed modification of the parts of the mouth, which enables the parasite to anchor itself, as it were, with living grapnels. There is hardly any animal which is not subject to the attacks of these tiresome mites, and even the hard-shelled tortoise itself is not free from them. They fix themselves so firmly with their barbed grapnels that, if they are roughly torn from their hold, they either leave their heads in the wound, or carry away part of the flesh. Under the microscope the head of any Ixodes forms a beautiful object, and is easily prepared by means of Canada balsam and pressure.

These creatures often swarm in thick woods, and attach themselves for the nonce to the leaves of shrubs, at no great height, waiting for the time when some animal may wander near and become their victim. Sometimes they swarm upon an animal to such an extent that they have been known to kill even a horse or an ox from sheer exhaustion. The French call the Ixodes of the dog, the LOUVETTE, and in America all the mites belonging to this group are known by the name of PIQUES.

These "ticks," as they are popularly called, are extremely annoying in tropical countries, where they swarm in every forest, and infest every living creature that passes by, provided its skin be sufficiently soft to be penetrated by their beaks. They are small and flat when they first settle themselves on their victim, but they suck the blood with such vehemence and industry, that they speedily swell and redden, until at last, when fully gorged, they are as large as broad beans, and as easily crushed as ripe gooseberries.

In these countries, after a walk in the forest, every one is obliged to undergo a thorough inspection from head to foot in order to rid himself of the ticks. When found, they must by no means be pulled away, as their barbed heads would then remain in the wound, and cause a festering sore. The proper method of detaching them is to touch them with oil, when they immediately begin to work their way out of their holding places, and may then be removed and killed. Sometimes a tick is only to be found by the pain which it causes. A dull aching pang, for example, shoots at intervals up the arm, and the experienced forester at once begins to look for a tick somewhere about the roots of the fingers. The creature in such a case is usually very small, not very much larger than a cheese-mite, but it still has strength enough to make its presence felt.

Even in the large forests, the ticks are numerous and unpleasant. In some of them, they are far too plentiful to be agreeable; and after a day's walk in the wood I have often been obliged to serve numbers of ticks with an oily notice of ejectment.

A CREATURE but too well known to millers and dealers in corn, is called FLOUR-MITE (*Acarus farinæ*). Although it is a very tiny creature, it contrives to travel over the loose flour with considerable speed. The well-known cheese-mite is closely allied to the Flour-mite. In these creatures the body is covered with numerous stout hairs, which are capable of movement, so that each hair must have at least two muscles, together with their tendons. Despite, therefore, of the minute size of these mites, their structure is not a jot less complicated than that of many larger beings, and possesses a wonderful series of organs of which the higher animals are destitute.

A little vesicle at the end of the foot is a beautiful object in the microscope, especially if the mite can be kept alive while imprisoned under the field of the instrument. In these creatures the females are larger than their mates. The eggs of this mite are oval, very white, and covered with a sort of brown network.

The BEETLE-MITE. This genus is a very large one, containing a great number of species.

Most persons who have been accustomed to see the common Watchman-beetle (*Geotrupes stercorarius*) in its wild state must have noticed the frequency with which the under part of the body is infested with certain pale yellow mites. This particular species is here represented. Sometimes the beetle is so covered with the mites that its whole body swarms with them; but, as a general fact, they confine themselves to the under surface. Many other insects are victims of mites belonging to the genus Gamasus, the humble-bee being very conspicuous in this respect.

Closely allied to the beetle-mite is the terrible RED MITE, so called by the bird-fanciers, in allusion to its color when gorged with blood. When hungry it is of a light yellow color, but when it has fed, the blood shows its ruddy hue through the transparent skin of the mite. It is a very small creature, and lives mostly in the crevices of the cage during the day, coming out to feed at night. I always used to destroy them by inserting insect-powder into the crevices of the cage, dusting the birds well with the same substance, and keeping a small camel's hair brush charged with oil, with which any stray mite could be at once killed.

The well-known SCARLET MITE, so plentiful on banks and in gardens, is covered with a soft down, which gives a very rich and pleasing depth to its color. This species is not of large size, but in the East Indies a species is found which is three or four times larger than our own Scarlet Mite. It yields a bright red dye, and is therefore called *Trombidium tinctórium*.

EXAMPLES of a different family, the Pediculidæ, are the HOG-TICK and the DOG-TICK. The former is found only upon swine, and not universally even upon those animals. It is of moderate dimensions, measuring about one line and a quarter in length. Its thorax is mostly brown, and its abdomen grayish-yellow.

Mr. Denny, in his "Monographia Anoplurorum," gives the following account of the Hog-tick :—

"*Hæmatopinus suis*. This species is found in great abundance on swine, but it does not appear so generally spread as might be expected from the dirty habits of the animals. It most frequently occurs on those fresh imported from Ireland, the Sister Isle. It was many months before I could obtain a single example. Here, in England, I had applied to both farmers and pig-butchers, neither of whom seemed to approve of the idea which I had conceived of *their* pigs being lousy, but referred me to those of the Emerald Isle as being sure to gratify my wishes.

"I accordingly visited a colony just arrived, where I most certainly met with a ready supply. But here they were confined almost entirely to lean animals; and wherever I found a pig fat and healthy, no game were to be seen. In walking, this species uses the claw and tibial tooth with great facility, which act as finger and thumb."

The Dog-tick is apt to be extremely troublesome, not only getting into the fur of the dog, but harboring in their bedding, and almost defying all attempts at destruction. White precipitate seems to be the best solid substance for this purpose, and a very weak solution of nitric acid answers well as a liquid. But, in both cases, the dog must be muzzled to prevent it from nibbling at its fur, and thus imbibing some of the poison. Its color is ashy-flesh, with a slight checkering. The skin is so transparent that the intestine can plainly be seen, of a dull red color. When gorged with blood, the creature becomes of a light scarlet. This species is also found on the ferret.

THE DEER-TICK and the HORSE-TICK refer to another family of these creatures. The Horse-tick is found both on the horse and ass, especially when fresh from pasture, and is very common under such circumstances. It is rather a pretty species, with a light chestnut head and thorax, and may be known by the squared thorax and the long club-like first joint of the antennæ.

The Long-horned Tick, or Deer-tick, is also a common species, and is parasitical on the common fallow deer, assembling in great numbers on the inner side of the thigh. The color of the head and thorax is something like that of the last-mentioned species, except that there is more red in it. Moreover, it can be distinguished by the antennæ, which have the second joint the longest and the third acute. The eyes, too, are large and prominent.

THE four creatures described in the following lines are parasitic on birds. The species called *Menopon pallidum* is unpleasantly familiar to poultry-keepers, swarming among the feathers to such a degree that the hands are often covered with these parasites when the fowls are plucked or even lifted up. They cling very tightly, and are not easily brushed away, as their bodies are smoothly polished, and offer scarcely any resistance. The color of this species is pale straw.

A parasite found on the common swan and other aquatic birds, such as the bean goose and Bewick's swan, is termed *Docophorus cygni*. It is colored after a rather peculiar fashion. The head, thorax, and legs are bright chestnut, smooth and shining; the abdomen is white, except the first segment, which is of the same color as the thorax. There is also a chestnut spot on the third segment, and a row of short, liver-colored bands runs down each side.

A parasite which is not very plentiful, but which is found on various birds, such as the rook, the raven, and the blackbird, is called *Goniódes falcicornis*. It has a hard, shelly surface, and is marked with numerous dark bars. The last example is the SICKLE-HORNED TICK, so called from the shape of its antennæ, which are rather large, flattish, and curved. It is a pretty species, its squared head being of a light chestnut color, and highly polished. The abdomen is broad, rather flat, and of a light tawny yellow, barred with deep red, and having the last segment of the same color as the head. It is parasitical upon the common peacock, and may almost invariably be found, after the death of the bird, congregated in tolerable numbers about the base of the beak.

MYRIAPODA.

IN accordance with the best systems of the present day, the MYRIAPODA are considered as a separate class.

Some writers have placed them at the end of the insects, on account of certain structural resemblances with certain insects in the larval state. There is also a strong resemblance to the Annelida, or Ringed Worms, which will be placed next in order; and, indeed, when we come to examine the lower forms of animal life, we find ourselves quite bewildered with their many relationships, and uncertain as to their true position in the scale of nature. Van der Hoeven, after reviewing some of the difficulties of systematic zoologists, makes the following pertinent remarks:—"Thus is the entire animal kingdom a *net everywhere connected*, and every attempt to arrange animals in a single ascending series must necessarily fail of effect."

The reader will remark that in the spiders the head and thorax are fused together into a single mass, the abdomen remaining separate. In the Myriapoda the reverse of this structure is seen, the head being perfectly distinct, while there is no outward mark to distinguish between the thorax and abdomen.

The Myriapoda are without even the rudiments of wings, and possess a great number of feet, not less than twelve pairs; and in some species there are more than forty pairs of legs. In allusion to their numerous feet, the Myriapoda are popularly called Hundred-legs, and their scientific title is even bolder, signifying ten thousand feet. To this class belong the well-known centipedes, so plentiful in our gardens, and the equally well-known millipedes, found under decaying wood and in similar localities.

In moderate climates none of the Myriapods attain to great dimensions; but in hot countries, and especially under the tropics, they become so large as to be positively formidable as well as repulsive. Even the common centipede of the garden is by no means an attractive being, and there are few persons who can handle one of those creatures without some feeling of disgust.

In all the Myriopada the feet are terminated by a single claw. Some species are totally blind, but those who possess visual organs have two masses or clusters of simple eyes, their number being variable, according to species or in the different stages of development in the same individual.

CHILOPODA.

THE first order of the Myriapods, called by Mr. Newport the Chilopoda, may be known by several characteristics. The head is broad and somewhat prominent, and the segments of the body are unequal, each having a single pair of legs. The mandibles are long, sickle-shaped, sharp, and prominent. The first tribe of the Chilopods has antennæ of great length, longer indeed than the body, very slender, and composed of many joints. The tarsi are also many-jointed, unequal, and very long. The eyes are prominent and rather globular.

The family to which the NOBLE CERMATIA (*Cermatia nobilis*) belongs is known by eight large bone-like plates or shields upon the back, looking very like the ridge tiles on the roof of an out-house. The members of the genus Cermatia, or Scutigera as it is sometimes called, are spread over the hotter parts of the world, and attain their greatest dimensions under the tropics. Specimens of these strange beings are found in the South of Europe, Madeira, many parts of Africa and Asia, Florida, New Holland, and Australia. The eyes of the Cermatia are unlike those of the generality of Myriapoda, the ocelli being crowded together, so that the facets assume a hexagonal form like those of the insects and some of the crustacea.

All the Cermatiæ are exceedingly active, running about on their long legs with an action that reminds the observer of the common harvest-spider. Indeed, the whole creature has very much the look of being composed of a number of harvest-spider's legs attached to the body of a centipede. The Cermatia is carnivorous in its habits, feeding upon insects and having a great predilection for spiders. A full-grown Cermatia will attack even one of the large and formidable spiders of the tropics, and, safe in its shelly mail, succeed in killing and devouring its foe. In the struggle it will probably lose a few legs; but the creature is in no wise fastidious about its proper complement of limbs, and loses six or seven legs with perfect indifference, behaving in this respect like the harvest-spider, the crane-fly, and other "lang-leggit" creatures, whose affection for their limbs seems to be in inverse ratio to their length.

Scolopendra formósa. (Lower figure.)   *Scolopendra lútea.* (Upper figure.)

The color of the noble Cermatia is pale brown, with a yellowish line running down its centre. The limbs are strongly marked with yellowish-brown, green, and rings of blue. It seldom exceeds two inches in length. This species is found in the East Indies and in the Mauritius.

A MYRIAPOD belonging to another family, termed the Lithobiidæ, is called *Lithobius rubriceps*. The members of this family may be known by the fifteen shields upon the back and their sharp, elongated angles. They are found in the open air, hiding under stones—a habit to which is due the title of the family and genus. The name Lithobius is composed of two Greek words, the former signifying a stone, and the latter to live.

In this species the head is large and squared, and of a deep red color. There are fourteen eyes on each side, and they are small and very black. The long antennæ are yellow, and the mandibles are of the same color, deepening into black at the points. The general color of the body is olive-brown, the green tinge being more conspicuous in some individuals than in others, and the legs are yellow. This is a small species, rather less than an inch and a half in length. This species inhabits the south of Spain.

More than twenty species of Lithobius are known, scattered over the greater part of the world, some being found even in comparatively cold countries. Several of them are very prettily colored, such as the Variegated Lithobius (*Lithobius variegatus*), which has a double row of dark spots along its body, and the Black-eyed Lithobius (*Lithobius melanops*), which is of a yellowish-green color, with an orange head, one joint of each leg of the same bright hue, and twelve large black eyes on either side of the head. The Forked Lithobius (*Lithobius forficatus*) is very common in this country, being found plentifully under stones and in similar localities. It is a quick, active creature, of a canary-red color, and is of moderate dimensions, measuring from an inch to an inch and a half in length. It has fifteen pairs of legs.

We now arrive at the true Scolopendræ, which, together with the allied genera, are popularly known by the name of Centipedes. The genus Scolopendra is a very large one, containing about sixty species, most of them inhabitants of the tropics, and many attaining a large size.

The great Scolopendræ are not only unpleasant and repulsive to the sight, but are really formidable creatures, being armed with fangs scarcely less terrible than the sting of the scorpion. These weapons are placed just below the mouth, and are formed from the second pair of feet, which are modified into a pair of strong claws, set horizontally in a manner resembling the falces of ordinary spiders, and terminated by a strong and sharp hook on each side. These hooks are perforated, and are traversed by a little channel leading from a poison gland, like that of the scorpion, so that the venomous secretion is forced into the wound by the very action of biting. These curious weapons cover the first pair of feet and the gnawing organs of the mouth. All the members of the order possess this remarkable modification of the feet, which has earned for them the title of Chilopoda, a term composed of two Greek words, the former signifying a beak, and the latter a foot.

Both the species of Scolopendra figured on the engraving are exotic, and of rather large dimensions. The *Scolopendra formosa* is well deserving of its specific title, which signifies beautiful, on account of the splendid coloring with which it is adorned while living, the feet being orange, with black teeth, and the edges of each segment being bright green. It is a native of the East Indies, and is about four inches in length. The second species is, as its name imports, of a yellowish color, with a deep orange-colored head and appendages. This is a native of the Caribbean Islands, and is of the same length as the preceding species.

Our next example is the GIANT CENTIPEDE, a creature that well deserves its name, sometimes attaining a foot or rather more in length.

This truly formidable being is a native of Venezuela, and possesses a pair of such powerful venom-fangs that its bite is nearly, if not quite, as dangerous as that of the viper.

As an example of the effects of the poisoned wound inflicted by these large centipedes, I may cite a passage from Williamson's valuable work on "Oriental Field Sports":—"Centipedes grow to nearly a foot long, and as thick as a man's little finger; their form is, indeed, flatter, or like tape. When young they are of a clay color, but become darker with age. They bite by means of a pair of strong forceps placed horizontally at their mouth, nearly as large as the hooked thorns on a blackberry bramble, causing much pain and inflammation, and often occasioning fever. Being from their shape so peculiarly capable of secreting themselves, they sometimes occasion very ludicrous accidents. I once saw a friend apply a flute to his mouth to play on it, but scarcely had he begun, when a large centipede fastened to his under lip, causing him to change his note very abruptly. Several have been bitten while smoking their hookahs; and I was myself once made to smart in putting on my gloves, a centipede having taken possession of one of the fingers.

"A very grave and respectable old gentleman, who was remarkably fond of starting an hypothesis and hunting it to death, and who would rather pay the piper than not have his dance out on all occasions, perceived a large centipede deliberately crawling up an old door at Bethsaron Gardens, near Chororinghee. The veteran assured the company that all venomous animals were in their nature inoffensive, and never wounded but when attacked. Experience having satisfied some present to the contrary, an argument arose, and the old gentleman, with much dignity, asserted that he would prove the validity of his position by placing his finger in the centipede's way. He did so, and received such a bite as occasioned a violent fever, from which a critical abscess under his arm-pit relieved him."

There are many of these fierce and venomous creatures scattered over the world, causing no small annoyance to the new-comers, who cannot for a long time look with indifference on a great centipede, some eight or ten inches long, running up the wall close to their heads, or traversing the floor within a short distance of their feet. Among military men the monotony of camp life is sometimes agreeably diversified by a centipede hunt, the creature being chased as eagerly as if it were a fox or a wolf, and neatly captured in a split bamboo, or between two sticks.

So extremely poisonous are the fangs of these myriapods, that they will even kill poultry without much difficulty, while the smaller creatures on which they prey die almost immediately under the bite. The force with which they can grasp is really terrible, the two hooked claws being driven into the flesh until they meet, and holding their position so firmly that the centipede will rather be torn asunder than loosen its grasp. The best way to assure ones self of the force and general structure of these fangs is to procure a specimen that has been preserved in spirits and dissect it, when the powerful muscles that work the poison-feet, the glands which supply the venom, and the perforated passage through which it is discharged into the wound, are easily made out.

The color of the Giant Centipede is bright rusty-red, with a deep green head and antennæ, and blackish or olive-colored feet. A closely allied species of similar dimensions is found in Jamaica.

The nest of these myriapods looks like a rounded object with an aperture on one side. One was found by Mr. Foxcroft while digging for beetles in Sierra Leone. It was formed in a reddish kind of earth, and many of these habitations were discovered in the same locality.

Two more examples of this large genus are the *Scolopendra angulata* and the *Scolopendra variegata*. The latter, a beautiful species, is a native of Demerara, and is generally about five inches in length when adult. Its color is rather rich and striking. The general hue of the upper surface is deep chestnut, and the front edge of the head segment, the hinder edge of the dorsal segments, and the lower surface are light orange. The antennæ are olive-green, and the feet are orange banded with olive.

The second species is found in the Island of Trinidad, and in color contrasts well with the preceding. Its length is not quite so great, measuring less than five inches. This creature is deep green, the lip and mandibles are reddish orange, and the feet are orange and green. It derives its specific name of *angulata* from the sharp angle on the sides of each segment.

WE now take our leave of the true Scolopendræ and pass to other genera.

Both the specimens which we will first describe are remarkable beings; one for its noisy nature, and the other for its phosphorescent power.

In the centipede called *Eucorybas crotalus*, the feet are modified into flat, plate-like appendages. As the centipede moves along it makes a clattering noise with these plates, and derives from this curious habit both its scientific names. The Corybantes were an ancient tribe to whom the education of Jupiter was intrusted when he was sought by his father, who wanted to eat him, and who, in order to cover the sound of his cries, continually danced and played the castanets around the infant. The specific name, *crotalus*, signifies a rattle, and is, therefore, appropriate to the creature.

This centipede is a native of Southern Africa, and is found about Natal. Its color is rusty-brown.

An allied species, but in no way conspicuous for its dimensions, is called *Arthronomalus longicornis*. It is, however, remarkable on another account. It has the power of giving out a tolerably strong phosphorescent light, which is only visible after dark, but is then very conspicuous, and has often caused the centipede to be mistaken for a glow-worm. It is not unfrequently found within peaches, apricots, plums, and similar fruits, when they are very ripe, and lies comfortably coiled up in the little space between the stone and the fruit, where the sweetest juices lie.

The color of this centipede is yellow; its head is deep rust color; its antennæ are very hairy and four times as long as the head segment. There are from fifty-one to fifty-five pairs of legs. Its length varies from two and a half inches to three inches.

Our next example, the *Gonibregmatus cumingii*, is remarkable for the enormous number of rings of which the creature is composed, and the consequent number of legs which are needed to carry it over the ground. Although this species is only from four to five inches, it has no less than three hundred and twenty-two legs. It is a native of the Philippine Islands, whence so many wonderful forms are brought.

The rather harsh generic name of this creature is composed of two Greek words, the former signifying an angle, and the latter the top of the head, and is given to the animal because the front edge of the head is formed into an acute angle. The general color of the species is ashen-gray, and the mandibles are black at their tips.

CHILOGNATA.

WE now come to a new group of Myriapoda, where the creatures have the power of rolling themselves up, more or less completely, like the hedgehog and the pill-woodlouse. A new species of this group has been termed *Zephrónia impressus*. It is a native of Borneo, and was found by Mr. A. R. Wallace. The general color of the creature is rusty-brown, inclining to red; the head is edged with blackish-brown, and the front segment is also edged with the same color. The surface is shining and polished, and all the segments, except that at the end of the body, are marked with deep longitudinal impressions. For this reason, I call the species "impressus." Its length is nearly two inches.

Before leaving these creatures, it may be as well to state that, during the earlier stages of their existence, these animals are much less perfect than when they have reached adult age; they have not their full complement of segments or limbs, nor an equal number of eyes. The metamorphosis, therefore, is complete, and serves to show the relationship between the Myriapoda and the insects.

Some species of this genus are remarkable for their beautiful markings, and the aspect which they present when rolled up for defence.

One of them, the Actæon Millepede, is a native of Madagascar, and was noted by the celebrated female traveller, Madame Ida Pfeiffer. The surface of its body is very polished and shining, and the general color is a livid yellow. A number of tiny puncturations are scattered rather sparely at the back of the head and between the eyes, but in the front they are more numerous, and along the sides they are nearly as thick as the little depressions on the end of a thimble. The species may also be known by the shape of the eighth to the eleventh segments included, which are curiously pointed, looking as if they had been snipped off diagonally with scissors.

The body of another species, termed *Zephrónia versicolor*, is very smooth and beautifully colored, being of a yellowish ground tint, boldly variegated with stripes and spots of deep black, so as to render it extremely conspicuous. The front of the head, the eyes, legs, and antennæ are pale green in the preserved specimens, and are thought to be darker during life. The head is marked with distinct points. This beautiful species is further remarkable from the fact that no two specimens ever seem to be colored exactly alike. On the upper ridge of the face there are from eight to ten little short spines. This species is a native of Ceylon.

These creatures are all natives of the hotter parts of the earth, but there exists an allied example in almost every garden, and certainly in every field throughout the greater part of Europe. This is the PILL-MILLEPEDE (*Glomeris marginâta*). It is found among moss and under stones, and, as it rolls itself up in a manner very similar to that which is employed by the armadillo-woodlouse, is often mistaken for that being. It may, however, be readily distinguished from that crustacean by the simple fact that the legs have their origin on a single line traversing the middle of the under surface, and that when the creature is walking, their extremities do not project beyond the edges of the shelly covering.

Like the armadillo-woodlouse, the Pill-millepede was formerly used in medicine, probably because it looks somewhat like a pill, and may be found among the old stock of druggists' shops, mixed with the veritable armadillo. Both these beings feed on the same substances, namely, decaying animal and vegetable matter. It seems to be rather a gregarious creature, as it is generally found in tolerable numbers in some favored locality.

We now come to another genus, termed Polydesmus. This term is composed of two Greek words, the former signifying "many," and the latter a "bundle," in allusion to the numerous groups of limbs arranged along the body. In all these creatures the body is covered with a hard skin, and the segments are flattened and lengthened at the sides. A handsome species, called *Polydesmus splendidus*, is found in India, and mostly attains the length of two inches. The color and general aspect of this species are rather striking. Independently of the very deep depression of the segments, which has a very strange effect, the color is bold and striking, being deep and very reddish-brown, diversified by an angular spot of bright yellow placed in the hinder angles of each segment. The body is smooth and slightly shining. When alarmed it is able to roll itself into a partial spiral, so as to present merely the hard shelly armor to the foe, and to shield the limbs within the coil.

To a new species belonging to this genus, I propose to give the specific name "granulatus," on account of the peculiar appearance of the body, which is thickly covered with very minute raised tubercles of a white color, such tubercles being called granules in scientific nomenclature. Perhaps I can give a better notion of the idea expressed by the word "granulated," by mentioning that it could be rightly applied to such substances as shagreen.

The general color of this species is rather dark drab, and it may be at once recognized by the peculiar form of the segments, which are flattened and elongated even more than usual in this genus, and are set at their extremities with three distinct teeth. The length of the specimen from which this description is taken is rather more than three inches.

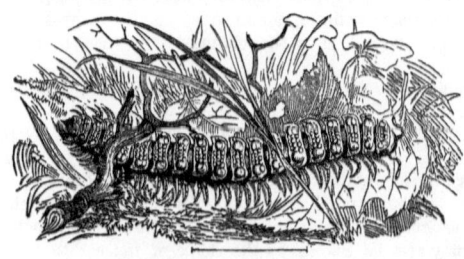

POLYDESMUS.—*Polydesmus complanatus.*

A species of this genus (*Polydesmus complanatus*), represented in the accompanying illustration, is found in Europe, and is not uncommon in gardens. It is about three-quarters of an inch in length, is very narrow, and has thirty-one pairs of feet. The genus is a very large one, and contains a great number of exotic species. The figure of the engraving is magnified.

Before passing to the next large family of myriapods, we may mention the pretty little PENCIL-TAIL (*Polyxenus lagúrus*), a tiny creature which seldom attains a greater length than the twelfth of an inch. It is found under the bark of trees, in clefts of walls, and in moss, and may be known by the twelve pairs of feet, the bunches of little scales on the sides, and the white pencil at the end of the tail.

THE members of the curious family Julidæ are very like those which have just been described, but may be known from them by the fact that the edges of the segments are not flattened and lengthened, but are continued in an unbroken circle. They feed mostly on decaying vegetable matters, but have been seen to eat dead earth-worms and mollusks.

They all exhale a peculiar and rather unpleasant odor, which is caused by a fluid secretion in certain little sacs along the sides, two on each ring. The little apertures through which this scented fluid exudes may be seen on examining the creatures closely, and by some of the earlier writers they were mistaken for spiracles, the sacs themselves being thought to be the breathing apparatus. The real spiracles may be seen on the under sides of the animal, close to the insertion of the feet. Like the preceding creatures, they can roll themselves up, but, on account of the length of their body, they can only assume a spiral form, as is shown by the left-hand figure in the illustration.

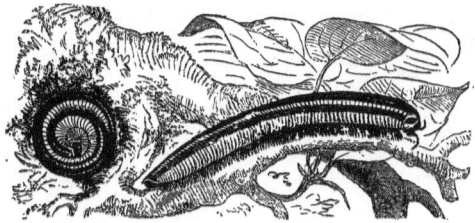

MILLEPEDE.—*Julus terrestris* (Magnified.)

The two figures shown in the accompanying illustration, represent the common MILLEPEDE of the garden. This little creature is very plentiful, and may be found under decaying wood, or below stones. Its movements are very curious. The little delicate feet, looking like white threads proceeding from below, move in a regularly graduated order, so that, as the creature glides along, a succession of waves seem to pass over its body. On being touched it immediately stops, and coils itself into a spiral form, lying necessarily on its side.

The development of the Julidæ is curious and interesting. In the early part of the spring, the female deposits sixty or seventy eggs in the earth, digging a hole expressly for their reception. Here they lie until they are hatched, which occurs in about three weeks' time, when the young Julidæ make their way into the world. They are then without any limbs, and retain the two halves of the egg-shell by means of a filament, which fastens them to the body. After a little while they gain three pairs of feet, and then are able to separate themselves from the egg-shell. At this period of their existence, they bear a great resemblance to the larvæ of some beetles. As they continue to grow, however, the number of segments and limbs increase, so that they gradually lose their resemblance to the beetle larvæ, and attain the shape and form of their parents.

The *Spirostreptes cinctátus* is a native of India, and sometimes attains considerable dimensions, reaching a length of nine inches. It is of a rusty red color, in some individuals inclining to yellowish clay, and has a drab ring round the middle of each segment. The legs also have a ring of the same color round the middle of each joint.

Our last example, the *Spirostreptes annulátipes*, is a creature of large size. This is also an Indian species, and somewhat resembles the preceding, except that its colors are much deeper; there is a narrow black ring round the middle of each segment, and each joint is broadly banded with the same color. There are seventy-five segments in this species, when it has reached full age.

ANNULATA.

A NEW class of animals now comes before us. These creatures are technically called Annulata, or sometimes Annelida, on account of the rings, or annuli, of which their bodies are composed. They may be distinguished from the Julidæ by the absence of true feet, although in very many species the place of feet is supplied by bundles of bristles, set along the sides. The respiration is carried on either by means of external gills, internal sacs, or even through the skin itself.

In most of the Annulata the body is long and cylindrical, but in some it is flattened and oval. The number of rings is very variable, even in the same species; so variable, indeed, that in some specimens of *Phyllodoce laminosa*, no less than five hundred rings have been counted, while others possess only three hundred.

SETIGERA.

THE group of worms which come first on our list is remarkable for the architectural powers of its members. In order to protect their soft-skinned body and delicate gills, they build for themselves a residence into which they exactly fit. This residence is in the form of a tube, and in some cases, as in the Serpulæ, is of a very hard shelly substance, and in some, as the Terebella, is soft and covered with grains of sand and fragments of shells.

The beautiful SERPULA is remarkable for its white shell, its exquisite fan-like branchiæ, and its brilliant operculum.

As may be seen by reference to the accompanying illustration, the shell of the Serpula is tolerably cylindrical, very hard, white, and moderately smooth on the exterior, though it is ridged at intervals, marking the different stages of its formation. The size of the tube increases with the growth of its inmate and architect, so that a perfect specimen is always very small at its origin, and much larger at its mouth. The Serpula is able to travel up and down this tube by the bundles of bristles, which project from the rings along the sides, and is able to retract itself with marvellous rapidity. It has no eyes, and yet is sensible of light. For example, if a Serpula be fully protruded, with its gill-fans extended to their utmost, and blazing in all its scarlet and white splendor, a hand moved between it and the window will cause it to disappear into its tube with a movement so rapid, that the eye cannot follow it. The figure in the illustration is of natural size.

The gills, whose exquisitely graceful form and delicate coloring have always attracted admiration, are affixed to the neck, as, if they were set at the opposite extremity of the body or along the sides, they would not obtain sufficient air from the small amount of water that could be contained in the tube.

The beautiful scarlet stopper ought also to be mentioned. Each set of gills is furnished with a tentacle-like appendage, one of which is small and thread-like, and the other expanded at its extremity into a conical operculum or stopper, marked with a number of ridges, which form a beautiful series of teeth around its circumference. The footstalk on which this stopper is mounted is a little longer than the gills, so that when the animal retreats into its tube the gills collapse and vanish, and the entrance of the tube is exactly closed by the conical stopper.

The Serpula is a lovely inhabitant of the aquarium, but has an inconvenient habit of dying, sometimes coming out of the tube for that purpose, and sometimes retreating to its farthest recesses, and there putrefying, to the great damage of the aquarium. There are several kinds of Serpula, some of which are only attached by the lower part of the tube, and hold the rest of that wonderful structure upright in the water; some, like the present species, intertwine their tubes very much like a handful of boiled macaroni; while others, such as the *Serpula triquetra*, form tubes which do not project at all, but are affixed to their supports throughout their entire length. This species makes a triangular tube. There are many interesting

SERPULA.—*Serpula contortuplicata.*

circumstances connected with the habits and structure of these lovely worms, but our failing space will not admit of a longer description.

We now come to another pretty tube-inhabiting annelid, which is called Sabella, because it lives in the sand and forms its tube of that substance. Several species of Sabella are found on the European coasts, the most common of which is the SHORE SABELLA (*Sabella alveolaria*), a little creature seldom exceeding three-quarters of an inch in length. As is the case with many of these worms, it has a thin tail-like appendage at the extremity of its body, which is doubled up within the tube. The head is furnished with a great number of little threadlike tentacles, which are very flexible, and under a good microscope are seen to have a groove running along the centre, and a double row of teeth along the edges, something like the snout of a saw-fish.

This is a useful species to the naturalist on account of its plentiful occurrence, and readiness to work while in captivity. If a Sabella be watched while it is building up its curious tube, it will be seen to choose the particles of sand with the greatest care, selecting and seeming to balance them with the tentacles, and cementing each in its place with a glutinous secretion, which has the property of setting while under water. If the creature can be induced to build its case against the side of a glass vessel the possessor has cause to be gratified, for the creature does not waste material, and will often make the glass answer for one side of its tube, thereby permitting the observer to watch its entire economy.

The skin of these worms is very tough. I remember once having to dissect the digestive organs for a lecture, and losing hour after hour in my endeavors to make a successful preparation. Just as the lecturer's servant came for the dissection, I had begun a fresh subject, and quite lost patience. So I gave the worm an angry tug with the forceps, when the whole skin of one side stripped off, leaving the digestive organs exposed as beautifully as if they had been carefully dissected.

To give the Sabella a variety of building materials, and to note which it accepts, is always an interesting amusement; for the worm is very fastidious, not to say capricious, in its choice, and always likes to have a stock of materials from which it may make its selection.

While wandering along sandy coasts, we frequently come across some moderately large tubes projecting from the sand, and rather conspicuous in the little puddles left by the receding tide. Round their mouth is usually a set of forked filaments which, like the tube itself, are composed of fragments of sand agglutinated together. The substance of this tube is very soft, but very tough, and will endure a tolerably hard pull without breaking. If the inhabitant of these tubes be sought, it will not be found without much labor, for the TEREBELLA retreats to the farther extremity at the least indication of danger; and as the tube is a foot or more in length, and is always conducted under stones or among rocks, it is not easily dislodged.

As in the case of the Sabella, this annelid performs its architectural labors by means of its tentacles, which are most wonderfully constructed, so as to be capable of extension or retraction, and at the same time can seize or throw away a particle of sand at any part of the tentacle. The method of working is very well given by Mr. T. Rymer Jones:—"If a specimen be dislodged from its tube, it swims by violent contortions in the water, after the manner of various marine annelids; the tentaculæ and the branchiæ are compressed and contracted about the head, like a brush; and as the animal is very soon exhausted by such unnatural exertions, it soon sinks to the bottom. Should a quantity of sand be now scattered from above, the tentaculæ, speedily relaxing, extend themselves in all directions to gather it up, sweeping the vessel quite clean, so that in a very short time not a particle is left behind that is within their reach, the whole having been collected to be employed in the construction of a new artificial dwelling, adapted to shelter the naked body of the architect.

"We will suppose a tube to have been partially constructed into the side of the aquarium, wherein a specimen is about to take up its permanent abode. During the earlier part of the day, the animal is found lurking in its interior, with only the extremities of the tentaculæ protruding beyond the orifice, and so it will remain till towards noon.

"But scarcely has the sun passed the meridian, than the creature begins to become restless; and towards four or five it will be seen to have risen upwards, the tentaculæ extending with the approach of evening, until after sunset, when they are in full activity. They are now spread out from the orifice of the tube like so many slender cords—each seizes on one or more grains of sand, and drags its burden to the summit of the tube, there to be employed according to the service required. Should any of the tentaculæ slip their hold, the same organs are again employed to search eagerly for the lost particle of sand, which is again seized and dragged towards its destination.

"Such operations are protracted during several hours, though so gradually as to be apparently of little effect. Nevertheless, on resuming inspection next morning, a surprising elongation of the tube will be discovered; or, perhaps, instead of a simple accession to its walls, the orifice will be surrounded by forking threads of sandy particles agglutinated together."

There are many species of Terebella. They have, to a considerable extent, the power of reproducing lost portions of the body; and it has been found that even the whole mass of plumy tentacles can be removed without much injury to the Terebella, which retreats to its tube, and after a while reproduces the whole of the missing organs.

The SHELL-BINDER is very plentiful on some coasts, especially those where the shells of various mollusks are found in profusion. The tube of this species is built almost entirely of little fragments of shell, and is of very great length—so long, indeed, and going so deeply into the sand and among the stones, that to procure a perfect specimen is almost an impossibility, except by some rare good fortune. As this creature makes its dwelling about midway between high and low water mark, it may sometimes be procured by setting to work as soon as the tide has retreated, and, with crowbar, pick, and shovel, making the best use of the few hours that can be given to the task. I have never yet succeeded in extracting an entire tube, though I have often tried to do so.

A species of Shell-binder is very common on the white mud of the lagoons of the Florida Reef. It is an interesting view, when gliding over the Reef in a boat, to look over in the shallow water and observe these creatures at work. They construct a tube about three-quarters of an inch diameter, and it projects about two inches above ground. Few objects of nature have arrested our attention with greater wonder than these tube-builders. Here we have a worm, of low organization, and, so far as intelligence is concerned, it might well be at the very foot of the animal scale. Here we have the creature picking up material around it to build a house. It not only picks up material, but it *selects*, as a stone-mason does, the most suitable. A singular circumstance is, that it builds its tube exclusively (its hard parts) of the little lime fronds of calcareous algæ—such as abound in the sand of the Reef. This algæ grows abundantly among the corals. The leaves, or fronds, are small, oval discs, when alive, covered by green vegetable tissue. The worm selects the lime parts and lays them neatly in courses, just as a stone-mason lays his wall. The worm occasionally places a bit of sea-weed in the courses, to aid in concealing the walls. These will be seen introduced in various parts of the tube, falling over and quite effectively breaking up the artificial aspect of the structure, which thus serves as a protective resemblance to the surrounding weed-covered objects. What are our thoughts, in view of this exhibition of "intelligence" in a base worm! If nothing more, it reminds us that human knowledge is finite. The worm goes a step further,—and what additional wonder do we not experience, when we see the creature hunt about for a bit of shell, an entirely different object, and bring it to the tube precisely as we have seen in the case of the trap-door spiders. Here the worm has a house. When he wishes to feed, he pushes his head against the shell door, which yields, and drops to its place when the worm retires. Once in the completed tube, the worm does not leave it entirely. Often the whole structure is concealed by a large piece of alga so fastened to the top that it falls over the structure.

Passing from the tube-inhabiting worms, we now come to those which are free and able to move about at pleasure.

No one who has walked on sandy coasts can have failed to notice the numerous worm-casts which appear in the sand, between high and low water, being most numerous where the sand is level, and becoming scarcer in proportion to the steepness of the slope. Sometimes, when a large, marshy flat makes its appearance, which is never entirely dry even at low water, these worm-casts become so numerous that the foot can hardly be placed between them; and even while the spectator is gazing on the wet sand, coil after coil of dark sand emerges from below, as if Michael Scott's familiars were trying to fulfil their task of making ropes from sea-sand.

These sandy coils are the casts of the Lug-worm, so valuable to fishermen as a bait, and which, when well settled upon the hook, and tipped with a mussel, prove most attractive to the whiting pout, rock cod, plaice, dabs, and other shore-loving fishes. At every low tide the fishermen's boys may be seen busily digging for Lug-worms, or Logs, as they generally term these annelids, and in a populous spot they will fill their square wooden pails in a wonderfully short time.

As a number of Lug-worms lie in a box, covered with sand, mud, and slime, twisting and writhing about in continual movement, they have by no means an attractive aspect, and might even be thought repulsive. But if a single worm be taken from the mass, washed, and placed in a vessel of clear sea-water, it assumes quite a different aspect, and becomes a really beautiful and interesting creature. Its color is very variable, but usually is dark green and carmine, some specimens being almost entirely of the latter hue. Others, again, are nearly brown, and some of a deep red.

Along the sides runs a double row of the wonderful bristles by means of which the creature is enabled to propel itself through the sand, and projecting from the back are thirteen pairs of light scarlet tufts, which, on examination, are found to be the gills of the worm. These are most beautiful organs, and when magnified are seen to be composed of many tufts, like the branches of a thick shrub.

The Lug-worm has some of the habits of the tube-making annelids, for, although it is perfectly free and able to move where it likes, it does not push its way through the sand at random, but forms a tunnel of moderate strength, through which it can pass and repass at pleasure. As it bores its way through the sand, it pours out a small quantity of the glutinous matter which has already been mentioned in the Terebella, and thus cements the sides of the tunnel together in a manner somewhat resembling the brickwork of a railway tunnel. Like that work of engineering skill, moreover, the tube of the Lug-worm cannot bear removal, breaking up when it is unsupported by the surrounding earth. It is, however, amply strong enough for its use, and will withstand the beatings of ordinary waves without yielding.

In the whole of the genus Arenicola there are no eyes nor jaws, and the head is not distinct. Several species of this genus are known.

The Great Eunice (*Eunice gigantea*) is another annelid closely allied to the Nereidæ. In this family the body is very long and composed of numerous segments. The proboscis has at least seven, and sometimes nine pairs of horny jaws. Sometimes it will attain a length of more than four feet, and comprise upwards of four hundred segments in its body, each segment furnished with its paddles, some seventeen hundred or more in number.

When in a living state, this is a most lovely creature, winding along its serpentine course with easy grace, and gleaming with all the colors of the rainbow as the sunbeams fall on its polished surface and active propellers.

An example of the beautiful genus Nereis is now given. The Nereidæ have both tentacles and eyes, and the proboscis is large, often being furnished with a single pair of horny jaws. In the typical genus the eyes are four, arranged in a sort of square, and the tentacles are four in number. The proboscis is thick, strong, and armed with two jaws.

The beautiful Nereids are found plentifully on European coasts, mostly hiding under stones and rocks, or hiding in the sand. They are well worthy of examination under the microscope; and, perhaps, the best method of making out the structure of these beautiful creatures is by taking a single segment and noticing its construction. On the back are seen certain tufts of different shapes in the various species, but all agreeing in being composed of numerous blood-vessels ramifying in a most complicated manner. These are the gills, or branchiæ, of the Nereis.

On each side are seen the organs of locomotion, sometimes consisting of a single, but mostly of a double, row of oars. Each oar is formed of a strong muscular footstalk, from the extremity of which proceeds a bundle of stiff bristles and a variously formed flap, which is technically called the "cirrus." If the bristles be examined separately, their wonderful forms cannot fail to attract admiration. They no longer appear as the simple hairs which the naked eye would assume them to be, but are transformed, as it were, into a very arsenal of destructive weapons, the barbed spear—the scimetar, the sabre, the sword-bayonet, and the cutlass, all being represented; while there is no lack of more peaceful instruments, such as the grapnel, the sickle, and the fish-hook.

The Nereids will live for a time in a shallow basin half filled with sea water, and are, therefore, valuable to those who really desire to study for themselves the beautiful forms with which they are surrounded, and which, but for the microscope, would ever be hidden from our eyes. The observer should not fail to examine the formidable proboscis with its terrible jaws. While the worm is at rest, this proboscis is retracted like the finger of a glove, and the jaws appear to be situated in the neck, where, indeed, they were once taken for a gizzard. But either by dissection or applying pressure in the right direction, the jaws can be drawn out, and are then found to be destructive weapons at the end of the proboscis. Many years ago, while examining, for the first time, a Nereis which I had found on the sea-shore, I took this structure for a gizzard, and find, on reference to my note-book, that a sketch of these internal jaws is marked with the title, "Gizzard of the above."

There are very many species of these interesting worms, among which we may mention the Glow-worm Nereis (*Nereis noctiluca*), a little species seldom more than an inch in length, but which is remarkable for its power of emitting phosphorescent light in a manner that reminds the observer of the luminous centipede already described. There is also the PEARLY NEREIS (*Nereis margaritacea*), so called on account of its pearly-white color. This is a much larger and more handsome species, measuring eight or ten inches in length when fully grown. Another species, the SHINING or IRIDESCENT NEREIS (*Nereis fulgens*), is remarkable from the fact that it constructs a tube of very thin silken texture transparent in itself, but often being studded with particles of sand. It seldom exceeds seven inches in length, and is of a deep orange-red color, with a blackish line running along the back. Like most of the Nereids, it is iridescent on the surface.

Another magnificent species, called the LAMINATED NEREIS (*Phyllodoce laminata*), deserves a passing notice. This fine specimen sometimes attains the length of two feet, and is certainly the finest example of the family that is to be found in the European seas. Its color is shining iridescent green, having a bluish tinge on the back; and changing gradually to a more leaf-green hue on the sides. There are no less than four hundred segments in the body of a full-grown Phyllodoce, and, consequently, eight hundred paddles and sets of bristles, by means of which it can swim through the water or crawl upon the sand with equal ease and grace.

As Mr. T. Rymer Jones well remarks: "The mechanism of this creature, its parts and their powers, are to be ranked among the more conspicuous and admirable works of creation, nor can they be contemplated without wonder. Issuing forth from its retreat, it swims by an undulating serpentine motion. Its unwieldy body, gradually withdrawn from its hiding-place, has its multiplied organs unfolded in regular order and arrangement, so that, whether intertwined or free, they never present any appearance of intricacy or confusion—each part performs its own proper functions, and the general effect is produced by the united exercise of the whole. When inactive, the lateral paddles are laid close over the back, but when in activity they spread widely out, acting like so many oars to aid the animal's course by their united impulse on the water.

"It is a pleasant thing to see a well-manned boat glide over the smooth surface of the sea, or to watch the long array of oars as silently they simultaneously dip and rise again, all flashing in the evening sunshine. But such a sight is but a paltry spectacle compared with that afforded by these gorgeous worms; four hundred pairs of oars, instinct with life, harmoniously respond in play, so active that the eye can scarcely trace their movements, save by the hues of iridescent splendor, violet and blue, and green and gold, the very rainbow's tints that indicate their course."

It is a remarkable fact, that in the Nereids their young are often produced by the simple process of breaking off a piece from the end of the body. The last ring but one becomes swollen and lengthens, and by degrees assumes the appearance of a young Nereid, with its eyes and antennæ. When it is sufficiently strong it is broken off, and goes forth to seek an independent life. Sometimes it happens that a second and a third are thus formed before the first is separated, and M. Milne-Edwards has seen a row of six young Nereids thus attached to their parent.

Many species of Nereids inhabit the sandy beaches of our coast, and offer an interesting field of study. The labors of naturalists attached to the Fishery Commission, at Wood's Holl, have resulted in a great amount of investigation.

The larger forms of this group of invertebrate animals are, many of them, of considerable beauty. The Sea Mouse (*Aphrodite*) is often taken on the hook by the fishermen off George's Banks. Its iridescent spines or hair-like covering render it extremely attractive. These are called Scale-bearing Annelids.

The form mentioned above is *Hermione hystrix*. *Lepidonotus squammatus* is a more common form, found in pools near shore. *L. sublevis* is another, familiar on the New England sea-shore. From their nature it is not likely these forms get to be designated by English terms. Their technical ones, however, are classical, the larger number being named from mythology.

The CIRRHATULUS derives its name from the numerous cirri, or thread-like appendages, which project from its sides, and which serve for legs as well as organs of respiration. These curious appendages arise from the alternate segments of the body, and are continued in two rows along the back almost to the very end of the body.

If placed under a good microscope, the transparent walls of these cirri permit the blood to be seen coursing through them. This is not, however, a very easy operation, as the creature is very timid, and when touched will contract the cirri into a shapeless bundle. When, however, the Cirrhatulus is quite at its ease, recumbent in its rocky home, it permits the cirri to lie flat on the ground, where they surround it like a mass of red worms continually writhing and twining throughout their length.

It is one of the light-hating creatures, always seeking a retreat under some stone or in a cleft of a marine rock; and it is, moreover, protected by a mass of sand, mud, and slime, which it collects around its body, so as effectually to disguise its shape. The length of the Cirrhatulus is about four inches, and its color is mostly red, with a tinge of brown.

The members of the genus SYLLIS may be easily distinguished by the shape of the tentacles, which are jointed in such a manner as to resemble closely the beads of a necklace. The number of the tentacles is always uneven, and this fact serves to separate them from an allied genus, where their number is even.

The SEA-MOUSE, a creature with a hairy coat, possesses beauties which never fail to strike even the unobservant eye of a casual passenger, as the wondrous hues of ruby, emerald, sapphire, and every imaginable gem, flash from the coat of this breathing rainbow. Each hair of the Sea-mouse is a living prism, and when held singly before the eyes is a most magnificent object in spite of its small dimensions, flinging out gleams of changing colors as it is moved in the fingers, or the direction of the light is changed. I have often thought that if Shakespeare had only known of the Aphrodite, he might have furnished Queen Mab with a still more fairy-like conveyance.

Yet the habits of the creature seem to be quite out of accordance with its exceeding beauty. When the sunlight falls on its surface, the many-hued hairs give forth a chromatic radiance which is almost painful to the eyes from its very intensity; and it would be but natural to conclude that the Aphrodite made its home in the sunniest spots, and welcomed the dawn with gladness. Such, however, is not the case; for this beautiful creature, which wears all the colors of the humming-bird and seems equally a child of the sun, passes its life under stones, shells, and similar localities at the muddy bottom of the sea.

The whole group of the Aphroditacea is separated from the rest of the order by reason of the curious mode in which its respiration is conducted. If the beautiful hairs be pushed aside, a series of scales will be seen upon the back, which are guarded by a covering of a loose felt-like substance, composed of interwoven hairs. This felt, if it may be so called, acts as a filter, which is very necessary, considering the muddy localities in which the creature lives, and permits the water to pass in a purified state to the breathing apparatus, which is set beneath the scales. These scales or plates move up and down, something like the gills of a fish, and by their alternating movements have the power of admitting the water and then expelling it in regular pulsations. If a recent specimen be examined, a considerable quantity of mud is always to be found entangled in the felt-like covering of the scales.

Sometimes this beautiful annelid attains a considerable size, reaching the length of five or even six inches. Generally, however, from three to four inches is the measurement. It is a slow-going, but very voracious creature, feeding even upon its own kind, and using its powerful proboscis as a means of capture.

In some species of this genus, the spines which edge the body are most marvellously formed. They are set upon projecting footstalks, and when not required for use can be drawn back into the body. Their shape, however, would seem to render such a proceeding dangerous, inasmuch as they are formed just like the many-barbed spears used by certain savage tribes. In the *Aphrodite hispida*, for example, they are just like doubly-barbed harpoons, and would wound the soft tissues of the body most severely when withdrawn. In order, therefore, to prevent this result, each spine is furnished with a sheath composed of two blades, which close upon the barbs when the weapon is withdrawn, and open again to allow its exit when it is protruded.

Another species, the PORCUPINE SEA-MOUSE, is easily distinguished from the preceding creature by the peculiar structure of the back, which is devoid of the felt-like substance, and the scales are consequently bare.

On account of the singular structure of the CHÆTOPTERUS, it has been placed in a family by itself, of which it constitutes the sole genus.

This remarkable worm is one of the tube-builders, and makes a dwelling of a tough parchment-like consistency, measuring eight or ten inches in length. It is found in the seas about the Antilles. As may be seen by reference to the engraving, in this curious being there is no distinct head, and no vestige of maxillæ, but the proboscis is furnished with a lip, to which are attached two small tentacles. "Then comes a disc with nine pairs of feet, then a pair of long silky bundles, like two wings. The gills, in the form of laminæ, are attached rather below than above, and predominate along the middle of the body." In the illustration the animal is represented in its natural size.

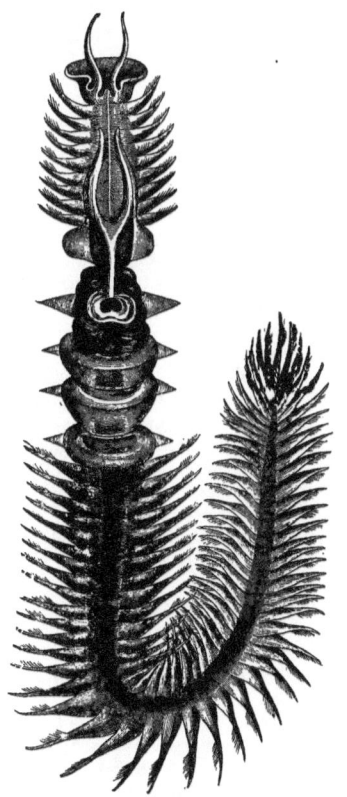

CHÆTOPTERUS.—*Chætópterus pergamentáceus.*

THE next family, of which the common EARTH-WORM is a very familiar example, is distinguished by the ringed body without any gills or feet, but with bristles arranged upon the rings for the purpose of progression.

In the well-known Earth-worm, the bristles are short and very stiff, and are eight in number on each ring, two pairs being placed on each side; so that, in fact, there are eight longitudinal rows of bristles on the body, four on the sides, and four below, which enable the creature to take a firm hold of the ground as it proceeds.

Except that the worm makes use of bristles, and the snake of the edges of its scales, the mode of progress is much the same in both cases. The whole body of the creature is very elastic, and capable of being extended or contracted to a wonderful degree. When it wishes to advance, it pushes forward its body, permits the bristles to hitch against the ground, and then, by contracting the rings together, brings itself forward, and is ready for another step. As in each full-grown Earth-worm there are at least one hundred and twenty rings, and each ring contains eight bristles, it may be imagined that the hold upon the ground is very strong.

As every one knows, the Earth-worm lives a very solitary life below ground, driving its little tunnels in all directions, and never seeing its friends, except at night, when it comes cautiously to the surface and searches for company. In the evening, if the observer be furnished with a "bull's-eye" lantern, and will examine the ground with a very gentle and cautious step, he will be sure to find many worms stretching themselves out of their holes, retaining for the most part their hold of the place of repose by a ring or two still left in the hole, and elongating themselves to an almost incredible extent. If, while thus employed, an Earth-worm be alarmed or touched, it springs back into its hole as if it had been a string of india-rubber that had been stretched and was suddenly released.

The worms have a curious habit of searching for various leaves and dragging them into their holes, the point downwards, and are always careful to select those particular leaves which they best like. As a general rule, they dislike evergreens; and the leaf which I have found to be most in favor is that of the primrose. I have often watched the worms engaged in this curious pursuit; and in the dusk of the evening it has a very strange effect to see a leaf moving over the ground as if by magic, the dull reddish-brown of the worm being quite invisible in the imperfect light.

The food of the Earth-worm is wholly of a vegetable nature, and consists of the roots of various plants, of leaves, and decaying vegetable substances. Many persons cherish a rooted fear of the Earth-worm, fancying that it lives in church-yards and feeds upon the dead. These fears are but idle prejudice, for the worm cares no more for the coffined dead than does the tiger for the full manger, or the ox for the bleeding gazelle. The corpse when once laid in the ground sinks into its dust by natural corruption, untouched by the imagined devourer.

The so-called worms that feed upon decaying animal substances are the larvæ of various flies and beetles, which are hatched from eggs laid by the parent; so that if the maternal insect be excluded, there cannot be any possibility of the larvæ. Moreover, neither the fly nor beetle could live at the depth in which a coffin is deposited in the earth; and if perchance one or two should happen to fall into the grave, they would be dead in half an hour, from the deprivation of air and the weight of the superincumbent soil.

Let, therefore, the poor Earth-worm be freed from causeless reproach; and though its form be not attractive, nor its touch agreeable, let it, at all events, be divested of the terrors with which it has hitherto been clothed.

The Earth-worm is a timid and retiring creature, living below the surface of the ground, and having a great objection to heat and light. Heat dries up the coat of mucus with which its body is covered, and which enables it to slide through the ground without retaining a particle of soil upon its surface. A very moderate amount of heat soon kills an Earth-worm; and if one of these annelids be placed in a spot where it cannot hide itself from the sun's rays, it soon dies, and either melts into a kind of soft jelly, or hardens into a thin strip of horny parchment.

The vexed question of its use to agriculture is too wide a subject to be treated at length in these pages; but we may safely come to two conclusions—first, that unless it were of some use it would never have been made; and secondly, that it will be wiser to find out wherein its use lies than to kill it first and then perhaps discover that its presence was absolutely needful and its absence injurious.

The Earth-worm is of no direct use to mankind, except, perhaps, as bait for the angler; and for this purpose they are easily obtained by the simple process of driving a garden-fork into the ground and shaking it about vigorously. The timid worms are very much alarmed at the tremulous earth, and come to the surface for the purpose of escaping, when they can be easily seized and captured.

SUCTORIA.

THE COMMON LEECH is almost as familiar as the earth-worm, and is one of a genus which furnish the blood-sucking creatures which are so largely used in surgery. It belongs to a large group of Annelida which have no projecting bristles to help them onward, and are, therefore, forced to proceed in a different manner.

All these Leeches are wonderfully adapted for the purpose to which they are applied, their mouths being supplied with sharp teeth to cut the vessels, and with a sucker-like disc, so that the blood can be drawn from its natural channels; while their digestive organs are little more than a series of sacs in which an enormous quantity of blood can be received and retained.

Every one who has had practical experience of Leeches, whether personally a sufferer or from seeing them applied to others, must have noticed the curious triangular wound which is made by the teeth. If the mouth of a Leech be examined, it will be seen to have three sets of minute and saw-like teeth, mounted on as many projections, which are set in the form of a triangle. The wound made by this apparatus is rather painful at the time, and is apt to be troublesome in healing, especially in the case of very thin-skinned persons, requiring the application of strong pressure and even the use of some powerful caustic.

At one meal the Leech will imbibe so large a quantity of blood that it will need no more food for a year, being able to digest by very slow degrees the enormous meal which it has taken. It is a very remarkable fact, that the blood remains within the Leech in a perfectly unchanged state—as fresh, as red, and as liquid as when it was first drawn—and even after the lapse of many months is found to have undergone no alteration.

The very great difficulty in inducing a Leech to make a second meal is well known, and can be well accounted for by the fact that it has already taken enough food to support existence during one-sixth of its whole life. In Europe this is almost impossible, as the time occupied in reducing the Leech to the requisite state of hunger is so long that it more than counterbalances the value of the creature itself. "Use up, and buy more," is the plan that is now pursued by the surgeon.

The Leeches that are used in England are mostly imported from Brittany, where they live in great numbers and constitute an important branch of commerce, being sold by millions annually. The Leech-gatherers take them in various ways. The simplest and most successful method is to wade into the water and pick off the Leeches as fast as they settle upon the bare legs. This plan, however, is by no means calculated to improve the health of the Leech-gatherer, who becomes thin, pale, and almost spectre-like, from the constant drain of blood, and seems to be a fit companion for the old worn-out horses and cattle that are occasionally driven into the Leech-ponds in order to feed these bloodthirsty annelids.

Another plan is to entangle the Leeches in a mass of reeds and rushes, and a third method is, to substitute pieces of raw meat for the legs of the man, and take off the Leeches as they gather round the spoil. This proceeding, however, is thought to injure the health of the Leeches, and is not held in much favor.

Those who keep Leeches, and desire that they should be preserved in a healthy state, will do well to line the sides of the vessel with clay, and to place a bundle of moss, equisetum, and similar materials, with the annelids. These creatures are invested with a coat of slime, and, as is often the case with such beings, are obliged frequently to change their skin. This operation is performed every four or five days, and is rather a troublesome one, unless the Leech be furnished with some such materials as have just been mentioned. Like the snake, when in the act of casting its slough, the Leech crawls among the stems of moss, and thus succeeds in rubbing off its cast garment.

The Leech lays its eggs in little masses, called cocoons, each of which contains, on the average, from six to sixteen eggs. These cocoons are placed in clay banks, and are of rather large size, being about three-quarters of an inch in length. In some parts of France, attempts are made to rear the Leeches; and it is found that these fastidious annelids will not lay their cocoons in small tanks, but require large reservoirs lined with clay and edged with weeds and other aquatic plants.

I may perhaps mention that some European waters contain other species of blood-sucking Leeches, which are found mostly in still or stagnant waters, and invariably gather to a spot where the mud is thick, soft, and plentiful. One summer, while bathing, I waded through some mud in order to pick some very fine dewberries that were overhanging the bank, and when I began to dress found that my feet were covered with Leeches of different sizes. I counted eighteen on one foot, and then found that their numbers were so great that I ceased to count them.

The common HORSE-LEECH, another example of this curious family, is plentiful in ditches and more sluggish rivers. This annelid is distinguished from the preceding by the character of its teeth, which are not nearly so numerous as in the medicinal leech, and much more blunt.

It is a carnivorous being, and feeds upon the common earth-worm, seizing it as it protrudes itself from the banks of the stream in which the Horse-leech resides. There is a popular prejudice against the Horse-leech, the wound which it makes being thought to be poisonous. This, however, is clearly erroneous, and the creature has evidently been confounded with another species, the BLACK LEECH (*Pseudobdella nigra*). The Horse-leech is much larger than the medicinal species, and may be known by its color, which is greenish-black, whereas that of the medicinal leech is green, with some longitudinal bands on the back, spotted with

SKATE SUCKER.—*Albiōne muricata.*

black at their edges and middle; the under surface yellowish-green edged, but not spotted with black.

The figure in the accompanying illustration represents the SKATE-SUCKER, so called because it is found adhering to several fishes, and is especially prevalent on the common skate and others of the ray tribe. Almost all the species of this genus are beset with the curious nodules upon the rings of the body, which give to the creatures so strange an aspect. In this genus, moreover, the portion containing the head is quite distinct and separated from the body by a sort of neck. Our figure is of natural size.

All these creatures have two modes of movement: they can crawl slowly along by means of moving their rings alternately, or they can proceed at a swifter pace by employing a similar mode of progress to that which is made use of by the larvæ of the geometrical moths. Being furnished with a sucker at either end, they first fix their hinder sucker against any object, and then extend the body well forwards. Having secured the sucker of the head, they loosen their hold of the posterior sucker, arch their bodies just like the looper-caterpillars, and so proceed.

Before bidding farewell to the Leeches, we must cast a casual glance at three remarkable members of this group.

The first is the BRANCHELLION, or BRANCHIOBDELLA, a flattish and not very large creature, which is notable for being parasitic upon the torpedo, and retaining its hold in spite of the electric powers of the fish. Another species of the same genus is found on the lobster.

The second of these creatures is the wonderful NEMERTES, a leech-like being not furnished with sucker, and attaining the extraordinary length of thirty or forty feet.

The last of these beings is the LAND-LEECH (*Hirudo zeylonica*), a terrible pest to those who travel through the forests, and often occurring in such vast numbers as even to endanger life.

INTERNAL WORMS; ENTOZOA.

WE now pass to the last members of this great class, the Entozoa, or Internal Worms, so called because they are all found in the systems of living animals. They have also, but wrongly, been termed Intestinal Worms, inasmuch as very many species inhabit the respiratory, or even the sensorial, organs, and are never found in the intestines. The Entozoa are very numerous, and are distributed throughout the world, inhabiting the interior of various living beings; and, indeed, their presence is so universal, that wherever an animal can live, there are Entozoa to be found within its structure.

To give an idea of the wide distribution of these strange beings, we will take one genus of Entozoa as exemplified by the specimens in the Museums, and note the various animals in which the members of that single genus have been found.

The restricted genus Ascaris is the type of its family, and many specimens are in the collection of the Museums. Those have been taken from the following animals: man, mole, dog, fox, cat, seal, wood-mouse, sow, horse, grizzly bear, heron, tortoise (several), toad, frog, ruffe, blenny, fishing-frog, barbel, cod (several), turbot, flounder, eel, goshawk, barn-owl, lapwing, red-wing, cormorant, and grouse.

The history of these remarkable beings is, for the most part, shrouded in mystery, and we know but little of their true habits and the manner of obtaining entrance into the various beings on which they live. It is, however, ascertained that the young of the Entozoa have a very different shape from that of their parents, and that they may obtain entrance into their future homes under the disguise of various forms.

To this curious sub-class belong many remarkable creatures, among which the following may be briefly mentioned. The GUINEA WORM (*Filaria medinensis*) is one of the most developed of this group of animals. It is a strange-looking creature, eight or ten feet in length, and not thicker than ordinary sewing-thread. It is found in many of the hotter parts of the world, especially in the country from which it takes its name. It is also found in America, though it does not appear to be plentiful, except in the Island of Curaçao.

The Guinea Worm is much dreaded by the inhabitants of the countries where it resides, on account of the pain and inconvenience which it occasions, and the great difficulty in destroying it. It mostly takes up its residence in the leg, and there grows to an inordinate length, causing much pain and swelling until the head of the worm makes its appearance. As soon as the sufferer perceives that the worm has made its appearance, he takes a small piece of wood or pasteboard, rolls the projecting end of the worm round it, and, after very cautiously drawing it out for a few inches, he winds up the loosened portion, and ties the stick close to the limb. Next day, he draws out a little more of the worm, and proceeds in a similar manner, until, in a fortnight or three weeks, the entire entozoon has been withdrawn and wound upon the roller.

This process is extremely simple, but demands the greatest care, as, if the worm should be broken, a most painful and even dangerous tumor is certain to arise. I have seen a moderately large specimen of the Guinea Worm extracted from an English sailor, into whose leg it had made an entrance, and who was quite incapable of work until his enemy had been destroyed. It is rather flat, like some kinds of silken thread, and is of a very pale brown color.

Passing by a few genera, such as Ascaris, Tricocephalus, and Strongylus, all of which are found in the human subject, we come to the large and important genus Tænia, which may be accepted as the type of all the Entozoa. The well-known Tape-worm (*Tænia solium*) belongs to this genus. It derives its name of *solium*, as also its French title of Ver Solitaire, from the supposed fact, that only one individual can infest the same person. It is, however, known that a few cases have occurred where the same individual has been afflicted with at least two specimens of Tape-worms. Those which belong to this genus may be known by the long, flat body, and the head with four suctorial spots, and almost invariably a circle of very small but very sharp hooks. The whole structure of the Tænia is interesting to those who study comparative anatomy, but too purely scientific to be described in these pages.

Another species of Tape-worm that is found in the common cat is termed *Tænia crassicollis*. An example of a Tape-worm that is found in several birds, such as the nightingale, blackcap, and the lark, is called, after the last-mentioned bird, Lark-worm (*Tænia platycephala*). Another species of this is found in the black-backed gull.

Tænia, Tape-worm, is a parasitic worm of flattened, tape-like form, living in the intestines of man and many of the lower animals. The long, tape-like creature is made up of many joined together, each joint being regarded as a distinct animal.

The *Tænia solium* is the more familiar form. In a tolerably healthy person it may remain parasitic during a long period.

The treatment for this pest is aimed to dislodge the head; for, if that is left, though all of the body, its numerous joints, be thrown off, the head is a nucleus or starting-point for more, and the disease becomes as troublesome as before. Once the head is dislodged, there is an end of the creature. Pomegranate root and the extract of the root of the male fern are regarded as the most efficient remedies.

There is another notable genus of Entozoa, commonly going by the name of Hydatids, on account of the large amount of liquid which is found within their cells. Within each large cyst, or cell, myriads of smaller cysts may be found, some in an early stage of progress, and others being further advanced, and containing other cysts within themselves.

I have seen some enormous Hydatids taken from the interior of a female monkey. They were so large and so full, that the owner of the animal thought that it had died from over-eating itself. On opening the creature, however, the stomach was found to contain very little food, and almost the whole cavity of the chest and abdomen was filled with huge cysts, which had encroached upon all the vital organs, pushed the heart on one side, enveloped the whole of one lung, and, in fact, had caused so strange a disturbance of the viscera, that the fact of the creature having supported life under such circumstances seemed almost incredible. The large cysts contained a vast number of smaller cells, and these again were filled with cysts of still less dimensions. A large quantity of fluid also existed, and floating in the liquid were found myriads of echinococci, very small, but with the characters exhibited perfectly well under the microscope.

We now come to our last examples of the Entozoa.

The RAY-WORM is, as its name imports, an inhabitant of fishes belonging to the ray tribe. The genus to which this creature belongs is a rather remarkable one, the head affording some curious characteristics. It is very large, in comparison with the body, and has four deep clefts set opposite to each other. Some of the species have only two of these clefts, which, in the Greek language, are called "bothria," but in all the genus either two or four of these clefts are to be found. The generic title literally signifies "furrow-headed."

One species of this genus (*Bothriocephalus latus*) is the well-known BROAD TAPE-WORM, which is so injurious to man whenever it takes up its abode within a human being. This creature will sometimes attain a length of twenty feet; and it may always be recognized by the peculiarly deep and opposite furrows on the head. Like all its kind, this Tape-worm increases with great rapidity, multiplying its joints by division, and in that manner extending itself to the great length which has already been mentioned.

The whole history of these strange creatures is very obscure, on account of the impossibility of watching them in the spots wherein they take their residence. They are all, as is generally the case with beings low in the scale of nature, exceedingly tenacious of life, and will survive treatment which would kill many beings apparently stronger and more capable of resistance than themselves. For example, several species of Entozoa have been found in a living state within meat after it has been cooked, especially in those spots where the heat had not penetrated sufficiently to destroy the natural ruddiness of its color. It is known that "measly" pork derives its peculiar aspect from the presence of Entozoa, and that many of these Entozoa, or, at all events, their eggs, are swallowed by human beings, within whom they make their homes, and whom they condemn to infinite worry, pain, and weakness.

Even immersion in alcohol does not seem to inflict any serious damage upon these creatures. Rudolphi found a specimen of an Ascaris quite lively in the throat of a cormorant that had been steeped in spirits of wine for nearly a fortnight. Even the severe process of being thoroughly dried appears to be quite as ineffectual to destroy these beings. A number of Ascarides that had been removed from a fish, and suffered to become quite dry, and were apparently nothing more than flat slips of membrane adhering to a board, began to revive as soon as they were wetted, and actually moved the moistened part of their bodies, while the remainder was still dry and adhering to the board.

THE TRICHINA is a parasitic nematoid worm, which infects the muscular tissue of the pig, the rat, and some other animals, and is liable to occur in man. It is usually a quiescent incysted parasite, occupying in great numbers, often, the voluntary muscles. The process of development of Trichinæ in the intestine, and the dispersal of their young throughout the system, produces in men and animals a severe and often fatal illness, known as *trichinosis*—severe intestinal irritation, with fever, are common symptoms. No remedy is found effectual in staying the ravages of this dreadful parasite, other than extreme cleanliness and caution.

The use of food imperfectly cooked is the great source of trouble. The fat of pork is not injurious, as the Trichinæ never infect that portion. The muscular parts, as in sausages, fresh pork, ham, and the lean parts of bacon, are liable to produce the disease. The parasite is so small that the butcher often cannot tell whether any given piece of pork is affected or not. Neither pickling nor smoking, as ordinarily practised, will destroy the life of the pest. The only protection is by very thorough cooking. The Trichinæ are killed by a temperature of 160° Far. Meat that is subjected a short time to this temperature is harmless. All parts of the piece of meat should be carefully heated at that point, as a ham may be at 160° on the surface, while inside the temperature may be much lower. The only safe rule is to boil until the meat is of a uniform color throughout.

RADIATA, OR ECHINODÉRMATA.

WE now arrive at a vast and comprehensive division of living beings, which have no joints whatever, and no limbs, and are called Radiata, because all their parts radiate from a common centre. The structure is very evident in some of these beings, but in others the formation is so exceedingly obscure, that it is only by anatomical investigation that their real position is discovered.

The highest forms in this division have been gathered together in the class Echinodermata. This word signifies Urchin-skinned, and is given to the animals comprising it because their skins are more or less furnished with spines, resembling those of the hedgehog. In these animals the radiate form is very plainly shown, some of them assuming a perfectly star-like shape, of which the common star-fishes of our coasts are familiar examples. In some of the Radiates, such as the sea-urchin, the whole body is encrusted with a chalky coat, while in others it is as soft and easily torn as if it were composed of mere structureless gelatine.

The mode of walking, or rather creeping, which is practised by these beings, is very interesting, and may be easily seen by watching the proceedings of a common star-fish when placed in a vessel of sea-water. At first it will be quite still, and lie as if dead, but by degrees the tips of the arms will be seen to curve slightly, and then the creature slides forward without any perceptible means of locomotion. If, however, it be suddenly taken from the water and reversed, the mystery is at once solved, and the walking apparatus is seen to consist of a vast number of tiny tentacles, each with a little round transparent head, and all moving slowly but continually from side to side, sometimes being thrust out to a considerable distance, and sometimes withdrawn almost wholly within the shell.

These are the "ambulacræ," or walking apparatus, and are among the most extraordinary means of progression in the animal kingdom. Each of these innumerable organs act as a sucker, its soft head being applied to any hard substance, and adhering thereto with tolerable firmness, until the pressure is relaxed and the sucker released. The suckers continually move forward, seize upon the ground, draw the body gently along, and then search for a new hold. As there are nearly two thousand suckers continually at work, some being protruded, others relaxed, and others still feeling for a holding-place, the progress of the creature is very regular and gliding, and hardly seems to be produced by voluntary motion.

The Echinoderms of our North American coast are not conspicuous for beauty, and, consequently, are not well known popularly. At this day it is probable that very few persons that visit the sea-shore and pick up one of the common sea-urchins that are seen there, would have the slightest conception what it is—indeed, most people would question its kingdom—whether vegetable or animal. In the absence of information, no wonder; but is it not an undesirable state of things, that the nature of the most common objects of our sea-shore should be so completely unfamiliar?

We will now proceed to our examples of these curious beings.

The HERMIT SIPUNCULUS (*Sipunculus bernhardus*) is a long, slender, worm-like being. It is a creature which is remarkable for the fact that it resides in the empty shells of mollusks, after the same fashion as is observed by the hermit crabs.

If taken out of the shell, the Sipunculus resembles a worm so closely, that it might easily be mistaken for an annelid; and, indeed, according to one of our best zoologists, it forms a link between these two great divisions, for in its person radiism sets and annulism begins. The end of the body, which is concealed within the shell, is capable of being enlarged into a bulb-like shape, which enables the creature to maintain a firm hold of its shelly retreat, and the other extremity is furnished with an external proboscis, at the end of which is a small circlet of tentacles.

Several species of this genus are eatable and held in great estimation by the Chinese, who catch them in a very ingenious manner. The EDIBLE SIPUNCULUS lives in holes in the sand, and always keeps the mouth of its burrow open. The Chinese fishermen arm themselves with a bundle of slender wooden rods, tapered to a point at one end, and having a little round knob at the extremity of the point. They proceed to the sands at low water, and drop one of these rods into each burrow, where they leave it for twelve hours. During this time the Sipunculus is sure to swallow the button, and as the elastic tissues contract it is unable to release itself; and when the tide has again retreated, the rod, with the Sipunculus attached, is drawn out of the burrow by the fishermen.

The species which we are now examining is very careful of its own comfort, and in order to make the entrance of its shell exactly suitable to its own size, it stops up the aperture with sand and similar substances.

A rather curious creature, notable for the long tuft-like appendage at its extremity, is the TAILED PRIAPULUS, a species which is found in the southern seas, and occasionally taken off the English coasts. A curious bundle of threads at its extremity is supposed to serve the purpose of respiratory organs. It has a retractile proboscis, but no tentacles round the mouth.

The SYRINX is distinguished from the Sipunculus by the proboscis, which in these creatures is rather short, and has an indented tentacular fold round the mouth. The generic name, Syrinx, is derived from the resemblance of the creature to the reed from which the ancient pipes were made. This species has a wonderful capacity for changing its shape. The SPOON-WORM is so called on account of the spoon-like appendage to the proboscis. Behind the proboscis are two shining, hook-like bristles. All the members of this genus are remarkable for the wonderful power of contraction and expansion possessed by the skin, and the extraordinary manner in which they can alter their shape. In consequence of this extremely contractile structure, the whole of the water contained in the body is spurted out as soon as a wound is made, and intestines are seen to be forced out after the water. One species of Thalassema is used as bait by fishermen.

All these species belong to the family Sipunculidæ.

The examples next described belong to the family HOLOTHURIDÆ, and are popularly known by the name of SEA CUCUMBERS, or SEA PUDDINGS. In these the body is mostly cylindrical, and is covered with a tough, leathery skin, upon which are placed a number of scattered chalky particles. The mouth is surrounded with a set of retractile tentacles.

Some species of this family are eaten by the Chinese, and a large trade is carried on in these strange products of the sea; the annual merchandise being worth about two hundred thousand pounds. The price of the TREPANGS (*Psolus phantapus*), as they are called, is very variable, according to the species, some kinds being comparatively cheap, and costing rather less than two pounds per hundredweight, while others will fetch thirty pounds for the same weight. There are, besides, "fancy prices" for some very scarce species of Trepang, which, however, are likely to be equalled in real value by the cheaper and commoner kinds.

They seem to be very unattractive creatures, black, wrinkled, and looking much as if they had been made out of the upper-leather of old shoes. They are, however, convertible into a rich and palatable soup, and are also stewed in various ways, taking, in fact, the same rank among the Chinese that turtle does with us. The Trepang is prepared for the market by being carefully opened and cleansed, laid in lime, and then dried, either in the sun or over wooden fires.

The Psolus is allied to a form which is often brought up on the fishermen's hooks on Georges' Banks. It is about six inches in length, composed on the exterior of a series of calcareous scales, arranged like those on fishes; these are of a brick-red color, sometimes of brilliant scarlet. One end, the mouth, is furnished with a flowing array of branching shrubbery tentacles, that are at the least disturbance entirely withdrawn, and concealed internally. This is one of the most desirable objects for the aquarium. On the under surface is a flat disc, provided with small tubes, which answer as propelling organs. These tubes have disc-shaped terminal parts, which adhere to surfaces it passes along. The same are seen in appropriate portions of the Sea Urchins. A few small species are found at Grand Menan, but the home of these forms is in tropical waters. The *Pentacta* is represented in our New England waters, and other forms, similar to the *Cucumarias*. The most notable of these forms is seen on the Florida Reef. A species of Holothuria inhabits the lagoons at Tortugas, measuring two feet in length. It is much like a great cucumber in appearance, though it is black and uninviting in aspect. These creatures are strewn along the bare places among the shrub corals of the lagoons in considerable numbers. Remove one of them and submerge it in a pail of water, the huge creature exhausts the oxygen very quickly. Another creature now protests there is not enough of the life-giving agent, and appears from the mouth of the great Holothurian, in the shape of a veritable fish, six or eight inches in length. Delicate, almost white, from its absence from light, it seems to have little faculty for swimming, though it is possessed of every ordinary requisite of fins. Careful as possible with this fish, we never could keep it alive an hour. Here is a singular case of commensalism. Chaucer is credited with the invention of the word commensal, as literally meaning eating at the same table. Naturalists have adopted this term to distinguish the cases like the present, where two creatures are intimately associated.

The Holothurias are prepared for food, and a large species, the Trepang of the Chinese, inhabiting the Pacific Ocean coral reefs, is similar to the great one just described. They are gathered on the reefs and "cured" there, when they are exported to Chinese ports in great quantities. Our friend Stimpson, of fame in these regions of the invertebrates, "cured" our smaller Holothuria, of Grand Menan, and pronounced it equal, at least, to the article of the Chinese markets, with which he was familiar. Our friend had personal experience in various other directions, to wit, in one case, testing the "smarting" powers of the tentacles of one of the great jelly fishes. Not content with the ordinary method of touching, he applied his tongue—with positive results as to potency.

A form of Holothurian is quite often thrown up on the beaches of New England after storms, which is very attractive from its pearly-white, soft, leathery exterior, and a beautiful pinkish blush on one side. This was called *Chirodota* by Gould, and is altered to the *Synaptas*.

An odd-looking little creature among them, called the PSOLINUS, is remarkable for the great length of the ambulacræ, which lift it well above the object on which it walks. Owing to this fact, it has quite an intelligent aspect as it crawls along, with its beautiful crown of tentacles expanded, and waving in the water. In these two curious genera, the ambulacræ are only distributed in the under surface, and in the present example are placed in three rows on a flattened disc, which occupies part of the under surface.

In the genus PENTACTE, the ambulacræ are placed in a series of parallel rows along the body, sometimes six, but mostly five in number.

It is a remarkable fact, that when one of the Holothuridæ is alarmed, or suffers from indigestion, or is affected in any way, it proceeds to an act which is the exact analogue of the Japanese custom of "happy despatch." Under any or either of these circumstances, it proceeds to disembowel itself, and does so with a completeness and promptitude that are almost incredible. It disgorges the whole of its interior, with all the complicated arrangements that render the Holothuridæ such singular beings to dissect, casts away all its viscera, its stomach, and even throws off the beautiful bell of tentacles.

Having done this, and reduced itself to the condition of an empty skin, which cannot eat because it has no mouth and no stomach, and will not walk, because it has no object for locomotion, it remains perfectly quiescent for some months. At the expiration of that period, a fresh set of tentacles begin to make their appearance; they are followed by other portions; and after a while, the animal is furnished with a completely new set of the important organs which it had cast away. It seems a singular cure for indigestion, but no one can deny its efficacy.

We now come to the SEA CUCUMBER, which has received its generic name from its great resemblance to that vegetable. The smaller species are appropriately named Sea Gherkins. The food of all these animals consists of marine mollusks and other small inhabitants of the sea. The complete but empty shells of several small mollusks have been found within the stomach of dissected specimens, proving that the creature must have swallowed the shell entire, and dissolved out its inhabitant by the process of digestion.

It may as well be mentioned that the only vestige of a skeleton in these creatures is a ring of chalky substance surrounding the beginning of the intestinal canal, and formed of ten pieces, five large and as many small. To this curious ring are attached the longitudinal muscles of the body, by which the creature can lengthen or shorten itself at will, the expansion and contraction of the body being due to a series of transverse muscular fibres. The longitudinal muscles are ten in number, and are arranged in five pairs.

Another example of the Cucumariæ is termed *Cucumaria lyatina*. It is remarkable for its beautiful mouth, which is adorned with a crown of tentacles.

A Cucumaria called *Synapta* is a more singular being. It derives its name from a Greek word signifying to seize hold of anything. This name is given to it because, when the hand is drawn over its surface, the skin is slightly arrested by some invisible agency.

On taking off part of the skin of the Synapta and placing it under a microscope, a most wonderful sight is disclosed. The skin is furnished with a number of little tubercles on which are set numbers of tiny spicules, which look as if they were anchors for a fairy fleet. They are of extremely minute dimensions, and are quite invisible without the aid of a microscope, but never fail to excite admiration when they are well exhibited. Perhaps the best method of bringing out their beautiful shapes is by using a parabolic condenser or a spotted lens, as then their translucent glassy forms shine out against a dark background.

These little objects are of exactly the same shape as the classic anchors of ancient times, and were it not for their extreme minuteness, the person who sees them for the first time is tempted to think that they have been manufactured by some ingenious impostor. But the hand of man is quite incapable of making these beautiful little objects, with their long shanks, their gracefully curved arms, and their sharply-pointed and regularly-serrated flukes.

Nor are the anchors the only wonders which so appropriately deck the skin of a marine animal. If the little prominences can be neatly placed under the microscope without being rubbed, each anchor is found to be affixed by the end of the shank to the end of a curiously-formed shield, made of the same translucent substance as the anchor itself, and pierced with a perfectly regular pattern like ladies' "cut-work" embroidery. These shields hold the anchor in such a way that, as the shield lies flatly upon the skin, the flukes of the anchor are held in the air. The object of this remarkable arrangement is not known.

There are several species of Synapta, all with the anchors and shields, but the pattern upon the shields is different in the various species, as is the shape of the anchor. These remarkable appendages have been compared by some authors to the little hooks on the calyx of the well-known burdock. Synaptas are abundant on the Reef, and one or more are found on the shores of New England.

WE now come to a new and beautiful family of this order, called Echinidæ, because they are covered with spines like the quills of the hedgehog. Popularly, they are known by the name of SEA-URCHINS, or SEA-EGGS. The general shape of these curious beings can be best learned by reference to our colored illustration, which in every respect is most true to nature.

In all these curious beings the upper parts are protected by a kind of shell always more or less dome-shaped, but extremely variable in form, as will be seen in the illustrations. The shell is one of the most marvellous structures in the animal kingdom, and the mechanical difficulties which are overcome in its formation are of no ordinary kind. In the case of the common SEA-EGG, the shell is nearly globular. Now, this shell increases in size with the age of the animal; and how a hollow spherical shell can increase regularly in size, not materially altering its shape, is a problem of extreme difficulty. It is, however, solved in the following manner:—

HOLOTHURIANS AND SEA STAR.

The shell is composed of a vast number of separate pieces, whose junction is evident when the interior of the shell is examined, but is almost entirely hidden by the projections upon the outer surface. These pieces are of a hexagonal or pentagonal shape, with a slight curve, and having mostly two opposite sides much longer than the others. As the animal grows, fresh deposits of chalky matter are made upon the edges of each plate, so that the plates increase regularly in size, still keeping their shape, and in consequence the dimensions of the whole shell increase, while the globular shape is preserved.

If a fresh and perfect specimen be examined, the surface is seen to be covered with short sharp spines set so thickly that the substance of the shell can hardly be seen through them. The structure of these spines is very remarkable, and under the microscope they present some most interesting details. Moreover, each spine is movable at the will of the owner, and works upon a true ball-and-socket joint, the ball being a round globular projection on the surface of the shell, and the socket sunk into the base of the spine. When the creature is dead and dried, the membrane which binds together the ball-and-socket joint becomes very fragile, so that at a slight touch the membrane is broken and the spines fall off.

Other peculiarities of structure will be noted in connection with the different species.

The common Sea-urchin is edible, and in some places is extensively consumed, fully earning its title of Sea-egg, by being boiled and eaten in the same manner as the eggs of poultry.

The fishing for these creatures in the Bay of Naples is graphically and quaintly described by Mr. R. Jones:—
"I had not swum very far from the beach before I found myself surrounded by some fifty or sixty human heads, the bodies belonging to which

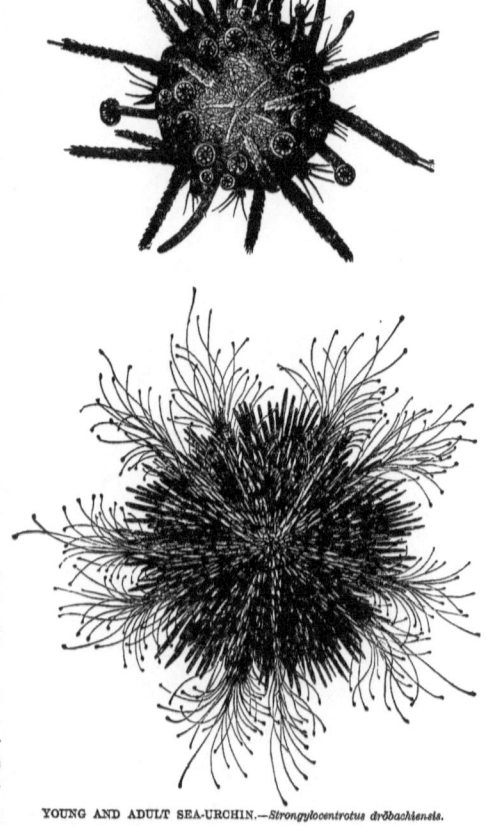

YOUNG AND ADULT SEA-URCHIN.—*Strongylocentrotus dröbachiensis.*

were invisible, and interspersed among these, perhaps, an equal number of pairs of feet sticking out of the water. As I approached the spot, the entire scene became sufficiently ludicrous and bewildering. . . . . Down went a head, up came a pair of heels—down went a pair of heels, up came a head; and as something like a hundred people were all diligently practising the same manœuvre, the strange vicissitude from heels to head and head to heels, going on simultaneously, was rather a puzzling spectacle."

After inquiry, it proved that these divers were engaged in fishing for Sea-urchins, which are especially valuable just before they deposit their eggs, the roe, as the aggregate egg masses are termed, being large and in as much repute as the "soft roe" of the herring.

These Sea-urchins are fond of burrowing into the sand, an operation which is conducted mostly by help of the movable spines. They will sink themselves entirely out of sight, but not without leaving a slight funnel-shaped depression in the sand, which is sufficient to guide a practised eye to their hiding-place.

The Sea-urchins are represented in New England by one species. It is found ensconced in pools among the rocks at low tide, being unaffected by the loss of water during the low tide which leaves them bare.

The Common Sea-urchin on our coast, bearing the heavy title of *Strongylocentrotus dröbachiensis* (see figure) is the only one quite familiar to Northern waters. It is exceedingly abundant in the tide pools and in the rocky cliffs, and is common in Alaska. A larger species is found in California. The Echinoderms are not very largely represented on the North American coast. In the warmer waters of the Florida Reef they are abundant.

The accompanying figure (No. 2) is about the average size of our species. The smaller figure (No. 1) represents the young as seen from the side of the mouth.

The genus to which another species, the PIPER-URCHIN, belongs, can always be recognized by the enormous comparative size of the tubercles sustaining the spines and the parallel rows of ambulacra.

The members of the genus Cidaris are mostly found in the hotter parts of the world, and are plentiful in the Indian Seas. The spines of several of the species have been made serviceable in the cause of education, being found to make excellent slate-pencils after being calcined. The missionaries have the credit of making this useful discovery.

The food of the Echini in general seems to consist of various substances, both of an animal and vegetable nature. Fragments of different sea-weeds have been found in the digestive cavity, as also certain portions of shells, which seem to prove that the Echinus had fed upon the mollusks, and broken their shells in pieces with its powerful jaws. The precise mode of feeding is not exactly ascertained; but it seems likely that the Echinus can seize its prey with any of its ambulacra, no matter on what portion of the body they may be situated, and pass it from one to the other until it reaches the mouth, which is placed in the centre of the open disc. Both univalve and bivalve mollusks appear to be eaten by the Echinus.

The creature which is represented in the accompanying illustration is appropriately named COMMON HEART-URCHIN, from its peculiar shape, and bears an evident resemblance to the heart-cockles already mentioned. Many species of Heart-urchins are found in a fossil state, and are especially common in the chalk formations.

The shell of this genus is slight and delicate, and is composed of very large plates, which, in consequence, are comparatively few in number. There is always a furrow of greater or less depth at the upper end. In the naked specimen the rows of pores through which the ambulacra pass are plainly perceptible, and even in the fossilized specimens, which have been buried in the earth for so many ages, these pores are still visible, and so plainly marked, that the genus and species of the dead shell can be made out with little less ease than if the animal were just taken out of the water.

The Heart-urchins are found in all parts of the world, and the European seas contain specimens of these curious beings. In the Mediterranean they are extremely plentiful, and mostly appear to live below the sand. They seem to feed on the animal substances that are mingled with the sand, for M. de Blainville found, on dissecting many specimens, that their digestive organs were always filled with fine sand. The walls of the digestive cavities are exceedingly delicate, and have been compared to the spider's web.

Another of these remarkable creatures, where the shell is formed into two points, is the FIDDLE HEART-URCHIN, so called from the fiddle-shaped mark upon the shell.

In some of the hotter parts of the world, such as the Indian seas, several species of Echinus are armed with sharp and slender spines, which are apt to pierce the bare foot of a bather, and to cause painful, and even dangerous wounds. Most of these Echini live in the crevices of rocks, but sometimes crawl over the sand, and inflict much suffering upon those who unwittingly place a foot upon them. Mr. F. D. Bennett, in his account of a "Whaling Voyage," had practical experience of these sharp spines:—"On one occasion, when searching for fish in the crevice of a coral rock, I felt a severe pain in my hand, and, upon withdrawing it, found my fingers covered with slender spines, evidently those of an Echinus, and of a gray color,

HEART-URCHIN.—*Perinopsis lyrifera*. (Natural size.)

elegantly banded with black. They projected from my fingers like well-planted arrows from a target, and their points, being barbed, could not be removed, but remained for some weeks imbedded as black specks in the skin.

"Its concealed situation did not permit me to examine this particular Echinus, but I subsequently noticed others of a similar nature fixed to the hollows in the rocks; they were equal in size to the *Echinus cidaris*, and their body was similarly depressed, but the spines were long, slender, and more vertically arranged, and their points finely serrated. Their color was jet-black. These animals adhered so firmly to the rocks, that they could not be detached without difficulty.

"When closely approached, they gave an irritable shrug to their spines, similar to that displayed by the porcupine or hedgehog. It was difficult to say if the hand had been brought in perfect contact with this Echinus before it was wounded by its weapons. In some experiments, I approached the spines with so much caution, that had they been the finest pointed needles in a fixed state, no injury could have been received from them; yet their points were always struck into my hand, rapidly and severely. The natives are well aware of the offensive character of these animals, and caution the stranger against handling them."

The same author mentions that a species of Cidaris is largely eaten by the South Sea Islanders, and that in various places on the sea-shore there are large heaps of its shells and spines, showing that feasts have been lately held in that locality.

THE curiously-formed Echinus which is shown in the illustration on next page is popularly called the CAKE-URCHIN, on account of its remarkably flattened form. It belongs to a family which are generally called Shield-urchins, from their flat, disc-like shapes. The shell is wonderfully flattened and slopes rapidly from the centre to the circumference. The general shape and arrangement of the plates from which the shell is built may be seen in the figure. The word "placenta" is Latin, signifying a flat cake, and is appropriately given to this species.

The development of the Echinus is so very remarkable, that it deserves a passing notice. This creature passes through a metamorphosis even more strange than that of the insect, and no one who was not acquainted with the animal could possibly recognize in the delicate framework of translucent spines the larval form of the globular Sea-urchin. At first, the little creatures are almost shapeless and globular, rolling about through the water in an uncertain kind of way. But by degrees they put forth a dome-like portion, from which proceed several slender calcareous rods, altogether making a figure that has been aptly compared to a skeleton French clock. In this state it was formerly known by the name of Plutens. As if to carry out the comparison still further, the first indication of change to its more perfect form is the development of a circular disc which will represent the face of the clock, upon which are traced certain lines that answer to the hands and figures. By rapid degrees, the disc expands and covers the gelatinous substance of the animal, and puts on hour by hour more of the Echinus as it loses its former skeleton shape. The latter becomes rapidly covered by and absorbed into the former, and in due time the framework of long, slender rods, which might also be well compared to an artist's easel, or the tripod stand of a theodolite, is converted into the well-known globular Echinus, with all its complicated apparatus of spines, pedicillariæ, and walking-organs.

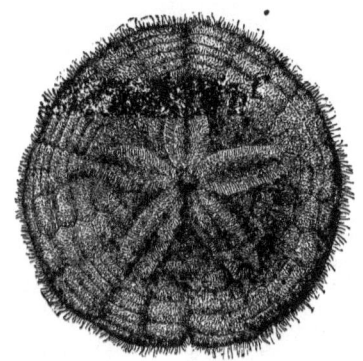

SHIELD-URCHIN.—*Echinarachnius parma.* (Natural size.)

The reader may perhaps have noticed that, on inspecting a common Echinus, especially from the interior, it exhibits in a very distinct manner its close alliance with the well-known star-fishes. Take, for example, a common five-finger star-fish out of the water, lay it on its back, and then gather all the five points together. Now, supposing the creature to be dead, strip the skin from the rays, leaving it only adherent down the centre, join the edges of the strips, and there is a very good imitation of the Sea-urchin.

The Cake-urchin is represented on the New England coast by one of about three inches diameter. This object is, perhaps, more puzzling to the average observer than any other. Its remarkable flatness is a stumbling-block to understanding it as an animal. Sand Cake is a name given it, and suggests its possible origin with the uninformed.

Two other curious members of this genus are the KEYHOLE-URCHIN and the WHEEL-URCHIN.

The latter, so called because of its wheel-like shape, is nearly as flat as a piece of money, and has a very slight elevation in the centre. It is remarkable for the very deep teeth into which one side of the disc is cut, giving the creature an aspect as if it were a cog-wheel in process of manufacture. The color of this species is mostly grayish-slate above, and dull white below. The under surface is veined over its whole extent, all the veinings radiating from the centre. The color of this species is, however, extremely variable. It is also called Rotula.

The second species might be well called the Keyhole-urchin. This remarkable creature, instead of being toothed at the edge like the preceding species, has its disc pierced with oblong apertures of a shape much resembling a keyhole. These apertures are rather variable in their shape, sometimes being merely pierced through the disc of the Urchin, and sometimes extending fairly to the edge. When full-grown, this is rather a large species, much resembling an ordinary pancake both in shape and dimensions. There are many species of Encope, most of which are inhabitants of the hotter seas, some being found in Southern America. The color of the Keyhole-urchin is dull gray. The whole family is a very remarkable one, and affords numerous points of interest to the careful observer.

STAR-FISHES; ASTERIADÆ.

LEAVING now the Echini, we pass to the next large group of Echinodermata, called scientifically Asteriadæ, and popularly known as Star-fishes. These creatures exhibit in the strongest manner the radiate form of body, the various organs boldly radiating from a common centre.

Many of these creatures are exceedingly common, so plentiful, indeed, as to be intensely hated by the fishermen. Of these, the common FIVE-FINGERS, ASTERIAS, BUTTHORN, or CROSS-FISH, is perhaps found in the greatest numbers. All Star-fishes are very wonderful beings, and well repay a close and lengthened examination of their habits, their development, and their anatomy. There are sufficient materials in a single Star-fish to fill a whole book as large as the present volume, and it is therefore necessary that our descriptions shall be but brief and compressed.

To begin with the ordinary habits of this creature.

Every one who has wandered by the sea-side has seen specimens of the common Five-fingers thrown on the beach, and perhaps may have passed it by as something too common-place to deserve notice. If it be taken up, it dangles helplessly from the hand, and appears to be one of the most innocuous beings on the face of the earth. Yet, this very creature has, in all probability, killed and devoured great numbers of the edible mollusks, and has either entirely or partially excited the anger of many an industrious fisherman.

To begin with the former delinquency. It is found that the Star-fish is a terrible foe to mollusks; and, although its body is so soft, and it is destitute of any jaws or levers, such as are employed by other mollusk-eating inhabitants of the sea, it can devour even the tightly-shut bivalves, however firmly they may close their valves. On looking at a Star-fish, it will be seen that its mouth is in the very centre of the rays, and it is through that simple-looking mouth that the Star-fish is able to draw its sustenance.

Even if it should come upon a mollusk which, like the oyster, is firmly attached to some object, it is by no means disconcerted, but immediately proceeds to action. Its first process is to lie upon its prey, folding its arms over it, so as to hold itself in the right position. It then applies the mouth closely to the victim, and deliberately begins to push out its stomach through the mouth, and wraps the mollusk in the folds of that organ. Some naturalists think that the Star-fish has the power of secreting some fluid which is applied to the shell, and causes the bivalve to unclose itself. But, whether this be the case or not, patience will always do her work, and in time the hapless mollusk surrenders itself to the devourer. In the case of smaller prey, the creature is taken wholly into the mouth, and there digested.

A very remarkable effect of the voracity of the Star-fish is often seen in specimens. It is not an unusual occurrence, that Star-fishes had managed to swallow entire a bivalve mollusk, and had dissolved out all the soft parts from the shell. This they were unable to throw out, as is the custom of Star-fishes, and, in consequence, the empty shell of the bivalve became a fixture within the body of the Star-fish.

The second delinquency of the Star-fish is achieved as follows:—By some wonderful power the Star-fish is enabled to detect prey at some distance, even though no organs of sight, hearing, or scent can be absolutely defined. When, therefore, the fishermen lower their baits into the sea, the Star-fishes and crabs often seize the hook, and so give the fisherman all the trouble of pulling up his line for nothing, baiting the hook afresh, and losing his time.

The fishermen always kill the Star-fish, in reprisal for its attack on their bait, and formerly were accustomed to tear it across and fling the pieces into the sea. This, however, is a very foolish plan of proceeding, for the Star-fish is wonderfully tenacious of life, and can bear the loss of one or all of its rays without seeming much inconvenienced. The two halves of the Asterias would simply heal the wound, put forth fresh rays, and, after a time, be transmuted into two perfect Star-fishes.

It often happens that the lounger on the sea-shore finds examples of this species with only four or even three rays, and, finding no vestige of a scar to mark the place whence the missing limb was torn, he is apt to fancy that he has found a new species which only possesses a small number of rays. The fact, however, is that the interval is immediately filled up by the creature; the rays on each side of the injury close up together, and all mark of a wound is soon obliterated. I have seen these strange beings with only one ray, proceeding quietly along without appearing to suffer any inconvenience from their loss.

The movements of the Star-fish are extremely graceful, the creature gliding onward with a beautifully smooth and regular motion. It always manages to accommodate itself to the surface over which it is passing, never bridging over even a slight depression, but exactly following all the inequalities of the ground. It can also pass through a very narrow opening, and does so by pushing one ray in front, and then folding the others back, so that they may afford no obstacle to the passage. It also has an odd habit of pressing the points of its rays upon the bottom of the sea, and raising itself in the middle, so as to resemble a five-legged stool. If the reader is desirous of keeping a few Star-fishes in an aquarium, the object may be easily accomplished by keeping them in a very cool place, as they are extremely impatient of heat, and soon die if the water becomes too warm. They also require that a supply of air be frequently pumped through the water in which they reside.

The bony apparatus, or skeleton, if it may be so called, of the Star-fish is a most beautiful and wondrous object. Without going into the tempting regions of anatomy, I may state that a few hours will be well bestowed in examining the structure of any of these beings. A very simple plan of doing so is to wash the creature well with fresh water, lest the salt should rust the scissors and scalpel, and then carefully look into the extraordinary array of tentacles, or ambulacra, on which the creature walks. Let it then be pinned to a flat piece of cork loaded with lead, and sunk about half an inch below the surface of clear fresh water. Slit up the skin along each ray, taking care to save a portion for the microscope, and turn the flaps aside.

In each ray will be seen the curious feathered and fern-like branches of the stomach, and under them lies the wondrous array of bone-like pieces of which the skeleton is made. Thousands upon thousands of pure white columns are ranked in double vistas, and are overarched by an elaborate structure of the same white material on the pillars. I know nothing that can compare with this sight for delicacy and beauty. Imagine a cathedral aisle half a mile in length, which is supported by a double row of white marble columns, and whose roof is formed of the same beautiful material; then, let all the pillars be bowed towards each other in pairs, so that their capitals rest against each other, and a dim idea will be formed of the wonderful structure of the Star-fish.

The piece of skin must be preserved in order to examine, with the aid of the microscope, the pedicillariæ and minute spiracles that stud its surface. A tolerably stout pair of scissors are required for the purpose of cutting the skin, as its substance is tough; and it is besides furnished with such an array of hard stony appendages, that the edge of a more delicate instrument would certainly be turned, and its blade run some risk of fracture.

Before we pass to the remaining examples of this family, a few words must be given to the development of this wonderful creature.

The eggs of the Star-fish are numerous, almost beyond the power of arithmetic to calculate, and thus keep up the needful supply of these creatures whose enemies are so numerous, and powers of escape so trifling. When first excluded, the eggs are not allowed to pass freely into the sea, but are protected for a time in a kind of cage or chamber formed by the parent by raising itself on the tips of its rays, as has already been mentioned. When hatched, the young are round and almost shapeless, bearing a very close resemblance to an imprisoned animalcule. They by degrees put forth their rays, the feet issue from the rays, and, after a while, they are enabled to shift for themselves, and are dismissed from their parental home.

The Butthorn is much like a species once thought to be very rare on the New England coast, but now known through the dredgings of the Fishery Commission to be abundant in certain localities in deep water. One haul off Portland, Me., during a summer we spent with Professor Baird as guest of the Fish Commission, produced a large number.

The common Star-fish of our American beaches is familiar enough, though as yet, like many another sea form, not understood. Though so diverse in shape, the Star-fish, Echinus, or Sea-chestnut, and the Holothurias, are closely allied as Echinoderms—spine-skinned animals.

Species like the Sun-star (*Solaster*) have been found, sparingly, by adhering to lines of fishermen on the fishing banks.

We now proceed to the examination of some of the more conspicuous species of Asteriadæ.

The common FIVE FINGERS, or CROSS-FISH, needs no more description than has already been given.

A pretty little species, called GIBBOUS STARLET, is notable for the manner in which the rays are connected by a membrane as far as their tips. Another species is the KNOTTY-CUSHION STAR, so called on account of the thick rounded rays.

IN the next examples we have several other curious forms of Star-fishes. The BIRD'S-FOOT SEA-STAR derives its name from its singular shape, which is not at all unlike that of a duck's foot, with its spread toes and connecting membrane. This beautiful species is very thin of texture, and has a pentagonal form, caused by the five rays and the connecting membrane. If the surface of this Star-fish be examined with a good magnifier, it will be found to be covered with tufts of very tiny spines arranged in a regular series, and forming a kind of pattern.

The colors of the Bird's-foot Star are positively splendid. Each ray is marked with a double line of bright scarlet, a narrow belt of the same color edges the connecting membrane, and the centre is also scarlet. The ground color is light yellow, and the contrast of these two beautiful colors has a remarkably splendid effect. This species is seldom seen in the shallow waters or above low-water mark, and is, as a general rule, taken with the dredge.

A boldly-rayed species, which looks something like the front view of a sunflower, is very common, and goes popularly by the appropriate name of SUN-STAR. It often attains to considerable dimensions, and is always a very conspicuous object from the glaring colors with which its surface is decorated, and the large amount of surface on which they can be displayed. The upper surface of this fine species is bright vermilion, and as it sometimes is eight or nine inches in diameter, it is a very brilliant object as it lies upon the rocks.

Should any reader be desirous of preserving this or any other of the Star-fishes for a cabinet, he may do so without difficulty, by taking a few precautions. The first process is to wash the Star-fish in plenty of fresh water, and it will be better to follow up this step by removing the whole of the stomach and its appendages. This may be done from the under surface of the rays; and it will, perhaps, be useful if a little cotton wool be judiciously inserted, so as to prevent the skin from collapsing during the process of drying. Star-fishes may be easily dried, either before the fire or in the sun, but in either case they must be carefully washed in fresh water; and if a fire be employed, as must be the case in wet or dull weather, the board on which the Star-fish is should not be placed very near the fire, and should be occasionally watched, so that any tendency to warping may be corrected.

In the EYED CRIBELLA, the eyes are rather blunt at their extremities, and are cleft nearly to the centre, so that there is no definite disc. This species is rather stiffer to the touch than the others. It must, however, be remarked that the consistency of the Star-fishes is extremely variable, even in the same species or the same individual. If, for example, a specimen of the common cross-fish be taken from the pool of water in which it is lying, a practised hand will at once know whether it is dead or alive. In the former case the creature is soft and flabby to the touch, yields readily to the impress of the fingers, and hangs down heavily like a mass of wet rag. If, on the contrary, any life should be left in the creature, the rays are tolerably firm and resisting to the touch, and when held by one ray it has altogether a firmer and more lively feeling about it. A simple but effectual mode of ascertaining whether a Star-fish be alive or dead; is to turn it on its back in some sea-water. If it be dead there will, of course, be no movement, but if the least particle of life be still latent in that body from which it can hardly be expelled, the ambulacra, or feet, are seen to put themselves in motion, some being thrust out while others are being withdrawn.

Our next examples are very curious species of Star-fish.

The BRITTLE-STARS (*Ophiocoma rosula*), of which there are several species, are very appropriately named, inasmuch as they are able to break up their rays in the most extraordinary manner, a capability which they mostly exercise when they feel alarmed. The generic name, Ophiocoma, is derived from two Greek words, the former signifying a serpent and the latter a lock of hair.

The whole of the Brittle-stars are curious and restless beings. They can never remain in the same attitude for the tenth part of a second, but are continually twining their long arms, as if they were indeed the serpents with which Medusa's head was surrounded. The least impurity in the water will cause these strange beings to break themselves to pieces in this extraordinary manner, but they never seem to disintegrate themselves with such rapidity as when they are touched, or otherwise alarmed.

The lamented Professor Forbes has left an admirably quaint description of this suicidal process. Having in vain attempted to secure a perfect specimen of a Brittle-star, he thought that he might achieve that object by having a pail of fresh water lowered into the sea, so that as soon as the dredge reached the surface of the sea it might be transferred to the bucket of fresh water, and all the inmates killed at once by the shock.

A fine specimen of the genus Luidia was then taken in the dredge. "As it does not generally break up before it is raised above the surface of the sea, cautiously and anxiously I sank my bucket to a level with the dredge's mouth, and proceeded, in the most gentle manner, to introduce Luidia to the purer element. Whether the cold element was too much for him, or the sight of the bucket too terrific, I know not; but in a moment he began to dissolve his corporation, and at every mesh of the dredge his fragments were seen escaping. In despair, I grasped the largest, and brought up the extremity of an arm with its terminating eye, the spinous eyelid of which opened and closed with something exceedingly like a wink of derision."

These Brittle-stars are, however, extremely capricious in their exercise of this curious power. It sometimes happens that, as in the instance so amusingly narrated, the creatures break themselves to pieces without any apparent provocation, while, in other cases, specimen after specimen may be taken, handled, killed, or wounded, without the loss of a ray. Even in the aquarium, they are equally uncertain in their habits, at one hour being entire and splendid specimens, and at the next being little but a solitary disc amid a ruined heap of broken arms.

The Brittle-stars are abundant in the warmer waters. When Dr. Gould published his "Report on Invertebrata of Massachusetts," 1841, his enumeration of *Echinodermata* embraced *Echinus granulatus*, Sea-egg, Sea-urchin; *Asterias*, four species. *A. rubens* being the common Star-fish, or Five-finger; and two species of *Ophiura*, which were visible only as brought up by dredging, or from the stomachs of fishes. The latter were not, as they are in the tropical waters, found crawling on the objects at low tide or in shallow waters. The coral shrubs, and dead and crumbling blocks of Meandrinas astreas, etc., are numerously inhabited by them. There are many others since discovered by the extensive dredging in deep and shallow waters on our coast.

THE WHITE SAND-STAR (*Ophiurus albidus*).

The word Ophiurus is of Greek origin, signifying snake-tail, and is therefore very appropriately given to these curious beings, whose slender arms twist and coil just like a handful of small serpents.

The Ophiuri are quite as voracious as the ordinary Star-fishes, and are able by means of the long arms to convey food to the mouth, which is placed in the central disc. The young of these Echinodermata are quite as curious as those of the sea-urchins, to which, indeed, they bear some resemblance. They have long been known to naturalists under the title of Easel animalcules, on account of their peculiar shape, their real origin not being suspected until later years. It is totally unlike the form which it attains when mature, and the relationship between the adult Star-fish and the Pluteus, as the larva is termed, has been well compared to the relation of an embroidery frame to the pictured canvas within.

THE wonderful creature which is called by the name of SHETLAND ARGUS is one of a most remarkable genus of Star-fishes, which are remarkable for the vast development of their arms.

Although the whole mass of arms is of so complicated a description, it will be found, on carefully examining the creature, to be formed by the simple process of twofold division. From the central disc spring five stout arms, each of which almost immediately divides into two smaller arms, and these again into two others; so that in a fine specimen the number of little arms or tendrils, if we may so call them, exceeds eighty thousand. All these organs are extremely flexible, and quite under the control of the animal, which is able to close or expand them at will. When the extremities of the arms are drawn together, it will be seen that the whole animal assumes the shape of a globular basket; and in consequence of this resemblance, the name of Basket-urchin, or Sea-basket, has been proposed for the creature.

It takes its food by means of these wonderful arms, using itself, in fact, like a living casting net, surrounding the prey with the spread arms, and inclosing it within their multitudinous lines. It has been known to embrace in this manner a fisherman's bait, and to allow itself to be drawn to the surface without losing its hold. It is one of the deep-sea Star-fishes, and is very seldom taken except by means of the dredge or line.

The structure of the Shetland Argus is most marvellously complicated, inasmuch as each of the numerous arms is composed of an enormous number of small joints, each exactly in its place, and so beautifully connected together, that they are as flexible as silken cords, and yet as perfectly under the command of their owner as if they were restricted to the original five from which they take their origin.

THE elegant and graceful Star-fish which is appropriately named the FEATHER-STAR, is a native of the English coasts, and has always attracted the attention of sea-side observers. It is not very readily seen, being one of the deep water species, but it may be captured by means of the dredge, and will live for some time in the marine aquarium. It is a very active being, combining in its own person the accomplishments of many different Star-fishes. For example, it can crawl with tolerable speed over the ground, can swim through the water with sufficient power to direct its course, can float about at will, driven by the tide, and will sometimes clasp pieces of floating wood, so as to be carried along by the waves without any fatigue.

Its habits while in the aquarium are very interesting, and have been well described by Mr. Gosse:—

"In captivity, the Feather-star sits upon the frond of a sea-weed or on a projecting angle of rock, which it grasps very firmly with its clawed filaments, so firmly that it is difficult to tear it from its hold. When violence is used, it catches hold of its support, or any other object within reach, with the tips of its arms, which it hooks down for the purpose, and with its pinnæ, so that it seems furnished with so many claws, the hard, stony nature of which is revealed by the creaking, scratching noise they make as they are forced from any hold, as if they were made of glass.

"I was surprised to observe that several of the arms were unsymmetrically short; and on examining these with a lens, saw distinctly that each had been broken off, and was renewed; the new part agreeing in structure and color with the rest, but the joints were much less in diameter; and this difference was strongly marked at the point of union, the first of the new joints being not more than one-third as wide as its predecessor. The appearance much reminded me of a lizard renewing its tail.

"In sitting, the Feather-star bends its arms with a sigmoid curve, the tips bending upwards. It waves them now and then, but not much, and remains long without moving from its hold. Though I repeatedly took it out of the water, removing it forcibly, it manifested no tendency to voluntary dislocation."

Perhaps, however, the strangest part of the Comatula's life is its early youth.

Every one who has the least smattering of geology is familiar with the fossils called Encrinites, and is well acquainted with them under the different popular names that they bear. They are, or rather were, Echinodermata set upon a long flexible stalk, and being constructed, like the Star-fishes, of an enormous number of joints. Popularly they are known by the name of Stone-lilies, or Screw-stones, and their disjointed members are very familiar under the title of St. Cuthbert's Beads. The number of joints in an adult Encrinite is almost incredible.

In the head only of one specimen, no less than one hundred and fifty thousand joints have been calculated to exist, exclusive of the numerous parts of which the stem is composed. These joints are frequently found separated from each other; and as they are perforated by a small hole through which a thread might be run, they were formerly strung together and used as rosaries. Encrinites were found very plentifully in many marbles, which, according to Dr. Buckland's energetic language, are as entirely made up of the petrified remains of Encrinites as a corn rick is of straws. These wonderful beings could hardly be dissected out of the stone by any exertion of human labor, but it is found that water will achieve a task at once too laborious and too delicate for human hands to undertake. It oftens happens that the abrupt faces of marble cliffs exposed to the weather, so that the annual rains are driven forciby upon them, and by their continual action wear away the soft surrounding substance of the stone, leaving the harder forms of the Encrinites as memorials of the time long passed away.

The Encrinites have long ago perished, but there are still some existing species of stalked Echinodermata, which are closely allied to them, and are still more nearly connected with the history of the Feather-star. These are termed Pentacrinites, because their joints are five-sided. Many fossil species of Pentacrinites are found, and are seen in positions which seem to prove that they must have been adherent by their bases to floating objects, and thus carried about from one place to another, like the barnacles, which have already described and figured.

The Feather-star is a great rarity. The *Comatula* is a stemmed form found sparingly in the waters off South Carolina. A large species is found off Greenland, and is occasionally brought from off the coast of Maine, near Eastport.

ONE living species of these strange creatures is still in existence. This being is appropriately called by the name of MEDUSA'S HEAD, as the many arms that wave about its summit bear some resemblance to the fabled head of Medusa, with its burden of venomous serpents.

It is not a very large species when compared with some of its fossil relatives, for the largest specimens hitherto discovered are only a few feet in length, and have a stem about as large as a common drawing-pencil. Several fossil species, on the contrary, are at least eleven or twelve feet in length, and measure a full inch across the stem. The Medusa's Head is the only species at present known, though it is probable that others may be yet discovered.

Euriale scutatum is a name applied to a very singular and always interesting form of Star-fish. A species found off Massachusetts Bay is named *Astrophyton agassizi*, Basket-fish, so called by old Governor Winthrop, of Massachusetts, who wrote an account of it with other natural productions, for the Proceedings of the Royal Society. Linnæus called it *Asterias caput medusæ*, a very good specific name—Medusa's Head. This has been regarded as very rare. An occasional instance of its being drawn up by fishermen, was all that was known until it was taken in quantities off Cape Cod in one locality.

A Star-fish, discovered by Mr. Thompson, and called by him *Pentacrinus europæus*, is, when full-grown, barely three-quarters of an inch in height, and with a stem no thicker than sewing silk. Without entering into the many and interesting details of structure, development, and the habits of this beautiful little creature, we need only observe that this being has been proved to be the young, or larval state of the Feather-star. During this stage of its existence, the young Comatula is affixed to its ever-lengthening stem, but when it has attained adult age, it leaves its footstalk and wanders freely through the ocean.

The reader will not fail to observe that herein the creature acts in precisely the opposite mode to that which is adopted by many beings which have already been described. In those marine animals of which the acorn-barnacle is a good example, the young enjoys freedom throughout its earlier stages, is furnished with certain organs which it afterwards loses, and does not settle down to one spot until it has attained adult age. In the case of the Comatula, the contrast between the two states of life is very strongly marked, the creature being of a more nomad nature than the rest of its kin, and in swimming, presenting a curious resemblance to the Medusa, the arms contracting and expanding in a manner that strongly reminds the observer of the pulsating disc of the acaleph.

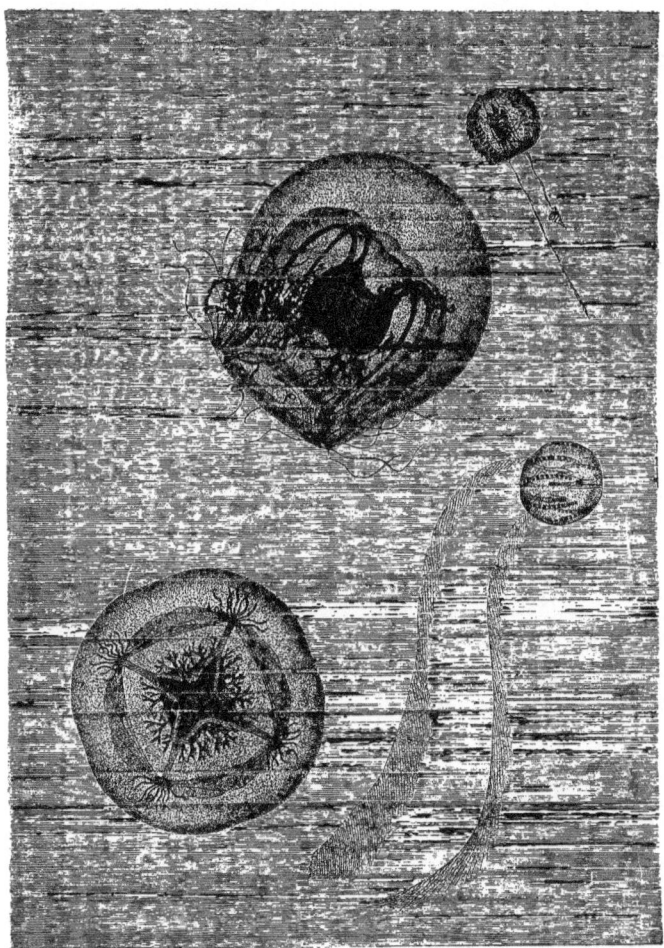

SWIMMING SEA-NETTLES.—*Acalepha.*

NETTLES; ACALEPHA.

WE now arrive at a large and important class of animals.

These beings, represented by some specimens in the accompanying illustration, are scientifically termed Acalepha, or Nettles, a word which may be freely rendered as Sea-nettles. The term is appropriate to many of the species which compose this large class, for a very great number of the Acalepha are possessed of certain poisoned weapons which pierce the skin, and irritate the nerves as if they were veritable stinging-nettles floating about in the sea. Popularly, they are known by the familiar term Jelly-fishes, because their structure is so gelatinous, mostly clear and transparent, but sometimes semi-opaque or colored with most beautiful tints.

The whole history of these remarkable animals is curious and interesting in the extreme, for not only do they exhibit some of the most graceful shapes and pleasing hues that can add beauty to a living being, but they also afford examples of the earlier forms of organs and members which in the higher animals attain their fullest development.

When they have attained their adult condition, they roam the seas freely, though in their earlier stages they are fixed to one spot and assume a shape quite unlike that of their parent.

The function of nutrition is carried on in these animals in a method sufficiently simple. They are furnished with a cavity, corresponding to the stomach of higher animals, in which the food is placed, and from which a number of diverging vessels convey the nutritive fluid to the rest of the body.

SIPHONOPHORA.

For convenience sake, this class is divided into three groups or orders, the first of which is called the Siphonophora, and includes the best organized members of the class. In them the shape of the body is irregular, and there is no central cavity. They are furnished with sucking organs, and move by means of a certain cavity into which water is received gently and from which it is expelled forcibly, or sometimes by means of little sacs or vesicles charged with air.

Owing to the vast number of species contained in this class, it is manifestly impossible to mention all the curious and interesting animals which it includes. Care, however, has been taken to select those species which afford the best types of their orders, and it will be found that almost every group of importance will find its representation in the following pages.

The present arrangement of the Jelly-fishes (1885), Hydroids so called, and Corals, is under the title *Cœlenterata*, constituting one of the great branches of the Animal Kingdom, the third in the scale, counting upwards from the lowest. See the classification and nomenclature tables at end of this volume. Three classes are recognized: *Hydrozoa*, *Actinozoa*, and *Ctenophora*. In the first and last of these classes are what are familiarly known as Jelly-fishes, while class Actinozoa embraces the Corals and other Sea Anemone forms.

The first class is Hydrozoa. The first Order embraces those forms called Hydroids. The fresh-water Hydra is a familiar example. The second Order, Discophora, embraces the great hemispherical jellies that inhabit our North.

Among the Hydroids, the first class of Cœlenterates, the Tubularia is familiar. It is not uncommon on our Atlantic shores. The plate on page 558 gives a very fine example of this Hydroid. A bunch of these creatures looks more like a group of beautiful pink-like flowers than any other marine form. The color is exquisite pink, while the stems are sober brown. They are found in our North American waters during the summer. The Discophores attain the largest size of all. Their popular names are Sea-nettles, in allusion to the stinging powers, Sea-bulbs, etc.

The bodies of these, though comparatively tough, are yet mostly water. A specimen weighing thirty-four pounds lost ninety per cent. on drying in the sun. These creatures are phosphorescent, glowing like living fire. We have seen the waters of the harbor of Havana one golden hue at night from their presence. The most common form in the Northern waters is the *Cyanea*, which attains a great size. Mrs. Agassiz records the following dimensions from personal measurement, taken from a specimen at Nahant. She says: "Encountering one day one of these huge Jelly-fishes, when out in a row-boat, we attempted to take a rough measurement of its dimensions on the spot. He was lying quietly near the surface, and did not seem in the least disturbed, but allowed the oar, eight feet in length, to be laid along its disc, which proved to be about seven feet in diameter. Backing the boat slowly along the line of the tentacles, which were floating at their utmost extension behind him, we then measured them in the same manner, and found them rather more than fourteen times the length of the oar, thus covering a space of some hundred and twelve feet."

TUBULARIAN HYDROIDS.—*Tubularia indivisa.*

This sounds so marvellous it may be taken as an exaggeration, but the facts are rather understated than otherwise. We may well regard such creatures with caution and dread for their stinging powers.

Class III., SIPHONOPHORA, embraces some of the most beautiful of the "Sea-jellies," or Medusæ, as they are called. The most notable, and surely the most beautiful of all, is the "Portuguese Man-'o'-War."

PHYSALIA. This class includes species of most diverse forms, yet closely allied. The essential parts, however, are not so varied—that is, the stomach and reproductive organs are a mass of soft flesh that hang from the floats. It is the upper and ornamental portion that varies. For example, see the difference between the beautiful bubble of the Physalia and the little oval floating raft of the Sallee Man with its low crest, and the crestless circular float of the Porpita. The latter, seen on the ocean as we have seen them in myriads, presents a pretty circular disc of the dimensions of a quarter dollar. This is a perfectly smooth float, of the same indigo and purple as the Physalia. On the under side is the fleshy mass of stomach and small tentacles. In some there is a delicate fringe on the periphery. These are the Porpitas. The Sallee Man, or Velella—meaning little boat—is more interesting from the curious form of its crest. On an oval float like that of the latter species, there stands an upright sail-like crest, of the thinnest isinglass-like substance when denuded, but when alive covered by the indigo-colored membrane seen in all. The twisted shape of this upright is pleasing for its beauty of form; and the denuded shells are exquisite in texture. These surely recall the "painted ships upon a painted ocean." The two forms are seen in company; and in some instances the Physalia is seen in great numbers also with them.

The Class IV., CTENOPHORA, embraces the highest forms of the *Medusæ*. Venus Girdle is an example. One of the prettiest of the Ctenophores is the *Bolina* of the New England coast—*Mnemiopsis*. These are the pretty transparent comb-bearing forms that float in great numbers on the waters during the hottest portion of the year. They are often strewn on the beaches in vast masses, their iridescent bodies, or combs, glowing brilliantly. The *Pleurobranchia rhodactyla* is a common, rather small, oval form of great attractiveness.

Beroe is another form, having no tentacles. It is of a delicate pink, which greatly enhances its beauty. The pretty oval forms of these creatures constitute one element in their beauty.

THE REMARKABLE creature, called by the popular name of SALLEE MAN, sometimes corrupted, in nautical fashion, into SALLYMAN, may be met in vast numbers, sometimes being crowded together in large masses, and of various sizes, though it seldom approaches land.

In this curious animal the body is membranous, oval, and very flat, and may be at once recognized by the cartilaginous crest which rises obliquely from its upper surface, and the numerous tubercles which depend from its lower surface and surround the mouth. This cartilaginous substance marks out the Sallee Man as possessing a somewhat higher organization than its merely gelatinous relatives, and it is therefore placed at the head of its order.

The Sallee Man, scientifically termed *Velella vulgaris*, is seldom seen on northern European coasts, although it sometimes happens to be driven, by stress of wind and waves, to regions more chilly than those in which it entered the world. It is thought with justice that the upright cartilage can act the part of a sail, and, by means of its diagonal setting, drive the creature through the sea. The exact direction of its movements is in all probability decided by the numerous tentacles which hang from its lower surface, and which, by contraction or extension, can become living rudders.

The Velella is very widely distributed, and is found in every sea except those that are subject to the cold influences of the poles:

There is an allied genus called RATARIA, in which the body is circular, and the row of tentacles round the mouth is single. The body is sustained by a flattened elevated cartilaginous plate, and possesses also a longitudinal crest above, muscular and movable.

The internal cartilage of the Velelladæ are sometimes found strewn in great numbers on the surface of the water. Sailors believe that the delicate substance of the creature has been destroyed by the hot sunbeams, but naturalists have now ascertained that the true cause of their destruction is to be found in the sea-lizard (*Glaucus*), which feeds upon these curious inhabitants of the ocean, and devours the whole body with the exception of the firm cartilaginous plates.

A CONSPICUOUS member of this class of animals is the celebrated PORTUGUESE MAN-OF-WAR (*Physalis pelagicus*).

This beautiful but most formidable acaleph is found in all the tropical seas, and never fails to attract the attention of those who see it for the first time. The general shape of this remarkable being is a bubble-like envelope filled with air, upon which is a membranous crest, and which has a number of long tentacles hanging from one end.

These tentacles can be protruded or withdrawn at will, and sometimes reach a considerable length. They are of different shapes, some being short, and only measuring a few inches in length, while the seven or eight central tentacles will extend to a distance of several feet.

These long tentacles are most formidably armed with stinging tentacles, too minute to be seen with the naked eye, but possessing venomous powers even more noxious than those of the common nettle. "It is in these appendages alone," writes Mr. D. Bennett, "that the stinging property of the Physalis resides. Every other part of the mollusk may be touched with impunity, but the slightest contact of the hand with the cables produces a sensation as painful and protracted as the stinging of nettles; while, like the effect of that vegetable poison, the skin of the injured part often presents a white elevation or wheal.

"Nor is the inconvenience confined to the hand; a dull aching pain usually proceeds up the arm and shoulder, and even extends to the muscles of the chest, producing an unpleasant feeling of anxiety and difficulty in respiration. Washing the injured part with water rather aggravates than relieves the pain, which is best remedied by friction with olive oil. The cables retain their urent property long after they have been detached from the animal, and their viscid secretion when received on a cloth retains the same virulent principle for many days, and communicates it to other objects."

It is most probable that these terrible appendages are employed for the purpose of procuring food, and that they serve to entangle and kill the creatures on which the Physalis lives. Several of these acalephs have been observed with the bodies of half decomposed fishes entangled among the short tentacles.

The colors of the Physalis are always beautiful, and slightly variable, both in tint and intensity. The delicate pink crest can be elevated or depressed at will, and is beautifully transparent, grooved vertically throughout its length. The general hue of its body is blue, taking a very deep tint at the pointed end, and fading into softer hues towards the tentacles. A general iridescence, however, plays over the body, which seems in certain lights to be formed of topaz, sapphire, or aquamarine. The short fringes are beautifully colored, the inner row being deep purple, and the outer row glowing crimson, as if formed of living carbuncle. The larger tentacles are nearly colorless, but are banded at very small intervals throughout their length, giving them the appearance of being jointed.

It is a common trick with sailors to induce a "green hand" to pick up a floating Physalis, and to make him buy a rather dear experience at the cost of several hours' smart.

The vesicular body seems to be permanently filled with air, the animal having no power of inflating or collapsing at will. Many of these beings may be found on the sea-shore, where they have been flung by a tempest, the tentacles all decayed, but the body still inflated with air.

This is one of the most familiar objects seen in tropical waters, and it is one of the most beautiful. A thin, bubble-like, pear-shaped float rests upon the water, colored a brilliant indigo shading to pink. Along its upper crest is a narrow ruffle of silvery-white. Delicate in the extreme is this gorgeous bubble. But on the under side, hanging in the water, is a jelly-like mass of flesh, from which depend in coils, of several feet in length, the tentacular organs. So low is this creature in the scale of life, it has no propelling power. The little sail-like ruffle along the crest of the bladder catches the breeze, and the tiny ship seems to sail before it. The adult length is about nine inches. Often these creatures are seen in great numbers, bedecking the ocean far and wide with their richly-colored floats.

THE DIPHYIDÆ.

While resident on the Florida Reef we had opportunity to observe the creatures, and often removed specimens to a pail of water for examination. A box with a bottom of window-glass, placed on the surface of the sea, renders objects visible with great clearness. Using this one day over a Physalia, we observed several small fishes swimming among its long curling tentacles. Here was a discovery, and a mystery. How is it that these little fishes should be exempt from harm, when it is known that the least touch of the tentacles causes instant death to other fishes? A power resides in the tentacles that these little fishes seem to be exceptionally proof against. They are so near the brilliant blue color of the Physalia that they seem to be a part of them. One would regard them as a natural accompaniment, so alike they are in color, and so completely protected are they from harm.

Since this discovery of so remarkable an association of animals of diverse habits, many others have been made in various parts of the world. The singular power possessed by the Physalia in common with many of the corals and jelly-fishes, may well be considered here. For a long time, naturally since corals and jelly-fishes have attracted attention of collectors, it has been observed that a sharp, unpleasant, stinging effect is produced by contact with these forms. Not until a few years since was this fully understood. We have before us photographs taken from the tentacles of the most powerful of these "stinging" creatures, those of the Physalia, just considered. On a light magnifying power the tentacles, the long, curling, extensible feelers, exhibit along their surfaces vast numbers of needle-point orifices, raised slightly from the surrounding surface. These are called *lasso cells*, because the little cell-shaped swellings contain each a veritable lasso-like, slender, tubular thread coiled up within, which is darted out instantly when needed. The interior anatomy of these organs is somewhat complicated, but it suffices us to know that the weapon thrown out is barbed, and though microscopic, yet penetrates the skin even of fishes, not only causing pain, but, from the deadly effects, seems to eject at the same instant a poisonous fluid. We have seen fishes swim up to the tentacles of a Physalia boldly, and very quickly turn over and die. At the same time, as we have seen, there are little fishes one would suppose equally vulnerable, quite at home within the dread portals, moving among and around the poisonous hanging mass.

The great power existing in the tentacles of the Physalia we may, perhaps, be allowed to illustrate by a personal relation. And we may find it proper at this place to say that our long residence on the Florida Reef naturally afforded opportunities for many observations of habits of marine animals not heretofore made public. In order to render such of the true value, we may be pardoned, perhaps, the frequent use of personal pronouns.

It was a common occurrence after the appearance of myriads of the beautiful Physalias, like minature glass ships upon the ocean, that the creatures, having no power to direct their course, would, after a considerable disturbance in the ocean, as after a gale, drift ashore in great numbers. The moat of the fortress was a point which caught many.

A lad of our family, indulging with others in a bath at this portion of the moat, inadvertently swam over a Physalia; its long slimy tentacles adhered to his chest and abdomen, and the shock of the millions of poisonous lassos that were thrown into his skin was such as to nearly prostrate him. Some soldiers at hand rescued him from the water, otherwise he, though an excellent swimmer, would have drowned. For several hours the most vigorous treatment of internal stimulants and external topical remedies was necessary, to keep him from sinking, with the vital powers wholly overcome. The treatment consisted in rubbing the parts with an abundantly strong soapy-water, warm, and the internal use of whiskey. For many months the marks of the tentacles were observed on his flesh, appearing like welts left after a severe lashing with a whip.

Our last example of this order belongs to a tolerably large family termed Diphyidæ, or double animals, because they are formed, as it were, of two animals, one fitting inside the other. Their general form is bell-shaped. In the present genus, both animals are similar, and of a somewhat pyramidal shape, and have a few points round the aperture.

The connection between the two portions of the DIPHYES seems to be very slight, inasmuch as the two halves are often found separated from each other. The progress of the animal is achieved simply by taking water slowly into the bells, and expelling it smartly, much after the fashion of the ordinary Medusæ.

Trailing from the interior of the bells may be seen a curiously-elongated appendage, studded with globules, which are, in fact, the offspring in different stages of development. A number of tiny discs set on footstalks are also distributed along this appendage, and save the power of adhesion to any object which they may happen to touch.

COMB-BEARERS; CTENOPHORA.

WE now come to a fresh order named Ctenophora, or comb-bearers, because their bodies are furnished with rows of flattened cilia, set in rows above each other something like the teeth of a comb. There are many members of this beautiful order to be found, of which the common CYDIPPE is an excellent example. In the accompanying illustration it is drawn of its natural size.

This lovely creature may easily be captured by the simple process of towing a gauze net over the side of a sailing boat. When removed from the water the net will be found studded with variously-sized knobs of transparent gelatine, not particularly attractive, and presenting no salient points whatever. Let, however, these apparently inanimate lumps of jelly be transferred to a vessel filled with sea-water, and then how different is their aspect!

Until the eye is accustomed to their shapes, they are not very easily seen, owing to their transparency and the similarity between their refractive powers and those of the water. I have often noticed persons looking at ss jars without discovering that a single living creature was within them, though each jar was tenanted by two or three of these beautiful creatures.

By degrees, however, they became plainly visible, the chief points of attraction being the eight bands of ever moving cilia that are drawn longitudinally over the body, and by means of which the creature performs its wonderful evolutions. The Cydippe is never still, but careers through the water with ceaseless movement, sometimes rising and falling in one spot, sometimes rolling over and over, sometimes spinning on its longer axis, but mostly pursuing a partly spiral course, turning slowly on itself as it proceeds through the water.

During these movements a faint iridescence plays over the whole body of the Cydippe, but its chief glories are concentrated upon the bands of cilia which are drawn over the body. On these the colors are too brilliant, and yet evanescent, for description. Miniature rainbows seem to ripple along these living belts; and as the Cydippe glides gracefully along, it appears to be encircled with many diadems of self-illumined jewelry. If examined by the microscope, the ciliæ of which the locomotive bands are composed are seen to bear some resemblance to very narrow Venetian blinds, each lath closing or opening in regular succession.

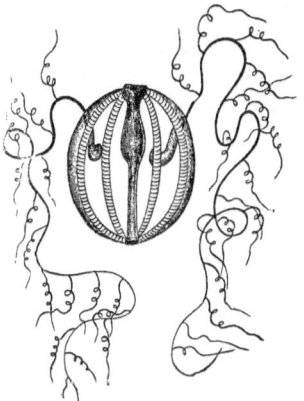

CYDIPPE.—*Cydippe pileus.*

Pendent from the body are further seen two long filaments, to which are attached a number of shorter and still finer threads, not unlike the hooks and snoods on a deep-sea line, and used, indeed, for a similar purpose. The Cydippe can protrude or retract these tentacles at will, and is continually throwing them out from the body or drawing them back again, so that they never seem to be exactly the same length, one being often three or four times as long as the other. The manner in which these tentacles trail after the creature is extremely graceful, and the observer cannot resist a feeling of wonder that they should avoid entanglement.

The tentacles are employed for the purpose of catching prey, the Cydippe having been observed in the very act of seizing and eating its food. The long threads arrested the object as soon as touched, and in a very short time they were drawn to the central mouth, and the prey softly lodged within. The smaller crustaceans appear to be the favorite food of the Cydippe. The vitality, or perhaps the irritability, of the cilia is very enduring, for they continue to act when the animal is cut into several parts, or even when a little piece is nipped off, and will carry the severed portions through the water quite merrily.

The development of the Cydippe is very interesting, the young being produced from minute vesicles, and passing through a series of stages before they assume their perfect form.

THE present illustration shows us a long, flat, riband-like creature edged with a delicate

VENUS' GIRDLE.—*Cestum veneris.* (One-half natural size.)

fringe of cilia. This curious being is called VENUS' GIRDLE, and from its beauty fully deserves the name.

This lovely creature is found in the Mediterranean, where it attains to the extraordinary length of five feet, the breadth being only two inches. Rightly, the words breadth and length ought to be transposed, as the development is wholly lateral. The mouth of the Venus' Girdle may be seen in the centre of the body, occupying a very small space, in proportion to the large dimensions of the creature to which it belongs. A very good idea of the appearance of the Venus' Girdle may be obtained by supposing a Cydippe two inches in length to be flattened and rolled out into a riband of five feet in length.

Owing to the great length and tenuity of this creature, it is seldom found quite entire, but it seems to care little for the loss of a foot or so of its substance.

DISC-BEARERS; DISCOPHORA.

WE now come to a very large order of acalephs, including all those beings which are so familiar under the title of JELLY FISHES, SLOBBERS, and similar euphonious names. They are all united under the name of Discophora, or disc-bearers, because they are furnished with a large umbrella-like disc, by means of which they are enabled to proceed through the water.

Each order is separated into several tribes, the first of which is termed Gymnophthalmata, or Naked-eyed Medusæ, because the little ocelli, or eye-specks, are either uncovered or altogether absent. The edge is either simple or branched. The name of Medusæ is given to these creatures on account of the long trailing filaments which depend from them like the snaky locks of Medusa from her head. In the Naked-eyed Medusæ, the circulating vessels may be seen radiating to the edge either simple or branched.

A good example of the family Sarsiadæ is the *Sarsia tubulosa*. This family contains several genera. All the Sarsiæ are pretty little creatures, and may be known by the four simple nutritive vessels and the egg-tubes placed in the footstalk. In this genus the umbrella is nearly hemispherical, and there are four tentacles set at the ends of the radiating vessels.

Though small, the Sarsiæ are interesting to the naturalist, on account of the curious method by which the young are produced, sprouting like buds from the footstalk, and presenting a very strange aspect as they project in different stages of development. In their first stage, the young Sarsiæ are nothing more than simple prominences upon the surface of the footstalk, and gradually increase in size, developing first one part and then another, until at last the little creatures are quite perfect, shake themselves free from the parent, and commence an independent existence.

There is a curious species of this genus, *Sarsia prolifera*, in which the base of every tentacle is supplied with a little bunch of young Medusæ, some just making their first appearance as mere lumps of gelatinous substance, some half-grown, and others nearly ready to free themselves from the parent stock.

THE members of the next family are known by their flattened discs and the egg-tubes running linearly along the vessels. The *Eudora undulosa* is a prominent species of this family. It is a rather curious creature which is devoid of footstalks and appendages, and has a disc almost as flat as a biscuit. In the pretty *Æquorea cyanea* the disc is rather more convex than in the preceding genus, the footstalk is very wide and expands into many lobes, with long and broad fringes; and the tentacles are very slender and variable in number. The present species inhabits the South seas.

A REALLY fine creature is the *Chrysaora lutea*. It belongs to the next tribe of the order, wherein the eye-specks are covered by certain flaps, and the circulating vessels united into a kind of network. This tribe is further divided into two families, in the first of which, the true Medusæ, solid food is received into a mouth; and in the second, there is no mouth, but nourishment is absorbed through the ends of branching vessels.

The CHRYSAORA belongs to the first of these families, and may be recognized by the long unfringed but furbelowed arms. A fine species belonging to this genus, *Chrysaora cyclonota*, was kept for some time by Mr. Gosse, and has afforded many useful hints to the students of Natural History. Experiments were made for the purpose of ascertaining the method of obtaining food, and it was discovered that the furbelowed arms as well as the tentacles are used for catching prey. A dead white-bait was first given to the Medusa, and, after having been caught by the tentacles and furbelows, was delivered to the former organs, the latter relinquishing their hold. Very gradually it was shifted towards the mouth of the footstalk, and there held for about an hour, when it was released and fell to the bottom of the vessel.

Thinking that the fish might have been too large a morsel for the Medusa, the experimenter next supplied the animal with a small piece of cooked meat. This was seized as the fish had been, and during the course of the night was conveyed into one of the four cavities of the footstalk. There it remained for about sixty hours, when it was rejected. On being examined, it was found to be perfectly white, but not in the least decomposed or having any putrescent smell.

A curious change then took place. "After I had kept this Chrysaora for about a week, its manners underwent a change. It no longer swam about freely in the water by means of its pumping contractions, nor was its appearance that of an umbrella. It began to turn itself inside out, and at length assumed this form permanently, its shape being that of an elegant vase or cup, with the rim turned over, and the tentacles depending loosely from it, the furbelows constituting a sort of foot.

"The latter were now put to a new use; the animal began habitually to rest near the bottom of the vessel, or upon the broad fronds of the *Iridæa*, which were growing in the water and preserving its purity, but occasionally it would rise midway to the surface and hang by one or two of the furbelows. A fold or two of the latter would come to the top of the water, and dilate upon the surface into a broad flat expansion, exactly like the foot of a swimming mollusk; from this the Medusa would hang suspended in an inverted position. All the other furbelows, and portions of this one that lay below the expansion, floated as usual through the water, except that on some occasions an accessory power was obtained by pressing a portion of another furbelow to the side of the glass and making it adhere just like the portion that was exposed to the surface of the air. The texture of the furbelows when thus stretched smooth was exquisitely delicate." This curious movement seemed to be a prelude to the production of eggs, which were seen in great numbers. As if its whole life powers were exhausted by this process, the creature soon became feeble and then died, its captive life having endured for almost three weeks.

An example of the typical genus of the Medusæ is the *Medusa aurita*. This is a sufficiently common species, and may be found plentifully on northern European shores, together with its kindred. There are few more beautiful sights than to stand on a pier head or lie in the stern sheets of a boat, and watch the Medusæ passing in shoals through the clear water, pulsating as if the whole being were but a translucent heart, trailing behind them their delicate fringes of waving cilia, and rolling gently over as if in excess of happiness. At night, the Medusæ put on new beauties, glowing with phosphorescent light like marine fire-flies, and giving to the ocean an almost unearthly beauty that irresistibly recalls to the mind the "sea of glass mingled with fire."

That scourge of the ocean, the VENOMOUS CYANÆA, though a harmless-looking creature, is, in truth, one of the few inhabitants of the sea that are to be feared by bathers on our favored shores; but its presence is so much to be dreaded that no one who has once suffered from the lash of its envenomed filaments will venture to bathe without keeping a careful watch on the surrounding water. I have twice undergone the torment occasioned by the contact of this creature, and know by experience the severity of its stroke.

At its first infliction, the pain is not unlike that caused by the common stinging-nettle, but rather sharper, and with more of a tingling sensation. Presently, however, it increases in violence, and then seems to attack the whole nervous system, occasionally causing a severe pain to dart through the body as if a rifle-bullet had passed in at one side and out at the other. Both the heart and lungs suffer spasmodically, and the victim occasionally feels as if he could not survive for another minute.

These symptoms last for ten or twelve hours before they fairly abate, and even after several days the very contact of the clothes is painful to the skin. The shooting pangs just mentioned are of longer duration, and I have felt them more than three months after the Cyanæa had stung me.

To the unaided eye the filaments which work such dread misery are most innocuous and feeble, being scarcely stronger than the gossamer floating in the air, and looking much as if the Medusa had broken away a spider's web, and were trailing the long threads behind it. The microscope, however, reveals a wondrous structure, which, though it cannot precisely compensate for the sufferings inflicted by these tentacles, can at all events endow them with an interest which would not otherwise be felt.

Lest any of my readers should become fellow-sufferers with myself, I advise them to be very careful when bathing after a strong south-west wind has prevailed, and if ever they see a tawny mass of membranes and fibres floating along, to retreat at once, and wait until it is at least a hundred yards away. Some may suppose that this advice is needlessly timid, but those who have once felt a single poison thread across their hand or foot, will recognize that discretion is by far the wisest part to be played whenever there is the least danger of being stung by the Cyanæa.

RHIZOSTOMA.—*Rhizostoma cuvieri.*

The last family, of which a small specimen is represented in the accompanying illustration, is easily known by the absence of a mouth. In the typical genus, RHIZOSTOMA, the footstalk is deeply scooped into semilunar orifices, and the eight cartilaginous arms are without fringes.

Before taking a final leave of these remarkable beings, it is needful that we should briefly notice the strange metamorphosis through which some of them pass before they assume their well-known form. Experiments were made on a species of Chrysaora, by Sir John Dalyell, with the following result:—When first sent into the world, the young Medusæ were little flat, worm-like creatures, too minute to be examined by any except the highest powers of the microscope. By degrees, these tiny beings settle down to one spot and affix themselves, the body lengthens, arms begin to be shown, and after a while the strange creature is developed into the being known as the *Hydra tuba*.

Satisfied, apparently, with its condition, the Hydra remains in the same spot for some time, and produces a number of young Hydras, which sprout like buds from its sides, and, when separated, resemble their parent. Here, we might naturally imagine to be the end of its history, for, with almost all animals, when a being is able to produce young, it is considered as having attained the utmost development of which it is capable. The Hydra, however, has yet other phases through which to pass. Towards spring, its body becomes much lengthened and wrinkled, so as to form a number of folds, just as if a series of threads had been tied tightly round it, one below the other. The upper rings now rapidly expand and the folds deepen, until the animal resembles a number of saucers regularly increasing in size, laid upon each other. The edges of each saucer are developed into two-cleft rays, and in this condition the animal proves to be the beautiful zoophyte discovered by M. Sars, and called the Strobila.

These are, indeed, strange vicissitudes in life, changes more marvellous than even those wrought by water and magic words, in the old days when Haroun Alraschid ruled the faithful. There is yet more to come. The uppermost and largest disc or saucer now lengthens its rays and assumes the form of an unmistakable, though shallow-disced Medusa. Its arms rapidly gain strength, the attachment becomes hourly weaker, until at last the whole disc is broken away, and floats into the wide sea in its new form.

How wonderful is this phenomenon, and how full of interest is the study of animate nature! Here we have a being which first enters into active existence in a shape like that of the infusorial animalcules; then changing into a hydra, and while in this state becoming the parent of a numerous offspring; then developing into a Strobila; and lastly, breaking up into a series of Medusæ.

ZOOPHYTES.

ACTINOIDA.

QUITTING the Acalephæ, we come to the vast class of Zoophytes, or animal plants, so called, because, though really belonging to the animal kingdom, many of them bear a singularly close resemblance to vegetable forms. In our beautiful oleograph, seven European species of this class are given, some to exhibit their forms as they appear when expanded, and the others to show the variety in colors. These seven species are: The THICK-PETALLED SEA-ROSE (*Thelia crassicornis*); the *Sagartia parasitica;* the SEA-PINK; the WIDOW; the RED-ANEMONE (*Sagartia rosea*); the WARTY-ANEMONE; and the GREEN-ANEMONE (*Anthea cereus*). As there exists a great similarity in the form and structure of these Sea-anemones, it will be sufficient to describe only some of them. The substance of these Zoophytes is always gelatinous and fleshy, and round the entrance to the stomach are set certain tentacles, used in catching prey and conveying it to the stomach. These tentacles are armed with myriads of offensive weapons contained in little capsules, and capable of being discharged with great force. Organs of sight, smell, taste, and hearing seem to be totally absent, though it is possible that an extended sense of touch may compensate the creature for these deficiencies.

Without entering further into the constitution of these singular beings, we will proceed to the examination of the various groups into which they have been divided.

IN the family of the Lucernariadæ, the tentacles are arranged in detached groups, a peculiarity whereby the creatures may easily be recognized. These organs are placed upon the outer edge of the membranous and expanded disc, in the centre of which is the squared mouth. They are mostly found adherent by a stem to some object, but they can swim with tolerable rapidity, their bodies pulsating like those of the Medusæ. None of them attain any great size, the largest being about one inch in height. Pink is their usual color.

Mr. Gosse, in his "Sea-anemones and Corals," remarks that the Lucernariadæ have closer affinities with the Medusæ than with the Actiniæ, on account of several structural peculiarities, among which may be mentioned the gelatinous texture, the expanded umbrella, the egg-sacs in the substance of the umbrella, and the squared mouth at the end of a free footstalk. I have, therefore, departed a little from the ordinary arrangement, and placed the Lucernariadæ immediately after the Acalephs, forming a kind of intermediate link between them.

THE highest form of true Zoophyte is, undoubtedly, that which is so familiar under the name of Sea-anemone—a name singularly inappropriate, inasmuch as the resemblance to an anemone is very far-fetched; while that to the chrysanthemum, daisy, or dandelion is very close. These creatures are called Actinoida, and are easily distinguished by having the stomach inclosed in a sac divided into compartments by radiating partitions. For convenience sake, this group is divided into two sub-orders, the first of which is the Actinaria, known by the number of tentacles (twelve or more), perforated above, and the radiating partitions sometimes depositing solid, chalky plates, commonly called "coral." The tribe Astræacea is known by the imperfect series of tentacles, and the family Actiniadæ by their circular arrangement.

SEA ANEMONES.

The beautiful OPELET, or Green-anemone, in the oleograph, may easily be recognized by the great length of its many tentacles, which wave, and twist, and twine, and curl like so many snakes. It has but little power of retracting the tentacles, and is, therefore, more conspicuous than many other species. It is tolerably hardy, enduring confinement well, but requiring food more often than is the case with the other Actiniæ. Like all other members of this order, the Opelet is able to arrest passing objects by means of the tentacles, and does so by the aid of a wonderful array of weapons unexampled in the animal kingdom.

If a portion of a tentacle be examined under a moderately powerful microscope, it will be seen to be studded with tiny cells, in each of which lies coiled a dark thread. On applying pressure to the cell, it suddenly discharges the coiled thread, which proves on closer examination to be a long, wiry dart, often of wondrously complex structure, and capable of penetrating into any soft substance with which it comes in contact. Elaborate accounts and drawings of these cells and their contained weapons may be found in Mr. Gosse's valuable "Sea-Anemones and Corals," a work to which I gladly refer my readers for many interesting details respecting the beautiful creatures on which we are at present engaged.

Though the human skin be a tougher and harder substance than the prey generally brought into contact with the tentacles, it yet can feel the effects of the individually minute but collectively potent weapons with which these delicate tentacles are armed. A finger which is touched by a tentacle is instantly conscious of being seized, as it were, and forced to adhere to the soft waving membrane which it could crush with a single effort. On most persons this adherence has no particular effect; but those who possess delicate skins, and a sensitive nervous system, are much worried by blisters and pustules occasioned by the assaults of these microscopical weapons. A young eel, measuring six inches in length, and half an inch in thickness, was killed in a few minutes by mere contact with the tentacles, and in a very short time was tucked quietly away in the creature's stomach. These weapons are most numerous at the tips of the tentacles, just where they are most needed.

THE SCOTTISH PEARLET (*Ilyanthus scoticus*). This is a member of a genus once thought very rare in Europe, but now necessarily expanded into a family, and found to contain a considerable number of species. Most of the Pearlets are able to crawl over solid bodies; some inhabit tubes; others are found burrowing in the sand; while nearly all are able to puff out the hinder part of the column with water.

Little is known respecting the history of the Scottish Pearlet, save that it is a very rare species, and has only been found in deep water. All the tentacles are very slender, and marked with a dark line.

The PUFFLETS are so called because they possess the power of puffing out the hinder part of the column until it assumes a somewhat globular shape. A European species of this genus, the PAINTED PUFFLET (*Edwardsia callimorpha*), appears to be one of the burrowers, its body being hidden beneath the sand, and the beautiful tentacles just protruding from the surface. None of the Pufflets have many tentacles.

WE may here briefly notice another example of the same family.

The VESTLET is one of those members of the family which inhabit tubes. All of them are remarkable from the fact that they possess no adherent base, but, as a compensation for this deficiency, are furnished with an adherent power upon the stem, enabling them to crawl freely over solid bodies. In this species, the tube is cylindrical, and very wide in comparison with the dimensions of the inhabitant; it is of tough, paper-like consistence, rather thick, and is composed of many layers of intertwining fibres, mixed with sand and mud. The ordinary length of the animal is six or seven inches, and the width of the flower-like plumes about an inch and a half. Mr. Gosse found that he was able to remove the creature from its opaque dwelling, and place it in a tube of glass, which the animal accepted as a useful substitute, without troubling itself to reconstruct another house.

THE beautiful creature called SEA-PINK, or PLUMOSE ANEMONE (*Actinoloba dianthus*), which is also shown in the oleograph, under the name of Plumose Anemone, is certainly the most magnificent of the European species.

It may be at once recognized by its bold cylindrical stem, firm and sturdy as the oak trunk, standing out bravely from the object to which it is affixed, and crowned with its lovely tufted tentacles, fringed and cut like the petals of the pink. Its color is extremely variable, being snowy-white, olive, red, orange, cream, or pale pink; and of all the varieties, the first is, in my eyes, the most beautiful. It is capable of much alteration in its general form, shrinking to a mere shapeless fleshy mass, and looking by no means a pleasing object; expanding itself to the fullest extent, or forming itself into many shapes, according to the caprice of the moment.

Fortunately for the owners of aquaria, the Plumose Anemone is hardy, and bears captivity well. It often separates itself into several parts, each of which becomes an independent being, and in some stages of this process looks as if two individuals had become fused together.

The pretty SNAKE-LOCKED ANEMONE, or WIDOW (*Sagartia viduata*), may be recognized in the colored engraving of the Sea-Anemones, by the long, slender stem, and the flexible, indistinctly-barred tentacles, with a dark line running down each side.

It is found on many shores of northern Europe, seeming to be rather local, but tolerably plentiful in the spots which it chooses for its residence. Though not adorned with brilliant colors, it is a remarkably pretty species, with its crown of delicate tentacles waving "like a thin blue cloud" upon the summit of its elongated stem. One of these Anemones has been known to produce some curious changes in its tentacles, at one time thickening them into knobs, and at another throwing out branches.

A widely-spread Anemone, with a circlet of pearl-like beads at the base of its tentacles, is well known under the name of BEADLET (*Actinia mesembryanthemum*).

It is a singularly hardy species, living mostly on the rocks that lie between high and low-water mark, and in some places collecting in extraordinary numbers. I remember on one occasion, after meeting a party of unsuccessful anemone-hunters, I filled their baskets in a quarter of an hour, though night had set in, and the only method of discovering the creatures was by the touch. It is, perhaps, more variable in color than any of the European Actiniæ, the body taking all imaginable hues, passing from bright scarlet to leaf-green, graduating from scarlet to crimson, from crimson to orange, from orange to yellow, and from yellow to green. The spherical beads around its mouth are more persistent in color than any other parts of the animal, being almost invariably a rich blue, just like a set of torquoises placed around the disc. These, however, are occasionally subject to change, and lose all color, looking like pearls rather than torquoises. Even the same individual is subject to change of color, being evidently influenced by various external conditions, such as light and shade, food, and the purity of the water in which it is placed.

In the aquarium it is wonderfully prolific, surrounding itself with many a brood of tiny young, whose minute forms are seen settled around their parent, opening their tentacles with a kind of competent air that has something of the ludicrous about it. The Beadlet is something of a wanderer, and will not only crawl slowly over the glassy sides of the aquarium, but, when it has reached the surface of the water, will invert itself so that the tentacles are downwards, make its base hollow, and float away, trusting itself freely to this shallow boat.

The GEM-PIMPLET, or Warty-Anemone (*Bunodes gemmacea*), may be recognized by the double series of large and small warty protuberances placed alternately on its body. There are six white bands on the stem, and the tentacles are thick, marked with white, oval spots. Like the preceding species, the Gem-Pimplet is not local, though gathering in considerable numbers in certain favored spots. Even when closed, with all the tentacles withdrawn, it may at once be known by the six bands of white which radiate from the orifice, and the great resemblance which its body bears to an echinus stripped of its spines.

THE Sea-Anemones are now (1885) embraced under the class II., ACTINOZOA,—and the Corals are included.

The term coral *insect* should now be abandoned. Yet we hear it used by persons who ought to know better. The idea that the coral animal is a separate creature which builds mechanically its blocks of coral, should be ignored, as the corals are, in most respects, the same as the shells of clams or other shell-fish, merely the lime skeletons of the soft animal, secreted, as our own bones are, for support to the soft parts. Thus, the corals are not anything like insects, and they are very much lower in the scale of life than the insects.

The genus *Actinia* includes a large number of soft-bodied creatures that do not secrete a lime, or hard covering, but have instead a thick, leather-like exterior. This creature may, for convenience, be compared to the naked mollusk, the garden-slug, while the coral, with its lime tube around it, may be regarded as corresponding to the garden-slug that has a shell, the garden-snail. This assumed analogy may help the reader to understand the subject. But, at all events, it must be understood that the corals which we admire so much for beauty of form and cells, are compound skeletons, which, like our bones, have been secreted by the animals that are now dead and decomposed. Corals are difficult to understand. We may readily get an idea of the corals that are made up of a congregation of stars, for each star is a sea-anemone, *Actinia* with its secreted skeleton around it. But when we are asked, how about the Meandrinas which have long, winding pits, we have to explain that the animals are compound, and

GREAT CRAMBACTIS.—*Crambactis arabica*. (A little diminished).

mouths are seen at intervals. It must be remembered that these animal forms are so low in the scale of life, it is not expected they will be found observing the strictest conventionalities.

We have, on the New England coast, a species of *Actinia* (*Metridium marginatum*) that grows to be about the size of a tea-cup. Its color is sometimes rather attractive, of a pinkish, but usually rather sombre hue. It is a pleasing object in the aquarium. It is found in shady pools among rocks at low water. Several beautiful forms are found in deep water. The warmer waters of the tropics teem with gorgeous colored species.

One example we remember as especially interesting, is a species that bears green, leaf-shaped appendages between the tentacles. One only we obtained, while resident on the Reef at Tortugas. This, or one very closely like it, is figured by Dana, in colors. It was about seven inches in length, and three inches diameter. Its body was columnar and fluted, of a delicate pinkish-white. Between the tentacles were organs that resembled oak-leaves, both in exquisite outline and in color. It was a most beautiful object. We kept it in our sea aquarium, until Mr. Barnum sent an agent to Tortugas for the purpose of gathering marine objects for his then new aquarium in New York. This *Actinia* we sent to Mr. B., and it proved an especial attraction for many months.

The GREAT CRAMBACTIS seen in the illustration recalls the latter Actinia in the respect that it has leaf-like appendages, but the latter are situated on the upper surface in this case, while those of the Reef specimen were under the tentacles, and quite like oak-leaves in shape and color.

LEAVING the sea-anemones, we now proceed to the next tribe, the Caryophylliaceæ, in which there are many tentacles, in two or more series, and the cells many-rayed. Many of these beings deposit a corallum.

The FUNGIA, or SEA-MUSHROOM, is so called from its great resemblance to a mushroom, the expanded disc and delicate lamellæ having a singularly fungine form. The hard corallum of this genus is not fixed, but the creature is protected from the violence of the waves by its habit of lying in clefts of rocks, or in the deep cavities of coral reefs, so that it enjoys free access of water, without the danger of being carried away by the currents or dashed ashore by the tempest.

CUP CORAL.—*Astroides calycularis*.

When young, however, the Fungiæ are affixed for a time, sometimes on rocks, and sometimes on the stony remains of their own kinds, being attached to a stem which gradually vanishes as the creature increases in age. While in this state, they bear some resemblance to the genus Caryophillia. Though all possessing the same general characteristics, they are not all circular, some being oval, and others bearing no small resemblance to slugs. The entire corallum is surrounded by the soft substance of the Fungia, which envelops it below as well as above. Most of the Fungia are found in the Indian seas, especially among the coral-beds.

WE now pass to another group of these curious beings. The ENDIVE CORAL is so called from the resemblance which its corallum bears to the crumpled leaves of that vegetable. The animal has no tentacles, and the cells are small, conical, and rather oblique. The corallum is fixed, sharply edged, and expanded from the base to the tip—a peculiarity which has earned it the specific title of Pavonia, or Peacock's-tail Coral. All the living members of this pretty genus are to be found in the East and West Indian seas.

The present illustration represents the CUP CORAL as it appears when the tentacles are fully expanded, and when they are closed. It is not a very large, but is a very pretty species, the color of its corallum being generally of a pure translucent white, sometimes tinged with a delicate rosy hue, while that of the living animal is pearly-white, variegated with rich chestnut, and the palest imaginable fawn.

RED CORAL AND EIGHT-ARMED CUTTLE.

It is mostly a deep-water species, not unfrequent on northern European coasts, and is seldom procured except by means of the dredge or grapnel. Sometimes, however, it is found near the coast, and at the equinoctial springtides may sometimes be procured from the rocks which are laid bare by the receding waters. Fortunately for the collectors, it is very gregarious in its habits; and when one specimen is found, others may generally be secured within a very short distance. It is a pretty inhabitant of the aquarium, and, as a general rule, may be induced to expand its long tendrils to their fullest extent, by placing a morsel of food upon the orifice. When properly managed it is tolerably hardy, but it does not brook inattention—shrinking up daily, and at last perishing hopelessly. When new to the mysteries of aquarium-management, I never could keep a Cup Coral more than a month.

In the family Oculinidæ, the corallum is branched and tree-like, and is here represented by the only known form, the TUFT CORAL. It is very rare. A remarkably fine specimen is figured by Mr. Gosse, who remarks that it was taken off Skye in the year 1852, entangled in the deep-sea-line of a fisherman. Another specimen, weighing six pounds, has been taken in a similar manner between the islands of Rum and Eig. As may be seen from the illustration, the corallum resembles a massive, thickly-branched tree. The individual corals are about half an inch in height and the same in diameter.

On the full-page illustration is seen a coral that has attained a singularly tree-like form, and, in consequence of this structure, has obtained the appropriate name of Dendrophyllia, or Tree-Coral.

The regular branched form of this coral can be seen by reference to the illustration, together with the manner in which the individuals are set on their common stem. The cells are rather deep, and the animals possess tentacles which are cleft longitudinally. It is a native of the hotter seas.

On the accompanying illustration we have some examples of those beings which we call Madrepores.

In the genus MADREPORA the animals are rather short, with twelve simple tentacles. The cells are deep, irregularly arranged upon the surface, and are crowded together towards the tips of the corallum, though they are scattered rather widely at its base. The cells are nearly cylindrical in their general shape and project outwards from a centre, something like the grains on an ear of wheat.

The genus Echinopora is distinguished by the peculiar arrangement of the cells, which are set only upon the upper surface of the coral. They are boldy radiated and rather irregular. All the true Madreporæ inhabit the hot seas, and are most plentiful under the tropics.

Also of great interest is the genus ASTRÆA, so called because the animals are sown over its surface like stars in the heavens. The cells are decidedly short, and the tentacles few in number. The genus is a very large one, including many recent and fossil specimens, many of which are familiar to us in the polished stones of which mantlepieces and other domestic ornaments are made. Owing to the vast number of the animals, and the rapidity with which they increase, the groups of Astræa often assume enormous dimensions; and in the secondary and tertiary rocks they frequently occur in such huge masses that whole rocks are composed of their remains.

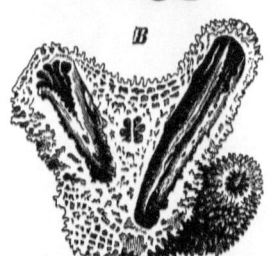

MADREPORA.—*Madrepora verrucosa*. A. Little tree in natural size. B. Cells in enlarged form.

In the accompanying illustration is seen a figure of that remarkable coral which is popularly called BRAIN-STONE, or BRAIN-CORAL, because the convolutions into which the corallum is moulded much resemble those of the human brain.

The animals of this genus are always united together in long waved series, each having a distinct mouth and series of very short tentacles. The cells are very shallow, and the valleys formed by their union are separated from each other by distinct ridges. The shape of this coral alters greatly with age, somewhat resembling the top-shells when young, but becoming rounded above when adult. The Brain-coral is found in several of the hot seas.

BRAIN-CORAL.—*Astræa pallida*. (One-half natural size.)

Among the Asteriadæ, as these creatures are called, in consequence of the star-like appearance of the polype or animal, the ORGAN-PIPE-CORAL is perhaps the most striking. It forms, as far as is yet known, the only example of the group to which it belongs, and which is called Tubuliporina, on account of the multiplied series of regular tubes from which it is formed. As will be seen by reference to our engraving, in this beautiful coral the tubes are arranged like the pipes of a church-organ, or the storied rows of basaltic columns of the Giant Causeway.

The color, too, is very pleasing, being a delicate pink, so that even the empty and lifeless corallum forms a really beautiful object. When living, however, it may fairly lay claim to the title of magnificent, for each tube is clothed, formed, and vivified by a light green polype, whose color contrasts beautifully with that of the structure which is raised by that soft and feeble body.

Two other species of true coral, such as are used so largely in the manufacture of ornaments, are termed *Corállium fecundum* and *Corállium nóbile*.

These beautiful zoophytes seem to be found only in the Mediterranean, where regular fisheries are established and the corals dragged from their recesses. The appliances, however, are very rude; and it is likely that more elaborate machinery would reap a rich harvest by permitting some selection to be made and by enabling the fishers to regulate the dimensions of the groups of coral branches. Although the stony centre is so thick and solid, the substance of the animal is quite delicate and membranous, enveloping the corallum like wetted goldbeater's skin.

A fan-like object is popularly called from its shape, the SEA-FAN (*Gorgónia flabellum*), and well deserves that title. In this genus the branching arms are united by a number of transparent pieces, which are, in fact, developments

ORGAN-PIPE CORAL. *Tubipara springa*. (Natural size.)

of the branches, are covered in a similar manner by the investing membrane, and bear the living polypes on their surface. The whole structure easily dries, and may be found in most curiosity shops, or in the dwelling-houses of mariners, who have brought home these remarkable objects as presents to their wives.

The Gorgónias, Sea-fans, Sea-feathers, Sea-whips, etc., belong to the Order *Halcyonoida* of this class. To this also belongs the Organ-pipe coral—the precious red coral of jewelry—the curious Sea-pens, Venellas, etc. The Sea-fans and Sea-feathers are abundant on the shoals of the Florida Reef. Acres of them may be seen, bending with the tide like so many land grasses or shrubs. Their colors are pretty and striking, while living, and some are of a beautiful red and sulphur-yellow when dead.

GORGONIA - *Gorgónia verrucosa*.

The GORGONIA VERRUCOSA figured above is a common example in other seas. The egg case of a shark is shown very prettily, with its coiling tendrils wound around the branches of the Gorgónia. The polyps of this species are shown plainly, while most others are too minute to show distinctly. The illustration is of natural size.

An allied species belonging to the same family (*Isis hippuris*) is formed in a very strange fashion. Its branches are composed of a number of strong joints, united together by horny rings, so that a certain amount of flexibility pervades its structure. Owing to this formation, it is sometimes called the HORN-PLANT, or SEA-SHRUB, titles surviving from the time when all the corals were thought to be vegetables, and the expanded polypes to be their flowers. They are always fixed by a base, and grow like trees, with their branches upwards. It is worthy of notice, that the Gorgoniæ are never bushy, and, for the most part, have their branches in the same plane.

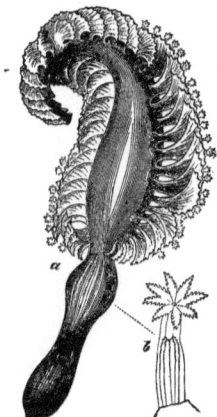

SEA PEN.—*Pennatula grisca.* (One-quarter natural size. b, Enlarged cell.)

In this illustration we have an example of a very interesting and extremely beautiful species.

The SEA-PEN is so named because its whole form bears the most remarkable resemblance to a quill-feather, consisting of a central shaft, from which a double row of "pinnæ" is developed at right angles, bearing the polype on their upper margin. As may be seen by the illustration, the whole form of this curious being is remarkably graceful, and it really seems as if it had been modelled upon a quill-feather plucked from the wing of some bird.

The Sea-pen is never attached to solid substances, but remains quite free in the ocean. It does not, however, swim, but is a helpless sort of being, and only kept in its proper position by the base being thrust into the mud or sand at the bottom of the sea. Some species of Sea-Pinnæ are phosphorescent, and present a magnificent sight in the darkness. It was once thought that the creature was able to swim by means of the webs, or pinnæ, which flapped like the fins of a fish, but it is now ascertained that no such power resides in these organs. The stem is of a rather soft consistency, strengthened by a bony centre, which reaches nearly to its tip.

An object of somewhat similar form, but considerably elongated, and with the pinnæ proportionately shorter, is called SEA-RUSH (*Virgularia mirabilis*), an animal belonging to a genus that can easily be distinguished from the preceding by several peculiarities. The pinnæ are short, deeply scooped above, and, with their bases, partly surround the central stem. The polypes are set only upon the edges of the pinnæ. There is an allied species belonging to the same family, called by the name of *Pavonaria*. In this remarkable genus, the general shape of the lengthened mass is four-sided, and the polypes are arranged in a somewhat spiral form on the stem, but only one side of its latter half. In temperate seas, the Sea-rushes do not grow to any great length; but under a tropical sun they reach great dimensions, some of them measuring more than a yard in length.

A very curious inhabitant of some seas, which is in the habit of encrusting all kinds of marine bodies, such as shells, stones, and stems of the large algæ, is popularly known under the name SEA-FINGER (*Alcyonium digitatum*). Its general mass runs into lobes, and is of a soft, spongy consistency, pierced with little holes, from which the polypes make their appearance when in health. When closely examined, the little holes or pores are seen to be formed of eight rays, in a kind of star-like pattern, and corresponding to the tentacles of the polypes which inhabit them. These little cells are placed at the ends of canals, which permeate the whole mass, and serve to unite into one common body the vast number of polypes which are thus aggregated together. When examined by the microscope, the substance of the polypidon is found to be filled with tiny particles of chalky matter, which serve to give consistency to the fabric, and add to its elasticity.

HYDROIDA.

WE now arrive at the order Hydroida, which are known by the internal cavity being simple, and the creature increasing by buds thrown out from the sides. The Tubulariadæ are the first family of these creatures. In the Tubulariadæ the buds grow from the base of the tentacles, and break off their attachments as soon as they have attained maturity. The buds, or young, are naked. The animals are sometimes naked, but are often inclosed in a horny, tubular covering, which we will term the polypidon. The first family is represented by its typical genus. The polypidon of this genus does not throw out branches, and the tentacles are delicate, thread-like, and arranged in two circles. The germs, or buds, are set on very short footstalks, and are gathered upon the bases of the lower tentacles.

Before leaving this interesting family of zoophytes, we must pause awhile, to cast a cursory glance at one or two of the more prominent examples.

The CLUB-ZOOPHYTE (*Clava multicornis*) has a large and rounded extremity, something like the head of a bludgeon, upon which are placed irregularly a number of thread-like tentacles.

The various species belonging to the genus Coryne are also worthy of notice. These conspicuous, though minute, zoophytes may be recognized by the globular tips of the tentacles. Sometimes the creatures are naked, and sometimes they are inclosed in a rude sort of tube. The word "Coryne" is Greek, and signifies a club. The head of each tentacle is most elaborately constructed, and adorned with very minute tentacles, each being furnished with a small bristle at its tip. These tentacles can be moved with tolerable rapidity, and are held in various attitudes, sometimes stretching out at right angles from the stem, but often bending upwards, with their heads directed towards a common centre, and have been happily compared to the bars of a turnstile or the weighted arms of a screw press.

Another genus is that which is appropriately named Eudendrium, from two Greek words, signifying a beautiful tree. As may be presumed from its name, it has a decidedly tree-like form, each twig terminating in a polype whose flower-like tentacles add in no slight degree to its beauty. It is found that, when in captivity, the Eudendrium is sadly apt to throw off all the lovely diadems with which it is crowned, but that it will in process of time supply the deficiency by new heads. Its reproduction is quite as remarkable as that of any creature which has hitherto been mentioned, but our failing space will not permit a detailed account.

In the Sertulariadæ, the buds are inclosed in vesicles, and do not break away when adult. They are placed in cup-like cells, which have no footstalks.

Any of the common Sertulariæ affords a good example of this family; and as they are easily procured, they are very valuable aids to those who wish to study the structure of these beautiful beings. Even the empty polypidon is not without its elegance, and is often made up into those flattened bouquets of so-called sea-weeds, which are sold in such quantities at sea-side bathing towns. But when the whole being is full of life and health, its multitudinous cells filled with the delicate polypes, each furnished with more than twenty tentacles all moving in the water, its beauty defies description. These little polypes are wonderfully active and suspicious. At the least alarm, they retreat into their cells as if withdrawn by springs, and when they again push out their tentacles, it is in a very wary and careful manner.

The reproduction of these beings is very curious, for it is known that they can be propagated by cuttings just like plants, as well as by cell vesicles, and that in the latter case the first stage of the young closely resembles that of the young medusæ already mentioned. They also reproduce by offshoots; and it is very likely that their capabilities in this respect are not limited even to these three methods.

The Campanulariæ, or Bell-zoophytes, may be distinguished from the last family by having the cells placed on footstalks.

The whole history of this creature is very interesting, but on account of failing space we must restrict ourselves to its chief peculiarities. Placed among the ordinary polype-cells may be seen, at certain times of the year, a few scattered egg-shape objects, some eight or ten usually being found on a branch. Within these cells are seen a small number of very minute living beings, which gradually develop themselves. A restless movement prevails towards the upper part, some slender tentacles make their appearance at the end, and at last the whole of the tip breaks loose, displaying itself as a tiny medusa.

This change is indeed a wonderful one, perhaps even more marvellous than the mutual transformations of hydra tuba and medusæ, inasmuch as the Campanularia and the medusa belong absolutely to separate classes; and that a medusa should spring from a zoophyte is hardly less surprising than that a perch should give birth to a human being.

These important discoveries were made simultaneously by Professor Van Beneden and Sir John Dalyell, and the former naturalist was able to observe a phenomenon which certainly seems to be the first step towards the return from the medusa into the zoophyte. Having isolated a specimen of the little medusæ, and made a careful drawing of it, he left it for about an hour, and on his return was surprised to find that the whole shape of the tiny being had altered. The convex disc had become concave, the tentacles were reversed, and the animal had changed the central footstalk of the medusa into the semblance of a zoophytic stem.

"My observations," remarks that accomplished naturalist, as quoted by Mr. T. R. Jones, "go no further; but although I have not seen the medusa give origin to a polype stem, I observed it up to the moment when it was about to form a new colony; and without fear of deceiving ourselves, we may form by analogy some idea of the changes which must necessarily occur. The Campanularia, in its medusa state, has only a single aperture, situated at the extremity of its central pedicle. We have already seen that its body becomes inverted like the finger of a glove, and that the marginal filaments become converted into true tentacles. The polype fixes itself by the extremity of its central appendage—that is, by what was previously its mouth; the back of the umbrella becomes depressed at the same time that the tentacles change their direction; and in the centre of the disc a new aperture is formed, which communicates with the central cavity, and becomes the permanent mouth, which is situated directly opposite to the original one.

"Being now fixed by its base, the body of the polype begins to grow; and as its external sheath becomes hardened, buds sprout at regular intervals from its surface. In a word, the growth of the polype resembles that of the hydra, with this difference, that in the latter there is no polype stem, and their buds sprout from another part of the body."

The name of Campanularia is given to this zoophyte in consequence of the bell-like form of its cells, and is derived from the Latin word "campana," a bell.

The delicate PLUMULARIA is so called on account of the feathery appearance of its polypidon. The cells are always small and the egg-vesicles are scattered. In some species the stem is composed of many parallel tubes, such as *Plumularia myriophyllum*, but in the present species it is quite simple. The egg-vesicles are rather widely scattered.

THE CORAL REEFS OF FLORIDA.

THE REEF PROPER OF THE FLORIDA STRAITS does not reach the surface, excepting in certain places, as follows: Carysport, where there is an iron pier lighthouse, Alligator Reef, Tennessee Reef, and a few shoals of less extent, but perhaps not less dangerous. These shoals give rise to heavy breakers, which show at most times in white caps. In a few places there is an accumulation of dead corals and *débris*, which brings the surface to a level with the water; then the dry land that is formed is called a key (cayo) or islet. The Dry Tortugas are so named, being originally of similar character.

Sombrero Key is an important example, on which is a fine lighthouse. Dove Key, the Sambos, and Sand Key, are others. Sand Key is situated at the entrance of the channel that leads to Key West harbor, and bears one of the most important lighthouses on the coast. This is about nine miles from Key West, southwesterly, and is the southernmost inhabited land of the United States.

Several safe anchorages are known, particularly at Key Largo, with from one to three fathoms of water.

The Bahamas are coral reefs and islands similar to the above.

The westernmost portion of the Florida Reef consists of several keys that barely rise above the sea, and are covered by fine white coral sand. Beach grasses have taken root, and even quite large trees are flourishing. A small bush, called bay cedar, is abundant, and covers some keys entirely.

As there is not extant a published account of these interesting islands, which are so intimately associated with what we have to say about the marine objects of the semi-tropical waters of North America, we feel sure that it will be acceptable to the reader to have a somewhat detailed account of them.

The Dry Tortugas, before the late conflict of 1861-'5, was little known to the average reader. The establishment of a military prison there soon made the name a terror to evil-doers, and a synonym for the dreadful. During the two years preceding the "conflict" it had been our fortune to reside at the Tortugas as United States surgeon. Fort Jefferson was then in progress of construction. It is an enormous work, involving many millions of money. During these two years the quiet life and delightful association with other officers of the post, and their families, interested in the same pursuits, rendered it an opportunity of exceptional excellence for the study of marine zoology.

The visitor to this region in years when the post was garrisoned would take the following course: Usually a stop, coming from the north, was made at Key West, the only important inhabited island then on the reef. From there a sail, usually by night, of sixty miles, brought one off Marquesas Keys and Rebecca Shoals. Daylight reveals in the western horizon a long row of castellated structures, impressing one as fairy castles, now illumined by the rays of the rising sun. The vessel now abruptly changes her course, to enter the peculiar winding channel that is so characteristic of the coral reef—five miles from the fortress. Anon there shoots forth a small cloud from the top of the work, and simultaneously rises the garrison ensign, followed at an interval by the booming sound of the sunrise gun.

In this delightful climate, even during the winter months, this scene is as enjoyable as it is novel. On all sides is the vast ocean. Not a sign else, save the four green-capped islets, slender white strips on the blue sea, with low green bushes on their surface. These now begin to be distinguished. Seven of those small islands, of sizes varying from a quarter of a mile to two miles in length, form a sort of irregular ring around a deep harbor. The intervening space is occupied by the solid reef that has been built up from the sea-bottom, and lies just under the surface, many miles in extent, the entire group being about circular and some seven miles in diameter. The water on this area varies in depth from one foot to twenty, and it is the abode of great numbers of the shoal-water corals, corallines, and algæ. In the centre, or nearly so, of the harbor, an islet of sand, formed like all the others on the solid coral basis of the reef, and about thirteen acres in extent. On this island, entirely covering it, is built Fort Jefferson, the largest structure of the kind in the United States.

Though these little islands look to us like mere sand-spits that any stout gale might demolish, they are grounded in the most endurable of material. The solid area of extended reef around them, just beneath the surface, is as firm as rock. Just at the edge of these islands, on the windward side, the waves break with great violence—the vast ocean depths are behind. The still waters within offer the safest anchorage, reached through the narrow, winding channels. The nature of coral reefs the world over is to grow in such shape as to inclose lagoons with more or less depth of water, which is usually sufficient to float the largest vessels. Hence the great value attached to coral islands in the great Indian Ocean, where passing vessels seek temporary shelter from storms.

The harbor within these islands is valuable for the navy in time of war, as otherwise the presence of a great fortification here is useless.

The important elements in the building up of these coral reefs are the Astrean Corals. These are not circumscribed in growth like many others, but are seemingly indefinite in boundary. Immense ledges are seen cropping out of the mud in shallow water.

The Brain Corals, so called from their resemblance to the brain, Meandrinas, from the meandering nature of their cells, exhibit a number of beautiful shapes, ranging from the most regular hemispheres to masses of indefinite shape and size. These, with the star corals, the astreas, as we have seen, form important elements in the building up of reefs.

In the coral regions of the West Indies and the Florida peninsula the islands are called cays, in English keys, a corruption from *cayo*, Spanish for an islet. The principal cay of Florida, or the only considerable one inhabited, was early called *Cayo hueso*. Bone Cay, or

Bone Islet, from the remarkably white appearance of its beaches, the white coral fragments thrown up by the sea, appeared like bleached bones to the first visitors, hence Bone Cay, now Key West.

It was the opinion of Professor Louis Agassiz that the entire peninsula of Florida had been built up from the sea-bottom by the several reef-building corals now living in the surrounding waters. This theory was seemingly verified by the discovery, in the interior of the state, of parallel ridges, which extend across the peninsula, and are the dead remains of species of corals that are seen living in the vicinity.

The process of reef-building is easily comprehended by observing the present living forms, their growth and decay. We will observe a single egg of a reef-building coral, an astrea. As it floats in the deep sea, its ultimate destination as a single object is to rest on some solid base, and there develop into a simple polyp, in its first stages resembling an actinia or sea-anemone. We have observed the development of these eggs in a glass of sea-water, and we may assume that on the sea-bottom the little animal flower is passing through the same phases of development. Soon we notice at the base of the polyp the first layers of a foundation wall. When finished the creature represents the perfect coral animal. It is like a sea-anemone inclosed within a tube of lime. In some respects it is like a clam or other shell-fish in its shell, a perfect animal. It has several ways of growth and extension into family groups, by eggs, and by development of buds out of its sides. If we take a piece of one of these reef-building corals in hand, we see that there are numerous stars, if it is an astrea, each star representing a single polyp, each a single animal; but the hard parts, that serve as skeletons, or that correspond somewhat to the shells of clams and other shell-fish, are closely united. Practically the young members of the family, the buds, stay at home, and build on to the old home the first house, and the result is an indefinite number of tenements united in one block. The great ledges of astrean corals seen in the waters of the Florida Reef, are thus built up. This is the principal element in the foundation of a coral island. We may now regard the sea-bottom covered to a certain extent with the outspreading ledge of these united stars. Among the numerous elements that must be recognized in reef-building are various species of burrowing shell-fish, and worms in great variety. These creatures kill the coral animals, and penetrate their limestone houses. Here we have the first steps in the building of the reef. The coral stars have secreted and deposited on the ocean bottom the masses of lime which form their houses; their enemies have destroyed them and penetrated their walls. The general *débris* of the ocean covers the broken walls. But the young of the coral animals are swimming in great numbers, ready to fasten upon any point. Myriads settle upon the old and dead ones, and found new houses; new blocks are built upon the old, and in time also yield to the inroads of their numerous enemies. The conflict thus goes on. The coral block of houses, solid material, becomes a compact mass, which rises gradually through this process of growth and decay, life and death, until this growing land has reached near the surface of the sea. In the shoaler water that now covers this coral-made land numerous small corals and algæ grow, objects that require shallow water in which to thrive. Here is manifested a wise provision. The larger corals cease to thrive because the water is not sufficient, then smaller species appear, which, with the soft corals, as gorgonias, sea-feathers, and fans, and masses of corallines, the latter being algæ or sea-weeds with solid lime bases, eventually quite bring the newly-made land to a point at the surface of the sea. Here we have an island, built up from the sea-bottom through the agency of living corals, their dead skeletons, algæ, and the accumulated *débris* of ocean. This island would be of little service were it to remain at the ocean level. Nature has provided for the extension of this land. The mangrove tree is found growing on the extreme ocean border. Its fruit drops into the sea. This fruit is so much in shape, size, and color like a cigar, one is quite sure to be deceived on viewing it. Myriads of this fruit float over the new-made land; one end being heavier inclines to touch bottom. During the still water, after the summer solstice, these fruits throw out roots, which find their way rapidly and strongly into the earth. Soon they have put out leaves and have become trees. The roots, instead of disappearing beneath the soil, remain to a certain extent exposed, so that when the tree has gotten to be a year old the roots are veritable flying buttresses. Remark how well adapted this plan is to finish the growth of the island, to bring it up to a safe height, when other elements shall be utilized. These flying buttresses catch all *débris* of the ocean, and hold it until a soil is formed. Now, birds come to roost here; they bring seeds, which are deposited in the excrement. Among these seeds are several kinds of great convolvuli, morning-glory plants, whose habits are to run on the ground like a pumpkin vine.

These great vines take root at intervals—many of them form resting-places for moving rubbish. Sand begins to collect. Innumerable agencies conspire to bring this low island to a greater height above water, when the land becomes dry; hence Dry Tortugas, in contradistinction to Wet Tortugas, or wet land that has not yet reached the point of being above water. Once the surface has become somewhat permanently dry, other seeds germinate, and grasses appear—the beach-grasses, whose rootlets catch and hold the sands. Eventually a considerable soil is formed. The visitation of sea birds brings guano, shrubs appear, and then great trees. Some of the older keys are heavily wooded with a variety of trees. By these processes it is supposed the larger portion of the State of Florida has been built up.

ROTIFERA.

ALTHOUGH the Rotifera, or Wheel Animalcules, are generally placed among the Infusoria, on account of their minute dimensions and aquatic habits, it is evident, from many peculiarities of their formation, that they deserve a much higher place, and in all probability constitute a class by themselves.

They are called Wheel Animalcules on account of a curious structure which is found upon many of their members, and which looks very like a pair of revolving wheels set upon the head. These so-called wheels are two disc-like lobes, the edges of which are fringed with cilia, which, when in movement, give to the creature an appearance as if it wore wheels on its head, like those of the fairy knight of ballad poetry. These wheels can be drawn into the body at will, or protruded to some little extent, and their object is evidently to procure food by causing currents of water to flow across the mouth. All, however, do not possess these appendages, but have a row of cilia, mostly broken into lobes, extending all round the upper portion of the body.

They have a well-defined muscular system, while their jaws are nearly, if not quite, as complicated as those of the echinus. Most of them can swim, some are able to attach themselves at will to any fixed objects, while others are fixed to one spot, from which they do not stir.

Distinct sexes have been discovered in several genera of Rotifers; and in those cases where the male has not been found, it is generally thought that the very small size and eccentric shape of the opposite sex may be the reason why it has not been discovered. In those instances where his existence has been indubitably ascertained, he is always a strange being, very unlike the female, very small, and what is even more strange, possessing neither jaws, throat, stomach, nor intestines. His life must therefore be very short, as is known to be the case with the male sex in many insects. It has been well suggested, that perhaps the males are only produced at certain times of the year, and are not, therefore, found so plentifully as their mates.

Fortunately for observers, the integuments of these animals are extremely transparent, so that it is possible to watch the whole of the vital processes, and to see the various functions carried on with as much ease as if the skin were of crystal. Their development is wonderfully rapid; for although but a few eggs are produced at one time, they are so quickly hatched, and the animal is so rapid in its growth, that Professor Ehrenberg calculated that in the genus Hydatica, although only three or four eggs are produced at a time, a single individual will be the progenitrix of nearly seventeen million descendants within the space of twenty-four days.

In this class the arrangement is very perplexing to systematic naturalists, and nothing is as yet settled about it.

These remarkable beings are mostly found in water that has become stagnant, but is partially purified by the presence of the Infusorians, which always swarm in such localities. There is, however, one very strange residence of the common Rotifer, namely, within the leaf-cells of the common bog-moss (*Sphagnum*). These cells are very large in proportion to the size of the leaf, are kept open by spiral threads coiled in their interior, and their walls are pierced with large apertures, so as to form a general communication throughout the whole mass of cells. Within these curious chambers the Rotifer is found, and is able to pass freely from one cell to another. They probably gain their admission in the egg state, and find sufficient moisture in the cells for their seeds.

The typical genus of this class is known by the name of Rotifer. In all the members of this genus the body is rather elongated, and furnished at the hinder end with a kind of telescopic tail, by means of which they can attach themselves at will to any object, and release themselves whenever they please. Sometimes they move their bodies gently about, while still grasping by the extremity of tail; sometimes they are nearly motionless, while they frequently rock themselves backwards and forwards so violently that they seem almost to be testing the strength of their hold.

These creatures can both swim and crawl, the former act of locomotion being achieved by the movement of the cilia, and the latter by creeping along after the fashion of the leech, the head and tail taking alternate hold of the object on which they are crawling.

The masticating apparatus is always conspicuous, whether the animal have the wheel protruded or withdrawn. It is situated behind the bases of the wheel-lobes, and looks, when the animal is at rest, something like a circular buckler, with a cross composed of double lines drawn over its surface. Even in the very young and undeveloped animals which are seen within the body of the parent, these jaws form the most conspicuous portions of their structure, and enable them to be recognized long before they are able to go out into their watery world and shift for themselves.

All the Rotifers have a marvellous fund of vitality, and survive under circumstances where animals less tenacious of life would die a thousand deaths. They have been thoroughly dried by means of chemical acid, wetted and restored to life, dried again, wetted again, and subjected to this treatment through many successive alternations, without perishing.

At first sight, this animal bears a strong resemblance to several of the Molluskoids; but a closer examination shows that the apparent tentacles are nothing more than extensions of the lobes on which the cilia are set, and the apparent cell is no cell at all, but a gelatinous secretion from the body. In one genus, however, a veritable tube is built up, composed of particles of solid matter, formed into little pellets by a special organ, and then deposited upon the edge of the tube. The organ which forms these pellets is set towards the front of the head, and on its under side, and looks like a little revolving disc.

RHIZOPODA.

THE whole arrangement of the beings which we are now about to examine is still very obscure, and the best zoologists of the present time have declared that any system which has been hitherto adopted can only be considered as provisional.

Some writers, for example, unite the Rhizopoda with the Infusoria, while others rank them among the Polyzoa; and others again consider them to be intermediate between the radiata and those simple forms of animal life which are appropriately named Protozoa. After taking into consideration the various systems that have been propounded by different authors, I have come to the conclusion that, at all events, as a provisional arrangement, the Rhizopoda ought to be ranked as a distinct class, and placed in the position which they here occupy.

The name Rhizopoda is of Greek origin, and literally signifies "root-footed." It is a very appropriate title, inasmuch as they put forth certain filamentous appendages from their bodies, which look very like the tender rootlets of plants, and serve a double purpose, namely, as organs of progression, and as instruments whereby they may catch their prey.

Some of these beings are quite unprotected, their soft gelatinous bodies being devoid of any covering; others are inclosed in a horny case, pierced with openings, through which the filaments can be projected; while the greater number of the known species are furnished with shells very similar in form to those of the mollusks, and in some cases wonderfully similar to the highly complicated dwelling of one of the highest mollusks, the pearly nautilus.

These minute though beautiful beings exist in numbers that are only rivalled by the sands of the sea for multitude; and the vast hosts of these creatures can be barely estimated even when we know that many large cities are built wholly of the dead skeletons of these microscopic beings, and that in a single ounce of sand from the Caribbean Sea nearly four millions of these shells have been discovered. The living species are not nearly so numerous as the fossil. They can be captured in various ways. If, for example, growing algæ be plucked, and placed in a glass vessel of sea-water, the Rhizopods will leave the algæ, and settle on the sides of the vessel. If they live in muddy substances, such as the "oyster-ooze," which is especially prolific in Rhizopod forms, the upper layer of mud should be taken off and stirred up in a vessel of clear sea-water, when the creatures will sink to the bottom of the vessel, and may easily be separated.

These modes are adopted for living specimens, but if the dead skeletons only are required, they can be procured in many ways. One of the simplest methods of finding Rhizopod shells is, to shake the dust out of sponges, and to examine it when laid thinly on black paper. An ordinary pocket magnifier is employed in the search, and the shells are readily seen against the black background. For removing them I always employ a single bristle, stuck into a handle—one taken from a shaving-brush is, perhaps, the best adapted to the purpose—and take up the shells singly by wetting the tip of the bristle.

There is also another method whereby the empty shells may be obtained in considerable numbers. The sand, mud, or other substance, in which they reside, should be well dried, heated, and then stirred into water. As the chambered cells of the Rhizopods will be filled with air, they will float on the surface of the water, and can be skimmed off without much difficulty.

The first sub-class of these beings is the Foraminifera, so called on account of the tiny openings, or *foramina*, with which the pretty shells are pierced. Sometimes, however, this shell is wanting, and its place supplied by a cover composed of matted sand-grains. The greater number of these creatures are formed by a succession of buds, each bud remaining in connection with that from which it sprung, and thus forming a composite body, which sometimes is rather complicated in its structure. Sometimes when the buds are merely arranged in a line, the result is a straight, rod-like form, divided into a series of joints, marking the spots where the buds have in their sequence issued from each other. If, on the other hand, each bud grows a little on one side of its predecessor, a spiral form is the result, and a nautilus-like shell is formed. The resemblance to this mollusk is further increased when each bud becomes rather larger than that from which it sprung.

The arrangements of the Foraminifera hitherto in use have mostly been founded upon the mode of growth; but Dr. Carpenter has clearly shown that this character is so extremely variable, that no reliance can be placed upon it. In a single genus, there is every gradation between the straight and the spiral forms; and, in many instances, a shell which commences in a spiral will end in a straight line.

As, therefore, the already existing systems have been shown to be based on false principles, and the arrangement which is to supplant them has not been fully decided upon, we will not occupy our space by insisting upon the characters by which the systems are established, but merely proceed to a brief description of the localities in which the various species may be found.

The greater number of the species are found in Europe, and are now known by the names of Dentalina, Polystomella, Rosalina, and Quinqueloculina. Some other species are to be found in Central America.

Another sub-class of Rhizopods is named Polycystina, and is notable for the singular structure of the shells, which are pierced in regular patterns, without orifices, and are often prolonged into curious spikes and projections that give them a most wonderful beauty when seen under a good microscope. They are, in general, smaller than the Foraminifera, and are found in the mud of various seas, especially those of the West Indian islands. The marvellous variety which is obtained by the carrying out of two principles, namely, the piercing of holes and the projection of spikes, is almost incredible; and the delicate tracery of the patterns thus produced is so artistic as to have been happily compared to the hollow ivory balls carved by the patient hands of Chinese artists.

There is one little creature, which is supposed by many physiologists to belong to the Rhizopoda, but whose position is very uncertain, and even its class not clearly ascertained. This is the NOCTILUCA, a tiny being, about as large as the head of a minikin pin, which is remarkable for its phosphorescent power. If a vessel be filled with sea-water, and brought into a dark room, the Noctiluca fills it with little sparklets of bluish light, which shine for an instant like stars in the firmament, and which can be induced to give out their momentary radiance by tapping the vessel, or even by a heavy footfall on the floor of the room.

Each of these little beings is furnished with a minute tail-like appendage, by means of which it is enabled to proceed through the water; and on certain favorable occasions they fill the sea with their luminous hosts, and cause each wave to become a breaking mass of liquid fire. A ship passing through the sea leaves a fiery wake behind her keel, and when the boatmen lift their oars from the sea, they appear to drop flames from the blades as they are raised, all dripping, into the air at every stroke.

Although so small as to be microscopic in their dimensions, they are yet large enough to be discerned by the unaided eye, and can therefore be isolated without difficulty and placed in the field of the microscope.

In the accompanying illustration will be seen an odd-looking object, which is considered as belonging to the Rhizopods, though not possessing any shell.

This creature, called AMŒBA, is remarkable for the fact that it really has no outline and no shape, for its body is continually altering its figure; so that the rounded object which was seen in the microscope but a few minutes before, will, in that short space of time, have protruded a number of elongations that look like fingers of a glove or the rays of a star-fish.

It can elongate itself to almost any extent, can then throw out its strange protrusions so as to resemble a club with a spiked head, or it can gather itself into a rude globular mass, as if pinched out of dough by a single squeeze of the hand, allowing the soft substance to protrude between the fingers. It has no particular stomach, but extemporizes that organ out of any part of its body with which its food happens to come in contact, literally pushing the food into its body and then digesting it without requiring any special apparatus for the purpose.

Some of the Lobose Rhizopods, as these creatures are called, are also furnished with a shelly or horny covering, such as the Arcellina, where the shield is cap-shaped, or the Difflugia, where it is pitcher-shaped, the animal protruding itself from that part which represents the mouth of the jug. Many physiologists suppose that

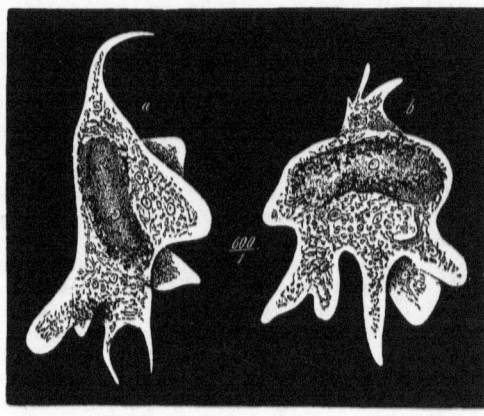

AMŒBA—*Amœba princeps.* 600 times enlarged. *a* and *b* show the same animal in changed form.

the Amœba is not a perfect being, but it is merely the larval state of some animal with a higher development, such as the Arcella and other shell-bearing Rhizopods.

The genus Perinidium may be known by the furrow that runs transversely around the body, and is furnished with cilia. The integument of the body is membranous. The Tripos Perinidium is remarkable for its power of shining by night. It may be recognized by the shelly case, which is concave, smooth, and is developed into three horns, two being long and the other comparatively short. The longer horns are in front. Its length is about 150th of an inch. The Kerona also belongs to this order, and is found in fresh water, where it may often be seen in considerable numbers. Besides the usual cilia, it is furnished with instruments of progression that enable it to climb and creep, and are formed like bristles or hooklets. Its length is rather variable, but is about equal to that of the Peridinium.

INFUSORIA.

WE now come to the Infusoria, creatures which are all of very minute dimensions, and respecting which there is great uncertainty prevailing. As with the preceding class, no definite system has yet been invented by which they can be arranged; and in many cases physiologists are undecided whether the tiny beings are veritable species, or whether they are but the larval forms of higher beings; while, in some cases, it cannot be precisely ascertained whether they belong to the animal or vegetable kingdom.

Without, therefore, occupying our space with disquisitions which would require a volume for their full elucidation, we will proceed at once to some of the more remarkable forms among these curious beings.

Two species of Infusoria, termed *Vorticella citrina* and *Stentor polymorphis*, may be found in soft water that has been allowed to remain in the open air, and in which any vegetable matter has been permitted to decay. Both these creatures are affixed by footstalks to some object on which they make their residence, and both agree in having a bell-like mouth, edged with a fringe of cilia.

These organs are set upon the edge of the mouth, and their object is indirectly to draw food into the system by creating certain currents in the surrounding water. When the cilia are exposed to a good microscope, they appear to be formed like the cogs of a little wheel, which is rotating with great rapidity; and it is not until a close examination detects the real cause of this appearance that its illusory nature is discovered. As in the case of the cilia attached to the higher animals, of which a notice has already been given, each fibril bends in regular succession, so as to produce the effect of waves upon the eye.

When the Infusoria are free, the continual movement of the cilia causes them to move with greater or lesser swiftness through the water, each fibril acting as a minute paddle, and having a distinct feathering movement, like that of an oar handled by a skilful rower. It is a most curious sight to observe the admirable manner in which they make their strokes, the flattened sides striking the water so as to give the greatest force to the blow, and the back stroke being made with the edge, so as to meet with the least possible resistance.

In the VORTICELLA, the footstalks on which the bell-like cup is seated are of considerable length, and capable of being shortened by being coiled into a spiral form. This is by no means an uncommon Infusorian, and is very liberal in displaying this capability. It is usually found associating in groups, so that there is hardly a stage in its life of which some example cannot be discovered. Though devoid of apparent organs of sense, this creature is marvellously timid, shrinking in a moment if the water be shaken, and tightening its coils until they resemble the spiral rings of a vine's tendril. It soon, however, recovers itself, and by slow degrees permits the spires to uncoil, and waves its fringed head boldly in the narrow prison to which it has been consigned. Sometimes the Vorticella breaks away from its footstalks, and is then carried rapidly through the water by the action of its ever-waving fringe of cilia.

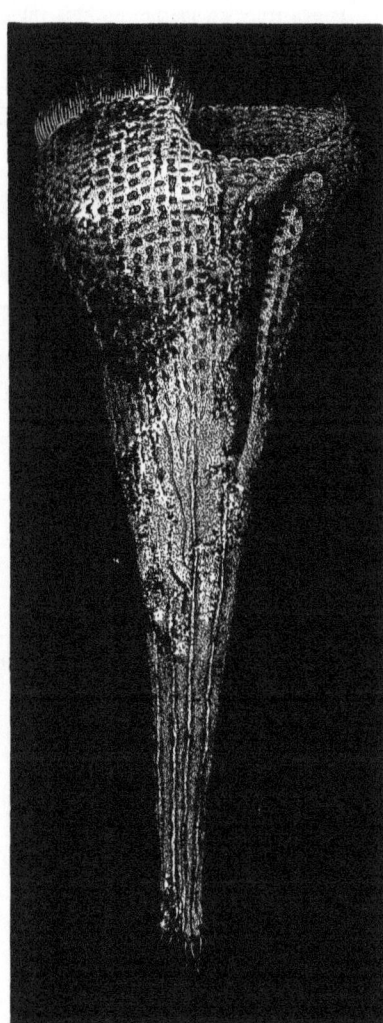

STENTOR.—*Stentor polymorphis.* (Two hundred times enlarged).

As is the case with many of its kindred, the Vorticella is able to increase its numbers by the simple process of splitting itself into two distinct beings, each of which is afterwards a complete and perfect being. At first, a single notch is seen upon the edge of the lip, but as time passes on the notch deepens, the cleft becomes more apparent, and in a wonderfully short space each half of the Vorticella is changed into a perfect individual, which in its turn is ready to divide and subdivide itself *ad infinitum*. It is a truly strange process, this subdivision, and forms one of the links that bind animals of a higher type of organization with these lowly, but not imperfect beings. Thus, therefore, the Vorticella never need die of old age, for it renews its youth, as it were, by this voluntary division, just as if a man of sixty were to split himself down his spine, and thus become two young men of thirty, or, by further subdivision, four lads of fifteen.

The figure in the illustration represents the STENTOR, so called because its general shape bears some resemblance to that of a speaking-trumpet. This is a comparatively large species, being visible to the naked eye, and readily distinguished by a practised observer. Sometimes it is found singly, either attached by its base or swimming boldly through the water; but in most instances it gathers itself round duck-weed, or floating sticks, and is produced in such numbers that its vast multitudes quite resemble a fringe of soft, filmy slime.

Like the preceding animal, the Stentor multiplies by self-division; but it is very likely that many other methods of increasing its numbers are employed. There is, for example, in these creatures, the remarkable phenomenon called "conjugation," which is almost identical with the same act as performed by some of the microscopic vegetables. If two free Infusoria of the same species—say, for example, the common Paramecium, that swarms so largely in stagnant waters—happen to meet at the proper season of the year, they adhere firmly to each other, as if they were magnets and iron, and go spinning about the water with no less speed than when each urged its single course. A vast number of very minute eggs are then produced by both of the individuals, but the further development of these eggs is not yet known. Sometimes, as in the Stentor, the Infusoria are fixed by their bases, and in such instances they bend their mouths towards each other, and so contrive to unite themselves in pairs.

PORIFERA.

E now arrive at a large class of beings, which, if they really do belong to the animal kingdom, and are not to be ranked among vegetables, are by common consent allowed to form the very lowest link in the animal chain.

The name Porifera is given to them because the whole of their surface is pierced with holes of various dimensions, the greater number being extremely minute, while others are of considerable dimensions. The well-known Turkey Sponge, so useful for the toilet, will afford a good example of the porous structure.

Yet no one can form an adequate idea of the living Sponge from the dry, dead skeleton which is sold under that name. Many of the species are decked with delicate colors, while all are truly beautiful creatures when viewed in full life and action. They are to be found widely distributed through the seas, and there is hardly a solid body on which a Sponge will not grow. Sponges are generally found hanging from the under sides of projecting rocks at some distance below the surface of the sea, or clinging to the roofs of submarine caverns. Some, however, are strong, sturdy, and branched, and stand boldly erect like the earth-plants which they so wonderfully resemble.

Even the living inhabitants of the sea are liable to become the resting-places of many a Sponge, and the crustacea are often forced to bear on their shells the additional burden of living Sponges and other zoophytes much more massive than their whole body.

The true living being which constitutes the Sponge is of a soft and almost gelatinous texture, to the unaided eye; and with the aid of the microscope is found to consist of an aggregation of separate bodies like those of the Amœbæ, some of which are furnished with long cilia. By the constant action of the cilia a current of water is kept up, causing the liquid to enter at the innumerable pores with which the surface is pierced, and to be expelled through the larger orifices. A Sponge in full action is a wonderful sight: the cilia drives the water in ceaseless torrents, whirling along all kinds of solid particles, arresting those which are useful for digestion, and rejecting those which it cannot assimilate.

The reader will at once see that a creature thus composed will stand in need of some solid framework on which the delicate fabric can be supported; and on examining a series of Sponges with the microscope, we find that it is mostly composed of a fibrous and rather horny network, strengthened with spiculæ of a hard mineral substance. The shape of the spiculæ is extremely variable, some being simple translucent bars, some looking much like rough flints rendered transparent, others star-shaped with several points, while the greater number resemble knotted clubs made of differently-colored glass, and having a lovely effect under the microscope.

In the genus Grantia, which is well known to marine zoologists as having furnished valuable information respecting the nutriment and reproduction of the Sponges, no horny network can be found, but its place is supplied by the singular form of the spiculæ, which are composed of three long-pointed spines arranged so as to form a star of three rays. These rays, on account of their shape, form an entangled mass, and answer the purpose of the ordinary horny framework. A new species, termed *Halichondria palmata*, inhabits the East Indies.

There are several European species of the genus Grantia. Some of them are hollow, and stand out with tolerable boldness from the objects on which they are set, while others are always found as whitish incrustations upon stones and other massive substances. Their structure is tolerably firm, and, on account of the absence of the horny framework, is not so elastic as are the generality of the Sponge tribe; and the texture is very close, but still porous. With a microscope of tolerably high power, magnifying from two to three hundred diameters, the layer of spiculæ can be readily made out, interlacing with each other in wonderful profusion, and so completely intermixed that a single spicula is scarcely ever separable from the general mass.

The shape of the species belonging to this genus is extremely variable, but in all the structure is remarkably simple, the wall being extremely thin, so that the ramifying canals are not needed, and the water is merely absorbed through the minute pores of the wall and expelled through the large orifice which forms the mouth of the sac. If the spiculæ of this or other Sponges be wanted in a separate state, the animal matter can be removed by heat; but a better, though slower process, is to immerse the specimen in strong nitric acid or liquor potassæ, according to the flinty or chalky nature of the spicules. When separated they may be mounted in two ways, namely, as dry and opaque objects, or in Canada balsam.

We must now briefly examine a rather important genus of Sponges, which has many representatives. It is a very extensive genus, and its members are variously shaped, all, however, agreeing in those salient points on which the group has been founded. They are all spongy, elastic, not slimy, and with a very porous surface.

One species is generally called the MERMAID'S GLOVE, because it is apt to spread into a form that bears a somewhat remote resemblance to a glove with extended fingers. It is certainly the largest of the European Sponges, sometimes attaining a height of two feet, and stretching out its branches boldly into the sea. The branches are rather flattened, and when full-grown are about an inch in width. They do not always remain separated throughout their whole extent, but are apt to coalesce in various parts, and sometimes to form rudely-shaped arches.

The color of this Sponge is generally of a pale straw-yellow, and to the touch its exterior is decidedly rough, on account of the myriads of spiculæ which slightly project from the surface. These spiculæ are needle-like, sometimes slightly curved and sometimes straight. Mostly they are pointed at both ends, but as they are fragile and snap asunder with the least violence, they often look as if they were only pointed at one end. They lie nearly parallel to each other, and many are so placed that their points are presented outwards. This Sponge is found in deep water in many parts of the European seas.

We now come to the large genus Halichondria.

The FUNNEL-SPONGE closely resembles an ordinary funnel. Its structure is very finely porous, and it is rather a pretty and elegant species. The spiculæ which contain the softer parts of the Funnel-sponge are long, slender, and sometimes curved. In most cases they are pointed at each end, but in others only one end is sharp, while the other is rounded, so that the spicule resembles a needle without an eye. They are rather loose, and either lie in bundles or crossing one another. The width of an ordinary specimen is about three inches, and its length is equal to its width.

The LING-HOOD has a shape which, when it is young, reminds the observer of the preceding species. It may, however, be readily distinguished from that Sponge by the thick coating of hair-like spines with which its surface is covered. It always becomes shallower by age, and is therefore extremely variable in its form. The edge is seldom so smooth and regular as that of the Funnel-sponge, being mostly cut into notches and the intervals developed into lobes.

One remarkable characteristic of this species is the very brittle exterior, which can be broken away by the fingers, and is found to consist almost wholly of flinty spicules, cemented together by the glutinous substance of the animal. Sometimes it almost loses the cup-like form, and spreads out like a fan, deriving therefrom the popular title of SEA-FAN. As the term, however, is applied to many other marine beings, it is not thought so useful as the name which has already been mentioned.

MANY of the zoophytes, especially the sertularia and its kin, are very liable to the gentle but irresistible attacks of this Sponge, which, settling upon them, increases rapidly and more or less envelops them in its own mass. Its mode of growth is always variable. Sometimes it follows all the branches of the zoophyte on which it is parasitic, causing it to resemble a tree thickly covered with ivy; while at other times it spreads out so widely, that it gathers all the branches together, covers them with its own substance, and forms them into a shapeless, spongy mass, from which a few ends of the branches vaguely protrude.

This Sponge does not, however, confine itself to zoophytes as resting-places, but settles upon stones, sticks, shells, and other objects. Its color is generally grayish-white, but it sometimes deepens its tint and becomes of a yellowish-brown. The texture of its substance is always rather coarse, but its softness differs greatly according to the object on which it has established itself and the locality in which it happens to be. The spiculæ are always short, rather curved, and are sharpened at one end and rounded at the other. About forty species of the genus Halichondria are found in our seas, several of which are remarkable for having the spiculæ knobbed at both ends.

The extraordinary object which is called by the appropriate name of NEPTUNE'S CUP is one of the most magnificent, as well as one of the most notable, of the Sponge-tribe. It hardly looks like a Sponge; and when a specimen is shown to persons who have no knowledge of the subject, they can hardly ever be made to believe that the exhibitor is not endeavoring to play a practical joke upon them.

The Neptune's Cup is of enormous dimensions, often measuring four feet in height and having a corresponding width. Its exterior is rough, gnarled, and knotted like the bark of some old tree; and if a portion were removed from the side, it might almost be mistaken for a piece of cork-tree bark. Many persons have imagined that the strangely-shaped object was made of the skin of an elephant's leg, and I have even heard a teacher telling her pupils that it was an old Roman wine-jar.

It is hardly possible to disabuse strangers of the notion that it is not the result of human ingenuity until they are allowed to lift it, and test personally its exceeding lightness. It is hollow, and is not at all unlike an old font that by some misfortune has been deprived of its base. Its capacity is enormous, and it would not only form a cup for Neptune, but even

Polyphemus himself might have filled its depths with the ruddy wine which he loved, and failed to empty the huge vessel at a draught.

The substance of this enormous Sponge is porous, rather stiff, and without much elasticity. It yields but slightly to pressure, and almost feels to the touch as if it had been made from cork.

This is one of the exotic Sponges, being found only in the hotter seas. In general shape it has some resemblance to the Funnel-sponge already described, but is of much coarser texture, and, save for its gigantic size, is not nearly so attractive.

IN the PUMICE-STONE SPONGE, we have an admirable example of the flinty structure developed to the utmost degree. The framework is wholly composed of flinty spiculæ, all fused together, and forming a highly porous mass, which at first sight resembles a madrepore rather than a true Sponge. It has not the least elasticity, but is as hard and as stiff as if it had been carved from stone. On account of its extreme porosity, it is very light, seeming to weigh not more than a piece of cork of the same size.

The whole surface, above and below, is plentifully sown with pores, which have a lovely effect under a magnifier, when the sunbeams fall on the glittering spiculæ of which the mass is composed. A number of the large apertures appear on both sides, and all converge towards the centre. The general shape of this remarkable Sponge is cup-like, but exceedingly shallow, and on the inside it is tolerably smooth, becoming rougher and deeply grooved on the outer surface. It has a peculiarly rough feel to the touch, almost exactly resembling the well-known rasping effect produced by rubbing pumice-stone upon the skin; and it is in consequence of this resemblance that it has gained its popular title.

The peculiarities of this very beautiful Sponge consist in the following distinctive characters, the most remarkable of which is its being formed entirely of silex, the reticulate structure of the mass being composed of transparent, glassy tubes, the silex forming the mass itself, and not, as in other instances, arranged as spiculæ in the horny membranes; consequently, it is perfectly rigid and sonorous when struck.

When viewed under a microscope of about seventy-five diameters, the net-like meshes are seen to be composed of beautiful glassy tubes, uniting one with the other in every direction, the external surface of the cylinders having a rugged aspect. The newest or last-formed portions appear to emanate from centres, and at certain distances from spherical knobs, from which straight tubes again arise, thus forming the net-like mass.

BEFORE taking leave of these interesting beings, we must glance rapidly at the method by which they distribute themselves so widely and increase with such marvellous rapidity.

It will be remembered that the soft animal matter of which the true Sponge is formed is composed of multitudinous bodies which closely resemble the Amœbæ, and many of which are furnished with thread-like cilia. In certain months of the year, which in moderate climates are generally found to be October and November, a vast number of very minute yellowish particles are to be seen studding the body of the Sponge. They are not often seen near the surface, but are gathered plentifully within its multitudinous cells. Small as are these yellow particles, they are formed of many eggs, or "gemmules," as they are called, of the Sponge, which gradually increase in size, and at last are expelled from the larger orifices, and thrown at random into the wide sea.

There they are, flimsy, minute, shelterless, feeble, and apparently helpless. Small, however, as they may be, they still possess the power of transporting themselves through the water by means of the cilia with which their bodies are abundantly studded. Their shape is very like that of a pear; and as they are wholly covered with cilia, except the narrow end, it is evident that their larger end must always be in front. They lead a free life for several days after their expulsion from the parental home; and even in this early stage some indications of the future framework are to be seen.

After the lapse of some little time, these gemmules meet with some object which affords them a suitable resting-place, and accordingly affix themselves to the spot, from which they never afterwards can move. The rounded body soon becomes flattened, as it adheres with a close grasp, and spreads itself into a nearly circular film. The cilia still exist on the upper surface of this film, but the effect of their action is then not to propel the Sponge, but to create a current of water which can pass over it.

As time passes on, the distinctive spiculæ become visible, and, after three weeks or a month have passed away, the spiculæ have been gathered into little bundles, which by their arrangement tend to preserve the shape of the Sponge and to keep the orifices open. The little being now spreads rapidly, by a process which much resembles the subdivision of the Infusoria, and the whole mass of the Sponge is evidently composed of a vast number of the Amœba-like bodies which have already been described. Thousands upon thousands of these gemmules are passed out into the sea from every Sponge that inhabits its waters; and the only wonder is, that, in consequence of such marvellously prolific properties, the Sponges do not swarm to such an extent as to fill the whole seas, and poison the entire earth with the odor of their decay.

THE editor of this edition would state here that though the Sponges were once regarded as forming the lowest branch of the animal kingdom, as stated in the original text of this work, they now constitute the second in the ascending scale, or next to the lowest, under the title BRANCH II.—PORIFERATA. As with other branches of the animal kingdom, we refer the reader who desires to gain knowledge of the present state of science as applied to this branch and that embracing the lowest animals, to the technical tables of classification and nomenclature of the Smithsonian Institution. Not long since, it is well known, they were so little understood that no one had quite the courage to say which kingdom they belonged to—whether of the animal or vegetable.

The present state of that section of science which refers to these low animals may be simply presented as follows: It is now known, as the result of much study and observation during the last ten or fifteen years, that the Sponges, in common with all other animals above them, are composed of myriads of cells, which perform each their respective offices in the animal economy. In some groups perpetuation by division of the body is observed. Yet in all these are specialized cells or eggs, for the purposes of reproduction.

In the first division of the animal kingdom, that embracing the lowest animals, and called BRANCH I.—PROTOZOA, it is observed that they differ by having only one simple cell; consequently they do not increase by means of eggs, but by division or segmentation. An analagous example is seen when vegetable roots are perpetuated by cuttings. This difference suggests to the naturalist two distinct divisions. Those animals having many cells are called collectively METAZOA, and the single-celled PROTOZOA.

Sponges are all aquatic; found in the ocean, and in fresh water to a very limited extent. They are all fixed, with very few exceptions, to some object near or on the bottom of the seas. The young, during a short period, are supplied with *cilia*, by which they move through the water until they become fixed. Myriads of floating microscopical plants and animals become their food by absorption through the pores and open channels so characteristic of these forms. The term *Poriferata* is selected to indicate this branch of animals from this prominent feature. The familiar vase form is characteristic of them. Some very beautiful examples are familiar.

The great Neptune's Cups are interesting forms, being complete vases in shape and construction, yet in their native element living animal structures.

Late authorities place Sponges as follows:

Class I.—CALCISPONGÆ. Lime Sponges, literally. Yet all do not have the lime spicules or skeleton framework of lime formation.

Order I. An American representative of this order is known through Mr. J. A. Ryder's observations, called *Camaraphysema*. It is a club-shaped mass, with a tough exterior.

Order II.—OLYNTHOIDEA. This order embraces those forms that have the framework of calcareous spicules. Some extremely curious forms of the latter are found, resembling artificial objects, as various forms of anchors, spears, "grains," etc.

Four sub-orders embrace the comparatively few species of this order.

Class II.—CARNEOSPONGIÆ. Most of the forms embraced here have the skeleton framework made up of horny or silicious spicules. Three orders embrace the Sponges of this class.

The III. Order, KERATOIDEA, includes the commercial Sponges; those having a horny framework.

Sub-order—*Sponginæ*. The genus Spongia embrace all the Sponges that are utilized in commerce. Six species are at present recognized, with varieties.

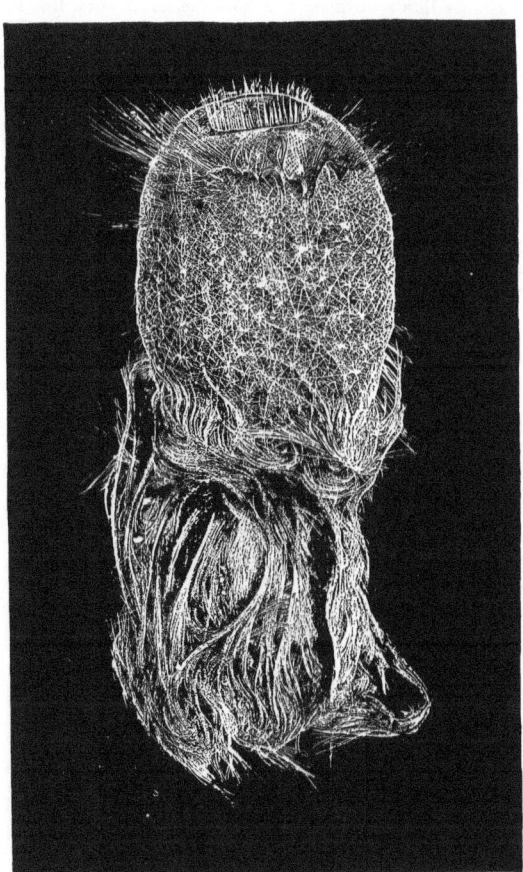

GLASS SPONGE.—*Hottenia carpenteri*. (Natural size.)

Three of the species are found in the Mediterranean and Red Sea, and three are native to Bahamas and the Florida Reef.

The Sponges of our coast are of the coarser kinds, yet of great practical value in the arts and household affairs.

It is an interesting fact that parties in the American localities have succeeded in raising Sponges from cuttings. This is done, of course, under the sea surface, as the moment the Sponge is removed from water it dies. The broad areas of lagoon on the Florida Reef will furnish profitable returns of this new product.

The well known "Dead Man's Finger" (*Chalinula oculata*), found frequently after storms on our beaches along the Atlantic coast, is of another order, called KEROTO-SILICOIDEA, on account of the union of the two kinds of spicules.

One of the sub-orders of this group embraces the species *Microciona prolifera*, abundant on pools at Cape Cod. Its color is a bright orange, and in this form, fresh from the sea, it will be remembered by many a visitor to the sea-shore after heavy gales.

Another sub-order embraces the familiar Crumb-of-Bread Sponge (*Halichondria panicea*). Another, *Suberites compacta*, is common on the south side of Cape Cod.

Some boring forms, small, but very destructive to shell-fish, are common on our shores. *Cliona sulphurea* is one notable species. No satisfactory explanation is yet forthcoming of its boring character. Another sub-order embraces the fresh-water Sponges. Two families, embracing ten species, represent the group at present.

It is said that the peculiar "cucumber odor," so called, is derived from the decay of these fresh-water Sponges.

THE RHIZOPODS.

Order II., SILICOIDEA, includes the highest of the class. The skeleton framework of these The *Tethia* and *Geodia* are prominent forms.

What are known as Glass Sponges are of this Order.

The *Hyalonema*, or Glass-rope Sponge, is found in the waters around the Philippine Islands. Though so much in appearance like spun glass, it is nevertheless of horn, like the nails and hoofs of animals. When burnt it has the same odor as horn. In 1860 naturalists first found the true nature of these objects; they, as well as the glass vases, were so artificial in appearance that they were taken as objects made up for a market.

The *Holtenia carpenteria*, seen in the engraving, is another of different shape. Of all, for beauty and singularity, the *Euplectella*, or Glass Vase—Venus-cup—is the most beautiful. A very fine example is here figured. This, it will be remembered, is the framework of a Sponge, just as the Sponges we use are frameworks of the Sponge. While alive all Sponges are quite heavy, some exceedingly so, with their flesh and the inclosed water. When lifted from the sea, an old black hat with many holes, is as good an illustration of the looks of a living Sponge as any. But what beautiful objects when denuded of flesh!

THE editor of this edition would, as in the case of the Branch of the Animal Kingdom II., *Poriferata*, treat of the First Branch where, according to late authorities, it properly belongs. For classification and nomenclature, according to modern authors, see tables of Smithsonian Institution.

Branch I.—PROTOZOA. First animals, is meant literally—or, looking at animal life in a descending series—the last, or lowest of animals. In our pages this latter view is adopted.

The simplest object that can be called an animal is embraced in this branch. All of this group can move, eat, and reproduce their kind. They move by little oars placed in all parts of the exterior; they eat by absorbing minute animal and vegetable substance into all parts of their bodies; and they are reproduced by the division of their forms indefinitely. Such creatures, we are ready to understand from their composition, are formed of single cells, whose parts are homogeneous, one bit being a representation of another or the whole.

GLASS VASE.—*Euplectella aspergillum*. (Two-thirds natural size.)

The *Protozoa* are mostly minute creatures—microscopic. There is some degree of rank observed among the *Protozoa*. The first class embraces the lowest, called *Monera*. One of these forms, called *Protomyxa*, is regarded as the simplest representative.

The *Rhizopods* form a second class, but the differences are extremely small.

The fresh-water Rhizopods of America have been treated in a magnificent work, with colored plates. In the first Order of this class, the well-known *Amœba* is placed. See the figure in accompanying engraving.

The Order *Radiolaria* embraces some very beautiful forms, radiated and resembling crystals of snow-flakes.

The Order *Reticularia* includes the... ... ... *Bulina*, etc. One of the most notable of these is the *Globigerina*, a wonderful little microscopic globe, from which radiate myriads of spine-like organs, giving the creature a resemblance to some of the *Echini*. The celebrated "*Globigerina* ooze," described by the naturalists of the "Challenger," is made up of this little animal in innumerable numbers, forming vast beds of mud on the ocean bottom.

The Class INFUSORIA includes certain more familiar forms called *Vorticella* and *Stentor*, etc. But the most notable is the *Noctiluca*, the largest of all, being visible to the naked eye. This is somewhat like a gooseberry in aspect. It is noted for its beautiful phosphorescence—the sea at times being wonderfully illuminated by myriads of it on the surface. It has been seen in this condition on the Coast of Maine and Massachusetts.

Many species of these "animalcules" exist, and many of their forms are as fantastic and beautiful as anything Nature produces.

THE FOOD FISHES

OF THE AMERICAN ATLANTIC COAST,

AND THE MARINE NATURAL HISTORY IN ITS RELATIONS TO THEIR INTEGRITY AS SUCH.

THE importance of the fisheries of the coast to the United States can scarcely be exaggerated. The amount of wholesome food yielded, the pecuniary value of the same in the various products incidental, the numbers of men and boys employed, furnishing profitable occupation, the stimulus to several important industries, as boat-building, with the various allied branches of manufacture, and the little less important item of furnishing a school for the merchant and national marine service.

But few years since, the bare suggestion that the abundant supply of food fishes, evidently swarming the ocean, and visiting our coasts in convenient times for their capture, would at any time possibly become reduced to a noticeable condition of scarcity, would have been received as idle speculation. With no thought that the time would ever come when any degree of circumspection would be necessary with reference to the economy of stock, the fisheries of our coasts have been carried on with a lavish hand, and, since the introduction of "trawls," with a reckless waste of material. There are, however, many legitimate channels through which an enormous and rapidly increasing draught of food fishes find their way to a market and consumption. The construction of railroads into remote parts of the country, and the possibility of using ice for the packing of fish, opens up large additional facilities for their consumption.

The discovery that fish can be made to supply a valuable oil by boiling and compression, and that the residue as well as the uncooked fish furnish a valuable manure, has constituted an additional source of consumption on an enormous scale.

It is not strange that such a wholesale consumption of fishes should materially lessen the supply, which, formerly, greatly exceeded the demand.

The first official notice taken of this state of things, with a view to adopting measures of relief, was instituted by the States of Massachusetts and Rhode Island; both being especially interested, as the alleged decrease is attributed to the localities on their borders. The cause assigned by the complainants was the use and multiplication of traps and pounds, which captured fish of all kinds in great numbers, and, as was supposed, in larger quantities than needful, and more than the natural fecundity of the fishes could replace yearly, and especially in view of the fact that these great catches were made during the season of spawning, destroying many fertile fishes, and preventing others from depositing their eggs.

Petitions were presented to the Legislatures of both these States, in the winter of 1869-70, asking that a law be passed prohibiting the use of fixed apparatus for capturing fishes, and the whole subject came before a committee of the Legislatures. The Massachusetts committee, of which Capt. Nathaniel Atwood, of Provincetown, was chairman, decided that there was no reasonable ground for complaint, and the committee was discharged. On the other hand, the Rhode Island committee, having given much greater personal attention to the subject, came to the conclusion that the prayers of the petitioners were well founded, and they reported in favor of a very stringent law prohibiting the further use of traps or "pounds," excepting within a limited district. In short, there was a radical difference between the findings of the two States' committees. The report of the Rhode Island committee, however, was not acted on by the Legislature. The question, unfortunately, became a political one; but there were some scientific men in the State, notably Mr. Samuel Powel, of Newport, a member of the State Senate, who urged a scientific investigation, as the question was one too little understood. Touching the shad fisheries, which are the more important in Connecticut, the latter State at this time united in some movements for reform in the methods of fishing. The question now began to assume an importance it deserved. It was considered that as the general United States government had absolute control of the rivers, harbors, and estuaries, the States really had no rights in this matter, and even if so, there naturally would be some conflicting measures that would be difficult of reconciliation. Then the interests of individuals whose rights might be themes of conflict would be difficult of adjustment. Accordingly, it seemed altogether best that a United States officer be commissioned for the purpose of taking entire charge of the question of coast fisheries, and by a fair adjustment reconcile all difficulties, first instituting an extensive series of investigations, to get at the exact cause of the apparent failure of the supply of food fishes. All depended now on determining the nature of the food sought by our food fishes, the growth of their spawn, and other circumstances bearing upon the solution of the problem in question. A bill was therefore introduced in Congress on the 9th of February, 1871, for the above purpose. This called for the "appointment by the President, by and with the consent," etc., "from among the civic officers of the government, one person of proved scientific and practical acquaintance with the fishes of the coast, to be commissioner of fish and fisheries." The various departments of the government were authorized to render all possible aid to the commissioner in the legitimate prosecution of his duties. This also included the examination of the Great Lakes with reference to their fisheries. An appropriation was made to meet the necessary expenses, and the President tendered the commission to Professor Spencer F. Baird, then assistant secretary of the Smithsonian.

The vicinity of Vineyard Sound being the one most referred to as exhibiting evidences of a decrease in the numbers of food fishes, Professor Baird proceeded to that point to commence his investigations, in June, 1871. He established his headquarters at Woods' Holl, a coast village about eighteen miles from New Bedford, and directly opposite the famous Holmes' Holl, now called Vineyard Haven. Prof. Baird now found it necessary to utilize any aid that had been promised him through the government officers. At New Bedford he secured the use of a revenue vessel attached to the custom-house. The Secretary of the Interior also directed the customs officer to detail the revenue cutter "Moccasin" for his services.

A systematic plan was now adopted, Professor Baird having the valuable assistance of Prof. Gill, and others connected with the Smithsonian Institution. The aids afforded by the Interior Department were supplemented by the free use of apparatus already belonging to the government. A paper was drawn up embracing the various points desirable to investigate systematically. Questions were framed, and these were included in a general circular which was distributed widely along the coast among sea-faring men.

The first serious difficulty in the accumulation of information is in the confusion of names of fishes. There are so many common designations for each species, differing at each remove of locality. The first question asks the local names of the fishes. The descriptive answers to questions are then made to correspond with the local name; otherwise great confusion arises at once. Then the geographical distribution of the species, its abundance at different periods of the year and in different seasons; its size, migration, and movements; relationship to its fellows, or to others; its food and peculiarities of reproduction, and many other things, altogether covering a list of eighty-eight questions.

As the history of species would not be complete without a knowledge of their associates in the sea, especially such as prey upon them, or, in turn, constitute their own food, it was regarded necessary to prosecute searching inquiries on these points. One important cause of scarcity of some food fishes alleged, is that of the great scarcity of certain species which heretofore deposited vast quantities of spawn in the rivers and estuaries, where the most of our food fishes resort for subsistence. The investigation on an adequate scale called for services of experts in various branches of marine zoology. Prof. Verrill and Mr. S. I. Smith, of Yale College, offered their services to conduct the experiments of the laboratory for invertebrate forms. The nature of these inquiries created very generally among naturalists a desire to take part in the results, and Prof. Baird very generously extended the most complete accommodations possible for them. The material collected on the dredging excursions, and by the numerous other methods, was large, and ample enough to afford not only individuals, but museums and similar institutions, with sets of specimens of marine invertebrate animals.

A photographer was employed to take views of the various species, in the different stages of growth and at different seasons. The temperature of the sea water was taken at various times of the day, and at stated depths.

Headquarters being established here at Woods' Holl, each season was devoted to the work of the commission, during the months of August, September, and October. Parties were commissioned, also, to continue certain work laid out, of importance in its relations to the cooler seasons as well as the warmer. Almost the first result of the marine dredging was the capture of new fishes, and some not hitherto known to visit our coast. Hosts of invertebrate forms were discovered at almost every haul of the dredge, gladdening the hearts of the zoologists and aiding the progress of the several branches of invertebrate forms.

Prof. Baird says at this time: "An interesting result of the work done during the summer of 1871 consisted in the great variety of the fishes obtained through the pounds and otherwise, many of them of kinds previously unknown on the New England coast. The total number actually taken and photographed amounted to one hundred and six species, of which twenty or more are not included in the great work of Dr. Storer on the 'Fishes of Massachusetts.'" Nine species are mentioned by various others as found in the waters of Vineyard Sound, but which were not secured, making one hundred and fifteen in all known to belong to that fauna.

Among the more interesting of the newly-recognized fishes is the species of Tunny, a kind of small horse-mackerel, the *Orcynus thunnina*, weighing about twenty pounds, and which, though well known in the Mediterranean, and in the warmer parts of the Atlantic, have never been known to visit the Atlantic side of America. Some five hundred of these were taken in one locality, called Menemsha Bight. Two species of the sword-fish family, never before seen in these waters, were also taken.

In connection with this work the algæ of the coast were carefully studied by several authors, and complete lists published by the commission. During the same summer, some investigations were carried on by Dr. Yarrow, U. S. A., on the coast of North Carolina. Also, during the same season, a deputy commissioner, Mr. Milner, made a tour of Lake Michigan, visiting every fishing locality, and reporting the results in connection with the general operations at Woods' Holl.

Two seasons' work now came to be the subject of a report of some of the "General Results of the Investigation." The objects to be gained, as authorized by Congress, were, first, to determine the facts as to the alleged decrease of the food fishes; secondly, if such a decrease be capable of substantiation, to ascertain the cause of the same; thirdly, to suggest methods for the restoration of the supply. A fourth object, incidental to the rest, was to work out the problems connected with the physical character of the seas adjacent to the fishing localities, and the natural history of the inhabitants of the water, whether vertebrate or invertebrate, and the associated vegetable life; as also to make copious and exhaustive collections of specimens, for the purpose of enriching the National Museum at Washington, and of furnishing duplicate specimens for distribution in series to such suitable collegiate and other cabinets as might be recommended for the purpose.

This research into the general natural history of the waters was considered legitimate, as, without a thorough knowledge of the subject, it would be impossible to determine, with precision, the causes that affect the abundance of animal life in the sea, and the methods for regulating it; and the records of these facts, accompanied by proper illustrative figures, it was believed, would be a very acceptable contribution to the cause of popular education, and supply a want that has long been felt in this country.

The direct operations of the commission required the use of extensive apparatus. It was found that the additional cost of procuring a large series of specimens of marine objects that could be distributed to all parts of our country and the Old World was so little, a liberal arrangement was at once commenced.

Most enlightened nations have devoted means to the same end, particularly the German government.

It was found, early in the investigations, that an alarming decrease in numbers of fish had occurred on the coast of New England, extending from Point Judith on the west to Monomoy on the east, including Narragansett Bay, Vineyard Sound, and Nantucket.

This fact, Prof. Baird says he has no hesitation in saying, has been established by his own investigations, as well as by evidence of those whose testimony was taken on the subject.

But few years since, this region was the scene of an extensive fishery, the most important on our coast; the number of southern or deep-sea species resorting to the inlets and bays to deposit their spawn being enormous.

The Scup, Black-fish, or Tautog, Striped Bass and Sea Bass, Sheep's-head King-fish, and Weak-fish. The appearance of these fish was extremely regular. Their arrival could be calculated with quite exactitude, as much so as with the migratory birds, varying only at times by changes of temperature, etc. There were a few that seem to have been independent of such fixed conditions, and whose movements were somewhat erratic, as the Mackerel, Bonito, Blue-fisn, etc.

The Scup, from the Indian name Mish-cûp-paûg, known as Porgy in the Southern States, and by the early English settlers as Bream, from its resemblance to the English fish of that name, is an important fish in this connection, as being the species that has suffered most in this wonderful decimation. Its technical name is *Stenotomus argyrops* (Linn.), Gill. Its southern range is to Cape Florida, and it is found in southern waters throughout the year, more abundantly in June and July.

The first run of these fish seen on our coast in the season is in May, when the largest are noticed; these proceed early in June to lay their spawn. These are from two to four pounds' weight, and about eighteen inches in length. It is thought that the spawning takes place in the eel-grass that spreads over the shoal waters of Narragansett Bay and Vineyard Sound. There is a regular interval noticed between the departure and arrival from point to point. For example, the Scup are taken at Montauk three weeks earlier than at Woods' Holl, and a week earlier at the latter place than at Hyannis, still farther east. They feed upon a great varity of marine animals, such as worms, crustacea, mollusks, etc., and they take a hook very readily as long as they remain.

The flesh of the Scup is very much prized by most persons. It is firm and flaky, and sweet.

Since the settlement of America, it has been the most important food fish taken in these waters, and the rapid diminution has caused great solicitude.

This fish is little known north of Cape Cod. Dr. Storer says they were introduced into Massachusetts Bay in 1833, and that they are taken only seldom. The great numbers of this species of fish once known to be in the vicinity of Vineyard Sound and Narragansett Bay are indicated by a record of schools seen in the latter place, where the water was nine feet and the fish so abundant they were crowded out of water. Six hundred barrels were taken at one haul of the seine, near Nantucket. At one time, in 1861, seven hundred barrels were let out of a trap, the market being glutted. Until within about eight years, one could take Scup anywhere from Point Judith to Cape Cod, almost as rapidly as a line with two baited hooks could be thrown over and hauled in.

The great reduction in numbers of these fish is estimated to cause the price of living among many families to be raised to a hundred dollars extra yearly.

The causes of the decrease of the numbers of food fishes are placed as follows: The decrease or disappearance of the food upon which the fish subsist, necessitates their departure to other localities.

Epidemic diseases, or peculiar atmospheric agencies, such as heat, cold, et cetera; destruction by other fishes; the agency of man, being manifested either by the pollution of waters, by the refuse of factories, or the excessive over-fishing, or the use of improper apparatus.

INDEX.

A.

Abaster, 136.
Aboma, 122.
Acalepha, 556.
Acanthias, 200.
Acanthonyx, 442.
Acanthopluis, 115.
Acanthopterygii, 212.
Acarina, 516.
Acarus, 517.
Acephala, 345.
Acerina, 219.
Achatina, 334, 335.
Acheta, 389.
Achirus, 265.
Achrochordus, 117.
Acipenser, 191, 192.
Aconthea, 413.
Acontiadæ, 69.
Acontias, 69.
Acontias, Blind, 70.
Acontias, Painted, 70.
Acrida, 391.
Acris, 154, 164.
Acrochorde, 116.
Acrosoma, 510.
Acrydæ, 391.
Actinaria, 568.
Actinia, 570, 571.
Actinolda, 568.
Actinoloba, 569.
Actinozoa, 557, 570.
Ada, 45.
Adder, 110.
Adder, Berg, 107.
Adder, Checkered, 132.
Adder, Das, 107.
Adder, Death, 115.
Adder, Deaf, 98.
Adder, Horned, 107.
Adder, Puff, 105.
Adder, Red, 98.
Adder, Sea-, 216.
Adder, Spotted, 132.
Adder, Water, 184.
Adippe, Fritillary, 411.
Admiral, Scarlet, 412.
Æcanthus, 391.
Ægeria, 419.
Ægle, Red-Spotted, 444.
Æquorea, 564.
Æsop, Couch's, 468.
Æthra, 444.
Agama, 87.
Agamas, 84.
Agate-Shell, 334.
Agelena, 507.
Agraulis, 411.
Agrion, 198.
Ahætulla, 138.
Ailanthus, 422.
Alausa, 277.
Albacore, Pacific, 246.
Albicore, 244.
Albione, 537.
Alca, 463.
Alcippe, 492.
Alcyonella, 370.
Alcyonidium, 369.
Alcyonium, 576.
Alecto, 367, 368.

VOL. III.

Aleutera, 289.
Alewife, 277.
Alima, 471.
Alligator, Young, 178.
Alligators, 35, 36.
Alloposus, 307.
Alopias, 202.
Aloponotus, 77.
Alosa, 276.
Alpheus, 467.
Alucita, 427.
Alytes, 159.
Amblyopsidæ, 275.
Amblyopsis, 276.
Amblystoma, 172, 175.
Amblystome, 175.
Amciva, 44.
Ammocœtes, 295.
Ammodyte, 110.
Ammodytes, 270, 271.
Amœba, 584, 593.
Amphibia gradientia, 170
Amphioxus, 296.
Amphipoda, 472.
Amphisbæna, 38.
Amphisbæna, White, 39.
Ampulla, bulla, 338.
Ampullaria, 336.
Anadas, 260.
Anableps, 285.
Anaconda, 124.
Anadia, Eyed, 55.
Anarrhichas, 257.
Anchorella, 486.
Anchovy, 276.
Ancistrodon, 97, 99.
Ancylus, 337.
Anemone, Green-, 568, 569.
Anemone, Plumose, 569.
Anemone, Red-, 568.
Anemone, Sea-, 557, 568.
Anemone, Snake-Locked, 570.
Anemone, Warty, 568, 570.
Angel-Fish, 204, 228.
Augler-Fish, 255.
Anguilla, 271, 272.
Anisopleura, 341.
Annulata, 527.
Anobium, 381.
Anodon, Rough, 137.
Anolis, Various, 80,81,82.
Anophthalmus, 375.
Ant-Lion, 395.
Ant, White, 393.
Anthea, 568.
Antennarius, 256.
Anthia, 374, 375.
Anthonomus, 385.
Anthalia, 398.
Anthrenus, 381.
Anthoceridæ, 419.
Aphaniptera, 432.
Aphrodite, 532, 533.
Aphrophora, 400, 429.
Apis, 403.
Apistos, 232.
Aplonote, 77.
Aplustrum, 338.
Aplysia, 339.

Aplysias, 338.
Apodidæ, 480.
Apogon, 224.
Aporrhais, 325.
Aprasia, 58.
Aprionodon, 196.
Aptera, 437.
Apus, 480.
Arachnida, 493.
Arachnida, Six-Eyed, 511
Araneidea, 493.
Arapaima, 240.
Ararambuya, 123.
Arcella, 585.
Archer Fish, 231.
Archippus, 411.
Architeuthis, 309.
Arctia, 423.
Arctopsis, 441.
Arcturus, Baffin's Bay, 477.
Arenicola, 531.
Argentine, 282.
Argonaut, 305.
Argonauta, 304.
Argonauts, 307.
Argulidæ, 485.
Argulus, Fish-, 485.
Argus, Shetland, 553.
Argyrophis, 70.
Arius, 431.
Arion, 335.
Aristolochia indica, 143.
Ark, Noah's, 351.
Ark-Shells, 351.
Arkys, 501, 502.
Armadillo, Pill-, 479.
Armed-Breast, 216.
Army-Worm, 434.
Aromochelys, 15.
Arrosoir, 355.
Arthronomalus, 523.
Arthropoda, 372.
Ascaris, 538.
Ascidians, 359.
Asilidæ, 435.
Asker, 172.
Asp, 109.
Aspidomorpha, 386.
Ass's Ear, 329.
Astacidæ, 465.
Astacus, 466.
Astarte, 347.
Asteriadæ, 549, 574.
Asterias, 549, 552, 554.
Astræa, 573, 574.
Astræacea, 568.
Astroides, 572.
Astrophyton, 554.
Ateuchus, 379.
Atherina, 261.
Atlantas, 342.
Atlas Beetle, 380.
Atractaspis, 149.
Attacus, 422.
Atun, 243.
Atypus, 496.
Auger, Spotted, 319.
Auxis, 246.
Avicularidæ, 347.

Axolotelos, 177.
Axolotl, 175.

B.

Back-Swimmers, 431.
Bacteria, 391.
Balænoptera, 491.
Balaninus, 384.
Balanus, 491.
Balistes, 288.
Balloon-Fish, 289.
Band-Fish, 263.
Barbel, 285.
Barca, 202.
Barideus, 385.
Barnacles, 488 to 491.
Barracoudas, 243.
Bascanium, 133.
Basket-Fish, 554.
Bass, Various, 218 to 220.
Bat-Fish, 256.
Batrachians, 149.
Batrachians, Crawling, 170.
Batrachians, Tailed, 179.
Batrachians, Tongued, 151.
Batrachidæ, 255.
Beadlet, 570.
Beania, 364.
Beauty, Camberwell, 413
Becker, 227.
Becuna, 242.
Bed-Bug, 430.
Bees, Various, 402 to 404.
Beetle, Bloody-Nose, 386.
Beetle, Cocktail, 373.
Beetle, Egyptian, 378.
Beetle, Musk, 373.
Beetle, Pellet, 379.
Beetle, Watchman-, 379.
Beetles, 373.
Beetles, Carrion, 377.
Beetles, Diamond, 383.
Beetles, Spring, 381.
Belemnites, 307, 310.
Bellows-Fish, 263.
Bembecidæ, 402.
Berœ, 559.
Berycidæ, 216.
Bicellaria, 364.
Bill-Fish, 292.
Bismore, 216.
Black-Fish, 220, 223.
Blattidæ, 388.
Bleak, 287.
Bleeding-Tooth Shell, 328.
Blennies, Various, 256 to 253.
Blennius, 257.
Blepsias, 233.
Blind-Fish, 275.
Blindworm, 63.
Blister-Fly, 382.
Blowers, 289.
Blue-Back, 277.
Blue-Fish, 247.
Blue-Tail, 62.
Boas, 118.
Boas, Various, 122, 123.

Bodian, 225.
Bombardier Beetle, 375.
Bombinator, 160.
Bombyx, 422.
Boide, 118, 136.
Boiga, 138.
Boiguacu, 123.
Bojobi, 123.
Bolina, 559.
Boltenia, 358.
Bombardier, 160.
Bombus, 403.
Bombycidæ, 434.
Bone-Dog, 195.
Bonito, 244, 246.
Bonnet Fleuk, 266.
Bonnet Head, 196.
Bonnet, Lady's, 332.
Bony-Fish, 278.
Boouslange, 137.
Boro Poloo, 422.
Bot-Fly, 436.
Bothriocephalus, 540.
Bothrs, 266.
Botryllus, Star-Shaped, 359.
Bounce, 194.
Bowerbankin, 368, 369.
Brachelytra, 377.
Brachinus, 375.
Brachiopoda, 344.
Brain-Stone, 574.
Braize, 227.
Bramolds, 249.
Branch-Horns, 481.
Branchellion, 538.
Branchiobdella, 538.
Branchiopoda, 480.
Bream, 286.
Bream, Sea-, 227, 228.
Breast-Plate, 362.
Brett, 266.
Brevortia, 278.
Brill, 266.
Brittle-Stars, 552.
Bubble-Shell, 338.
Buccinums, 330, 331.
Bucephalus, 138.
Buckle, Roaring, 317.
Buckwheat-Nose, 129.
Bufo, 163, 164.
Bug-Fish, 278.
Bug, Harvest-, 516.
Bug, Mealy, 429.
Bugong, 410.
Bugula, 364.
Bulimi, 334.
Bulimus, Lemon, 334.
Bulina, 594.
Bull-Frog Shell, 316.
Bull-Frogs, 152 to 155.
Bull-Head, Armed, 236.
Bullhead, 234.
Bungarus, 140, 141.
Bunkers, 278.
Bunodes, 570.
Buprestidæ, 380.
Burying Beetles, 377.
Bushmaster, 96.
Buskia, 369.
Butt, 266.

INDEX.

Butter-Fish, 258.
Butterflies, and Varieties, 404 to 415.
Butterfly-Fish, 257.
Butthorn, 549.

C.

Caberea, 364.
Cactus, 429.
Caddis-Flies, 397.
Cæcilia, 180.
Calamaries, 307.
Calandra, 384.
Calcispongæ, 591.
Calico, 265.
Caligus, 485.
Callianassa, 464, 478.
Callimorpha, 424.
Callionymus, 253.
Callorhynchus, 193.
Calocaris, 405.
Calpidium, 362.
Calydna, 414.
Calyptræa, 332.
Camaleao, 77.
Camara, 453.
Camaraphysema, 591.
Campanulariæ, 577.
Camposcia, 440.
Cancer, 444.
Cantharidæ, 382.
Canthocamptus, 484.
Caouane, 23.
Caprella, 476.
Capeuna, 225.
Carabidæ, 375.
Caramurn, 184.
Carangidæ, 250.
Caraux, 249.
Carbasea, 365.
Carcharias, 203.
Cardiadæ, 352.
Cardinal Beetle, 382.
Cardium, 352.
Caret, 23.
Cariuaria, 342.
Carneospongiæ, 592.
Carp, 284.
Carphophiops, 136.
Caryophylliaceæ, 572.
Cassididæ, 386.
Cassis, 320, 331.
Castnia, 420.
Cat, Sea, 193.
Cat, Duck-Billed, 192.
Cat-Thresher, 277.
Catagramma, 405, 413.
Catenicella, 362.
Catocala, 425.
Caudisona, 104.
Cayman, Spectacled, 38.
Caymans, 35.
Cecrops, 486.
Celestus, 67.
Cellepora, 366.
Cellularia, 363.
Centipede, Giant, 522.
Centorhynchus, 385.
Centriscidæ, 263.
Centrocymnus, 196.
Centronotus, 258.
Centropristis, 220.
Centropyx, 45.
Cephalophora, 312.
Cephalopoda, 304.
Cephaloptera, 212.
Cepola, 263.
Cerambys, 373.
Cerambyx, 285.
Cerapus, 475.
Cercopidæ, 429.
Cercosaura, 55.
Cerithium, 325.
Cermatia, 522.
Cerxartes, 107, 108.
Cerberus, 117.
Cerithiadæ, 325.
Cerura, 423.
Cestracion, 203.
Cestum, 563.
Cetorhinus, 202.
Cethosia, 412.
Cetochilus, 484.

Vol. III.

Chad, 228.
Chætodons, Various, 229, 230.
Chætopterus, 534.
Chain, Little, 362.
Chalcis, 54.
Chalgua, Achagual, 193.
Chalimus, 486.
Chalinula, 592.
Chamæsaura, 55.
Chameleon, 82, 90.
Charcharinus, 198.
Chariua, 136.
Charioteer, 230.
Charr, 282.
Cheirotes, 39.
Chelifer, 514.
Chelmo, 230.
Chelodines, 18, 19.
Chelonia, 20.
Chelonura, 20.
Chelopus, 14.
Chelydes, 17.
Chelydra, 16.
Chelys, 17.
Chersæa, 109.
Chersydrus, 116.
Chetodonts, 228.
Chigoe, 432.
Chilabothrus, 125.
Chilocorus, 386.
Chilodactyle, Banded, 231.
Chilognata, 524.
Chilopoda, 520.
Chimæra, 193.
Chiragra, 441.
Chirocolo, 55.
Chirodota, 543.
Chitons, Various, 332, 333
Chittul, 116.
Chlamydosaurus, 85.
Chlorion, 401.
Chogsett, 222.
Chondracanthus, 486.
Chondropterygii, 190.
Chorinus, 441.
Chrysaora, 564, 566.
Chrysalis-Shell, 335.
Chrysemys, 14.
Chrysochroa, 380.
Chrysomela, 386.
Chrysomelidæ, 386.
Chrysophora, 380.
Chrysophrys, 228.
Chub, 242, 287, 294.
Chunk-Head, 98.
Chuss, 260.
Chydorus, 482.
Cicincelidæ, 373.
Cirripedes, 488.
Cicada, 390, 428.
Cicigna, 68.
Cicindela, 374.
Cidaris, 546.
Cimbex, 398.
Cimex, 430.
Ciniflo, 506.
Cinosternum, 15.
Cirrhatulus, 532.
Cirrhitidæ, 231.
Cistudo, 12.
Cladocera, 481.
Clam, and Species, 345, 347, 352, 354.
Clam, Cracker, 211.
Claquerrs, 72.
Clava, 577.
Clavellina, 358, 359.
Clear-Wings, 419 420.
Cleg, 434.
Cleodora, 342.
Cleodorus, 344.
Clifden Nonpareil, 425.
Cliona, 592.
Clione, 344.
Clotho, 107, 504.
Club-Shell, Great, 325.
Clubiona, 504.
Clubs, 325.
Clupea, 276, 277, 278.
Clytus, 385.
Cnethocampa, 434.
Cobbler-Fish, 249.

Cobra, African, 148.
Cobra Di Capello, 141.
Coccinella, 386.
Coccus, 429, 430.
Cochineal Insect, 429.
Cock-Paidle, 253.
Cockles, 352, 353.
Cockchaffer, Common, 377.
Cockroach, 388.
Cocktails, 377.
Cod-Fish, 267, 260.
Cœlenterata, 557.
Cœlotes, 507.
Coffer-Fish, 288.
Coleoptera, 373, 377, 386.
Colorado Potato Beetle, 385.
Coluber, 130, 132, 133.
Colubrinæ, 125.
Comatula, 554.
Comb-Bearers, 562.
Comb, Venus', 314.
Conch, 330.
Conch, Queen, 331.
Conch-Shell, 315.
Conchoderma, 492.
Conchologist, 328.
Cone-Shells, 321.
Cones, 304, 321, 330.
Conidæ, 321.
Conis, 321.
Conger, 272.
Conocephalus, 125.
Conotrachelus, 385.
Contia, 136.
Cooter, 13.
Copepoda, 480.
Copepodœ, 484.
Copris, 379.
Coqs de Mer, 453.
Cora-Mota, 262.
Coral Reefs of Florida, 578.
Coralline, Bird's-Head, 364.
Corallium, 574.
Corals, 557, 572 to 575.
Cordonnier, 249.
Cordyles, 50, 51.
Coregonus, 282.
Cornuda, 199.
Coronella, 126.
Coronulas, 492.
Corophilus, 154.
Corophium, Long-Horned, 476.
Correganus, 281.
Corvina, 241.
Corybantes, 529.
Coryne, 487, 577.
Coryphænas, 248.
Coryphene, 248.
Cottidæ, 234.
Cotton-Mouth, 98.
Cottus, 234, 235.
Cow-Fish, 288.
Cowrey, Orange, 304.
Cowries, 322 to 324.
Crab, Angular, 451.
Crab, Armed, 452.
Crab, Bearded, 456.
Crab, Black-, Toulouse, 448.
Crab, Brassy, 445.
Crab, Calling, 450.
Crab, Death's-Head, 456.
Crab, Domed, 448.
Crab, Crested, 452.
Crab, Edible, 443.
Crab, Edible, of American, 447.
Crab, Fighting, 449.
Crab, Flattened Mud-, 448.
Crab, Floating, 451.
Crab, Frog-, Toothed, 457.
Crab, Gonty, 441.
Crab, Great Burrowing, 464.
Crab, Green, 445.
Crab, Oar-Foot, 458.
Crab, Hairy, 444, 456.

Crab, Harper-, 441.
Crab, Herald, 442.
Crab, Heraldic, 442.
Crab, Hermit-, 458.
Crab, Hermit-, Crafty, 459.
Crab, Keeled, 454.
Crab, Lady-, 450.
Crab, Long-Tailed Molucca, 487.
Crab, Long-Snouted, 442.
Crab, Mask, 454.
Crab, Montagu's, 444.
Crab, Nipper-, 447.
Crab, Noduled, 457.
Crab, Nut-, Pennant's, 454.
Crab, Oceanic Swimming, 446.
Crab, Painted, 451.
Crab, Pea-, 449.
Crab, Polished, 454.
Crab, Porcelain-, Long-Horned, 462.
Crab, Porcelain-, Broad-Claw, 461.
Crab, Porcupine-, 456.
Crab, Purse-, 460.
Crab, Racing, 450.
Crab, Ram's-Horn, 442.
Crab, Red, 464.
Crab, Robber-, 460.
Crab, Sand-, 450.
Crab, Saucepan-, 488.
Crab, Scallop-, 455.
Crab, Sentinel, 447.
Crab, Seven-Spined, 454.
Crab, Shore-, 445.
Crab, Spider, 439.
Crab, Spider-, Four-Horned, 441.
Crab, Spider-, Great, 440, 441.
Crab, Spider-, Hornback, 442.
Crab, Spider-, Three-Spined, 442.
Crab, Spirit, 451.
Crab, Spotted, 445.
Crab, Stone-, Northern, 456.
Crab, Strawberry-, 443.
Crab, Thorn-Claw, 442.
Crab, Tortoise, 453.
Crab, Urania, 453.
Crab, Velvet Fiddler, 446.
Crab, Violet-, of Jamaica, 448.
Crab, Wooley, 455.
Crabs, Fiddler, 450.
Crabs, Spider-, 441, 452.
Crabs, Swift-Footed, 440.
Crabs, Swimming, 445.
Crabes Honteux, 453.
Crambactis, 571, 572.
Cramp-Fish, 206.
Cranchia, 308.
Crane-Flies, 433.
Crangon, 467.
Craspedocephalus, 96.
Craw-Fish, 466.
Cray-Fish, Sea, 464.
Cray-Fish, 466.
Creophilus, 377.
Crevalli, 247.
Cribella, Eyed, 551.
Crickets, 389.
Cricket, Western, 154.
Criocarcinus, 442.
Crisia, 367.
Cristatella, 360.
Cronkers, 241.
Crocodiles, 28 to 30.
Crocodilus, 31, 32, 33, 34, 35.
Crooner, 236.
Cross-Fish, 549, 551.
Crotalidæ, 95, 103.
Crotalus, 100, 103, 104.
Crustacea, 438.
Crustacea, Sessile-Eyed, 472.
Crustacea, Sword-Tailed, 487.

Crustacea, Ten-Footed, 307.
Crustaceans, Mouth-Footed, 460.
Crustaceans, Stalk-Eyed, 438.
Crustaceans, Stomapod, 470.
Crustaceans, Ten-Legged, 438.
Crytoblepharus, 57.
Cryptobranchus, 178.
Cryptophialus, 492.
Cryptopods, 453.
Cryptopodia, 443.
Ctenobranchia, 330.
Ctenolabrus, 222.
Ctenophora, 557, 559, 562.
Cuckoo Flies, 400.
Cuckoo-Spit, 429.
Cucujo, 380.
Cucumaria, 544.
Cucumber, Sea, 544.
Cucumbus, Sea, 542.
Culebra de Agua, 124.
Cunners, 222.
Cupularia, 367.
Curculio, 385.
Curucnen, 96.
Cusk, 269.
Cuttle-Fishes, 311.
Cuttles, Eight-Armed, 306.
Cyamus, 477.
Cyanæa, Venomous, 565.
Cyanea, 557.
Cyclopidæ, 484.
Cyclops, 480, 484.
Cyclopterus, 254.
Cydippe, 562.
Cylichna, 338.
Cymba, 322.
Cymbulia, 343.
Cynips, 399.
Cynoscion, 241.
Cynthia, 358, 359.
Cypræa, 322.
Cypræa, 330.
Cypridinidæ, 483.
Cyprina, 347.
Cyprinus, 285.
Cypris, 483.
Cyrestis, 412.
Cystignathus, 158, 159.
Cythere, 483.
Cytherea, 353.

D.

Dab, Common, 266.
Daboia, 104, 105.
Dace, 287.
Dactylethra, 150.
Dactylocera, 475.
Daddy Long-Legs, 418.
Dalader, 431.
Delatias, 204.
Danais, 411.
Daphnia, 481.
Dasee-Worm, 422.
Dasychira, 424.
Dasypeltidæ, 137.
Date-Shell, Finger, 351.
Date-Shell, Fork-Tailed, 350.
Dead Man's Finger, 592.
Deal-Fish, 258.
Decacera, 307.
Decapoda, 307, 438.
Deer-Swallower, 124.
Delma, 58.
Dendrapsis, 140.
Dendronotus, 340, 342.
Dendrophidæ, 137.
Dendrophyllia, 573.
Dendrosaura, 89.
Dentalina, 584.
Dentalium, 332.
Dermatochelys, 22.
Dermestes, 381.
Desmognath, 172.
Desmotenthis, 308.
Devil-Fish, 210, 310.

INDEX.

Devil, Mud, 178.
Devil, Sea, 210, 231.
Deweeboraloowah, 226.
Dhab, 59.
Diachoris, 365.
Diactor, 432.
Diadophis, 136.
Diamond-Back, 14.
Dibranchiata, 305.
Dichelertium, 485.
Dido, 411.
Dinetopus, 364.
Diodon, 280.
Dipera, 433.
Diphyes, 561.
Diphyidæ, 561.
Diplodus, 227.
Diploglossus, Sagra's, 67.
Dipsas, 140.
Disaulax, 385.
Disc-Bearers, 564.
Discoboli, 253.
Discophora, 557, 564.
Discopora, 368.
Distaff Shell 316.
Doclea, 441.
Docophorus, 519.
Dogania, 20, 21.
Dog-Fishes, 179, 193 to 208.
Dolabella, 339.
Dolium, 319.
Doliums, 331.
Dolomedes, 499.
Dolphin-Shell, 329.
Dolphins, 248.
Dominula, 424.
Doris, 340, 342.
Dorsc, 370.
Dory, 219.
Dory, John, 248.
Doryphora, 385.
Doto, 341.
Dove Shell, 321.
Dragon-Flies, 394, 395.
Dragon, Flying, 84.
Dragon, Fringed, 84.
Dragon, Great, 45.
Dragonet, Common, 252.
Drassus, 504.
Dreissena, 350.
Dromia, 455, 456.
Dromicus, 137.
Drone-Fly, 436.
Drum-Fish, 241.
Dryadidæ, 134.
Drylophidæ, 130.
Dumb Rattle, 98.
Dung-Beetle, 370.
Dynastes, 380.
Dysdera, 511.
Dyticus, 376.

E.

Ear-Shells, 329, 330.
Earwigs, 387.
Earth-Worm, 534.
Echeneis, 248.
Echinarachnius, 548.
Echinidæ, 544.
Echinopora, 573.
Echinodermata, 541.
Echinorhinus, 204.
Echinus, 547, 552.
Echis, 100.
Edwardsia, 569.
Eel, Lamprey, 294.
Eel, Mud, 186.
Eel-Pout, 208.
Eels, 270 to 275.
Eft, 172.
Eft, Spotted, 175.
Egg-Shells, 324.
El Adda, 59.
Elaps, 134, 149, .
Elateridæ, 381.
Elodone, 307.
Elephant-Fish, 193.
Elysia, 342.
Emmets, King of, 39.
Emydosauri, 28.
Emys, 10, 11, 14.
Encrinites, 554.

VOL. III.

Endendrium, 577.
Ensis, 354.
Engraulis, 276.
Entomophaga, 399.
Entomostraca, 480.
Entoza, 538.
Eolis, 341, 342.
Epeira, 509.
Ephemera, 396.
Epibulus, Sly, 222.
Epicalia, 405.
Epinephilus, 224.
Eques, 240.
Equites, 406.
Erato, 409.
Eromias, Namaqua, 49.
Erosus, 499.
Eretmochelys, 23.
Ergasilidæ, 485.
Ergatis, 506.
Erichthus, 470.
Eripus, 510.
Eristalis, 355, 436.
Erpeton, 117.
Erycinidæ, 414.
Eschara, 366, 367.
Eucorybas, 523.
Eudora, 564.
Eumces, 60, 61.
Eunice, 531.
Euplæa, 410.
Euplectella, 596.
Euriale, 554.
Eurygone, 44.
Eutropia, 134.
Evat, 172.
Exocœtus, 279.

F.

Fan-Fish, 251.
Fan-Foot, 71.
Farancia, 136.
Farciminaria, 364.
Fat-Back, 278.
Father Lasher, 235.
Feather-Star, 553.
Fer-De-Lance, 95.
Fiddler, Marbled, 446.
Fierasfers, 270.
Fig, Little, 316.
Filaria, 538.
Filc-Fishes, 288.
Filc-Shell, 346.
Fire-Fish, Red, 232.
Fireflies, 380, 381.
Fish-Louse, 478.
Fishes, 188.
Fishes, Flat, 265.
Fishing-Frog, 255.
Fissurellidæ, 351.
Fistularia, 264.
Five-Fingers, 549,551,552
Flea Water, 481.
Ficas, 432.
Fier, 367.
Fleuk, 266.
Flies, 436.
Flounders, 266.
Flustra, 360.
Flying-Fish, 278.
Foot, Pelican's, 325.
Foraminifera, 588, 594.
Forest-Fly, 437.
Forester, Green, 419.
Fox, 284.
Fox-Fish, 252.
Fox, Sea, 202.
Fredericella, 371.
Frog-Fishes, 255.
Frog-Hopper, 420.
Frog-Shell, 316.
Frogs, 151 to 169.
Frost-Fish, 269.
Fulgoridæ, 428.
Fusus, 319, 317.·

G.

Gaobua, 262.
Gadus, 268, 260.
Gad-Fly, 434.
Galathea, 463.
Galene, 445.

Galeocerdo, 196.
Galeodes, 512.
Galeorhinus, 200.
Gall Insects, 399.
Galliwasps, 67, 69.
Gammasus, 518.
Gammarus, 475.
Gaper Shell, 355.
Gar-Fish, 279.
Garteracantha, 510.
Gasperean, 277.
Gasteropoda, Inopercu- late, 333.
Gasterosteus, 213, 215.
Gastrochæna, 356.
Gastrophilus, 436.
Gavial, 29.
Gebin, 464, 465.
Gecarcinus, 448.
Geckos, 71 to 74.
Geissosauri, 55.
Gelasinus, 449, 450.
Gemellaria, 364.
Geodia, 593.
Geometridæ, 426.
Geotrupes, 370.
Gerrhosaurus,Bibron's,51
Gherkins, Sea, 544.
Gilt-Head, 228.
Gini-Maha, 282.
Glass-Eye, 219.
Glass Vase, 593.
Glaucus, 341.
Globe-Fish, Prickly, 280.
Globigerina, 504.
Glomeris, 524.
Glossina, 435.
Glove, Mermaid's, 588.
Glow-Worm, 381.
Glycimeris, 354.
Gnat, Common, 433.
Goa, 32.
Goat-Fishes, 227.
Gobies, 251, 252.
Gold-Fish, 285.
Gongylus, 60.
Gonibrogmatus, 524.
Goniodes, 519.
Gonoleptes, 512.
Gonyocephale, 84.
Goody, 242.
Gooseberry-Fly, 398.
Goose-Mussel, 489.
Gopher, 6, 13.
Gore-Bill, 270.
Gorgonia, 367, 574, 575.
Gorgonias, 574.
Gossamer, 502.
Gowdie, 252.
Grammatophore, 86.
Grancio, 448.
Grantia, 588.
Grapsus, 451.
Grasshoppers, 391.
Grayling, 281, 282.
Green-Fish, 223.
Greenbone, 258, 279.
Gribble, 478.
Ground Beetles, 375.
Grouper, Red, 224.
Grunts, 227.
Gryllotalpa, 389, 390.
Gryllus, 389.
Guana, 78.
Guffer, 258.
Gunnel, Spotted, 258.
Gurnards, 231, 234 to 238.
Gymnetrus, Oared, 258.
Gymnodontes, 289.
Gymnophthalmata, 564.
Gymnophthalmidæ, 57.
Gymnostoma, 344.
Gymnotus, 274.
Gynæcia, 405.
Gyrinophilus, 172.
Gyrinus, 376.

H.

Haddock, 270.
Hæmulon, 225.
Hag-Fish, Glutinous, 295
Hair-Tail, Silvery, 243.

Haje, 148.
Hakes, 269, 270.
Halcyonoida, 574.
Haldea, 136.
Halibuts, 266, 267.
Halichondria, 588, 592.
Haliotis, 329, 330.
Hamadryas, 140, 142.
Hammer-Head, 199.
Hammer-Heads, 196.
Hammer-Shell, 347, 348.
Hard-Head, 278.
Hardim, 86.
Hare, Sea, 338.
Harpa, 318.
Harpalus, 374, 375.
Harp-Shells, 317.
Harp, Ventricose, 318.
Harvest-Bug, 516.
Harvest-Man, 511.
Hatchet-Shell, 339.
Hatteria, 77, 78.
Head-Fish, 290.
Hector, 407.
Hedgehog, Sea, 289.
Hedgehog-Shell, 314.
Heerwurm, 434.
Helbut, 267.
Helices, 338.
Helicidæ, 333.
Heliconia, 409.
Heliceps, 405, 414.
Helix, 334, 338.
Hellbender, 178.
Helmet, Ruddy, 320.
Helmet-Shells, 320.
Heloderma, 44.
Hemerobiidæ, 395.
Hemidactyle, Spotted, 72
Hemirrhamphus, 279.
Hemmatoccrus, 431.
Hepialus, 420.
Heptangius, 203.
Hera, 424.
Hercules Beetle, 380.
Hermione, 532.
Hermit-Screw, Flem-
 ing's, 474.
Herpeton, 117.
Herrings, 276 to 278.
Herrings, King of, 193.
Hetæra, 414.
Heterocera, 415.
Heterodactylus, 55.
Heterodon, 120.
Heteropoda, 342.
Heteroptera, 430.
Heteroteuthis, 308.
Hidden Feet, 433.
Hinnites, 347.
Hinulia, 61.
Hippa, Asiatic, 457.
Hippobosea, 437.
Hippocampus, 291.
Hippocrepia, 363.
Hippoglossus, 266.
Hippolyte, 408.
Hirudo, 538.
Histioteuthis, 308.
Historidæ, 377.
Hoe, 195, 202, 203.
Hoc-Choke Cover Clip, 265.
Hœmatopinus, 518.
Hog-Fish, 227.
Holocanthus, 230.
Holocentrum, 217.
Holothuria, 543.
Holtenia, 592, 593.
Homolidæ, 457.
Homoptera, 427.
Hoplostethus, 216.
Hoppers, 420.
Horn-Fish, 219, 279.
Hornels, 270.
Hornera, 367.
Hornet, 402.
Hornet-Fly, Banded,435.
Hornsman, 107.
Horse-Couch, 331.
Horse Crevallé, 250.
Horseman, 240.
Horseshoe Animals, 363.

Hortalia, 120.
Hound, Smooth, 196, 200.
Hoverer-Flies, 436.
Huenia, 442.
Hyalea, 342.
Hyalonema, 593.
Hyas, 441.
Hydatids, 539.
Hydra, 557, 566.
Hydroida, 576.
Hydroids, 557, 558.
Hydromedusa, 18.
Hydrophis, 115, 116.
Hydrozoa, 557.
Hyla, 154, 164.
Hylaplesia, 169.
Hylaplesura, 168.
Hylas, 419.
Hymenoptera, 397.
Hypoderma, 437.
Hypogymna, 423.
Hypsirhina, 120.

I.

Ibaeus, Spotted, 463.
Ichneumons, 398, 399.
Idmonea, 367.
Iguanas, 75 to 78.
Iguanas of the Old World, 84.
Iguanidæ, 79.
Ilyanthus, 569.
Ilybius, 376.
Infundibulata, 362.
Infusoria, 585, 594.
Insect-Eaters, 399.
Insects, 372.
Internal Worms, 538.
Invertebrata, 298.
Isabelita, 228.
Isis, 575.
Isogomphodon, 196.
Isoplcura, 332.
Isopoda, 477.
Issc, 424.
Isurus, 201, 203.
Itoniza, 510.
Ithonia, 409.
Ivory-Shell, Spotted, 319
Ixn, 454.
Ixodes, 516.

J.

Jacare, 37, 38.
Jaeculina, 486.
Janthina, 330.
Jelly-Fishes, 556,557,564.
Jew-Fish, 220.
Julius, 526.
Jumper-Fish, 257.
Jumpers, 472.
Jungia, 572.

K.

Kakaan, 225.
Karoo Bokadam, 117.
Katuka, 104.
Katydid, 391.
Keratoidea, 592.
Keroto-Silicoidon, 592.
King-Fish, 242, 249.
King, Sand, 132.
Kingston, 205.
Kite, 266.
Kleg, 270.
Klip-Fish, 268.
Knee-Pans, Little. 330.
Knights, 406.

L.

Lebia, 374.
Labrus, 222, 223.
Lac Insect, 430.
Lacerta, 47, 48.
Lachesis, 97.
Ladies' Slipper, 364.
Ladybirds, 386.
Lafayette-Fish, 241.
Lambrus, Spine-Armed, 443.

INDEX.

Lamellicorn Beetles, 377.
Lamna, 203.
Lamp-Shell, 344.
Lampern, 295.
Lampreys, 293, 295.
Lampris, 249.
Lamproglena, 486.
Lancelet, 296.
Langaha, 139.
Lantern-Flies, 428.
Lark-Worm, 539.
Latilidæ, 222.
Launce, Sand, 270.
Leaf Insect, 392.
Leather-Jackets, 288.
Lebia, 375.
Leeches, 535 to 538.
Leconia, Spotted, 453.
Leous, 492.
Lepidonotus, 532.
Lepidoptera, 404.
Lepidopus, 243.
Lepidosiren, 181, 184.
Lepidosteus, 293.
Lepralia, 366.
Leptalis, 409.
Leptocephalus, 275, 342.
Leptocircus, 408.
Leptoglossæ, 71.
Leptopodia, 439.
Lernæa, 204, 487.
Lernæocera, 487.
Lernæodiscus, 486.
Lernæopoda, 487.
Lerneadiæ, 486.
Lernentoma, 486.
Leucania, 434.
Leuciscus, 287.
Leucosia, 453.
Lialis, 58.
Libellula, 395.
Lice, 437.
Lightning-Colored Shell, 321.
Lima, 346.
Limacina, 344.
Limax, 335, 338.
Limulus, 337.
Limpets, 331, 332, 337.
Limulus, 488.
Ling, 269.
Ling-Hood, 589.
Linyphia, 508.
Lip-Fishes, 221.
Liparis, 254, 255.
Lisette, 283.
Lissotriton, 175.
Lithobius, 521, 522.
Lithodes, 456.
Lithodomus, 347.
Lithosiidæ, 424.
Litorina, 327.
Lizard, Big Water, 179.
Lizard, Sea, 341.
Lizard, Spine-Backed, of New Guinea, 67.
Lizards, Band-Tailed, 50.
Lizards, Basket-, 51.
Lizards, Nocturnal, 71.
Lizards, Rough-Tailed, 71.
Lizards, Serpent-Eyed, 40
Lizards, Slender-Tongued, 40, 71.
Lizards, Spine-Foot, 40.
Lizards, Thick-Tongued, 71.
Lizards, True, 89.
Lizards, True, 40, 45, 49.
Lizards, Varieties of, 39 to 52, 55, 67, 71, 83 to 86.
Lobe-Bearer, 339.
Lobiger, 330.
Lobster-Louse, 485.
Lobsters, 463 to 465.
Locust, Migratory, 390.
Locust, Sea, 232.
Lodia, 136.
Loligo, 311.
Longicorn Beetles, 385.
Long-Nose, 279.
Long-Tail, 202.
Lophius, 242.
Lophobranchii, 290.
VOL. III.

Lopholatilus, 222.
Lophopus, 370.
Loricata, 28.
Lota, 286.
Louvette, 517.
Lucernariadæ, 568.
Lucinia, 347.
Lucioperca, 218.
Luidia, 502.
Lug-Worm, 530.
Lump-Fish, 253.
Lump-Sucker, 253, 254.
Lumper, 258.
Lunulites, 366.
Luth, 21.
Lutjanus, 227.
Lycænidæ, 415.
Lycosa, 497, 499.
Lynceidæ, 482.
Lyrie, 236.
Lysimnia, 410.
Lytta, 382.

M.

Mabouya, 62.
Mackerel Guide, 279.
Mackerels, 244, 247, 250.
Macrochelys, 20.
Macroglossa, 419.
Mactra, 353, 354.
Madrepora, 573.
Magilus, 318.
Maiadæ, 441.
Maigre, 241.
Mail-Shells, 332.
Malaclemys, 11.
Malacodermmys, 14.
Malleus, 347.
Malthea, 256.
Man-Eater, 202.
Manta, 210.
Manticora, 374.
Mantis, 392.
Marginella, 322.
Marmignatto, 508.
Marpesia, 412.
Martesia, 356.
Mastacembelus, 264.
Mastigure, Egyptian, 88.
Matamata, 17.
Matlametlo, 153.
May-Fly, 396.
Meal-Worm, 383.
Meandrinas, 579.
Meantia, 185.
Mechanitis, 405, 409.
Mecistops, 29, 30.
Medusa's Head, 554.
Medusæ, 559, 564, 565.
Megalochile, Eared, 87.
Melania, 336.
Meleagrina, 349.
Membraniporn, 365.
Menhaden, 278.
Menobranchus, 179.
Menopoma, 178.
Menopon, 519.
Menipea, 363.
Menticirrus, 242.
Merlangus, 268, 270.
Merluelus, 270.
Mesosemia, 405.
Metapoceros, 78.
Metazon, 591.
Metœcus, 474.
Micippa, 442.
Micraspis, 386.
Micrhybina, 168.
Microchiera, 440.
Microciona, 592.
Micrognator, 399.
Microlepidotus, 51.
Microperca, 220.
Mictidæ, 431.
Midamus, 410.
Migranes, 453.
Millepedes, 524, 526.
Miller's Thumb, 234.
Mineralogist, 328.
Minnow, 287.
Misipsa, 414.
Mites, 516 to 518.
Mitre, Bishop's, 321.

Mnemiopsis, 559.
Moccasins, 97, 98.
Mocoa, 61.
Modiolus, 347.
Moina, 482.
Mola, 290.
Mollusca, 302, 345.
Mullusks, Four-Gilled, 311.
Mollusks, Headless, 345.
Mollusks, Naked-Gilled, 340.
Mollusks, Rearward-Gilled, 338.
Mollusks, Shore, 327.
Mollusks, Wing-Footed, 342.
Moloch, 89.
Monedula, 401.
Monera, 593.
Money-Spinners, 508.
Monitors, 40 to 42.
Monk-Fish, 205.
Monochirus, 265.
Moon-Fishes, 250.
Moray, 273.
Mordellidæ, 382.
Morella, 118.
Morgay, 193.
Mormolyce, 375.
Moroteuthis, 309.
Morpho, 415.
Morrhua, 268, 270.
Morris, Anglesey, 275.
Moss-Bunker, 278.
Moths, 415 to 427.
Mud-Borer, 464.
Mud-Burrower, 464.
Mud-Eel, 186.
Mud-Fish, 181.
Mud Puppies, 179.
Mugger, 32.
Mugil, 261, 262.
Mullets, Three-Banded, 224, 226, 227, 261.
Mullus, 226.
Muræna, 272, 273.
Murænopsis, 179, 180.
Murex, 314, 315, 331.
Muscidæ, 436.
Mushroom, Sea-, 572.
Musk Beetle, 385.
Mussels, 350, 351.
Mustelus, 196, 200.
Mutilla, 401.
Mya, 353, 354, 355.
Myctiris, Long-Armed, 449.
Mygale, 499.
Myliobatis, 211.
Mypsis, 470.
Myriapoda, 520.
Myripristis, 217.
Myrmarachna, 500.
Myrmeleon, 396.
Myxine, 295.

N.

Nakoo, 29.
Nalla Whallagee Pam, 115.
Narara, 78.
Naseus, 260.
Naticella-Shell, 324.
Natricidæ, 127.
Natter, Sand-, 110.
Natterjack, 163.
Nautilus, 305, 311.
Necrophagœ, 377.
Necrophorus, 377.
Necturus, 179, 186.
Needle-Fish, 292.
Needle-Shell, Spotted, 319.
Nematus, 398.
Nemertes, 538.
Nemoptera, Coa, 396.
Neoptolemus, 415.
Nepidæ, 431.
Neptune's Boat, 322.
Neptune's Cup, 589, 591.
Neptunus, 447.
Nereis, 531, 532.

Neritas, 324.
Neritina, 325.
Nettles, 556, 557.
Neuroptera, 393.
Newts, 172.
Nicothoë, 485.
Nigger, 398.
Nipper, 223.
Noah's Ark, 351.
Noble, 236.
Noctiluca, 584, 594.
Noctuidæ, 424, 434.
Nomada, 403.
Nops, 511.
Notamia, 364.
Notodelphys, 484.
Notoneeta, 431.
Nowd, 236.
Nucleobranchiata, 342.
Nudibranchs, 340.
Numb-Fish, 206.
Nurse, 196.
Nursia, 454.
Nyctisaura, 71.
Nymphalidæ, 411.

O.

Obisium, 514.
Octopus, 306, 307.
Ocypode, 450.
Ocypus, 377.
Odynerus, 402.
Œstrus, 436, 487.
Old Wife, 222.
Oliva, 320.
Olivas, 320, 330, 331.
Olynthoidea, 591.
Ommastrephes, 307, 309, 311.
Oncuidophorus, 45.
Oniscus, 479.
Onychodactylus, 175.
Onychophis, 70.
Opahs, 249.
Opelet, 569.
Opheosaurus, 54.
Ophibolus, 131, 132, 133.
Ophidin, 93.
Ophiocephalus, 262, 263.
Ophiocoma, 552.
Ophiomore, 68.
Ophiops, 49.
Ophiura, 552.
Opisthobranchiata, 338.
Orange-Tip, 408.
Orchestia, 478.
Orcynus, 245, 246.
Orcocephale, Marine, 78.
Oreosoma, 288.
Orgyia, 423.
Ornithoptera, 406.
Orthoptera, 388.
Osmerus, 281.
Osmia, 383.
Ostracion, 288.
Ostreæoda, 480, 483.
Ostrea, 346.
Otiothops, 511.
Ouanlibi, 220.
Oular Carron, 116.
Ovules, 330.
Oyster-Fish, 255.
Oysters, 345 to 348.

P.

Pachyglossæ, 71.
Paddle-Fish, 192.
Pagellus, 228.
Pagurus, 458.
Palæmon, 468.
Palinurus, 463, 464.
Paludicella, 371.
Pam, Horatta, 109.
Pammelas, 250.
Pandora, 227.
Panella, 487.
Pan-Fish, 488.
Panopœa, 354.
Paper Sailors, 307.
Papilio, 405, 407, 408.
Paramecium, 587.

Parrot-Fish, Tesselated, 224.
Parthenope, Spinose, 443.
Passerita, 139.
Patellas, 330.
Pavonaria, 576.
Pavonin, 572.
Pear-Shells, 316, 330.
Pearl, 266.
Pearlet, Scottish, 569.
Pecten, 348.
Pectinatella, 370.
Pedda-Poda, 119.
Pedicellina, 369.
Pediculidæ, 518.
Pegasus, 290.
Pelamis, 115, 116.
Pelonæa, 358.
Pelor, 233.
Peltocephala, 485.
Pemphegus, 430.
Penæus, 469.
Pencil-Tail, 525.
Penella, 480.
Pennatula, 576.
Pentacrinus, 554.
Pentacte, 543.
Pepsis, 401.
Perca, 218.
Perch, 223.
Perch, Climbing, 260.
Perch-Sucker, 486.
Perches, Various, 217 to 220, 223.
Percophis, Brazilian, 239.
Perimela, Toothed, 444.
Perinidium, 585.
Perinopsis, 547.
Peristethus, 237.
Periwinkle, 327.
Perlidæ, 396.
Perlous, 203.
Pete, 57.
Petricolidæ, 356.
Petromyzon, 294.
Phasgonuren, 391.
Phasiuidæ, 391.
Pheasant-Shells, 328.
Philodromus, 503.
Philodryas, 136.
Pholas, 356.
Phoio, 409.
Phorus, 328.
Photinus, 381.
Phoxichilidium, 487.
Phronima, 474.
Phrynosoma, 83.
Phrynus, 514.
Phycis, 269.
Phylactolæmata, 369.
Phyllodoce, 527, 532.
Phyllopoda, 480.
Phyllopteryx, Horse-Like, 291.
Phyllosome, Club-Horned, 470.
Phylloxera, Grape, 430.
Phyllure, White's, 74.
Physalia, 559, 561.
Physanoptera, 393.
Piddocks, 356.
Pieris, 408, 409.
Pigeons, Sea, 338.
Pikes, Various, 219, 279, 293.
Pilchard, 276.
Pilot Fishes, 246, 250.
Pimmus, 444.
Pimplet, Gem-, 570.
Pinna, 349.
Pinnotheres, 350, 449.
Pipa, 150.
Pipe-Fishes, 292.
Piques, 517.
Pirai, 282.
Pirarucu, 240.
Pirate, River, 279.
Piraya, 282.
Pisces, 189.
Pissodes, 385.
Pityophis, 130.
Plaice, 266.
Planes, 451.
Planorbis, 337.

D

INDEX.

Platessa, 266.
Platycephalus, 235.
Platyoniculus, 450.
Plectodera, 386.
Plectognathi, 287.
Plestiodons, 61, 62.
Plethodon, 172.
Pleurobranchia, 559.
Pleuronectes, 266.
Plonæa, 358.
Plum Gouger, 385.
Plumatella, 371, 476.
Plumularia, 578.
Plateaus, 548.
Poacher, Sea, 236.
Podophthalmata, 438.
Pœcilopoda, 485.
Pogge, 236.
Poghagen, 278.
Pogonias, 241.
Pogy, 278.
Polewig, 252.
Polish Scarlet Grain, 430.
Pollack, 270.
Pollux, 413.
Polybius, 447.
Polycera, 560.
Polychrus, 77.
Polycystina, 584.
Polydesmus, 525.
Polyphemus, 482.
Polyprion, Couch's, 220.
Polypus, 306.
Polystomella, 584, 594.
Polyxenus, 525.
Polyzoa, 361.
Polyzoa, Marine, 362.
Pomacanthus, 228.
Pomadasys, 227.
Pomatomus, 247.
Pompano, 247, 250.
Pompilus, 401.
Porbeagles, 201, 203.
Porcellana, 462.
Porcupine-Fish, 289.
Porgee, 227.
Porifera, 587.
Poriferata, 591, 593.
Porpita, 559.
Portuguese Man-o'-War, 559, 560.
Portunus, 446.
Poseidon, 406.
Postillons, 165.
Pouch-Shell, 337.
Pout, 270.
Prawns, 467, 468.
Praying Insects, 392.
Priam, 407.
Priapulus, Tailed, 542.
Prickleﬁsh, 213.
Pride, Sand, 295.
Pristidurus, 195.
Pristiophorus, 206.
Pristipoma, 225.
Pristis, 206.
Proach, Lucky, 235.
Proteida, 179.
Protealius, 408.
Proteus, 185, 186.
Protolepas, 492.
Protomyxa, 593.
Protonopsis, 179.
Protozoa, 591, 593.
Psammophylax, 126.
Pseudemys, 14.
Pseudis, 152.
Pseudobdella, 537.
Pseudophidia, 180.
Pseudopus, 52.
Pseudoscorpiones, 512.
Pseudotriton, 170.
Psolius, 543.
Psolus, 542.
Pterocircoza, Eyed, 391.
Pteraclis, 249.
Pteropods, 242 to 244.
Pteroceroda, 330.
Pterotracheaea, 342.
Ptilinus, 381.
Ptiniidæ, 381.
Ptyodactylus, 72.
Puffers, 289.
Puffers, 569.

VOL. III.

Pulex, 432.
Pulmonata, 338.
Pupas, 338.
Puppies, 178, 179.
Purple, Common, 318.
Purple-Shells, 330.
Purpura, 318.
Purpuras, 330.
Pustulopora, 367.
Pycnogonum, 487.
Pygopus, 57.
Pyrochroa, 382.
Pyrophorus, 381.
Pyrosoma, 359.
Pyrula, 316.
Pyrulas, 330.
Pythons, 118 to 120.
Pyxis, 8, 9.

Q.

Quahog, 347, 353.
Quinqueloculina, 584.
Quoit-Fishes, 253.

R.

Rabbit-Fish, 193.
Racers, 130, 131.
Radiata, 541.
Radiolaria, 594.
Rain, 212.
Raidæ, 206.
Raja, 207, 209.
Rana, 154, 158.
Ranellas, 331.
Ranina, 457.
Rataria, 559.
Rat-Fish, 193.
Rattlesnakes, 99, 103, 104.
Ray-Worm, 539.
Rays, 206 to 212.
Razor-Shells, 355.
Red-Throat, Four-Streaked, 225.
Rodavius, 431.
Reef Proper of the Florida Straits, 578.
Regalecus, 258.
Regenia, 40, 41.
Remoras, 248.
Reniceps, 196.
Reptiles, 3.
Reptiles, Blind, 70.
Reptiles, Mailed, 28.
Reptiles, Scaled, 40.
Reptiles, Shielded, 5.
Retepora, 366.
Reticularia, 594.
Rhinaster, 517.
Rhinoceros, 517.
Scorpion, Book-, 514.
Rhinochilus, 136.
Rhinophryne, 169.
Rhizopoda, 380.
Rhizopods, 580, 593.
Rhizostoma, 566.
Rhopalura, 405.
Rhynchites, 383.
Rhynchocephalia, 78.
Rhyssa, 398, 400.
Ribbon-Fish, 258.
Ripiphorus, 382.
River Jack, 107.
River Pirate, 279.
Roach, 242, 286.
Robin Huss, 193.
Roccus, 220.
Rock-Fish, 251, 268.
Roller, Oak-Leaf, 426.
Rosalina, 584.
Rose-Buds, 330.
Rossia, 308.
Rotifera, 581.
Rougegueule, 225.
Rove-Beetles, 377.
Rudder-Fish, 249.
Ruffe, 219.
Rutelidæ, 380.

S.

Sabella, 528.
Saddle-Shell, 346.
Sadis, 240.

Sagartia, 568, 570.
Sagittas, 342.
Sail-Fish, 202, 251.
Sailors' Choice, 227.
Salamanders, Various, 170, 172, 175, 176.
Salarias, 257.
Salicornaria, 362.
Sallyman, 559.
Salmo, 280, 282, 283.
Salmon, 279.
Salmon, Jack, 219.
Salpa, 360.
Salticus, 452, 500.
Salvadora, 135.
Sand-Hopper, 472.
Sand-Screw, Kroyer's, 473.
Sand-Skipper, 472.
Sand-Star, White, 552.
Sauger, 219.
Sardu, 240.
Sarpedon, 407.
Sarpo, 255.
Sarsia, 564.
Satyridæ, 414.
Saura, 40.
Sauroplis, 52.
Savala, 249.
Saw-Belly, 277.
Saw-Fishes, 206.
Saw-Flies, True, 397.
Saw-Tail, 195.
Scabbard-Fish, 243.
Scad, 250.
Scalaria, 304.
Scale-Fin, 243.
Scale-Insect, 429.
Scallops, 348.
Scaphlopus, 154.
Scaphiorhynchops, 192.
Scaphopoda, 342.
Scarabaeus, 378, 379.
Scaritidæ, 375.
Schnap-Sticker, 125.
Scholtopusic, 52, 57.
Schizodactylus, 390.
Scicuridæ, 240, 241.
Sciara, 434.
Sciccus, 59.
Scleranchius, 510.
Sclerodermi, 287.
Scolia, 401.
Scoliodon, 196.
Scolopendra, 521 to 523.
Scolytus, 384.
Scomber, 244, 245.
Scomberomorus, 247.
Scopelus, 282, 283.
Scorpaenas, 231 to 233.
Scorpion, Book-, 514.
Scorpion-Fishes, 231.
Scorpion-Flies, 395.
Scorpion, Sea, 231, 234.
Scorpion-Shells, 313.
Scorpion, Water, 481.
Scorpions, True, 514.
Screws, 475, 476.
Screw-Stones, 554.
Scrubicularia, 354.
Sculpins, 234, 232.
Scup, or Scupping, 227.
Scutibranchia, 328.
Scyllaridæ, 463.
Scyllium, 195.
Scytodes, 511.
Sea-Adder, 216.
Sea-Basket, 553.
Sea-Bream, King of, 227.
Sea-Bulbs, 557.
Sea-Cat, 193, 256.
Sea-Dace, 218.
Sea-Devils, 210, 212.
Sea-Dragon, 200.
Sea-Eggs, 544, 552.
Sea-Fan, 574.
Sea-Feathers, 574.
Sea-Finger, 576.
Sea-Fox, 202.
Sea-Hare, 338.
Sea-Horse, 290.
Sea-Jellies, 559.
Sea-Mats, 365.
Sea-Mouse, 532, 533.

Sea-Needle, 279.
Sea-Owl, 253.
Sea-Pens, 575.
Sea-Pigeons, 338.
Sea-Pink, 568, 569.
Sea-Puddings, 542.
Sea-Rose, Thick-Petalled, 568.
Sea-Shrub, 575.
Sea-Star, Bird's Foot, 551.
Sea-Surgeon, 259.
Sea-Toad, 441.
Sea-Whips, 574.
Sea-Wolf, 256.
Scopaard, 233.
Sogestrium, 511.
Selenaria, 366.
Semotilus, 204.
Senoculata, 511.
Sepia, 306, 310.
Sepiola, 308.
Sops, 68.
Septoglossæ, 40.
Serialaria, 368.
Serpent, Water, 124.
Serpents, False, 180.
Serpents, Poisonous, 104.
Serpents, River, 115.
Serpents, Sea, 115.
Serpula, 367, 527, 528.
Serranus, 220.
Serranus, Ruddy, 220.
Serrosalmo, 284.
Sertulariadæ, 577.
Sosiadæ, 419.
Setigera, 527.
Shad, 276.
Shan, or Shanny, 257.
Shark, Fresh-Water, 279.
Shark-Sucker, 486.
Sharks, 193 to 205.
Sharp-Nose, 196.
Sharplin, 213.
Sheat-Fish, 287.
Sheepshead, 227.
Shell-Binder, 529.
Shell, Fountain, 313.
Shield-Tail, Philippine, 71.
Ship-Worm, 356.
Shore-Hopper 473.
Shovel-Fish, 192.
Shovel-Head, 196.
Shrimps, Various, 464, 466, 467, 469, 470, 471, 475, 476, 478, 480, 481.
Shuttle, Weaver's, 324.
Sida, 482.
Silicoidea, 593.
Siliquaria, 326.
Siliquas, 354.
Silk-Worms, 422.
Sillago, Indian, 239.
Silurus, 286, 287.
Single-Thorn, Japanese, 210.
Siphon, 480.
Siphonophora, 557, 559.
Siphonostoma, 485.
Sipunculus, 541, 542.
Siren, 179, 180.
Sitaris, 383.
Sitophilus, 385.
Sivado, 409.
Skate-Sucker, 537.
Skates, 208 to 210, 212.
Skinks, 57 to 61, 69.
Skip-Jack, 246.
Skippaups, 278.
Skippers, 415.
Skittle-Dog, 195.
Slaters, 470.
Slobbers, 564.
Slow-Worm, 63.
Slugs, 335.
Smelts, 261, 281, 282.
Smeltie, 270.
Smerinthus, 416, 419.
Snails, 337, 330, 333, 334, 336 to 338.
Snails, Sea, 254, 324.
Snake, Bead, 134, 148.

Snake, Black, 126, 130, 133, 135, 137.
Snake, Black and Red, Little, 136.
Snake, Brown, 136.
Snake, Brown, of America, 125.
Snake, Carpet, 118.
Snake, Chain, Kennicott's, 133.
Snake, Chicken, 132.
Snake, Clawed, 70.
Snake, Coach-Whip, 134.
Snake, Congo, 179.
Snake, Copper-Head, 98, 99.
Snake, Coral, 125.
Snake, Corn-, 130.
Snake, Diamond, of Australia, 118.
Snake, Emerald Whip, 136.
Snake-Fish, 263.
Snake, Fox, 130.
Snake, Garter, 134.
Snake, Glass, 53.
Snake, Gopher, 133.
Snake, Gray, of Jamaica, 137.
Snake, Green, 135, 136.
Snake, Ground, 136.
Snake, Hog-Nose, 129.
Snake, Horn, 136.
Snake, House, 132.
Snake, Indigo, 133.
Snake, Javelin, 69.
Snake, King, Scarlet, 131.
Snake, Labarri, 149.
Snake, Milk, 132.
Snake, Pine, 130.
Snake, Red-Bellied, 136.
Snake, Ribbon, 134.
Snake, Ring-Necked, 136.
Snake, Ringed, 127.
Snake, Rock, 119, 120.
Snake, Sachem, 132.
Snake, Say's, 132.
Snake, Scarlet, 136.
Snake-Shell, 326.
Snake, Silver, 70.
Snake, Spitting, 148.
Snake-Stone, 146.
Snake, Thorny-, 115.
Snake, Thunder, 131.
Snake, Tree-, Golden, 139.
Snake, Water, Red-Bellied, 134.
Snake, Worm, 136.
Snake, Yellow, of Jamaica, 125.
Snakes, 93.
Snakes, Sea-, 390.
Snakes, Viperine, 95.
Snakes, Wood-, 139.
Snap-Beetles, 381.
Snappers, 20, 227.
Snipe, Sea, 263.
Solaster, 551.
Soldiers and Sailors, 381.
Sole, Land, 335.
Solen, 265.
Solenmyas, 354.
Solen, 354.
Solenette, 265.
Solens, 354.
Soles, 265.
Solpuga, 513.
Sonora, 136.
Spade-Foot, Solitary, 154.
Spanish Flies, 382.
Sparassus, 503.
Sparidæ, 227.
Sparling, 282.
Spectres, 392.
Spelerpes, 172.
Sphænops, 60.
Sphaerodactyle, Banded, 74.
Sphingidæ, 416.
Sphinx, 417, 418.
Sphyræna 242.
Sphyrnias, 199.

INDEX.

Spider, Cardinal-, 507.
Spider Crab, 439, 441.
Spider, Crab, Great, 494.
Spider, Cross-, 509.
Spider, Garden-, 509, 510.
Spider, Harvest-, 511.
Spider, Pirate-, 499.
Spider-Shells, 313.
Spider, Tarantula-, 497.
Spider, Water, 505.
Spiders, Crab-, 501.
Spiders, Hunting, 452, 499.
Spiders, Sea-, 439.
Spiders, Trap Door, 496.
Spiders, True, 493.
Spiders, Wolf-, 497, 498.
Spiders, Zebra, 452.
Spike-Shell, 342.
Spilotes, 133.
Spindle-Shells, 312, 316, 317.
Spine-Belly, Striped,289.
Spine-Foot, Cape, 49.
Spio, 400.
Spiralidæ, 311.
Spirling, 282.
Spirostreptes, 526.
Spirula, 307.
Spitters, 72.
Spondylus, 347.
Sponge, Crumb-of-Bread, 592.
Sponge, Fresh Water, 370.
Sponge, Funnel, 589.
Sponge, Glass-Rope, 593.
Sponge, Pumice-Stone, 590.
Sponge, Turkey, 587.
Sponges, Fresh-Water, 592.
Sponges, Glass, 593.
Sponges, Lime, 591.
Spongia, 592.
Spout-Shell, 325.
Sprugh-Slange, 148.
Spyrnius, 199.
Squamata, 40.
Squamipinnes, 228.
Squatina, 205.
Squeteague, 241.
Squilla, 471.
Squinado, 442.
Squirt, Sea, 358.
Stag Beetle, 378.
Staircase-Shell, 327.
Staphylinus, 373, 377.
Star-Fishes, 549, 552.
Star-Gazers, 238, 284.
Staurotheutis, 307.
St. Cuthbert's Beads, 554
Stellio, 86, 88.
Stenorhynchus, 439.
Stentor, 586, 594.
Sticklebacks, 212 to 216.
Stidostedium, 219.
Sting-A-Ree, 211.
Stomapoda, 460.
Stomatia, 330.
Stone-Flies, 396.
Stone-Lillies, 554.
Stone-Suckers, 294.
Stone-Toter, 294.
Storeria, 136.
Strepsiptera, 404.
Strobilosaura, 75.
Stromateus, 241.
Strombs, 313.
Strombus, 313, 331, 459.
Strongylocentrotus, 545, 546.
Strongylus, 539.

Sturgeons, 190 to 192.
Suberites, 592.
Suckauhock, 353.
Suckers, 254.
Suckers, Stone-, 294.
Sucking-Fishes, 247, 248.
Suctoria, 585.
Sun-Fish, 202, 249, 289.
Sun, Shooter, 115.
Surmullets, 226, 227.
Swift, New Zealand, 420.
Swine-Fishes, Checkered, 221, 256.
Swingle-Tail, 202.
Sword-Fishes, 250, 251.
Swordlick, 258.
Symnista, 458.
Synaptas, 543.
Syngnathus, 291, 292.
Syntethys, 359.
Syrphidæ, 436.

T.

Tabanus, 434.
Tachydrome, 52.
Tadpole, Development of, 156.
Tænia, 275, 539.
Tantilla, 136.
Taonius, 308.
Tapayaxin, Crowned, 83, 370.
Tape-Worm, 539.
Taraguira, Six-Lined,45.
Tarantula, 496.
Tarentola, Cape, 73.
Tautog, 223.
Tectibranchiata, 338.
Togouaria, 507.
Tegnexins, 43, 45.
Teidæ, 45.
Tejus, 43.
Telephoridæ, 381.
Tellinidæ, 354.
Tench, 285.
Tenebrio, 383.
Terebella, 529.
Terebratula, 344.
Teredo, 357.
Termes, 393.
Terrapins, 9 to 16.
Testacella, 336.
Testudo, 6, 7.
Tethia, 592.
Tetrabranchiata, 311.
Tetragnathon, 509.
Tetraodon, 289.
Teuthidæ, 307, 308.
Tic-Polonga, 104.
Ticks, 437, 516, 518, 519.
Tiedemania, 344.
Tiger Beetles, 373, 374.
Tile-Fishes, 222, 289.
Tiligugu, 69.
Tiliqua, 57, 69.
Timarcha, 386.
Tinea, 426.
Tinker, 210.
Tittlebat, 213.
Thais, 408.
Thalassinidæ, 464.
Thalassochelys, 23.
Thecla, 405.
Thecosomata, 344.
Thelia, 568.
Theridion, 508.
Thoas, 408.
Thomisus, 501, 502.
Thorntail, Unicorn, 260.
Thracia, 354.
Thresher, 202.

Thunderbolts, 310.
Thymallus, 281, 282, 283.
Thyodamas, 411.
Thyraitos, 243.
Toad-Fish, 255, 256.
Toads, 150, 159 to 164,166.
Tobacco-Pipe Fish, 264.
Tommy Logge, 234.
Tooth-Shells, 332.
Top-Shells, 328.
Topo, 199.
Tops, 328, 329.
Torpedo, 206, 207.
Tortoise Beetles, 386.
Tortoise, Blanding's, 14.
Tortoise, Box, 12.
Tortoise, Chicken, 10.
Tortoise, Common Land, 7.
Tortoise, Gopher, 6.
Tortoise, Indian, 7.
Tortoise, Lettered, 9.
Tortoise, Mud, 15.
Tortoise, Muhlenberg's, 14.
Tortoise, Mungofa, 6.
Tortoise, Painted, 14.
Tortoise, Quaker, 10.
Tortoise, Snake, 18.
Tortoise, Speckled, 14.
Tortoise, Wood, 14.
Tortoises, 5.
Tortoises, Land, 13.
Tortoises, True, 6.
Tortrix, 125, 400, 426.
Tracheliastes, 486.
Trachinus, 239.
Trachurus, 245.
Trachycephalus, 167.
Trachynotus, 247, 250.
Trachypterus, 258.
Trachysaurus, 61.
Trepangs, 542.
Triacis, 196.
Tribolonotus, 67.
Trichina, 540.
Trichinus, 243, 244.
Trichoptera, 397.
Tricocephalus, 539.
Tricondyla, 374.
Tridacna, 352.
Trigger-Fishes, 288, 289.
Trigla, 235, 236.
Triodons, 289.
Trionyx, 19, 20.
Tritonidæ, 340.
Tritons, 172, 174, 315.
Trochus, 327, 329.
Trombidium, 518.
Tropidonotus, 128, 134.
Trough-Shells, 353.
Trouts, 241, 281, 282.
Trumpet-Fish, 263.
Trumpet, Sea, 315.
Trumpets, 330.
Truncatella, 327.
Trunk-Fishes, 287, 288.
Tsetse, 435.
Tuatera, 78.
Tubipora, 574.
Tubularia, 557, 558.
Tubulariadæ, 576.
Tubulipora, 367, 368.
Tubuliporina, 574.
Tudes, 199.
Tulip-Shell, Great, 316.
Tumble-Bug, 379.
Tun-Shell, Apple, 319.
Tuna, 246.
Tunicata, 358.
Tunnies, 244, 245.
Turbo Versicolor, 329.
Turbots, 265, 266.

Turnip-Fly, 398.
Turritella, Common, 326.
Turtle, Box, 13.
Turtle, Hawk's-Bill, 23.
Turtle, Hieroglyphic, 14.
Turtle, Green, 25.
Turtle, Leathery, 21.
Turtle, Loggerhead, 23.
Turtle, Map, 14.
Turtle, Musk, 15.
Turtle, Rhinoceros, 23.
Turtle, Salt Marsh, 14.
Turtle, Small Box, 15.
Turtle, Small Mud, 15.
Turtle, Snapping, 17, 19.
Turtles, 21.
Turtles, Pond, 13.
Turtles, Soft, 19.
Tusseh, 422.
Tusser, 422.
Tweeg, 178.
Typhline, 70.
Typhlops, 70.
Tyrse, 20.

U.

Ular, Sawa, 120.
Umbrella, Indian, 339.
Unionidæ, 347.
Urania, 420.
Uranoscopus, 239.
Urchin, Basket, 553.
Urchin, Cake-, 547.
Urchin-Fish, 289.
Urchin-Fish, Hairy, 289.
Urchin, Heart-, Common, 546.
Urchin, Heart-, Fiddle, 546.
Urchin, Keyhole, 548.
Urchin, Piper-, 546.
Urchin, Sea-, 552.
Urchin, Shield-, 548.
Urchin, Wheel-, 548.
Urchins, Sea-, 544.
Urocerus, 308, 400.
Uroctea, 505.
Uronastix, 88.
Uropeltis, 71.

V.

Vaagmar, Riband-Shaped, 258.
Vanessa, 412.
Varans, 40, 41.
Veined-White, 408.
Velella, 559.
Venduce, 282.
Venellas, 573.
Venus, 347.
Venus Cup, 593.
Venus' Girdle, 559, 563.
Venus-Shells, 353.
Ver Palmiste, 384.
Ver Solitaire, 539.
Vermetus, 326.
Vespa, 402.
Vestlet, 569.
Viper, 110.
Viper, Black, 130.
Viper, Blowing, 129.
Viper, Death, 115.
Viper, Horned, 108.
Viver, Plumed, 107.
Viper, Red, 98.
Viper, Water, 98.
Vipera, 106, 109, 110.
Viperidæ, 104.
Viperina, 95.
Vipers, 104.
Virginia, 136.

Virgularia, 576.
Volucella Flies, 436.
Volutes, 304,321, 322, 330
Volutidæ, 321.
Vorticella, 586, 594.

W.

Walking-Fish, 256.
Walking-Stick Insect, 391.
Wasps, Wood, 400 to 402.
Water Beetles, 375, 376.
Water Boatman, 431.
Water-Dog, 179.
Water Hog-Louse, 470.
Watering Pot-Shell, 355.
Wax Insects, 428.
Weak-Fish, 241.
Weaver, Great, 239.
Weevils, 383 to 385.
Wenona, Lead-Colored, 136.
Wentletraps, 326, 327.
Whale-Louse, 477.
Wheel Animalcules, 581.
Wheel-Bug, 431.
Whelks, 316, 317.
Whirligigs, 376.
Whirlwig Beetles, 376.
White-Bait, 278.
White-Fish, 278, 281.
Whiting, 270.
Whiting, Bermuda, 242.
Wide-Gab, 255.
Widow, 568, 570.
Willow-Fly, 396.
Window Pane, 266.
Window-Shell, Chinese, 346.
Wing-Shells, 347.
Wire-Worm, 424.
Wood-Borers, 404.
Woodcocks, 314.
Woodlice, 479.
Woodlouse, 74.
Woolly Bear, 423.
Worm, Guinea, 538.
Worm-Shell, 326.
Wrasses, 222, 223.
Wreck-Fish, 220.

X.

Xenobalanus, 402.
Xenocerus, 385.
Xenopus, 150.
Xiphidæ, 251.
Xiphosoma, 122.
Xiphosura, 487.
Xiphosurus, 81.

Y.

Yakare, 28.
Yellow Sally, 396.
Yellow-Tail, 278.
Yponomeutidæ, 426.

Z.

Zenidæ, 248.
Zeonia, 414.
Zephronia, 524.
Zeus, 249.
Zoarces, 254.
Zonurus, 50.
Zoophyte, Club-, 577.
Zoophytes, 508.
Zoophytes, Bell-, 577
Zootoca, 46.
Zoxymus, 445.
Zwarte Slang, 126.
Zygobranchia, 330.

VOL. III.   F

www.ingramcontent.com/pod-product-compliance
Lightning Source LLC
Chambersburg PA
CBHW021219300426
44111CB00007B/353